Principles of GNSS, Inertial, and Multisensor Integrated Navigation Systems

Second Edition

DISCLAIMER OF WARRANTY

The technical descriptions, procedures, and computer programs in this book have been developed with the greatest of care and they have been useful to the author in a broad range of applications; however, they are provided as is, without warranty of any kind. Artech House, Inc. and the author and editors of the book titled *Principles of GNSS, Inertial, and Multisensor Integrated Navigation Systems Second Edition* make no warranties, expressed or implied, that the equations, programs, and procedures in this book or its associated software are free of error, or are consistent with any particular standard of merchantability, or will meet your requirements for any particular application. They should not be relied upon for solving a problem whose incorrect solution could result in injury to a person or loss of property. Any use of the programs or procedures in such a manner is at the user's own risk. The editors, author, and publisher disclaim all liability for direct, incidental, or consequent damages resulting from use of the programs or procedures in this book or the associated software.

Accompanying software can be found:
https://us.artechhouse.com/Assets/downloads/Groves_005.zip

Principles of GNSS, Inertial, and Multisensor Integrated Navigation Systems

Second Edition

Paul D. Groves

ARTECH HOUSE

BOSTON | LONDON
artechhouse.com

Library of Congress Cataloging-in-Publication Data
A catalog record for this book is available from the U.S. Library of Congress.

British Library Cataloguing in Publication Data
A catalog record for this book is available from the British Library.

ISBN-13: 978-1-60807-005-3

Cover design by Vicki Kane and Igor Valdman

© 2013 Paul D. Groves

All rights reserved. Printed and bound in the United States of America. No part of this book may be reproduced or utilized in any form or by any means, electronic or mechanical, including photocopying, recording, or by any information storage and retrieval system, without permission in writing from the publisher.

All terms mentioned in this book that are known to be trademarks or service marks have been appropriately capitalized. Artech House cannot attest to the accuracy of this information. Use of a term in this book should not be regarded as affecting the validity of any trademark or service mark.

10 9 8 7 6 5 4 3 2 1

Contents

Preface	xvii
Acknowledgments	xix

CHAPTER 1
Introduction   1

1.1	Fundamental Concepts	1
1.2	Dead Reckoning	5
1.3	Position Fixing	7
	1.3.1 Position-Fixing Methods	7
	1.3.2 Signal-Based Positioning	12
	1.3.3 Environmental Feature Matching	14
1.4	The Navigation System	15
	1.4.1 Requirements	16
	1.4.2 Context	17
	1.4.3 Integration	18
	1.4.4 Aiding	18
	1.4.5 Assistance and Cooperation	19
	1.4.6 Fault Detection	20
1.5	Overview of the Book	20
	References	22

CHAPTER 2
Coordinate Frames, Kinematics, and the Earth   23

2.1	Coordinate Frames	23
	2.1.1 Earth-Centered Inertial Frame	25
	2.1.2 Earth-Centered Earth-Fixed Frame	26
	2.1.3 Local Navigation Frame	27
	2.1.4 Local Tangent-Plane Frame	28
	2.1.5 Body Frame	28
	2.1.6 Other Frames	29
2.2	Attitude, Rotation, and Resolving Axes Transformations	30
	2.2.1 Euler Attitude	33
	2.2.2 Coordinate Transformation Matrix	35
	2.2.3 Quaternion Attitude	40
	2.2.4 Rotation Vector	42
2.3	Kinematics	43

	2.3.1	Angular Rate	44
	2.3.2	Cartesian Position	46
	2.3.3	Velocity	48
	2.3.4	Acceleration	50
	2.3.5	Motion with Respect to a Rotating Reference Frame	51
2.4	Earth Surface and Gravity Models	53
	2.4.1	The Ellipsoid Model of the Earth's Surface	54
	2.4.2	Curvilinear Position	57
	2.4.3	Position Conversion	61
	2.4.4	The Geoid, Orthometric Height, and Earth Tides	64
	2.4.5	Projected Coordinates	65
	2.4.6	Earth Rotation	66
	2.4.7	Specific Force, Gravitation, and Gravity	67
2.5	Frame Transformations	72
	2.5.1	Inertial and Earth Frames	73
	2.5.2	Earth and Local Navigation Frames	74
	2.5.3	Inertial and Local Navigation Frames	75
	2.5.4	Earth and Local Tangent-Plane Frames	76
	2.5.5	Transposition of Navigation Solutions	77
	References	78

CHAPTER 3

Kalman Filter-Based Estimation — 81

3.1	Introduction	82
	3.1.1 Elements of the Kalman Filter	82
	3.1.2 Steps of the Kalman Filter	84
	3.1.3 Kalman Filter Applications	86
3.2	Algorithms and Models	87
	3.2.1 Definitions	87
	3.2.2 Kalman Filter Algorithm	91
	3.2.3 System Model	96
	3.2.4 Measurement Model	100
	3.2.5 Kalman Filter Behavior and State Observability	103
	3.2.6 Closed-Loop Kalman Filter	106
	3.2.7 Sequential Measurement Update	107
3.3	Implementation Issues	109
	3.3.1 Tuning and Stability	109
	3.3.2 Algorithm Design	111
	3.3.3 Numerical Issues	113
	3.3.4 Time Synchronization	114
	3.3.5 Kalman Filter Design Process	117
3.4	Extensions to the Kalman Filter	117
	3.4.1 Extended and Linearized Kalman Filter	118
	3.4.2 Unscented Kalman Filter	121
	3.4.3 Time-Correlated Noise	123
	3.4.4 Adaptive Kalman Filter	124

		3.4.5 Multiple-Hypothesis Filtering	125
		3.4.6 Kalman Smoothing	129
3.5	The Particle Filter		131
	References		135

CHAPTER 4
Inertial Sensors — 137

4.1	Accelerometers	139
	4.1.1 Pendulous Accelerometers	140
	4.1.2 Vibrating-Beam Accelerometers	142
4.2	Gyroscopes	142
	4.2.1 Optical Gyroscopes	143
	4.2.2 Vibratory Gyroscopes	146
4.3	Inertial Measurement Units	149
4.4	Error Characteristics	151
	4.4.1 Biases	152
	4.4.2 Scale Factor and Cross-Coupling Errors	154
	4.4.3 Random Noise	155
	4.4.4 Further Error Sources	157
	4.4.5 Vibration-Induced Errors	159
	4.4.6 Error Models	160
	References	161

CHAPTER 5
Inertial Navigation — 163

5.1	Introduction to Inertial Navigation	164
5.2	Inertial-Frame Navigation Equations	168
	5.2.1 Attitude Update	168
	5.2.2 Specific-Force Frame Transformation	170
	5.2.3 Velocity Update	171
	5.2.4 Position Update	172
5.3	Earth-Frame Navigation Equations	172
	5.3.1 Attitude Update	173
	5.3.2 Specific-Force Frame Transformation	174
	5.3.3 Velocity Update	174
	5.3.4 Position Update	175
5.4	Local-Navigation-Frame Navigation Equations	176
	5.4.1 Attitude Update	176
	5.4.2 Specific-Force Frame Transformation	178
	5.4.3 Velocity Update	179
	5.4.4 Position Update	179
	5.4.5 Wander-Azimuth Implementation	180
5.5	Navigation Equations Optimization	183
	5.5.1 Precision Attitude Update	183
	5.5.2 Precision Specific-Force Frame Transformation	187
	5.5.3 Precision Velocity and Position Updates	188

		5.5.4 Effects of Sensor Sampling Interval and Vibration	189
		5.5.5 Design Tradeoffs	195
	5.6	Initialization and Alignment	195
		5.6.1 Position and Velocity Initialization	196
		5.6.2 Attitude Initialization	196
		5.6.3 Fine Alignment	200
	5.7	INS Error Propagation	203
		5.7.1 Short-Term Straight-Line Error Propagation	204
		5.7.2 Medium- and Long-Term Error Propagation	209
		5.7.3 Maneuver-Dependent Errors	212
	5.8	Indexed IMU	214
	5.9	Partial IMU	215
		References	216

CHAPTER 6

Dead Reckoning, Attitude, and Height Measurement — 217

	6.1	Attitude Measurement	217
		6.1.1 Magnetic Heading	218
		6.1.2 Marine Gyrocompass	222
		6.1.3 Strapdown Yaw-Axis Gyro	223
		6.1.4 Heading from Trajectory	225
		6.1.5 Integrated Heading Determination	226
		6.1.6 Accelerometer Leveling and Tilt Sensors	226
		6.1.7 Horizon Sensing	227
		6.1.8 Attitude and Heading Reference System	228
	6.2	Height and Depth Measurement	229
		6.2.1 Barometric Altimeter	230
		6.2.2 Depth Pressure Sensor	231
		6.2.3 Radar Altimeter	232
	6.3	Odometry	233
		6.3.1 Linear Odometry	234
		6.3.2 Differential Odometry	238
		6.3.3 Integrated Odometry and Partial IMU	239
	6.4	Pedestrian Dead Reckoning Using Step Detection	240
	6.5	Doppler Radar and Sonar	245
	6.6	Other Dead-Reckoning Techniques	249
		6.6.1 Correlation-Based Velocity Measurement	249
		6.6.2 Air Data	249
		6.6.3 Ship's Speed Log	250
		References	250

CHAPTER 7

Principles of Radio Positioning — 255

	7.1	Radio Positioning Configurations and Methods	255
		7.1.1 Self-Positioning and Remote Positioning	255
		7.1.2 Relative Positioning	257

		7.1.3	Proximity	258
		7.1.4	Ranging	260
		7.1.5	Angular Positioning	269
		7.1.6	Pattern Matching	271
		7.1.7	Doppler Positioning	274
	7.2	Positioning Signals		276
		7.2.1	Modulation Types	276
		7.2.2	Radio Spectrum	277
	7.3	User Equipment		279
		7.3.1	Architecture	279
		7.3.2	Signal Timing Measurement	280
		7.3.3	Position Determination from Ranging	282
	7.4	Propagation, Error Sources, and Positioning Accuracy		287
		7.4.1	Ionosphere, Troposphere, and Surface Propagation Effects	287
		7.4.2	Attenuation, Reflection, Multipath, and Diffraction	288
		7.4.3	Resolution, Noise, and Tracking Errors	290
		7.4.4	Transmitter Location and Timing Errors	292
		7.4.5	Effect of Signal Geometry	292
		References		297

CHAPTER 8
GNSS: Fundamentals, Signals, and Satellites — 299

8.1	Fundamentals of Satellite Navigation			300
	8.1.1	GNSS Architecture		300
	8.1.2	Signals and Range Measurement		303
	8.1.3	Positioning		307
	8.1.4	Error Sources and Performance Limitations		309
8.2	The Systems			312
	8.2.1	Global Positioning System		312
	8.2.2	GLONASS		313
	8.2.3	Galileo		313
	8.2.4	Beidou		314
	8.2.5	Regional Systems		314
	8.2.6	Augmentation Systems		314
	8.2.7	System Compatibility		316
8.3	GNSS Signals			317
	8.3.1	Signal Types		318
	8.3.2	Global Positioning System		320
	8.3.3	GLONASS		323
	8.3.4	Galileo		324
	8.3.5	Beidou		326
	8.3.6	Regional Systems		326
	8.3.7	Augmentation Systems		327
8.4	Navigation Data Messages			327
	8.4.1	GPS		327
	8.4.2	GLONASS		328

		8.4.3	Galileo	329
		8.4.4	SBAS	329
		8.4.5	Time Base Synchronization	329
	8.5	Satellite Orbits and Geometry		330
		8.5.1	Satellite Orbits	330
		8.5.2	Satellite Position and Velocity	332
		8.5.3	Range, Range Rate, and Line of Sight	339
		8.5.4	Elevation and Azimuth	344
		References		345

CHAPTER 9

GNSS: User Equipment Processing and Errors — 349

9.1	Receiver Hardware and Antenna		350
	9.1.1	Antennas	350
	9.1.2	Reference Oscillator	351
	9.1.3	Receiver Front End	352
	9.1.4	Baseband Signal Processor	355
9.2	Ranging Processor		367
	9.2.1	Acquisition	367
	9.2.2	Code Tracking	372
	9.2.3	Carrier Tracking	377
	9.2.4	Tracking Lock Detection	384
	9.2.5	Navigation-Message Demodulation	385
	9.2.6	Carrier-Power-to-Noise-Density Measurement	386
	9.2.7	Pseudo-Range, Pseudo-Range-Rate, and Carrier-Phase Measurements	387
9.3	Range Error Sources		389
	9.3.1	Ephemeris Prediction and Satellite Clock Errors	390
	9.3.2	Ionosphere and Troposphere Propagation Errors	391
	9.3.3	Tracking Errors	395
	9.3.4	Multipath, Nonline-of-Sight, and Diffraction	401
9.4	Navigation Processor		407
	9.4.1	Single-Epoch Navigation Solution	409
	9.4.2	Filtered Navigation Solution	413
	9.4.3	Signal Geometry and Navigation Solution Accuracy	424
	9.4.4	Position Error Budget	429
	References		431

CHAPTER 10

GNSS: Advanced Techniques — 437

10.1	Differential GNSS		437
	10.1.1	Spatial and Temporal Correlation of GNSS Errors	438
	10.1.2	Local and Regional Area DGNSS	439
	10.1.3	Wide Area DGNSS and Precise Point Positioning	440
	10.1.4	Relative GNSS	441

10.2	Real-Time Kinematic Carrier-Phase Positioning and Attitude Determination		442
	10.2.1	Principles of Accumulated Delta Range Positioning	443
	10.2.2	Single-Epoch Navigation Solution Using Double-Differenced ADR	446
	10.2.3	Geometry-Based Integer Ambiguity Resolution	447
	10.2.4	Multifrequency Integer Ambiguity Resolution	449
	10.2.5	GNSS Attitude Determination	450
10.3	Interference Rejection and Weak Signal Processing		451
	10.3.1	Sources of Interference, Jamming, and Attenuation	452
	10.3.2	Antenna Systems	452
	10.3.3	Receiver Front-End Filtering	453
	10.3.4	Extended Range Tracking	454
	10.3.5	Receiver Sensitivity	455
	10.3.6	Combined Acquisition and Tracking	456
	10.3.7	Vector Tracking	456
10.4	Mitigation of Multipath Interference and Nonline-of-Sight Reception		458
	10.4.1	Antenna-Based Techniques	459
	10.4.2	Receiver-Based Techniques	460
	10.4.3	Navigation-Processor-Based Techniques	461
10.5	Aiding, Assistance, and Orbit Prediction		462
	10.5.1	Acquisition and Velocity Aiding	463
	10.5.2	Assisted GNSS	464
	10.5.3	Orbit Prediction	465
10.6	Shadow Matching		465
	References		467

CHAPTER 11

Long- and Medium-Range Radio Navigation   473

11.1	Aircraft Navigation Systems		473
	11.1.1	Distance Measuring Equipment	474
	11.1.2	Range-Bearing Systems	479
	11.1.3	Nondirectional Beacons	480
	11.1.4	JTIDS/MIDS Relative Navigation	481
	11.1.5	Future Air Navigation Systems	481
11.2	Enhanced Loran		481
	11.2.1	Signals	482
	11.2.2	User Equipment and Positioning	484
	11.2.3	Error Sources	487
	11.2.4	Differential Loran	488
11.3	Phone Positioning		488
	11.3.1	Proximity and Pattern Matching	489
	11.3.2	Ranging	490
11.4	Other Systems		491
	11.4.1	Iridium Positioning	491

11.4.2	Marine Radio Beacons	492
11.4.3	AM Radio Broadcasts	492
11.4.4	FM Radio Broadcasts	493
11.4.5	Digital Television and Radio	493
11.4.6	Generic Radio Positioning	494
References		495

CHAPTER 12

Short-Range Positioning 499

12.1	Pseudolites	499
	12.1.1 In-Band Pseudolites	500
	12.1.2 Locata and Terralite XPS	500
	12.1.3 Indoor Messaging System	501
12.2	Ultrawideband	501
	12.2.1 Modulation Schemes	502
	12.2.2 Signal Timing	503
	12.2.3 Positioning	504
12.3	Short-Range Communications Systems	506
	12.3.1 Wireless Local Area Networks (Wi-Fi)	506
	12.3.2 Wireless Personal Area Networks	507
	12.3.3 Radio Frequency Identification	508
	12.3.4 Bluetooth Low Energy	508
	12.3.5 Dedicated Short-Range Communication	509
12.4	Underwater Acoustic Positioning	509
12.5	Other Positioning Technologies	512
	12.5.1 Radio	512
	12.5.2 Ultrasound	512
	12.5.3 Infrared	512
	12.5.4 Optical	513
	12.5.5 Magnetic	513
	References	513

CHAPTER 13

Environmental Feature Matching 517

13.1	Map Matching	519
	13.1.1 Digital Road Maps	520
	13.1.2 Road Link Identification	521
	13.1.3 Road Positioning	526
	13.1.4 Rail Map Matching	527
	13.1.5 Pedestrian Map Matching	528
13.2	Terrain-Referenced Navigation	530
	13.2.1 Sequential Processing	531
	13.2.2 Batch Processing	532
	13.2.3 Performance	535
	13.2.4 Laser TRN	535

		13.2.5	Sonar TRN	536
		13.2.6	Barometric TRN	537
		13.2.7	Terrain Database Height Aiding	537
	13.3	Image-Based Navigation		538
		13.3.1	Imaging Sensors	539
		13.3.2	Image Feature Comparison	541
		13.3.3	Position Fixing Using Individual Features	543
		13.3.4	Position Fixing by Whole-Image Matching	546
		13.3.5	Visual Odometry	546
		13.3.6	Feature Tracking	548
		13.3.7	Stellar Navigation	548
	13.4	Other Feature-Matching Techniques		550
		13.4.1	Gravity Gradiometry	551
		13.4.2	Magnetic Field Variation	552
		13.4.3	Celestial X-Ray Sources	552
		References		552

CHAPTER 14

INS/GNSS Integration — 559

	14.1	Integration Architectures		560
		14.1.1	Correction of the Inertial Navigation Solution	562
		14.1.2	Loosely Coupled Integration	566
		14.1.3	Tightly Coupled Integration	567
		14.1.4	GNSS Aiding	569
		14.1.5	Deeply Coupled Integration	571
	14.2	System Model and State Selection		573
		14.2.1	State Selection and Observability	574
		14.2.2	INS State Propagation in an Inertial Frame	577
		14.2.3	INS State Propagation in an Earth Frame	582
		14.2.4	INS State Propagation Resolved in a Local Navigation Frame	584
		14.2.5	Additional IMU Error States	589
		14.2.6	INS System Noise	590
		14.2.7	GNSS State Propagation and System Noise	593
		14.2.8	State Initialization	594
	14.3	Measurement Models		596
		14.3.1	Loosely Coupled Integration	598
		14.3.2	Tightly Coupled Integration	602
		14.3.3	Deeply Coupled Integration	606
		14.3.4	Estimation of Attitude and Instrument Errors	614
	14.4	Advanced INS/GNSS Integration		615
		14.4.1	Differential GNSS	615
		14.4.2	Carrier-Phase Positioning	616
		14.4.3	GNSS Attitude	618
		14.4.4	Large Heading Errors	619

		14.4.5	Advanced IMU Error Modeling	621
		14.4.6	Smoothing	622
		References		622

CHAPTER 15

INS Alignment, Zero Updates, and Motion Constraints — 627

- 15.1 Transfer Alignment — 627
  - 15.1.1 Conventional Measurement Matching — 629
  - 15.1.2 Rapid Transfer Alignment — 631
  - 15.1.3 Reference Navigation System — 633
- 15.2 Quasi-Stationary Alignment — 634
  - 15.2.1 Coarse Alignment — 634
  - 15.2.2 Fine Alignment — 637
- 15.3 Zero Updates — 638
  - 15.3.1 Stationary-Condition Detection — 638
  - 15.3.2 Zero Velocity Update — 639
  - 15.3.3 Zero Angular Rate Update — 640
- 15.4 Motion Constraints — 641
  - 15.4.1 Land Vehicle Constraints — 641
  - 15.4.2 Pedestrian Constraints — 643
  - 15.4.3 Ship and Boat Constraint — 644
  - References — 644

CHAPTER 16

Multisensor Integrated Navigation — 647

- 16.1 Integration Architectures — 647
  - 16.1.1 Cascaded Single-Epoch Integration — 648
  - 16.1.2 Centralized Single-Epoch Integration — 651
  - 16.1.3 Cascaded Filtered Integration — 652
  - 16.1.4 Centralized Filtered Integration — 654
  - 16.1.5 Federated Filtered Integration — 655
  - 16.1.6 Hybrid Integration Architectures — 658
  - 16.1.7 Total-State Kalman Filter Employing Prediction — 659
  - 16.1.8 Error-State Kalman Filter — 661
  - 16.1.9 Primary and Reversionary Moding — 663
  - 16.1.10 Context-Adaptive Moding — 665
- 16.2 Dead Reckoning, Attitude, and Height Measurement — 666
  - 16.2.1 Attitude — 667
  - 16.2.2 Height and Depth — 673
  - 16.2.3 Odometry — 674
  - 16.2.4 Pedestrian Dead Reckoning Using Step Detection — 677
  - 16.2.5 Doppler Radar and Sonar — 680
  - 16.2.6 Visual Odometry and Terrain-Referenced Dead Reckoning — 682
- 16.3 Position-Fixing Measurements — 682
  - 16.3.1 Position Measurement Integration — 683
  - 16.3.2 Ranging Measurement Integration — 685

		16.3.3	Angular Measurement Integration	690

	16.3.3	Angular Measurement Integration	690
	16.3.4	Line Fix Integration	694
	16.3.5	Handling Ambiguous Measurements	695
	16.3.6	Feature Tracking and Mapping	697
	16.3.7	Aiding of Position-Fixing Systems	698
	References		699

CHAPTER 17

Fault Detection, Integrity Monitoring, and Testing — 701

17.1	Failure Modes		702
	17.1.1	Inertial Navigation	702
	17.1.2	Dead Reckoning, Attitude, and Height Measurement	702
	17.1.3	GNSS	703
	17.1.4	Terrestrial Radio Navigation	703
	17.1.5	Environmental Feature Matching and Tracking	704
	17.1.6	Integration Algorithm	704
	17.1.7	Context	705
17.2	Range Checks		705
	17.2.1	Sensor Outputs	705
	17.2.2	Navigation Solution	706
	17.2.3	Kalman Filter Estimates	706
17.3	Kalman Filter Measurement Innovations		706
	17.3.1	Innovation Filtering	707
	17.3.2	Innovation Sequence Monitoring	709
	17.3.3	Remedying Biased State Estimates	711
17.4	Direct Consistency Checks		712
	17.4.1	Measurement Consistency Checks and RAIM	713
	17.4.2	Parallel Solutions	715
17.5	Infrastructure-Based Integrity Monitoring		719
17.6	Solution Protection and Performance Requirements		720
17.7	Testing		724
	17.7.1	Field Trials	724
	17.7.2	Recorded Data Testing	725
	17.7.3	Laboratory Testing	725
	17.7.4	Software Simulation	725
	References		726

CHAPTER 18

Applications and Future Trends — 729

18.1	Design and Development	729
18.2	Aviation	731
18.3	Guided Weapons and Small UAVs	733
18.4	Land Vehicle Applications	733
18.5	Rail Navigation	734
18.6	Marine Navigation	735
18.7	Underwater Navigation	737

18.8	Spacecraft Navigation		737
18.9	Pedestrian Navigation		738
18.10	Other Applications		739
18.11	Future Trends		740
	References		741

List of Symbols	743
Acronyms and Abbreviations	751
About the Author	757
DVD Contents	759

Preface

The main aims of this book are as follows:

- To describe, both qualitatively and mathematically, global navigation satellite systems (GNSS), inertial navigation, and many other navigation and positioning technologies, focusing on their principles of operation, their performance characteristics, and how they may be integrated together;
- To provide a clear and accessible introduction to navigation systems suitable for those with no prior knowledge;
- To review the state of the art in navigation and positioning, introducing new ideas, as well as presenting established technology.

This book is aimed at professional engineers and scientists in industry, academia, and government, and at students, mainly at the master's and Ph.D. levels. This book covers navigation of air, land, sea, underwater, and space vehicles, both piloted and autonomous, together with pedestrian navigation. It is also relevant to other positioning applications, including mobile mapping, machine control, and vehicle testing.

This book begins with a basic introduction to the main principles of navigation and a summary of the different technologies. This is followed by a mathematical grounding in coordinate frames, attitude representations, multiframe kinematics, Earth modeling, and Kalman filter-based estimation. The different navigation and positioning technologies are then described. For each topic, the basic principles are explained before going into detail. The book goes beyond GNSS and inertial navigation to describe terrestrial radio navigation, short-range positioning, environmental feature matching, and dead reckoning techniques, such as odometry, pedestrian dead reckoning (PDR), and Doppler radar/sonar. The Global Positioning System (GPS) and the other GNSS systems are described together. The final chapters describe inertial navigation system (INS)/GNSS and multisensory integration; INS alignment, zero updates, and motion constraints; fault detection, integrity monitoring, and testing; and navigation applications.

The emphasis throughout is on providing an understanding of how navigation systems work, rather than on engineering details. This book focuses on the physical principles on which navigation systems are based, how they generate a navigation solution, how they may be combined, the origins of the error sources, and their mitigation. Later chapters build on material covered in earlier chapters, with comprehensive cross-referencing.

The second edition is more than 50% larger than the first, providing the opportunity to devote more space to the underlying principles and explore more topics in detail. Eight chapters are new or substantially rewritten, and the remaining chapters have all been revised and expanded. Subjects covered in more depth include map matching, image-based navigation, attitude determination, deeply coupled INS/GNSS integration, acoustic positioning, PDR, GNSS operation in poor reception environments, and a number of terrestrial and short-range radio positioning techniques, including ultrawideband (UWB) positioning. New topics include the unscented Kalman filter and particle filter, GNSS shadow matching, motion constraints, context, cooperation/collaboration, partial inertial measurement units (IMUs), system design, and testing.

An accompanying DVD has also been introduced. This DVD contains worked examples (in a Microsoft Excel format), problems, and MATLAB software, as well as eleven appendices containing additional material.

Acknowledgments

I would like to thank the team at Artech House and the many people who have given me helpful comments and suggestions for the book. Particular thanks go to those who have commented on drafts of this new edition, including Ramsey Faragher, Simon Julier, Naomi Li, Sherman Lo, Bob Mason, Philip Mattos, Washington Ochieng, Alex Parkins, Andrey Soloviev, Toby Webb, Paul Williams, and Artech House's anonymous reviewer; and to those who commented on the draft of the first edition (listed therein). I would like to thank QinetiQ for letting me reuse material I wrote for the "Principles of Integrated Navigation" course. This is marked by footnotes as QinetiQ copyright and appears in Chapters 2, 3, and 5. Finally, I would like to thank my family, friends, and colleagues for their patience and support.

CHAPTER 1
Introduction

What is meant by "navigation"? What is the difference between "position" and "location"? How do global navigation satellite systems (GNSS), such as the Global Positioning System (GPS), work? What is an inertial navigation system (INS)? This chapter introduces the basic concepts of navigation technology, compares the main technologies, and provides a qualitative overview of the material covered in the body of the book.

Section 1.1 introduces the fundamental concepts of navigation and positioning and defines the scope of the book. Sections 1.2 and 1.3 introduce the different navigation techniques and technologies, covering dead reckoning and position fixing, respectively. Section 1.4 then considers the navigation system as a whole, discussing requirements, context, integration, aiding, assistance, cooperation, and fault detection. Finally, Section 1.5 presents an overview of the rest of the book and the accompanying CD.

1.1 Fundamental Concepts

There is no universally agreed definition of *navigation*. The *Concise Oxford Dictionary* [1] defines navigation as "any of several methods of determining or planning a ship's or aircraft's position and course by geometry, astronomy, radio signals, etc." This encompasses two concepts. The first concept is the determination of the position and velocity of a moving body with respect to a known reference point, sometimes known as the science of navigation. The second concept is the planning and maintenance of a course from one location to another, avoiding obstacles and collisions. This is sometimes known as the art of navigation and may also be known as guidance, pilotage, or routing, depending on the vehicle. A navigation technique is thus a method for determining position and velocity or a course or both. It may be either manual or automatic. This book is concerned only with the science of navigation, the determination of position and velocity, and focuses on automatic techniques.

Positioning is the determination of the position of a body and is thus a subset of navigation. However, navigation is also one of a number of applications of positioning. Others include surveying, mapping, tracking, surveillance, machine control, construction, vehicle testing, Earth sciences, intelligent transportation systems (ITS), and location-based services (LBS).

Positioning techniques may be categorized in three ways. The first way is into real-time and postprocessed techniques. Postprocessed techniques typically determine position hours or days after the measurements are made. However, navigation requires real-time positioning, whereby the position is calculated as soon as possible after making the measurements. Real-time positioning may also be subdivided into continuous positioning, as required for navigation, and instantaneous positioning for applications requiring position at a single point in time.

The second way of classifying positioning is whether the object of interest is fixed or movable. The positioning of fixed objects is known as static positioning, whereas the positioning of movable objects is mobile, dynamic, or kinematic positioning. Navigation thus requires mobile positioning, which may be further divided into techniques that only directly determine a position solution and those that also measure velocity. Velocity is needed for navigation. However, it may be derived from a rapidly updated position or using a different technique, so both types of mobile positioning are relevant.

The final categories are self-positioning and remote positioning. Most navigation applications use self-positioning, whereby the position is calculated at the object whose position is to be determined. In remote positioning, the position is calculated elsewhere and the cooperation of the object tracked is not necessarily required, which is useful for covert surveillance. However, for navigation, a communication link is needed to send the position and velocity and/or guidance instructions to the moving body. Examples include the radar surveillance systems used by air traffic control and vessel traffic services.

Figure 1.1 summarizes the different positioning categories and some of their applications. This book focuses on continuous real-time mobile self-positioning. However, much of the information presented is also relevant to the other classes of positioning.

A *navigation system*, sometimes known as a navigation aid, is a device that determines position and velocity automatically. Similarly, a positioning system determines position. An *integrated navigation system* determines position and velocity using more than one technology. This may also be called a *hybridized positioning system*.

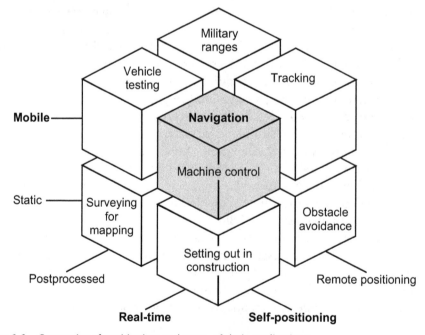

Figure 1.1 Categories of positioning and some of their applications.

1.1 Fundamental Concepts

A *navigation sensor* is a device used to measure a property from which the navigation system computes its outputs; examples include accelerometers, gyroscopes, and radio navigation receivers.

The output of a navigation system or technique is known as the *navigation solution*. It comprises the position and velocity of the navigating object. Some navigation systems also provide some or all of the attitude (including heading), acceleration, and angular rate. Similarly, the position solution is just the position of the object. For navigation of cars, trains, ships, and outdoor pedestrians, the vertical component of position and velocity is not required, enabling two-dimensional positioning techniques that only operate in the horizontal plane to be used. Other applications, such as air, space, underwater, and indoor pedestrian navigation, require three-dimensional positioning.

For navigation, it is assumed that the *user*, which may be a person or computer software (e.g., route guidance) is part of the object to be positioned. Thus, the user's navigation solution is the same as that of the object. The parts of the navigation system located on the object to be positioned (sometimes the entire system) are known as *user equipment*.

The terms position and *location* are nominally interchangeable, but are normally used to denote two different concepts. Thus, position is expressed quantitatively as a set of numerical coordinates, whereas location is expressed qualitatively, such as a city, street, building, or room. A navigation system will calculate a position, whereas a person, signpost, or address will describe a location. A map or geographic information system (GIS) matches locations to positions, so it is a useful tool for converting between the two. Figure 1.2 illustrates this.

Some authors use the term *localization* instead of positioning, particularly for short-range applications. The two are essentially interchangeable, although "localization" is also used to describe techniques that constrain the position solution to a particular area, such as a street or room, instead of determining coordinates.

All navigation and positioning techniques are based on one of two fundamental methods: position fixing and dead reckoning. Position fixing uses identifiable external

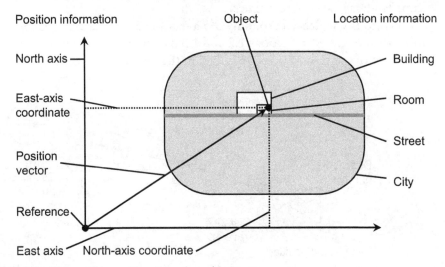

Figure 1.2 The position and location of an object.

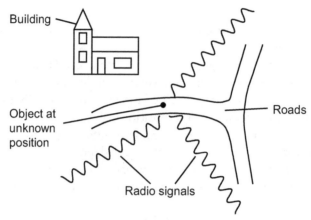

Figure 1.3 Examples of information available for position fixing.

information to determine position directly. This may be signals or environmental features; Figure 1.3 illustrates some examples. Signals are usually transmitted by radio (e.g., GNSS), but may also be acoustic, ultrasound, optical, or infrared. Environmental features include buildings or parts thereof, signs, roads, rivers, terrain height, sounds, smells, and even variations in the magnetic and gravitational fields.

Position may be inferred directly by matching the signals receivable and/or features observable at a given location with a database. Alternatively, more distant landmarks at known positions may be selected and their distance and/or direction from the user measured. A landmark may be a transmitter (or receiver) of signals or an environmental feature. A landmark installed specifically for navigation is known as an *aid to navigation* (AtoN).

Dead reckoning measures the distance and direction traveled. Therefore, if the initial position is known, the current position may be determined as shown in Figure 1.4. A dead-reckoning system, such as an INS, may be self-contained aboard the navigating vehicle, requiring no external infrastructure. However, environmental features may also be used for dead reckoning by comparing measurements of the same landmark at different times.

Figure 1.5 depicts a taxonomy of navigation and positioning, showing how the methods introduced in Sections 1.2 and 1.3 may be classified. The figure also includes examples of the technologies described in the rest of the book.

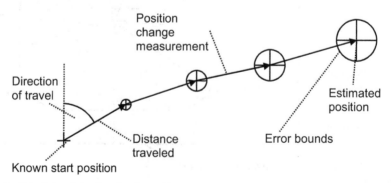

Figure 1.4 Principle of dead reckoning.

1.2 Dead Reckoning

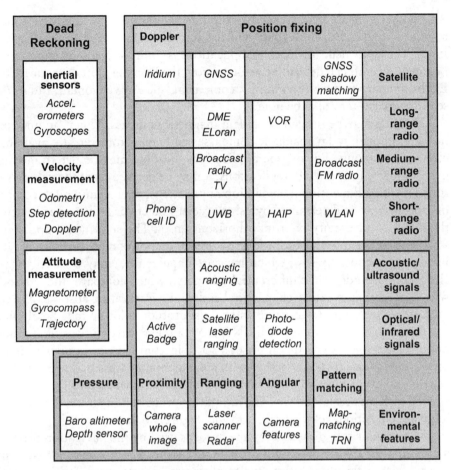

Figure 1.5 A taxonomy of navigation and positioning technology.

1.2 Dead Reckoning

Dead reckoning (possibly derived from "deduced reckoning") either measures the change in position or measures the velocity and integrates it. This is added to the previous position in order to obtain the current position as shown in Figure 1.4. The speed or distance traveled is measured in body-aligned axes, so a separate attitude solution is required to obtain the direction of travel with respect to the environment. For two-dimensional navigation, a heading measurement is sufficient, whereas for three-dimensional navigation, a full three-component attitude measurement is needed. Where the attitude is changing, the smaller the step size in the position calculation, the more accurate the navigation solution will be. The calculations were originally performed manually, severely limiting the data rate, but are now done by computer.

Traditional distance and velocity measurement methods include counting paces, using a pacing stick, and spooling a knotted rope off the back of a ship—hence, the use of the knot as a unit of speed by the maritime community. Today, pace counting can be automated using a pedometer, while more sophisticated pedestrian dead reckoning (PDR) techniques using accelerometers also determine the step length.

An odometer measures distance by counting the rotations of a wheel. Today, it is standard equipment on all road vehicles, but the technique dates back to Roman times. The equivalent for marine applications is a ship's electromagnetic speed log or sonar. Aircraft can determine velocity from the Doppler shift of radar reflections. Environmental feature tracking by comparing successive camera, radar, or laser scanner images may also be used.

Heading may be measured using a magnetic compass. This is an ancient technology, although today magnetic compasses and magnetometers are available with electronic readouts. For marine applications, heading may be determined using a gyrocompass, and for land applications, it may be derived from the vehicle's trajectory. For three-dimensional navigation applications, the roll and pitch components of attitude may be determined by using accelerometers or a tilt sensor to determine the direction of gravity or from a horizon sensor. The sun, moon, and stars may also be used to determine attitude if the time and approximate position are known. Finally, gyroscopes (gyros), which measure angular rate, may be used to measure changes in attitude, while differential odometry, which compares the left and right wheel speeds, can measure changes in heading. By integrating absolute and relative attitude sensors, a more accurate and robust attitude solution may be obtained.

An inertial navigation system (INS) is a complete three-dimensional dead-reckoning navigation system. It comprises a set of inertial sensors, known as an *inertial measurement unit* (IMU), together with a navigation processor. The inertial sensors usually comprise three mutually orthogonal accelerometers and three gyroscopes aligned with the accelerometers. The navigation processor integrates the IMU outputs to give the position, velocity, and attitude. Figure 1.6 illustrates this.

The angular rate measured by the gyros is used by the navigation processor to maintain the INS's attitude solution. The accelerometers, however, measure specific force, which is the acceleration due to all forces except for gravity. Thus, the measurements produced by stationary accelerometers comprise the reaction to gravity. In a strapdown INS, the accelerometers are aligned with the navigating body, so the attitude solution is used to transform the specific force measurement into the resolving axes used by the navigation processor. A gravity model is then used to obtain the acceleration from the specific force using the position solution. Integrating the acceleration produces the velocity solution and integrating the velocity gives

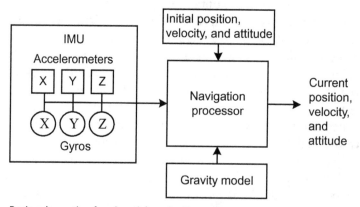

Figure 1.6 Basic schematic of an inertial navigation system.

the position solution. The position, velocity, and attitude must be initialized before a navigation solution can be computed.

Overall navigation performance can vary by several orders of magnitude, depending on the quality of the inertial sensors. Inertial sensors are available for a few dollars or euros, but are not accurate enough for navigation. INSs that exhibit a horizontal position error drift of less than 1,500m in the first hour cost around $100,000 (€80,000) each and are used in military aircraft and commercial airliners. Intermediate quality sensors are suitable for use as part of an integrated navigation system.

The principal advantages of inertial navigation and other dead-reckoning techniques, compared to position fixing, are continuous operation, a high update rate, low short-term noise, and the provision of attitude, angular rate, and acceleration as well as position and velocity. The main drawbacks are that the position solution must be initialized and the position error grows with time because the errors in successive distance and direction measurements accumulate. In an integrated navigation system, position-fixing measurements may be used to correct the dead-reckoning navigation solution and also calibrate the dead-reckoning sensor errors.

1.3 Position Fixing

This section describes and compares the main position-fixing methods and then summarizes the main signal-based positioning systems and environmental feature-matching techniques.

1.3.1 Position-Fixing Methods

There are five main position-fixing methods: proximity, ranging, angular positioning, pattern matching, and Doppler positioning [2]. Each is described in turn, followed by a discussion of the common issues.

Basic proximity is the simplest method. If a radio signal is received, the receiver position is taken to be the transmitter position. Similarly, if a nearby environmental feature, such as a building, is identified, the position is assumed to be that of the feature. Thus, the closer the user is to the landmark, the more accurate proximity positioning is. Very short-range radio signals, such as Bluetooth and radio frequency identification (RFID), and indoor features are thus suited to it. If multiple landmarks are used, an average of their positions may be taken.

A more advanced version of proximity positioning is containment intersection. It uses the same Boolean measurements: a landmark is either observed or not observed. However, a containment zone is defined for each landmark, representing the area within which a radio signal may be received or an environmental feature observed. If a landmark is observed, the position is localized to that landmark's containment zone. With multiple landmarks, the position is localized to the intersection of the containment zones and the center of this intersection may be taken as the position fix. Figure 1.7 illustrates this.

Figure 1.8 shows how a position fix may be obtained in two dimensions using ranging. Each measurement defines a circular line of position (LOP) of radius equal to the measured range between the user and a landmark and centered at the known

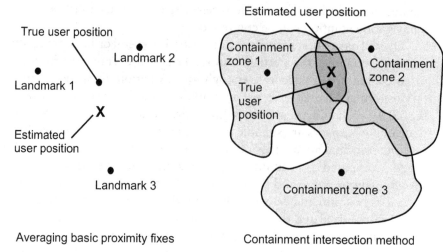

Figure 1.7 Basic and advanced proximity positioning using multiple landmarks.

position of that landmark. The user may be located anywhere on that LOP. Where two LOPs are available, the user position lies where they intersect. However, a pair of circles intersects at two points, only one of which is the correct position. Often, prior information can be used to determine which one this is. Otherwise, a third range measurement is required.

In three dimensions, each range measurement defines a spherical surface of position (SOP), centered at the landmark. Two of these SOPs intersect to form a circular LOP, while three spherical SOPs intersect at two points. Thus, three or four range measurements are required to obtain a unique position fix, depending on what additional information is available. However, if the user and all of the landmarks are within the same plane, it is only possible to determine the components of position within that plane, not in the perpendicular direction. Consequently, it is difficult to obtain vertical position from a long- or medium-range terrestrial ranging system.

The range can be determined from a signal transmitted from a landmark to the user equipment and/or vice versa by measuring the time of flight (TOF) of the signal and multiplying it by the speed of light or sound (as appropriate). Accurate TOF measurement requires time synchronization of the transmitter and receiver. In a two-way ranging system, such as distance measuring equipment (DME), signals are transmitted in both directions, cancelling out most of the time synchronization error.

For one-way ranging between transmitters at known locations and a receiver at an unknown location, the transmitter clocks are synchronized with each other and the receiver clock offset is treated as an additional unknown in the position solution. Determining this additional unknown parameter requires a ranging measurement from an additional transmitter. This technique is known as passive ranging and is how GNSS works. Alternatively, a reference receiver at a known position may be used to measure the transmitter clock offsets.

Where the landmark is an environmental feature, the range must be measured using an active sensor. This transmits a modulated signal to the landmark, where it is reflected, and then measures the round-trip time of the returned signal. Radar, sonar, or laser ranging is typically used.

1.3 Position Fixing

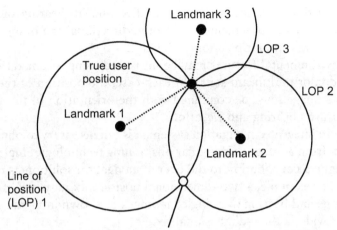

Figure 1.8 Positioning by ranging in two dimensions.

A bearing is the angle within the horizontal plane between the line of sight to an object and a known direction, usually true or magnetic north. When the direction of north is known, a two-dimensional position fix may be obtained by measuring the bearing to two landmarks at known positions, as shown in Figure 1.9. Each measurement defines a straight LOP in the direction of the bearing which passes through the landmark. The two LOPs intersect at a single point which is the user position. When there is no reference direction, a curved LOP may be determined by measuring the difference in directions of two landmarks, in which case, three landmarks are required for a two-dimensional position fix.

Angular positioning may be extended to three dimensions by measuring the elevation angle to one of the landmarks, where the elevation is the angle between the line of sight to the object and a horizontal plane. For a given angular measurement accuracy, the accuracy of the position fix will degrade with distance from the landmarks. Note also that, because of the curvature of the Earth's surface, bearings and elevations measured at the landmark and at the user will not be equal and opposite.

The angle of arrival (AOA) of a radio signal may be determined either by direction finding or from nonisotropic transmissions. In direction finding, a directional antenna system with a steerable reception pattern is used to measure the bearing at the receiver. Any signal may be used; it does not have to be designed for positioning. A nonisotropic transmission comprises a signal broadcast whose modulation varies

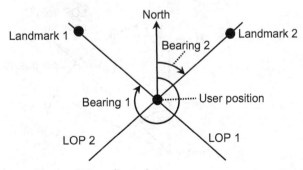

Figure 1.9 Angular positioning in two dimensions.

with direction, enabling the receiver to determine its bearing and/or elevation at the transmitter. Examples include VHF omnidirectional radiorange (VOR) and Nokia high-accuracy indoor positioning (HAIP).

Environmental features may be measured using a camera, laser scanner, imaging radar, or multibeam sonar. In each case, the position of the feature within the sensor's image must be combined with the orientation of the sensor to determine the feature's bearing and elevation.

In an integrated navigation system, it is not necessary to obtain a complete position fix from a ranging or angular positioning technology. Single measurements can still make a contribution to the overall navigation solution as discussed in Section 1.4.3. For example, a two-dimensional position fix may be obtained by measuring the range and bearing of a single landmark as shown in Figure 1.10. Adding elevation provides a three-dimensional fix.

Positioning using landmarks requires them to be identified. A signal can normally be identified by demodulating it. Digital signals usually include a transmitter identification, while analog signals can be identified using the frequency and/or repetition rate. An environmental feature must be identified by comparing an image of it with stored information. Enough detail must be captured to uniquely identify it. In practice, this usually requires careful selection of features with unique characteristics and the input of an approximate position solution to limit the size of database that must be searched to obtain a match. Even so, positioning using environmental features is normally more processor intensive than signal-based positioning.

In pattern matching, a database is maintained of measurable parameters that vary with position. Examples include the terrain height, received signal strengths from multiple wireless local area network (WLAN) access points, the environmental magnetic field, and the determination of which GNSS signals are obstructed by buildings. Values measured at the current unknown user position are compared with stored values at a series of candidate positions, typically arranged in a grid pattern. Whichever candidate position gives the best match is then the position solution. If several neighboring candidates give good matches, the position can be determined by interpolation. As with feature matching for landmark identification, the input of an approximate position solution limits the size of the database to be searched. To improve the chances of obtaining a unique position solution from pattern matching, multiple measured parameters may be combined into a *location signature* and matched with the database together.

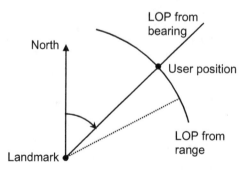

Figure 1.10 Two-dimensional positioning from the range and bearing of a single landmark.

1.3 Position Fixing

In some cases, such as terrain height, there is insufficient information to obtain an unambiguous position fix from measurements made at one location. However, if the navigation system is moving, measurements may be made at multiple positions, collectively known as a *transect*. Using dead reckoning to determine the relative positions of the transect points enables the measurements to be combined into a location signature, which is then compared with the database. Figure 1.11 illustrates this. Note that the transect and database point spacing will generally be different, requiring interpolation to match them.

The final position-fixing method is Doppler positioning, which requires relative motion between the transmitter and receiver of a signal. By measuring the signal's Doppler shift, the component of relative velocity along the line of sight is obtained, from which an approximately conical surface of position may be determined. This is used for Iridium positioning.

Height can be computed from pressure measurements using a barometric altimeter (baro). A pressure sensor may also be used to measure depth underwater. A radar altimeter (radalt) measures the height above the terrain, so it can be used to determine an aircraft's height where the terrain height is known.

All position-fixing methods require data, such as the position of landmarks, feature identification information, and pattern-matching data. This data may be preloaded into the user equipment. However, it then needs to be kept up-to-date, while a lot of data storage may be required to navigate over a large area. Some databases, particularly in older systems, only cover the host vehicle's planned route or a series of positions along that route known as waypoints.

A navigation system may also build its own landmark database using a technique known as simultaneous localization and mapping (SLAM), whereby it explores the environment, observing features several times and using dead reckoning to measure the distance traveled. New signals and environmental features may be added to an existing database using the same approach.

Many signal-based self-positioning systems include the transmitter positions in the signals transmitted. However, this can introduce a delay between first receiving a signal and computing a position from it. A separate data link may be used to provide the necessary information on demand; this is known as *assistance* and is discussed in Section 1.4.5.

Position fixing is essential for determining absolute position and the errors are independent of the distance traveled. However, it relies on the availability of suitable

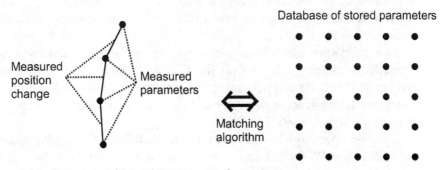

Figure 1.11 Pattern matching using a transect of measurements.

signals or environmental features. Without them, it does not work. The availability of a position solution is boosted by combining multiple position-fixing technologies and/or using dead reckoning to bridge gaps.

1.3.2 Signal-Based Positioning

Radio was first used for navigation in the 1920s. Low- and medium-frequency transmitters were used for direction finding while 75-MHz marker beacons were used to delineate airways using simple proximity. The first truly global radio navigation system, with worldwide coverage and continuous availability, albeit a relatively poor accuracy, was Omega. This achieved global coverage in the early 1970s and operated until 1997.

Today's terrestrial radio positioning systems fall into two main categories. The first comprises the survivors of a generation of radio navigation systems that were developed largely in the 1940s and 1950s and continued to evolve into the 1980s. These include DME and VOR, used for aircraft navigation; various beacons used for direction finding; and, in some countries, an updated version of the Loran (Long-range navigation) system, used for marine navigation. These systems were developed specifically for navigation and are long-range with transmitter coverage radii of hundreds of kilometers (up to 3,000 km for Loran).

The second category comprises techniques developed in the 1990s and 2000s to exploit existing communications and broadcasting signals for positioning purposes. Mobile phone signals, WLANs or Wi-Fi, wireless personal area networks (WPANs), such as Bluetooth and Zigbee, RFID, ultrawideband (UWB) communications, television signals, and broadcast radio are all used. UWB systems designed specifically for positioning have also been developed. Although broadcast signals can typically be received up to 100 km away and some mobile phone signals up to 35 km away, most of these technologies are short-range with coverage radii of tens of meters. Only some of these positioning techniques require the cooperation of the network operator. Signals that are used for positioning without the cooperation of the operator are known as signals of opportunity (SOOP, SOP, or SOO). Ranging using SOOP requires determination of the transmitter timing. When the transmitter clock is stable and the transmission pattern is regular, transmitter timing may be determined using a calibration process. Otherwise, a reference station at a known location must be used.

Terrestrial position fixes may also be obtained using other types of signal. Acoustic signals are used for underwater ranging over a few kilometers. Ultrasound, infrared, and optical signals may be used for short-range positioning, typically within a single room.

The world's first satellite navigation system was the U.S. Transit System, designed primarily for shipping. The first satellite was launched in 1961 and the system operated until 1996. Doppler positioning was used to obtain a two-dimensional position fix every 1–2 hours, accurate to about 25m (for a single fix). Russia implemented a similar system, known as Tsikada.

The first operational prototype satellite of the U.S. Global Positioning System was launched in 1978 and initial operational capability (IOC) of the full system was

1.3 Position Fixing

declared in 1993. Global'naya Navigatsionnaya Sputnikovaya Sistema (GLONASS) is operated by Russia and was developed in parallel to GPS. A third satellite navigation system, Galileo, is under development by the European Union and other partners with IOC planned for 2016. In addition, regional systems are being deployed by China, India, and Japan, with the Chinese Beidou system being expanded to provide global coverage by 2020. These systems, collectively known as global navigation satellite systems, operate under the same principle.

Each global GNSS constellation is designed to incorporate 24 or more satellites. This ensures that signals from at least four satellites are available at any location, the minimum required for the user equipment to derive a three-dimensional position fix and calibrate its clock offset by passive ranging. Figure 1.12 illustrates the basic concept. In practice, there are usually more satellites in view from a given constellation, although many receivers use signals from multiple constellations. This enables the position accuracy to be improved and faults to be identified by comparing measurements.

GNSS offers a basic positioning accuracy of a few meters. Differential techniques can improve this by making use of base stations at known locations to calibrate some of the errors. Carrier-phase positioning techniques can give centimeter accuracy for real-time navigation and can also be used to measure attitude. However, they are much more sensitive to interference, signal interruptions, and satellite geometry than basic positioning.

GNSS provides three-dimensional positioning, whereas most terrestrial technologies are limited to horizontal positioning because of their signal geometry. GNSS also provides higher accuracy than the terrestrial systems, except for UWB, and is the only current position fixing technology to offer global coverage. However, GNSS signals are weak and thus vulnerable to incidental interference, deliberate jamming, and attenuation by obstacles such as buildings, foliage, and mountains. Long-range terrestrial systems, such as DME and enhanced Loran, provide a backup to GNSS for safety-critical and mission-critical applications, while the short-range systems provide coverage of indoor and dense urban environments that GNSS signals struggle to penetrate. Thus, by making use of more than one type of signal for positioning, the availability and robustness of the navigation solution is maximized. Position may be determined using a combination of different types of signals without computing a

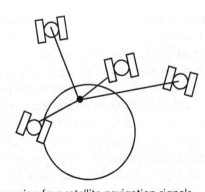

Figure 1.12 Passive ranging using four satellite navigation signals.

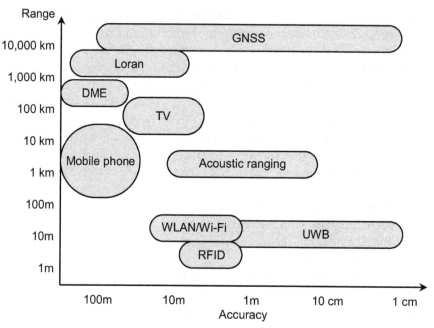

Figure 1.13 Range and accuracy of signal-based positioning technologies.

separate position solution from each technology. Figure 1.13 summarizes the ranges and accuracies of the different positioning technologies.

1.3.3 Environmental Feature Matching

Humans and other animals naturally navigate using environmental features. These features may be compared with maps, pictures, written directions, or memory in order to determine position. Features must either be static or move in a predictable way. Historically, LOPs were obtained from manually-identified distant terrestrial landmarks by angular positioning using a theodolite and magnetic compass. Image-based positioning techniques have now been developed that automate this process, while using a stereo camera, radar, laser scanner, or sonar also enables ranging to be performed. Pattern-matching techniques can also directly infer the user position from an image.

The sun, the moon, and the stars can also be used as landmarks. For example, the highest elevation angle of the sun above the horizon at an equinox is equal to the latitude (a measure of the north-south position). More generally, the position may be calculated from the elevations of two or more stars, together with the time at a known location. These elevations were historically measured using a sextant. Today, a star imager automates the whole process. Accurate time is needed to determine the longitude (east-west position). This has been practical on transoceanic voyages since the 1760s, following major advances in timing technology by John Harrison [3].

Terrain-referenced navigation (TRN) determines the user position from the height of the terrain below. Figure 1.14 illustrates this for different types of host vehicle. For an aircraft, a radalt or laser scanner is used to measure the height above terrain, which is differenced with the vehicle height from the navigation solution to obtain

1.4 The Navigation System

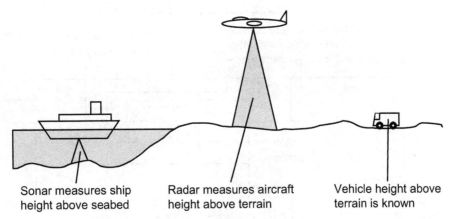

Figure 1.14 The concept of terrain-referenced navigation.

the terrain height. A ship or submarine uses sonar to measure the depth of the terrain below the vessel, while a land vehicle may infer terrain height directly from its own height solution. In each case, a series of measurements is compared with a terrain height database using pattern matching to determine the host vehicle position. Radalt-based techniques for aircraft navigation have been developed since the 1950s and are accurate to about 50m. TRN works best over hilly and mountainous terrain and will not give position fixes over flat terrain.

Map-matching techniques use the fact that land vehicles generally travel on roads or rails and pedestrians do not walk through walls to constrain the drift of a dead-reckoning solution and/or correct errors in a position-fixing measurement. They follow the navigation solution on a map and apply corrections where it strays outside the permitted areas. Map matching is a key component of car navigation and combines aspects of both the proximity and pattern-matching positioning methods. Maps can also be used to infer height from a horizontal position solution.

Other environmental features that may be used for position fixing include anomalies in the Earth's magnetic or gravity field and pulsars. Position may also be determined by using a heterogeneous mix of different types of feature.

All position-fixing techniques that use environmental features rely on pattern matching, either directly or for identifying landmarks. Pattern matching occasionally produces false matches, resulting in erroneous or ambiguous position fixes. Therefore, fault detection and recovery techniques should be always be implemented.

1.4 The Navigation System

The requirements that a navigation system must meet will vary between applications. Its operating context should inform the system design and may contribute additional information to the navigation solution. When multiple positioning technologies are used, their outputs should be combined to produce an optimal integrated navigation solution and one technology may be used to aid another. A communications link can be used to provide additional information to assist the navigation system, while direct communication between navigation systems at different locations enables them to cooperate (or collaborate). Finally, the provision of a reliable navigation solution

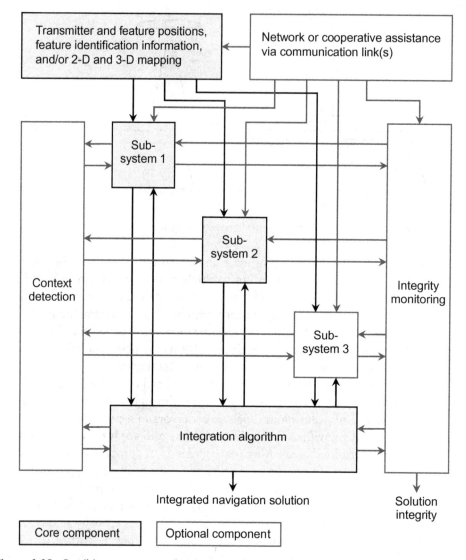

Figure 1.15 Possible components of an integrated navigation system.

requires faults to be detected and corrected where possible. This section introduces and discusses each topic in turn. Figure 1.15 shows how these different functions interact within an integrated navigation system.

1.4.1 Requirements

Different navigation applications have very different requirements in terms of accuracy, update rate, reliability, budget, size, and mass, and whether an attitude solution is required as well as position and velocity. For example, high-value, safety-critical assets, such as airliners and ships, require a guarantee that the navigation solution is always within the error bounds indicated, known as integrity, and require a high level of solution availability. However, the accuracy requirements are relatively modest and there is a large budget. For military applications, a degree of risk is accepted, but the navigation system must be stealthy and able to operate in an electronic warfare environment; the accuracy requirements vary. For personal navigation and road vehicle applications, the key drivers are typically cost, size, weight, and power consumption. Consequently, different combinations of navigation sensors are suited for different applications.

Different requirements lead to different positioning philosophies. For high-value applications, a system is designed to meet a specific set of requirements and the user equipment and infrastructure supplied accordingly. For lower-value applications, a philosophy of making the best use of whatever information happens to be available is often adopted. Thus, the user equipment often comprises sensors and radios that were originally introduced for other purposes and positioning is based on whatever motion, signals, and environmental features they detect. Performance then tends to be dependent on the context.

1.4.2 Context

Context is the environment in which a navigation system operates and the behavior of its host vehicle or user. This can contribute additional information to the navigation solution and is best illustrated with some examples. Land vehicles remain close to the terrain, while ships and boats remain on the water, so one dimension of the position solution is essentially known. The facts that cars drive on roads, trains travel on rails, and pedestrians do not walk through walls may be used to constrain the position solution.

Every vehicle or person has a maximum speed, acceleration, and angular rate, which varies with direction (e.g., the forward component of velocity is normally the largest). There are also relationships between speed and maximum turn rate. This can be used by a navigation system to optimally weight new and older measurements to minimize noise while remaining responsive to dynamics. The vertical and transverse motion constraints imposed by traveling on wheels can be used to reduce the number of sensors required for dead reckoning or constrain the error growth of an INS, while PDR depends inherently on the characteristics of human walking.

The environment is also important. In indoor, urban, and open environments, different radio signals are available and their error characteristics vary. Pedestrian and vehicle behavior also changes. A car typically travels more slowly, stops more, and turns more in an urban environment compared to an open environment. Different radio signals and environmental features are available for aircraft navigation, depending on the aircraft's height and whether it is traveling over land or sea. Finally, most radio signals do not propagate underwater.

A navigation system design should therefore be matched to its context. However, the context can change, particularly for devices, such as smartphones, which move between indoor and outdoor environments and can be stationary, on a pedestrian, or in a vehicle. For best performance, a navigation system should therefore be able to detect its operating context and adapt accordingly; this is *context-adaptive* or *cognitive* positioning.

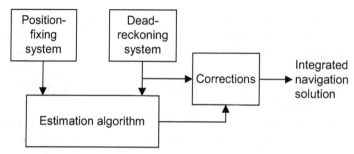

Figure 1.16 A typical position-fixing and dead-reckoning integration architecture.

1.4.3 Integration

An integrated navigation system comprises two or more subsystems based on different navigation technologies. These may be just position-fixing technologies or a mixture of position fixing and dead reckoning. The integration of position-fixing systems is more robust where the individual measurements from each system, such as ranges and bearing, are input to a common position estimation algorithm. This is known as *measurement-domain integration* and has the advantage that a position-fixing system can contribute to the integrated navigation solution even when it has insufficient information to calculate an independent position solution. It is also easier to characterize the measurement errors, ensuring optimal weighting within the navigation solution. The alternative *position-domain integration* inputs the position solutions from the different systems.

Figure 1.16 shows a typical architecture for integrating position-fixing and dead-reckoning systems, such as GNSS and INS. This exploits their very different error characteristics by using the dead-reckoning system to provide the integrated navigation solution as it operates continuously. The measurements from the position-fixing system are then used by an estimation algorithm, usually based on the Kalman filter, to apply corrections to the dead-reckoning system's navigation solution and calibrate its sensor errors.

1.4.4 Aiding

There are a number of ways in which a position-fixing or dead-reckoning system may be aided using either the integrated navigation solution or another positioning technology. Dead reckoning requires an initialization of its position and velocity solution and may also require attitude initialization. Navigation solution corrections, estimated by the integration algorithm, may be fed back at regular intervals. Where the sensor errors are estimated by the integration algorithm, these may also be fed back to the dead-reckoning system and used to correct the sensor outputs.

Any position-fixing system that uses pattern matching, either for position determination or to identify environmental features used as landmarks, requires an approximate position solution to be input in order to limit its search area. This can also help a signal-based position-fixing system to search for signals. Thus, a position fix may be tiered, with one technology used to obtain a coarse position and another used to provide a more precise position using position aiding from the first system.

Transect-based pattern-matching techniques, such as TRN, require a velocity solution in order to combine parameters measured at different positions into a single location signature that may be matched with the database. Velocity aiding can also be used to help increase the sensitivity of radio positioning systems and compensate for the effects of vehicle motion in two-way ranging systems.

1.4.5 Assistance and Cooperation

Assistance is the use of a separate communications link to provide the navigation system with information about the signals and environmental features available for positioning. This can include the positions of transmitters and other landmarks, signal characteristics, such as frequencies and modulation information, feature identification information, and pattern-matching data.

As an alternative to storing the relevant data within the navigation system, assistance can provide more up-to-date information and reduce the system's data storage requirements as information is then only required for the current location and surrounding area. As an alternative to downloading the information from the positioning systems themselves, assistance can enable a position fix to be obtained more quickly or when the reception of the positioning signals is poor.

Assistance data may be provided by a commercial service provider, such as a mobile phone operator or road traffic information service. This is known as *network assistance* and incurs a subscription charge, although the positioning data is typically included with the main service.

Alternatively, nearby users may exchange assistance information directly over a short-range communications link. This is an example of *cooperative positioning*, also known as *collaborative* or *peer-to-peer positioning*. Participants in a cooperative positioning system may be a group of military, security, or emergency service personnel, a fleet of vehicles of any type, or even members of the public. An individual's smartphone could also cooperate with his or her car, or receive information from a train, ferry, or aircraft.

Cooperative positioning is not limited to the exchange of data obtainable from positioning signals or service providers. Participants may also synchronize their clocks and exchange information that they have gathered themselves. Examples include:

- Availability and quality of signals and/or environmental features;
- Transmitter clock offset and position information for signals of opportunity;
- Positions of environmental features and associated identification information;
- Terrain height;
- Calibration parameters for barometric height.

Cooperative positioning can also incorporate relative positioning, whereby participants measure their relative positions using proximity, ranging, and/or angular positioning. This enables participants to make use of signals and features that they cannot observe directly and is particularly useful where there is insufficient

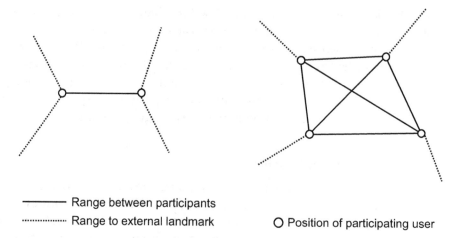

Figure 1.17 Two-dimensional cooperative positioning incorporating relative range measurements.

information available to determine a stand-alone position solution. Figure 1.17 shows some examples.

1.4.6 Fault Detection

To guarantee a reliable navigation solution, it is necessary to detect any faults that may occur, whether they lie within the user equipment hardware, software, or database, or in external components, such as radio signals. This is known as *integrity monitoring* and can be provided at various levels. *Fault detection* simply informs the user that a fault is present; *fault isolation* identifies where the fault has occurred and produces a new navigation solution without data from the faulty component; and *fault exclusion* additionally verifies that the new navigation solution is fault-free. User-based integrity monitoring can potentially detect a fault from any source, provided there is sufficient redundant information (i.e., more than the minimum number of measurements needed to determine position). However, faults in radio navigation signals can be more effectively detected by base stations at known locations, with alerts then transmitted to the navigation system user; this is known as infrastructure-based integrity monitoring. For safety-critical applications, such as civil aviation, the integrity monitoring system must be formally certified to ensure it meets a number of performance requirements.

1.5 Overview of the Book

This section briefly summarizes the contents of the remaining 17 chapters and the accompanying CD, and then discusses some of the conventions used.

Chapters 2 and 3 provide the mathematical grounding for the book. Chapter 2 introduces coordinate frames, attitude representations, multiframe kinematics, Earth modeling, and frame transformations. Chapter 3 describes Kalman filter-based estimation, the core statistical tool used to maintain an optimal real-time navigation solution from position-fixing and dead-reckoning systems.

Chapters 4 to 13 describe the individual navigation technologies, beginning with dead reckoning. Chapter 4 describes the principles and properties of inertial sensors, and Chapter 5 shows how they may be used to obtain an inertial navigation solution. Chapter 6 describes a range of dead-reckoning, attitude, and height measurement technologies, including compasses, altimeters, odometry, PDR using step detection, Doppler radar, and sonar.

Chapters 7 to 12 describe radio positioning. Chapter 7 introduces the main principles, including configurations and methods, signals, user equipment, signal propagation, and error sources. Chapters 8 to 10 are devoted to GNSS, beginning with the fundamentals, systems, signals, satellite orbits, and geometry. The antenna, receiver hardware, ranging processor, error sources, and navigation processor are then described, followed by a review of advanced techniques for enhancing accuracy and robustness. Chapter 11 describes long- and medium-range radio navigation systems, including DME, enhanced Loran, and mobile phone positioning. Chapter 12 describes short-range radio positioning technologies, including pseudolites, UWB, and WLAN, together with acoustic positioning.

Chapter 13 describes position-fixing and dead-reckoning techniques based on environmental feature matching, including map matching, terrain-referenced navigation, and image-based navigation.

Chapters 14 to 16 describe integrated navigation. Chapter 14 focuses on INS/GNSS integration, covering the loosely coupled, tightly coupled, and deeply coupled integration architectures. Chapter 15 describes INS alignment, the application of zero updates when the system is stationary, and context-dependent motion constraints. Chapter 16 then covers multisensor integrated navigation, reviewing the different architectures and describing the integration of dead-reckoning, attitude, height, and position-fixing measurements.

Chapter 17 describes fault detection and integrity monitoring, including a summary of common failure modes, a review of the different methods of fault detection, and a discussion of integrity certification. Navigation system testing is also discussed. Finally, Chapter 18 discusses how the technology described in the preceding chapters may be deployed to meet the requirements of a wide range of navigation applications and discusses future trends. Lists of key symbols and acronyms complete the book.

The accompanying CD includes appendices, worked examples, problems and exercises, and some MATLAB INS/GNSS simulation software. Appendices A and B provide background material on vectors, matrices, statistics, probability, and random processes. Appendices C to I provide additional topics on the Earth, state estimation, inertial navigation, GNSS and other radio positioning techniques, environmental feature matching, INS/GNSS integration, and multisensor integration. Appendix J discusses the software simulation of all types of navigation systems, with an emphasis on GNSS and inertial navigation. Finally, Appendix K describes some historical navigation and positioning technology. The worked examples are also provided as Microsoft Excel files to enable interaction and modification.

Like many fields, navigation does not always adopt consistent notation and terminology. Here, a consistent notation has been adopted throughout the book, with common alternatives indicated where appropriate. The most commonly used conventions have generally been adopted, with some departures to avoid clashes and aid clarity.

Scalars are italicized and may be either upper or lower case. Vectors are lowercase bold and matrices are uppercase bold, with the corresponding scalar used to indicate their individual components. The vector (or cross) product is denoted by $\wedge$ and Dirac notation (i.e., $\dot{x}, \ddot{x}$, and so on) is generally used to indicate time derivatives. All equations presented assume base SI units: the meter, second, and radian. Other units used include the degree ($1° = \pi/180$ rad), the hour (1 hour = 3,600 seconds), and the g unit, describing acceleration due to gravity (1g = 9.80665 m s^{-2}).

Unless stated otherwise, all uncertainties and error bounds quoted are ensemble 1σ standard deviations, which correspond to a 68% confidence level where a Gaussian (normal) distribution applies. This convention is adopted because integration and other estimation algorithms model the 1σ error bounds.

Despite everyone's best efforts, most books contain errors and information can become out of date. A list of updates and corrections is therefore provided online. This can be accessed via the CD menu.

Problems and exercises for this chapter are on the accompanying CD.

References

[1] *The Concise Oxford Dictionary*, 9th ed., Oxford, U.K.: Oxford University Press, 1995.
[2] Bensky, A., *Wireless Positioning Technologies and Applications*, Norwood, MA: Artech House, 2008.
[3] Sobel, D., *Longitude*, London, U.K.: Fourth Estate, 1996.

CHAPTER 2
Coordinate Frames, Kinematics, and the Earth

This chapter provides the mathematical and physical foundations of navigation. Section 2.1 introduces the concept of a coordinate frame and how it may be used to represent an object, reference, or set of resolving axes. The main coordinate frames used in navigation are described. Section 2.2 explains the different methods of representing attitude, rotation, and resolving axes transformations, and shows how to convert between them. Section 2.3 defines the angular rate, Cartesian position, velocity, and acceleration in a multiple coordinate frame environment where the reference frame or resolving axes may be rotating; it then introduces the centrifugal and Coriolis pseudo-forces. Section 2.4 shows how the Earth's surface is modeled and defines latitude, longitude, and height. It also describes projected coordinates and Earth rotation, introduces specific force, and explains the difference between gravity and gravitation. Finally, Section 2.5 presents the equations for transforming between different coordinate frame representations.

2.1 Coordinate Frames

The science of navigation describes the position, orientation, and motion of objects. An object may be a piece of navigation equipment, such as a GNSS antenna or an INS. It may be a vehicle, such as an aircraft, ship, submarine, car, train, or satellite. It may also be a person, animal, mobile computing device, or high-value asset.

To describe the position and linear motion of an object, a specific point on that object must be selected. This is known as the *origin* of that object. It may be the center of mass of that object, the geometrical center, or an arbitrarily convenient point, such as a corner. For radio positioning equipment, the phase center of the antenna is a suitable origin as this is the point at which the radio signals appear to arrive. A point at which the sensitive axes of a number of dead-reckoning sensors intersect is also a suitable origin.

To describe the orientation and angular motion of an object, a set of three axes must also be selected. These axes must be noncoplanar and should also be mutually perpendicular. Suitable axis choices include the normal direction of motion of the object, the vertical direction when the object is at rest, the sensitive axis of an inertial or other dead-reckoning sensor, and an antenna's boresight (the normal to its plane and usually also the direction of maximum sensitivity).

However, the position, orientation, and motion of an object are meaningless on their own. Some form of reference is needed, relative to which the object may be described. The reference is also defined by an origin and a set of axes. Suitable origins include the center of the Earth, the center of the solar system, and convenient local landmarks. Suitable axes include the north, east, and vertical directions; the Earth's axis of rotation and vectors within the equatorial plane; the alignment of a local road grid; the walls of a building; and a line joining two landmarks. Another object may also act as the reference.

The origin and axes of either an object or a reference collectively comprise a *coordinate frame*. When the axes are mutually perpendicular, the coordinate frame is orthogonal and has six degrees of freedom. These are the position of the origin, o, and the orientation of the axes, x, y, and z. They must be expressed with respect to another frame to define them. Figure 2.1 illustrates this with the superscripts denoting the frames to which the origins and axes apply. A convention is adopted here of using Greek letters to denote generic coordinate frames and Roman letters to denote specifically defined frames.

In the *right-handed convention*, the x-, y-, and z-axes are always oriented such that if the thumb and first two fingers of the right hand are extended perpendicularly, the thumb is the x-axis, the first finger is the y-axis, and the second finger is the z-axis. The opposite convention is left-handed and is rarely used. All coordinate frames considered here are both orthogonal and follow the right-handed convention. In formal terms, their axes may be described as orthogonal right-handed basis sets.

A coordinate frame may be used to describe either an object or a reference. The two concepts are actually interchangeable. In a two-frame problem, defining which one is the object frame and which one is the reference frame is arbitrary and tends to be a matter of conceptual convenience. It is equally valid to describe the position and orientation of frame α with respect to frame β as it is to describe frame β with respect to frame α. This is a principle of relativity: the laws of physics appear the same for all observers. In other words, describing the position of a road with respect to a car conveys the same information as the position of the car with respect to the road.

Any navigation problem thus involves at least two coordinate frames. These are the object frame, describing the body whose position and/or orientation is desired, and the reference frame, describing a known body, such as the Earth, relative to which the object position and/or orientation is desired. However, many navigation problems involve more than one reference frame or even more than one object frame. For example, inertial sensors measure motion with respect to inertial space, whereas a typical navigation system user wants to know their position with respect to the Earth. It is not sufficient to model motion with respect to the Earth while ignoring its rotation, as is typically done in simple mechanics problems; this can cause significant errors. Reference frame rotation also impacts GNSS positioning

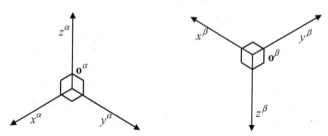

Figure 2.1 Two orthogonal coordinate frames. (From: [1]. ©2002 QinetiQ Ltd. Reprinted with permission.)

as it affects the apparent signal propagation speed. Thus, for accurate navigation, the relationship between the different coordinate frames must be properly modeled.

Any two coordinate frames may have any relative orientation, known as attitude. This may be represented in a number of different ways, as described in Section 2.2. However, within each representation, the attitude of one frame with respect to the other comprises a unique set of numbers.

A pair of coordinate frames may also have any relative position, velocity, acceleration, angular rate, and so forth. However, these quantities comprise vectors which may be resolved into components along any set of three mutually-perpendicular axes. For example, the position of frame α with respect to frame β may be described using the α-frame axes, the β-frame axes, or the axes of a third frame, γ. In practical terms, the position of a car with respect to a local road grid could be resolved about the axes of the car body frame; the road grid frame; or north, east, and down. Here, a superscript is used to denote the axes in which a quantity is expressed, known as the resolving frame. Note that it is not necessary to define the origin of the resolving frame. The position, velocity, acceleration, and angular rate in a multiple coordinate frame problem are defined in Section 2.3.

A coordinate frame definition comprises a set of rules, known as a *coordinate system*, and a set of measurements that enable known objects to be described with respect to that frame using the coordinate system. A coordinate frame may be considered a realization of the corresponding coordinate system using the measurements. Frames that are different realizations of the same coordinate system will differ slightly. Historically, nations performed their own realizations. However, international realizations, coordinated by the International Earth Rotation and Reference Systems Service (IERS), are increasingly being adopted. For more information on frame realization, the reader is directed to geodesy texts (see Selected Bibliography).

The remainder of this section defines the main coordinate systems used in navigation: Earth-centered inertial (ECI), Earth-centered Earth-fixed (ECEF), local navigation, local tangent-plane, and body frames. A brief summary of some other types of coordinate frame completes the section.

2.1.1 Earth-Centered Inertial Frame

In physics, any coordinate frame that does not accelerate or rotate with respect to the rest of the Universe is an *inertial frame*. An *Earth-centered inertial frame*, denoted by the symbol i, is nominally centered at the Earth's center of mass and oriented with respect to the Earth's spin axis and the stars. This is not strictly an inertial frame as the Earth experiences acceleration in its orbit around the Sun, its spin axis slowly moves, and the galaxy rotates. However, these effects are smaller than the measurement noise exhibited by navigation sensors, so an ECI frame may be treated as a true inertial frame for all practical purposes.

Figure 2.2 shows the origin and axes of an ECI frame and the rotation of the Earth with respect to space. The z-axis always points along the Earth's axis of rotation from the frame's origin at the center of mass to the true north pole (not the magnetic pole). The x- and y-axes lie within the equatorial plane, but do not rotate with the Earth. The y-axis points 90° ahead of the x-axis in the direction of the Earth's rotation. Note that a few authors define these axes differently.

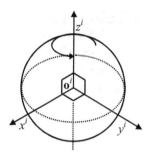

Figure 2.2 Origin and axes of an Earth-centered inertial frame. (*From:* [1]. ©2002 QinetiQ Ltd. Reprinted with permission.)

To complete the definition of the coordinate system, it is also necessary to specify the time at which the inertial frame axes coincide with those of the corresponding Earth-centered Earth-fixed frame. There are three common solutions. The first solution is simply to align the two coordinate frames when the navigation solution is initialized. The second solution is to align the coordinate frames at midnight, noting that a number of different time bases may be used, such as local time, Coordinated Universal Time (UTC), International Atomic Time (TAI), or GPS time. The final solution, used within the scientific community, is to define the x-axis as the direction from the Earth to the Sun at the vernal equinox, which is the spring equinox in the northern hemisphere. This is the same as the direction from the center of the Earth to the intersection of the Earth's equatorial plane with the Earth-Sun orbital plane (ecliptic). This version of an ECI frame is sometimes known as celestial coordinates.

A problem with realizing an ECI frame in practice is determining where the center of the Earth is with respect to known points on the surface. Instead, the origin of an ECI frame is taken as the center of an ellipsoidal representation of the Earth's surface (Section 2.4.1), which is close to the true center of mass.

A further problem, in which a precise realization of the coordinate frame is needed, is polar motion. The spin axis actually moves with respect to the solid Earth, with the poles roughly following a circular path of radius 15m. One solution is to adopt the IERS Reference Pole (IRP) or Conventional Terrestrial Pole (CTP), which is the average position of the pole surveyed between 1900 and 1905. The inertial coordinate system that adopts the center of an ellipsoidal representation of the Earth's surface as its origin, the IRP/CTP as its z-axis, and the x-axis based on the Earth-Sun axis at vernal equinox is known as the Conventional Inertial Reference System (CIRS).

Inertial frames are important in navigation because inertial sensors measure motion with respect to a generic inertial frame. An inertial reference frame and resolving axes also enables the simplest form of navigation equations to be used, as shown in later chapters.

2.1.2 Earth-Centered Earth-Fixed Frame

An *Earth-centered Earth-fixed frame*, commonly abbreviated to Earth frame, is similar to an Earth-centered inertial frame, except that all axes remain fixed with respect to the Earth. The two coordinate systems share a common origin, the center of the ellipsoid modeling the Earth's surface (Section 2.4.1), which is roughly at the center of mass. An ECEF frame is denoted by the symbol e.

2.1 Coordinate Frames

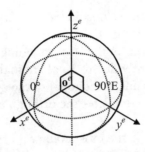

Figure 2.3 Origin and axes of an Earth-centered Earth-fixed frame. (*From:* [1]. ©2002 QinetiQ Ltd. Reprinted with permission.)

Figure 2.3 shows the origin and axes of an ECEF frame. The z-axis is the same as that of the corresponding ECI frame. It always points along the Earth's axis of rotation from the center to the north pole (true not magnetic). The x-axis points from the center to the intersection of the equator with the IERS Reference Meridian (IRM) or Conventional Zero Meridian (CZM), which defines 0° longitude. The y-axis completes the right-handed orthogonal set, pointing from the center to the intersection of the equator with the 90° east meridian. Again, note that a few authors define these axes differently. The ECEF coordinate system using the IRP/CTP and the IRM/CZM is also known as the Conventional Terrestrial Reference System (CTRS), and some authors use the symbol t to denote it.

The Earth-centered Earth-fixed coordinate system is important in navigation because the user wants to know his or her position relative to the Earth, so its realizations are commonly used as both a reference frame and a resolving frame.

2.1.3 Local Navigation Frame

A *local navigation frame*, local level navigation frame, or geodetic, geographic, or topocentric frame is denoted by the symbol n (some authors use g or l). Its origin is the object described by the navigation solution. This could be part of the navigation system itself or the center of mass of the host vehicle or user.

Figure 2.4 shows the origin and axes of a local navigation frame. The axes are aligned with the topographic directions: north, east, and vertical. In the convention used here, the z-axis, also known as the down (D) axis, is defined as the normal to

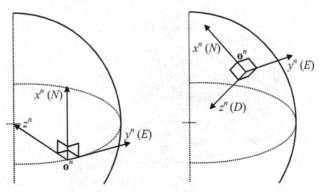

Figure 2.4 Origin and axes of a local navigation frame.

the surface of the reference ellipsoid (Section 2.4.1) in the direction pointing towards the Earth. Simple gravity models (Section 2.4.7) assume that the gravity vector is coincident with the z-axis of the corresponding local navigation frame. True gravity deviates from this slightly due to local anomalies. The x-axis, or north (N) axis, is the projection in the plane orthogonal to the z-axis of the line from the user to the north pole. The y-axis completes the orthogonal set by pointing east and is known as the east (E) axis.

North, east, down is the most common order of the axes in a local navigation coordinate system and will always be used here. However, there are other forms in use. The combination x = east, y = north, z = up is common, while x = north, y = west, z = up and x = south, y = west, z = down are also used, noting that the axes must form a right-handed set.

The local navigation coordinate system is important in navigation because the user wants to know his or her attitude relative to the north, east, and down directions. For position and velocity, it provides a convenient set of resolving axes, but is not used as a reference frame.

A major drawback of local navigation frames is that there is a singularity at each pole because the north and east axes are undefined there. Thus, navigation equations mechanized using this frame are unsuitable for use near the poles. Instead, an alternative frame should be used with conversion of the navigation solution to the local navigation frame at the end of the processing chain.

In a multibody problem, each body will have its local navigation frame. However, only one is typically of interest in practice. Furthermore, the differences in orientation between the local navigation frames of objects in close proximity are usually negligible.

2.1.4 Local Tangent-Plane Frame

A *local tangent-plane frame*, denoted by l (some authors use t), has a fixed origin with respect to the Earth, usually a point on the surface. Like the local navigation frame, its z-axis is aligned with the vertical (pointing either up or down). Its x- and y-axes may be also aligned with the topographic directions (i.e., north and east), in which case it may be known as a local geodetic frame or topocentric frame. However, the x- and y-axes may be also aligned with an environmental feature, such as a road or building. As with the other frames, the axes form a right-handed orthogonal set. Thus, this frame is Earth-fixed, but not Earth-centered. This type of frame is used for navigation within a localized area. Examples include aircraft landing and urban and indoor positioning.

A planar frame, denoted by p, is used for two-dimensional positioning; its third dimension is neglected. It may comprise the horizontal components of the local tangent-plane frame or may be used to express projected coordinates (Section 2.4.5).

2.1.5 Body Frame

A *body frame*, sometimes known as a vehicle frame, comprises the origin and orientation of the object described by the navigation solution. The origin is thus coincident with that of the corresponding local navigation frame. However, the axes remain fixed

2.1 Coordinate Frames

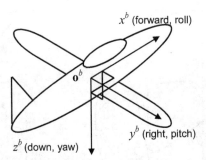

Figure 2.5 Body frame axes. (*From:* [1]. ©2002 QinetiQ Ltd. Reprinted with permission.)

with respect to the body. Here, the most common convention is adopted, whereby x is the forward axis, pointing in the usual direction of travel; z is the down axis, pointing in the usual direction of gravity; and y is the right axis, completing the orthogonal set. For angular motion, the body-frame axes are also known as roll, pitch, and yaw. Roll motion is about the x-axis, pitch motion is about the y-axis, and yaw motion is about the z-axis. Figure 2.5 illustrates this. A right-handed corkscrew rule applies, whereby if the axis is pointing away, then positive rotation about that axis is clockwise.

A body frame is essential in navigation because it describes the object that the navigation solution refers to. Inertial sensors and other dead-reckoning sensors measure the motion of a body frame and most have a fixed orientation with respect to that frame.

The symbol b is used to denote the body frame of the primary object of interest. The body frame origin may be within a navigation sensor or it may be the center of mass of the host vehicle as this simplifies the kinematics in a control system. Many navigation problems involve multiple objects, each with their own body frame, for which alternative symbols must be used. Examples include a for an antenna; c for a camera's imaging sensor; f for front wheels or an environmental feature; r for rear wheels, a reference station, or a radar transponder; s for a satellite; and t for a transmitter. For multiple satellites, transmitters, or environmental features, frames can be denoted by numbers.

2.1.6 Other Frames

A *wander-azimuth frame*, w (some authors use n), is a variant of a local navigation frame and shares the same origin and z-axis. However, the x- and y-axes are displaced from north and east by an angle, ψ_{nw} (some authors use α), known as the wander angle. Figure 2.6 illustrates this. The wander angle varies as the frame moves with respect to the Earth and is always known, so transformation of the navigation solution to a local navigation frame is straightforward. A wander-azimuth frame avoids the polar singularity of a local navigation frame, so is commonly used to mechanize inertial navigation equations. It is discussed further in Section 5.3.5.

Another variant of a local navigation frame is a *geocentric frame*. This differs in that the z-axis points from the origin to the center of the Earth instead of along the normal to the ellipsoid. The x-axis is defined in the same way as the projection of the line to the north pole in the plane orthogonal to the z-axis.

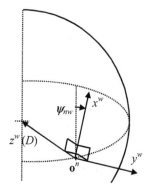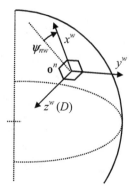

Figure 2.6 Axes of the wander-azimuth frame.

In navigation systems with directional sensors, such as an IMU, odometer, radar, Doppler sonar, and imaging sensors, the sensitive axis of each sensor may be considered to have its own body frame, known as a *sensor frame* or *instrument frame*. Thus, an IMU could be considered as having a coordinate frame for each accelerometer and gyro. However, it is generally simpler to assume that each sensor has a known orientation with respect to the navigation system body frame, particularly in cases, such as most IMUs, where the sensitive axes of the instruments are nominally aligned with the body frame. Departures from this (i.e., the instrument mounting misalignments) are then treated as a set of perturbations that must be accounted for when modeling the errors of the system.

For some sensors, such as accelerometers and odometers, a lever arm transformation (Section 2.5.5) must be performed to translate measurements from the sensor frame origin to the system body frame origin. For inertial navigation, this transformation is usually performed within the IMU (see Section 4.3).

In calculating the motion of satellites, orbital coordinate frames, denoted by o, are used. An orbital frame is an inertial frame with its origin at the Earth's center of mass, but its axes tilted with respect to the ECI frame so that satellite moves in the xy plane. More details may be found in Section 8.5.2.

A line-of-sight (LOS) frame is essentially a body frame with a zero-bank constraint (see Section 2.2.1). It is defined with its x-axis along the boresight from the sensor to the target, its y-axis in the horizontal plane, pointing to the right when looking along boresight, and its z-axis completing the orthogonal set, such that it points down when the boresight is in the horizontal plane.

2.2 Attitude, Rotation, and Resolving Axes Transformations

Attitude describes the orientation of the axes of one coordinate frame with respect to those of another. One way of representing attitude is the rotation required to align one set of axes with another. Figure 2.7 illustrates this in two dimensions; a clockwise rotation of frame γ through angle ψ, with respect to frame β, is required to align the axes of frame γ with those of frame β. Alternatively, frame β could be rotated through an angle of $-\psi$ with respect to frame γ to achieve the same axis alignment. Unless a third frame is introduced, the two rotations are indistinguishable. It is not necessary for the frame origins to coincide.

2.2 Attitude, Rotation, and Resolving Axes Transformations

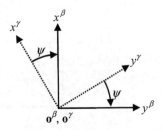

Figure 2.7 Rotation of the axes of frame γ to align with those of frame β.

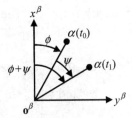

Figure 2.8 Rotation of the line βα with respect to the axes of frame β.

Consider now a line of fixed length, $r_{\beta\alpha}$, from the origin of frame β to a point, α, that is free to rotate about the origin of frame β. Figure 2.8 shows the position of the line at times t_0 and t_1.

At time t_0, the position of α with respect to the origin of frame β and resolved about the axes of that frame may be described by

$$\begin{aligned} x^{\beta}_{\beta\alpha}(t_0) &= r_{\beta\alpha} \cos\phi \\ y^{\beta}_{\beta\alpha}(t_0) &= r_{\beta\alpha} \sin\phi \end{aligned}, \quad (2.1)$$

where the superscript β denotes the frame of the resolving axes.

At time t_1, the line has rotated through an angle ψ, so the position of α is described by

$$\begin{aligned} x^{\beta}_{\beta\alpha}(t_1) &= r_{\beta\alpha} \cos(\phi+\psi) \\ y^{\beta}_{\beta\alpha}(t_1) &= r_{\beta\alpha} \sin(\phi+\psi) \end{aligned}. \quad (2.2)$$

Using trigonometric identities, it may be shown that the coordinates describing the position of α at the two times are related by

$$\begin{pmatrix} x^{\beta}_{\beta\alpha}(t_1) \\ y^{\beta}_{\beta\alpha}(t_1) \end{pmatrix} = \begin{pmatrix} \cos\psi & -\sin\psi \\ \sin\psi & \cos\psi \end{pmatrix} \begin{pmatrix} x^{\beta}_{\beta\alpha}(t_0) \\ y^{\beta}_{\beta\alpha}(t_0) \end{pmatrix}. \quad (2.3)$$

Note that the matrix describing the rotation is a function only of the angle of rotation, not the original orientation of the line.

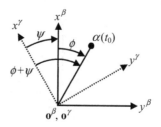

Figure 2.9 Orientation of the line $\beta\alpha$ with respect to the axes of frames β and γ.

Figure 2.9 depicts the orientation of the line $\beta\alpha$ at time t_0 with respect to frames β and γ. The position of α with respect to the origin of frame β, but resolved about the axes of frame γ is thus

$$x^\gamma_{\beta\alpha}(t_0) = r_{\beta\alpha} \cos(\phi + \psi)$$
$$y^\gamma_{\beta\alpha}(t_0) = r_{\beta\alpha} \sin(\phi + \psi) \quad . \quad (2.4)$$

Applying trigonometric identities again, it may be shown that the coordinates describing the position of α resolved about the two sets of axes are related by

$$\begin{pmatrix} x^\gamma_{\beta\alpha} \\ y^\gamma_{\beta\alpha} \end{pmatrix} = \begin{pmatrix} \cos\psi & -\sin\psi \\ \sin\psi & \cos\psi \end{pmatrix} \begin{pmatrix} x^\beta_{\beta\alpha} \\ y^\beta_{\beta\alpha} \end{pmatrix}. \quad (2.5)$$

Note that the matrix describing the coordinate transformation is a function only of the angle of rotation required to align one set of resolving axes with the other. Comparing this with (2.3), it can be seen that the rotation matrix of (2.3) is identical to the coordinate transformation matrix of (2.5). This is because the rotation of an object with respect to a set of resolving axes is indistinguishable from an equal and opposite rotation of the resolving axes with respect to the object.

Consequently, a coordinate transformation matrix may be used to describe a rotation and is thus a valid way of representing attitude. Conversely, a coordinate transformation (without a change in reference frame) may be represented as rotation. As the magnitude of the vector does not change, transforming a vector from one set of resolving axes to another may be thought of as applying a rotation in space to that vector.

Extending this to three dimensions, the coordinate transformation matrix is simply expanded from a 2×2 matrix to a 3×3 matrix. The 3-D extension of the rotation angle is more complex. It may be expressed as three successive scalar rotations, known as Euler angles, about defined axes. Alternatively, it may be expressed as a single scalar rotation about a particular axis that must be defined; this is represented either as a set of quaternions or as a rotation vector.

This section presents detailed descriptions of Euler angles and the coordinate transformation matrix, basic descriptions of quaternions and the rotation vector, and the equations for converting between these different attitude representations.

When combining successive rotations or axes transformations, it is essential that they are applied in the correct order, regardless of the method used to represent

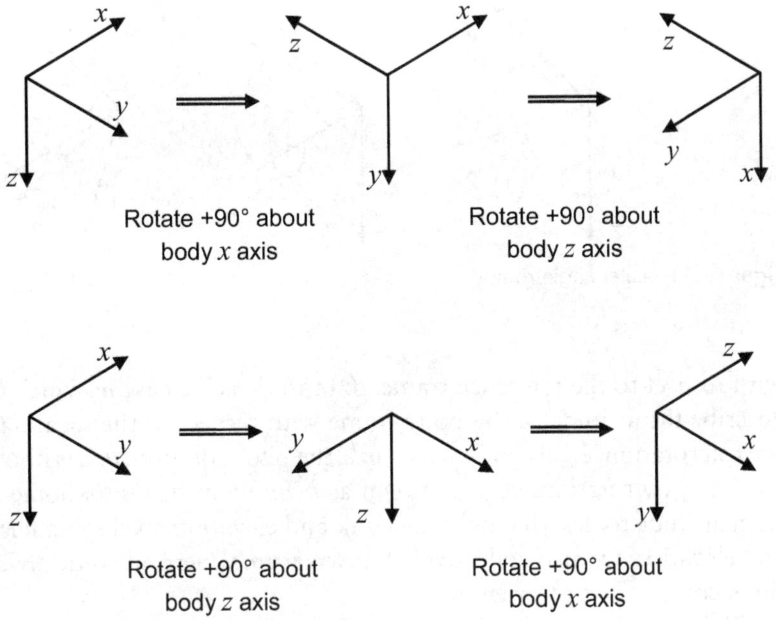

Figure 2.10 Noncommutivity of rotations.

them. This is because the order of rotations or transformations determines the final outcome. In formal terms, they do not commute. For example, a 90° rotation about the *x*-axis followed by a 90° rotation about the *z*-axis leads to a different orientation from a 90° *z*-axis rotation followed by a 90° *x*-axis rotation. This applies regardless of whether the rotations are made about the axes of the object's body frame or of a reference frame. Figure 2.10 illustrates this.

2.2.1 Euler Attitude

Euler angles (pronounced as "oiler") are the most intuitive way of describing an attitude, particularly that of a body frame with respect to the corresponding local navigation frame. The attitude is broken down into three successive rotations, with each rotation about an axis orthogonal to that of its predecessor and/or successor. Figure 2.11 illustrates this for the rotation of the axes of a coordinate frame from alignment with frame β to alignment with frame α, via alignments with two intermediate frames, ψ and θ.

The first rotation, through the angle $\psi_{\beta\alpha}$, is the *yaw* rotation. This is performed about the common *z*-axis of the β frame and the first intermediate frame. Thus, the *x*- and *y*-axes are rotated but the *z*-axis is not. Next, the *pitch rotation*, through $\theta_{\beta\alpha}$, is performed about the common *y*-axis of the first and second intermediate frames. This rotates the *x*- and *z*-axes. Finally, the *roll rotation*, through $\phi_{\beta\alpha}$, is performed about the common *x*-axis of the second intermediate frame and the α frame. This rotates the *y*- and *z*-axes.

It is convenient to represent the orientation of an object frame with respect to a reference frame using the Euler angles describing the rotation from the reference frame resolving axes to those of the object frame. Thus, the roll, pitch, and yaw Euler rotations, $\phi_{\beta\alpha}$, $\theta_{\beta\alpha}$, and $\psi_{\beta\alpha}$, describe the orientation of the object frame, α,

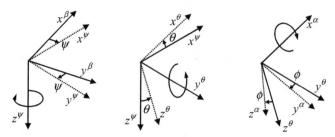

Figure 2.11 Euler angle rotations.

with respect to the reference frame, β. In the specific case in which the Euler angles describe the attitude of the body frame with respect to the local navigation frame, the roll rotation, ϕ_{nb}, is known as *bank*, the pitch rotation, θ_{nb}, is known as *elevation*, and the yaw rotation, ψ_{nb}, is known as *heading* or *azimuth*. Some authors use the term attitude to describe only the bank and elevation, excluding heading. The bank and elevation are also collectively known as *tilts*. Here, attitude always describes all three components of orientation.

Euler angles can also be used to transform a vector, $\mathbf{x} = (x, y, z)$, from one set of resolving axes, β, to a second set, α. As with rotation, the transformation occurs in three stages. First, the yaw step transforms the x and y components of the vector by performing a rotation through the angle $\psi_{\beta\alpha}$, but leaves the z component unchanged. The resulting vector is resolved about the axes of the first intermediate frame, denoted by ψ:

$$\begin{aligned} x^\psi &= x^\beta \cos\psi_{\beta\alpha} + y^\beta \sin\psi_{\beta\alpha} \\ y^\psi &= -x^\beta \sin\psi_{\beta\alpha} + y^\beta \cos\psi_{\beta\alpha} \\ z^\psi &= z^\beta \end{aligned} \quad (2.6)$$

Note that this resolving axes rotation is in the opposite direction to that in the earlier example described by (2.4).

Next, the pitch step transforms the x and z components of the vector by performing a rotation through $\theta_{\beta\alpha}$. This results in a vector resolved about the axes of the second intermediate frame, denoted by θ:

$$\begin{aligned} x^\theta &= x^\psi \cos\theta_{\beta\alpha} - z^\psi \sin\theta_{\beta\alpha} \\ y^\theta &= y^\psi \\ z^\theta &= x^\psi \sin\theta_{\beta\alpha} + z^\psi \cos\theta_{\beta\alpha} \end{aligned} \quad (2.7)$$

Finally, the roll step transforms the y and z components by performing a rotation through $\phi_{\beta\alpha}$. This produces a vector resolved about the axes of the α frame as required:

$$\begin{aligned} x^\alpha &= x^\theta \\ y^\alpha &= y^\theta \cos\phi_{\beta\alpha} + z^\theta \sin\phi_{\beta\alpha} \\ z^\alpha &= -y^\theta \sin\phi_{\beta\alpha} + z^\theta \cos\phi_{\beta\alpha} \end{aligned} \quad (2.8)$$

The Euler rotation from frame β to frame α may be denoted by the vector

$$\psi_{\beta\alpha} = \begin{pmatrix} \phi_{\beta\alpha} \\ \theta_{\beta\alpha} \\ \psi_{\beta\alpha} \end{pmatrix}, \tag{2.9}$$

noting that the Euler angles are listed in the reverse order to that in which they are applied. The order in which the three rotations are carried out is critical as each is performed in a different coordinate frame. If they are performed in a different order (e.g., with the roll first), the orientation of the axes at the end of the transformation is generally different. In formal terms, the three Euler rotations do not commute.

The Euler rotation $(\phi_{\beta\alpha} + \pi, \pi - \theta_{\beta\alpha}, \psi_{\beta\alpha} + \pi)$ gives the same result as the Euler rotation $(\phi_{\beta\alpha}, \theta_{\beta\alpha}, \psi_{\beta\alpha})$. Consequently, to avoid duplicate sets of Euler angles representing the same attitude, a convention is adopted of limiting the pitch rotation, θ, to the range $(-90° \leq \theta \leq 90°)$. Another property of Euler angles is that the axes about which the roll and yaw rotations are made are usually not orthogonal, although both are orthogonal to the axis about which the pitch rotation is made.

To reverse an Euler rotation, either the original operation must be reversed, beginning with the roll, or a different transformation must be applied. Simply reversing the sign of the Euler angles does not return to the original orientation, thus[*]

$$\begin{pmatrix} \phi_{\alpha\beta} \\ \theta_{\alpha\beta} \\ \psi_{\alpha\beta} \end{pmatrix} \neq \begin{pmatrix} -\phi_{\beta\alpha} \\ -\theta_{\beta\alpha} \\ -\psi_{\beta\alpha} \end{pmatrix}. \tag{2.10}$$

Similarly, successive rotations cannot be expressed simply by adding the Euler angles:

$$\begin{pmatrix} \phi_{\beta\gamma} \\ \theta_{\beta\gamma} \\ \psi_{\beta\gamma} \end{pmatrix} \neq \begin{pmatrix} \phi_{\beta\alpha} + \phi_{\alpha\gamma} \\ \theta_{\beta\alpha} + \theta_{\alpha\gamma} \\ \psi_{\beta\alpha} + \psi_{\alpha\gamma} \end{pmatrix}. \tag{2.11}$$

A further difficulty is that the Euler angles exhibit a singularity at $\pm 90°$ pitch, where the roll and yaw become indistinguishable. Because of these difficulties, Euler angles are rarely used for attitude computation.[†]

2.2.2 Coordinate Transformation Matrix

The *coordinate transformation matrix*, or rotation matrix, is a 3×3 matrix, denoted as $\mathbf{C}_\alpha^\beta$ (some authors use $\mathbf{R}$ or $\mathbf{T}$). A vector may be transformed in one step from one

---

[*]This and subsequent paragraphs are based on material written by the author for QinetiQ, so comprise QinetiQ copyright material.
[†]End of QinetiQ copyright material.

set of resolving axes to another by premultiplying it by the appropriate coordinate transformation matrix. Thus, for an arbitrary vector, **x**,

$$\mathbf{x}^\beta = \mathbf{C}_\alpha^\beta \mathbf{x}^\alpha, \tag{2.12}$$

where the superscript of **x** denotes the resolving axes. The lower index of the matrix represents the "from" coordinate frame and the upper index the "to" frame. The rows of a coordinate transformation matrix are in the "to" frame, whereas the columns are in the "from" frame.

When the matrix is used to represent attitude, it is more common to use the upper index (the "to" frame) to represent the reference frame, β, and the lower index (the "from" frame) to represent the object frame, α. This is because the rotation of an object's orientation with respect to a set of resolving axes is equal and opposite to the rotation of the resolving axes with respect to the object. The first case corresponds to attitude and the second to coordinate transformation. However, many authors represent attitude as a reference frame to object frame transformation, $\mathbf{C}_\beta^\alpha$.

Figure 2.12 shows the role of each element of the coordinate transformation matrix in transforming the resolving axes of a vector from frame α to frame β.

It can be shown [2] that the coordinate transformation matrix elements are the product of the unit vectors describing the axes of the two frames, which, in turn, are equal to the cosines of the angles between the axes:

$$\mathbf{C}_\alpha^\beta = \begin{pmatrix} \mathbf{u}_{\beta x} \cdot \mathbf{u}_{\alpha x} & \mathbf{u}_{\beta x} \cdot \mathbf{u}_{\alpha y} & \mathbf{u}_{\beta x} \cdot \mathbf{u}_{\alpha z} \\ \mathbf{u}_{\beta y} \cdot \mathbf{u}_{\alpha x} & \mathbf{u}_{\beta y} \cdot \mathbf{u}_{\alpha y} & \mathbf{u}_{\beta y} \cdot \mathbf{u}_{\alpha z} \\ \mathbf{u}_{\beta z} \cdot \mathbf{u}_{\alpha x} & \mathbf{u}_{\beta z} \cdot \mathbf{u}_{\alpha y} & \mathbf{u}_{\beta z} \cdot \mathbf{u}_{\alpha z} \end{pmatrix} = \begin{pmatrix} \cos \mu_{\beta x, \alpha x} & \cos \mu_{\beta x, \alpha y} & \cos \mu_{\beta x, \alpha z} \\ \cos \mu_{\beta y, \alpha x} & \cos \mu_{\beta y, \alpha y} & \cos \mu_{\beta y, \alpha z} \\ \cos \mu_{\beta z, \alpha x} & \cos \mu_{\beta z, \alpha y} & \cos \mu_{\beta z, \alpha z} \end{pmatrix}, \tag{2.13}$$

where $\mathbf{u}_i$ is a unit vector describing axis i and $\mu_{i,j}$ is the resultant angle between axes i and j. Hence, the term direction cosine matrix (DCM) is often used to describe these matrices.

Coordinate transformation matrices are easy to manipulate. As (2.13) shows, to reverse a rotation or coordinate transformation, the transpose of the matrix, denoted by the superscript, T (see Section A.2 in Appendix A on the CD), is used. Thus,

$$\mathbf{C}_\beta^\alpha = \left(\mathbf{C}_\alpha^\beta\right)^\mathrm{T}. \tag{2.14}$$

$$\mathbf{C}_\alpha^\beta = \begin{array}{|c|c|c|} \hline \alpha_x \to \beta_x & \alpha_y \to \beta_x & \alpha_z \to \beta_x \\ \hline \alpha_x \to \beta_y & \alpha_y \to \beta_y & \alpha_z \to \beta_y \\ \hline \alpha_x \to \beta_z & \alpha_y \to \beta_z & \alpha_z \to \beta_z \\ \hline \end{array}$$

Figure 2.12 The coordinate transformation matrix component functions.

To perform successive transformations or rotations, the coordinate transformation matrices are simply multiplied:

$$\mathbf{C}_\alpha^\gamma = \mathbf{C}_\beta^\gamma \mathbf{C}_\alpha^\beta. \qquad (2.15)$$

However, as with any matrix multiplication, the order is critical, so

$$\mathbf{C}_\alpha^\gamma \neq \mathbf{C}_\alpha^\beta \mathbf{C}_\beta^\gamma. \qquad (2.16)$$

This reflects the fact that rotations themselves do not commute as shown in Figure 2.10.

Performing a transformation and then reversing the process must return the original vector or matrix, so

$$\mathbf{C}_\alpha^\beta \mathbf{C}_\beta^\alpha = \mathbf{I}_3, \qquad (2.17)$$

where $\mathbf{I}_n$ is the $n \times n$ identity or unit matrix. Thus, coordinate transformation matrices are orthonormal (see Section A.3 in Appendix A on the CD).

A coordinate transformation matrix can also be used to transform a matrix to which specific resolving axes apply. Consider a matrix, $\mathbf{M}$, used to transform a vector $\mathbf{a}$ into a vector $\mathbf{b}$. If $\mathbf{a}$ and $\mathbf{b}$ may be resolved about axes α or β, the transformation may be written as

$$\mathbf{b}^\alpha = \mathbf{M}^\alpha \mathbf{a}^\alpha \qquad (2.18)$$

or

$$\mathbf{b}^\beta = \mathbf{M}^\beta \mathbf{a}^\beta. \qquad (2.19)$$

Thus, the rows and columns of $\mathbf{M}$ must be resolved about the same axes as $\mathbf{a}$ and $\mathbf{b}$. Applying (2.12) to (2.18) gives

$$\mathbf{C}_\beta^\alpha \mathbf{b}^\beta = \mathbf{M}^\alpha \mathbf{C}_\beta^\alpha \mathbf{a}^\beta. \qquad (2.20)$$

Premultiplying by $\mathbf{C}_\alpha^\beta$, applying (2.17), and substituting the result into (2.19) give

$$\mathbf{M}^\beta = \mathbf{C}_\alpha^\beta \mathbf{M}^\alpha \mathbf{C}_\beta^\alpha. \qquad (2.21)$$

where the left-hand coordinate transformation matrix transforms the rows of $\mathbf{M}$ and the right-hand matrix transforms the columns. When the resolving frame of only the rows or only the columns of a matrix are to be transformed, respectively, only the left-hand or the right-hand coordinate transformation matrix is applied.

Although a coordinate transformation matrix has nine components, the requirement to meet (2.17) means that only three of these are independent. Thus, it has the same number of independent components as Euler attitude. A set of Euler angles is converted to a coordinate transformation matrix by first representing each of the rotations of (2.6)–(2.8) as a matrix and then multiplying, noting that with matrices, the first operation is placed on the right. Thus, for coordinate transformations, Euler angles are converted to a coordinate transformation matrix using

$$C_\beta^\alpha = \begin{pmatrix} 1 & 0 & 0 \\ 0 & \cos\phi_{\beta\alpha} & \sin\phi_{\beta\alpha} \\ 0 & -\sin\phi_{\beta\alpha} & \cos\phi_{\beta\alpha} \end{pmatrix} \begin{pmatrix} \cos\theta_{\beta\alpha} & 0 & -\sin\theta_{\beta\alpha} \\ 0 & 1 & 0 \\ \sin\theta_{\beta\alpha} & 0 & \cos\theta_{\beta\alpha} \end{pmatrix} \begin{pmatrix} \cos\psi_{\beta\alpha} & \sin\psi_{\beta\alpha} & 0 \\ -\sin\psi_{\beta\alpha} & \cos\psi_{\beta\alpha} & 0 \\ 0 & 0 & 1 \end{pmatrix}$$

$$= \begin{bmatrix} \cos\theta_{\beta\alpha}\cos\psi_{\beta\alpha} & \cos\theta_{\beta\alpha}\sin\psi_{\beta\alpha} & -\sin\theta_{\beta\alpha} \\ \begin{pmatrix} -\cos\phi_{\beta\alpha}\sin\psi_{\beta\alpha} \\ +\sin\phi_{\beta\alpha}\sin\theta_{\beta\alpha}\cos\psi_{\beta\alpha} \end{pmatrix} & \begin{pmatrix} \cos\phi_{\beta\alpha}\cos\psi_{\beta\alpha} \\ +\sin\phi_{\beta\alpha}\sin\theta_{\beta\alpha}\sin\psi_{\beta\alpha} \end{pmatrix} & \sin\phi_{\beta\alpha}\cos\theta_{\beta\alpha} \\ \begin{pmatrix} \sin\phi_{\beta\alpha}\sin\psi_{\beta\alpha} \\ +\cos\phi_{\beta\alpha}\sin\theta_{\beta\alpha}\cos\psi_{\beta\alpha} \end{pmatrix} & \begin{pmatrix} -\sin\phi_{\beta\alpha}\cos\psi_{\beta\alpha} \\ +\cos\phi_{\beta\alpha}\sin\theta_{\beta\alpha}\sin\psi_{\beta\alpha} \end{pmatrix} & \cos\phi_{\beta\alpha}\cos\theta_{\beta\alpha} \end{bmatrix}, \quad (2.22)$$

while the reverse conversion is

$$\begin{aligned} \phi_{\beta\alpha} &= \arctan_2\left(C^\alpha_{\beta 2,3}, C^\alpha_{\beta 3,3}\right) \\ \theta_{\beta\alpha} &= -\arcsin C^\alpha_{\beta 1,3} \\ \psi_{\beta\alpha} &= \arctan_2\left(C^\alpha_{\beta 1,2}, C^\alpha_{\beta 1,1}\right) \end{aligned}, \quad (2.23)$$

noting that four-quadrant (360°) arctangent functions must be used where $\arctan_2(a, b)$ is equivalent to $\arctan(a/b)$. These conversions are used in the MATLAB functions, Euler_to_CTM and CTM_to_Euler, included on the accompanying CD.

For converting between attitude representations (e.g., between $\mathbf{\psi}_{nb}$ and C_b^n), the following is normally used

$$C_\alpha^\beta = \begin{bmatrix} \cos\theta_{\beta\alpha}\cos\psi_{\beta\alpha} & \begin{pmatrix} -\cos\phi_{\beta\alpha}\sin\psi_{\beta\alpha} \\ +\sin\phi_{\beta\alpha}\sin\theta_{\beta\alpha}\cos\psi_{\beta\alpha} \end{pmatrix} & \begin{pmatrix} \sin\phi_{\beta\alpha}\sin\psi_{\beta\alpha} \\ +\cos\phi_{\beta\alpha}\sin\theta_{\beta\alpha}\cos\psi_{\beta\alpha} \end{pmatrix} \\ \cos\theta_{\beta\alpha}\sin\psi_{\beta\alpha} & \begin{pmatrix} \cos\phi_{\beta\alpha}\cos\psi_{\beta\alpha} \\ +\sin\phi_{\beta\alpha}\sin\theta_{\beta\alpha}\sin\psi_{\beta\alpha} \end{pmatrix} & \begin{pmatrix} -\sin\phi_{\beta\alpha}\cos\psi_{\beta\alpha} \\ +\cos\phi_{\beta\alpha}\sin\theta_{\beta\alpha}\sin\psi_{\beta\alpha} \end{pmatrix} \\ -\sin\theta_{\beta\alpha} & \sin\phi_{\beta\alpha}\cos\theta_{\beta\alpha} & \cos\phi_{\beta\alpha}\cos\theta_{\beta\alpha} \end{bmatrix} \quad (2.24)$$

and

$$\begin{aligned} \phi_{\beta\alpha} &= \arctan_2\left(C^\beta_{\alpha 3,2}, C^\beta_{\alpha 3,3}\right) \\ \theta_{\beta\alpha} &= -\arcsin C^\beta_{\alpha 3,1} \\ \psi_{\beta\alpha} &= \arctan_2\left(C^\beta_{\alpha 2,1}, C^\beta_{\alpha 1,1}\right) \end{aligned}. \quad (2.25)$$

Again, four-quadrant arctangent functions must be used. Example 2.1 on the CD illustrates the conversion of the coordinate transformation matrix to and from Euler angles and is editable using Microsoft Excel.

When the coordinate transformation matrix and Euler angles represent a small angular perturbation for which the small angle approximation is valid, (2.22) becomes

$$\mathbf{C}_\beta^\alpha \approx \begin{pmatrix} 1 & \psi_{\beta\alpha} & -\theta_{\beta\alpha} \\ -\psi_{\beta\alpha} & 1 & \phi_{\beta\alpha} \\ \theta_{\beta\alpha} & -\phi_{\beta\alpha} & 1 \end{pmatrix} = \mathbf{I}_3 - [\mathbf{\psi}_{\beta\alpha} \wedge] \qquad (2.26)$$

and (2.24) becomes

$$\mathbf{C}_\alpha^\beta \approx \mathbf{I}_3 + [\mathbf{\psi}_{\beta\alpha} \wedge] \qquad (2.27)$$

where $[\mathbf{\psi}_{\beta\alpha} \wedge]$ denotes the skew-symmetric matrix of the Euler angles (see Section A.3 in Appendix A on the CD). Note that under the small angle approximation, $\mathbf{\psi}_{\alpha\beta} \approx -\mathbf{\psi}_{\beta\alpha}$.

One of the eigenvalues (see Section A.6 in Appendix A on the CD) of a coordinate transformation matrix is 1 (the other two are complex and have a magnitude of 1). Consequently, there exist vectors that remain unchanged following the application of a coordinate transformation matrix, or its transpose. These vectors are of the form $k\mathbf{e}_{\beta\alpha}^{\alpha/\beta}$, where k is any scalar and $\mathbf{e}_{\beta\alpha}^{\alpha/\beta}$ is the unit vector describing the axis of the rotation that the coordinate transformation matrix can be used to represent. As this vector is unchanged by (2.12), the axis of rotation is the same when resolved in the axes of the two frames transformed between. Thus,

$$\mathbf{e}_{\beta\alpha}^\alpha = \mathbf{e}_{\beta\alpha}^\beta = \mathbf{e}_{\beta\alpha}^{\alpha/\beta}. \qquad (2.28)$$

Note that this rotation-axis vector takes a different value when resolved in the axes of any other frame. It may be obtained by solving

$$\begin{aligned} \mathbf{e}_{\beta\alpha}^{\alpha/\beta} &= \mathbf{C}_\beta^\alpha \mathbf{e}_{\beta\alpha}^{\alpha/\beta} \\ \mathbf{e}_{\beta\alpha}^{\alpha/\beta\mathrm{T}} \mathbf{e}_{\beta\alpha}^{\alpha/\beta} &= 1 \end{aligned}. \qquad (2.29)$$

This has two solutions with opposite signs. It is conventional to select the solution

$$\mathbf{e}_{\beta\alpha}^{\alpha/\beta} = \frac{1}{2 \sin \mu_{\beta\alpha}} \begin{pmatrix} C_{\beta 2,3}^\alpha - C_{\beta 3,2}^\alpha \\ C_{\beta 3,1}^\alpha - C_{\beta 1,3}^\alpha \\ C_{\beta 1,2}^\alpha - C_{\beta 2,1}^\alpha \end{pmatrix}, \qquad (2.30)$$

where $\mu_{\beta\alpha}$ is the magnitude of the rotation. This is given by

$$\begin{aligned} \mu_{\beta\alpha} &= \arcsin\left(\tfrac{1}{2}\sqrt{\left(C_{\beta 2,3}^\alpha - C_{\beta 3,2}^\alpha\right)^2 + \left(C_{\beta 3,1}^\alpha - C_{\beta 1,3}^\alpha\right)^2 + \left(C_{\beta 1,2}^\alpha - C_{\beta 2,1}^\alpha\right)^2} \right) \\ &= \arccos\left[\tfrac{1}{2}\left(C_{\beta 1,1}^\alpha + C_{\beta 2,2}^\alpha + C_{\beta 3,3}^\alpha - 1\right)\right] \end{aligned}. \qquad (2.31)$$

Note that $\mathbf{e}_{\alpha\beta}^{\beta/\alpha} = -\mathbf{e}_{\beta\alpha}^{\alpha/\beta}$. The axis of rotation and scalar multiples thereof are the only vectors which are invariant to a coordinate transformation (except where $\mathbf{C}_{\beta}^{\alpha} = \mathbf{I}_3$).

2.2.3 Quaternion Attitude

A rotation may be represented using a *quaternion*, which is a hyper-complex number with four components:

$$\mathbf{q} = (q_0, q_1, q_2, q_3),$$

where q_0 is a function only of the magnitude of the rotation and the other three components are functions of both the magnitude and the axis of rotation. Some authors number the components 1 to 4, with the magnitude component either at the beginning as q_1 or at the end as q_4. Thus, care must be taken to ensure that a quaternion is interpreted correctly.

As with coordinate transformation matrices, the axis of rotation is the same in both the "to" and the "from" coordinate frames of the rotation. As with the other attitude representations, only three components of the attitude quaternion are independent. It is defined as

$$\mathbf{q}_{\beta}^{\alpha} = \begin{pmatrix} \cos(\mu_{\beta\alpha}/2) \\ e_{\beta\alpha,1}^{\alpha/\beta} \sin(\mu_{\beta\alpha}/2) \\ e_{\beta\alpha,2}^{\alpha/\beta} \sin(\mu_{\beta\alpha}/2) \\ e_{\beta\alpha,3}^{\alpha/\beta} \sin(\mu_{\beta\alpha}/2) \end{pmatrix}, \tag{2.32}$$

where $\mu_{\beta\alpha}$ and $\mathbf{e}_{\beta\alpha}^{\alpha/\beta}$ are the magnitude and axis of rotation as defined in Section 2.2.2. Conversely,

$$\mu_{\beta\alpha} = 2\arccos(q_{\beta\alpha 0}), \qquad \mathbf{e}_{\beta\alpha}^{\alpha/\beta} = \frac{\mathbf{q}_{\beta 1:3}^{\alpha}}{|\mathbf{q}_{\beta 1:3}^{\alpha}|}, \tag{2.33}$$

where $(\mathbf{q}_{1:3} = q_1, q_2, q_3)$.

With only four components, the quaternion attitude representation is more computationally efficient for some processes than the coordinate transformation matrix. It also avoids the singularities inherent in Euler angles. However, manipulation of quaternions is not intuitive, so their use makes navigation equations more difficult to follow, increasing the chances of mistakes being made. Consequently, discussion of quaternions in the main body of the book is limited to their transformations to and from the other attitude representations. More details on quaternion properties and methods may be found in Section E.6 of Appendix E on the CD.

2.2 Attitude, Rotation, and Resolving Axes Transformations

A quaternion attitude is converted to and from the corresponding coordinate transformation matrix using [3]

$$\mathbf{C}_\beta^\alpha = \begin{pmatrix} q_{\beta 0}^{\alpha\,2} + q_{\beta 1}^{\alpha\,2} - q_{\beta 2}^{\alpha\,2} - q_{\beta 3}^{\alpha\,2} & 2(q_{\beta 1}^\alpha q_{\beta 2}^\alpha + q_{\beta 3}^\alpha q_{\beta 0}^\alpha) & 2(q_{\beta 1}^\alpha q_{\beta 3}^\alpha - q_{\beta 2}^\alpha q_{\beta 0}^\alpha) \\ 2(q_{\beta 1}^\alpha q_{\beta 2}^\alpha - q_{\beta 3}^\alpha q_{\beta 0}^\alpha) & q_{\beta 0}^{\alpha\,2} - q_{\beta 1}^{\alpha\,2} + q_{\beta 2}^{\alpha\,2} - q_{\beta 3}^{\alpha\,2} & 2(q_{\beta 2}^\alpha q_{\beta 3}^\alpha + q_{\beta 1}^\alpha q_{\beta 0}^\alpha) \\ 2(q_{\beta 1}^\alpha q_{\beta 3}^\alpha + q_{\beta 2}^\alpha q_{\beta 0}^\alpha) & 2(q_{\beta 2}^\alpha q_{\beta 3}^\alpha - q_{\beta 1}^\alpha q_{\beta 0}^\alpha) & q_{\beta 0}^{\alpha\,2} - q_{\beta 1}^{\alpha\,2} - q_{\beta 2}^{\alpha\,2} + q_{\beta 3}^{\alpha\,2} \end{pmatrix}, \quad (2.34)$$

$$\begin{aligned} q_{\beta 0}^\alpha &= \tfrac{1}{2}\sqrt{1 + C_{\beta 1,1}^\alpha + C_{\beta 2,2}^\alpha + C_{\beta 3,3}^\alpha} = \tfrac{1}{2}\sqrt{1 + C_{\alpha 1,1}^\beta + C_{\alpha 2,2}^\beta + C_{\alpha 3,3}^\beta} \\ q_{\beta 1}^\alpha &= \frac{C_{\beta 2,3}^\alpha - C_{\beta 3,2}^\alpha}{4 q_{\beta 0}^\alpha} = \frac{C_{\alpha 3,2}^\beta - C_{\alpha 2,3}^\beta}{4 q_{\beta 0}^\alpha} \\ q_{\beta 2}^\alpha &= \frac{C_{\beta 3,1}^\alpha - C_{\beta 1,3}^\alpha}{4 q_{\beta 0}^\alpha} = \frac{C_{\alpha 1,3}^\beta - C_{\alpha 3,1}^\beta}{4 q_{\beta 0}^\alpha} \\ q_{\beta 3}^\alpha &= \frac{C_{\beta 1,2}^\alpha - C_{\beta 2,1}^\alpha}{4 q_{\beta 0}^\alpha} = \frac{C_{\alpha 2,1}^\beta - C_{\alpha 1,2}^\beta}{4 q_{\beta 0}^\alpha} \end{aligned} \quad . \quad (2.35)$$

In cases where $q_{\beta 0}^\alpha$ is close to zero, (2.35) should be replaced by

$$\begin{aligned} q_{\beta 1}^\alpha &= \tfrac{1}{2}\sqrt{1 + C_{\beta 1,1}^\alpha - C_{\beta 2,2}^\alpha - C_{\beta 3,3}^\alpha} = \tfrac{1}{2}\sqrt{1 + C_{\alpha 1,1}^\beta - C_{\alpha 2,2}^\beta - C_{\alpha 3,3}^\beta} \\ q_{\beta 0}^\alpha &= \frac{C_{\beta 2,3}^\alpha - C_{\beta 3,2}^\alpha}{4 q_{\beta 1}^\alpha} = \frac{C_{\alpha 3,2}^\beta - C_{\alpha 2,3}^\beta}{4 q_{\beta 1}^\alpha} \\ q_{\beta 2}^\alpha &= \frac{C_{\beta 2,1}^\alpha + C_{\beta 1,2}^\alpha}{4 q_{\beta 1}^\alpha} = \frac{C_{\alpha 1,2}^\beta + C_{\alpha 2,1}^\beta}{4 q_{\beta 1}^\alpha} \\ q_{\beta 3}^\alpha &= \frac{C_{\beta 3,1}^\alpha + C_{\beta 1,3}^\alpha}{4 q_{\beta 1}^\alpha} = \frac{C_{\alpha 1,3}^\beta + C_{\alpha 3,1}^\beta}{4 q_{\beta 1}^\alpha} \end{aligned} \quad . \quad (2.36)$$

The transformation between quaternion and Euler attitude is [3]

$$\begin{aligned} \phi_{\beta\alpha} &= \arctan_2\left[2\left(q_{\beta 0}^\alpha q_{\beta 1}^\alpha + q_{\beta 2}^\alpha q_{\beta 3}^\alpha\right), \left(1 - 2 q_{\beta 1}^{\alpha\,2} - 2 q_{\beta 2}^{\alpha\,2}\right)\right] \\ \theta_{\beta\alpha} &= \arcsin\left[2\left(q_{\beta 0}^\alpha q_{\beta 2}^\alpha - q_{\beta 1}^\alpha q_{\beta 3}^\alpha\right)\right] \\ \psi_{\beta\alpha} &= \arctan_2\left[2\left(q_{\beta 0}^\alpha q_{\beta 3}^\alpha + q_{\beta 1}^\alpha q_{\beta 2}^\alpha\right), \left(1 - 2 q_{\beta 2}^{\alpha\,2} - 2 q_{\beta 3}^{\alpha\,2}\right)\right] \end{aligned} \quad , \quad (2.37)$$

where four-quadrant arctangent functions must be used, and

$$q_{\beta 0}^{\alpha} = \cos\left(\frac{\phi_{\beta\alpha}}{2}\right)\cos\left(\frac{\theta_{\beta\alpha}}{2}\right)\cos\left(\frac{\psi_{\beta\alpha}}{2}\right) + \sin\left(\frac{\phi_{\beta\alpha}}{2}\right)\sin\left(\frac{\theta_{\beta\alpha}}{2}\right)\sin\left(\frac{\psi_{\beta\alpha}}{2}\right)$$

$$q_{\beta 1}^{\alpha} = \sin\left(\frac{\phi_{\beta\alpha}}{2}\right)\cos\left(\frac{\theta_{\beta\alpha}}{2}\right)\cos\left(\frac{\psi_{\beta\alpha}}{2}\right) - \cos\left(\frac{\phi_{\beta\alpha}}{2}\right)\sin\left(\frac{\theta_{\beta\alpha}}{2}\right)\sin\left(\frac{\psi_{\beta\alpha}}{2}\right)$$

$$q_{\beta 2}^{\alpha} = \cos\left(\frac{\phi_{\beta\alpha}}{2}\right)\sin\left(\frac{\theta_{\beta\alpha}}{2}\right)\cos\left(\frac{\psi_{\beta\alpha}}{2}\right) + \sin\left(\frac{\phi_{\beta\alpha}}{2}\right)\cos\left(\frac{\theta_{\beta\alpha}}{2}\right)\sin\left(\frac{\psi_{\beta\alpha}}{2}\right)$$

$$q_{\beta 3}^{\alpha} = \cos\left(\frac{\phi_{\beta\alpha}}{2}\right)\cos\left(\frac{\theta_{\beta\alpha}}{2}\right)\sin\left(\frac{\psi_{\beta\alpha}}{2}\right) - \sin\left(\frac{\phi_{\beta\alpha}}{2}\right)\sin\left(\frac{\theta_{\beta\alpha}}{2}\right)\cos\left(\frac{\psi_{\beta\alpha}}{2}\right)$$

(2.38)

Example 2.1 on the CD also illustrates the conversion of quaternion attitude to and from the coordinate transformation matrix and Euler forms.

2.2.4 Rotation Vector

The final method of representing attitude discussed here is the rotation vector [4]. This is a three-component vector, ρ (some authors use σ), and is simply the product of the axis-of-rotation unit vector and the magnitude of the rotation. Thus,

$$\rho_{\beta\alpha} = \mu_{\beta\alpha} e_{\beta\alpha}^{\alpha/\beta}. \tag{2.39}$$

Conversely,

$$\mu_{\beta\alpha} = |\rho_{\beta\alpha}|, \qquad e_{\beta\alpha}^{\alpha/\beta} = \frac{\rho_{\beta\alpha}}{|\rho_{\beta\alpha}|}. \tag{2.40}$$

Like quaternion attitude, manipulation of rotation vectors is not intuitive, so coverage in this book is limited. More details on rotation vector methods in navigation may be found in [2].

The transformation between a rotation vector and quaternion attitude is:

$$q_{\beta 0}^{\alpha} = \cos\left(\frac{|\rho_{\beta\alpha}|}{2}\right), \qquad q_{\beta 1:3}^{\alpha} = \sin\left(\frac{|\rho_{\beta\alpha}|}{2}\right)\frac{\rho_{\beta\alpha}}{|\rho_{\beta\alpha}|}, \tag{2.41}$$

$$\rho_{\beta\alpha} = \frac{2\arccos(q_{\beta 0}^{\alpha})}{\sqrt{1 - q_{\beta 0}^{\alpha\,2}}} q_{\beta 1:3}^{\alpha}. \tag{2.42}$$

A rotation vector is converted to a coordinate transformation matrix using

$$\begin{aligned}C_{\beta}^{\alpha} &= \exp[-\rho_{\beta\alpha}\wedge] \\ &= I_3 - \frac{\sin|\rho_{\beta\alpha}|}{|\rho_{\beta\alpha}|}[\rho_{\beta\alpha}\wedge] + \frac{1-\cos|\rho_{\beta\alpha}|}{|\rho_{\beta\alpha}|^2}[\rho_{\beta\alpha}\wedge]^2.\end{aligned} \tag{2.43}$$

From (2.30) and (2.39), the reverse transformation is

$$\boldsymbol{\rho}_{\beta\alpha} = \frac{\mu_{\beta\alpha}}{2\sin\mu_{\beta\alpha}} \begin{pmatrix} C^{\alpha}_{\beta 2,3} - C^{\alpha}_{\beta 3,2} \\ C^{\alpha}_{\beta 3,1} - C^{\alpha}_{\beta 1,3} \\ C^{\alpha}_{\beta 1,2} - C^{\alpha}_{\beta 2,1} \end{pmatrix}, \qquad (2.44)$$

where $\mu_{\beta\alpha}$ is defined in terms of $\mathbf{C}^{\alpha}_{\beta}$ by (2.31).

The transformation from a rotation vector to the corresponding Euler attitude is

$$\phi_{\beta\alpha} = \arctan_2\left[\left(\frac{\sin|\boldsymbol{\rho}_{\beta\alpha}|}{|\boldsymbol{\rho}_{\beta\alpha}|}\rho_{\beta\alpha 1} + \frac{1-\cos|\boldsymbol{\rho}_{\beta\alpha}|}{|\boldsymbol{\rho}_{\beta\alpha}|^2}\rho_{\beta\alpha 2}\rho_{\beta\alpha 3}\right), \frac{\rho^2_{\beta\alpha 3} - \cos|\boldsymbol{\rho}_{\beta\alpha}|(\rho^2_{\beta\alpha 1} + \rho^2_{\beta\alpha 2})}{|\boldsymbol{\rho}_{\beta\alpha}|^2}\right]$$

$$\theta_{\beta\alpha} = \arcsin\left[\frac{\sin|\boldsymbol{\rho}_{\beta\alpha}|}{|\boldsymbol{\rho}_{\beta\alpha}|}\rho_{\beta\alpha 2} - \frac{1-\cos|\boldsymbol{\rho}_{\beta\alpha}|}{|\boldsymbol{\rho}_{\beta\alpha}|^2}\rho_{\beta\alpha 1}\rho_{\beta\alpha 3}\right],$$

$$\psi_{\beta\alpha} = \arctan_2\left[\left(\frac{\sin|\boldsymbol{\rho}_{\beta\alpha}|}{|\boldsymbol{\rho}_{\beta\alpha}|}\rho_{\beta\alpha 3} + \frac{1-\cos|\boldsymbol{\rho}_{\beta\alpha}|}{|\boldsymbol{\rho}_{\beta\alpha}|^2}\rho_{\beta\alpha 1}\rho_{\beta\alpha 2}\right), \frac{\rho^2_{\beta\alpha 1} - \cos|\boldsymbol{\rho}_{\beta\alpha}|(\rho^2_{\beta\alpha 2} + \rho^2_{\beta\alpha 3})}{|\boldsymbol{\rho}_{\beta\alpha}|^2}\right]$$

$$(2.45)$$

noting that the rotation vector and Euler angles are the same in the small angle approximation. In general, the reverse transformation is complicated, so it is better performed via the quaternion attitude or coordinate transformation matrix.

Rotation vectors are useful for interpolating attitudes as they are the only form of attitude that enables rotations to be linearly interpolated. For example, if the frame γ is at the orientation where a proportion k of the rotation from frame β to frame α has completed, the rotation vectors describing the relative attitudes of the three frames are related by

$$\begin{aligned}\boldsymbol{\rho}_{\beta\gamma} &= k\boldsymbol{\rho}_{\beta\alpha} \\ \boldsymbol{\rho}_{\gamma\alpha} &= (1-k)\boldsymbol{\rho}_{\beta\alpha}\end{aligned}. \qquad (2.46)$$

Note that noncolinear rotation vectors neither commute nor combine additively.

2.3 Kinematics

In navigation, the linear and angular motion of one coordinate frame must be described with respect to another. Kinematics is the study of the motion of objects without consideration of the causes of that motion. This is in contrast to dynamics, which studies the relationship between the motion of objects and its causes.

Most kinematic quantities, such as position, velocity, acceleration, and angular rate, involve three coordinate frames:

- The frame whose motion is described, known as the *object frame*, α;
- The frame with which that motion is respect to, known as the *reference frame*, β;
- The set of axes in which that motion is represented, known as the *resolving frame*, γ.

The object frame, α, and the reference frame, β, must be different; otherwise, there is no motion. The resolving frame, γ, may be either the object frame, the reference frame, or a third frame. Its origin need not be defined; only the orientation of its axes is required. Note also that the choice of resolving frame does not affect the magnitude of a vector.

To describe these kinematic quantities fully, all three frames must be explicitly stated. Most authors do not do this, potentially causing confusion. Here, the following notation is used for Cartesian position, velocity, acceleration, and angular rate:

$$\mathbf{x}^{\gamma}_{\beta\alpha}$$

where the vector, $\mathbf{x}$, describes a kinematic property of frame α with respect to frame β, expressed in the frame γ axes. Note that, for attitude, only the object frame, α, and reference frame, β, are involved; there is no resolving frame.

In this section, the angular rate, Cartesian (as opposed to curvilinear) position, velocity, and acceleration are described in turn, correctly accounting for any rotation of the reference frame and resolving frame. Motion with respect to a rotating reference frame and the ensuing centrifugal and Coriolis pseudo-forces are then described.

2.3.1 Angular Rate

The *angular rate vector*, $\boldsymbol{\omega}^{\gamma}_{\beta\alpha}$, is the rate of rotation of the α-frame axes with respect to the β-frame axes, resolved about the γ-frame axes. Figure 2.13 illustrates the directions of the angular rate vector and the corresponding rotation that it represents. The rotation is within the plane perpendicular to the angular rate vector. Some authors use the notation p, q, and r to denote the components of angular rate about, respectively, the x-, y-, and z-axes of the resolving frame, so $\boldsymbol{\omega}^{\gamma}_{\beta\alpha} = (p^{\gamma}_{\beta\alpha}, q^{\gamma}_{\beta\alpha}, r^{\gamma}_{\beta\alpha})$.

The object and reference frames of an angular rate may be transposed simply by reversing the sign:

$$\boldsymbol{\omega}^{\gamma}_{\beta\alpha} = -\boldsymbol{\omega}^{\gamma}_{\alpha\beta}. \tag{2.47}$$

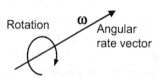

Figure 2.13 Angular rate rotation and vector directions.

2.3 Kinematics

Angular rates resolved about the same axes may simply be added, provided the object frame of one angular rate is the same as the reference frame of the other; thus

$$\omega^{\gamma}_{\beta\alpha} = \omega^{\gamma}_{\beta\delta} + \omega^{\gamma}_{\delta\alpha}. \qquad (2.48)$$

The resolving axes may be changed simply by premultiplying by the relevant coordinate transformation matrix:

$$\omega^{\delta}_{\beta\alpha} = \mathbf{C}^{\delta}_{\gamma}\omega^{\gamma}_{\beta\alpha}. \qquad (2.49)$$

Note that the magnitude of the angular rate, $|\omega^{\gamma}_{\beta\alpha}|$, is independent of the resolving axes, so may be written simply as $\omega_{\beta\alpha}$. However, the magnitude of the angular acceleration, $|\dot{\omega}^{\gamma}_{\beta\alpha}|$, does depend on the choice of resolving frame.

The skew-symmetric matrix of the angular rate vector is also commonly used:

$$\mathbf{\Omega}^{\gamma}_{\beta\alpha} = \left[\omega^{\gamma}_{\beta\alpha} \wedge \right] = \begin{pmatrix} 0 & -\omega^{\gamma}_{\beta\alpha 3} & \omega^{\gamma}_{\beta\alpha 2} \\ \omega^{\gamma}_{\beta\alpha 3} & 0 & -\omega^{\gamma}_{\beta\alpha 1} \\ -\omega^{\gamma}_{\beta\alpha 2} & \omega^{\gamma}_{\beta\alpha 1} & 0 \end{pmatrix}, \qquad (2.50)$$

where the resolving frame, γ, of the vector $\omega^{\gamma}_{\beta\alpha}$ applies to both the rows and the columns of its skew-symmetric matrix $\mathbf{\Omega}^{\gamma}_{\beta\alpha}$. Therefore, from (2.21), skew-symmetric matrices transform as

$$\mathbf{\Omega}^{\delta}_{\beta\alpha} = \mathbf{C}^{\delta}_{\gamma}\mathbf{\Omega}^{\gamma}_{\beta\alpha}\mathbf{C}^{\gamma}_{\delta}. \qquad (2.51)$$

The time derivative of a coordinate transformation matrix is defined as [5, 6]

$$\dot{\mathbf{C}}^{\alpha}_{\beta}(t) = \lim_{\delta t \to 0}\left(\frac{\mathbf{C}^{\alpha}_{\beta}(t+\delta t) - \mathbf{C}^{\alpha}_{\beta}(t)}{\delta t}\right). \qquad (2.52)$$

If the object frame, α, is considered to be rotating with respect to a stationary reference frame, β, the coordinate transformation matrix at time $t+\delta t$ may be written as

$$\mathbf{C}^{\alpha}_{\beta}(t+\delta t) = \mathbf{C}^{\alpha(t+\delta t)}_{\alpha(t)}\mathbf{C}^{\alpha}_{\beta}(t). \qquad (2.53)$$

The rotation of the object frame over the interval t to $t+\delta t$ is infinitesimal, so may be represented by the small angle $\mathbf{\psi}_{\alpha(t)\alpha(t+\delta t)}$. Therefore, from (2.26),

$$\begin{aligned}\mathbf{C}^{\alpha}_{\beta}(t+\delta t) &= \left(\mathbf{I}_3 - \left[\mathbf{\psi}_{\alpha(t)\alpha(t+\delta t)} \wedge\right]\right)\mathbf{C}^{\alpha}_{\beta}(t) \\ &= \left(\mathbf{I}_3 - \delta t\left[\omega^{\alpha}_{\beta\alpha} \wedge\right]\right)\mathbf{C}^{\alpha}_{\beta}(t) \\ &= \left(\mathbf{I}_3 - \delta t\mathbf{\Omega}^{\alpha}_{\beta\alpha}\right)\mathbf{C}^{\alpha}_{\beta}(t)\end{aligned} \qquad (2.54)$$

Substituting this into (2.52) gives

$$\dot{C}^\alpha_\beta = -\Omega^\alpha_{\beta\alpha} C^\alpha_\beta. \tag{2.55}$$

If the above steps are repeated under the assumption that the β frame is rotating and the α frame is stationary, the result $\dot{C}^\alpha_\beta = -C^\alpha_\beta \Omega^\beta_{\beta\alpha}$ is obtained. However, applying (2.51) and (2.17) shows that these results are equivalent. From (2.47), the general result is

$$\begin{aligned} \dot{C}^\alpha_\beta &= -C^\alpha_\beta \Omega^\beta_{\beta\alpha} = C^\alpha_\beta \Omega^\beta_{\alpha\beta} \\ &= -\Omega^\alpha_{\beta\alpha} C^\alpha_\beta = \Omega^\alpha_{\alpha\beta} C^\alpha_\beta \end{aligned}. \tag{2.56}$$

The inverse relationship is

$$\begin{aligned} \Omega^\alpha_{\alpha\beta} &= \dot{C}^\alpha_\beta C^\beta_\alpha, & \Omega^\alpha_{\beta\alpha} &= -\dot{C}^\alpha_\beta C^\beta_\alpha \\ \Omega^\beta_{\alpha\beta} &= C^\beta_\alpha \dot{C}^\alpha_\beta, & \Omega^\beta_{\beta\alpha} &= -C^\beta_\alpha \dot{C}^\alpha_\beta \end{aligned}. \tag{2.57}$$

The time derivative of the Euler attitude may be expressed in terms of the angular rate using [5]

$$\begin{pmatrix} \dot{\phi}_{\beta\alpha} \\ \dot{\theta}_{\beta\alpha} \\ \dot{\psi}_{\beta\alpha} \end{pmatrix} = \begin{pmatrix} 1 & \sin\phi_{\beta\alpha}\tan\theta_{\beta\alpha} & \cos\phi_{\beta\alpha}\tan\theta_{\beta\alpha} \\ 0 & \cos\phi_{\beta\alpha} & -\sin\phi_{\beta\alpha} \\ 0 & \sin\phi_{\beta\alpha}/\cos\theta_{\beta\alpha} & \cos\phi_{\beta\alpha}/\cos\theta_{\beta\alpha} \end{pmatrix} \omega^\alpha_{\beta\alpha}. \tag{2.58}$$

The inverse relationship is

$$\omega^\alpha_{\beta\alpha} = \begin{pmatrix} 1 & 0 & -\sin\theta_{\beta\alpha} \\ 0 & \cos\phi_{\beta\alpha} & \sin\phi_{\beta\alpha}\cos\theta_{\beta\alpha} \\ 0 & -\sin\phi_{\beta\alpha} & \cos\phi_{\beta\alpha}\cos\theta_{\beta\alpha} \end{pmatrix} \begin{pmatrix} \dot{\phi}_{\beta\alpha} \\ \dot{\theta}_{\beta\alpha} \\ \dot{\psi}_{\beta\alpha} \end{pmatrix}. \tag{2.59}$$

2.3.2 Cartesian Position

As Figure 2.14 shows, the *Cartesian position* of the origin of frame α with respect to the origin of frame β, resolved about the axes of frame γ, is $\mathbf{r}^\gamma_{\beta\alpha} = (x^\gamma_{\beta\alpha}, y^\gamma_{\beta\alpha}, z^\gamma_{\beta\alpha})$, where x, y, and z are the components of position in the x, y, and z axes of the γ frame. Cartesian position differs from curvilinear position (Section 2.4.2) in that the resolving axes are independent of the position vector. It is also known as the Euclidean position.

The object and reference frames of a Cartesian position may be transposed simply by reversing the sign:

$$\mathbf{r}^\gamma_{\beta\alpha} = -\mathbf{r}^\gamma_{\alpha\beta}. \tag{2.60}$$

2.3 Kinematics

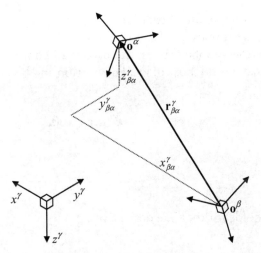

Figure 2.14 Position of the origin of frame α with respect to the origin of frame β in frame γ axes.

Similarly, two positions with common resolving axes may be subtracted if the reference frames are common or added provided the object frame of one matches the reference frame of the other:

$$\begin{aligned} \mathbf{r}_{\beta\alpha}^{\gamma} &= \mathbf{r}_{\delta\alpha}^{\gamma} - \mathbf{r}_{\delta\beta}^{\gamma} \\ &= \mathbf{r}_{\beta\delta}^{\gamma} + \mathbf{r}_{\delta\alpha}^{\gamma} \end{aligned} \qquad (2.61)$$

This may also be used to transform a position from one reference frame to another and holds for time derivatives.

Position may be resolved in a different frame by applying a coordinate transformation matrix:[*]

$$\mathbf{r}_{\beta\alpha}^{\delta} = \mathbf{C}_{\gamma}^{\delta} \mathbf{r}_{\beta\alpha}^{\gamma}. \qquad (2.62)$$

Note that[†]

$$\mathbf{r}_{\alpha\beta}^{\alpha} = -\mathbf{C}_{\beta}^{\alpha} \mathbf{r}_{\beta\alpha}^{\beta}. \qquad (2.63)$$

and

$$\begin{aligned} \mathbf{r}_{\beta\alpha}^{\beta} &= \mathbf{C}_{\delta}^{\beta} \left(\mathbf{r}_{\beta\delta}^{\delta} + \mathbf{r}_{\delta\alpha}^{\delta} \right) \\ &= \mathbf{r}_{\beta\delta}^{\beta} + \mathbf{C}_{\delta}^{\beta} \mathbf{r}_{\delta\alpha}^{\delta} \end{aligned} \qquad (2.64)$$

The magnitude of the Cartesian position, $\left| \mathbf{r}_{\beta\alpha}^{\gamma} \right|$, is independent of the resolving axes, so it may be written simply as $r_{\beta\alpha}$. However, the magnitude of its time derivative,

---

[*]This and subsequent paragraphs are based on material written by the author for QinetiQ, so comprise QinetiQ copyright material.

[†]End of QinetiQ copyright material.

$|\dot{\mathbf{r}}^\gamma_{\beta\alpha}|$, depends on the rate of rotation of the resolving frame with respect to the reference and object frames (see Section 2.3.3).

Considering specific frames, the origins of commonly realized ECI and ECEF frames coincide, as do those of local navigation and body frames for the same object. Therefore,

$$\mathbf{r}^\gamma_{ie} = \mathbf{r}^\gamma_{nb} = 0, \tag{2.65}$$

and

$$\mathbf{r}^\gamma_{ib} = \mathbf{r}^\gamma_{eb} = \mathbf{r}^\gamma_{in} = \mathbf{r}^\gamma_{en}, \tag{2.66}$$

which also holds for the time derivatives.

2.3.3 Velocity

Velocity is defined as the rate of change of the position of the origin of an object frame with respect to the origin and axes of a reference frame. This may, in turn, be resolved about the axes of a third frame. Thus, the velocity of frame α with respect to frame β, resolved about the axes of frame γ, is[‡]

$$\mathbf{v}^\gamma_{\beta\alpha} = \mathbf{C}^\gamma_\beta \dot{\mathbf{r}}^\beta_{\beta\alpha} \tag{2.67}$$

A velocity is thus registered if the object frame, α, moves with respect to the β-frame origin, or the reference frame, β, moves with respect to the α-frame origin. However, the velocity is defined not only with respect to the origin of the reference frame, but with respect to the axes as well. Therefore, a velocity is also registered if the reference frame, β, rotates with respect to the α-frame origin. For example, if an observer is spinning on an office chair, surrounding objects will be moving with respect to the axes of a chair-fixed reference frame. This is important in navigation as many of the commonly used reference frames rotate with respect to each other.

Figure 2.15 illustrates the three types of motion that register a velocity. No velocity is registered if the object frame rotates. Rotation of the resolving axes, γ, with respect to the reference frame, β, has no impact on the magnitude of the velocity.

It should be noted that the velocity, $\mathbf{v}^\gamma_{\beta\alpha}$, is not equal to the time derivative of the Cartesian position, $\mathbf{r}^\gamma_{\beta\alpha}$, where there is rotation of the resolving frame, γ, with respect to the reference frame, β. From (2.62) and (2.67),

$$\begin{aligned}\dot{\mathbf{r}}^\gamma_{\beta\alpha} &= \dot{\mathbf{C}}^\gamma_\beta \mathbf{r}^\beta_{\beta\alpha} + \mathbf{C}^\gamma_\beta \dot{\mathbf{r}}^\beta_{\beta\alpha} \\ &= \dot{\mathbf{C}}^\gamma_\beta \mathbf{r}^\beta_{\beta\alpha} + \mathbf{v}^\gamma_{\beta\alpha}\end{aligned}. \tag{2.68}$$

Rotation between the resolving axes and the reference frame is important in navigation because a local navigation frame rotates with respect to an ECEF frame as the origin of the former moves with respect to the Earth.

[‡]This paragraph, up to this point, is based on material written by the author for QinetiQ, so comprises QinetiQ copyright material.

2.3 Kinematics

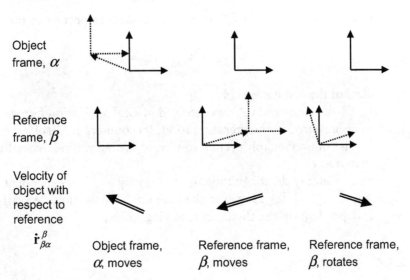

Figure 2.15 Motion causing a velocity to register.

Unlike with Cartesian position, the reference and object frames cannot be interchanged by reversing the sign unless there is no angular motion between them. The correct relationship is

$$\mathbf{v}_{\alpha\beta}^{\gamma} = -\mathbf{v}_{\beta\alpha}^{\gamma} - \mathbf{C}_{\alpha}^{\gamma}\dot{\mathbf{C}}_{\beta}^{\alpha}\mathbf{r}_{\beta\alpha}^{\beta}, \tag{2.69}$$

although

$$\mathbf{v}_{\alpha\beta}^{\gamma}\big|_{\dot{\mathbf{C}}_{\beta}^{\alpha}=0} = -\mathbf{v}_{\beta\alpha}^{\gamma}. \tag{2.70}$$

Similarly, addition of velocities is not valid if the reference frames are rotating with respect to each other. Thus,

$$\mathbf{v}_{\beta\alpha}^{\gamma} \neq \mathbf{v}_{\beta\delta}^{\gamma} + \mathbf{v}_{\delta\alpha}^{\gamma}, \tag{2.71}$$

although

$$\mathbf{v}_{\beta\alpha}^{\gamma}\big|_{\dot{\mathbf{C}}_{\beta}^{\delta}=0} = \mathbf{v}_{\beta\delta}^{\gamma} + \mathbf{v}_{\delta\alpha}^{\gamma}. \tag{2.72}$$

Velocity may be transformed from one resolving frame to another using the appropriate coordinate transformation matrix:

$$\mathbf{v}_{\beta\alpha}^{\delta} = \mathbf{C}_{\gamma}^{\delta}\mathbf{v}_{\beta\alpha}^{\gamma}. \tag{2.73}$$

Commonly-realized ECI and ECEF frames have a common origin, as do body and local navigation frames of the same object. Therefore,

$$\mathbf{v}_{ie}^{\gamma} = \mathbf{v}_{nb}^{\gamma} = 0, \quad \mathbf{v}_{ib}^{\gamma} = \mathbf{v}_{in}^{\gamma}, \quad \mathbf{v}_{eb}^{\gamma} = \mathbf{v}_{en}^{\gamma}. \tag{2.74}$$

However, because an ECEF frame rotates with respect to an inertial frame,

$$\mathbf{v}_{ib}^{\gamma} \neq \mathbf{v}_{eb}^{\gamma}, \quad \mathbf{v}_{in}^{\gamma} \neq \mathbf{v}_{en}^{\gamma}, \tag{2.75}$$

regardless of the resolving axes.

The Earth-referenced velocity resolved in local navigation frame axes, $\mathbf{v}_{eb}^{n}$ or $\mathbf{v}_{en}^{n}$, is often abbreviated in the literature to $\mathbf{v}^{n}$. Its counterpart resolved in ECEF frame axes, $\mathbf{v}_{eb}^{e}$, is commonly abbreviated to $\mathbf{v}^{e}$, and the inertial-referenced velocity, $\mathbf{v}_{ib}^{i}$, is abbreviated to $\mathbf{v}^{i}$.

Speed is simply the magnitude of the velocity and is independent of the resolving axes, so $v_{\beta\alpha} = |\mathbf{v}_{\beta\alpha}^{\gamma}|$. However, the magnitude of the time derivative of velocity, $|\dot{\mathbf{v}}_{\beta\alpha}^{\gamma}|$, is dependent on the choice of resolving frame.

2.3.4 Acceleration

Acceleration is defined as the second time derivative of the position of the origin of one frame with respect to the origin and axes of another frame. Thus, the acceleration of frame α with respect to frame β, resolved about the axes of frame γ, is[‡]

$$\mathbf{a}_{\beta\alpha}^{\gamma} = \mathbf{C}_{\beta}^{\gamma}\ddot{\mathbf{r}}_{\beta\alpha}^{\beta}. \tag{2.76}$$

The acceleration is the force per unit mass on the object applied from the reference frame. Its magnitude is necessarily independent of the resolving frame. It is not the same as the time derivative of $\mathbf{v}_{\beta\alpha}^{\gamma}$ or the second time derivative of $\mathbf{r}_{\beta\alpha}^{\gamma}$. These depend on the rotation of the resolving frame, γ, with respect to the reference frame, β:

$$\dot{\mathbf{v}}_{\beta\alpha}^{\gamma} = \dot{\mathbf{C}}_{\beta}^{\gamma}\dot{\mathbf{r}}_{\beta\alpha}^{\beta} + \mathbf{a}_{\beta\alpha}^{\gamma}, \tag{2.77}$$

$$\begin{aligned}\ddot{\mathbf{r}}_{\beta\alpha}^{\gamma} &= \ddot{\mathbf{C}}_{\beta}^{\gamma}\mathbf{r}_{\beta\alpha}^{\beta} + \dot{\mathbf{C}}_{\beta}^{\gamma}\dot{\mathbf{r}}_{\beta\alpha}^{\beta} + \dot{\mathbf{v}}_{\beta\alpha}^{\gamma} \\ &= \ddot{\mathbf{C}}_{\beta}^{\gamma}\mathbf{r}_{\beta\alpha}^{\beta} + 2\dot{\mathbf{C}}_{\beta}^{\gamma}\dot{\mathbf{r}}_{\beta\alpha}^{\beta} + \mathbf{a}_{\beta\alpha}^{\gamma}.\end{aligned} \tag{2.78}$$

From (2.56) and (2.62),

$$\ddot{\mathbf{C}}_{\beta}^{\gamma}\mathbf{r}_{\beta\alpha}^{\beta} = \left(\mathbf{\Omega}_{\beta\gamma}^{\gamma}\mathbf{\Omega}_{\beta\gamma}^{\gamma} - \dot{\mathbf{\Omega}}_{\beta\gamma}^{\gamma}\right)\mathbf{r}_{\beta\alpha}^{\gamma}, \tag{2.79}$$

while from (2.68), (2.56), and (2.62),

$$\begin{aligned}\dot{\mathbf{C}}_{\beta}^{\gamma}\dot{\mathbf{r}}_{\beta\alpha}^{\beta} &= -\mathbf{\Omega}_{\beta\gamma}^{\gamma}\mathbf{C}_{\beta}^{\gamma}\dot{\mathbf{r}}_{\beta\alpha}^{\beta} = \mathbf{\Omega}_{\beta\gamma}^{\gamma}\left(\dot{\mathbf{C}}_{\beta}^{\gamma}\mathbf{r}_{\beta\alpha}^{\beta} - \dot{\mathbf{r}}_{\beta\alpha}^{\gamma}\right) \\ &= -\mathbf{\Omega}_{\beta\gamma}^{\gamma}\mathbf{\Omega}_{\beta\gamma}^{\gamma}\mathbf{r}_{\beta\alpha}^{\gamma} - \mathbf{\Omega}_{\beta\gamma}^{\gamma}\dot{\mathbf{r}}_{\beta\alpha}^{\gamma}.\end{aligned} \tag{2.80}$$

[‡]This paragraph, up to this point, is based on material written by the author for QinetiQ, so comprises QinetiQ copyright material.

2.3 Kinematics

Substituting these into (2.78) gives

$$\ddot{\mathbf{r}}^\gamma_{\beta\alpha} = -\Omega^\gamma_{\beta\gamma}\Omega^\gamma_{\beta\gamma}\mathbf{r}^\gamma_{\beta\alpha} - 2\Omega^\gamma_{\beta\gamma}\dot{\mathbf{r}}^\gamma_{\beta\alpha} - \dot{\Omega}^\gamma_{\beta\gamma}\mathbf{r}^\gamma_{\beta\alpha} + \mathbf{a}^\gamma_{\beta\alpha}. \tag{2.81}$$

The first three terms of this are related to the centrifugal, Coriolis, and Euler pseudo-forces described in Section 2.3.5 [7].

As with velocity, addition of accelerations is not valid if the reference frames are rotating with respect to each other:

$$\mathbf{a}^\gamma_{\beta\alpha} \neq \mathbf{a}^\gamma_{\beta\delta} + \mathbf{a}^\gamma_{\delta\alpha}, \tag{2.82}$$

Similarly, an acceleration may be resolved about a different set of axes by applying the appropriate coordinate transformation matrix:

$$\mathbf{a}^\delta_{\beta\alpha} = \mathbf{C}^\delta_\gamma \mathbf{a}^\gamma_{\beta\alpha}. \tag{2.83}$$

2.3.5 Motion with Respect to a Rotating Reference Frame

In navigation, it is convenient to describe the motion of objects with respect to a rotating reference frame, such as an ECEF frame. Newton's laws of motion state that, with respect to an inertial reference frame, an object will move at constant velocity unless acted upon by a force. This does not apply with respect to a rotating frame.

Consider an object that is stationary with respect to a reference frame that is rotating at a constant rate. With respect to an inertial frame, the same object is moving in a circle centered about the axis of rotation of the rotating frame (assuming that axis is fixed with respect to the inertial frame). As the object is moving in a circle with respect to inertial space, it must be subject to a force. If the position of the object, α, with respect to inertial frame, i, is described by

$$\begin{aligned} x^i_{i\alpha} &= r\cos\omega_{i\beta}t \\ y^i_{i\alpha} &= r\sin\omega_{i\beta}t \end{aligned}, \tag{2.84}$$

where $\omega_{i\beta}$ is the angular rate and t is time, then the acceleration is

$$\begin{pmatrix} \ddot{x}^i_{i\alpha} \\ \ddot{y}^i_{i\alpha} \end{pmatrix} = -\omega^2_{i\beta} r \begin{pmatrix} \cos\omega_{i\beta}t \\ \sin\omega_{i\beta}t \end{pmatrix} = -\omega^2_{i\beta} \begin{pmatrix} x^i_{i\alpha} \\ y^i_{i\alpha} \end{pmatrix}. \tag{2.85}$$

Thus, the acceleration is towards the axis of rotation. This is *centripetal acceleration* and the corresponding force is *centripetal force*. A person on a carousel (roundabout) must be subject to a centripetal force in order to remain on the carousel.

With respect to the rotating reference frame, however, the acceleration of the object is zero. The centripetal force is still present. Therefore, from the perspective of the rotating frame, there must be another force that is equal and opposite to the centripetal force. This is the *centrifugal force* and is an example of a pseudo-force,

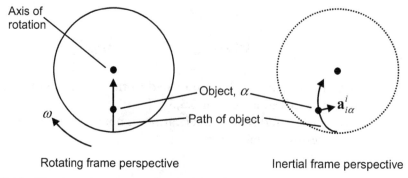

Figure 2.16 Object moving at constant velocity with respect to a rotating frame.

also known as a virtual force or a fictitious force. It arises from the use of a rotating reference frame rather than from a physical process. However, it can behave like a real force. A person on a carousel who is subject to insufficient centripetal force will appear to be pulled off the carousel by the centrifugal force.

Consider now an object that, with respect to the rotating reference frame, is moving towards the axis of rotation at a constant velocity. With respect to the inertial frame, the object is moving in a curved path and must therefore be accelerating. Figure 2.16 illustrates this. The object's velocity, with respect to inertial space, along the direction of rotation reduces as it approaches the axis of rotation. Therefore, it must be subject to a retarding force along the direction opposing rotation as well as to the centripetal force. With respect to the rotating frame, the object is moving at constant velocity, so it must have zero acceleration. Therefore, there must be a second pseudo-force that opposes the retarding force. This is the *Coriolis force* [7].

The Coriolis acceleration is always in a direction perpendicular to the object's velocity with respect to the rotating reference frame. Figure 2.17 presents some examples. If an object is set in motion, but no force is applied to maintain a constant velocity with respect to a rotating reference frame, its velocity will be constant with respect to an inertial frame. Therefore, with respect to the rotating frame, its path will appear to be curved by the action of the Coriolis force. This may be demonstrated experimentally by throwing or rolling a ball on a carousel.

Consider an object that is stationary with respect to an inertial frame. With respect to a rotating frame, the object is ascribing a circle centered at the rotation axis and is thus subject to centripetal acceleration. However, all objects described with respect to a rotating reference frame are subject to a centrifugal acceleration, in this case equal and opposite to the centripetal acceleration. However, there is no contradiction because the object is moving with respect to the rotating frame and is thus subject to a Coriolis acceleration. The centrifugal and Coriolis pseudo-accelerations sum to the centripetal acceleration required to describe the object's motion with respect to the rotating frame.

Consider the motion of an object frame, α, with respect to a rotating reference frame, β. The pseudo-acceleration, $\mathbf{a}_{\beta\alpha}^{P\beta}$, is obtained by subtracting the difference in inertially referenced accelerations of the object and reference from the total acceleration. Thus,

$$\mathbf{a}_{\beta\alpha}^{P\beta} = \mathbf{a}_{\beta\alpha}^{\beta} - \mathbf{a}_{i\alpha}^{\beta} + \mathbf{a}_{i\beta}^{\beta}, \qquad (2.86)$$

2.4 Earth Surface and Gravity Models

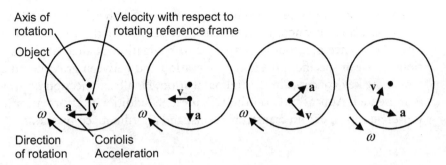

Figure 2.17 Examples of Coriolis acceleration.

where i is an inertial frame. Applying (2.76) and then (2.61),

$$\begin{aligned} \mathbf{a}_{\beta\alpha}^{P\beta} &= \ddot{\mathbf{r}}_{\beta\alpha}^{\beta} - \mathbf{C}_i^{\beta}\left(\ddot{\mathbf{r}}_{i\alpha}^{i} - \ddot{\mathbf{r}}_{i\beta}^{i}\right) \\ &= \ddot{\mathbf{r}}_{\beta\alpha}^{\beta} - \mathbf{C}_i^{\beta}\ddot{\mathbf{r}}_{\beta\alpha}^{i} \end{aligned} \quad (2.87)$$

From (2.29),

$$\mathbf{r}_{\beta\alpha}^{\beta} = \mathbf{C}_i^{\beta}\mathbf{r}_{\beta\alpha}^{i}. \quad (2.88)$$

Differentiating this twice,

$$\ddot{\mathbf{r}}_{\beta\alpha}^{\beta} = \ddot{\mathbf{C}}_i^{\beta}\mathbf{r}_{\beta\alpha}^{i} + 2\dot{\mathbf{C}}_i^{\beta}\dot{\mathbf{r}}_{\beta\alpha}^{i} + \mathbf{C}_i^{\beta}\ddot{\mathbf{r}}_{\beta\alpha}^{i}. \quad (2.89)$$

Substituting this into (2.87),

$$\mathbf{a}_{\beta\alpha}^{P\beta} = \ddot{\mathbf{C}}_i^{\beta}\mathbf{r}_{\beta\alpha}^{i} + 2\dot{\mathbf{C}}_i^{\beta}\dot{\mathbf{r}}_{\beta\alpha}^{i}. \quad (2.90)$$

Applying (2.79) and (2.80) and rearranging,

$$\mathbf{a}_{\beta\alpha}^{P\beta} = -\mathbf{\Omega}_{i\beta}^{\beta}\mathbf{\Omega}_{i\beta}^{\beta}\mathbf{r}_{\beta\alpha}^{\beta} - 2\mathbf{\Omega}_{i\beta}^{\beta}\dot{\mathbf{r}}_{\beta\alpha}^{\beta} - \dot{\mathbf{\Omega}}_{i\beta}^{\beta}\mathbf{r}_{\beta\alpha}^{\beta}, \quad (2.91)$$

where the first term is the centrifugal acceleration, the second term is the Coriolis acceleration, and the final term is the Euler acceleration. The Euler force is the third pseudo-force and arises when the reference frame undergoes angular acceleration with respect to inertial space.

2.4 Earth Surface and Gravity Models

For most applications, a position solution with respect to the Earth's surface is required. Obtaining this requires a reference surface to be defined with respect to the center and axes of the Earth. A set of coordinates for expressing position with respect to that surface, the latitude, longitude, and height, must then be defined. For mapping, a method of projecting these coordinates onto a flat surface is required. To transform inertially referenced measurements to Earth referenced, the Earth's rotation must also be defined. This section addresses each of these issues in turn, showing how the science of geodesy is applied to navigation. It then explains the distinctions between specific force and acceleration and between gravity and gravitation, which are key concepts in inertial navigation. Finally, a selection of gravity models is presented. Appendix C on the CD presents additional information on position representations, datum transformations, and coordinate conversions.

2.4.1 The Ellipsoid Model of the Earth's Surface

An Earth-centered Earth-fixed coordinate frame enables the user to navigate with respect to the center of the Earth. However, for most practical navigation problems, the user wants to know his or her position relative to the Earth's surface. The first step is to define that surface in an ECEF frame. The Earth's surface is an oblate spheroid. *Oblate* means that it is wider at its equatorial plane than along its axis of rotational symmetry, while *spheroid* means that it is close to a sphere. Unfortunately, the Earth's surface is irregular. Modeling it accurately within a navigation system is not practical, requiring a large amount of data storage and more complex navigation algorithms. Therefore, the Earth's surface is approximated to a regular shape, which is then fitted to the true surface of the Earth at mean sea level.

The model of the Earth's surface used in most navigation systems is an oblate ellipsoid of revolution. Figure 2.18 depicts a cross-section of this reference ellipsoid, noting that this and subsequent diagrams exaggerate the flattening of the Earth. The ellipsoid exhibits rotational symmetry about the north-south (z^e) axis and mirror symmetry over the equatorial plane. It is defined by two radii. The equatorial radius, R_0, or the length of the semi-major axis, a, is the distance from the center to any point on the equator, which is the furthest part of the surface from the center. The polar radius, R_P, or the length of the semi-minor axis, b, is the distance from the center to either pole, which are the nearest points on the surface to the center.

The ellipsoid is commonly defined in terms of the equatorial radius and either the (primary or major) eccentricity of the ellipsoid, e, or the flattening of the ellipsoid, f. These are defined by

$$e = \sqrt{1 - \frac{R_P^2}{R_0^2}}, \qquad f = \frac{R_0 - R_P}{R_0}, \qquad (2.92)$$

and are related by

$$e = \sqrt{2f - f^2}, \qquad f = 1 - \sqrt{1 - e^2}. \qquad (2.93)$$

The Cartesian position of a point, S, on the ellipsoid surface is $\mathbf{r}_{eS}^e = (x_{eS}^e, y_{eS}^e, z_{eS}^e)$. The distance of that point from the center of the Earth is known as the geocentric radius and is simply

$$r_{eS}^e = |\mathbf{r}_{eS}^e| = \sqrt{{x_{eS}^e}^2 + {y_{eS}^e}^2 + {z_{eS}^e}^2}. \qquad (2.94)$$

2.4 Earth Surface and Gravity Models

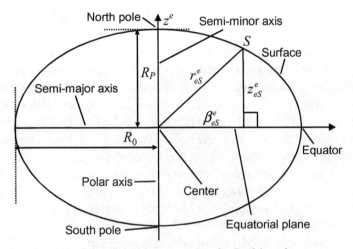

Figure 2.18 Cross-section of the ellipsoid representing the Earth's surface.

It is useful to define the magnitude of the projection of $\mathbf{r}^e_{eS}$ into the equatorial plane as β^e_{eS}. Thus,

$$\beta^e_{eS} = \sqrt{x^{e\,2}_{eS} + y^{e\,2}_{eS}}. \tag{2.95}$$

The cross-section of the ellipsoid shown in Figure 2.18 is the vertical plane containing the vector $\mathbf{r}^e_{eS}$. Thus, z^e_{eS} and β^e_{eS} are constrained by the ellipse equation:

$$\left(\frac{\beta^e_{eS}}{R_0}\right)^2 + \left(\frac{z^e_{eS}}{R_P}\right)^2 = 1. \tag{2.96}$$

Substituting in (2.95) defines the surface of the ellipsoid:

$$\left(\frac{x^e_{eS}}{R_0}\right)^2 + \left(\frac{y^e_{eS}}{R_0}\right)^2 + \left(\frac{z^e_{eS}}{R_P}\right)^2 = 1. \tag{2.97}$$

As well as providing a reference for determining position, the ellipsoid model is also crucial in defining a local navigation frame (Section 2.1.3), as the down direction of this frame is defined as the normal to the ellipsoid, pointing to the equatorial plane. Note that the normal to an ellipsoid does not intersect the ellipsoid center unless it passes through the poles or the equator.

Realizing the ellipsoid model in practice requires the positions of a large number of points on the Earth's surface to be measured. There is no practical method of measuring position with respect to the center of the Earth, noting that the center of an ellipsoid is not necessarily the center of mass. Consequently, position has been measured by surveying the relative positions of a number of points, a process known as triangulation. This has been done on national, regional, and continental bases, providing a host of different ellipsoid models, or geodetic datums, that provide a good fit to the Earth's surface across the area of interest, but a poor fit elsewhere in the world [8].

The advent of satellite navigation has enabled the position of points across the whole of the Earth's surface to be measured with respect to a common reference, the satellite constellation, leading to the development of global ellipsoid models. The two main standards are the World Geodetic System 1984 (WGS 84) [9] and the International Terrestrial Reference Frame (ITRF) [10]. Both of these datums have their origin at the Earth's center of mass and define rotation using the IRP/CTP.

WGS 84 was developed by the Defense Mapping Agency, now the National Geospatial-Intelligence Agency (NGA), as a standard for the U.S. military and is a refinement of predecessors WGS 60, WGS 66, and WGS 72. Its use for GPS and in most INSs led to its adoption as a global standard for navigation systems. WGS 84 was originally realized with 1691 Transit position fixes, each accurate to 1–2m, and was revised in the 1990s using GPS measurements and ITRF data [11]. As well as defining an ECEF coordinate frame and an ellipsoid, WGS 84 provides models of the Earth's geoid (Section 2.4.4) and gravity field (Section 2.4.7) and a set of fundamental constants. WGS 84 defines the ellipsoid in terms of the equatorial radius and the flattening. The polar radius and eccentricity may be derived from this. The values are listed in Table 2.1.

The ITRF is maintained by the IERS and is the datum of choice for the scientific community, particularly geodesists. It is based on a mixture of measurements from satellite laser ranging, lunar laser ranging, very long baseline interferometry (VLBI) and GPS. It is used in association with the Geodetic Reference System 1980 (GRS80) ellipsoid, also described in Table 2.1, which differs by less than a millimeter from the WGS84 ellipsoid.

ITRF is more precise than WGS 84, although the revision of the latter in the 1990s brought the two into closer alignment and WGS 84 is now considered to be a realization of the ITRF. Galileo uses a realization of the ITRF known as the Galileo Terrestrial Reference Frame (GTRF). GLONASS uses the PZ-90.02 datum, which has an origin offset from that of the ITRF by about 0.4 m. Similarly, Beidou uses the China Geodetic Coordinate System 2000 (CGCS 2000), also nominally aligned with the ITRF.

All datums must be regularly updated to account for plate tectonic motion, which causes the position of all points on the surface to move by a few centimeters each year with respect to the center of the Earth. Section C.1 of Appendix C on the CD presents more information on datums, including the transformation of coordinates between datums.

Table 2.1 Parameters of the WGS84 and GRS80 Ellipsoids

Parameter	WGS84 Value	GRS80 Value
Equatorial radius, R_0	6,378,137.0m	6,378,137.0m
Polar radius, R_P	6,356,752.31425m	6,356,752.31414m
Flattening, f	1/298.257223563	1/298.257222101
Eccentricity, e	0.0818191908425	0.0818191910428

2.4 Earth Surface and Gravity Models

2.4.2 Curvilinear Position

Position with respect to the Earth's surface is described using three mutually orthogonal coordinates, aligned with the axes of a local navigation frame. The distance from the body described to the surface along the normal to that surface is the *height* or *altitude*. The north-south axis coordinate of the point on the surface where that normal intersects is the *latitude*, and the coordinate of that point in the east-west axis is the *longitude*. Each of these is defined in detail later. Because the orientation of all three axes with respect to the Earth varies with location, the latitude, longitude, and height are collectively known as *curvilinear* or *ellipsoidal position*.

Connecting all points on the ellipsoid surface of the same latitude produces a circle centered about the polar (north-south) axis; this is known as a *parallel* and has radius β^e_{eS}. Similarly, the points of constant longitude on the ellipsoid surface define a semi-ellipse, running from pole to pole, known as a *meridian*. A parallel and a meridian always intersect at 90°. Planes containing a parallel or a meridian are known as parallel sections and meridian sections, respectively.

Traditionally, latitude was measured by determining the local vertical with a plumb bob and the Earth's axis of rotation from the motion of the stars. However, this *astronomical latitude* has two drawbacks. First, due to local gravity variation, multiple points along a meridian can have the same astronomical latitude [8]. Second, as a result of polar motion, the astronomical latitude of any point on the Earth varies slightly with time.

The *geocentric latitude*, Φ, illustrated in Figure 2.19, is the angle of intersection of the line from the center to a point on the surface of the ellipsoid with the equatorial plane. For all types of latitude, the convention is that latitude is positive in the northern hemisphere and negative in the southern hemisphere. By trigonometry, the geocentric latitude of a point S on the surface is given by

$$\tan \Phi_S = \frac{z^e_{eS}}{\beta^e_{eS}} = \frac{z^e_{eS}}{\sqrt{x^{e\,2}_{eS} + y^{e\,2}_{eS}}}, \qquad \sin \Phi_S = \frac{z^e_{eS}}{r^e_{eS}}. \qquad (2.98)$$

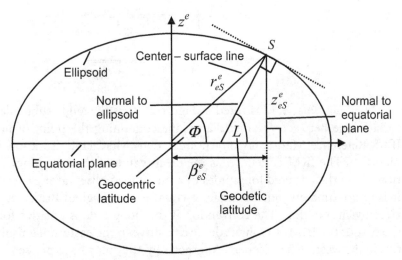

Figure 2.19 Geocentric and geodetic latitude.

The *geodetic latitude*, L, also shown in Figure 2.19, is the angle of intersection of the normal to the ellipsoid with the equatorial plane. This is sometimes known as the ellipsoidal latitude. The symbol ϕ is also commonly used. Geodetic latitude is a rationalization of astronomical latitude, retaining the basic principle, but removing the ambiguity. It is the standard form of latitude used in terrestrial navigation. As the geodetic latitude is defined by the normal to the surface, it can be obtained from the gradient of that surface. Thus, for a point S on the surface of the ellipsoid,

$$\tan L_S = -\frac{\partial \beta^e_{eS}}{\partial z^e_{eS}}. \tag{2.99}$$

Differentiating (2.96) and then substituting (2.92) and (2.95),

$$\frac{\partial \beta^e_{eS}}{\partial z^e_{eS}} = -\frac{z^e_{eS} R_0^2}{\beta^e_{eS} R_P^2} = -\frac{z^e_{eS}}{(1-e^2)\beta^e_{eS}}. \tag{2.100}$$

Thus,

$$\tan L_S = \frac{z^e_{eS}}{(1-e^2)\beta^e_{eS}} = \frac{z^e_{eS}}{(1-e^2)\sqrt{x^{e\,2}_{eS} + y^{e\,2}_{eS}}}. \tag{2.101}$$

Substituting in (2.98) gives the relationship between the geodetic and geocentric latitudes:

$$\tan \Phi_S = (1-e^2)\tan L_S. \tag{2.102}$$

For a body, b, which is not on the surface of the ellipsoid, the geodetic latitude is given by the coordinates of the point, $S(b)$, where the normal to the surface from that body intersects the surface. Thus,

$$\tan L_b = \frac{z^e_{eS(b)}}{(1-e^2)\sqrt{x^{e\,2}_{eS(b)} + y^{e\,2}_{eS(b)}}}.$$
$$\tan L_b \neq \frac{z^e_{eb}}{(1-e^2)\sqrt{x^{e\,2}_{eb} + y^{e\,2}_{eb}}} \tag{2.103}$$

The *longitude*, λ, illustrated in Figure 2.20, is the angle subtended in the equatorial plane between the meridian plane containing the point of interest and the IERS Reference Meridian/Conventional Zero Meridian, also known as the prime meridian. The IRM is defined as the mean value of the zero longitude determinations from the adopted longitudes of a number of observatories around the world. It is approximately, but not exactly, equal to the original British zero meridian at Greenwich, London. The convention is that longitude is positive for meridians to the east of the IRM, so longitudes are positive in the eastern hemisphere and negative in the western hemisphere. Alternatively, they may be expressed between 0° and 360° or 0 and 2π rad. Note that some authors use the symbol λ for latitude and l,

2.4 Earth Surface and Gravity Models

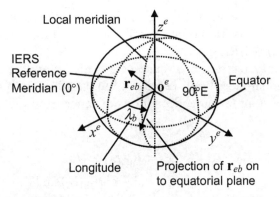

Figure 2.20 Illustration of longitude.

L, or ϕ for longitude. By trigonometry, the longitude of a point S on the surface and of any body, b, is given by

$$\tan \lambda_S = \frac{y^e_{eS}}{x^e_{eS}}, \qquad \tan \lambda_b = \frac{y^e_{eb}}{x^e_{eb}}. \tag{2.104}$$

Note that longitude is undefined at the poles, exhibiting a singularity similar to that of Euler angles at ±90° pitch. Significant numerical computation errors can occur when attempting to compute longitude very close to the north or south pole.

At this point, it is useful to define the radii of curvature of the ellipsoid. The radius of curvature for north-south motion, R_N (some authors use M or ρ), is known as the meridian radius of curvature. It is the radius of curvature of a meridian, a cross-section of the ellipsoid surface in the north-down plane, at the point of interest. This is the same as the radius of the best-fitting circle to the meridian ellipse at the point of interest. The meridian radius of curvature varies with latitude and is smallest at the equator, where the geocentric radius is largest, and largest at the poles. It is given by

$$R_N(L) = \frac{R_0(1 - e^2)}{(1 - e^2 \sin^2 L)^{3/2}} \tag{2.105}$$

The rate of change of geodetic latitude for a body traveling at unit velocity along a meridian is $1/R_N$.

The radius of curvature for east-west motion, R_E (some authors use N or ν), is known as the transverse radius of curvature, normal radius of curvature, or prime vertical radius of curvature. It is the radius of curvature of a cross-section of the ellipsoid surface in the east-down plane at the point of interest. This is the vertical plane perpendicular to the meridian plane and is not the plane of constant latitude. The transverse radius of curvature varies with latitude and is smallest at the equator. It is also equal to the length of the normal from a point on the surface to the polar axis. It is given by

$$R_E(L) = \frac{R_0}{\sqrt{1 - e^2 \sin^2 L}}. \tag{2.106}$$

The rate of change of the angle subtended at the rotation axis for a body traveling at unit velocity along the surface normal to a meridian (which is not the same as a parallel) is $1/R_E$.

The transverse radius of curvature is also useful in defining the parallels on the ellipsoid surface. From (2.92), (2.96), (2.101), and (2.106), the radius of the circle of constant latitude, β^e_{eS}, and its distance from the equatorial plane, z^e_{eS}, are given by

$$\begin{aligned} \beta^e_{eS} &= R_E(L_S)\cos L_S \\ z^e_{eS} &= (1-e^2)R_E(L_S)\sin L_S \end{aligned}. \quad (2.107)$$

The rate of change of longitude for a body traveling at unit velocity along a parallel is $1/\beta^e_{eS}$.

Figure 2.21 shows the meridian and transverse radii of curvature and the geocentric radius as a function of latitude and compares them with the equatorial and polar radii. Note that the two radii of curvature are the same at the poles, where the north-south and east-west directions are undefined. Both radii of curvature are calculated by the MATLAB function, Radii_of_curvature, on the CD.

The radius of curvature in an arbitrary direction described by the azimuth ψ_{nu} is

$$R = \left(\frac{\cos^2 \psi_{nu}}{R_N} + \frac{\sin^2 \psi_{nu}}{R_E}\right)^{-1}. \quad (2.108)$$

The *geodetic height* or altitude, h, sometimes known as the ellipsoidal height or altitude, is the distance from a body to the ellipsoid surface along the normal to that ellipsoid, with positive height denoting that the body is outside the ellipsoid. This is illustrated in Figure 2.22. By trigonometry, the height of a body, b, is given by

$$h_b = \frac{z^e_{eb} - z^e_{eS(b)}}{\sin L_b}. \quad (2.109)$$

Substituting in (2.107),

$$h_b = \frac{z^e_{eb}}{\sin L_b} - (1-e^2)R_E(L_b). \quad (2.110)$$

The curvilinear position of a body, b, may be expressed in vector form as $\mathbf{p}_b = (L_b, \lambda_b, h_b)$. Note that only the object frame is specified as an ECEF reference frame and local navigation frame resolving axes are implicit in the definition of curvilinear position.

At a height h_b above the ellipsoid, the meridian and transverse radii of curvature are, respectively, $R_N(L_b) + h_b$ and $R_E(L_b) + h_b$. Similarly, the radius of curvature within the parallel plane is $(R_E(L_b) + h_b)\cos L_b$. The velocity along a curve divided by the radius of curvature of that curve is equal to the time derivative of the angle

2.4 Earth Surface and Gravity Models

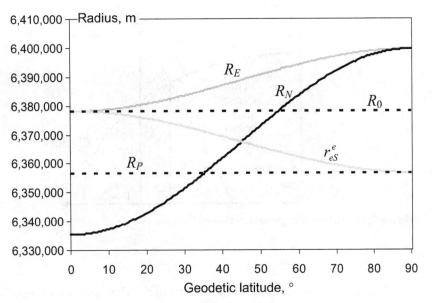

Figure 2.21 Variation of meridian and transverse radii of curvature and geocentric radius with latitude.

subtended. Therefore, the time derivative of curvilinear position is the following linear function of the Earth-referenced velocity in local navigation frame axes:

$$\dot{L}_b = \frac{v^n_{eb,N}}{R_N(L_b) + h_b}$$
$$\dot{\lambda}_b = \frac{v^n_{eb,E}}{(R_E(L_b) + h_b)\cos L_b} \quad . \quad (2.111)$$
$$\dot{h}_b = -v^n_{eb,D}$$

This enables curvilinear position to be integrated directly from velocity without having to use the Cartesian position as an intermediary.

2.4.3 Position Conversion

Using (2.95), (2.104), (2.107), and (2.110), the Cartesian ECEF position may be obtained from the curvilinear position using

$$x^e_{eb} = (R_E(L_b) + h_b)\cos L_b \cos \lambda_b$$
$$y^e_{eb} = (R_E(L_b) + h_b)\cos L_b \sin \lambda_b \quad . \quad (2.112)$$
$$z^e_{eb} = [(1-e^2)R_E(L_b) + h_b]\sin L_b$$

Example 2.2 on the CD illustrates this and is editable using Microsoft Excel. It is also included in the MATLAB function, NED_to_ECEF, also on the CD.

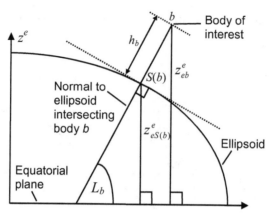

Figure 2.22 Height and geodetic latitude of a body, b.

The curvilinear position is obtained from the Cartesian ECEF position by implementing the inverse of the previous [12]:

$$\tan L_b = \frac{z_{eb}^e \left[R_E(L_b) + h_b \right]}{\sqrt{x_{eb}^{e\,2} + y_{eb}^{e\,2}} \left[(1 - e^2) R_E(L_b) + h_b \right]}$$

$$\tan \lambda_b = \frac{y_{eb}^e}{x_{eb}^e}, \qquad (2.113)$$

$$h_b = \frac{\sqrt{x_{eb}^{e\,2} + y_{eb}^{e\,2}}}{\cos L_b} - R_E(L_b)$$

where a four-quadrant arctangent function must be used for longitude. Note that, because R_E is a function of latitude, the latitude and height must be solved iteratively as Section C.2.1 of Appendix C on the CD explains in more detail. When a previous curvilinear position solution is available, it should be used to initialize the calculation. Otherwise, the convergence of the iterative process may be sped up by initializing the geodetic latitude with the geocentric latitude, Φ_b, given by

$$\Phi_b = \arctan\left(\frac{z_{eb}^e}{\sqrt{x_{eb}^{e\,2} + y_{eb}^{e\,2}}} \right). \qquad (2.114)$$

In polar regions, (2.113) is replaced by

$$\tan\left(\frac{\pi}{2} - L_b \right) = \frac{\sqrt{x_{eb}^{e\,2} + y_{eb}^{e\,2}} \left[(1 - e^2) R_E(L_b) + h_b \right]}{z_{eb}^e \left[R_E(L_b) + h_b \right]}$$

$$\tan \lambda_b = \frac{y_{eb}^e}{x_{eb}^e}, \qquad (2.115)$$

$$h_b = \frac{z_{eb}^e}{\sin L_b} - (1 - e^2) R_E(L_b)$$

2.4 Earth Surface and Gravity Models

The following approximate closed-form latitude solution is accurate to within 1 cm for positions close to the Earth's surface [13]:

$$\tan L_b \approx \frac{z_{eb}^e \sqrt{1-e^2} + e^2 R_0 \sin^3 \zeta_b}{\sqrt{1-e^2}\left(\sqrt{x_{eb}^{e2} + y_{eb}^{e2}} - e^2 R_0 \cos^3 \zeta_b\right)}, \qquad (2.116)$$

where

$$\tan \zeta_b = \frac{z_{eb}^e}{\sqrt{1-e^2}\sqrt{x_{eb}^{e2} + y_{eb}^{e2}}}. \qquad (2.117)$$

Section C.2 of Appendix C on the CD presents an iterative version of this. It also describes further iterated solutions and two closed-form exact solutions. All methods are included in Example 2.2 on the CD, while the Borkowski closed-form exact solution is included in the MATLAB function, ECEF_to_NED, on the CD.

Great care should be taken when Cartesian and curvilinear positions are mixed within a set of navigation equations to ensure that the curvilinear position computation is performed with sufficient precision. Otherwise, a divergent position solution could result.

Small perturbations to the position may be converted between Cartesian and curvilinear representation using

$$\delta \mathbf{r}_{eb}^e \approx \mathbf{C}_n^e \mathbf{T}_p^{r(n)} \delta \mathbf{p}_b, \qquad \delta \mathbf{p}_b \approx \mathbf{T}_{r(n)}^p \mathbf{C}_e^n \delta \mathbf{r}_{eb}^e \qquad (2.118)$$

where

$$\mathbf{T}_{r(n)}^p = \frac{\partial \mathbf{p}_b}{\partial \mathbf{r}_{eb}^n} = \begin{pmatrix} \frac{1}{R_N(L_b) + h_b} & 0 & 0 \\ 0 & \frac{1}{(R_E(L_b) + h_b)\cos L_b} & 0 \\ 0 & 0 & -1 \end{pmatrix}, \qquad (2.119)$$

$$\mathbf{T}_p^{r(n)} = \frac{\partial \mathbf{r}_{eb}^n}{\partial \mathbf{p}_b} = \begin{pmatrix} R_N(L_b) + h_b & 0 & 0 \\ 0 & (R_E(L_b) + h_b)\cos L_b & 0 \\ 0 & 0 & -1 \end{pmatrix}. \qquad (2.120)$$

These are particularly useful for converting error standard deviations.

Section C.3 of Appendix C on the CD describes the normal vector representation of curvilinear position, which avoids the longitude singularity at the poles.

Finally, although most navigation systems now use the WGS 84 datum, many maps are based on national and regional datums. This is partly for historical reasons and partly because it is convenient to map features using datums that move with the tectonic plates. Consequently, it may be necessary to transform curvilinear or Cartesian position from one datum to another. The datums may use different origins, axis alignments, and scalings as well as different radii of curvature. Datum transformations are described in Section C.1 of Appendix C on the CD. No conversion between WGS 84 and ITRF position is needed as the differences between the two datums are less than the uncertainty bounds.

2.4.4 The Geoid, Orthometric Height, and Earth Tides

The gravity potential is the potential energy required to overcome gravity (see Section 2.4.7). As water will always flow from an area of higher gravity potential to an area of lower gravity potential, mean sea level, which is averaged over the tide cycle, maintains a surface of approximately equal gravity potential (differences arise due to permanent ocean currents). The geoid is a model of the Earth's surface that has a constant gravity potential; it is an example of an equipotential surface. The geoid is generally within 1m of mean sea level [13]. Note that, over land, the physical surface of the Earth, known as the terrain, is generally above the geoid. The gravity vector at any point on the Earth's surface is thus perpendicular to the geoid, not the ellipsoid or the terrain, although, in practice, the difference is small.

As the Earth's gravity field varies with location, the geoid can differ from the ellipsoid by up to 100m. The height of the geoid with respect to the ellipsoid is denoted N; this is known as the *geoid–ellipsoid separation*. The current WGS 84 geoid model is known as the Earth Gravitational Model 2008 (EGM 08) and has 4,730,400 (= 2,160 × 2,190) coefficients defining the geoid height, N, and gravitational potential as a spherical harmonic function of geodetic latitude and longitude [14]. A geoid model is also known as a vertical datum.

The height of a body above the geoid is known as the *orthometric height* or *orthometric altitude* and is denoted as H. The height or altitude above mean sea level (AMSL) is also commonly used. The orthometric height of the terrain is known as *elevation*. The orthometric height is related to the geodetic height by

$$H_b \approx h_b - N(L_b, \lambda_b). \tag{2.121}$$

This is not exact because the geodetic height is measured normal to the ellipsoid, whereas the orthometric height is measured normal to the geoid. Figure 2.23 illustrates the two heights, the geoid, ellipsoid, and terrain.

For many applications, orthometric height is more useful than geodetic height. Maps tend to express the height of the terrain and features with respect to the geoid, making orthometric height critical for aircraft approach, landing, and low-level flight. It is also important in civil engineering, for example, to determine the direction of flow of water. Thus, a navigation system will often need to incorporate a geoid model to convert between geodetic and orthometric height.

It is well known that lunar gravitation causes ocean tides. However, it also causes tidal movement of the Earth's crust and there are tidal effects due to solar gravitation. Together, these are known as solid Earth tides and cause the positions of the terrain and features thereon to vary with respect to the geoid and ellipsoid with an amplitude of about half a meter. The vertical displacement is largest, but there is also horizontal displacement. There are multiple oscillations with varying

2.4 Earth Surface and Gravity Models

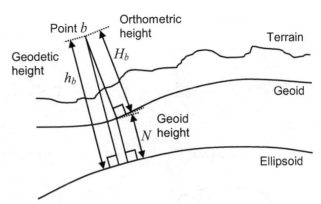

Figure 2.23 Height, geoid, ellipsoid, and terrain. (*After:* [12].)

periods that contribute to the solid Earth tides, with the largest components having approximately diurnal (~24 hour) and semidiurnal (~12 hour) periods.

Solid Earth tides predominantly affect positioning using GNSS and other satellite-based techniques. An appropriate correction is applied to obtain a time-invariant position solution [15]. However, for most navigation applications, solid Earth tides are neglected.

2.4.5 Projected Coordinates

Projected coordinates provide a way of representing the ellipsoid as a flat surface. This is essential for printing maps on paper or displaying them on a flat screen. A projection converts geodetic latitude and longitude to and from planar Cartesian coordinates. The projection may be arbitrary, but more commonly represents a straight line from a focal point or line, through the surface of the ellipsoid, to the corresponding point on the 2-D surface. The 2-D surface may be represented in 3-D space as a plane. Alternatively, it may be wrapped into a cylinder or cone. Projections are thus categorized as cylindrical, conical, or planar. Figure 2.24 illustrates some examples [13, 16].

The aspect denotes the orientation of the 2-D surface. A cylindrical or conical projection has a normal aspect if its axis of rotational symmetry is aligned with the north-south axis of the ellipsoid, a transverse aspect if its axis is within the equatorial plane, and an oblique aspect otherwise. A planar projection has a normal aspect if it is perpendicular to the equatorial plane, a polar aspect if it is parallel, and an oblique aspect otherwise [16]. Aspects are indicated in Figure 2.24.

All projections distort the shape of large-scale features as geometry on a flat surface is fundamentally different from that on a curved surface. Different classes of projections preserve some features of geometry and distort others. A conformal projection preserves the shape of small-scale features, an equal-area projection preserves areas, an equidistant projection preserves distances along at least one line, and an azimuthal projection preserves the angular relationship of all features with respect to the center of the projection [16].

A transverse Mercator projection is a conformal transverse cylindrical projection, commonly used by national mapping agencies. Examples include the Universal

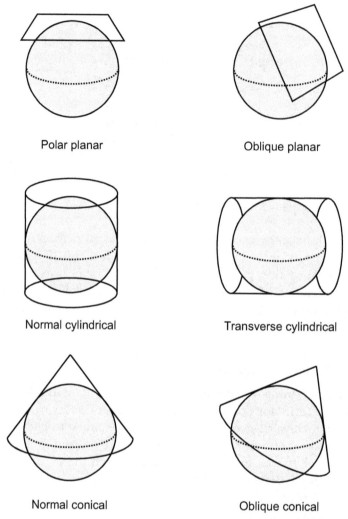

Figure 2.24 Example projections.

Transverse Mercator (UTM) system, Gauss-Krueger zoned system, many U.S. state planes, and the U.K. National Grid. Section C.4 of Appendix C on the CD describes the projection in more detail and presents formulae for converting between latitude and longitude and transverse Mercator projected coordinates.

2.4.6 Earth Rotation

The ECI and ECEF coordinate systems are defined such that the Earth rotates, with respect to space, clockwise about their common z-axis, shown in Figure 2.25. Thus, the Earth-rotation vector resolved in an ECI or ECEF frame is given by

$$\omega_{ie}^i = \omega_{ie}^e = \begin{pmatrix} 0 \\ 0 \\ \omega_{ie} \end{pmatrix}. \tag{2.122}$$

2.4 Earth Surface and Gravity Models

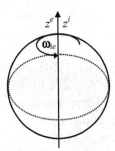

Figure 2.25 Earth rotation in an ECI or ECEF frame. (*From:* [1]. © 2002 QinetiQ Ltd. Reprinted with permission.)

The Earth-rotation vector resolved into local navigation frame axes is a function of geodetic latitude:

$$\boldsymbol{\omega}_{ie}^n = \begin{pmatrix} \omega_{ie} \cos L_b \\ 0 \\ -\omega_{ie} \sin L_b \end{pmatrix}. \tag{2.123}$$

The period of rotation of the Earth with respect to space is known as the sidereal day and is about 23 hours, 56 minutes, 4 seconds. This differs from the 24-hour mean solar day as the Earth's orbital motion causes the Earth–Sun direction with respect to space to vary, resulting in one more rotation than solar day each year (note that 1/365 of a day is about 4 minutes). The rate of rotation is not constant and the sidereal day can vary by several milliseconds from day to day. There are random changes due to wind and seasonal changes as ice forming and melting alters the Earth's moment of inertia. There is also a long term reduction of the Earth rotation rate due to tidal friction [12].

For navigation purposes, a constant rotation rate is assumed, based on the mean sidereal day. The WGS 84 value of the Earth's angular rate is $\omega_{ie} = 7.292115 \times 10^{-5}$ rad s^{-1} [9].

The velocity of the Earth's surface due to Earth rotation is given by

$$\mathbf{v}_{iS}^e = \boldsymbol{\omega}_{ie}^e \wedge \mathbf{r}_{eS}^e, \qquad \mathbf{v}_{iS}^n = \mathbf{C}_e^n \big(\boldsymbol{\omega}_{ie}^e \wedge \mathbf{r}_{eS}^e \big). \tag{2.124}$$

See Section 2.5.5 for how this is obtained. The maximum speed is 465 m s^{-1} at the equator.

2.4.7 Specific Force, Gravitation, and Gravity

Specific force is the nongravitational force per unit mass on a body, sensed with respect to an inertial frame. It has no meaning with respect to any other frame, although it can be resolved in any axes. *Gravitation* is the fundamental mass attraction force; it does not incorporate any centripetal components.*

---

*This and subsequent paragraphs are based on material written by the author for QinetiQ, so comprise QinetiQ copyright material.

Specific force is what people and instruments sense. Gravitation is not sensed because it acts equally on all points, causing them to move together. Other forces are sensed as they are transmitted from point to point. The sensation of weight is caused by the forces opposing gravity.† This reaction to gravity is known as the restoring force on land, buoyancy at sea, and lift in the air.

During freefall, the specific force is zero so there is no sensation of weight. Conversely, under zero acceleration when the specific force is equal and opposite to the acceleration due to gravitation, the reaction to gravitation is sensed as weight. Figure 2.26 illustrates this with a mass in freefall and a mass suspended by a spring. In both cases, the gravitational force on the mass is the same. However, in the suspended case, the spring exerts an equal and opposite force.

A further example is provided by the upward motion of an elevator, illustrated by Figure 2.27. As the elevator accelerates upward, the specific force is higher and the occupants appear to weigh more. As the elevator decelerates, the specific force is lower than normal and the occupants feel lighter. In a windowless elevator, this can create the illusion that the elevator has overshot the destination floor and is dropping down to correct for it.*

Thus, specific force, **f**, varies with acceleration, **a**, and the acceleration due to the gravitational force, **γ**, as

$$\mathbf{f}_{ib}^{\gamma} = \mathbf{a}_{ib}^{\gamma} - \boldsymbol{\gamma}_{ib}^{\gamma} \tag{2.125}$$

Specific force is the quantity measured by accelerometers. The measurements are made in the body frame of the accelerometer triad; thus, the sensed specific force is $\mathbf{f}_{ib}^{b}$.

As a prelude to defining gravity, it is useful to consider an object that is stationary with respect to a rotating frame, such as an ECEF frame. This has the properties

$$\mathbf{v}_{eb}^{e} = 0, \qquad \mathbf{a}_{eb}^{e} = 0. \tag{2.126}$$

From (2.67) and (2.76), and applying (2.66),

$$\dot{\mathbf{r}}_{ib}^{e} = \dot{\mathbf{r}}_{eb}^{e} = 0, \qquad \ddot{\mathbf{r}}_{ib}^{e} = \ddot{\mathbf{r}}_{eb}^{e} = 0. \tag{2.127}$$

The inertially referenced acceleration in ECEF frame axes is given by (2.81), noting that $\dot{\boldsymbol{\Omega}}_{ie}^{e} = 0$ as the Earth rate is assumed constant:

$$\mathbf{a}_{ib}^{e} = \boldsymbol{\Omega}_{ie}^{e}\boldsymbol{\Omega}_{ie}^{e}\mathbf{r}_{ib}^{e} + 2\boldsymbol{\Omega}_{ie}^{e}\dot{\mathbf{r}}_{ib}^{e} + \ddot{\mathbf{r}}_{ib}^{e}. \tag{2.128}$$

Applying (2.127),

$$\mathbf{a}_{ib}^{e} = \boldsymbol{\Omega}_{ie}^{e}\boldsymbol{\Omega}_{ie}^{e}\mathbf{r}_{eb}^{e}. \tag{2.129}$$

---

†End of QinetiQ copyright material.

*This and subsequent paragraphs are based on material written by the author for QinetiQ, so comprise QinetiQ copyright material.

2.4 Earth Surface and Gravity Models

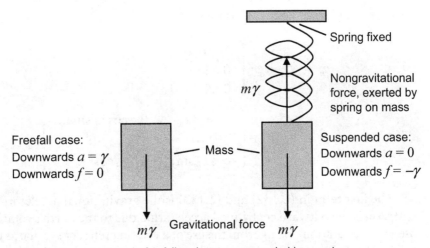

Figure 2.26 Forces on a mass in freefall and a mass suspended by a spring.

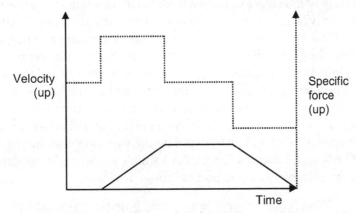

Figure 2.27 Velocity and specific force of an elevator moving up (*From:* [1]. © 2002 QinetiQ Ltd. Reprinted with permission.)

Substituting this into the specific force definition, (2.125), gives

$$\mathbf{f}_{ib}^e = \mathbf{\Omega}_{ie}^e \mathbf{\Omega}_{ie}^e \mathbf{r}_{eb}^e - \boldsymbol{\gamma}_{ib}^e. \tag{2.130}$$

The specific force sensed when stationary with respect to an Earth frame is the reaction to what is known as the acceleration due to *gravity*, which is thus defined by[†]

$$\mathbf{g}_b^\gamma = -\mathbf{f}_{ib}^\gamma \big|_{\mathbf{a}_{eb}^\gamma=0, \mathbf{v}_{eb}^\gamma=0}. \tag{2.131}$$

Therefore, from (2.130), the acceleration due to gravity is

$$\mathbf{g}_b^\gamma = \boldsymbol{\gamma}_{ib}^\gamma - \mathbf{\Omega}_{ie}^\gamma \mathbf{\Omega}_{ie}^\gamma \mathbf{r}_{eb}^\gamma, \tag{2.132}$$

[†]End of QinetiQ copyright material.

noting from (2.122) and (2.123) that

$$\mathbf{g}_b^e = \boldsymbol{\gamma}_{ib}^e + \omega_{ie}^2 \begin{pmatrix} 1 & 0 & 0 \\ 0 & 1 & 0 \\ 0 & 0 & 0 \end{pmatrix} \mathbf{r}_{eb}^e$$

$$\mathbf{g}_b^n = \boldsymbol{\gamma}_{ib}^n + \omega_{ie}^2 \begin{pmatrix} \sin^2 L_b & 0 & \cos L_b \sin L_b \\ 0 & 1 & 0 \\ \cos L_b \sin L_b & 0 & \cos^2 L_b \end{pmatrix} \mathbf{r}_{eb}^n$$
(2.133)

The first term in (2.132) and (2.133) is the gravitational acceleration. The second term is the outward centrifugal acceleration due to the Earth's rotation; this is a pseudo-acceleration arising from the use of a rotating reference frame as discussed in Section 2.3.5. Figure 2.28 illustrates the two components of gravity. From an inertial frame perspective, a centripetal acceleration, (2.129), is applied to maintain an object stationary with respect to the rotating Earth. It is important not to confuse gravity, **g**, with gravitation, **γ**. At the Earth's surface, the total acceleration due to gravity is about 9.8 m s^{-2}, with the centrifugal component contributing up to 0.034 m s^{-2}. In orbit, the gravitational component is smaller and the centrifugal component is larger. However, an inertial reference frame is normally used for orbital applications.

The centrifugal component of gravity can be calculated exactly at all locations, but calculation of the gravitational component is more complex. For air applications, it is standard practice to use an empirical model of the surface gravity, $\mathbf{g}_0$, and apply a simple scaling law to calculate the variation with height.[‡]

The WGS 84 datum [9] provides a simple model of the acceleration due to gravity at the ellipsoid as a function of latitude:

$$g_0(L) \approx 9.7803253359 \frac{(1 + 0.001931853 \sin^2 L)}{\sqrt{1 - e^2 \sin^2 L}} \text{ m s}^{-2}.$$
(2.134)

This is known as the Somigliana model. Note that it is a gravity model, not a gravitational model. The geoid (Section 2.4.4) defines a surface of constant gravity potential. However, the acceleration due to gravity is obtained from the gradient of the gravity potential, so it is not constant across the geoid. Although the true gravity vector is perpendicular to the geoid (not the terrain), it is a reasonable approximation for most navigation applications to treat it as perpendicular to the ellipsoid. Thus,

$$\mathbf{g}_0^\gamma(L) \approx g_0(L) \mathbf{u}_{nD}^\gamma,$$
(2.135)

where $\mathbf{u}_{nD}^\gamma$ is the down unit vector of a local navigation frame.

The gravitational acceleration at the ellipsoid can be obtained from the acceleration due to gravity by subtracting the centrifugal acceleration. Thus,

$$\boldsymbol{\gamma}_0^\gamma(L) = \mathbf{g}_0^\gamma(L) + \boldsymbol{\Omega}_{ie}^\gamma \boldsymbol{\Omega}_{ie}^\gamma \mathbf{r}_{eS}^\gamma(L).$$
(2.136)

[‡]This paragraph, up to this point, is based on material written by the author for QinetiQ, so comprises QinetiQ copyright material.

2.4 Earth Surface and Gravity Models

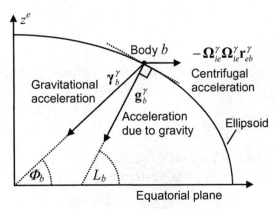

Figure 2.28 Gravity, gravitation and centrifugal acceleration.

From (2.107), the geocentric radius at the surface is given by

$$r_{eS}^e(L) = R_E(L)\sqrt{\cos^2 L + (1-e^2)^2 \sin^2 L}. \tag{2.137}$$

The gravitational field varies roughly as that for a point mass, so gravitational acceleration can be scaled with height as

$$\gamma_{ib}^\gamma \approx \frac{(r_{eS}^e(L_b))^2}{(r_{eS}^e(L_b) + h_b)^2} \gamma_0^\gamma(L_b). \tag{2.138}$$

For heights less than about 10 km, the scaling can be further approximated to $(1 - 2h_b/r_{eS}^e(L_b))$. The acceleration due to gravity, **g**, may then be recombined using (2.132). As the centrifugal component of gravity is small, it is reasonable to apply the height scaling to **g** where the height is small and/or poor quality accelerometers are used. Alternatively, a more accurate set of formulae for calculating gravity as a function of latitude and height is given in [9]. An approximation for the variation of the down component with height is

$$g_{b,D}^n(L_b, h_b) \approx g_0(L_b)\left\{1 - \frac{2}{R_0}\left[1 + f(1 - 2\sin^2 L_b) + \frac{\omega_{ie}^2 R_0^2 R_P}{\mu}\right]h_b + \frac{3}{R_0^2}h_b^2\right\}, \tag{2.139}$$

where μ is the Earth's gravitational constant and its WGS 84 value [9] is $3.986004418 \times 10^{14}$ m^{3}s^{-2}. The north component of gravity varies with height as [17]

$$g_{b,N}^n(L_b, h_b) \approx -8.08 \times 10^{-9} h_b \sin 2L_b \text{ m s}^{-2}. \tag{2.140}$$

This model is used in the MATLAB function, Gravity_NED, on the accompanying CD. Example 2.3 on the CD comprises calculations of the acceleration due to gravity at different latitudes and heights.

When working in an inertial reference frame, only the gravitational acceleration is required. This can be calculated directly at varying height using [18]

$$\gamma_{ib}^i = -\frac{\mu}{|\mathbf{r}_{ib}^i|^3}\left\{\mathbf{r}_{ib}^i + \frac{3}{2}J_2\frac{R_0^2}{|\mathbf{r}_{ib}^i|^2}\begin{Bmatrix}\left[1-5\left(r_{ib,z}^i/|\mathbf{r}_{ib}^i|\right)^2\right]r_{ib,x}^i \\ \left[1-5\left(r_{ib,z}^i/|\mathbf{r}_{ib}^i|\right)^2\right]r_{ib,y}^i \\ \left[3-5\left(r_{ib,z}^i/|\mathbf{r}_{ib}^i|\right)^2\right]r_{ib,z}^i\end{Bmatrix}\right\}, \quad (2.141)$$

where J_2 is the Earth's second gravitational constant and takes the value 1.082627×10^{-3} [9]. Resolved about ECEF-frame axes, it becomes

$$\gamma_{ib}^e = -\frac{\mu}{|\mathbf{r}_{eb}^e|^3}\left\{\mathbf{r}_{eb}^e + \frac{3}{2}J_2\frac{R_0^2}{|\mathbf{r}_{eb}^e|^2}\begin{Bmatrix}\left[1-5\left(r_{eb,z}^e/|\mathbf{r}_{eb}^e|\right)^2\right]r_{eb,x}^e \\ \left[1-5\left(r_{eb,z}^e/|\mathbf{r}_{eb}^e|\right)^2\right]r_{eb,y}^e \\ \left[3-5\left(r_{eb,z}^e/|\mathbf{r}_{eb}^e|\right)^2\right]r_{eb,z}^e\end{Bmatrix}\right\}. \quad (2.142)$$

These models are used in the MATLAB functions, Gravitation_ECI and Gravity_ECEF, on the CD.

Much higher precision may be obtained using a spherical harmonic model, such as the 4,730,400-coefficient EGM 2008 gravity model [14]. Further precision is given by a gravity anomaly database, which comprises the difference between the measured and modeled gravity fields over a grid of locations. Gravity anomalies tend to be largest over major mountain ranges and ocean trenches.

2.5 Frame Transformations

An essential feature of navigation mathematics is the capability to transform quantities between different coordinate frames. This section summarizes the equations for expressing the attitude of one frame with respect to another and transforming Cartesian position, velocity, acceleration, and angular rate between references to inertial, Earth, and local navigation frames, and between ECEF and local tangent-plane frames. The section concludes with the equations for transposing a navigation solution between different objects.

Cartesian position, velocity, acceleration, and angular rate referenced to the same frame transform between resolving axes simply by applying the coordinate transformation matrix (2.12):*

$$\mathbf{x}_{\beta\alpha}^\gamma = \mathbf{C}_\delta^\gamma \mathbf{x}_{\beta\alpha}^\delta \quad \mathbf{x} \in \mathbf{r}, \mathbf{v}, \mathbf{a}, \boldsymbol{\omega} \quad \gamma, \delta \in i, e, n, l, b. \quad (2.143)$$

---

*This and subsequent paragraphs are based on material written by the author for QinetiQ, so comprise QinetiQ copyright material.

2.5 Frame Transformations

Therefore, these transforms are not presented explicitly for each pair of frames.†
The coordinate transformation matrices involving the body frame—that is,

$$C_b^\beta, C_\beta^b \quad \beta \in i, e, n, l$$

describe the attitude of that body with respect to a reference frame. The body attitude with respect to a new reference frame may be obtained simply by multiplying by the coordinate transformation matrix between the two reference frames:

$$C_b^\delta = C_\beta^\delta C_b^\beta \qquad C_\delta^b = C_\beta^b C_\delta^\beta \qquad \beta, \delta \in i, e, n, l. \tag{2.144}$$

Transforming Euler, quaternion, or rotation vector attitude to a new reference frame is more complex. One solution is to convert to the coordinate transformation matrix representation, transform the reference, and then convert back.

2.5.1 Inertial and Earth Frames

The center and z-axes of commonly-realized Earth-centered inertial and Earth-centered Earth-fixed coordinate frames are coincident. The x- and y-axes are coincident at time t_0 and the frames rotate about the z-axes at ω_{ie} (see Section 2.4.6). Thus,*

$$C_i^e = \begin{pmatrix} \cos\omega_{ie}(t-t_0) & \sin\omega_{ie}(t-t_0) & 0 \\ -\sin\omega_{ie}(t-t_0) & \cos\omega_{ie}(t-t_0) & 0 \\ 0 & 0 & 1 \end{pmatrix}$$

$$C_e^i = \begin{pmatrix} \cos\omega_{ie}(t-t_0) & -\sin\omega_{ie}(t-t_0) & 0 \\ \sin\omega_{ie}(t-t_0) & \cos\omega_{ie}(t-t_0) & 0 \\ 0 & 0 & 1 \end{pmatrix}. \tag{2.145}$$

Positions referenced to the two frames are the same, so only the resolving axes need to be transformed:

$$\mathbf{r}_{eb}^e = C_i^e \mathbf{r}_{ib}^i, \quad \mathbf{r}_{ib}^i = C_e^i \mathbf{r}_{eb}^e. \tag{2.146}$$

Velocity and acceleration transformation is more complex:

$$\begin{aligned} \mathbf{v}_{eb}^e &= C_i^e \left(\mathbf{v}_{ib}^i - \mathbf{\Omega}_{ie}^i \mathbf{r}_{ib}^i \right) \\ \mathbf{v}_{ib}^i &= C_e^i \left(\mathbf{v}_{eb}^e + \mathbf{\Omega}_{ie}^e \mathbf{r}_{eb}^e \right) \end{aligned}, \tag{2.147}$$

---

†End of QinetiQ copyright material.
*This and subsequent paragraphs are based on material written by the author for QinetiQ, so comprise QinetiQ copyright material.

$$\begin{aligned}\mathbf{a}_{eb}^{e} &= \mathbf{C}_{i}^{e}\left(\mathbf{a}_{ib}^{i} - 2\mathbf{\Omega}_{ie}^{i}\mathbf{v}_{ib}^{i} + \mathbf{\Omega}_{ie}^{i}\mathbf{\Omega}_{ie}^{i}\mathbf{r}_{ib}^{i}\right) \\ \mathbf{a}_{ib}^{i} &= \mathbf{C}_{e}^{i}\left(\mathbf{a}_{eb}^{e} + 2\mathbf{\Omega}_{ie}^{e}\mathbf{v}_{eb}^{e} + \mathbf{\Omega}_{ie}^{e}\mathbf{\Omega}_{ie}^{e}\mathbf{r}_{eb}^{e}\right)\end{aligned}. \quad (2.148)$$

Angular rates transform as[†]

$$\boldsymbol{\omega}_{eb}^{e} = \mathbf{C}_{i}^{e}\left(\boldsymbol{\omega}_{ib}^{i} - \begin{pmatrix} 0 \\ 0 \\ \omega_{ie} \end{pmatrix}\right), \quad \boldsymbol{\omega}_{ib}^{i} = \mathbf{C}_{e}^{i}\left(\boldsymbol{\omega}_{eb}^{e} + \begin{pmatrix} 0 \\ 0 \\ \omega_{ie} \end{pmatrix}\right). \quad (2.149)$$

Example 2.4 on the CD illustrates the position, velocity, acceleration, and angular rate transformations, and is editable using Microsoft Excel. The MATLAB functions, ECEF_to_ECI and ECI_to_ECEF, on the CD implement the position, velocity, and attitude transformations.

Note that accurate timing is critical for conversion between ECI and ECEF frames. For example, if there is a 1-ms offset between the time bases used to specify t and t_0, a position error of up to 0.465m will occur when transforming between frames. Great caution should therefore be exercised in using the ECI frame where accurate timing is not available.

2.5.2 Earth and Local Navigation Frames

The relative orientation of commonly-realized Earth and local navigation frames is determined by the geodetic latitude, L_b, and longitude, λ_b, of the body frame whose center coincides with that of the local navigation frame:

$$\begin{aligned}\mathbf{C}_{e}^{n} &= \begin{pmatrix} -\sin L_b \cos \lambda_b & -\sin L_b \sin \lambda_b & \cos L_b \\ -\sin \lambda_b & \cos \lambda_b & 0 \\ -\cos L_b \cos \lambda_b & -\cos L_b \sin \lambda_b & -\sin L_b \end{pmatrix} \\ \mathbf{C}_{n}^{e} &= \begin{pmatrix} -\sin L_b \cos \lambda_b & -\sin \lambda_b & -\cos L_b \cos \lambda_b \\ -\sin L_b \sin \lambda_b & \cos \lambda_b & -\cos L_b \sin \lambda_b \\ \cos L_b & 0 & -\sin L_b \end{pmatrix}\end{aligned}. \quad (2.150)$$

Conversely, the latitude and longitude may be obtained from the coordinate transformation matrices using

$$\begin{aligned} L_b &= \arctan\left(-C_{e3,3}^{n}/C_{e1,3}^{n}\right) = \arctan\left(-C_{n3,3}^{e}/C_{n3,1}^{e}\right) \\ \lambda_b &= \arctan_2\left(-C_{e2,1}^{n}, C_{e2,2}^{n}\right) = \arctan_2\left(-C_{n1,2}^{e}, C_{n2,2}^{e}\right) \end{aligned}. \quad (2.151)$$

Position, velocity, and acceleration referenced to a local navigation frame are meaningless as the center of the corresponding body frame coincides with the

---

[†]End of QinetiQ copyright material.

2.5 Frame Transformations

navigation frame center. The resolving axes of Earth-referenced position, velocity, and acceleration are simply transformed using (2.143). Thus,

$$\mathbf{r}_{eb}^n = \mathbf{C}_e^n \mathbf{r}_{eb}^e, \qquad \mathbf{r}_{eb}^e = \mathbf{C}_n^e \mathbf{r}_{eb}^n$$
$$\mathbf{v}_{eb}^n = \mathbf{C}_e^n \mathbf{v}_{eb}^e, \qquad \mathbf{v}_{eb}^e = \mathbf{C}_n^e \mathbf{v}_{eb}^n. \qquad (2.152)$$
$$\mathbf{a}_{eb}^n = \mathbf{C}_e^n \mathbf{a}_{eb}^e, \qquad \mathbf{a}_{eb}^e = \mathbf{C}_n^e \mathbf{a}_{eb}^n$$

Angular rates transform as

$$\begin{aligned}\boldsymbol{\omega}_{nb}^n &= \mathbf{C}_e^n \left(\boldsymbol{\omega}_{eb}^e - \boldsymbol{\omega}_{en}^e\right) \\ &= \mathbf{C}_e^n \boldsymbol{\omega}_{eb}^e - \boldsymbol{\omega}_{en}^n\end{aligned}, \qquad \boldsymbol{\omega}_{eb}^e = \mathbf{C}_n^e \left(\boldsymbol{\omega}_{nb}^n + \boldsymbol{\omega}_{en}^n\right), \qquad (2.153)$$

noting that a solution for $\boldsymbol{\omega}_{en}^n$ is obtained in Section 5.4.1 The velocity, acceleration, and angular rate transformations are illustrated by Example 2.5 on the CD. The MATLAB functions, ECEF_to_NED and NED_to_ECEF, on the CD implement the velocity and attitude transformations.

2.5.3 Inertial and Local Navigation Frames

The inertial-local navigation frame coordinate transformation matrices are obtained by multiplying (2.145) and (2.150):[*]

$$\mathbf{C}_i^n = \begin{pmatrix} -\sin L_b \cos(\lambda_b + \omega_{ie}(t-t_0)) & -\sin L_b \sin(\lambda_b + \omega_{ie}(t-t_0)) & \cos L_b \\ -\sin(\lambda_b + \omega_{ie}(t-t_0)) & \cos(\lambda_b + \omega_{ie}(t-t_0)) & 0 \\ -\cos L_b \cos(\lambda_b + \omega_{ie}(t-t_0)) & -\cos L_b \sin(\lambda_b + \omega_{ie}(t-t_0)) & -\sin L_b \end{pmatrix}$$

$$\mathbf{C}_n^i = \begin{pmatrix} -\sin L_b \cos(\lambda_b + \omega_{ie}(t-t_0)) & -\sin(\lambda_b + \omega_{ie}(t-t_0)) & -\cos L_b \cos(\lambda_b + \omega_{ie}(t-t_0)) \\ -\sin L_b \sin(\lambda_b + \omega_{ie}(t-t_0)) & \cos(\lambda_b + \omega_{ie}(t-t_0)) & -\cos L_b \sin(\lambda_b + \omega_{ie}(t-t_0)) \\ \cos L_b & 0 & -\sin L_b \end{pmatrix}.$$

(2.154)

Earth-referenced velocity and acceleration in navigation frame axes transform to and from their inertial frame inertial reference counterparts as[†]

$$\mathbf{v}_{eb}^n = \mathbf{C}_i^n \left(\mathbf{v}_{ib}^i - \boldsymbol{\Omega}_{ie}^i \mathbf{r}_{ib}^i\right)$$
$$\mathbf{v}_{ib}^i = \mathbf{C}_n^i \mathbf{v}_{eb}^n + \mathbf{C}_e^i \boldsymbol{\Omega}_{ie}^e \mathbf{r}_{eb}^e, \qquad (2.155)$$

---

[*]This and subsequent paragraphs are based on material written by the author for QinetiQ, so comprise QinetiQ copyright material.
[†]End of QinetiQ copyright material.

$$\begin{aligned}
\mathbf{a}_{eb}^n &= \mathbf{C}_i^n\left(\mathbf{a}_{ib}^i - 2\mathbf{\Omega}_{ie}^i\mathbf{v}_{ib}^i + \mathbf{\Omega}_{ie}^i\mathbf{\Omega}_{ie}^i\mathbf{r}_{ib}^i\right) \\
\mathbf{a}_{ib}^i &= \mathbf{C}_n^i\left(\mathbf{a}_{eb}^n + 2\mathbf{\Omega}_{ie}^n\mathbf{v}_{eb}^n\right) + \mathbf{C}_e^i\mathbf{\Omega}_{ie}^e\mathbf{\Omega}_{ie}^e\mathbf{r}_{eb}^e
\end{aligned} \qquad (2.156)$$

Angular rates transform as

$$\begin{aligned}
\boldsymbol{\omega}_{nb}^n &= \mathbf{C}_i^n\left(\boldsymbol{\omega}_{ib}^i - \boldsymbol{\omega}_{in}^i\right) & \boldsymbol{\omega}_{ib}^i &= \mathbf{C}_n^i\left(\boldsymbol{\omega}_{nb}^n + \boldsymbol{\omega}_{in}^n\right) \\
&= \mathbf{C}_i^n\left(\boldsymbol{\omega}_{ib}^i - \boldsymbol{\omega}_{ie}^i\right) - \boldsymbol{\omega}_{en}^n, & &= \mathbf{C}_n^i\left(\boldsymbol{\omega}_{nb}^n + \boldsymbol{\omega}_{en}^n\right) + \boldsymbol{\omega}_{ie}^i
\end{aligned} \qquad (2.157)$$

Example 2.6 on the CD illustrates the velocity, acceleration, and angular rate transformations. Again, timing accuracy is critical for accurate frame transformations.

2.5.4 Earth and Local Tangent-Plane Frames

The orientation with respect to an ECEF frame of a local tangent-plane frame whose axes are aligned with north, east, and down may be determined using the geodetic latitude, L_l, and longitude, λ_l, of the local tangent-plane origin:

$$\begin{aligned}
\mathbf{C}_e^l &= \begin{pmatrix} -\sin L_l \cos \lambda_l & -\sin L_l \sin \lambda_l & \cos L_l \\ -\sin \lambda_l & \cos \lambda_l & 0 \\ -\cos L_l \cos \lambda_l & -\cos L_l \sin \lambda_l & -\sin L_l \end{pmatrix} \\
\mathbf{C}_l^e &= \begin{pmatrix} -\sin L_l \cos \lambda_l & -\sin \lambda_l & -\cos L_l \cos \lambda_l \\ -\sin L_l \sin \lambda_l & \cos \lambda_l & -\cos L_l \sin \lambda_l \\ \cos L_l & 0 & -\sin L_l \end{pmatrix}
\end{aligned} \qquad (2.158)$$

The origin and orientation of a local tangent-plane frame with respect to an ECEF frame are constant. Therefore, the velocity, acceleration, and angular rate may be transformed simply by rotating the resolving axes:

$$\begin{aligned}
\mathbf{v}_{lb}^l &= \mathbf{C}_e^l\mathbf{v}_{eb}^e, & \mathbf{v}_{eb}^e &= \mathbf{C}_l^e\mathbf{v}_{lb}^l \\
\mathbf{a}_{lb}^l &= \mathbf{C}_e^l\mathbf{a}_{eb}^e, & \mathbf{a}_{eb}^e &= \mathbf{C}_l^e\mathbf{a}_{lb}^l \\
\boldsymbol{\omega}_{lb}^l &= \mathbf{C}_e^l\boldsymbol{\omega}_{eb}^e, & \boldsymbol{\omega}_{eb}^e &= \mathbf{C}_l^e\boldsymbol{\omega}_{lb}^l
\end{aligned} \qquad (2.159)$$

The Cartesian position transforms as

$$\begin{aligned}
\mathbf{r}_{lb}^l &= \mathbf{C}_e^l\left(\mathbf{r}_{eb}^e - \mathbf{r}_{el}^e\right) \\
\mathbf{r}_{eb}^e &= \mathbf{r}_{el}^e + \mathbf{C}_l^e\mathbf{r}_{lb}^l
\end{aligned}, \qquad (2.160)$$

where $\mathbf{r}_{el}^e$ is the Cartesian ECEF position of the l-frame origin, obtained from L_l and λ_l using (2.112).

2.5 Frame Transformations

2.5.5 Transposition of Navigation Solutions

Sometimes, there is a requirement to transpose a navigation solution from one position to another on a vehicle, such as between an INS and a GNSS antenna, between an INS and the center of gravity, or between a reference and an aligning INS. Here, the equations for transposing position, velocity, and attitude from describing the b frame to describing the B frame are presented.*

Let the orientation of frame B with respect to frame b be $\mathbf{C}_b^B$ and the position of frame B with respect to frame b in frame b axes be $\mathbf{l}_{bB}^b$, which is known as the *lever arm* or *moment arm*. Note that the lever arm is mathematically identical to the Cartesian position with B as the object frame and b as the reference and resolving frames. Figure 2.29 illustrates this.

Attitude transformation is straightforward:

$$\mathbf{C}_\beta^B = \mathbf{C}_b^B \mathbf{C}_\beta^b, \qquad \mathbf{C}_B^\beta = \mathbf{C}_b^\beta \mathbf{C}_B^b. \tag{2.161}$$

Cartesian position may be transposed using

$$\mathbf{r}_{\beta B}^\gamma = \mathbf{r}_{\beta b}^\gamma + \mathbf{C}_b^\gamma \mathbf{l}_{bB}^b. \tag{2.162}$$

Precise transformation of latitude, longitude, and height requires conversion to Cartesian position and back. However, if the small angle approximation is applied to $1/R$, where R is the Earth radius, a simpler form may be used:

$$\begin{pmatrix} L_B \\ \lambda_B \\ h_B \end{pmatrix} \approx \begin{pmatrix} L_b \\ \lambda_b \\ h_b \end{pmatrix} + \begin{pmatrix} 1/(R_N(L_b) + h_b) & 0 & 0 \\ 0 & 1/[(R_E(L_b) + h_b)\cos L_b] & 0 \\ 0 & 0 & -1 \end{pmatrix} \mathbf{C}_b^n \mathbf{l}_{bB}^b. \tag{2.163}$$

The velocity transposition is obtained by differentiating (2.162) and substituting it into (2.67):

$$\mathbf{v}_{\beta B}^\gamma = \mathbf{v}_{\beta b}^\gamma + \mathbf{C}_\beta^\gamma \dot{\mathbf{C}}_b^\beta \mathbf{l}_{bB}^b, \tag{2.164}$$

assuming $\mathbf{l}_{bB}^b$ is constant. Substituting (2.56),†

$$\mathbf{v}_{\beta B}^\gamma = \mathbf{v}_{\beta b}^\gamma + \mathbf{C}_b^\gamma \left(\boldsymbol{\omega}_{\beta b}^b \wedge \mathbf{l}_{bB}^b \right). \tag{2.165}$$

Similarly, the acceleration transposition is

$$\mathbf{a}_{\beta B}^\gamma = \mathbf{a}_{\beta b}^\gamma + \mathbf{C}_b^\gamma \left[\boldsymbol{\omega}_{\beta b}^b \wedge \left(\boldsymbol{\omega}_{\beta b}^b \wedge \mathbf{l}_{bB}^b \right) + \left(\dot{\boldsymbol{\omega}}_{\beta b}^b \wedge \mathbf{l}_{bB}^b \right) \right]. \tag{2.166}$$

Problems and exercises for this chapter are on the accompanying CD.

---

*This and subsequent paragraphs are based on material written by the author for QinetiQ, so comprise QinetiQ copyright material.

†End of QinetiQ copyright material.

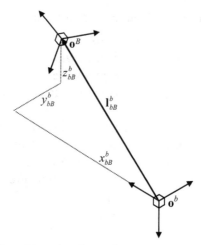

Figure 2.29 The lever arm from frame *b* to frame *B*.

References

[1] Groves, P. D., "Principles of Integrated Navigation," Course Notes, QinetiQ Ltd., 2002.
[2] Grewal, M. S., L. R. Weill, and A. P. Andrews, *Global Positioning Systems, Inertial Navigation, and Integration*, 2nd ed., New York: Wiley, 2007.
[3] Kuipers, J. B., *Quaternions and Rotation Sequences*, Princeton, NJ: Princeton University Press, 1999.
[4] Bortz, J. E., "A New Mathematical Formulation for Strapdown Inertial Navigation," *IEEE Trans. on Aerospace and Electronic Systems*, Vol. AES-7, No. 1, 1971, pp. 61–66.
[5] Farrell, J. A., *Aided Navigation: GPS with High Rate Sensors*, New York: McGraw-Hill, 2008.
[6] Rogers, R. M., *Applied Mathematics in Integrated Navigation Systems*, Reston: VA, AIAA, 2000.
[7] Feynman, R. P., R. B. Leighton, and M. Sands, *The Feynman Lectures on Physics, Volume 1*, Reading, MA: Addison-Wesley, 1963.
[8] Ashkenazi, V., "Coordinate Systems: How to Get Your Position Very Precise and Completely Wrong," *Journal of Navigation*, Vol. 39, No.2, 1986, pp. 269–278.
[9] Anon., *Department of Defense World Geodetic System 1984*, National Imagery and Mapping Agency (now NGA), TR8350.2, Third Edition, 1997.
[10] Boucher, C., et al., *The ITRF 2000*, International Earth Rotation and Reference Systems Service Technical Note, No. 31, 2004.
[11] Malys, S., et al., "Refinements to the World Geodetic System, 1984," *Proc. ION GPS-97*, Kansas, MO, September 1997, pp. 915–920.
[12] Misra, P., and P. Enge, *Global Positioning System Signals, Measurements, and Performance*, 2nd ed., Lincoln, MA: Ganga-Jamuna Press, 2006.
[13] Iliffe, J., and R. Lott, *Datums and Map Projections for Remote Sensing, GIS and Surveying*, 2nd ed., Edinburgh, U.K.: Whittles Publishing, 2008.
[14] Petit, G., and B. Luzum, (eds.), *IER Conventions (2010)*, IERS Technical Note No. 36, Frankfurt am Main, Germany: Verlag des Bundesamts für Kartographie und Geodäsie, 2010.
[15] Galati, S. R., *Geographic Information Systems Demystified*, Norwood, MA: Artech House, 2006.

[16] NGA, *WGS 84 Earth Gravitational Model*, http://earth-info.nga.mil/GandG/wgs84/gravitymod/, accessed February 21, 2010.

[17] Jekeli, C., *Inertial Navigation Systems with Geodetic Applications*, Berlin, Germany: de Gruyter, 2000.

[18] Britting, K. R., *Inertial Navigation Systems Analysis*, New York: Wiley, 1971.

Selected Bibliography

Bomford, G., Geodesy, Fourth Edition, London, UK: Clarendon Press, 1980.

Smith, J. R., Introduction to Geodesy: The History and Concepts of Modern Geodesy, New York: Wiley, 1997.

Torge, W., Geodesy, Berlin, Germany: de Gruyter, 2001.

CHAPTER 3
Kalman Filter-Based Estimation

A state estimation algorithm determines the values of a number of parameters of a system, such as its position and velocity, from measurements of the properties of that system. The Kalman filter forms the basis of most state estimation algorithms used in navigation systems. Its uses include maintaining an optimal satellite navigation solution, integration of GNSS user equipment with other navigation sensors, and alignment and calibration of an INS. State estimation is key to obtaining the best possible navigation solution from the various measurements available. A Kalman filter uses all the measurement information input to it over time, not just the most recent set of measurements.

This chapter provides an introduction to the Kalman filter and a review of how it may be adapted for practical use in navigation applications. Section 3.1 provides a qualitative description of the Kalman filter, with the algorithm and mathematical models introduced in Section 3.2. Section 3.3 discusses the practical application of the Kalman filter, while Section 3.4 reviews some more advanced estimation techniques, based on the Kalman filter, that are relevant to navigation problems. These include the extended Kalman filter (EKF), commonly used in navigation applications, the unscented Kalman filter (UKF), and the Kalman smoother, which can give improved performance in postprocessed applications. Finally, Section 3.5 provides a brief introduction to the particle filter. In addition, Appendix D on the CD describes least-squares estimation, summarizes the Schmidt-Kalman filter, and provides further information on the particle filter, while Appendix B on the CD provides background information on statistical measures, probability, and random processes.

Examples of the Kalman filter's applications in navigation are presented within Chapters 9 and 14 to 16, while the MATLAB software on the accompanying CD includes Kalman-filter based estimation algorithms for GNSS positioning and INS/GNSS integration. For a more formalized and detailed treatment of Kalman filters, there are many applied mathematics books devoted solely to this subject [1–6].

At this point, it is useful to introduce the distinction between systematic and random errors. A systematic error is repeatable and can thus be predicted from previous occurrences using a Kalman filter or another estimation algorithm. An example is a bias, or constant offset, in a measurement. A random error is nonrepeatable; it cannot be predicted. In practice, an error will often have both systematic and random components. An example is a bias that slowly varies in an unpredictable way. This can also be estimated using a Kalman filter.

3.1 Introduction

The Kalman filter is an estimation algorithm, rather than a filter. The basic technique was invented by R. E. Kalman in 1960 [7] and has been developed further by numerous authors since. It maintains real-time estimates of a number of parameters of a system, such as its position and velocity, that may continually change. The estimates are updated using a stream of measurements that are subject to noise. The measurements must be functions of the parameters estimated, but the set of measurements at a given time need not contain sufficient information to uniquely determine the values of the parameters at that time.

The Kalman filter uses knowledge of the deterministic and statistical properties of the system parameters and the measurements to obtain optimal estimates given the information available. It is a Bayesian estimation technique. It is supplied with an initial set of estimates and then operates recursively, updating its working estimates as a weighted average of their previous values and new values derived from the latest measurement data. By contrast, nonrecursive estimation algorithms derive their parameter estimates from the whole set of measurement data without prior estimates. For real-time applications, such as navigation, the recursive approach is more processor efficient, as only the new measurement data need be processed on each iteration. Old measurement data may be discarded.

To enable optimal weighting of the data, a Kalman filter maintains a set of uncertainties in its estimates and a measure of the correlations between the errors in the estimates of the different parameters. This is carried forward from iteration to iteration alongside the parameter estimates. It also accounts for the uncertainties in the measurements due to noise.

This section provides a qualitative description of the Kalman filter and the steps forming its algorithm. Some brief examples of Kalman filter applications conclude the section. A quantitative description and derivation follow in Section 3.2.

3.1.1 Elements of the Kalman Filter

Figure 3.1 shows the five core elements of the Kalman filter: the state vector and covariance, the system model, the measurement vector and covariance, the measurement model, and the algorithm.

The *state vector* is the set of parameters describing a system, known as *states*, which the Kalman filter estimates. Each state may be constant or time varying. For most navigation applications, the states include the components of position or position error. Velocity, attitude, and navigation sensor error states may also be estimated. Beware that some authors use the term state to describe the whole state vector rather than an individual component.

Associated with the state vector is an *error covariance matrix*. This represents the uncertainties in the Kalman filter's state estimates and the degree of correlation between the errors in those estimates. The correlation information within the error covariance matrix is important for three reasons. First, it enables the error distribution of the state estimates to be fully represented. Figure 3.2 illustrates this for north and east position estimates; when the correlation is neglected, the accuracy is overestimated in one direction and underestimated in another. Second, there is not

3.1 Introduction

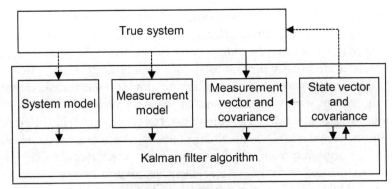

Figure 3.1 Elements of the Kalman filter. (*From:* [8]. © 2002 QinetiQ Ltd. Reprinted with permission.)

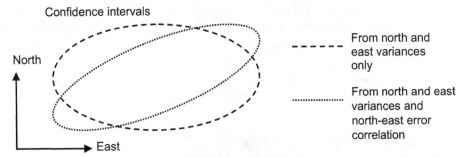

Figure 3.2 Example position error ellipses with and without error correlation.

always enough information from the measurements to estimate the Kalman filter states independently. The correlation information enables estimates of linear combinations of those states to be maintained while awaiting further measurement information. Finally, correlations between errors can build up over the intervals between measurements. Modeling this can enable one state to be determined from another (e.g., velocity from a series of positions). A Kalman filter is an iterative process, so the initial values of the state vector and covariance matrix must be set by the user or determined from another process.

The *system model*, also known as the process model or time-propagation model, describes how the Kalman filter states and error covariance matrix vary with time. For example, a position state will vary with time as the integral of a velocity state; the position uncertainty will increase with time as the integral of the velocity uncertainty; and the position and velocity estimation errors will become more correlated. The system model is deterministic for the states as it is based on known properties of the system.[*]

A state uncertainty should also be increased with time to account for unknown changes in the system that cause the state estimate to go out of date in the absence of new measurement information. These changes may be unmeasured dynamics or

---

[*]This paragraph, up to this point, is based on material written by the author for QinetiQ, so comprises QinetiQ copyright material.

random noise on an instrument output. For example, a velocity uncertainty must be increased over time if the acceleration is unknown. This variation in the true values of the states is known as *system noise* or process noise, and its assumed statistical properties are usually defined by the Kalman filter designer.

The *measurement vector* is a set of simultaneous measurements of properties of the system which are functions of the state vector. Examples include the set of range measurements from a radio navigation system and the difference in navigation solution between an INS under calibration and a reference navigation system. This is the information from which all of the state estimates are derived after initialization. Associated with the measurement vector is a *measurement noise covariance* matrix which describes the statistics of the noise on the measurements. For many applications, new measurement information is input to the Kalman filter at regular intervals. In other cases, the time interval between measurements can be irregular.

The *measurement model* describes how the measurement vector varies as a function of the true state vector (as opposed to the state vector estimate) in the absence of measurement noise. For example, the velocity measurement difference between an INS under calibration and a reference system is directly proportional to the INS velocity error. Like the system model, the measurement model is deterministic, based on known properties of the system.*

The *Kalman filter algorithm* uses the measurement vector, measurement model, and system model to maintain optimal estimates of the state vector.

3.1.2 Steps of the Kalman Filter

The Kalman filter algorithm consists of two phases, system propagation and measurement update, which together comprise up to 10 steps per iteration. These are shown in Figure 3.3. Steps 1–4 form the system-propagation phase and steps 5–10 the measurement-update phase. Each complete iteration of the Kalman filter corresponds to a particular point in time, known as an epoch.

The purpose of the system-propagation, or time-propagation, phase is to predict forward the state vector estimate and error covariance matrix from the time of validity of the last measurement set to the time of the current set of measurements using the known properties of the system. So, for example, a position estimate is predicted forward using the corresponding velocity estimate. This provides the Kalman filter's best estimate of the state vector at the current time in the absence of new measurement information. The first two steps calculate the deterministic and noise parts of the system model. The third step, *state propagation*, uses this to bring the state vector estimate up to date. The fourth step, *covariance propagation*, performs the corresponding update to the error covariance matrix, increasing the state uncertainty to account for the system noise.

In the measurement-update, or correction, phase, the state vector estimate and error covariance are updated to incorporate the new measurement information. Steps 5 and 6, respectively, calculate the deterministic and noise parts of the measurement model. The seventh step, *gain computation*, calculates the Kalman gain matrix. This

---
*This paragraph, up to this point, is based on material written by the author for QinetiQ, so comprises QinetiQ copyright material.

3.1 Introduction

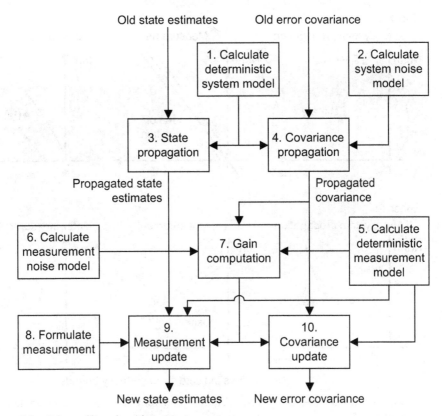

Figure 3.3 Kalman filter algorithm steps.

is used to optimally weight the correction to the state vector according to the uncertainty of the current state estimates and how noisy the measurements are. The eighth step formulates the measurement vector. The ninth step, the *measurement update*, updates the state estimates to incorporate the measurement data weighted with the Kalman gain. Finally, the *covariance update* updates the error covariance matrix to account for the new information that has been incorporated into the state vector estimate from the measurement data.

Figure 3.4 illustrates qualitatively how a Kalman filter can determine a position solution from successive incomplete measurements. At epoch 1, there is a 2-D position estimate with a large uncertainty. The measurement available at this epoch is a single line of position (LOP). This could be from a range measurement using a distant transmitter or from a bearing measurement. The measurement only provides positioning information along the direction perpendicular to the LOP. A unique position fix cannot be obtained from it. Implementing a Kalman filter measurement update results in the position estimate moving close to the measurement LOP. There is a large reduction in the position uncertainty perpendicular to the measurement LOP, but no reduction along the LOP.

At epoch 2, the Kalman filter system-propagation phase increases the position uncertainty to account for possible movement of the object. The measurement available at epoch 2 is also a single LOP, but in a different direction to that of the first measurement. This provides positioning information along a different direction to

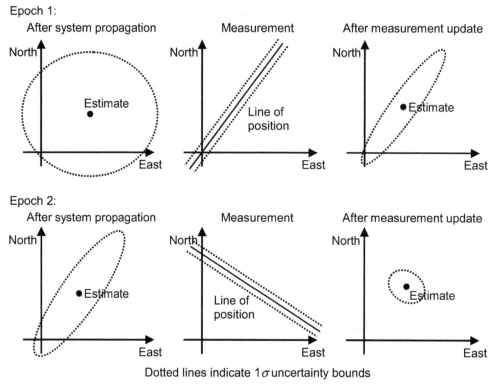

Figure 3.4 Kalman filter 2-D position determination from two successive incomplete measurements.

the first measurement. Consequently, implementing the Kalman filter measurement update results in a position estimate with a small uncertainty in both directions.

3.1.3 Kalman Filter Applications

Kalman filter-based estimation techniques have many applications in navigation. These include GNSS and terrestrial radio navigation, GNSS signal monitoring, INS/GNSS and multisensor integration, and fine alignment and calibration of an INS.

For stand-alone GNSS navigation, the states estimated are the user antenna position and velocity, and the receiver clock offset and drift. The measurements are the line-of-sight ranging measurements of each satellite signal made by the receiver. The GNSS navigation filter is described in Section 9.4.2. For terrestrial radio navigation, the height and vertical velocity are often omitted due to insufficient signal geometry, while the clock states may be omitted if the ranging measurements are two-way or differenced across transmitters (see Chapter 7). A single navigation filter may process both GNSS and terrestrial radio navigation measurements as discussed in Chapter 16.

GNSS signal monitoring uses the same measurements as GNSS navigation. However, the user antenna position and velocity are accurately known and a high precision receiver clock is used, so the time-correlated range errors may be estimated as Kalman filter states. With a network of monitor stations at different locations, the different contributing factors to the range errors may all be estimated as separate states.

For most INS/GNSS and multisensor integration architectures, the errors of the constituent navigation systems, including position and velocity errors, are estimated. In some architectures, the navigation solution itself is also estimated. The measurements processed vary with the type of integration implemented. Examples include position measurements, ranging measurements and sensor measurements. INS/GNSS integration techniques are described in Chapter 14, with multisensor integration described in Chapter 16.

For alignment and calibration of an INS, the states estimated are position, velocity, and attitude errors, together with inertial instrument errors, such as accelerometer and gyro biases. The measurements are the position, velocity, and/or attitude differences between the aligning-INS navigation solution and an external reference, such as another INS or GNSS.[*] More details are given in Section 5.6.3 and Chapters 14 and 15.

3.2 Algorithms and Models

This section presents and derives the Kalman filter algorithm, system model, and measurement model, including open- and closed-loop implementations and a discussion of Kalman filter behavior and state observability. Prior to this, error types are discussed and the main Kalman filter parameters defined. Although a Kalman filter may operate continuously in time, discrete-time implementations are most common as these are suited to digital computation. Thus, only the discrete-time version is presented here.[*]

3.2.1 Definitions

The time variation of all errors modeled within a discrete-time Kalman filter is assumed to fall into one of three categories: systematic errors, white noise sequences, and Gauss-Markov sequences. These are shown in Figure 3.5. *Systematic errors* are assumed to be constant—in other words, 100% time-correlated, though a Kalman filter's estimates of these quantities may vary as it obtains more information about them.

A *white noise sequence* is a discrete-time sequence of mutually uncorrelated random variables from a zero-mean distribution. Samples, w_i, have the property

$$E(w_i w_j) = \begin{matrix} \sigma_w^2 & i = j \\ 0 & i \neq j \end{matrix}, \quad (3.1)$$

where E is the expectation operator and σ_w^2 is the variance. A white noise process is described in Section B.4.2 of Appendix B on the CD. Samples from a band-limited white noise process may be treated as a white noise sequence provided the sampling

---

[*]This paragraph, up to this point, is based on material written by the author for QinetiQ, so comprises QinetiQ copyright material.

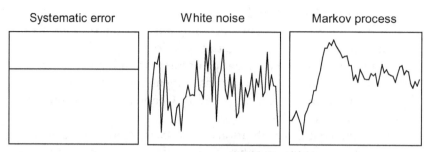

Figure 3.5 Example systematic error, white noise, and Markov process.

rate is much less than the double-sided noise bandwidth. The variance of a white noise sequence, obtained by integrating a white noise process over the time interval τ_w, is

$$\sigma_w^2 = \tau_w S_w, \qquad (3.2)$$

where S_w is the power spectral density (PSD) of the white noise process. This is the variance per unit bandwidth. In general, the PSD is a function of frequency. However, for band-limited white noise, the PSD is constant within the white noise bandwidth, which must significantly exceed $1/\tau_w$ for (3.2) to apply. In a Kalman filter, white noise is normally assumed to have a Gaussian (or normal) distribution (see Section B.3.2 in Appendix B on the CD).

A *Gauss-Markov sequence* is a quantity that varies with time as a linear function of its previous values and a white noise sequence. When the properties of a Gauss-Markov sequence are known, it can be modeled in a Kalman filter. It typically varies slowly compared to the update interval. A first-order Gauss-Markov sequence may be represented as a linear function only of its previous value and noise. A Markov process is the continuous-time equivalent of a Markov sequence. A first-order Gauss-Markov process, x_{mi}, may be described by

$$\frac{\partial x_{mi}}{\partial t} = -\frac{x_{mi}}{\tau_{mi}} + w_i, \qquad (3.3)$$

where t is time and τ_{mi} is the correlation time. It is often known as an exponentially correlated Markov process as it has an exponentially decaying auto-correlation function. Markov processes and sequences are described in more detail in Section B.4.3 of Appendix B on the CD. In a Kalman filter, they are normally assumed to have Gaussian distributions.

A principal assumption of Kalman filter theory is that the errors of the modeled system are systematic, white noise, or Gauss-Markov processes. They may also be linear combinations or integrals thereof. For example, a random walk process is integrated white noise, while a constant acceleration error leads to a velocity error that grows with time. Error sources modeled as states are assumed to be systematic, Markov processes, or their integrals. All noise sources are assumed to be white, noting that Markov processes have a white noise component. Real navigation system errors do not fall neatly into these categories, but, in many cases, can be approximated to them, provided the modeled errors adequately overbound their real counterparts. A good analogy is that you can fit a square peg into a round hole if you make the hole sufficiently large.

3.2 Algorithms and Models

The set of parameters estimated by a Kalman filter, known as the *state vector*, is denoted by **x**. The Kalman filter estimate of the state vector is denoted $\hat{\mathbf{x}}$, with the caret, ^, also used to indicate other quantities calculated using the state estimates.[*] Estimating absolute properties of the system, such as position, velocity, and attitude, as states is known as a *total-state* implementation. Estimation of the errors in a measurement made by the system, such as INS position, velocity, and attitude, as states is known as an *error-state* implementation. However, a state vector may comprise a mixture of total states and error states.

Note that it is not always sufficient for a Kalman filter only to estimate those states required to directly determine or correct the navigation solution. Significant systematic error sources and Markov processes that impact the states or measurements must be added to the state vector to prevent corruption of the navigation states. This is because a Kalman filter assumes that all error sources that are not modeled as states are white noise. The addition of these extra states is sometimes known as augmentation.

The *state vector residual*, $\delta \mathbf{x}$, is the difference between the true state vector and the Kalman filter estimates thereof. Thus,[†]

$$\delta \mathbf{x} = \mathbf{x} - \hat{\mathbf{x}}. \quad (3.4)$$

In an error-state implementation, the state vector residual represents the errors remaining in the system after the Kalman filter estimates have been used to correct it. The errors in the state estimates are obtained simply by reversing the sign of the state residuals.

The *error covariance matrix*, **P**, defines the expectation of the square of the deviation of the state vector estimate from the true value of the state vector. Thus,[‡]

$$\mathbf{P} = \mathrm{E}\!\left((\hat{\mathbf{x}} - \mathbf{x})(\hat{\mathbf{x}} - \mathbf{x})^{\mathrm{T}}\right) = \mathrm{E}(\delta \mathbf{x}\, \delta \mathbf{x}^{\mathrm{T}}). \quad (3.5)$$

The **P** matrix is symmetric (see Section A.3 of Appendix A on the CD). The diagonal elements are the variances of each state estimate, while their square roots are the uncertainties. Thus,

$$P_{ii} = \sigma_i^2, \quad (3.6)$$

where σ_i is the uncertainty of the ith state estimate. The off-diagonal elements of **P**, the covariances, describe the correlations between the errors in the different state estimates. They may be expressed as

$$P_{ij} = P_{ji} = \sigma_i \sigma_j \rho_{i,j} \quad (3.7)$$

---

[*]This paragraph, up to this point, is based on material written by the author for QinetiQ, so comprises QinetiQ copyright material.

[†]This and subsequent paragraphs are based on material written by the author for QinetiQ, so comprise QinetiQ copyright material.

[‡]End of QinetiQ copyright material.

where $\rho_{i,j}$ is the correlation coefficient, defined in Section B.2.1 of Appendix B on the CD. Note that $\rho_{i,j} = 1$ where $i = j$.

The errors in the estimates of different states can become significantly correlated with each other where there is insufficient information from the measurements to estimate those states independently. It is analogous to having a set of simultaneous equations where there are more unknowns than equations. This subject is known as observability and is discussed further in Section 3.2.5.

In an error-state implementation, all state estimates are usually given an initial value of zero. In a total-state implementation, the states may be initialized by the user, by a coarse initialization process, or with the estimates from the previous time the host equipment was used. The initialization values of the covariance matrix are generally determined by the Kalman filter designer and are normally selected cautiously. Thus, the state initialization values are a priori estimates, while the initial covariance matrix values indicate the confidence in those estimates.

In the continuous-time Kalman filter system and measurement models, the state vector and other parameters are shown as functions of time, t. In the discrete-time Kalman filter, the subscript k is used to denote the epoch or iteration to which the state the state vector and other parameters apply. Therefore, $\mathbf{x}_k \equiv \mathbf{x}(t_k)$.

It is necessary to distinguish between the state vector and error covariance after complete iterations of the Kalman filter and in the intermediate step between propagation and update. Thus, the time-propagated state estimates and covariance are denoted $\hat{\mathbf{x}}_k^-$ and $\mathbf{P}_k^-$ (some authors use $\hat{\mathbf{x}}_k(-)$ and $\mathbf{P}_k(-)$, $\hat{\mathbf{x}}_{k|k-1}$ and $\mathbf{P}_{k|k-1}$, or $\hat{\mathbf{x}}(k|k-1)$ and $\mathbf{P}(k|k-1)$). Their counterparts following the measurement update are denoted $\hat{\mathbf{x}}_k^+$ and $\mathbf{P}_k^+$ (some authors use $\hat{\mathbf{x}}_k(+)$ and $\mathbf{P}_k(+)$, $\hat{\mathbf{x}}_{k|k}$ and $\mathbf{P}_{k|k}$, or $\hat{\mathbf{x}}(k|k)$ and $\mathbf{P}(k|k)$).

The measurement vector, $\mathbf{z}$ (some authors use $\mathbf{y}$), is a set of measurements of the properties of the system described by the state vector. This could be a set of range measurements or the difference between two navigation systems' position and velocity solutions. It comprises a deterministic function, $\mathbf{h}(\mathbf{x})$, and noise, $\mathbf{w}_m$ (many authors use $\mathbf{v}$, while some use $\mathbf{\mu}$ or $\mathbf{w}$). Thus,[†]

$$\mathbf{z} = \mathbf{h}(\mathbf{x}) + \mathbf{w}_m. \quad (3.8)$$

The *measurement innovation*, $\delta\mathbf{z}^-$ (some authors use $\mathbf{\mu}$ or $\mathbf{r}$), is the difference between the true measurement vector and that computed from the state vector estimate prior to the measurement update:[‡]

$$\delta\mathbf{z}^- = \mathbf{z} - \mathbf{h}(\hat{\mathbf{x}}^-). \quad (3.9)$$

For example, it could be the difference between an actual set of range measurements and a set predicted using a Kalman filter's position estimate. The *measurement residual*, $\delta\mathbf{z}^+$, is the difference between the true measurement vector and that computed from the updated state vector:

[†]This and subsequent paragraphs are based on material written by the author for QinetiQ, so comprise QinetiQ copyright material.
[‡]End of QinetiQ copyright material.

3.2 Algorithms and Models

$$\delta z^+ = z - h(\hat{x}^+). \quad (3.10)$$

Beware that some authors use the term residual to describe the innovation.

The measurement innovations and residuals are a mixture of state estimation errors and measurement errors that are uncorrelated with the state estimates, such as the noise on a set of range measurements. The standard Kalman filter assumes that these measurement errors form a zero-mean distribution, normally assumed to be Gaussian, that is uncorrelated in time, and models their standard deviations with the *measurement noise covariance matrix*, **R**. This defines the expectation of the square of the measurement noise. Thus,

$$\mathbf{R} = E(\mathbf{w}_m \mathbf{w}_m^T). \quad (3.11)$$

The diagonal terms of **R** are the variances of each measurement, and the off-diagonal terms represent the correlation between the different components of the measurement noise. The **R** matrix is also symmetric. For most navigation applications, the noise on each component of the measurement vector is independent so **R** is a diagonal matrix. The rest of the Kalman filter notation is defined as it is used.[*]

3.2.2 Kalman Filter Algorithm

With reference to Figure 3.3, the discrete-time Kalman filter algorithm comprises the following steps:[†]

1. Calculate the transition matrix, $\mathbf{\Phi}_{k-1}$.
2. Calculate the system noise covariance matrix, $\mathbf{Q}_{k-1}$.
3. Propagate the state vector estimate from $\hat{\mathbf{x}}_{k-1}^+$ and $\hat{\mathbf{x}}_k^-$.
4. Propagate the error covariance matrix from $\mathbf{P}_{k-1}^+$ to $\mathbf{P}_k^-$.
5. Calculate the measurement matrix, $\mathbf{H}_k$.
6. Calculate the measurement noise covariance matrix, $\mathbf{R}_k$.
7. Calculate the Kalman gain matrix, $\mathbf{K}_k$.
8. Formulate the measurement, $\mathbf{z}_k$.
9. Update the state vector estimate from $\hat{\mathbf{x}}_k^-$ to $\hat{\mathbf{x}}_k^+$.
10. Update the error covariance matrix from $\mathbf{P}_k^-$ to $\mathbf{P}_k^+$.[‡]

The Kalman filter steps do not have to be implemented strictly in this order, provided that the dependencies depicted in Figure 3.3 are respected. Although many Kalman filters simply alternate the system-propagation and measurement-update phases, other processing cycles are possible as discussed in Section 3.3.2.

---

[*] This paragraph, up to this point, is based on material written by the author for QinetiQ, so comprises QinetiQ copyright material.

[†] This and subsequent paragraphs are based on material written by the author for QinetiQ, so comprise QinetiQ copyright material.

[‡] End of QinetiQ copyright material.

The first four steps comprise the system-propagation phase of the Kalman filter, also known as the system-update, system-extrapolation, prediction, projection, time-update, or time-propagation phase. The system model is derived in Section 3.2.3.

Step 1 is the calculation of the *transition matrix*, $\mathbf{\Phi}_{k-1}$ (a few authors use $\mathbf{F}_{k-1}$). This defines how the state vector changes with time as a function of the dynamics of the system modeled by the Kalman filter. For example, a position state will vary as the integral of a velocity state. The rows correspond to the new values of each state and the columns to the old values.

The transition matrix is different for every Kalman filter application and is derived from a linear system model as shown in Section 3.2.3. It is nearly always a function of the time interval, τ_s, between Kalman filter iterations and is often a function of other parameters. When these parameters vary over time, the transition matrix must be recalculated on every Kalman filter iteration. Note that, in a standard Kalman filter, the transition matrix is never a function of any of the states; otherwise, the system model would not be linear.

Example A is a Kalman filter estimating position and velocity along a single axis in a nonrotating frame. The state vector and transition matrix are

$$\mathbf{x}_A = \begin{pmatrix} r^i_{ib,x} \\ v^i_{ib,x} \end{pmatrix}, \quad \mathbf{\Phi}_A = \begin{pmatrix} 1 & \tau_s \\ 0 & 1 \end{pmatrix} \qquad (3.12)$$

as position is the integral of velocity. Example B is a Kalman filter estimating 2-D position, again in a nonrotating frame. Its state vector and transition matrix are

$$\mathbf{x}_B = \begin{pmatrix} r^i_{ib,x} \\ r^i_{ib,y} \end{pmatrix}, \quad \mathbf{\Phi}_B = \begin{pmatrix} 1 & 0 \\ 0 & 1 \end{pmatrix} \qquad (3.13)$$

as the transition matrix is simply the identity matrix where all states are independent. Examples 3.1 and 3.2 on the CD, both of which are editable using Microsoft Excel, comprise numerical implementations of a complete Kalman filter cycle based on Examples A and B, respectively.

Step 2 is the calculation of the *system noise covariance matrix*, $\mathbf{Q}_{k-1}$, also known as the process noise covariance matrix. It defines how the uncertainties of the state estimates increase with time due to unknown changes in the true values of those states, such as unmeasured dynamics and instrument noise. These changes are treated as noise sources in the Kalman filter's system model. The system noise is always a function of the time interval between iterations, τ_s. Depending on the application, it may be modeled as either time-varying or as constant (for a given time interval). The system noise covariance is a symmetric matrix and is often approximated to a diagonal matrix.

In Example A, system noise arises from changes in the velocity state over time. Example B does not include a velocity state, so system noise arises from changes in the two position states. System noise covariance matrices for these examples are presented in Section 3.2.3.

Step 3 comprises the propagation of the state vector estimate through time using

$$\hat{\mathbf{x}}_k^- = \mathbf{\Phi}_{k-1}\hat{\mathbf{x}}_{k-1}^+. \tag{3.14}$$

Step 4 is the corresponding error covariance propagation. The standard form is

$$\mathbf{P}_k^- = \mathbf{\Phi}_{k-1}\mathbf{P}_{k-1}^+\mathbf{\Phi}_{k-1}^T + \mathbf{Q}_{k-1}. \tag{3.15}$$

Note that the first $\mathbf{\Phi}$ matrix propagates the rows of the error covariance matrix, while the second, $\mathbf{\Phi}^T$, propagates the columns. Following this step, each state uncertainty should be either larger or unchanged.

The remaining steps in the Kalman filter algorithm comprise the measurement-update or correction phase. The measurement model is derived in Section 3.2.4.

Step 5 is the calculation of the *measurement matrix*, $\mathbf{H}_k$ (some authors use $\mathbf{M}_k$, while $\mathbf{G}_k$ or $\mathbf{A}_k$ is sometimes used in GNSS navigation filters). This defines how the measurement vector varies with the state vector. Each row corresponds to a measurement and each column to a state. For example, the range measurements from a radio navigation system vary with the position of the receiver. In a standard Kalman filter, each measurement is assumed to be a linear function of the state vector. Thus,

$$\mathbf{h}(\mathbf{x}_k, t_k) = \mathbf{H}_k\mathbf{x}_k. \tag{3.16}$$

In most applications, the measurement matrix varies, so it must be calculated on each iteration of the Kalman filter. In navigation, $\mathbf{H}_k$ is commonly a function of the user kinematics and/or the geometry of transmitters, such as GNSS satellites.

In Examples A and B, the measurements are, respectively, single-axis position and 2-D position, plus noise. The measurement models and matrices are thus

$$z_A = r_{ib,x}^i + w_m, \qquad \mathbf{H}_A = \begin{pmatrix} 1 & 0 \end{pmatrix} \tag{3.17}$$

and

$$\mathbf{z}_B = \begin{pmatrix} r_{ib,x}^i + w_{m,x} \\ r_{ib,y}^i + w_{m,y} \end{pmatrix}, \qquad \mathbf{H}_B = \begin{pmatrix} 1 & 0 \\ 0 & 1 \end{pmatrix}. \tag{3.18}$$

Measurement updates using these models are shown in Examples 3.1 and 3.2 on the CD.

Step 6 is the calculation of the measurement noise covariance matrix, $\mathbf{R}_k$. Depending on the application, it may be assumed constant, modeled as a function of dynamics, and/or modeled as a function of signal-to-noise measurements.

Step 7 is the calculation of the *Kalman gain matrix*, $\mathbf{K}_k$. This is used to determine the weighting of the measurement information in updating the state estimates. Each row corresponds to a state and each column to a measurement. The Kalman gain depends on the error covariance matrices of both the true measurement vector, $\mathbf{z}_k$, and that predicted from the state estimates, $\mathbf{H}_k\hat{\mathbf{x}}_k^-$, noting that the diagonal elements of the matrices are the squares of the uncertainties. From (3.8), (3.9), and (3.10), the error covariance of the true measurement vector is

$$E\left((\mathbf{z}_k - \mathbf{H}_k\mathbf{x}_k)(\mathbf{z}_k - \mathbf{H}_k\mathbf{x}_k)^\mathrm{T}\right) = \mathbf{R}_k, \qquad (3.19)$$

and, from (3.5), the error covariance of the measurement vector predicted from the state vector is

$$E\left((\mathbf{H}_k\hat{\mathbf{x}}_k^- - \mathbf{H}_k\mathbf{x}_k)(\mathbf{H}_k\hat{\mathbf{x}}_k^- - \mathbf{H}_k\mathbf{x}_k)^\mathrm{T}\right) = \mathbf{H}_k\mathbf{P}_k^-\mathbf{H}_k^\mathrm{T}. \qquad (3.20)$$

The Kalman gain matrix is

$$\mathbf{K}_k = \mathbf{P}_k^-\mathbf{H}_k^\mathrm{T}\left(\mathbf{H}_k\mathbf{P}_k^-\mathbf{H}_k^\mathrm{T} + \mathbf{R}_k\right)^{-1}. \qquad (3.21)$$

where $(\ )^{-1}$ denotes the inverse of a matrix. Matrix inversion is discussed in Section A.4 of Appendix A on the CD. Some authors use a fraction notation for matrix inversion; however, this can leave the order of matrix multiplication ambiguous. Note that, as the leading $\mathbf{H}_k$ matrix of (3.20) is omitted in the "numerator" of the variance ratio, the Kalman gain matrix transforms from measurement space to state space as well as weighting the measurement information. The correlation information in the off-diagonal elements of the $\mathbf{P}_k^-$ matrix couples the measurement vector to those states that are not directly related via the $\mathbf{H}_k$ matrix.

In Example A, the measurement is scalar, simplifying the Kalman gain calculation. If the covariance matrices are expressed as

$$\mathbf{P}_{A,k}^- = \begin{pmatrix} \sigma_r^2 & P_{rv} \\ P_{rv} & \sigma_v^2 \end{pmatrix}, \qquad R_{A,k} = \sigma_z^2, \qquad (3.22)$$

substituting these and (3.17) into (3.21) gives a Kalman gain of

$$\mathbf{K}_{A,k} = \begin{pmatrix} \sigma_r^2 \\ P_{rv} \end{pmatrix} \frac{1}{\sigma_r^2 + \sigma_z^2}. \qquad (3.23)$$

Note that the velocity may be estimated from the position measurements provided the prior position and velocity estimates have correlated errors.

Step 8 is the formulation of the measurement vector, $\mathbf{z}_k$. In some cases, such as radio navigation range measurements and Examples A and B, the measurement vector components are already present in the system modeled by the Kalman filter. In other cases, $\mathbf{z}_k$ must be calculated as a function of other system parameters. An example is the navigation solution difference between a system under calibration and a reference system.

3.2 Algorithms and Models

For many applications, the measurement innovation, $\delta\mathbf{z}_k^-$, may be calculated directly by applying corrections derived from the state estimates to those parameters of which the measurements are a function. For example, the navigation solution of an INS under calibration may be corrected by the Kalman filter state estimates prior to being differenced with a reference navigation solution.

Step 9 is the update of the state vector with the measurement vector using

$$\begin{aligned}\hat{\mathbf{x}}_k^+ &= \hat{\mathbf{x}}_k^- + \mathbf{K}_k\left(\mathbf{z}_k - \mathbf{H}_k\hat{\mathbf{x}}_k^-\right) \\ &= \hat{\mathbf{x}}_k^- + \mathbf{K}_k\delta\mathbf{z}_k^-\end{aligned} \quad (3.24)$$

The measurement innovation, $\delta\mathbf{z}_k^-$, is multiplied by the Kalman gain matrix to obtain a correction to the state vector estimate.

Step 10 is the corresponding update of the error covariance matrix with

$$\mathbf{P}_k^+ = (\mathbf{I} - \mathbf{K}_k\mathbf{H}_k)\mathbf{P}_k^-. \quad (3.25)$$

As the updated state vector estimate is based on more information, the updated state uncertainties are smaller than before the update. Note that for an application where Φ_{k-1} and Q_{k-1} are both zero, the Kalman filter is the same as a recursive least-squares estimator (see Section D.1 of Appendix D on the CD).

Figure 3.6 summarizes the data flow in a Kalman filter.

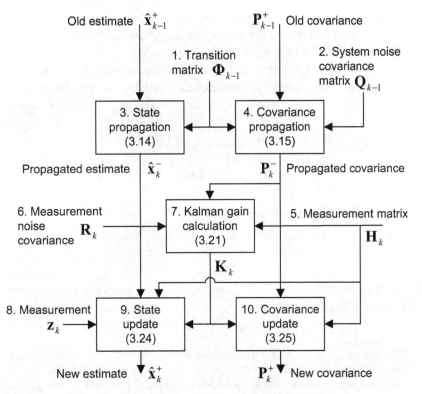

Figure 3.6 Kalman filter data flow.

The algorithm presented here is for an *open-loop implementation* of the Kalman filter, whereby all state estimates are retained in the Kalman filter algorithm. Section 3.2.6 describes the closed-loop implementation, whereby state estimates are fed back to correct the system.[*]

3.2.3 System Model

To propagate the state vector estimate, $\hat{\mathbf{x}}$, and error covariance, $\mathbf{P}$, forward in time, it is necessary to know how those states vary with time. This is the function of the system model. This section shows how the Kalman filter system propagation equations, (3.14) and (3.15), may be obtained from a model of the state dynamics, an application of linear systems theory.

An assumption of the Kalman filter is that the time derivative of each state is a linear function of the other states and of white noise sources. Thus, the true state vector, $\mathbf{x}(t)$, at time, t, of any Kalman filter is described by the following dynamic model:[†]

$$\dot{\mathbf{x}}(t) = \mathbf{F}(t)\mathbf{x}(t) + \mathbf{G}(t)\mathbf{w}_s(t), \qquad (3.26)$$

where $\mathbf{w}_s(t)$ is the continuous system noise vector (many authors use $\mathbf{w}$ and some use $\boldsymbol{\omega}$ or $\mathbf{v}$), $\mathbf{F}(t)$ is the system matrix (some authors use $\mathbf{A}$), and $\mathbf{G}(t)$ is the continuous system noise distribution matrix. The system noise vector comprises a number of independent random noise sources, each assumed to have a zero-mean symmetric distributions, such as the Gaussian distribution. $\mathbf{F}(t)$ and $\mathbf{G}(t)$ are always known functions. To determine the system model, these functions must be derived from the known properties of the system.[‡]

In Example A, the two-state Kalman filter estimating position and velocity along a single axis, the acceleration is not estimated so it must be represented as system noise. Thus, the state vector and system noise for this example are

$$\mathbf{x}_A = \begin{pmatrix} r^i_{ib,x} \\ v^i_{ib,x} \end{pmatrix}, \qquad w_{s,A} = a^i_{ib,x}. \qquad (3.27)$$

The state dynamics are simply

$$\dot{r}^i_{ib,x} = v^i_{ib,x}, \qquad \dot{v}^i_{ib,x} = a^i_{ib,x}. \qquad (3.28)$$

Substituting (3.27) and (3.28) into (3.16) gives the system matrix and system noise distribution matrix:

---

[*] This paragraph, up to this point, is based on material written by the author for QinetiQ, so comprises QinetiQ copyright material.

[†] This and subsequent paragraphs are based on material written by the author for QinetiQ, so comprise QinetiQ copyright material.

[‡] End of QinetiQ copyright material.

3.2 Algorithms and Models

$$\mathbf{F}_A = \begin{pmatrix} 0 & 1 \\ 0 & 0 \end{pmatrix}, \qquad \mathbf{G}_A = \begin{pmatrix} 0 \\ 1 \end{pmatrix}, \qquad (3.29)$$

noting that, in this case, neither matrix is a function of time.

To obtain an estimate, the expectation operator, E, is applied. The expectation value of the true state vector, $\mathbf{x}(t)$, is the estimated state vector, $\hat{\mathbf{x}}(t)$. The expectation value of the system noise vector, $\mathbf{w}_s(t)$, is zero as the noise is assumed to be of zero mean. $\mathbf{F}(t)$ and $\mathbf{G}(t)$ are assumed to be known functions and thus commute with the expectation operator. Hence, taking the expectation of (3.26) gives[*]

$$E(\dot{\mathbf{x}}(t)) = \frac{\partial}{\partial t}\hat{\mathbf{x}}(t) = \mathbf{F}(t)\hat{\mathbf{x}}(t). \qquad (3.30)$$

Solving (3.30) gives the state vector estimate at time t as a function of the state vector estimate at time $t - \tau_s$:

$$\hat{\mathbf{x}}(t) = \lim_{n \to \infty} \prod_{i=1}^{n} \exp\left(\mathbf{F}\left(t - \frac{i}{n}\tau_s\right)\frac{\tau_s}{n}\right)\hat{\mathbf{x}}(t - \tau_s), \qquad (3.31)$$

noting (A.17) in Appendix A on the CD. When $\mathbf{F}$ may be treated as constant over the interval $t - \tau_s$ to t, the approximation

$$\hat{\mathbf{x}}(t) \approx \exp(\mathbf{F}(t)\tau_s)\hat{\mathbf{x}}(t - \tau_s) \qquad (3.32)$$

may be made, noting that this is exact where $\mathbf{F}$ is actually constant [9].

In the discrete Kalman filter, the state vector estimate is modeled as a linear function of its previous value, coupled by the transition matrix, $\mathbf{\Phi}_{k-1}$, repeating (3.14):[†]

$$\hat{\mathbf{x}}_k^- = \mathbf{\Phi}_{k-1}\hat{\mathbf{x}}_{k-1}^+.$$

The discrete and continuous forms of the Kalman filter are equivalent, with $\hat{\mathbf{x}}_k \equiv \hat{\mathbf{x}}(t_k)$ and $\hat{\mathbf{x}}_{k-1} \equiv \hat{\mathbf{x}}(t_k - \tau_s)$. So, substituting (3.32) into (3.14),

$$\mathbf{\Phi}_{k-1} \approx \exp(\mathbf{F}_{k-1}\tau_s), \qquad (3.33)$$

where, assuming data is available at times $t_{k-1} = t_k - \tau_s$ and t_k, but not at intervening intervals, the system matrix, $\mathbf{F}_{k-1}$, can be calculated either as $\tfrac{1}{2}(\mathbf{F}(t_k - \tau_s) + \mathbf{F}(t_k))$ or by taking the mean of the parameters of $\mathbf{F}$ at times $t_k - \tau_s$ and t_k and making a single calculation of $\mathbf{F}$. In general, (3.33) cannot be computed directly; the exponent of the

---

[*]This paragraph, up to this point, is based on material written by the author for QinetiQ, so comprises QinetiQ copyright material.
[†]This and subsequent paragraphs are based on material written by the author for QinetiQ, so comprise QinetiQ copyright material.

matrix is not the matrix of the exponents of its components.[‡] Numerical methods are available [10], but these are computationally intensive where the matrices are large. Therefore, the transition matrix is usually computed as a power-series expansion of the system matrix, $\mathbf{F}$, and propagation interval, τ_s:

$$\mathbf{\Phi}_{k-1} = \sum_{r=0}^{\infty} \frac{\mathbf{F}_{k-1}^r \tau_s^r}{r!} = \mathbf{I} + \mathbf{F}_{k-1}\tau_s + \tfrac{1}{2}\mathbf{F}_{k-1}^2 \tau_s^2 + \tfrac{1}{6}\mathbf{F}_{k-1}^3 \tau_s^3 + \cdots. \tag{3.34}$$

The Kalman filter designer must decide where to truncate the power-series expansion, depending on the likely magnitude of the states, the length of the propagation interval, and the available error margins. With a shorter propagation interval, a given accuracy may be attained with a shorter truncation. Different truncations may be applied to different terms and exact solutions may be available for some elements of the transition matrix. In some cases, such as Example A, $\mathbf{F}^2$ is zero, so the first-order solution, $\mathbf{I} + \mathbf{F}_{k-1}\tau_s$, is exact.

The true state vector can be obtained as a function of its previous value, $\mathbf{x}_{k-1}$, by integrating (3.26) between times $t_k - \tau_s$ and t_k under the approximation that $\mathbf{F}(t)$ and $\mathbf{G}(t)$ are constant over the integration interval and substituting (3.33):

$$\mathbf{x}_k = \mathbf{\Phi}_{k-1}\mathbf{x}_{k-1} + \mathbf{\Gamma}_{k-1}\mathbf{w}_{s,k-1}, \tag{3.35}$$

where $\mathbf{w}_{s,k-1}$ is the discrete system noise vector and $\mathbf{\Gamma}_{k-1}$ is the discrete system noise distribution matrix, such that

$$\mathbf{\Gamma}_{k-1}\mathbf{w}_{s,k-1} = \int_{t_k-\tau_s}^{t_k} \exp(\mathbf{F}_{k-1}(t_k - t'))\mathbf{G}_{k-1}\mathbf{w}_s(t')\,\mathrm{d}t', \tag{3.36}$$

Note that, as system noise is introduced throughout the propagation interval, it is subject to state propagation via $\mathbf{F}$ for the remainder of that propagation interval. The system noise distribution matrix, $\mathbf{G}_{k-1}$, is calculated in a similar manner to $\mathbf{F}_{k-1}$, either as $\tfrac{1}{2}(\mathbf{G}(t_k - \tau_s) + \mathbf{G}(t_k))$ or by taking the mean of the parameters of $\mathbf{G}$ at times $t_k - \tau_s$ and t_k and making a single calculation of $\mathbf{G}$.

From (3.5), the error covariance matrix before and after the time propagation, and after the measurement update, is

$$\begin{aligned} \mathbf{P}_{k-1}^+ &= \mathrm{E}\!\left[(\hat{\mathbf{x}}_{k-1}^+ - \mathbf{x}_{k-1})(\hat{\mathbf{x}}_{k-1}^+ - \mathbf{x}_{k-1})^{\mathrm{T}}\right] \\ \mathbf{P}_k^- &= \mathrm{E}\!\left[(\hat{\mathbf{x}}_k^- - \mathbf{x}_k)(\hat{\mathbf{x}}_k^- - \mathbf{x}_k)^{\mathrm{T}}\right] \\ \mathbf{P}_k^+ &= \mathrm{E}\!\left[(\hat{\mathbf{x}}_k^+ - \mathbf{x}_k)(\hat{\mathbf{x}}_k^+ - \mathbf{x}_k)^{\mathrm{T}}\right] \end{aligned} \tag{3.37}$$

Subtracting (3.35) from (3.14),

$$\hat{\mathbf{x}}_k^- - \mathbf{x}_k = \mathbf{\Phi}_{k-1}\!\left(\hat{\mathbf{x}}_{k-1}^+ - \mathbf{x}_{k-1}\right) - \mathbf{\Gamma}_{k-1}\mathbf{w}_{s,k-1}. \tag{3.38}$$

[‡]End of QinetiQ copyright material.

3.2 Algorithms and Models

The errors in the state estimates are uncorrelated with the system noise, so

$$\mathrm{E}\big[(\hat{\mathbf{x}}_k^{\pm} - \mathbf{x}_k)\mathbf{w}_s^{\mathrm{T}}(t)\big] = 0, \quad \mathrm{E}\big[\mathbf{w}_s(t)(\hat{\mathbf{x}}_k^{\pm} - \mathbf{x}_k)^{\mathrm{T}}\big] = 0. \tag{3.39}$$

Therefore, substituting (3.38) and (3.39) into (3.37) gives

$$\mathbf{P}_k^- = \boldsymbol{\Phi}_{k-1}\mathbf{P}_{k-1}^+\boldsymbol{\Phi}_{k-1}^{\mathrm{T}} + \mathrm{E}\big[\boldsymbol{\Gamma}_{k-1}\mathbf{w}_{s,k-1}\mathbf{w}_{s,k-1}^{\mathrm{T}}\boldsymbol{\Gamma}_{k-1}^{\mathrm{T}}\big]. \tag{3.40}$$

Defining the system noise covariance matrix as

$$\mathbf{Q}_{k-1} = \mathrm{E}\big[\boldsymbol{\Gamma}_{k-1}\mathbf{w}_{s,k-1}\mathbf{w}_{s,k-1}^{\mathrm{T}}\boldsymbol{\Gamma}_{k-1}^{\mathrm{T}}\big] \tag{3.41}$$

gives the covariance propagation equation, (3.15)

$$\mathbf{P}_k^- = \boldsymbol{\Phi}_{k-1}\mathbf{P}_{k-1}^+\boldsymbol{\Phi}_{k-1}^{\mathrm{T}} + \mathbf{Q}_{k-1}.$$

Note that some authors define $\mathbf{Q}$ differently. Substituting (3.36) into (3.41) gives the system noise covariance in terms of the continuous system noise:

$$\mathbf{Q}_{k-1} = \mathrm{E}\left[\int_{t_k-\tau_s}^{t_k}\int_{t_k-\tau_s}^{t_k} \exp(\mathbf{F}_{k-1}(t_k - t'))\mathbf{G}_{k-1}\mathbf{w}_s(t')\mathbf{w}_s^{\mathrm{T}}(t'')\mathbf{G}_{k-1}^{\mathrm{T}}\exp(\mathbf{F}_{k-1}^{\mathrm{T}}(t_k - t''))\,\mathrm{d}t'\,\mathrm{d}t''\right]. \tag{3.42}$$

If the system noise is assumed to be white, applying (B.102) from Section B.4.2 of Appendix B on the CD gives

$$\mathbf{Q}_{k-1} = \int_{t_k-\tau_s}^{t_k} \exp(\mathbf{F}_{k-1}(t_k - t'))\mathbf{G}_{k-1}\mathbf{S}_{s,k-1}\mathbf{G}_{k-1}^{\mathrm{T}}\exp(\mathbf{F}_{k-1}^{\mathrm{T}}(t_k - t'))\,\mathrm{d}t', \tag{3.43}$$

where $\mathbf{S}_{s,k-1}$ is a diagonal matrix comprising the single-sided PSDs of the components of the continuous system noise vector, $\mathbf{w}_s(t)$.

The system noise covariance is usually approximated. The simplest version is obtained by neglecting the time propagation of the system noise over an iteration of the discrete-time filter, giving

$$\mathbf{Q}_{k-1} \approx \mathbf{Q}'_{k-1} = \mathbf{G}_{k-1}\mathrm{E}\left(\int_{t_k-\tau_s}^{t_k}\int_{t_k-\tau_s}^{t_k}\mathbf{w}_s(t')\mathbf{w}_s^{\mathrm{T}}(t'')\,\mathrm{d}t'\,\mathrm{d}t''\right)\mathbf{G}_{k-1}^{\mathrm{T}} \tag{3.44}$$

in the general case or

$$\mathbf{Q}_{k-1} \approx \mathbf{Q}'_{k-1} = \mathbf{G}_{k-1}\mathbf{S}_{s,k-1}\mathbf{G}_{k-1}^{\mathrm{T}}\tau_s \tag{3.45}$$

where white noise is assumed. This is known as the impulse approximation and, like all approximations, should be validated against the exact version prior to use.

Alternatively, (3.15) and (3.42) to (3.43) may be approximated to the first order in $\Phi_{k-1}Q'_{k-1}\Phi^T_{k-1}$, giving

$$P^-_k \approx \Phi_{k-1}\left(P^+_{k-1} + \tfrac{1}{2}Q'_{k-1}\right)\Phi^T_{k-1} + \tfrac{1}{2}Q'_{k-1}. \qquad (3.46)$$

Returning to Example A, if the acceleration is approximated as white Gaussian noise, the exact system noise covariance matrix is

$$Q_A = \begin{pmatrix} \tfrac{1}{3}S_a\tau_s^3 & \tfrac{1}{2}S_a\tau_s^2 \\ \tfrac{1}{2}S_a\tau_s^2 & S_a\tau_s \end{pmatrix}, \qquad (3.47)$$

where S_a is the PSD of the acceleration. This accounts for the propagation of the system noise onto the position state during the propagation interval. If the propagation interval is sufficiently small, the system noise covariance may be approximated to

$$Q_A \approx Q'_A = \begin{pmatrix} 0 & 0 \\ 0 & S_a\tau_s \end{pmatrix}. \qquad (3.48)$$

In Example B, the two states have no dependency through the system model. Therefore, the exact system noise covariance is simply

$$Q_B = \begin{pmatrix} S_{vx}\tau_s & 0 \\ 0 & S_{vy}\tau_s \end{pmatrix}, \qquad (3.49)$$

where S_{vx} and S_{vy} are the PSDs of the velocity in the x- and y-axes, respectively. Calculations of Q_A and Q_B are shown in Examples 3.1 and 3.2, respectively, on the CD.

Time-correlated system noise is discussed in Section 3.4.3.

3.2.4 Measurement Model

To update the state vector estimate with a set of measurements, it is necessary to know how the measurements vary with the states. This is the function of the measurement model. This section presents the derivation of the Kalman filter measurement-update equations, (3.21), (3.24), and (3.25), from the measurement model.

In a standard Kalman filter, the measurement vector, $z(t)$, is modeled as a linear function of the true state vector, $x(t)$, and the white noise sources, $w_m(t)$. Thus,

$$z(t) = H(t)x(t) + w_m(t), \qquad (3.50)$$

where $H(t)$ is the measurement matrix and is determined from the known properties of the system. For example, if the state vector comprises the position error of a dead-reckoning system, such as an INS, and the measurement vector comprises the difference between the dead-reckoning system's position solution and that of a positioning system, such as GNSS, then the measurement matrix is simply the identity matrix.

3.2 Algorithms and Models

If the measurements are taken at discrete intervals, (3.50) becomes

$$\mathbf{z}_k = \mathbf{H}_k \mathbf{x}_k + \mathbf{w}_{mk}. \tag{3.51}$$

Given this set of measurements, the new optimal estimate of the state vector is a linear combination of the measurement vector and the previous state vector estimate. Thus,

$$\hat{\mathbf{x}}_k^+ = \mathbf{K}_k \mathbf{z}_k + \mathbf{L}_k \hat{\mathbf{x}}_k^-, \tag{3.52}$$

where $\mathbf{K}_k$ and $\mathbf{L}_k$ are weighting matrices to be determined. Substituting in (3.51),

$$\hat{\mathbf{x}}_k^+ = \mathbf{K}_k \mathbf{H}_k \mathbf{x}_k + \mathbf{K}_k \mathbf{w}_{mk} + \mathbf{L}_k \hat{\mathbf{x}}_k^-. \tag{3.53}$$

A Kalman filter is an unbiased estimation algorithm, so the expectations of the errors in both the new and previous state vector estimates, $\hat{\mathbf{x}}_k^+ - \mathbf{x}_k$, and $\hat{\mathbf{x}}_k^- - \mathbf{x}_k$ are zero. The expectation of the measurement noise, $\mathbf{w}_{mk}$, is also zero. Thus, taking the expectation of (3.53) gives

$$\mathbf{L}_k = \mathbf{I} - \mathbf{K}_k \mathbf{H}_k. \tag{3.54}$$

Substituting this into (3.52) gives the state vector update equation [repeating (3.24)]:

$$\begin{aligned}\hat{\mathbf{x}}_k^+ &= \hat{\mathbf{x}}_k^- + \mathbf{K}_k \left(\mathbf{z}_k - \mathbf{H}_k \hat{\mathbf{x}}_k^- \right) \\ &= \hat{\mathbf{x}}_k^- + \mathbf{K}_k \delta \mathbf{z}_k^-\end{aligned}.$$

Substituting (3.51) into (3.24) and subtracting the true state vector,

$$\hat{\mathbf{x}}_k^+ - \mathbf{x}_k = (\mathbf{I} - \mathbf{K}_k \mathbf{H}_k)(\hat{\mathbf{x}}_k^- - \mathbf{x}_k) + \mathbf{K}_k \mathbf{w}_{mk}. \tag{3.55}$$

The error covariance matrix after the measurement update, $\mathbf{P}_k^+$, is then obtained by substituting this into (3.37), giving

$$\mathbf{P}_k^+ = \mathrm{E}\begin{bmatrix} (\mathbf{I} - \mathbf{K}_k \mathbf{H}_k)\mathbf{P}_k^-(\mathbf{I} - \mathbf{K}_k \mathbf{H}_k)^\mathrm{T} + \mathbf{K}_k \mathbf{w}_{mk}(\hat{\mathbf{x}}_k^- - \mathbf{x}_k)^\mathrm{T}(\mathbf{I} - \mathbf{K}_k \mathbf{H}_k)^\mathrm{T} \\ +(\mathbf{I} - \mathbf{K}_k \mathbf{H}_k)(\hat{\mathbf{x}}_k^- - \mathbf{x}_k)\mathbf{w}_{mk}^\mathrm{T}\mathbf{K}_k^\mathrm{T} + \mathbf{K}_k \mathbf{w}_{mk}\mathbf{w}_{mk}^\mathrm{T}\mathbf{K}_k^\mathrm{T} \end{bmatrix}. \tag{3.56}$$

The error in the state vector estimates is uncorrelated with the measurement noise so,[†]

$$\mathrm{E}\left[(\hat{\mathbf{x}}_k^- - \mathbf{x}_k)\mathbf{w}_{mk}^\mathrm{T}\right] = 0, \qquad \mathrm{E}\left[\mathbf{w}_{mk}(\hat{\mathbf{x}}_k^- - \mathbf{x}_k)^\mathrm{T}\right] = 0. \tag{3.57}$$

[†]This and subsequent paragraphs are based on material written by the author for QinetiQ, so comprise QinetiQ copyright material.

$\mathbf{K}_k$ and $\mathbf{H}_k$ commute with the expectation operator, so substituting (3.57) and (3.11) into (3.56) gives‡

$$\mathbf{P}_k^+ = (\mathbf{I} - \mathbf{K}_k\mathbf{H}_k)\mathbf{P}_k^-(\mathbf{I} - \mathbf{K}_k\mathbf{H}_k)^\mathrm{T} + \mathbf{K}_k\mathbf{R}_k\mathbf{K}_k^\mathrm{T}, \qquad (3.58)$$

noting that the measurement noise covariance matrix, $\mathbf{R}_k$, is defined by (3.11). This equation is known as the Joseph form of the covariance update.

There are two methods for determining the weighting function, $\mathbf{K}_k$, the minimum variance method [1, 2, 4], used here, and the maximum likelihood method [1, 3, 5]. Both give the same result. The criterion for optimally selecting $\mathbf{K}_k$ by the minimum variance method is the minimization of the error in the estimate, $\hat{\mathbf{x}}_k^+$. The variances of the state estimates are given by the diagonal elements of the error covariance matrix. It is therefore necessary to minimize the trace of $\mathbf{P}_k^+$ (see Section A.2 in Appendix A on the CD) with respect to $\mathbf{K}_k$:

$$\frac{\partial}{\partial \mathbf{K}_k}\left[\mathrm{Tr}(\mathbf{P}_k^+)\right] = 0. \qquad (3.59)$$

Substituting in (3.58) and applying the matrix relation (A.42) from Appendix A on the CD gives

$$-2(\mathbf{I} - \mathbf{K}_k\mathbf{H}_k)\mathbf{P}_k^-\mathbf{H}_k^\mathrm{T} + 2\mathbf{K}_k\mathbf{R}_k = 0. \qquad (3.60)$$

Rearranging this gives (3.21)

$$\mathbf{K}_k = \mathbf{P}_k^-\mathbf{H}_k^\mathrm{T}\left(\mathbf{H}_k\mathbf{P}_k^-\mathbf{H}_k^\mathrm{T} + \mathbf{R}_k\right)^{-1}.$$

As explained in [2], this result is independent of the units and/or scaling of the states.

By substituting (3.21) into (3.58), the error covariance update equation may be simplified to (3.25):

$$\mathbf{P}_k^+ = (\mathbf{I} - \mathbf{K}_k\mathbf{H}_k)\mathbf{P}_k^-.$$

This may also be computed as

$$\mathbf{P}_k^+ = \mathbf{P}_k^- - \mathbf{K}_k(\mathbf{H}_k\mathbf{P}_k^-), \qquad (3.61)$$

which is more efficient where the measurement vector has fewer components than the state vector.

An alternative form of measurement update, known as sequential processing, is described in Section 3.2.7.

Returning to the simple example at the beginning of the subsection, a Kalman filter estimates INS position error using the INS–GNSS position solution difference as

---

‡End of QinetiQ copyright material.

the measurement, so the measurement matrix, **H**, is the identity matrix. The problem may be simplified further if all components of the measurement have independent noise of standard deviation, σ_z, and the state estimates are uncorrelated and each have an uncertainty of σ_x. This is denoted Example C and may be expressed as

$$\mathbf{H}_{C,k} = \mathbf{I}_3, \qquad \mathbf{R}_{C,k} = \sigma_z^2 \mathbf{I}_3, \qquad \mathbf{P}_{C,k}^- = \sigma_x^2 \mathbf{I}_3. \tag{3.62}$$

Substituting this into (3.21), the Kalman gain matrix for this example is

$$\mathbf{K}_{C,k} = \frac{\sigma_x^2}{\sigma_x^2 + \sigma_z^2} \mathbf{I}_3. \tag{3.63}$$

From (3.24) and (3.25), the state estimates and error covariance are then updated using

$$\begin{aligned}\hat{\mathbf{x}}_{C,k}^+ &= \frac{\sigma_z^2 \hat{\mathbf{x}}_{C,k}^- + \sigma_x^2 \mathbf{z}_{C,k}}{\sigma_x^2 + \sigma_z^2} \\ \mathbf{P}_{C,k}^+ &= \frac{\sigma_z^2}{\sigma_x^2 + \sigma_z^2} \mathbf{P}_{C,k}^- = \frac{\sigma_x^2 \sigma_z^2}{\sigma_x^2 + \sigma_z^2} \mathbf{I}_3\end{aligned}. \tag{3.64}$$

Suppose the measurement vector input to the Kalman filter, **z**, is computed from another set of measurements, **y**. For example, a position measurement might be converted from range and bearing to Cartesian coordinates. If the measurements, **y**, have noise covariance, $\mathbf{C}_y$, the Kalman filter measurement noise covariance is determined using

$$\mathbf{R} = \left(\frac{d\mathbf{z}}{d\mathbf{y}} \bigg|_{\mathbf{y}=\hat{\mathbf{y}}} \right) \mathbf{C}_y \left(\frac{d\mathbf{z}}{d\mathbf{y}} \bigg|_{\mathbf{y}=\hat{\mathbf{y}}} \right)^{\mathrm{T}}, \tag{3.65}$$

where vector differentiation is described in Section A.5 of Appendix A on the CD. Note that **z** will typically not be a linear function of **y** as such transformations may be required where the original measurements, **y**, are not a linear function of the state vector, **x**. Nonlinear estimation is discussed in Sections 3.4.1, 3.4.2, and 3.5.

3.2.5 Kalman Filter Behavior and State Observability

Figure 3.7 shows how the uncertainty of a well-observed state estimate varies during the initial phase of Kalman filter operation, where the state estimates are converging with their true counterparts. Note that the state uncertainties are the root diagonals of the error covariance matrix, **P**. Initially, when the state uncertainties are large, the Kalman gain will be large, weighting the state estimates towards the new measurement data. The Kalman filter estimates will change quickly as they converge with the true values of the states, so the state uncertainty will drop rapidly. However, assuming a constant measurement noise covariance, **R**, this causes the Kalman gain to drop, weighting the state estimates more towards their previous values. This

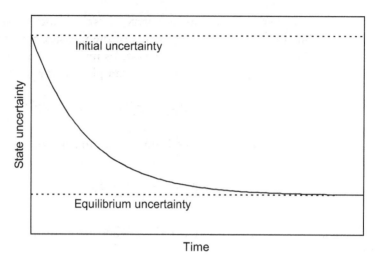

Figure 3.7 Kalman filter state uncertainty during convergence.

reduces the rate at which the states change, so the reduction in the state uncertainty slows. Eventually, the Kalman filter will approach equilibrium, whereby the decrease in state uncertainty with each measurement update is matched by the increase in uncertainty due to system noise. At equilibrium, the state estimates may still vary, but the level of confidence in those estimates, reflected by the state uncertainty, will be more or less fixed.

The rate at which a state estimate converges, if at all, depends on the *observability* of that state. There are two types of observability: deterministic, also known as geometric, and stochastic. *Deterministic observability* indicates whether there is sufficient measurement information, in the absence of noise, to independently determine all of the states, a condition known as full observability.

The Kalman filter's measurement model is analogous to a set of simultaneous equations where the states are the unknowns to be found, the measurements are the known quantities, and the measurement matrix, $\mathbf{H}$, provides the coefficients of the states. Therefore, on a single iteration, the Kalman filter cannot completely observe more states than there are components of the measurement vector, merely linear combinations of those states. However, if the measurement matrix changes over time or there is a time-dependent relationship between states through the transition matrix, $\boldsymbol{\Phi}$, then it is possible, over time, to observe more states than there are measurement components. The error covariance matrix, $\mathbf{P}$, records the correlations between the state estimates as well as their uncertainties. A good example in navigation is determination of velocity from the rate of change of position.

To determine whether the state vector $\mathbf{x}_1$ can be fully observed from a set of measurement vectors $\mathbf{z}_1, \mathbf{z}_2, \ldots, \mathbf{z}_k$, an observability matrix, $\mathbf{O}_{1:k}$ is defined by [2, 6, 11]

$$\begin{pmatrix} \mathbf{z}_1 \\ \mathbf{z}_2 \\ \vdots \\ \mathbf{z}_k \end{pmatrix} = \mathbf{O}_{1:k} \mathbf{x}_1 + \begin{pmatrix} \mathbf{w}_1 \\ \mathbf{w}_2 \\ \vdots \\ \mathbf{w}_k \end{pmatrix}, \qquad (3.66)$$

where the noise vectors, $\mathbf{w}_i$, comprise both measurement and system noise. Thus, from (3.35) and (3.51),

$$\mathbf{O}_{1:k} = \begin{pmatrix} \mathbf{H}_1 \\ \mathbf{H}_2\mathbf{\Phi}_1 \\ \vdots \\ \mathbf{H}_k\mathbf{\Phi}_{k-1}\cdots\mathbf{\Phi}_2\mathbf{\Phi}_1 \end{pmatrix}. \tag{3.67}$$

where $\mathbf{x}_1$ is fully observable, an estimate may be obtained by applying the expectation operator to (3.66), assuming the noise distributions are zero mean. Thus,

$$\hat{\mathbf{x}}_1 = \left(\mathbf{O}_{1:k}^\mathrm{T}\mathbf{O}_{1:k}\right)^{-1}\mathbf{O}_{1:k}^\mathrm{T} \begin{pmatrix} \mathbf{z}_1 \\ \mathbf{z}_2 \\ \vdots \\ \mathbf{z}_k \end{pmatrix}. \tag{3.68}$$

This only has a solution where the observability matrix has a pseudo-inverse, which requires $\mathbf{O}_{1:k}^\mathrm{T}\mathbf{O}_{1:k}$ to be nonsingular (see Section A.4 of Appendix A on the CD). This requires $\mathbf{O}_{1:k}$ to be of rank n, where n is the number of elements of the state vector, $\mathbf{x}$. The rank of a matrix is equal to the number of rows of the largest square submatrix (not necessarily continuous) with a nonzero determinant.

When $\mathbf{O}_{1:k}$ is of rank m, where $m < n$, there are m observable linear combinations of the states and $n - m$ unobservable combinations. The state vector is thus partially observable. A vector, $\mathbf{y}$, comprising m observable linear combinations of the states may be determined using [11]

$$\mathbf{y} = \mathbf{T}\mathbf{O}_m\mathbf{x}, \tag{3.69}$$

where $\mathbf{O}_m$ comprises m linearly independent rows of $\mathbf{O}_{1:k}$ and $\mathbf{T}$ is an arbitrary nonsingular $m \times m$ matrix.

An alternative method of determining full observability is to calculate the information matrix, $\mathbf{Y}$

$$\mathbf{Y} = \mathbf{H}_1^\mathrm{T}\mathbf{R}_1^{-1}\mathbf{H}_1 + \sum_{i=2}^{k}\left[\left(\prod_{j=1}^{i-1}\mathbf{\Phi}_{i-j}^\mathrm{T}\right)\mathbf{H}_i^\mathrm{T}\mathbf{R}_i^{-1}\mathbf{H}_i\left(\prod_{j=1}^{i-1}\mathbf{\Phi}_j\right)\right], \tag{3.70}$$

where $\prod_{j=1}^{n}\mathbf{\Phi}_j = \mathbf{\Phi}_n\cdots\mathbf{\Phi}_2\mathbf{\Phi}_1$. This is positive definite (see Section A.6 of Appendix A on the CD) where the state vector is fully observable [1, 6].

The observability of many parameters is dynamics dependent. For example, the attitude errors and accelerometer biases of an INS are not separately observable at constant attitude, but they are after a change in attitude as this changes the relationship between the states in the system model. Observation of many higher-order gyro and accelerometer errors requires much higher dynamics. However, if two states have the same effect on the measurements and vary with time and dynamics in the same way, they will never be separately observable, so should be combined to avoid wasting processing resources.

Given that a state, or linear combination of states, is deterministically observable, the rate of convergence depends on the *stochastic observability*. This depends on the measurement sampling rate, the magnitude and correlation properties of the measurement noise, and the level of system noise. The higher the sampling rate (subject to correlation time constraints) and the lower the measurement and system noise, the greater the stochastic observability.

Conversely, system and measurement noise can mask the effects of those states that only have a small impact on the measurements, making those states effectively unobservable. For a state that is stochastically unobservable, the equilibrium state uncertainty will be similar to the initial uncertainty or may even be larger.

The combined observability of states and their combinations may be studied by analyzing the normalized error covariance matrix, $\mathbf{P}'_k$, after k Kalman filter (or covariance propagation and update) cycles. This is defined by

$$P'_{k,ij} = \frac{P_{k,ij}}{\sigma_{0,i}\sigma_{0,j}}, \qquad (3.71)$$

where $\sigma_{0,i}$ is the initial uncertainty of the ith state. When a diagonal element of $\mathbf{P}'_k$ is close to zero, the corresponding state is strongly observable, whereas if it is close to unity (or larger), the state is weakly observable. As discussed above, linear combinations of weakly observable states may be strongly observable. These may be identified by calculating the eigenvalues and eigenvectors of $\mathbf{P}'_k$ (see Section A.6 of Appendix A on the CD). The eigenvectors corresponding to the smallest eigenvalues indicate the most strongly observed linear combinations of normalized states (i.e., $x_i/\sigma_{0,i}$).

When a Kalman filter is well designed, a reduction in the state uncertainty, as defined by the error covariance matrix, will be accompanied by a reduction in the corresponding state residual. Thus, the Kalman filter is convergent. However, poor design can result in state uncertainties much smaller than the corresponding state residuals or even residuals growing as the uncertainties drop, a phenomenon known as divergence. Section 3.3 discusses the causes of these problems and how they may be mitigated in a practical Kalman filter design.

3.2.6 Closed-Loop Kalman Filter

A linear system model is an assumption of the standard Kalman filter design. However, in many navigation applications, such as integration, alignment, and calibration of an INS, the true system model is not linear (i.e., the time differential of the state vector varies with terms to second order and higher in the state vector elements). One solution is to use a modified version of the Kalman filter algorithm, such as an extended Kalman filter (Section 3.4.1) or an unscented Kalman filter (Section 3.4.2). However, it is often possible to neglect the higher-order terms in the system model and still obtain a practically useful Kalman filter. The larger the values of the states that contribute to the neglected terms, the poorer a given linearity approximation will be.

A common technique for getting the best performance out of an error-state Kalman filter with a linearity approximation applied to the system model is the *closed-loop implementation*. Here, the errors estimated by the Kalman filter are fed back every iteration, or at regular intervals, to correct the system itself, zeroing the Kalman filter states in the process. This feedback process keeps the Kalman filter states small, minimizing the effect of neglecting higher order products of states in the system model. Conversely, in the *open-loop implementation*, when there is no feedback, the states will generally get larger as time progresses.[†]

The best stage in the Kalman filter algorithm to feed back the state estimates is immediately after the measurement update. This produces zero state estimates at the start of the state propagation, (3.14), enabling this stage to be omitted completely. The error covariance matrix, **P**, is unaffected by the feedback process as the same amount is added to or subtracted from both the true and estimated states, so error covariance propagation, (3.15), is still required.

The closed-loop and open-loop implementations of the Kalman filter may be mixed such that some state estimates are fed back as corrections, whereas others are not. This configuration is useful for applications where feeding back states is desirable, but some states cannot be fed back as there is no way of applying them as corrections to the system. In designing such a Kalman filter, care must be taken in implementing the state propagation as for some of the fed-back states, $\mathbf{x}_k^-$ may be nonzero due to coupling with nonfed-back states through the system model.

When a full closed-loop Kalman filter is implemented (i.e., with feedback of every state estimate at every iteration), $\mathbf{H}_k \hat{\mathbf{x}}_k^-$ is zero, so the measurement, $\mathbf{z}_k$, and measurement innovation, $\delta \mathbf{z}_k^-$, are the same.

In navigation, closed-loop Kalman filters are common for the integration, alignment, and calibration of low-grade INS and may also be used for correcting GNSS receiver clocks.[‡]

3.2.7 Sequential Measurement Update

The sequential measurement-update implementation of the Kalman filter, also known as the scalar measurement update or sequential processing, replaces the vector measurement update, (3.21), (3.24), and (3.25), with an iterative process using only one component of the measurement vector at a time. The system propagation is unchanged from the standard Kalman filter implementation. For each measurement, denoted by the index j, the Kalman gain is calculated and the state vector estimate and error covariance matrix are updated before moving onto the next measurement. The notation $\hat{\mathbf{x}}_k^j$ and $\mathbf{P}_k^j$ is used to respectively denote the state vector estimate and error covariance that have been updated using all components of the measurement vector up to and including the jth. If the total number of measurements is m,

$$\begin{aligned} \hat{\mathbf{x}}_k^0 &\equiv \hat{\mathbf{x}}_k^-, & \mathbf{P}_k^0 &\equiv \mathbf{P}_k^- \\ \hat{\mathbf{x}}_k^m &\equiv \hat{\mathbf{x}}_k^+, & \mathbf{P}_k^m &\equiv \mathbf{P}_k^+ \end{aligned}$$ (3.72)

[†]This and subsequent paragraphs are based on material written by the author for QinetiQ, so comprise QinetiQ copyright material.
[‡]End of QinetiQ copyright material.

When the components of the measurement vector are statistically independent, the measurement noise covariance matrix, $\mathbf{R}_k$, will be diagonal. In this case, the Kalman gain calculation for the jth measurement is

$$\mathbf{k}_k^j = \frac{\mathbf{P}_k^{j-1}\mathbf{H}_{k,j}^{\mathrm{T}}}{\mathbf{H}_{k,j}\mathbf{P}_k^{j-1}\mathbf{H}_{k,j}^{\mathrm{T}} + R_{k,j,j}}, \qquad (3.73)$$

where $\mathbf{H}_{k,j}$ is the jth row of the measurement matrix, so $\mathbf{H}_{k,j}\mathbf{P}_k^j\mathbf{H}_{k,j}^{\mathrm{T}}$ is a scalar. Note that $\mathbf{k}_k^j$ is a column vector. The sequential measurement update equations are then

$$\begin{aligned}\hat{\mathbf{x}}_k^j &= \hat{\mathbf{x}}_k^{j-1} + \mathbf{k}_k^j\left(z_{k,j} - \mathbf{H}_{k,j}\hat{\mathbf{x}}_k^{j-1}\right) \\ &= \hat{\mathbf{x}}_k^{j-1} + \mathbf{k}_k^j\delta\bar{z}_{k,j}\end{aligned}, \qquad (3.74)$$

$$\mathbf{P}_k^j = \mathbf{P}_k^{j-1} - \mathbf{k}_k^j\left(\mathbf{H}_{k,j}\mathbf{P}_k^{j-1}\right), \qquad (3.75)$$

noting that the jth component of the measurement innovation, $\delta\bar{z}_{k,j}$, must be calculated after the $j-1$th step of the measurement update has been performed. Because no matrix inversion is required to calculate the Kalman gain, the sequential form of the measurement update is always more computationally efficient (see Section 3.3.2) where the components of the measurement are independent.

When the measurement noise covariance, $\mathbf{R}_k$, is not diagonal, indicating measurement noise that is correlated between measurements at a given epoch, a sequential measurement update may still be performed. However, it is first necessary to reformulate the measurement into statistically independent components using

$$\begin{aligned}\mathbf{z}_k' &= \mathbf{T}_k\mathbf{z}_k \\ \mathbf{R}_k' &= \mathbf{T}_k\mathbf{R}_k\mathbf{T}_k^{\mathrm{T}}, \\ \mathbf{H}_k' &= \mathbf{T}_k\mathbf{H}_k\end{aligned} \qquad (3.76)$$

where the transformation matrix, $\mathbf{T}_k$, is selected to diagonalize $\mathbf{R}_k$ using Cholesky factorization as described in Section A.6 of Appendix A on the CD. The measurement update is then performed with $\mathbf{z}_k'$, $\mathbf{R}_k'$, and $\mathbf{H}_k'$ substituted for $\mathbf{z}_k$, $\mathbf{R}_k$, and $\mathbf{H}_k$ in (3.73) to (3.75). Calculation of the transformation matrix requires inversion of an $m \times m$ matrix, as is required for the conventional Kalman gain calculation. Therefore, with correlated measurement components, the sequential measurement update can only provide greater computational efficiency if the same transformation matrix is used at every epoch, k, and it is relatively sparse.

A hybrid of the sequential and conventional measurement updates may also be performed whereby the measurement vector is divided into a number of subvectors, which are then used to update the state estimates and error covariance sequentially. This can be useful where there is noise correlation within groups of measurements, but not between those groups.

3.3 Implementation Issues

This section discusses the implementation issues that must be considered in designing a practical Kalman filter. These include tuning and stability, efficient algorithm design, numerical issues, and synchronization. An overall design process is also recommended. Detection of erroneous measurements and biased state estimates is discussed in Chapter 17.

3.3.1 Tuning and Stability

The tuning of a Kalman filter is the selection by the designer or user of values for three matrices. These are the system noise covariance matrix, $\mathbf{Q}_k$, the measurement noise covariance matrix, $\mathbf{R}_k$, and the initial values of the error covariance matrix, $\mathbf{P}_0^+$. It is important to select these parameters correctly. If the values selected are too small, the actual errors in the Kalman filter estimates will be much larger than the state uncertainties obtained from $\mathbf{P}$. Conversely, if the values selected are too large, the reported uncertainties will be too large.[*] These can cause an external system that uses the Kalman filter estimates to apply the wrong weighting to them.

However, the critical parameter in Kalman filtering is the ratio of the error and measurement noise covariance matrices, $\mathbf{P}_k^-$ and $\mathbf{R}_k$, as they determine the Kalman gain, $\mathbf{K}_k$. Figure 3.8 illustrates this. If **P/R** is too small, the Kalman gain will be too small and state estimates will converge with their true counterparts more slowly than necessary. The state estimates will also be slow to respond to changes in the system. Conversely, if **P/R** is too large, the Kalman gain will be too large. This will bias the filter in favor of more recent measurements, which may result in unstable or biased state estimates due to the measurement noise having too great an influence on them. Sometimes, the state estimates can experience positive feedback of the measurement noise through the system model, causing them to rapidly diverge from their truth counterparts.[*]

In an ideal Kalman filter application, tuning the noise models to give consistent estimation errors and uncertainties will also produce stable state estimates that track their true counterparts. However, in practice, it is often necessary to tune the filter to give 1σ state uncertainties substantially larger (two or three times is typical) than the corresponding error standard deviations in order to maintain stability. This is because the Kalman filter's model of the system is only an approximation of the real system.

There are a number of sources of approximation in a Kalman filter. Smaller error states are often neglected due to observability problems or processing-capacity limitations. The system and/or measurement models may have to be approximated to meet the linearity requirements of the Kalman filter equations. The stochastic properties of slowly time-varying states are often oversimplified. Nominally constant states may also vary slowly with time (e.g., due to temperature or pressure changes). Finally, the Kalman filter assumes that all noise sources are white, whereas, in practice, they

---

[*]This paragraph, up to this point, is based on material written by the author for QinetiQ, so comprises QinetiQ copyright material.

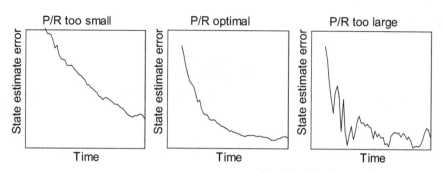

Figure 3.8 Kalman filter error propagation for varying **P/R** ratio.

will exhibit some time correlation due to band-limiting effects. Therefore, to overcome the limitations of the Kalman filter model, sufficient noise must be modeled to overbound the real system's behavior. By analogy, to fit the square peg of the real world problem into the round hole of the Kalman filter model, the hole must be widened to accommodate the edges of the peg.

A further issue is that allowing state uncertainties to become very small can precipitate numerical problems (see Section 3.3.3). Therefore, it is advisable to model system noise on all states and ensure that $\mathbf{Q}_k$ is positive definite (see Section A.6 of Appendix A on the CD). Alternatively, lower limits to the state uncertainties may be maintained.

For most applications, manufacturers' specifications and laboratory test data may be used to determine suitable initial values for the error covariance matrix. The same approach may be adopted for the system noise covariance matrices in cases in which the system model is a good representation of the truth and the system noise is close to white. In other cases, the system noise may be highly colored or dominated by the compensation of modeling approximations, in which case a more empirical approach will be needed, making use of test data gathered in typical operational environments. It may also be necessary to model the system noise as a function of vibration and/or user dynamics.

Similarly, when the measurement noise is close to being white, manufacturer's specifications or simple laboratory variance measurements may be used. However, it is often necessary to exaggerate **R** in order to account for time correlation in the measurement noise due to band-limiting or synchronization errors, while measurement noise can also vary with vibration and user dynamics. For radio navigation, the measurement noise is also affected by signal reception conditions.

In tuning a Kalman filter, it can be difficult to separate out the effects of measurement noise from those of the system noise and modeling limitations. Therefore, a good tuning philosophy is to fix $\mathbf{P}_0^+$, together with whichever of $\mathbf{Q}_k$ and $\mathbf{R}_k$ is easier to define analytically, then vary the remaining tuning matrix by trial and error to find the smallest value that gives stable state estimates. If this does not give satisfactory performance, the other tuning parameters can also be varied. Automatic real-time tuning techniques are discussed in Section 3.4.4.

Tuning a Kalman filter is essentially a tradeoff between convergence rate and stability. However, it is important to note that the convergence rate can also affect

3.3 Implementation Issues

Table 3.1 Multiplications and Additions Required by Kalman Filter Processes

Kalman Filter Process	Equation	Multiplications Required	Additions Required
System-propagation phase			
State propagation	(3.14)	n^2	$n(n-1)$
Covariance propagation	(3.15)	$2n^3$	$n^2(2n-1)$
System noise distribution matrix computation	(3.41)	$n(n+1)l$	$n^2(l-1)$
Measurement-update phase (vector implementation)			
Kalman gain calculation	(3.21)	$2mn^2 = 2m^2n$	$mn(m+n-2)+m^2$
Matrix inversion		$(3/2)m^3 - (1/2)m$	$\sim m^3$
State vector update	(3.24)	$2mn$	$2nm$
Covariance update	(3.25)	$mn^2 + n^3$	$n^2(n+m-1)$
	or (3.61)	$2mn^2$	$mn(2n-1)$
Measurement-update phase (sequential implementation, assuming diagonal **R**)			
Kalman gain calculation	(3.73)	$2mn^2 + 2mn$	$m(n^2-n+1)$
State vector update	(3.74)	$2mn$	$2nm$
Covariance update	(3.75)	$2mn^2$	$mn(2n-1)$

the long-term accuracy, as this is reached once the convergence rate matches the rate at which the true states change due to noise effects. For some Kalman filtering applications, integrity monitoring techniques (Chapter 17) can be used to detect and remedy state instability, in which case the tuning may be selected to optimize convergence.*

3.3.2 Algorithm Design

The processing load for implementation of a Kalman filter depends on the number of components of the state vector, n, measurement vector, m, and system noise vector, l, as shown in Table 3.1. When the number of states is large, the covariance propagation and update require the largest processing capacity. However, when the measurement vector is larger than the state vector, the Kalman gain calculation has the largest impact on processor load for the vector implementation of the measurement update. Therefore, implementing a sequential measurement update can significantly reduce the processor load when there are a large number of uncorrelated measurement components at each epoch.

In moving from a theoretical to a practical Kalman filter, a number of modifications can be made to improve the processing efficiency without significantly impacting on performance. For example, many elements of the transition, $\mathbf{\Phi}_k$, and measurement, $\mathbf{H}_k$, matrices are zero, so it is more efficient to use sparse matrix multiplication routines that only multiply the nonzero elements. However, there is a tradeoff between processing efficiency and algorithm complexity, with more complex algorithms taking longer to develop, code, and debug.*

---

*This paragraph, up to this point, is based on material written by the author for QinetiQ, so comprises QinetiQ copyright material.

Another option takes advantage of the error covariance matrix, $\mathbf{P}_k$, being symmetric about the diagonal. By computing only the diagonal elements and either the upper or lower triangle, the computational effort required to propagate and update the covariance matrix may be almost halved.

Sparse matrix multiplication cannot be used for the matrix inversion within the Kalman gain calculation, while its use in updating the covariance, (3.25), is limited to computing $\mathbf{K}_k\mathbf{H}_k$ and cases in which $\mathbf{H}_k$ has some columns that are all zeros. Consequently, the measurement-update phase of the Kalman filter will always require more computational capacity than the system-propagation phase. The interval between measurement updates may be limited by processing power. It may also be limited by the rate at which measurements are available or by the correlation time of the measurement noise. In any case, the measurement-update interval can sometimes be too large to calculate the transition matrix, $\mathbf{\Phi}_k$, over. This is because the system propagation interval, τ_s, must be sufficiently small for the system matrix, $\mathbf{F}$, to be treated as constant and the power-series expansion of $\mathbf{F}\tau_s$ in (3.34) to converge. However, the different phases of the Kalman filter do not have to be iterated at the same rate. The system propagation may be iterated at a faster rate than the measurement update, reducing the propagation interval, τ_s. Similarly, if a measurement update cannot be performed due to lack of valid data, the system propagation can still go ahead. The update rate for a given measurement stream should not be faster than the system-propagation rate.

The Kalman filter equations involving the covariance matrix, $\mathbf{P}$, impose a much higher computational load than those involving the state vector, $\mathbf{x}$. However, the accuracy requirement for the state vector is higher, particularly for the open-loop Kalman filter, requiring a shorter propagation interval to maximize the transition matrix accuracy. Therefore, it is sometimes more efficient to iterate the state vector propagation, (3.14), at a higher rate than the error covariance propagation, (3.15).

When the measurement update interval that processing capacity allows is much greater than the noise correlation time of the measurement stream, the noise on the measurements can be reduced by time averaging. In this case, the measurement innovation, $\delta \mathbf{z}^-$, is calculated at a faster rate and averaged measurement innovations are used to update the state estimates, $\hat{\mathbf{x}}$, and covariance, $\mathbf{P}$, at the rate allowed by the processing capacity. When the measurements, $\mathbf{z}$, rather than the measurement innovations, are averaged, the measurement matrix, $\mathbf{H}$, must be modified to account for the state propagation over the averaging interval [12]. Measurement averaging is also known as prefiltering.

Altogether, a Kalman filter algorithm may have four different iteration rates for the state propagation, (3.14), error covariance propagation, (3.15), measurement accumulation, and measurement update, (3.21), (3.24), and (3.25). Figure 3.9 presents an example illustration. Furthermore, different types of measurement input to the same Kalman filter, such as position and velocity or velocity and attitude, may be accumulated and updated at different rates.[*]

---

[*]This paragraph, up to this point, is based on material written by the author for QinetiQ, so comprises QinetiQ copyright material.

3.3 Implementation Issues

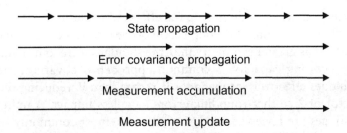

Figure 3.9 Example Kalman filter iteration rates.

3.3.3 Numerical Issues

When a Kalman filter is implemented on a computer, the precision is limited by the number of bits used to store and process each parameter. The fewer bits used, the larger the rounding errors on each computation will be. Thus, double-precision (64-bit) arithmetic is more robust than single precision (32 bit), which is more robust than 16-bit arithmetic, used in early implementations.

The effect of rounding errors on the state estimates can be accounted for by increasing the system noise covariance, **Q**, and, in many cases, is corrected by the Kalman filter's measurement update process. However, there are no corresponding corrections to the error covariance matrix, **P**. The longer the Kalman filter has been running and the higher the iteration rate, the greater the distortion of the matrix. This distortion manifests as breakage of the symmetry about the diagonal and can even produce negative diagonal elements, which represent imaginary uncertainty. Small errors in the **P** matrix are relatively harmless. However, large **P**-matrix errors distort the Kalman gain matrix, **K**. Gains that are too small produce unresponsive state estimates while gains that are too large can produce unstable, oscillatory state estimates. If an element of the Kalman gain matrix is the wrong sign, a state estimate is liable to diverge away from truth. Extreme covariance matrix distortion can also cause software crashes. Thus, the Kalman filter implementation must be designed to minimize computational errors in the error covariance matrix. In particular, **P** must remain positive definite (i.e., retain a positive determinant and positive eigenvalues).

There is a particular risk of numerical problems at the first measurement update following initialization in cases where the initial uncertainties are very large and the measurement noise covariance is small. This is because there can be a very large change in the error covariance matrix, with the covariance update comprising the multiplication of very large numbers with very small numbers. If problems occur, the initial state uncertainties should be set artificially small. As long as the values used are still larger than those expected after convergence, the state uncertainties will be corrected as the Kalman filter converges [4].

In general, rounding errors may be reduced by scaling the Kalman filter states so that all state uncertainties are of a similar order of magnitude in numerical terms, effectively reducing the dynamic range of the error covariance matrix. Rescaling of the measurement vector may also be needed to reduce the dynamic range of the $\mathbf{H}_k \mathbf{P}_k^- \mathbf{H}_k^T + \mathbf{R}_k$ matrix that is inverted to calculate the Kalman gain. Scaling is essential where fixed-point, as opposed to floating-point, arithmetic is used.

Another way of minimizing the effects of rounding errors is to modify the Kalman filter algorithm. The Joseph form of the covariance update replaces (3.25) with (3.58). It has greater symmetry than the standard form, but requires more than twice the processing capacity. A common approach is covariance factorization. These techniques effectively propagate $\sqrt{P}$ rather than P, reducing the dynamic range by a factor of 2 so that rounding errors have less impact. A number of factorization techniques are reviewed in [3, 5, 6], but the most commonly used is the Bierman-Thornton or UDU method [3, 13].

Ad hoc methods of stabilizing the error covariance matrix include forcibly maintaining symmetry by averaging the P-matrix with its transpose after each system propagation and measurement update, and applying minimum values to the state uncertainties.

3.3.4 Time Synchronization

Different types of navigation system exhibit different data lags between the time at which sensor measurements are taken, known as the *time of validity*, and the time when a navigation solution based on those measurements is output. There may also be a communication delay between the navigation system and the Kalman filter processor. When Kalman filter measurements compare the outputs of two different navigation systems, it is important to ensure that those outputs correspond to the same time of validity. Otherwise, differences in the navigation system outputs due to the time lag between them will be falsely attributed by the Kalman filter to the states, corrupting the estimates of those states. The greater the level of dynamics encountered, the larger the impact of a given time-synchronization error will be. Poor time synchronization can be mitigated by using very low gains in the Kalman filter; however, it is better to synchronize the measurement data.

Data synchronization requires the outputs from the faster responding system, such as an INS, to be stored. Once an output is received from the slower system, such as a GNSS receiver, an output from the faster system with the same time of validity is retrieved from the store and used to form a synchronized measurement input to the Kalman filter. Figure 3.10 illustrates the architecture. It is usually better to interpolate the data in the store rather than use the nearest point in time. Data-lag compensation is more effective where all data is time-tagged, enabling precise synchronization. When time tags are unavailable, data lag compensation may operate using an assumed average time delay, provided this is known to within about 10 ms and the actual lag does not vary by more than about ±100 ms.[*] It is also possible to estimate the time lag of one data stream with respect to another as a Kalman filter state (see Section I.6 of Appendix I on the CD).

The system-propagation phase of the Kalman filter usually uses data from the faster responding navigation system. Consequently, the state estimates may be propagated to a time ahead of the measurement time of validity. The optimal solution is to postmultiply the measurement matrix, H, by a transition matrix, Φ, that propagates from the state time of validity, t_s, to the measurement time of validity, t_m. Thus,

---

[*]This paragraph, up to this point, is based on material written by the author for QinetiQ, so comprises QinteQ copyright material.

3.3 Implementation Issues

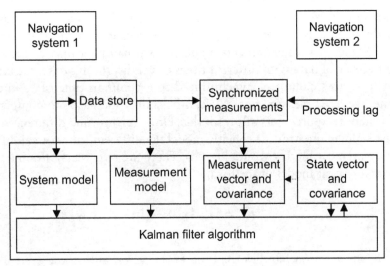

Figure 3.10 Data synchronization (open-loop Kalman filter).

$$\mathbf{H}(t_s) = \mathbf{H}(t_m)\mathbf{\Phi}(t_m, t_s). \tag{3.77}$$

Another option is simply to limit the extent of the system propagation to the measurement time of validity. However, the simplest solution is simply to increase the measurement noise covariance, **R**, to account for the effects of the timing offset, noting that these may be time correlated. The choice of solution is a tradeoff between performance impact, processing load, and complexity and will depend on the requirements for the application concerned.

For some high-integrity applications, there may a lag of tens of seconds between the measurement time of validity and its application. This is to enable multi-epoch fault detection tests (see Chapter 17) to be performed on the measurement stream prior to accepting the measurements. In this case, (3.77) should be used to link the measurements to the current states within the measurement model.

A similar problem is the case where the measurement vector comprises measurements with different times of validity. Again, the simplest option is to increase **R**, while an optimal approach is to postmultiply the relevant rows of the measurement matrix by transition matrices that propagate between the state and measurement times of validity. Another optimal approach is to divide the measurement vector up and perform a separate measurement update for each time of validity, interspersed with a Kalman filter system propagation.

When the closed-loop correction of the navigation system(s) under calibration by the Kalman filter is used, data-delay compensation introduces a delay in applying the corrections to the Kalman filter measurement stream. Further delays are introduced by the time it takes to process the Kalman filter measurement update and communicate the correction. Figure 3.11 illustrates this. As a result of these lags, one or more uncorrected measurement set may be processed by the Kalman filter, causing the closed-loop correction to be repeated. Overcorrection of a navigation system can cause instability with the navigation solution oscillating about the truth. The optimal solution to this problem is to apply corrections to the measurement innovations or the data store in Figure 3.10. However, a simpler solution is to down-weight the Kalman gain, **K**, either directly or via the measurement noise covariance, **R**.

Sometimes measurements input to a Kalman filter may be the sum, average, or difference of data with different times of validity. In this case, the measurement and state vectors cannot be synchronized to a common time of validity. For summed and averaged measurements, **R** can be simply increased to compensate if the performance requirements allow for this. However, for time-differenced measurements, the Kalman filter must explicitly model the different times of validity, otherwise the measurement matrix would be zero. There are two ways of doing this. Consider a measurement model of the form

$$\mathbf{z}(t) = \mathbf{H}(t)\big(\mathbf{x}(t) - \mathbf{x}(t - \tau)\big) + \mathbf{w}_m(t). \tag{3.78}$$

One solution handles the time propagation within the system model by augmenting the state vector at time t with a replica valid at time $t - \tau$. These additional states are known as delayed states. The combined state vector, transition matrix, system noise covariance matrix, and measurement matrix, denoted by the superscript C thus become

$$\mathbf{x}^C(t) = \begin{pmatrix} \mathbf{x}(t) \\ \mathbf{x}(t - \tau) \end{pmatrix}, \qquad \mathbf{\Phi}^C(t, t - \tau) = \begin{pmatrix} \mathbf{\Phi}(t, t - \tau) & 0 \\ \mathbf{I} & 0 \end{pmatrix},$$

$$\mathbf{Q}^C(t, t - \tau) = \begin{pmatrix} \mathbf{Q}(t, t - \tau) & 0 \\ 0 & 0 \end{pmatrix}, \qquad \mathbf{H}^C(t) = \begin{pmatrix} \mathbf{H}(t) & -\mathbf{H}(t) \end{pmatrix} \tag{3.79}$$

where $\mathbf{\Phi}(t, t - \tau)$ is the continuous-time transition matrix for the state vector between times $t - \tau$ and t, noting that $\mathbf{\Phi}(t_k, t_k - \tau_s) = \mathbf{\Phi}_{k-1}$. Similarly, $\mathbf{Q}(t, t - \tau)$ is the continuous-time system noise covariance matrix. This enables the standard Kalman filter measurement model, (3.50), to be used. In practice, only those components of $\mathbf{x}(t - \tau)$ to which the measurement matrix directly couples need be included in the state vector.

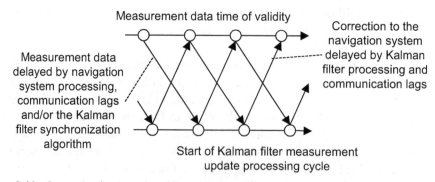

Figure 3.11 Processing lag in a closed-loop Kalman filter.

The other solution incorporates the time propagation of the state vector between epochs within the measurement model by replacing the measurement matrix with

$$\begin{aligned} \mathbf{H}'(t) &= \mathbf{H}(t)\big[\mathbf{I} - \mathbf{\Phi}(t-\tau,t)\big] \\ &= \mathbf{H}(t)\big[\mathbf{I} - \mathbf{\Phi}(t,t-\tau)^{-1}\big] \end{aligned} \quad (3.80)$$

and retaining the conventional single-epoch state vector. This imposes a lower processing load than the first method but neglects the effect of the system noise between times $t - \tau$ and t. Therefore, it should be validated against the first method before use. Both methods are extendable to measurement averaging and summing over multiple epochs.

3.3.5 Kalman Filter Design Process

A good design philosophy [14] for a Kalman filter is to first select as states all known errors or properties of the system which are modelable, observable, and contribute to the desired output of the overall system, generally a navigation solution. This is sometimes known as the *truth model*. System and measurement models should then be derived based on this state selection.

A software simulation should be developed, containing a version of the Kalman filter in which groups of states may be deselected and different phases of the algorithm run at different rates. With all states selected and all Kalman filter phases run at the fastest rate, the filter should be tuned and tested to check that it meets the requirements. Processor load need not be a major consideration at this stage.

Assuming the requirements are met, simulation runs should then be conducted with different groups of Kalman filter states de-selected and their effects modeled as system noise. Runs should also be conducted with phases of the Kalman filter run at a range of slower rates. Combinations of these configurations should also be investigated. Those changes that have the least effect on Kalman filter performance for a given reduction in processor load should then be implemented in turn until the computational load falls within the available processing capacity.

The reduced Kalman filter, sometimes known as the *design model*, should then be carefully retuned and assessed by simulation and trials to verify its performance.

3.4 Extensions to the Kalman Filter

The derivation of the Kalman filter algorithm is based on a number of assumptions about the properties of the states estimated and noise sources accounted for. However, these assumptions do not always apply to real navigation systems. This section looks at how the basic Kalman filter technique may be extended to handle a nonlinear measurement or system model, time-correlated noise, unknown system or measurement noise standard deviations, and non-Gaussian measurement distributions. In addition, Kalman smoothing techniques, which take advantage of the extra information available in postprocessed applications, are discussed.

3.4.1 Extended and Linearized Kalman Filter

In a standard Kalman filter, the measurement model is assumed to be linear (i.e., the measurement vector, z, is a linear function of the state vector, x). This is not always the case for real systems. In some applications, such as most INS alignment and calibration problems, a linear approximation of the measurement model is useful, though this can introduce small errors. However, for applications processing ranging measurements, such as a GNSS navigation filter, the measurement model is highly nonlinear.*

The system model is also assumed to be linear in the standard Kalman filter (i.e., $\dot{\mathbf{x}}$ is a linear function of x). Closed-loop correction of the system using the state estimates (Section 3.2.6) can often be used to maintain a linear approximation in the system model. However, it is not always possible to perform the necessary feedback to the system. An example of this is total-state INS/GNSS integration (see Section 14.1.1), where the absolute position, velocity, and attitude are estimated rather than the errors therein.

A nonlinear version of the Kalman filter is the *extended Kalman filter*. In an EKF, the system matrix, **F**, and measurement matrix, **H**, can be replaced in the state propagation and update equations by nonlinear functions of the state vector, respectively, **f(x)** and **h(x)**. It is common in navigation applications to combine the measurement-update phase of the EKF with the system-propagation phase of the standard Kalman filter. The reverse combination may also be used, though it is rare in navigation.

The system dynamic model of the EKF is

$$\dot{\mathbf{x}}(t) = \mathbf{f}(\mathbf{x}(t), t) + \mathbf{G}(t)\mathbf{w}_s(t), \tag{3.81}$$

where the nonlinear function of the state vector, **f**, replaces the product of the system matrix and state vector and the other terms are as defined in Section 3.2.3. The state vector propagation equation is thus

$$\hat{\mathbf{x}}_k^- = \hat{\mathbf{x}}_{k-1}^+ + \int_{t_k - \tau_s}^{t_k} \mathbf{f}(\hat{\mathbf{x}}(t'), t') \, dt', \tag{3.82}$$

replacing (3.14). When **f** may be assumed constant over the propagation interval, this simplifies to

$$\hat{\mathbf{x}}_k^- = \hat{\mathbf{x}}_{k-1}^+ + \mathbf{f}(\hat{\mathbf{x}}_{k-1}^+, t_k)\tau_s, \tag{3.83}$$

In the EKF, it is assumed that the error in the state vector estimate is much smaller than the state vector, enabling a linear system model to be applied to the state vector residual:

$$\delta\dot{\mathbf{x}}(t) = \mathbf{F}(t)\delta\mathbf{x}(t) + \mathbf{G}(t)\mathbf{w}_s(t). \tag{3.84}$$

---

*This paragraph, up to this point, is based on material written by the author for QinetiQ, so comprises QinetiQ copyright material.

3.4 Extensions to the Kalman Filter

The conventional error covariance propagation equation, (3.15), may thus be used with the system matrix linearized about the state vector estimate using

$$\mathbf{F}_{k-1} = \left.\frac{\partial \mathbf{f}(\mathbf{x},t_k)}{\partial \mathbf{x}}\right|_{\mathbf{x}=\hat{\mathbf{x}}_{k-1}^+}. \tag{3.85}$$

Provided the propagation interval, $\tau_s = t_k - t_{k-1}$, is sufficiently small for the approximation $\mathbf{f}(\mathbf{x},t_k) \approx \mathbf{f}(\mathbf{x},t_{k-1})$ to be valid, the transition matrix is calculated using (3.33):

$$\mathbf{\Phi}_{k-1} \approx \exp(\mathbf{F}_{k-1}\tau_s),$$

which is solved using a power-series expansion as in the conventional Kalman filter.

The measurement model of the EKF is

$$\mathbf{z}(t) = \mathbf{h}(\mathbf{x}(t),t) + \mathbf{w}_m(t), \tag{3.86}$$

where $\mathbf{h}$ is a nonlinear function of the state vector. The state vector is then updated with the true measurement vector using

$$\begin{aligned}\hat{\mathbf{x}}_k^+ &= \hat{\mathbf{x}}_k^- + \mathbf{K}_k\left[\mathbf{z}_k - \mathbf{h}(\hat{\mathbf{x}}_k^-,t_k)\right] \\ &= \hat{\mathbf{x}}_k^- + \mathbf{K}_k\delta\mathbf{z}_k^-\end{aligned}, \tag{3.87}$$

replacing (3.24), where from (3.9) and (3.86), the measurement innovation is

$$\begin{aligned}\delta\mathbf{z}_k^- &= \mathbf{z}_k - \mathbf{h}(\hat{\mathbf{x}}_k^-,t_k) \\ &= \mathbf{h}(\mathbf{x}_k,t_k) - \mathbf{h}(\hat{\mathbf{x}}_k^-,t_k) + \mathbf{w}_{mk}\end{aligned}. \tag{3.88}$$

Once the state vector estimate has converged with its true counterpart, the measurement innovations will be small, so they can legitimately be modeled as a linear function of the state vector where the full measurements cannot. Thus,

$$\delta\mathbf{z}_k^- \approx \mathbf{H}_k\delta\mathbf{x}_k^- + \mathbf{w}_{mk}, \tag{3.89}$$

where

$$\mathbf{H}_k = \left.\frac{\partial \mathbf{h}(\mathbf{x},t_k)}{\partial \mathbf{x}}\right|_{\mathbf{x}=\hat{\mathbf{x}}_k^-} = \left.\frac{\partial \mathbf{z}(\mathbf{x},t_k)}{\partial \mathbf{x}}\right|_{\mathbf{x}=\hat{\mathbf{x}}_k^-}. \tag{3.90}$$

A consequence of this linearization of $\mathbf{F}$ and $\mathbf{H}$ is that the error covariance matrix, $\mathbf{P}$, and Kalman gain, $\mathbf{K}$, are functions of the state estimates. This can occasionally cause stability problems and the EKF is more sensitive to the tuning of the P-matrix initialization than a standard Kalman filter.[*]

---

[*]This paragraph, up to this point, is based on material written by the author for QinetiQ, so comprises QinetiQ copyright material.

For the EKF to be valid, the values of **F** and **H** obtained by, respectively, linearizing the system and measurement models about the state vector estimate must be very close to the values that would be obtained if they were linearized about the true state vector. Figures 3.12 and 3.13 show examples of, respectively, valid and invalid linearization of single-state system and measurement models.

One way to assess whether the system model linearization is valid is to test for the condition:

$$\left.\frac{\partial \mathbf{f}(\mathbf{x},t_k)}{\partial \mathbf{x}}\right|_{\mathbf{x}=\hat{\mathbf{x}}_{k-1}^{+}+\Delta \mathbf{x}_{k-1}^{+i}} \approx \left.\frac{\partial \mathbf{f}(\mathbf{x},t_k)}{\partial \mathbf{x}}\right|_{\mathbf{x}=\hat{\mathbf{x}}_{k-1}^{+}-\Delta \mathbf{x}_{k-1}^{+i}}, \qquad \Delta x_{k-1,j}^{+i} = \begin{array}{ll} \sqrt{P_{k-1,i,i}^{+}} & j = i \\ 0 & j \neq i \end{array} \qquad (3.91)$$

for each state, i. This determines whether there is significant variation in the gradient of the system function, $\mathbf{f}$, over the uncertainty bounds of the state vector estimate.

Similarly, the validity of the measurement model linearization may be assessed by testing for the condition

$$\left.\frac{\partial \mathbf{h}(\mathbf{x},t_k)}{\partial \mathbf{x}}\right|_{\mathbf{x}=\hat{\mathbf{x}}_{k}^{-}+\Delta \mathbf{x}_{k}^{-i}} \approx \left.\frac{\partial \mathbf{h}(\mathbf{x},t_k)}{\partial \mathbf{x}}\right|_{\mathbf{x}=\hat{\mathbf{x}}_{k}^{-}-\Delta \mathbf{x}_{k}^{-i}}, \qquad \Delta x_{k,j}^{-i} = \begin{array}{ll} \sqrt{P_{k,i,i}^{-}} & j = i \\ 0 & j \neq i \end{array}, \qquad (3.92)$$

again, for each state, i. This determines whether the gradient of the measurement function, $\mathbf{h}$, varies significantly over the uncertainty bounds of the state vector estimate. A more sophisticated approach is described in [15].

When the above conditions are not met, the error covariance computed by the EKF will tend to be overoptimistic, which can eventually lead to divergent state estimates. This is most likely to happen at initialization, when the error covariance

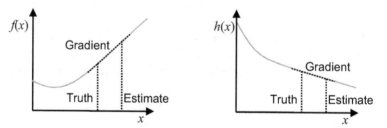

Figure 3.12 Example of valid system and measurement model linearization using an EKF (gradients are the same for the true and estimated values of x).

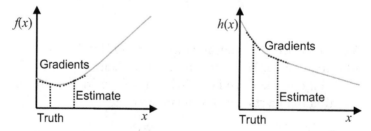

Figure 3.13 Example of invalid system and measurement model linearization using an EKF (gradients are different for the true and estimated values of x).

matrix, **P**, is normally at its largest. When the validity of the system model linearization is marginal, the system noise covariance, **Q**, may be increased to compensate. Similarly, the measurement noise covariance, **R**, may be increased where the validity of the measurement model linearization is marginal. However, if either linearization is clearly invalid, a higher-order approach must be used, such as the unscented Kalman filter (Section 3.4.2), the iterated EKF, or the second-order EKF [2, 6]. A new approach, designed to handle multiple classes of second-order problem, is the intelligent method for recursive estimation (IMRE) Kalman filter [16].

An alternative to the EKF that maintains an error covariance and Kalman gain that are independent of the state estimates is the linearized Kalman filter. This takes the same form as the EKF with the exception that the system and measurement models are linearized about a predetermined state vector, $\mathbf{x}^P$:

$$\mathbf{F}_{k-1} = \left.\frac{\partial \mathbf{f}(\mathbf{x},t_k)}{\partial \mathbf{x}}\right|_{\mathbf{x}=\mathbf{x}^P_{k-1}}, \quad \mathbf{H}_k = \left.\frac{\partial \mathbf{h}(\mathbf{x},t_k)}{\partial \mathbf{x}}\right|_{\mathbf{x}=\mathbf{x}^P_k}. \qquad (3.93)$$

For this to be valid, the above values of **F** and **H** must be very close to those that would be obtained if the system and measurement models were linearized about the true state vector. Therefore, it is not generally suited to cases where the EKF is invalid. A suitable application is guided weapons, where the approximate trajectory is known prior to launch and the Kalman filter is estimating the navigation solution.

3.4.2 Unscented Kalman Filter

The unscented Kalman filter, also known as the sigma-point Kalman filter, is a nonlinear adaptation of the Kalman filter that does not require the gradients of the system function, **f**, and measurement function, **h**, to be approximately constant over the uncertainty bounds of the state estimate [6, 17].

The UKF relies on the *unscented transformation* from an n-element state vector estimate, $\hat{\mathbf{x}}$, and its error covariance matrix, **P**, to a set of $2n$ parallel state vectors, known as sigma points. The transform is reversible as the mean and variance of the sigma points are the state vector estimate and error covariance matrix, respectively. There are a number of different types of unscented transformation [6]; the root-covariance type is used here. Like the standard Kalman filter and the EKF, the UKF assumes that all states and noise sources have distributions that can be described using only a mean and covariance (e.g., the Gaussian distribution).

For applications where only the system model is significantly nonlinear, the system-propagation phase of the UKF may be combined with the measurement-update phase of a conventional Kalman filter or EKF. Similarly, when only the measurement model is significantly nonlinear, a UKF measurement update may be combined with the system propagation of an EKF or conventional Kalman filter. In navigation, the UKF system propagation is useful for applications where there are large attitude uncertainties, while the UKF measurement update is useful where there is ranging between a transmitter and receiver a short distance apart.

The first step in the system-propagation phase of the UKF is to obtain the square root of the error covariance matrix, $\mathbf{S}^+_{k-1}$, by using Cholesky factorization (see Section A.6 of Appendix A on the CD) to solve

$$P_{k-1}^+ = S_{k-1}^+ S_{k-1}^{+\,T}. \tag{3.94}$$

Next, the sigma points are calculated using:

$$x_{k-1}^{+(i)} = \begin{matrix} \hat{x}_{k-1}^+ + \sqrt{n}S_{k-1,:,i}^+ & i \le n \\ \hat{x}_{k-1}^+ - \sqrt{n}S_{k-1,:,(i-n)}^+ & i > n \end{matrix}, \tag{3.95}$$

where the subscript :,i denotes the ith column of the matrix.

Each sigma point is then propagated through the system model using

$$x_k^{-(i)} = x_{k-1}^{+(i)} + f\left(x_{k-1}^{+(i)}, t_k\right)\tau_s, \tag{3.96}$$

where f is assumed constant over the propagation interval; otherwise, (3.82) must be used. The propagated state estimate and its error covariance are then calculated using

$$\hat{x}_k^- = \frac{1}{2n}\sum_{i=1}^{2n} x_k^{-(i)}, \tag{3.97}$$

$$P_k^- = \frac{1}{2n}\sum_{i=1}^{2n}\left(x_k^{-(i)} - \hat{x}_k^-\right)\left(x_k^{-(i)} - \hat{x}_k^-\right)^T + Q_{k-1}, \tag{3.98}$$

assuming that the system noise may be propagated linearly through the system model. Otherwise, the sigma point state vectors are augmented to incorporate system noise terms.

The measurement-update phase of the UKF begins by generating new sigma points using

$$x_k^{-(i)} = \begin{matrix} \hat{x}_k^- + \sqrt{n}S_{k,:,i}^- & i \le n \\ \hat{x}_k^- - \sqrt{n}S_{k,:,(i-n)}^- & i > n \end{matrix}, \quad P_k^- = S_k^- S_k^{-\,T}. \tag{3.99}$$

This step may be omitted to save processing capacity, using the sigma points from the system propagation phase, instead. However, there is some degradation in performance.

The sigma point and mean measurement innovations are calculated using

$$\begin{aligned} \delta z_k^{-(i)} &= z_k - h\left(\hat{x}_k^{-(i)}, t_k\right) \\ \delta z_k^- &= \frac{1}{2n}\sum_{i=1}^{2n} \delta z_k^{-(i)} \end{aligned}. \tag{3.100}$$

The covariance of the measurement innovations is given by

$$C_{\delta z,k}^- = \frac{1}{2n}\sum_{i=1}^{2n}\left(\delta z_k^{-(i)} - \delta z_k^-\right)\left(\delta z_k^{-(i)} - \delta z_k^-\right)^T + R_k, \tag{3.101}$$

assuming that the measurement noise may be propagated linearly through the measurement model. Otherwise, the sigma point state vectors are augmented to incorporate measurement noise terms.

Finally, the Kalman gain, state vector update and error covariance update of the UKF are

$$\mathbf{K}_k = \left[\frac{1}{2n} \sum_{i=1}^{2n} \left(\mathbf{x}_k^{-(i)} - \hat{\mathbf{x}}_k^- \right)\left(\delta \mathbf{z}_k^{-(i)} - \delta \mathbf{z}_k^- \right)^{\mathrm{T}} \right] \left(\mathbf{C}_{\delta z,k}^- \right)^{-1}, \quad (3.102)$$

$$\hat{\mathbf{x}}_k^+ = \hat{\mathbf{x}}_k^- + \mathbf{K}_k \delta \mathbf{z}_k^-, \quad (3.103)$$

$$\mathbf{P}_k^+ = \mathbf{P}_k^- - \mathbf{K}_k \mathbf{C}_{\delta z,k}^- \mathbf{K}_k^{\mathrm{T}}. \quad (3.104)$$

The system-propagation phase of the UKF may be combined with the measurement-update phase of the standard Kalman filter or EKF, or vice versa.

3.4.3 Time-Correlated Noise

In Kalman filtering, it is assumed that all measurement errors, $\mathbf{w}_m$, are time uncorrelated; in other words, the measurement noise is white. In practice this is often not the case. For example, Kalman filters in navigation often input measurements output by another Kalman filter, a loop filter, or another estimation algorithm. There may also be time-correlated variation in the lever arm between navigation systems. A Kalman filter attributes the time-correlated parts of the measurement innovations to the states. Consequently, correlated measurement noise can potentially corrupt the state estimates.

There are three main ways to account for time-correlated measurement noise in a Kalman filter. The optimal solution is to estimate the time-correlated noise as additional Kalman filter states. However, this may not be practical due to observability or processing capacity limitations. The second, and simplest, option is to reduce the gain of the Kalman filter. The measurement update interval may be increased to match the measurement noise correlation time; the assumed measurement noise covariance, $\mathbf{R}$, may be increased; or the Kalman gain, $\mathbf{K}$, down-weighted. Measurement averaging may be used in conjunction with an increased update interval, provided the averaged measurement is treated as a single measurement for statistical purposes. These gain-reduction techniques will all increase the time it takes the Kalman filter to converge and the uncertainty of the estimates at convergence. The third method of handling time-correlated noise is to a use a Schmidt-Kalman filter with uncertain measurement noise parameters [18]. This effectively increases the error covariance matrix, $\mathbf{P}$, to model the time-correlated noise and is described in Section D.2 of Appendix D on the CD.

Another assumption of Kalman filters is that the system noise, $\mathbf{w}_s$, is not time correlated. However, the system often exhibits significant systematic and other time-correlated errors that are not estimated as states due to observability or processing power limitations, but that affect the states that are estimated. These errors must be accounted for.[†]

[†]This and subsequent paragraphs are based on material written by the author for QinetiQ, so comprise QinetiQ copyright material.

When the correlation times are relatively short, these system errors may be modeled as white noise. However, the white noise must overbound the correlated noise, affecting the Kalman filter's convergence properties. For error sources correlated over more than a minute or so, a white noise approximation does not effectively model how the effects of these error sources propagate with time.‡ The solution is to use a Schmidt-Kalman filter with uncertain system noise parameters [18]. Details are provided in Section D.2 of Appendix D on the CD.

Another Kalman filter formulation, designed for handling time-correlated GNSS measurement errors is described in [19].

3.4.4 Adaptive Kalman Filter

For most applications, the Kalman filter's system noise covariance matrix, $\mathbf{Q}$, and measurement noise covariance matrix, $\mathbf{R}$, are determined during the development phase by laboratory measurements of the system, simulation and trials. However, there are some cases where this cannot be done. For example, if an INS/GNSS integration algorithm or INS calibration algorithm is designed for use with a range of different inertial sensors, the system noise covariance will not be known in advance of operation. Similarly, if a transfer alignment algorithm (Section 15.1) is designed for use on different aircraft and weapon stores without prior knowledge of the flexure and vibration environment, the measurement noise covariance will not be known in advance.*

In other cases, the optimum Kalman filter tuning might vary over time as the context varies. For example, a GNSS navigation filter in a mobile device that may be stationary, on a walking pedestrian, or in a car, would require a different system noise model in each case. Similarly, the accuracy of GNSS ranging measurements varies with the signal-to-noise level, and multipath environment.

For both applications where the optimum tuning is unknown and applications where it varies, an adaptive Kalman filter may be used to estimate $\mathbf{R}$ and/or $\mathbf{Q}$ as it operates. There are two main approaches, innovation-based adaptive estimation (IAE) [20, 21] and multiple-model adaptive estimation (MMAE) [22].

The IAE method calculates the system noise covariance, $\mathbf{Q}$, the measurement noise covariance, $\mathbf{R}$, or both from the measurement innovation statistics. The first step is the calculation of the covariance of the last n measurement innovations, $\mathbf{C}$:

$$\tilde{\mathbf{C}}^-_{\delta z,k} = \frac{1}{n} \sum_{j=k-n}^{k} \delta \mathbf{z}_j^- \delta \mathbf{z}_j^{-\mathrm{T}}. \tag{3.105}$$

This can be used to compute $\mathbf{Q}$ and/or $\mathbf{R}$:

$$\begin{aligned} \tilde{\mathbf{Q}}_k &= \mathbf{K}_k \tilde{\mathbf{C}}^-_{\delta z,k} \mathbf{K}_k^{\mathrm{T}} \\ \tilde{\mathbf{R}}_k &= \tilde{\mathbf{C}}^-_{\delta z,k} - \mathbf{H}_k \mathbf{P}_k^- \mathbf{H}_k^{\mathrm{T}} \end{aligned}. \tag{3.106}$$

---

‡End of QinetiQ copyright material.

*This paragraph, up to this point, is based on material written by the author for QinetiQ, so comprises QinetiQ copyright material.

3.4 Extensions to the Kalman Filter

Initial values of $\mathbf{Q}$ and $\mathbf{R}$ must be provided for use while the first set of measurement innovation statistics is compiled. These should be selected cautiously. Minimum and maximum values should also be imposed to stabilize the filter in the event of faults.

The MMAE method uses a bank of parallel Kalman filters with different values of the system and/or measurement noise covariance matrices, $\mathbf{Q}$ and $\mathbf{R}$. Different initial values of the error covariance matrix, $\mathbf{P}$, may also be used. Each of the Kalman filter hypotheses, denoted by the index i, is allocated a probability as follows [3, 4]:

$$p_{k,i} = \frac{\Lambda_{k,i}}{\sum_{j=1}^{l}\Lambda_{k,j}},$$

$$\Lambda_{k,i} = \frac{p_{k-1,i}}{\sqrt{(2\pi)^m |\mathbf{H}_{k,i}\mathbf{P}_{k,i}^{-}\mathbf{H}_{k,i}^{T} + \mathbf{R}_{k,i}|}} \exp\left[-\tfrac{1}{2}\delta\mathbf{z}_{k,i}^{-\mathrm{T}}\left(\mathbf{H}_{k,i}\mathbf{P}_{k,i}^{-}\mathbf{H}_{k,i}^{T} + \mathbf{R}_{k,i}\right)^{-1}\delta\mathbf{z}_{k,i}^{-}\right]$$

(3.107)

where m is the number of components of the measurement vector, l is the number of filter hypotheses, and Λ is a likelihood. Note that the matrix inversion is already performed as part of the Kalman gain calculation. The filter hypothesis with the smallest normalized measurement innovations is most consistent with the measurement stream, so is allocated the largest probability.

Over time, the probability of the best filter hypothesis will approach unity while the others approach zero. To make best use of the available processing capacity, weak hypotheses should be deleted and the strongest hypothesis periodically subdivided to refine the filter tuning and allow it to respond to changes in the system.

The overall state vector estimate and error covariance are obtained as follows:

$$\hat{\mathbf{x}}_k^+ = \sum_{i=1}^{l} p_{k,i}\hat{\mathbf{x}}_{k,i}^+, \qquad (3.108)$$

$$\mathbf{P}_k^+ = \sum_{i=1}^{l} p_{k,i}\left[\mathbf{P}_{k,i}^+ + \left(\hat{\mathbf{x}}_{k,i}^+ - \hat{\mathbf{x}}_k^+\right)\left(\hat{\mathbf{x}}_{k,i}^+ - \hat{\mathbf{x}}_k^+\right)^{\mathrm{T}}\right], \qquad (3.109)$$

noting that the error covariance matrix must account for the spread in the state vector estimates of the filter hypotheses as well as the error covariance of each hypothesis.

Comparing the IAE and MMAE adaptive Kalman filter techniques, the latter is more computationally intensive, as a bank of Kalman filters must be processed instead of just one. However, in an IAE Kalman filter, the system noise covariance, measurement noise covariance, error covariance, and Kalman gain matrices may all be functions of the state estimates, whereas they are independent in the MMAE filter bank (assuming conventional Kalman filters rather than EKFs). Consequently, the MMAE is less prone to filter instability.

3.4.5 Multiple-Hypothesis Filtering

An assumption of the standard Kalman filter is that the measurements have unimodal distributions (e.g., Gaussian), enabling the measurement vector to be modeled as a mean, **z**, and covariance, **R**. However, this is not the case for every navigation system. Ranging systems can produce bimodal position measurements where there are insufficient signals for a unique fix, while some feature-matching techniques (Chapter 13) can produce a fix in the form of a highly irregular position distribution. To process these measurements in a Kalman filter-based estimation algorithm, they must first be expressed as a sum of Gaussian distributions, known as hypotheses, each with a mean, z_i, a covariance, R_i, and also a probability, p_i. A probability score, p_0, should also be allocated to the null hypothesis, representing the probability that none of the other hypotheses are correct. The probability scores sum to unity:

$$\sum_{i=0}^{n_k} p_{k,i} = 1, \quad (3.110)$$

where n_k is the number of hypotheses and k denotes the Kalman filter iteration as usual.

There are three main methods of handling multiple-hypothesis measurements using Kalman filter techniques: best fix, weighted fix, and multiple-hypothesis filtering. The best-fix method is a standard Kalman filter that accepts the measurement hypothesis with the highest probability score and rejects the others. It should incorporate a prefiltering algorithm that rejects all of the measurement hypotheses where none is dominant. This method has the advantage of simplicity and can be effective where one hypothesis is clearly dominant on most iterations.

Weighted-fix techniques input all of the measurement hypotheses, weighted according to their probabilities, but maintain a single set of state estimates. An example is the probabilistic data association filter (PDAF) [23, 24], which is predominantly applied to target tracking problems. The system-propagation phase of the PDAF is the same as for a standard Kalman filter. In the measurement-update phase, the Kalman gain calculation is performed for each of the measurement hypotheses:

$$\mathbf{K}_{k,i} = \mathbf{P}_k^- \mathbf{H}_k^T \left[\mathbf{H}_k \mathbf{P}_k^- \mathbf{H}_k^T + \mathbf{R}_{k,i} \right]^{-1}. \quad (3.111)$$

The state vector and error covariance matrix are then updated using

$$\begin{aligned} \hat{\mathbf{x}}_k^+ &= \hat{\mathbf{x}}_k^- + \sum_{i=1}^{n_k} p_{k,i} \mathbf{K}_{k,i} \left(\mathbf{z}_{k,i} - \mathbf{H}_k \hat{\mathbf{x}}_k^- \right) \\ &= \hat{\mathbf{x}}_k^- + \sum_{i=1}^{n_k} p_{k,i} \mathbf{K}_{k,i} \delta \mathbf{z}_{k,i}^- \\ &= \sum_{i=1}^{n_k} p_{k,i} \hat{\mathbf{x}}_{k,i}^+ \end{aligned} \quad (3.112)$$

$$\mathbf{P}_k^+ = \left[\mathbf{I} - \left(\sum_{i=1}^{n_k} p_{k,i} \mathbf{K}_{k,i} \right) \mathbf{H}_k \right] \mathbf{P}_k^- + \sum_{i=1}^{n_k} p_{k,i} \left(\hat{\mathbf{x}}_{k,i}^+ - \hat{\mathbf{x}}_k^+ \right) \left(\hat{\mathbf{x}}_{k,i}^+ - \hat{\mathbf{x}}_k^+ \right)^T, \quad (3.113)$$

3.4 Extensions to the Kalman Filter

where

$$\begin{aligned}\hat{\mathbf{x}}_{k,i}^+ &= \hat{\mathbf{x}}_k^- + \mathbf{K}_{k,i}\left(\mathbf{z}_{k,i} - \mathbf{H}_k\hat{\mathbf{x}}_k^-\right) \\ &= \hat{\mathbf{x}}_k^- + \mathbf{K}_{k,i}\delta\mathbf{z}_{k,i}^-\end{aligned}. \tag{3.114}$$

Note that, where the measurement hypotheses are widely spread compared to the prior state uncertainties, the state uncertainty (root diagonals of $\mathbf{P}$) can be larger following the measurement update; this cannot happen in a standard Kalman filter.

Compared to the best-fix technique, the PDAF has the advantage that it incorporates all true measurement hypotheses, but the disadvantage that it also incorporates all of the false hypotheses. It is most suited to applications where false hypotheses are not correlated over successive measurement sets or the truth is a combination of overlapping Gaussian measurement hypotheses.

Where false measurement hypotheses are time correlated, a multiple-hypothesis Kalman filter (MHKF) enables multiple state vector hypotheses to be maintained

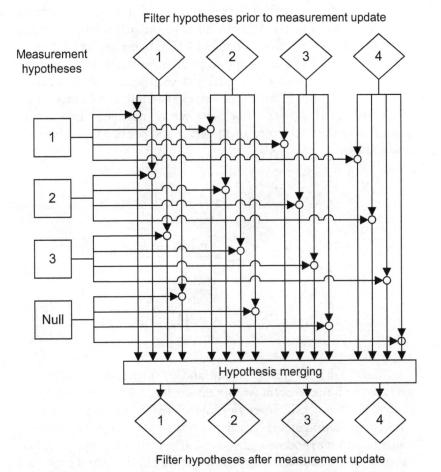

Figure 3.14 Multiple-hypothesis Kalman filter measurement update ($l = 4$, $n_k = 3$).

in parallel using a bank of Kalman filters. The technique was originally developed for target tracking [25], so is often known as multiple-hypothesis tracking (MHT). As the true hypothesis is identified over a series of filter cycles, the false measurement hypotheses are gradually eliminated from the filter bank. Like the MMAE filter, the MHKF maintains a set of l state vector and error covariance matrix hypotheses that are propagated independently through the system model using the conventional Kalman filter equations. Each of these hypotheses has an associated probability score.

For the measurement update phase, the filter bank is split into $(n_k + 1)l$ hypotheses, combining each state vector hypothesis with each measurement hypothesis and the null measurement hypothesis. Figure 3.14 shows the principle. A conventional Kalman filter update is then performed for each hypothesis and a probability score allocated that multiplies the probabilities of the state and measurement hypotheses. The new hypotheses must also be scored for consistency between the state vector and measurement hypotheses; a probability weighting similar to that used for the MMAE [see (3.107)] is suitable. Following this, the probability scores must be renormalized, noting that the scores for the null measurement hypotheses should remain unchanged.

It is clearly impractical for the number of state vector hypotheses to increase on each iteration of the Kalman filter, so the measurement update process must conclude with a reduction in the number of hypotheses to l. This is done by merging hypotheses. The exact approach varies between implementations, but, generally, similar hypotheses are merged with each other and the weakest hypotheses, in terms of their probability scores, are merged into their nearest neighbor. Hypotheses with probability scores below a certain minimum may simply be deleted. A pair of hypotheses, denoted by indices α and β, are merged into a new hypothesis, denoted by γ, using

$$p_{k,\gamma} = p_{k,\alpha} + p_{k,\beta}, \qquad (3.115)$$

$$\hat{\mathbf{x}}_{k,\gamma}^+ = \frac{p_{k,\alpha}\hat{\mathbf{x}}_{k,\alpha}^+ + p_{k,\beta}\hat{\mathbf{x}}_{k,\beta}^+}{p_{k,\gamma}}, \qquad (3.116)$$

$$\mathbf{P}_{k,\gamma}^+ = \sum_{i=\alpha,\beta} p_{k,i}\left[\mathbf{P}_{k,i}^+ + \left(\hat{\mathbf{x}}_{k,i}^+ - \hat{\mathbf{x}}_{k,\gamma}^+\right)\left(\hat{\mathbf{x}}_{k,i}^+ - \hat{\mathbf{x}}_{k,\gamma}^+\right)^T\right]. \qquad (3.117)$$

The overall state vector estimate and error covariance can either be the weighted average of all the hypotheses, obtained using (3.108) and (3.109), or the highest-probability hypothesis, depending on the needs of the application. When closed-loop correction (Section 3.2.6) is used, it is not possible to feed back corrections from the individual filter hypotheses as this would be contradictory; the closed-loop feedback must come from the filter bank as a whole. The corrections fed back to the system must also be subtracted from all of the state vector hypotheses to maintain constant differences between the hypotheses. Thus, the state estimates are not zeroed at feedback, so state vector propagation using (3.14) must take place in the same manner as for the open-loop Kalman filter.

The iterative Gaussian mixture approximation of the posterior (IGMAP) method [26], which can operate with either a single or multiple hypothesis state vector, combines the fitting of a set of Gaussian distributions to the measurement probability distribution and the measurement-update phase of the estimation algorithm into a single iterative process. By moving the approximation as a sum of Gaussian distributions from the beginning to the end of the measurement-update cycle, the residuals of the approximation process are reduced, producing more accurate state estimates. The system-propagation phase of IGMAP is the same as for a conventional Kalman filter or MHKF. However, IGMAP does require more processing capacity than a PDAF or MHKF.

The need to apply a Gaussian approximation to the measurement noise and system noise distributions can be removed altogether by using a Monte Carlo estimation algorithm, such as a particle filter (Section 3.5). However, this imposes a much higher processing load than Kalman filter-based estimation.

3.4.6 Kalman Smoothing

The Kalman filter is designed for real-time applications. It estimates the properties of a system at a given time using measurements of the system up to that time. However, for applications such as surveillance and testing, where the properties of a system are required after the event, a Kalman filter effectively throws away half the measurement data as it does not use measurements taken after the time of interest.[*]

The Kalman smoother is the extension of the Kalman filter that uses measurement information from after the time at which state estimates are required as well as before that time. This leads to more accurate state estimates for nonreal-time applications. There are two main methods, the forward-backward filter [2, 27], and the Rauch, Tung, and Striebel (RTS) method [4, 28].

The forward-backward filter comprises two Kalman filters, a forward filter and a backward filter. The forward filter is a standard Kalman filter. The backward filter is a Kalman filter algorithm working backward in time from the end of the data segment to the beginning. The two filters are treated as independent, so the backward filter must not be initialized with the final solution of the forward filter. The smoothed estimates are obtained simply by combining the estimates of the two filters, weighted according to the ratio of their error covariance matrices:[*]

$$\hat{\mathbf{x}}_k^+ = \left(\mathbf{P}_{f,k}^{+\,-1} + \mathbf{P}_{b,k}^{+\,-1}\right)^{-1}\left(\mathbf{P}_{f,k}^{+\,-1}\hat{\mathbf{x}}_{f,k}^+ + \mathbf{P}_{b,k}^{+\,-1}\hat{\mathbf{x}}_{b,k}^+\right)$$
$$\mathbf{P}_k^+ = \left(\mathbf{P}_{f,k}^{+\,-1} + \mathbf{P}_{b,k}^{+\,-1}\right)^{-1}$$

(3.118)

---

[*]This paragraph, up to this point, is based on material written by the author for QinetiQ, so comprises QinetiQ copyright material.

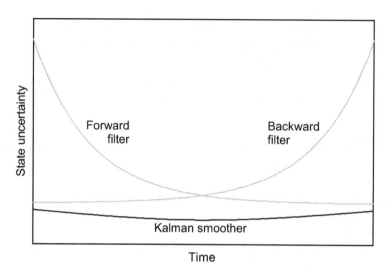

Figure 3.15 Forward-backward Kalman smoother state uncertainty.

where the subscripts f and b refer to the forward and backward filters, respectively.* The index, k, refers to the same point in time for both filters, so the backward filter must count backward. Figure 3.15 shows how the state uncertainty varies with time for the forward, backward and combined filters. It is only necessary to store the state vectors and error covariance matrices and perform the matrix inversion at the points of interest. Note that it is not necessary to run the forward filter beyond the last point of interest and the backward filter beyond the first point of interest.

In the RTS method, a conventional Kalman filter runs forward in time, but storing the state vector, $\mathbf{x}$, and the error covariance matrix, $\mathbf{P}$, after each system propagation and measurement update. The transition matrix, $\mathbf{\Phi}$, is also stored. Once the end of the data set is reached, smoothing begins, starting at the end and working back to the beginning. The smoothing gain on each iteration, $\mathbf{A}_k$, is given by

$$\mathbf{A}_k = \mathbf{P}_k^+ \mathbf{\Phi}_k^T \left(\mathbf{P}_{k+1}^-\right)^{-1}. \tag{3.119}$$

The smoothed state vector, $\hat{\mathbf{x}}_k^s$, and error covariance, $\mathbf{P}_k^s$, are then given by

$$\begin{aligned}\hat{\mathbf{x}}_k^s &= \hat{\mathbf{x}}_k^+ + \mathbf{A}_k\left(\hat{\mathbf{x}}_{k+1}^s - \hat{\mathbf{x}}_{k+1}^-\right) \\ \mathbf{P}_k^s &= \mathbf{P}_k^+ + \mathbf{A}_k\left(\mathbf{P}_{k+1}^s - \mathbf{P}_{k+1}^-\right)\mathbf{A}_k^T\end{aligned}. \tag{3.120}$$

When the smoothed solution is required at all points, the RTS method is more efficient, whereas the forward-backward method is more efficient where a smoothed solution is only required at a single point.

Kalman smoothing can also be used to provide a quasi-real-time solution by making use of information from a limited period after the time of interest. A continuous solution is then output at a fixed lag. This can be useful for tracking applications, such as logistics, security, and road-user charging, that require bridging of GNSS outages.

---

*This paragraph, up to this point, is based on material written by the author for QinetiQ, so comprises QinetiQ copyright material.

3.5 The Particle Filter

This section provides an introduction to the particle filter [29, 30], a nonlinear non-Gaussian Bayesian estimation technique, using the terminology and notation of Kalman filter-based estimation. Further information is available in standard texts [6, 31–33], noting that much of the particle filtering literature assumes a background in advanced statistics.

In state estimation, the measurements, noise sources, and the state estimates themselves are all probability distributions; the exact values are unknown. In Kalman filter-based estimation, all distributions are modeled using the mean and covariance. This is sufficient for modeling Gaussian distributions but is not readily applicable to all types of navigation system. Pattern-matching systems produce inherently non-Gaussian measurement distributions, which are often multimodal. Ranging and angular positioning using environmental features can produce multimodal measurements where the landmark identity is ambiguous. An INS can also have a non-Gaussian error distribution where the attitude uncertainty is too large for the small-angle approximation to be applied.

A particle filter is a type of sequential Monte Carlo estimation algorithm (some authors equate the two). As such, the state estimates are represented as a set of discrete state vectors, known as particles, which are spread throughout their joint probability distribution. Figure 3.16 illustrates this for a bivariate Gaussian distribution. This requires at least an order of magnitude more processing power than the mean and covariance representation used in Kalman filter-based estimation. However, it has the key advantage that any shape of probability distribution may be represented, as illustrated by Figure 3.17. There is no need to approximate the distribution to a multivariate Gaussian (or sum thereof in the case of a MHKF).

The more particles used, the more accurately the probability distribution of the state estimates is represented. Similarly, the more complex the distribution, the greater the number of particles required to represent it to a given accuracy. Most particle filters deploy at least a thousand particles and some use a million or more. The mean, $\hat{\mathbf{x}}_k^+$, and covariance, $\mathbf{P}_k^+$, of the state estimate at epoch k are given by

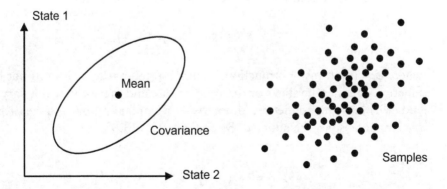

Figure 3.16 Representation of two states with a bivariate Gaussian distribution using a mean and covariance (left) and a set of particles (right).

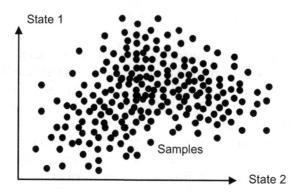

Figure 3.17 Representation of two states with a non-Gaussian distribution using a set of particles.

$$\hat{\mathbf{x}}_k^\pm = \sum_{i=1}^N p_{X,k}^{\pm(i)} \hat{\mathbf{x}}_k^{(i)}, \quad \mathbf{P}_k^\pm = \sum_{i=1}^N p_{X,k}^{\pm(i)} \left(\hat{\mathbf{x}}_k^{(i)} - \hat{\mathbf{x}}_k^\pm \right) \left(\hat{\mathbf{x}}_k^{(i)} - \hat{\mathbf{x}}_k^\pm \right)^\mathrm{T}, \quad (3.121)$$

where $\hat{\mathbf{x}}_k^{(i)}$ and $p_{X,k}^{\pm(i)}$ are, respectively, the state vector estimate and probability of the ith particle, N is the number of particles; the superscripts, $-$ and $+$, denote before and after the measurement update, respectively; and

$$\sum_{i=1}^N p_{X,k}^{\pm(i)} = 1. \quad (3.122)$$

A particle filter has three phases, shown in Figure 3.18. The system-propagation and measurement-update phases are equivalent to their Kalman filter counterparts, while the resampling phase has no Kalman filter equivalent.

Each particle is propagated through the system model separately. The first step, performed independently for each particle, is to sample the discrete system noise vector, $\mathbf{w}_{s,k-1}^{(i)}$, from a distribution, common to all particles. This distribution may be constant or vary with time and/or other known parameters. However, it does not vary with the estimated states. The particle's state vector estimate is then propagated using

$$\hat{\mathbf{x}}_k^{(i)} = \boldsymbol{\varphi}_{k-1}\left(\hat{\mathbf{x}}_{k-1}^{(i)}, \mathbf{w}_{s,k-1}^{(i)} \right), \quad (3.123)$$

where $\boldsymbol{\varphi}_{k-1}$ is a transition function, common to all particles. It need not be a linear function of either the states or the system noise and may or may not vary with time and other known parameters. Alternatively, a similar approach to the EKF and UKF system models may be adopted. By analogy with (3.82),

$$\begin{aligned} \hat{\mathbf{x}}_k^{(i)} &= \hat{\mathbf{x}}_{k-1}^{(i)} + \int_{t_k-\tau_s}^{t_k} \mathbf{f}\left(\hat{\mathbf{x}}_{k-1}^{(i)}, \mathbf{w}_{s,k-1}^{(i)}, t' \right) dt' \\ &\approx \hat{\mathbf{x}}_{k-1}^{(i)} + \mathbf{f}\left(\hat{\mathbf{x}}_{k-1}^{(i)}, \mathbf{w}_{s,k-1}^{(i)}, t_k \right) \tau_s \end{aligned} \quad (3.124)$$

3.5 The Particle Filter

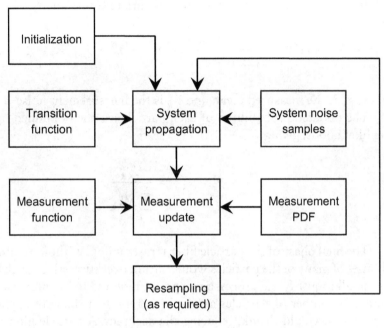

Figure 3.18 Phases of the particle filter.

If the system model is linear, this simplifies to

$$\hat{\mathbf{x}}_k^{(i)} = \hat{\mathbf{x}}_{k-1}^{(i)} + \boldsymbol{\Phi}_{k-1}\hat{\mathbf{x}}_{k-1}^{(i)} + \boldsymbol{\Gamma}_{k-1}\mathbf{w}_{s,k-1}^{(i)}. \tag{3.125}$$

The propagation of multiple state vectors through a model of the system, each with a different set of randomly sampled noise sources is a type of Monte Carlo simulation (see Section J.5 of Appendix J on the CD). This is why the particle filter is classified as a Monte Carlo estimation algorithm.

The system propagation phase of a particle filter changes the state estimates of each particle but leaves the probabilities unchanged, thus $p_{\mathbf{X},k}^{-(i)} \equiv p_{\mathbf{X},k-1}^{+(i)}$. By contrast, the measurement update phase changes the probabilities but not the state estimates. The first step in the measurement update is to obtain a prediction of the measurement vector from each particle's state vector estimate. Thus,

$$\hat{\mathbf{z}}_k^{(i)} = \mathbf{h}(\hat{\mathbf{x}}_k^{(i)}), \tag{3.126}$$

where $\mathbf{h}$ is the deterministic measurement function, as defined by (3.8) and used in the EKF and UKF; it need not be linear.

Next, the predicted measurements are compared with the probability distribution function obtained from the actual measurement process, $\tilde{f}_{\mathbf{Z},k}$, to obtain the relative likelihood of each particle. This is multiplied by the prior probability to obtain the absolute likelihood. Thus,

$$\Lambda_{\mathbf{X},k}^{(i)} = p_{\mathbf{X},k}^{-(i)} \tilde{f}_{\mathbf{Z},k}(\hat{\mathbf{z}}_k^{(i)}). \tag{3.127}$$

If the measurement distribution is m-variate Gaussian, its PDF is

$$\tilde{f}_{\mathbf{z},k}\left(\hat{\mathbf{z}}_k^{(i)}\right) = \frac{1}{(2\pi)^{m/2}|\mathbf{R}_k|^{1/2}} \exp\left[-\tfrac{1}{2}\left(\hat{\mathbf{z}}_k^{(i)} - \tilde{\mathbf{z}}_k\right)^{\mathrm{T}} \mathbf{R}_k^{-1}\left(\hat{\mathbf{z}}_k^{(i)} - \tilde{\mathbf{z}}_k\right)\right], \quad (3.128)$$

where $\tilde{\mathbf{z}}_k$ is the measured mean and $\mathbf{R}_k$ is the measurement noise covariance.

The updated probabilities of each particle are then obtained by renormalizing the likelihoods, giving

$$p_{\mathrm{X},k}^{+(i)} = \frac{\Lambda_{\mathrm{X},k}^{(i)}}{\sum_{j=1}^{n} \Lambda_{\mathrm{X},k}^{(j)}}. \quad (3.129)$$

The final phase of the particle filter is resampling. Without resampling, the probabilities of many of the particles would shrink over successive epochs until they were too small to justify the processing capacity required to maintain them. At the same time, the number of particles available to represent the core of the state estimate distribution would shrink, reducing the accuracy. A particle filter is most efficient when the particles have similar probabilities. Therefore, in the resampling phase, low-probability particles are deleted and high-probability particles are duplicated. The independent application of system noise to each particle ensures that duplicate particles become different at the next system propagation phase of the particle filter. Section D.3.1 of Appendix D on the CD describes resampling in more detail. The most commonly used resampling algorithms allocate equal probability (i.e., 1/N) to the resampled particles.

Resampling makes the particle filter more receptive to new measurement information, but it also adds noise to the state estimation process, degrading the accuracy. Therefore, it is not desirable to perform it on every filter cycle. Resampling can be triggered after a fixed number of cycles or based on the effective sample size, $N_{\textit{eff}}$, given by

$$N_{\textit{eff}} = \left[\sum_{i=1}^{N}\left(p_{\mathrm{X},k}^{+(i)}\right)^2\right]^{-1}, \quad (3.130)$$

dropping below a certain threshold, such as N/2 or 2N/3.

To initialize a particle filter, it is necessary to generate a set of particles by sampling randomly from the initial distribution of the states. For a uniform or Gaussian distribution, this is straightforward (see Sections J.4.1 and J.4.3 of Appendix J on the CD). For more complex distributions, importance sampling must be used as described in Section D.3.2 of Appendix D on the CD.

Hybrid filters that combine elements of the particle filter and the Kalman filter may also be implemented. These are discussed in Section D.3.3 of Appendix D on the CD.

Problems and exercises for this chapter are on the accompanying CD.

References

[1] Jazwinski, A. H., *Stochastic Processes and Filtering Theory*, San Diego, CA: Academic Press, 1970.

[2] Gelb, A., (ed.), *Applied Optimal Estimation*, Cambridge, MA: MIT Press, 1974.

[3] Maybeck, P. S., *Stochastic Models, Estimation and Control*, Vols. 1–3, New York: Academic Press, 1979–1983.

[4] Brown, R. G., and P. Y. C. Hwang, *Introduction to Random Signals and Applied Kalman Filtering*, 3rd ed., New York: Wiley, 1997.

[5] Grewal, M. S., and A. P. Andrews, *Kalman Filtering: Theory and Practice*, 2nd ed., New York: Wiley, 2000.

[6] Simon, D., *Optimal State Estimation*, New York: Wiley, 2006.

[7] Kalman, R. E., "A New Approach to Linear Filtering and Prediction Problems," *ASME Transactions, Series D: Journal of Basic Engineering*, Vol. 82, 1960, pp. 35–45.

[8] Groves, P. D., "Principles of Integrated Navigation," Course Notes, QinetiQ Ltd., 2002.

[9] Kailath, T., *Linear Systems*, Englewood Cliffs, NJ: Prentice-Hall, 1980.

[10] Golub, G. H., and C. F. Van Loan, *Matrix Computations*, Baltimore, MD: Johns Hopkins University Press, 1983.

[11] Farrell, J. A., *Aided Navigation: GPS with High Rate Sensors*, New York: McGraw-Hill, 2008.

[12] Rogers, R. M., *Applied Mathematics in Integrated Navigation Systems*, Reston, VA: AIAA, 2000.

[13] Bierman, G. L., *Factorization Methods for Discrete Sequential Estimation*, Mathematics in Science and Engineering, Vol. 128, New York: Academic Press, 1977.

[14] Stimac, L. W., and T. A. Kennedy, "Sensor Alignment Kalman Filters for Inertial Stabilization Systems," *Proc. IEEE PLANS*, Monterey, CA, March 1992, pp. 321–334.

[15] Xing, Z., and D. Gebre-Egziabher, "Comparing Non-Linear Filters for Aided Inertial Navigators," *Proc. ION ITM*, Anaheim, CA, January 2009, pp. 1048–1053.

[16] Draganov, A., L. Haas, and M. Harlacher, "The IMRE Kalman Filter—A New Kalman Filter Extension for Nonlinear Applications," *Proc. IEEE/ION PLANS*, Myrtle Beach, SC, April 2012, pp. 428-440.

[17] Julier, S. J., and J. K. Uhlmann, "A New Extension of the Kalman Filter to Nonlinear Systems," *Proc. AeroSense: The 11th Int. Symp. on Aerospace/Defence Sensing, Simulation and Controls*, SPIE, 1997.

[18] Schmidt, S. F., "Application of State Space Methods to Navigation Problems," in *Advanced in Control Systems*, Vol. 3, C. T. Leondes, (ed.), New York: Academic Press, 1966.

[19] Petovello, M. G., et al., "Consideration of Time-Correlated Errors in a Kalman Filter Applicable to GNSS," *Journal of Geodesy*, Vol. 83, No. 1, 2009, pp. 51-56 and Vol. 85, No. 6, 2011, pp. 367–368.

[20] Mehra, R. K., "Approaches to Adaptive Filtering," *IEEE Trans. on Automatic Control*, Vol. AC-17, 1972, pp. 693–698.

[21] Mohammed, A. H., and K. P. Schwarz, "Adaptive Kalman Filtering for INS/GPS," *Journal of Geodesy*, Vol. 73, 1999, pp. 193–203.

[22] Magill, D. T., "Optimal Adaptive Estimation of Sampled Stochastic Processes," *IEEE Trans. on Automatic Control*, Vol. AC-10, 1965, pp. 434–439.

[23] Bar-Shalom, Y., and T. E. Fortmann, *Tracking and Data Association*, New York: Academic Press, 1988.

[24] Dezert, J., and Y. Bar-Shalom, "Joint Probabilistic Data Association for Autonomous Navigation," *IEEE Trans. on Aerospace and Electronic Systems*, Vol. 29, 1993, pp. 1275–1285.

[25] Reid, D. B., "An Algorithm for Tracking Multiple Targets," *IEEE Trans. on Automatic Control*, Vol. AC-24, 1979, pp. 843–854.

[26] Runnalls, A. R., P. D. Groves, and R. J. Handley, "Terrain-Referenced Navigation Using the IGMAP Data Fusion Algorithm," *Proc. ION 61st AM*, Boston, MA, June 2005, pp. 976–987.

[27] Fraser, D. C., and J. E. Potter, "The Optimum Linear Smoother as a Combination of Two Optimum Linear Filters," *IEEE Trans. on Automatic Control*, Vol. 7, 1969, pp. 387–390.

[28] Rauch, H. E., F. Tung, and C. T. Striebel, "Maximum Likelihood Estimates of Linear Dynamic Systems," *AIAA Journal*, Vol. 3, 1965, pp. 1445–1450.

[29] Gordon, N. J., D. J. Salmond, and A. F. M. Smith, "A Novel Approach to Nonlinear/Non-Gaussian Bayesian State Estimation," *Proc. IEE Radar Signal Process.*, Vol. 140, 1993, pp. 107–113.

[30] Gustafsson, F., et al., "Particle Filters for Positioning, Navigation and Tracking," *IEEE Trans. on Signal Processing*, Vol. 50, 2002, pp. 425–437.

[31] Ristic, B., S. Arulampalam, and N. Gordon, *Beyond the Kalman Filter: Particle Filters for Tracking Applications*, Norwood, MA: Artech House, 2004.

[32] Doucet, A., N. de Freitas, and N. Gordon, (eds.), *Sequential Monte Carlo Methods in Practice*, New York: Springer, 2001.

[33] Doucet, A., and A. M. Johansen, "A Tutorial on Particle Filtering and Smoothing: Fifteen Years Later," in *Oxford Handbook of Nonlinear Filtering*, C. Crisan and B. Rozovsky, (eds.), Oxford, U.K.: OUP, 2011, pp. 656-704.

CHAPTER 4
Inertial Sensors

Inertial sensors comprise accelerometers and gyroscopes, commonly abbreviated to gyros. An *accelerometer* measures specific force and a *gyroscope* measures angular rate, both without an external reference. Devices that measure the velocity, acceleration, or angular rate of a body with respect to features in the environment are not inertial sensors.

Most types of accelerometer measure specific force along a single sensitive axis. Similarly, most types of gyro measure angular rate about a single axis. An *inertial measurement unit* (IMU) combines multiple accelerometers and gyros, usually three of each, to produce a three-dimensional measurement of specific force and angular rate. An IMU is the sensor for an inertial navigation system, described in Chapter 5, which produces an independent three-dimensional navigation solution. New designs of INS all employ a strapdown architecture, whereby the inertial sensors are fixed with respect to the navigation system casing. Lower-grade IMUs are also used in attitude and heading reference systems (AHRSs), described in Section 6.1.8; for pedestrian dead reckoning using step detection, discussed in Section 6.4; and can be used for context detection, discussed in Section 16.1.10. Gyrocompasses, described in Section 6.1.2, also use gyroscope technology. Finally, inertial sensors have many uses outside navigation as reviewed in [1].

This chapter describes the basic principles of accelerometer, gyro, and IMU technology, compares the different types of sensor, and reviews the error sources. Inertial sensor technology is reviewed in [1, 2].

Most accelerometers either are pendulous or use vibrating beams. Both technologies share the same basic principle and are described in Section 4.1, while Section E.1 of Appendix E on the CD introduces time-domain switching (TDS) accelerometers and inertial sensing using cold-atom interfereometry. There are three main types of gyro technology: optical, vibratory, and spinning mass, each of which is based on a different physical principle. These are described in Section 4.2 and in Section E.2 of Appendix E on the CD, while Section E.3 of Appendix E discusses angular rate measurement using accelerometers. The size, mass, performance, and cost of inertial sensors vary by several orders of magnitude, both within and between the different technologies. In general, higher-performance sensors are larger and more massive as well as more costly.

Current inertial sensor development is focused mainly on microelectromechanical systems (MEMS) technology. This enables small and light quartz and silicon sensors to be mass-produced at a low cost using etching techniques with several sensors on a single wafer [3]. MEMS sensors also exhibit much greater shock tolerance than conventional mechanical and optical designs, enabling them to be used in gun-launched guided munitions [4]. However, most MEMS sensors offer relatively poor performance. Micro-optical-electromechanical systems (MOEMS) technology replaces the capacitive pickoff of many MEMS sensors with an optical readout, offering potential improvements in performance [5], but was still at the research stage at the time of this writing.

The IMU regulates the power supplies to the inertial sensors, converts their outputs to engineering units, and transmits them on a data bus. It also calibrates out many of the raw sensor errors. The IMU functions are discussed in Section 4.3, while Section 4.4 discusses the error behavior of the calibrated accelerometers and gyros.

There is no universally agreed definition of high-, medium-, and low-grade IMUs and inertial sensors. One author's medium grade can be another's high or low grade. IMUs, INSs, and inertial sensors may be grouped into five broad performance categories: marine, aviation, intermediate, tactical, and consumer.

The highest grades of inertial sensors discussed here are used in military ships, submarines, some intercontinental ballistic missiles, and some spacecraft. A *marine-grade* INS can cost in excess of a $1 million (Euros €800,000) and offers a navigation-solution drift of less than 1.8 km in a day. Early systems offering that level of performance were very large, with a diameter of about a meter; current systems are much smaller. Indexing (Section 5.8) is sometimes used to achieve the required performance with lower-grade sensors.

Aviation-grade, or navigation-grade, INSs used in U.S. military aircraft are required to meet the Standard Navigation Unit (SNU) 84 standard, specifying a maximum horizontal position drift of ~1.5 km in the first hour of operation. These INSs are also used in commercial airliners and in military aircraft worldwide. They cost around $100,000 (€80,000) and have a standard size of 178×178×249 mm. An *intermediate-grade IMU*, about an order of magnitude poorer in performance terms, is used in small aircraft and helicopters and costs $20,000–$50,000 (€16,000–€40,000).

A *tactical-grade IMU* can only be used to provide a useful stand-alone inertial navigation solution for a few minutes. However, an accurate long-term navigation solution can be obtained by integrating it with a positioning system, such as GPS. These systems typically cost between $2,000 and $30,000 (€1,600 and €25,000) and are typically used in guided weapons and unmanned air vehicles (UAVs). Most are less than a liter in volume. Tactical grade covers a wide span of sensor performance, particularly for gyros.

The lowest grade of inertial sensors is known as *consumer grade* or *automotive grade*. They are often sold as individual accelerometers and gyros, rather than as IMUs and, without calibration, are not accurate enough for inertial navigation, even when integrated with other navigation systems, but can be used in an AHRS, for PDR using step detection, and for context detection. They are typically used in pedometers, antilock braking systems (ABSs), active suspension, and airbags. Accelerometers cost around a dollar or a euro while gyro prices start at about $10 (€8) [6]. Individual sensors can be as small as 5×5×1 mm.

The extent of calibration and other processing applied within the IMU can affect performance dramatically, particularly for MEMS sensors [7]. Sometimes, the same MEMS inertial sensors are sold at consumer grade without calibration and tactical grade with calibration. The term "low cost" is commonly applied to both consumer grade and tactical grade, spanning a very wide price range.

The range of inertial sensors from consumer to marine grade spans six orders of magnitude of gyro performance, but only three orders of magnitude of accelerometer performance. This is partly because gyro performance has more impact on navigation solution drift over periods in excess of about 40 minutes as explained in Section 5.7.2.

4.1 Accelerometers

Figure 4.1 shows a simple accelerometer. A proof mass is free to move with respect to the accelerometer case along the accelerometer's sensitive axis, restrained by springs, which are sometimes referred to as the *suspension*. A pickoff measures the position of the mass with respect to the case. When an accelerating force along the sensitive axis is applied to the case, the proof mass will initially continue at its previous velocity, so the case will move with respect to the mass, compressing one spring and stretching the other. Stretching and compressing the springs alters the forces that they transmit to the proof mass from the case. Consequently, the case will move with respect to the mass until the acceleration of the mass due to the asymmetric forces exerted by the springs matches the acceleration of the case due to the externally applied force. The resultant position of the mass with respect to the case is proportional to the acceleration applied to the case. By measuring this with a pickoff, an acceleration measurement is obtained. The exception to this is acceleration due to the gravitational force. Gravitation acts on the proof mass directly, not via the springs and applies the same acceleration to all components of the accelerometer, so there is no relative motion of the mass with respect to the case. Therefore, all accelerometers sense specific force, the nongravitational acceleration, not the total acceleration (see Section 2.4.7).

The object frame for accelerometer measurements is the accelerometer case, while the reference frame is inertial space, and measurements are resolved along the sensitive axes of the accelerometers. Thus, an IMU containing an accelerometer triad measures the specific force of the IMU body with respect to inertial space in body axes, the vector $\mathbf{f}_{ib}^{b}$.

The accelerometer shown in Figure 4.1 is incomplete. The proof mass needs to be supported in the axes perpendicular to the sensitive axis, and damping is needed to limit oscillation of the proof mass. However, all accelerometer designs are based on the basic principle shown. Practical accelerometers used in strapdown

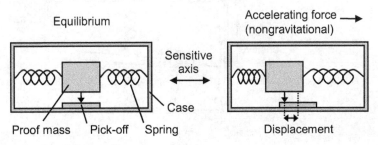

Figure 4.1 A simple accelerometer.

navigation systems currently follow either a pendulous or a vibrating-beam design, both of which are discussed later. Pendulous designs have been around for decades, while vibrating-beam accelerometers originated in the 1980s. Both types of accelerometer may be built using either conventional mechanical construction or MEMS technology. MEMS accelerometers of either design may be built with sensitive axes either in the plane of the device or perpendicular to that plane, enabling a three-axis accelerometer triad and associated electronics to be etched onto a single silicon chip [3, 8]. Although most MEMS accelerometers offer consumer- or tactical-grade performance, intermediate-grade designs have been developed [9].

Section E.1 of Appendix E on the CD describes a new type of MEMS accelerometer, the time-domain switching accelerometer. Section E.1.2 then describes cold-atom interferometry, which can measure specific force to a much higher precision than conventional technologies. Another type of accelerometer, the pendulous integrating gyro accelerometer (PIGA), can also exhibit very high precision but is not suited to strapdown use. In addition, research has been conducted into a number of novel accelerometer designs making use of optical and MEMS techniques [1].

The operating range of an accelerometer is typically quoted in terms of the acceleration due to gravity, abbreviated to 'g,' where 1 g = 9.80665 m s^{-2} [10]. Note that the actual acceleration due to gravity varies with location (see Section 2.4.7). Humans can only tolerate sustained accelerations of a few g, up to about 10g with specialist training and equipment. However, parts of the body commonly undergo very brief accelerations of up to 10g during normal walking (see Section 6.4). Thus, for navigation, accelerometers must have an operational range of at least ±10g. A greater range is needed in high-vibration applications and for some guided weapons and unmanned aircraft. Mechanical accelerometers typically have a range of ±100g [1]. However, many MEMS accelerometers have a much smaller range. Those designed for use in inclinometers (or tilt sensors) only have a range of ±2g. MEMS accelerometers are typically available with a variety of operating ranges; these are often directly proportional to the quantization error (see Section 4.4.3).

4.1.1 Pendulous Accelerometers

Figure 4.2 shows a mechanical open-loop pendulous accelerometer. The proof mass is attached to the case via a pendulous arm and hinge, forming a pendulum. This leaves the proof mass free to move along the sensitive axis while supporting it in the

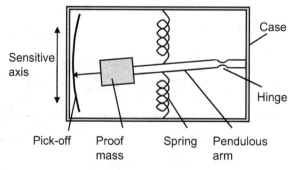

Figure 4.2 Mechanical open-loop pendulous accelerometer.

other two axes. A pair of springs or a single spring is used to transmit force from the case to the pendulum along the sensitive axis while the hinge provides damping. Further damping may be obtained by filling the case with oil.

Although the open-loop design produces a practical accelerometer, its performance is severely limited by three factors. First, the resolution of the pickoff, typically a variable resistor, is relatively poor. Second, the force exerted by a spring is only approximately a linear function of its compression or extension, displaying hysteresis as well as nonlinearity. Finally, the sensitive axis is perpendicular to the pendulous arm so, as the pendulum moves, the sensitive axis moves with respect to the case. This results in both nonlinearity of response along the desired sensitive axis and sensitivity to orthogonal specific force.

To resolve these problems, precision accelerometers use a closed-loop, or *force-feedback*, configuration [1, 2]. In a force-feedback accelerometer, a torquer is used to maintain the pendulous arm at a constant position with respect to the case, regardless of the specific force to which the accelerometer is subject. The pickoff detects departures from the equilibrium position, and the torquer is adjusted to return the pendulum to that position. In a force-feedback accelerometer, the force exerted by the torquer, rather than the pickoff signal, is proportional to the applied specific force. Figure 4.3 depicts a mechanical force-feedback accelerometer. The torquer comprises an electromagnet mounted on the pendulum and a pair of permanent magnets of opposite polarity mounted on either side of the case. The diagram shows a capacitive pickoff, comprising four capacitor plates, mounted such that two capacitors are formed between the case and pendulum. As the pendulum moves, the capacitance of one pair of plates increases while that of the other decreases. Alternatively, an inductive or optical pickoff may be used.

The closed-loop configuration ensures that the sensitive axis remains aligned with the accelerometer case, while the torquer offers much greater dynamic range and linearity than the open-loop accelerometer's springs and pickoff. However, a drawback is that the pendulum is unrestrained when the accelerometer is unpowered, risking damage in transit, particularly where the case is gas-filled rather than oil-filled. The design of the hinge, pendulous arm, proof mass, torquer, pickoff system, and control electronics all affect performance. By varying the component quality, a range of different grades of performance can be offered at different prices.

Both open-loop and closed-loop pendulous MEMS accelerometers are available, with the latter using an electrostatic, rather than magnetic, torquer. The pickoff may

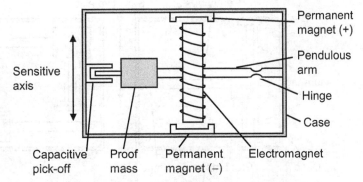

Figure 4.3 Mechanical force-feedback pendulous accelerometer. (*After:* [1].)

be capacitive, as described above, or a resistive element mounted on the hinge, whose resistance varies as it is stretched and compressed.

4.1.2 Vibrating-Beam Accelerometers

The *vibrating-beam accelerometer* (VBA) or resonant accelerometer retains the proof mass and pendulous arm from the pendulous accelerometer. However, the proof mass is supported along the sensitive axis by a vibrating beam, largely constraining its motion with respect to the case. When a force is applied to the accelerometer case along the sensitive axis, the beam pushes or pulls the proof mass, causing the beam to be compressed in the former case and stretched in the latter. The beam is driven to vibrate at its resonant frequency by the accelerometer electronics. However, compressing the beam decreases the resonant frequency, whereas tensing it increases the frequency. Therefore, by measuring the resonant frequency, the specific force along the sensitive axis can be determined.

Performance is improved by using a pair of vibrating beams, arranged such that one is compressed while the other is stretched. They may support either a single proof mass or two separates masses; both arrangements are shown in Figure 4.4. Two-element tuning-fork resonators are shown, as these are more balanced than single-element resonators. Larger-scale VBAs all use quartz elements as these provide a sharp resonance peak. MEMS VBAs have been fabricated out of both quartz and silicon.

The VBA is an inherently open-loop device. However, the proof mass is essentially fixed; there is no variation in the sensitive axis with respect to the casing.

4.2 Gyroscopes

This section describes the principles of optical and vibratory gyroscopes. There are two main types of optical gyro. The ring laser gyro (RLG) originated in the 1960s [11] as a high-performance technology, while the interferometric fiber-optic gyro (IFOG) was developed in the 1970s [12] as a lower-cost solution. Now, the performance ranges overlap with IFOGs available at tactical, intermediate, and aviation

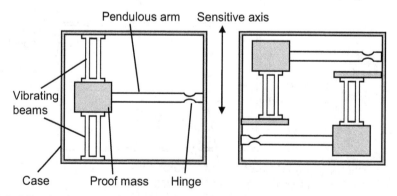

Figure 4.4 Vibrating beam accelerometers.

4.2 Gyroscopes

grades. A resonant fiber-optic gyro (RFOG) and a micro-optic gyro (MOG) have also been developed [2].

Practical vibratory gyros were developed in the 1980s. All MEMS gyros operate on the vibratory principle, but larger vibratory gyros are also available and the technique spans the full performance range.

The third main angular-rate-sensing technology is spinning-mass gyros. These use conservation of angular momentum to sense rotation. A motor spins a mass about one axis. If a torque is then applied about a perpendicular axis, the spinning mass rotates about the axis perpendicular to both the spinning and the applied torque. Details are presented in Section E.2 of Appendix E on the CD. Spinning-mass gyros have largely been superseded by optical and vibratory gyros.

Cold-atom interferometry, described in Section E.1.2 of Appendix E on the CD, offers the potential of much higher precision than current gyroscope technology. A number of other gyroscope technologies, including nuclear magnetic resonance (NMR), flueric sensors, and angular accelerometers, have also been researched [1]. NMR gyro technology is now being developed on a chip scale [13]. Angular rate can also be sensed using accelerometers as described in Section E.3 of Appendix E on the CD.

The object frame for gyro measurements is the gyro case, while the reference frame is inertial space, and measurements are resolved along the sensitive axes of the gyros. Thus, an IMU containing a gyro triad measures the angular rate of the IMU body with respect to inertial space in body axes, the vector $\boldsymbol{\omega}_{ib}^{b}$.

Manned vehicles typically rotate at up to 3 rad s^{-1} [14]. However, a gun-launched guided shell can rotate at up to 120 rad s^{-1} [4]. Thus, the gyro operating-range requirement varies with the application, while different technologies offer differing performance.

4.2.1 Optical Gyroscopes

Optical gyroscopes work on the principle that, in a given medium, light travels at a constant speed in an inertial frame. If light is sent in both directions around a non-rotating closed-loop waveguide made of mirrors or optical fiber, the path length is the same for both beams. However, if the waveguide is rotated about an axis perpendicular to its plane, then, from the perspective of an inertial frame, the reflecting surfaces are moving further apart for light traveling in the same direction as the

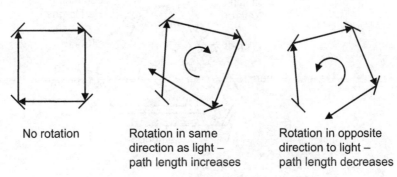

No rotation   Rotation in same direction as light – path length increases   Rotation in opposite direction to light – path length decreases

Figure 4.5 Effect of closed-loop waveguide rotation on path length.

rotation and closer together for light traveling in the opposite direction. Thus, rotating the waveguide in the same direction as the light path increases the path length and rotating it in the opposite direction decreases the path length. This is known as the *Sagnac effect*. Figure 4.5 illustrates it. By measuring the changes in path length, the angular rate of the waveguide with respect to inertial space can be determined. Note that, from the perspective of the rotating frame, the path length remains unchanged, but the speed of light changes. Optical gyros can typically measure angular rates in excess of ±20 rad s^{-1} [1].

4.2.1.1 Ring Laser Gyro

Figure 4.6 shows a ring laser gyro. A closed-loop tube with at least three arms is filled with a helium-neon gas mixture; this is known as a laser cavity. A high-reflectivity mirror is place at each corner. Finally, a cathode and anode are used to apply a high potential difference across the gas, generating an electric field.

A gas atom can absorb energy from the electric field, producing an excited state of the atom. Excited states are unstable, so the atom will eventually return to its normal state, known as the ground state, by emitting the excess energy as a photon. There is some variation in the potential energies of the ground and excited states, so the wavelengths of the spontaneously emitted photons are distributed over a resonance curve. The excited-state atoms can also be stimulated to emit photons by other photons in the laser cavity that are within the resonance curve. A photon produced by stimulated emission has the same wavelength, phase, and trajectory as the stimulating photon; this is known as coherence.

Photons of the same wavelength within the laser cavity interfere with each other. When there are an integer number of wavelengths within the length of the laser cavity, the interference is constructive. This is known as a resonant mode. Otherwise, the interference is destructive. For a practical laser, the resonant modes of the cavity must have a narrower bandwidth than the resonance of the atom transition, and there should be more than one cavity mode within the atomic resonance curve. The laser will then adopt a lasing mode whereby the photons adopt the wavelength of the cavity mode closest to the atom resonance peak.

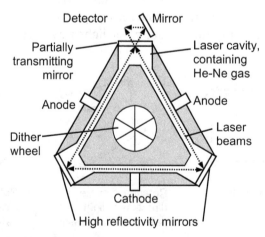

Figure 4.6 A typical ring laser gyro.

4.2 Gyroscopes

A ring laser has two lasing modes, one in each direction. If the laser cavity does not rotate, both modes have the same wavelength. However, if the laser cavity is rotated about an axis perpendicular to its plane, the cavity length is increased for the lasing mode in the direction of rotation and decreased for the mode in the opposite direction. Consequently, the lasing mode in the direction of rotation exhibits an increase in wavelength and decrease in frequency, while the converse happens for the other mode. In a ring laser gyro, one of the cavity mirrors is partially transmitting, enabling photons from both lasing modes to be focused on a detector, where they interfere. The beat frequency of the two modes is given by [1]

$$\Delta f \approx \frac{4A\omega_\perp}{\lambda_0}, \qquad (4.1)$$

where λ_0 is the wavelength of the nonrotating laser, A is the area enclosed by the RLG's light paths in the absence of rotation, and $\omega_\perp$ is the angular rate about an axis perpendicular to the plane of the laser cavity.

Because of scattering within the laser cavity, there is coupling between the clockwise and counterclockwise laser modes. At low angular rates, this prevents the wavelengths of the two laser modes from diverging, a process known as lock-in. Thus, a basic ring laser gyro is unable to detect low angular rates. To mitigate this problem, most RLGs implement a dithering process, whereby the laser cavity is subject to low-amplitude, high-frequency angular vibrations about the sensitive axis with respect to the gyro case. Alternatively, the Kerr effect may be used to vary the refractive index within part of the cavity. This constantly changes the lock-in region in terms of the gyro case angular rate, which is the quantity to be measured [1, 2].

Most RLG triads contain three separate instruments. However, a few designs comprise a single laser cavity with lasing modes in three planes.

4.2.1.2 Interferometric Fiber-Optic Gyro

Figure 4.7 shows the main elements of an *interferometric fiber-optic gyro*, often abbreviated to just fiber-optic gyro (FOG) [1, 2, 15]. A broadband light source is divided using beam splitters into two equal portions that are then sent through a fiber-optic coil in opposite directions. The beam splitters combine the two beams at the detector, where the interference between them is observed. Two beam splitters, rather than one, are used so that both light paths include an equal number of transmissions and reflections. When the fiber-optic coil is rotated about an axis

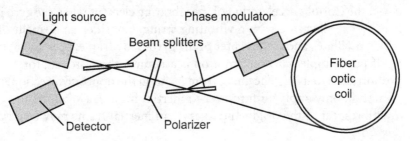

Figure 4.7 Interferometric fiber-optic gyro.

perpendicular to its plane, a phase change, ϕ_c, is introduced between the two light paths, given by

$$\phi_c \approx \frac{8\pi NA\omega_\perp}{\lambda_0 c_c}, \quad (4.2)$$

where λ_0 is the wavelength of the light source, which does not change; A is the area enclosed by the coil; N is the number of turns in the coil; c_c is the speed of light within the coil; and $\omega_\perp$ is the angular rate as before.

A phase modulator is placed on the entrance to the coil for one light path and the exit for other. This introduces a time-dependent phase shift, such that light arriving at the detector simultaneously via the two paths is subject to different phase shifts. The phase-shift difference between the two paths, $\phi_p(t)$, is also time variant. By synchronizing the duty cycle of the detector with the phase modulator, samples can be taken at a particular value of ϕ_p. The intensity of the signal received at the detector is then

$$I_d = I_0\big[1 + \cos\big(\phi_c(\omega_\perp) + \phi_p(t)\big)\big], \quad (4.3)$$

where I_0 is a constant. The scale factor of the intensity as a function of the rotation induced phase shift is

$$\frac{\partial I_d}{\partial \phi_c} = -I_0 \sin\big(\phi_c(\omega_\perp) + \phi_p(t)\big). \quad (4.4)$$

This is highly nonlinear and, without the phase modulator, gives zero scale factor for small angular rates. The sensitivity of the IFOG is optimized by selecting ϕ_p at the sampling time to maximize the scale factor. Best performance is obtained with closed-loop operation, whereby ϕ_p at the sampling time is constantly varied to keep the scale factor at its maximum value. The gyro sensitivity is also optimized by maximizing the coil diameter and number of turns. IFOGs are more reliable than both RLGs and spinning-mass gyros.

4.2.2 Vibratory Gyroscopes

A *vibratory gyroscope* comprises an element that is driven to undergo simple harmonic motion. The vibrating element may be a string, beam, pair of beams, tuning fork, ring, cylinder, or hemisphere. All operate on the same principle, which is to detect the Coriolis acceleration of the vibrating element when the gyro is rotated. This is easiest to illustrate with a vibrating string. Consider an element of the string, a, which oscillates about the center of the gyro body frame, b, at an angular frequency, ω_v. If pure simple harmonic motion is assumed, the restoring force on a is directly proportional to its displacement from b and in the opposite direction. Newton's laws of motion only apply with respect to inertial frames. Therefore, the acceleration of a with respect to the origin and axes of an inertial frame may be described as

$$\mathbf{a}_{ia}^b = -\omega_v^2 \mathbf{r}_{ba}^b, \quad (4.5)$$

where the resolving axes of the gyro body frame have been used for convenience.

In practice, the restoring force will depend on the direction of the displacement, while a will also be acted upon by the gravitational force, which will be counteracted by the restoring force at equilibrium. Furthermore, the vibration will be damped in those directions where it is not driven and heavily damped in directions where motion is constrained. Thus, the acceleration of a becomes

$$\mathbf{a}_{ia}^b = -\mathbf{K}\mathbf{r}_{ba}^b - \mathbf{L}\mathbf{v}_{ba}^b + \boldsymbol{\gamma}_{ia}^b, \qquad (4.6)$$

where $\mathbf{K}$ and $\mathbf{L}$ are nominally symmetric matrices describing the coefficients of the restoring and damping forces, respectively.

The gyro body can rotate with respect to inertial space. Therefore, applying (2.86) and (2.91) and substituting in (4.6), the equation of motion of the string element with respect to the gyro body is

$$\mathbf{a}_{ba}^b = -\boldsymbol{\Omega}_{ib}^b\boldsymbol{\Omega}_{ib}^b\mathbf{r}_{ba}^b - 2\boldsymbol{\Omega}_{ib}^b\mathbf{v}_{ba}^b - \dot{\boldsymbol{\Omega}}_{ib}^b\mathbf{r}_{ba}^b - \mathbf{K}\mathbf{r}_{ba}^b - \mathbf{L}\mathbf{v}_{ba}^b - \mathbf{f}_{ib}^b, \qquad (4.7)$$

where the first term is the centrifugal acceleration, the second term is the Coriolis acceleration, and the third term is the Euler acceleration. These are discussed in Section 2.3.5. Note that the motion of the string is also sensitive to the specific force of the gyro body (the gravitational acceleration of a and b are the same). If the vibration rate is set sufficiently high, (4.7) may be approximated to

$$\mathbf{a}_{ba}^b \approx -2\boldsymbol{\Omega}_{ib}^b\mathbf{v}_{ba}^b - \mathbf{K}\mathbf{r}_{ba}^b - \mathbf{L}\mathbf{v}_{ba}^b. \qquad (4.8)$$

The Coriolis acceleration instigates simple harmonic motion along the axis perpendicular to both the driven vibration and the projection of the angular rate vector, $\boldsymbol{\omega}_{ib}^b$, in the plane perpendicular to the driven vibration. The amplitude of this motion is proportional to the angular rate. Rotation about the vibration axis does not produce a Coriolis acceleration. In practice, the motion of the vibrating element is constrained along one of the axes perpendicular to the driven vibration, so only rotation about this input axis leads to significant oscillation in the output axis, mutually perpendicular to the input and driven axes. Figure 4.8 illustrates this.

How the output vibration is detected depends on the gyro architecture [1, 2]. For string and single-beam gyros, the vibration of the element itself must be detected. In double-beam and tuning-fork gyros, the two elements are driven in antiphase,

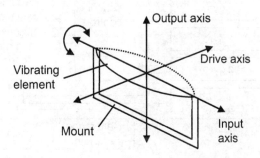

Figure 4.8 Axes of a vibrating gyro.

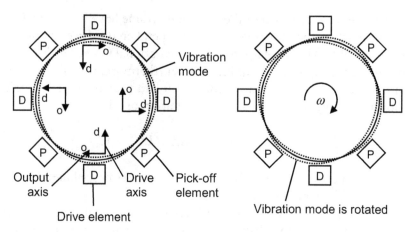

Figure 4.9 Vibration modes of ring, cylinder, and hemispherical vibratory gyros.

so their Coriolis-induced vibration is also in antiphase. This induces an oscillating torsion in the stem, which may be detected directly or via a pair of pickoff tines. Ring, cylinder, and hemispherical resonators have four drive units placed at right angles and four detectors at intermediate points. When the gyro is not rotating, the detectors are at the nodes of the vibration mode, so no signal is detected. When angular rate is applied, the vibration mode is rotated about the input axis. Figure 4.9 illustrates this.

Most vibratory gyros are low-cost, low-performance devices, often using MEMS technology [3] and with quartz giving better performance than silicon. The exception is the hemispherical resonator gyro (HRG), which can offer aviation grade performance. The HRG is light and compact and operates in a vacuum, so it has become popular for space applications [8].

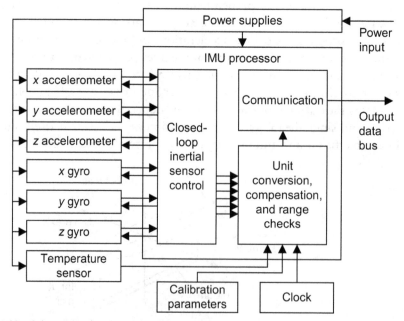

Figure 4.10 Schematic of an inertial measurement unit.

Operating ranges for MEMS gyros can be anything from ±3 rad s^{-1} to ±120 rad s^{-1} [1, 4], depending on the model.

4.3 Inertial Measurement Units

Figure 4.10 shows the main elements of a typical inertial measurement unit: accelerometers and gyroscopes, the IMU processor, a calibration-parameters store, a temperature sensor, and associated power supplies.

The accelerometers and gyroscopes without the other elements are sometimes known as an inertial sensor assembly (ISA) [14]. Most IMUs have three accelerometers and three single-degree-of-freedom gyroscopes, mounted with orthogonal sensitive axes. However, some IMUs incorporate additional inertial sensors in a skewed configuration to protect against single sensor failure; this is discussed in Section 17.4. Additional MEMS sensors can also be used to aid bias calibration (see Section 4.4.1). IMUs with fewer than six sensors are known as partial IMUs and are sometimes used for land navigation as described in Section 5.9. All-accelerometer IMUs are discussed in Section E.3 of Appendix E on the CD.

The IMU processor performs unit conversion on the inertial sensor outputs, provides compensation for the known errors of the inertial sensors, and performs range checks to detect sensor failure. It may also incorporate closed-loop force feedback or rebalance control for the accelerometers and/or gyros. Unit conversion transforms the inertial sensor outputs from potential difference, current, or pulses into units of specific force and angular rate. Many IMUs integrate the specific force and angular rate over the sampling interval, τ_i, producing

$$\mathbf{v}_{ib}^b(t) = \int_{t-\tau_i}^{t} \mathbf{f}_{ib}^b(t')\,dt', \qquad \boldsymbol{\alpha}_{ib}^b(t) = \int_{t-\tau_i}^{t} \boldsymbol{\omega}_{ib}^b(t')\,dt'. \tag{4.9}$$

These are often referred to as "delta-v"s and "delta-θ"s. However, this can be misleading: the delta-θs, $\boldsymbol{\alpha}_{ib}^b$, are attitude increments, but the delta-vs, $\mathbf{v}_{ib}^b$, are not velocity increments. The IMU outputs specific forces and angular rates, or their integrals, in the form of integers, which can be converted to SI units using scaling factors in the IMU's documentation. Output rates typically vary between 100 and 1,000 Hz.

Some IMUs sample the sensors at a higher rate than they output data. Samples may be simply summed over the data output interval or they may be combined as described in Section 5.5.4, minimizing the coning and sculling errors.

Inertial sensors exhibit constant errors that can be calibrated in the laboratory and stored in memory, enabling the IMU processor to correct the sensor outputs. Calibration parameters generally comprise accelerometer and gyro biases, scale factor and cross-coupling errors, and gyro g-dependent biases (see Section 4.4). These errors vary with temperature, so the calibration is performed at a range of temperatures and the IMU is equipped with a temperature sensor. However, the temperature within each individual sensor does not necessarily match the ambient temperature of the IMU, so some high-performance IMUs implement temperature control instead [1]. The cost of calibration may be minimized by applying the same set of calibration coefficients to a whole production batch of sensors. However, best performance is obtained by calibrating each sensor or IMU individually, noting that IMU-level calibration is needed to fully capture the cross-coupling errors. A Kalman filter may be used to obtain the calibration coefficients from the measurement data [16, 17]. This process is known as laboratory calibration, to distinguish it from the in-run calibration discussed later.

A further source of accelerometer errors that the IMU processor can compensate is the *size effect*. To compute a navigation solution for a single point in space, the IMU's angular rate and specific force measurements must also apply to a single reference point, sometimes known as the center of percussion. However, in practice, the size of the inertial sensors demands that they are placed a few centimeters apart (generally less for MEMS sensors). Figure 4.11 illustrates this. For the gyros, this does not present a problem. However, rotation of an accelerometer about the reference point causes it to sense a centrifugal force that is not observed at the reference point, while angular acceleration causes it to sense a Euler force. Both of these virtual forces are described in Section 2.3.5. From (2.91), the pseudo-accelerations of the accelerometers with respect to the reference point are given by

$$\begin{pmatrix} \mathbf{a}_{bx}^{bP} & \mathbf{a}_{by}^{bP} & \mathbf{a}_{bz}^{bP} \end{pmatrix} = -(\boldsymbol{\Omega}_{ib}^{b}\boldsymbol{\Omega}_{ib}^{b} + \dot{\boldsymbol{\Omega}}_{ib}^{b})\begin{pmatrix} \mathbf{r}_{bx}^{b} & \mathbf{r}_{by}^{b} & \mathbf{r}_{bz}^{b} \end{pmatrix}, \quad (4.10)$$

where $\mathbf{r}_{bx}^{b}$, $\mathbf{r}_{by}^{b}$, and $\mathbf{r}_{bz}^{b}$ are, respectively, the displacements of the x-, y-, and z-axis accelerometers from the IMU reference point that are known and constant, noting that a pendulous accelerometer measures acceleration at the proof mass, not at the hinge. From (2.86), the resulting error in the measurement of the specific force at the reference point is therefore

$$\delta \mathbf{f}_{ib,\text{size}}^{b} = \begin{pmatrix} a_{ix,x}^{b} - a_{ib,x}^{b} \\ a_{iy,y}^{b} - a_{ib,y}^{b} \\ a_{iz,z}^{b} - a_{ib,z}^{b} \end{pmatrix} = -\begin{pmatrix} a_{bx,x}^{bP} \\ a_{by,y}^{bP} \\ a_{bz,z}^{bP} \end{pmatrix}$$

$$= \begin{bmatrix} -\left(\omega_{ib,y}^{b}{}^{2} + \omega_{ib,z}^{b}{}^{2}\right)x_{bx}^{b} + \left(\omega_{ib,x}^{b}\omega_{ib,y}^{b} - \ddot{\omega}_{ib,z}^{b}\right)y_{bx}^{b} + \left(\omega_{ib,x}^{b}\omega_{ib,z}^{b} + \ddot{\omega}_{ib,y}^{b}\right)z_{bx}^{b} \\ -\left(\omega_{ib,z}^{b}{}^{2} + \omega_{ib,x}^{b}{}^{2}\right)y_{by}^{b} + \left(\omega_{ib,x}^{b}\omega_{ib,y}^{b} + \ddot{\omega}_{ib,z}^{b}\right)x_{by}^{b} + \left(\omega_{ib,y}^{b}\omega_{ib,z}^{b} - \ddot{\omega}_{ib,x}^{b}\right)z_{by}^{b} \\ -\left(\omega_{ib,x}^{b}{}^{2} + \omega_{ib,y}^{b}{}^{2}\right)z_{bz}^{b} + \left(\omega_{ib,x}^{b}\omega_{ib,z}^{b} - \ddot{\omega}_{ib,y}^{b}\right)x_{bz}^{b} + \left(\omega_{ib,y}^{b}\omega_{ib,z}^{b} + \ddot{\omega}_{ib,x}^{b}\right)y_{bz}^{b} \end{bmatrix}.$$

$$(4.11)$$

Where the reference point is defined as the intersection of the sensitive axes of the three accelerometers, $y_{bx}^{b} = z_{bx}^{b} = x_{by}^{b} = z_{by}^{b} = x_{bz}^{b} = y_{bz}^{b} = 0$, so the size effect error simplifies to

$$\delta \mathbf{f}_{ib,\text{size}}^{b} = -\begin{bmatrix} \left(\omega_{ib,y}^{b}{}^{2} + \omega_{ib,z}^{b}{}^{2}\right)x_{bx}^{b} \\ \left(\omega_{ib,z}^{b}{}^{2} + \omega_{ib,x}^{b}{}^{2}\right)y_{by}^{b} \\ \left(\omega_{ib,x}^{b}{}^{2} + \omega_{ib,y}^{b}{}^{2}\right)z_{bz}^{b} \end{bmatrix}. \quad (4.12)$$

4.4 Error Characteristics

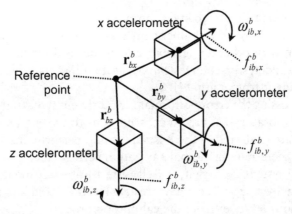

Figure 4.11 Accelerometer mounting relative to the IMU reference point.

The size effect correction applied by the IMU's processor is simply $-\delta \mathbf{f}_{ib,\text{size}}^b$.

Note that not all IMUs apply calibration to the sensor outputs or even correct for the size effect. This includes some IMUs with a temperature sensor.

As discussed in Section 4.4.5, inertial sensors are sensitive to vibration (e.g., from a propulsion system), transmitted both mechanically and as sound waves. The extent to which vibration is transmitted from the environment to the sensors depends on their packaging, their mounting within the IMU, and the mounting of the IMU itself. This will vary with both the frequency and direction of the vibration. Many IMUs therefore incorporate vibration isolators, which also protect the components from shock. These isolators must be designed to limit the transmission of vibrations at frequencies (and harmonics thereof) close to either the mechanical resonances of the sensors or the computational update rates of the IMU [1, 14].

4.4 Error Characteristics

All types of accelerometer and gyro exhibit biases, scale factor and cross-coupling errors, and random noise to a certain extent. Higher-order errors and angular rate-specific force cross-sensitivity may also occur, depending on the sensor type. Each of these errors is discussed in turn, followed by a discussion on vibration-induced errors and a summary of error modeling.

Each systematic error source has four components: a fixed contribution, a temperature-dependent variation, a run-to-run variation, and an in-run variation. The fixed contribution is present each time the sensor is used and is corrected by the IMU processor using the laboratory calibration data. The temperature-dependent component can also be corrected by the IMU using laboratory calibration data. When this is not corrected, the sensor will typically exhibit variation of its systematic errors over the first few minutes of operation while the sensor is warming-up to its normal operating temperature.

The run-to-run variation of each error source results in a contribution to the total error, which is different each time the sensor is used, but remains constant within any run. It cannot be corrected by the IMU processor, but it can be calibrated by the INS alignment and/or integration algorithms each time the IMU is used, as described in Section 5.6.3 and Chapters 14 and 15. Finally, the in-run variation contribution to the error source slowly changes during the course of a run. It cannot be corrected by the IMU or by an alignment process. In theory, it can be corrected through integration with other navigation sensors, but is difficult to observe in practice. In addition, sudden step changes can occur if an IMU is subject to a large shock, such as launching it from a gun [4].

In discussing the error performance of different types and grades of inertial sensor here, the laboratory-calibrated contributions to the error sources, corrected within the IMU, are neglected as the postcalibration performance of the inertial sensors is relevant in determining inertial navigation performance and designing an integrated navigation system. Note that, as well as the run-to-run and in-run variations contributing to each error source, there are also residual fixed and temperature-dependent contributions left over from the calibration process.

4.4.1 Biases

The *bias* is a constant error exhibited by all accelerometers and gyros. It is independent of the underlying specific force and angular rate. Figure 4.12 illustrates this. In most cases, the bias is the dominant term in the overall error of an inertial instrument. It is sometimes called the g-independent bias to distinguish it from the g-dependent bias discussed in Section 4.4.4.

The accelerometer and gyro biases of an IMU, following sensor calibration and compensation, are denoted by the vectors $\mathbf{b}_a = (b_{a,x}, b_{a,y}, b_{a,z})$ and $\mathbf{b}_g = (b_{g,x}, b_{g,y}, b_{g,z})$, respectively. IMU errors are always expressed in body axes, so the superscript b may be omitted. When the accelerometers and gyros form orthogonal triads, $b_{a,x}$ is the bias of the x-axis accelerometer (i.e., sensitive to specific force along the body frame x-axis), $b_{g,y}$ is the bias of the y-axis gyro, and so forth. For skewed-sensor configurations, the IMU biases may still be expressed as three-component vectors, but the components do not correspond to individual instruments.

It is sometimes convenient to split the biases into static, $\mathbf{b}_{as}$ and $\mathbf{b}_{gs}$, and dynamic, $\mathbf{b}_{ad}$ and $\mathbf{b}_{gd}$, components, where

$$\begin{aligned} \mathbf{b}_a &= \mathbf{b}_{as} + \mathbf{b}_{ad} \\ \mathbf{b}_g &= \mathbf{b}_{gs} + \mathbf{b}_{gd} \end{aligned}. \tag{4.13}$$

The static component, also known as the fixed bias, turn-on bias, or bias repeatability, comprises the run-to-run variation of each instrument bias plus the residual fixed bias remaining after sensor calibration. It is constant throughout an IMU operating period, but varies from run to run. The dynamic component, also known as the in-run bias variation or bias instability, varies over periods of the order of a minute and also incorporates the residual temperature-dependent bias remaining after sensor calibration. The dynamic bias is typically about 10% of the static bias.

Accelerometer and gyro biases are not usually quoted in SI units. Units of milli-g (mg) or micro-g (μg), where 1g = 9.80665 m s^{-2}, are used for accelerometer biases. For gyro biases, degrees per hour (° hr^{-1} or deg/hr) are used where 1 ° hr^{-1} = 4.848 × 10^{-6} rad s^{-1}, except for very poor-quality gyros where degrees per second are used. Table 4.1 gives typical accelerometer and gyro biases for different grades of IMU [1, 8].

Pendulous accelerometers span most of the performance range, while VBAs exhibit biases of 0.1 mg upward. MEMS accelerometers using both technologies

4.4 Error Characteristics

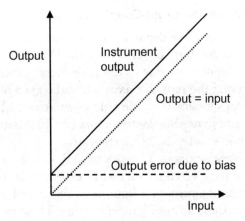

Figure 4.12 Sensor input versus output with a bias error.

exhibit the largest biases, ranging from 0.3 mg to more than 10 mg. Ring laser gyros exhibit biases as low as 0.001 ° hr^{-1}. However, low-cost RLGs can exhibit biases up to 10 ° hr^{-1}. IFOGs typically exhibit biases between 0.01 and 100 ° hr^{-1}, while vibratory-gyro biases range from 1 ° hr^{-1} to 1 ° s^{-1}. Uncalibrated MEMS accelerometers and gyros can exhibit large temperature-dependent biases, varying by several degrees per second or milli-g over the sensor's range of operating temperatures [18].

MEMS sensors manufactured in the same batch can exhibit similar bias characteristics. Consequently, by using two sensors, mounted with their sensitive axes in opposing directions, and differencing their outputs, much of the bias may be cancelled out, reducing its impact by an order of magnitude [18]. A set of twelve sensors may be combined using

$$\tilde{\mathbf{f}}_{ib}^b = \frac{1}{2} \begin{pmatrix} \tilde{f}_{ib,x}^{b,+} - \tilde{f}_{ib,x}^{b,-} \\ \tilde{f}_{ib,y}^{b,+} - \tilde{f}_{ib,y}^{b,-} \\ \tilde{f}_{ib,z}^{b,+} - \tilde{f}_{ib,z}^{b,-} \end{pmatrix}, \quad \tilde{\boldsymbol{\omega}}_{ib}^b = \frac{1}{2} \begin{pmatrix} \tilde{\omega}_{ib,x}^{b,+} - \tilde{\omega}_{ib,x}^{b,-} \\ \tilde{\omega}_{ib,y}^{b,+} - \tilde{\omega}_{ib,y}^{b,-} \\ \tilde{\omega}_{ib,z}^{b,+} - \tilde{\omega}_{ib,z}^{b,-} \end{pmatrix}, \quad (4.14)$$

where the superscripts + and −, respectively, denote the positively and negatively aligned sensors.

Table 4.1 Typical Accelerometer and Gyro Biases for Different Grades of IMU

IMU Grade	Accelerometer Bias		Gyro Bias	
	mg	m s^{-2}	° hr^{-1}	rad s^{-1}
Marine	0.01	10^{-4}	0.001	5×10^{-9}
Aviation	0.03–0.1	3×10^{-4}–10^{-3}	0.01	5×10^{-8}
Intermediate	0.1–1	10^{-3}–10^{-2}	0.1	5×10^{-7}
Tactical	1–10	0.01–0.1	1–100	5×10^{-6}–5×10^{-4}
Consumer	>3	>0.03	>100	>5×10^{-4}

4.4.2 Scale Factor and Cross-Coupling Errors

The *scale factor error* is the departure of the input-output gradient of the instrument from unity following unit conversion by the IMU. Figure 4.13 illustrates this. The accelerometer output error due to the scale factor error is proportional to the true specific force along the sensitive axis, while the gyro output error due to the scale factor error is proportional to the true angular rate about the sensitive axis. The accelerometer and gyro scale factor errors of an IMU are denoted by the vectors $\mathbf{s}_a = (s_{a,x}, s_{a,y}, s_{a,z})$ and $\mathbf{s}_g = (s_{g,x}, s_{g,y}, s_{g,z})$, respectively.

Cross-coupling errors in all types of IMUs arise from the misalignment of the sensitive axes of the inertial sensors with respect to the orthogonal axes of the body frame due to manufacturing limitations as illustrated in Figure 4.14. Hence, some authors describe these as misalignment errors. These make each accelerometer sensitive to the specific force along the axes orthogonal to its sensitive axis and each gyro sensitive to the angular rate about the axes orthogonal to its sensitive axis. The axes misalignment also produces additional scale factor errors, but these are typically two to four orders of magnitude smaller than the cross-coupling errors. In vibratory sensors, cross-coupling errors can also arise due to the cross-talk between the individual sensors. In consumer-grade MEMS sensors, the cross-coupling errors of the sensor itself, sometimes known as cross-axis sensitivity, can exceed those due

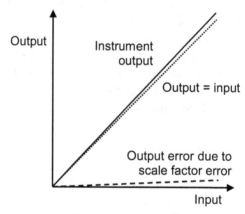

Figure 4.13 Scale factor error. (*From:* [19]. ©2002 QinetiQ Ltd. Reprinted with permission.)

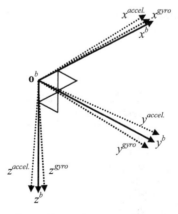

Figure 4.14 Misalignment of accelerometer and gyro sensitive axes with respect to the body frame.

4.4 Error Characteristics

to mounting misalignment. The notation $m_{a,\alpha\beta}$ is used to denote the cross-coupling coefficient of β-axis specific force sensed by the α-axis accelerometer, while $m_{g,\alpha\beta}$ denotes the coefficient of β-axis angular rate sensed by the α-axis gyro.

The scale factor and cross-coupling errors for a nominally orthogonal accelerometer and gyro triad may be expressed as the following matrices:

$$\mathbf{M}_a = \begin{pmatrix} s_{a,x} & m_{a,xy} & m_{a,xz} \\ m_{a,yx} & s_{a,y} & m_{a,yz} \\ m_{a,zx} & m_{a,zy} & s_{a,z} \end{pmatrix} \quad \mathbf{M}_g = \begin{pmatrix} s_{g,x} & m_{g,xy} & m_{g,xz} \\ m_{g,yx} & s_{g,y} & m_{g,yz} \\ m_{g,zx} & m_{g,zy} & s_{g,z} \end{pmatrix}. \tag{4.15}$$

The total specific force and angular rate measurement errors due to the scale factor and cross-coupling errors are then $\mathbf{M}_a \mathbf{f}_{ib}^b$ and $\mathbf{M}_g \boldsymbol{\omega}_{ib}^b$, respectively. Scale factor and cross-coupling errors are unitless and typically expressed in parts per million (ppm) or as a percentage. Some manufacturers quote the axis misalignments instead of the cross-coupling errors, noting that the latter is the sine of the former.

Where the cross-coupling errors arise only from axis misalignment, 3 of the 12 components may be eliminated by defining the body-frame axes in terms of the sensitive axes of the inertial sensors. One convention is to define the body-frame z-axis as the sensitive axis of the z gyro and the body-frame y-axis such that the sensitive axis of the y gyro lies in the yz plane. This eliminates $m_{g,zx}$, $m_{g,zy}$, and $m_{g,yx}$.

For most inertial sensors, the scale factor and cross-coupling errors are between 10^{-4} and 10^{-3} (100–1,000 ppm). There are two main exceptions. Some uncalibrated consumer-grade MEMS sensors exhibit scale factor errors as high as 0.1 (10%) and cross-coupling errors of up to 0.02 (2%). Ring laser gyros exhibit low scale factor errors, typically between 10^{-6} and 10^{-4} (1–100 ppm).

The lowest-cost sensors can exhibit significant scale factor asymmetry, whereby the scale factor errors are different for positive and negative readings. Figure 4.15 illustrates this.

4.4.3 Random Noise

All inertial sensors exhibit random noise from a number of sources. Electrical noise limits the resolution of inertial sensors, particularly MEMS sensors, where the signal

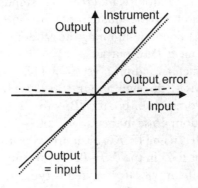

Figure 4.15 Scale factor asymmetry.

is very weak. Pendulous accelerometers exhibit noise due to mechanical instabilities, while the residual lock-in effects of an RLG, after dithering is applied, manifest as noise [1]. VBAs and vibratory gyros can exhibit high-frequency resonances. In addition, vibration from RLG dither motors and spinning-mass gyros can induce accelerometer noise [20]. The random noise on each IMU sample is denoted by the vectors $\mathbf{w}_a = (w_{a,x}, w_{a,y}, w_{a,z})$ and $\mathbf{w}_g = (w_{g,x}, w_{g,y}, w_{g,z})$ for the accelerometers and gyros, respectively.

The spectrum of accelerometer and gyro noise for frequencies below 1 Hz is approximately white, so the standard deviation of the average specific force and angular rate noise varies in inverse proportion to the square root of the averaging time. Inertial sensor noise is thus usually quoted in terms of the root PSD. The customary units are $\mu g/\sqrt{Hz}$ for accelerometer random noise, where $1~\mu g/\sqrt{Hz} = 9.80665 \times 10^{-6}$ m s$^{-1.5}$, and °/$\sqrt{hr}$ or °/hr/$\sqrt{Hz}$ for gyro random noise, where 1 °/$\sqrt{hr} = 2.909 \times 10^{-4}$ rad s$^{-0.5}$ and 1 °/hr/$\sqrt{Hz} = 4.848 \times 10^{-6}$ rad s$^{-0.5}$. The standard deviations of the random noise samples are obtained by multiplying the corresponding root PSDs by the root of the sampling rate or dividing them by the root of the sampling interval. White random noise cannot be calibrated and compensated as there is no correlation between past and future values.

MEMS sensors can also exhibit significant high-frequency noise [21]. Within the IMU body frame, this noise averages out over the order of a second, so passing the sensor outputs through inertial navigation equations (Chapter 5) will eliminate most of the effects of this noise. However, if the IMU is rotating, the noise will not average out to the same extent within the frame used to compute the inertial navigation solution. Consequently, caution should be exercised in selecting these sensors for highly-dynamic applications.

Applying lowpass filtering directly to the sensor or IMU outputs reduces the high-frequency noise regardless of the dynamics. Methods using wavelet filtering techniques [22] or an artificial neural network (ANN) [23] can give better performance than conventional lowpass filtering. However, all of these techniques both introduce time lags and reduce the effective sensor bandwidth. One solution to the latter problem is to vary the passband of the filter in real time according to the level of dynamics [24].

The accelerometer and gyro random noise are sometimes described as random walks, which can be a cause of confusion. Random noise on the specific force measurements is integrated to produce a random-walk error on the inertial velocity solution. Similarly, random noise on the angular rate measurements is integrated to produce an attitude random-walk error. The standard deviation of a random-walk process is proportional to the square root of the integration time. The same random-walk errors are obtained by summing the random noise on integrated specific force and attitude increment IMU outputs.

The accelerometer random-noise root PSD varies from about 20 $\mu g/\sqrt{Hz}$ for aviation-grade IMUs, through about 100 $\mu g/\sqrt{Hz}$ for tactical-grade IMUs using pendulous accelerometers or quartz VBAs to 80–1,000 $\mu g/\sqrt{Hz}$ for MEMS sensors. RLGs exhibit random noise in the range 0.001–0.02 °/$\sqrt{hr}$, depending on the grade. Tactical-grade IMUs using IFOGs or quartz vibratory gyros typically exhibit a gyro random noise root PSD in the 0.03–0.1 °/$\sqrt{hr}$ range. The root PSD for the random noise of MEMS silicon vibratory gyros is typically 0.06–2 °/$\sqrt{hr}$ and can increase with the input angular rate [7].

For consumer-grade MEMS accelerometers and gyros, many manufacturers quote the standard deviation of the total noise (white and high frequency) at the sensor output rate instead of providing PSD information. At output rates of 1 kHz or more, noise levels of 2.5–10 mg for accelerometers and 0.3–1 °/s for gyros are typical.

A further source of noise is the quantization of the IMU data-bus outputs. This rounds the sensor output to an integer multiple of a constant, known as the quantization level, as shown in Figure 4.16. Word lengths of 16 bits are typically used for the integrated specific force and attitude increment outputs of a tactical-grade IMU, v_{ib}^b and α_{ib}^b, giving quantization levels of the order of 10^{-4} m s^{-1} and 2×10^{-6} rad, respectively. The IMU's internal processor generally operates to a higher precision, so the residuals are carried over to the next iteration. Consequently, the standard deviation of the quantization noise averaged over successive IMU outputs varies in inverse proportion to the number of samples, rather than the square root, until the IMU internal quantization limit is reached.

For consumer-grade sensors, a shorter word length of 8 to 12 bits is typically used, so quantization errors can be higher at around 10^{-3} m s^{-1} and 2×10^{-5} rad for v_{ib}^b and α_{ib}^b, respectively. However, the quantization level is typically slightly less than the quoted noise standard deviation. Quantization residuals are not normally carried over in these sensors, so the standard deviation of the average quantization noise over successive IMU outputs is inversely proportion to the square root of the number of samples. Although the quantization error has a linear distribution, its average (with sufficient samples) has a Gaussian distribution due to the central limit theorem.

4.4.4 Further Error Sources

Accelerometer and gyros exhibit further error characteristics depending on the sensor design.

Vibratory gyros, spinning-mass gyros, and some designs of IFOG exhibit sensitivity to specific force, known as the *g-dependent bias*. The sensitivity of vibratory gyros to specific force is shown in (4.7). The coefficient of the g-dependent bias is around 1 °/hr/g (4.944×10^{-5} rad m^{-1} s) for an IFOG and 10–200 °/hr/g for an uncalibrated vibratory gyro [1]. Gyros can be sensitive to accelerations along all three axes, so the g-dependent bias for a gyro triad comprises the 3×3 matrix, $\mathbf{G}_g$.

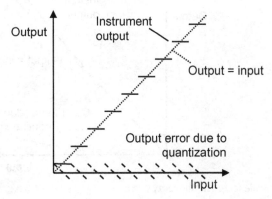

Figure 4.16 Effect of quantization on sensor output.

Inertial sensors can exhibit scale factor nonlinearity, sometimes just called nonlinearity, whereby the scale factor varies with the specific force or angular rate. The ensuing error may be modeled as a power series of terms proportional to the square, cube, fourth power, and so forth of the true angular rate or specific force measured by the sensor. Figure 4.17 illustrates this.

Some instrument specifications provide standard deviations of the quadratic and cubic coefficients of the nonlinearity power series (e.g., the K_2 and K_3 terms of a VBA [25]). However, the nonlinearity is normally expressed as the variation of the scale factor over the operating range of the sensor. This does not describe the shape of the scale factor variation, which can range from linear to irregular and need not be symmetric about the zero point. The scale factor nonlinearity ranges from 10^{-5} for some RLGs, through 10^{-4} to 10^{-3} for most inertial sensors, to 10^{-2} for MEMS gyros. The largest departures from scale factor linearity typically occur at the maximum angular rates or specific forces that the sensor will measure, often known as full scale (FS).

Open-loop sensors, including some MEMS accelerometers, and vibratory gyros can also exhibit variation of the cross-coupling errors (including the sign) as a function of the specific force or angular rate that they are measuring. This is because the direction of the sensitive axis changes slightly in response to sensed specific force or angular rate. The resulting measurement errors are known as anisoinertia errors. The accelerometer anisoinertia errors are proportional to the products of the specific force along two orthogonal axes. For pendulous accelerometers, the sensitive and pendulum axes product dominates. Similarly, the gyro anisoinertia errors are proportional to the product of the angular rate about two orthogonal axes. This phenomenon is also partially responsible for scale factor nonlinearity.

MEMS sensors often exhibit errors due to their operating ranges being exceeded, in which case the sensor simply outputs its largest possible positive or negative reading. Human motion can easily exceed the maximum ranges of typical smartphone accelerometers and gyros. It is therefore important to match the sensors to the application.

Errors can also arise when the bandwidth of the sensor is exceeded. The maximum bandwidth is half the update rate. However, some sensors have a lower bandwidth

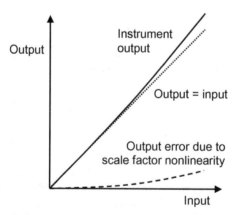

Figure 4.17 Scale factor nonlinearity.

than this, due to either damping of the sensor hardware or filtering to reduce noise. Thus, the bandwidth of the sensor must be matched to that of the motion to be measured. This is particularly important for high-vibration environments, such as aircraft, for which a bandwidth of more than 50 Hz is required to minimize coning and sculling errors (see Section 5.5.4).

Further higher-order systematic errors occur depending on the sensor type. Pendulous accelerometers can exhibit hysteresis errors, which depend on whether the measured specific force is increasing or decreasing [1]. Vibratory gyros can be sensitive to angular acceleration about the input and output axes and to the square of the angular rate about the output axis as well as the input axis, as shown in (4.7). Vibratory gyros can also exhibit g-dependent scale factor errors and large errors during their first few seconds of operation as they settle [7].

Foot-mounted IMUs can exhibit larger than expected errors due to the high specific force that occurs each time the foot hits the ground. Errors can arise through a mixture of nonlinearity, cross-axis sensitivity, and operating range limits.

Finally, in smartphones, inertial sensor measurements are typically accessed through the phone's operating system. This can cause a range of problems, including variable lags, missing or repeated measurements, reduced update rates, and increased quantization.

4.4.5 Vibration-Induced Errors

In a vibration environment, the motion will interact with the sensor scale factor and cross-coupling errors to produce oscillating sensor errors. Over time, these will average to zero. However any asymmetry and/or nonlinearity of the scale factor and cross-coupling errors will result in a component of the vibration-induced sensor error that does not cancel out over time. This is known as a vibration rectification error (VRE) and behaves like a bias that varies with the amplitude of the vibration. Asymmetric damping within the sensor can also lead to a VRE [26]. Note that vibratory gyros are sensitive to linear as well as angular vibration [7]. VREs in MEMS accelerometers can also vary with the underlying specific force [7].

Synchronized vibration along the sensitive and pendulum axes of a pendulous accelerometer also interacts with the anisoinertia error to produce a bias-like error known as the vibropendulous error [1].

A further problem is that oscillating errors on different sensor outputs will interact through inertial navigation equations, producing coning and sculling errors as described in Section 5.5.4.

Where the frequency of the external vibration is close to one of the inertial sensor's resonant frequencies or the update rate of a processing cycle, it will take a long time for the vibration-induced errors to cancel, resulting in a time-varying bias at the beat frequency between the vibration and the resonance or processing cycle.

MEMS sensors are particularly vulnerable to vibration-induced errors for several reasons. The scale factor and cross-coupling error nonlinearity and asymmetry are relatively large. Lowpass filtering to mitigate high frequency noise can exaggerate coning and sculling effects, while interaction between the vibration and the high-frequency sensor noise can produce time-correlated errors if the frequencies are close.

4.4.6 Error Models

The following equations show how the main error sources contribute to the accelerometer and gyro outputs:

$$\tilde{\mathbf{f}}_{ib}^b = \mathbf{b}_a + (\mathbf{I}_3 + \mathbf{M}_a)\mathbf{f}_{ib}^b + \mathbf{w}_a, \quad (4.16)$$

$$\tilde{\boldsymbol{\omega}}_{ib}^b = \mathbf{b}_g + (\mathbf{I}_3 + \mathbf{M}_g)\boldsymbol{\omega}_{ib}^b + \mathbf{G}_g \mathbf{f}_{ib}^b + \mathbf{w}_g, \quad (4.17)$$

where $\tilde{\mathbf{f}}_{ib}^b$ and $\tilde{\boldsymbol{\omega}}_{ib}^b$ are the IMU-output specific force and angular rate vectors, $\mathbf{f}_{ib}^b$ and $\boldsymbol{\omega}_{ib}^b$ are the true counterparts, and $\mathbf{I}_3$ is the identity matrix. These are implemented in the MATLAB function, IMU_model, on the accompanying CD. The total accelerometer and gyro errors are

$$\begin{aligned}\delta\mathbf{f}_{ib}^b &= \tilde{\mathbf{f}}_{ib}^b - \mathbf{f}_{ib}^b \\ \delta\boldsymbol{\omega}_{ib}^b &= \tilde{\boldsymbol{\omega}}_{ib}^b - \boldsymbol{\omega}_{ib}^b\end{aligned}. \quad (4.18)$$

Example 4.1 on the CD shows the variation of the specific force error exhibited by a single accelerometer and is editable using Microsoft Excel.

Where estimates of the biases, scale factor and cross-coupling errors, and gyro g-dependent errors are available, corrections may be applied:

$$\begin{aligned}\hat{\mathbf{f}}_{ib}^b &= (\mathbf{I}_3 + \hat{\mathbf{M}}_a)^{-1}(\tilde{\mathbf{f}}_{ib}^b - \hat{\mathbf{b}}_a) \\ &\approx (\mathbf{I}_3 + \hat{\mathbf{M}}_a)^{-1}\tilde{\mathbf{f}}_{ib}^b - \hat{\mathbf{b}}_a\end{aligned}, \quad (4.19)$$

$$\begin{aligned}\hat{\boldsymbol{\omega}}_{ib}^b &= (\mathbf{I}_3 + \hat{\mathbf{M}}_g)^{-1}(\tilde{\boldsymbol{\omega}}_{ib}^b - \hat{\mathbf{b}}_g - \hat{\mathbf{G}}_g \hat{\mathbf{f}}_{ib}^b) \\ &\approx (\mathbf{I}_3 + \hat{\mathbf{M}}_g)^{-1}\tilde{\boldsymbol{\omega}}_{ib}^b - \hat{\mathbf{b}}_g - \hat{\mathbf{G}}_g \hat{\mathbf{f}}_{ib}^b\end{aligned}, \quad (4.20)$$

where the carat, ^, is used to denote an estimate and, applying a power-series expansion,

$$(\mathbf{I}_3 + \hat{\mathbf{M}}_{a/g})^{-1} = \mathbf{I}_3 + \sum_r \binom{-1}{r} \hat{\mathbf{M}}_{a/g}^r \approx \mathbf{I}_3 - \hat{\mathbf{M}}_{a/g} + \hat{\mathbf{M}}_{a/g}^2. \quad (4.21)$$

The approximate versions of (4.19) and (4.20) neglect products of IMU errors. A similar formulation is used for applying the laboratory calibration within the IMU processor, noting that, in that case, the error measurements are functions of temperature.

Problems and exercises for this chapter are on the accompanying CD.

References

[1] Titterton, D. H., and J. L. Weston, *Strapdown Inertial Navigation Technology*, 2nd ed., Stevenage, U.K.: IEE, 2004.

[2] Lawrence, A., *Modern Inertial Technology*, 2nd ed., New York: Springer-Verlag, 2001.

[3] Kempe, V., *Inertial MEMS Principles and Practices*, Cambridge, U.K.: Cambridge University Press, 2011.

[4] Karnick, D., et al., "Honeywell Gun-Hard Inertial Measurement Unit (IMU) Development," *Proc. ION NTM*, San Diego, CA, January 2007, pp. 718–724.

[5] Norgia, M., and S. Donati, "Hybrid Opto-Mechanical Gyroscope with Injection-Interferometer Readout," *Electronics Letters*, Vol. 37, No. 12, 2001, pp. 756–758.

[6] El-Sheimy, N., and X. Niu, "The Promise of MEMS to the Navigation Community," *Inside GNSS*, March–April 2007, pp. 46–56.

[7] Pethel, S. J., "Test and Evaluation of High Performance Micro Electro-Mechanical System Based Inertial Measurement Units," *Proc. IEEE/ION PLANS*, San Diego, CA, April 2006, pp. 772–794.

[8] Barbour, N. M., "Inertial Navigation Sensors," *Advances in Navigation Sensors and Integration Technology*, NATO RTO Lecture Series-232, London, U.K., October 2003, paper 2.

[9] Zwahlen, P., et al., "Breakthrough in High Performance Inertial Navigation Grade Sigma-Delta MEMS Accelerometer," *Proc. IEEE/ION PLANS*, Myrtle Beach, SC, April 2012, pp. 15–19.

[10] Tennent, R. M., *Science Data Book*, Edinburgh, U.K.: Oliver & Boyd, 1971.

[11] Macek, W. M., and D. T. M. Davis, "Rotation Rate Sensing with Traveling-Wave Ring Lasers," *Applied Physics Letters*, Vol. 2, No. 5, 1963, pp. 67–68.

[12] Vali, V., and R. W. Shorthill, "Fiber Ring Interferometer," *Applied Optics*, Vol.15, No. 15, 1976, pp. 1099–1100.

[13] Donely, E. A., "Nuclear Magnetic Resonance Gyroscopes," *Proc. IEEE Sensors 2010*, Waikoloa, HI, November 2010, pp. 17–22.

[14] Grewal, M. S., L. R. Weill, and A. P. Andrews, *Global Positioning Systems, Inertial Navigation, and Integration*, 2nd ed., New York: Wiley, 2007.

[15] Matthews, A., "Utilization of Fiber Optic Gyros in Inertial Measurement Units," *Navigation: JION*, Vol. 27, No. 1, 1990, pp. 17–38.

[16] Fountain, J. R., "Silicon IMU for Missile and Munitions Applications," *Advances in Navigation Sensors and Integration Technology*, NATO RTO Lecture Series-232, London, U.K., October 2003, paper 10.

[17] Rogers, R. M., *Applied Mathematics in Integrated Navigation Systems*, Reston, VA: AIAA, 2000.

[18] Yuksel, Y., N. El-Sheimy, and A. Noureldin, "Error Modeling and Characterization of Environmental Effects for Low Cost Inertial MEMS Units," *Proc. IEEE/ION PLANS*, Palm Springs, CA, May 2010, pp. 598–612.

[19] Groves, P. D., "Principles of Integrated Navigation," Course Notes, QinetiQ Ltd., 2002.

[20] Woolven, S., and D. B. Reid, "IMU Noise Evaluation for Attitude Determination and Stabilization in Ground and Airborne Applications," *Proc. IEEE PLANS*, Las Vegas, NV, April 1994, pp. 817–822.

[21] Fountain, J. R., "Characteristics and Overview of a Silicon Vibrating Structure Gyroscope," *Advances in Navigation Sensors and Integration Technology*, NATO RTO Lecture Series-232, London, U.K., October 2003, paper 8.

[22] Shalard, J., A. M. Bruton, and K. P. Schwarz, "Detection and Filtering of Short Term ($1/f^\gamma$) Noise in Inertial Sensors," *Navigation: JION*, Vol. 46, No. 2, 1999, pp. 97–107.

[23] El-Rabbany, A., and M. El-Diasty, "An Efficient Neural Network Model for De-noising of MEMS-Based Inertial Data," *Journal of Navigation*, Vol. 57, No. 3, 2004, pp. 407–415.

[24] De Agostino, M., "A Multi-Frequency Filtering Procedure for Inertial Navigation," *Proc. IEEE/ION PLANS*, Monterey, CA, May 2008, pp. 115–121.

[25] Le Traon, O., et al., "The VIA Vibrating Beam Accelerometer: Concept and Performances," *Proc. IEEE PLANS*, Palm Springs, CA, 1998, pp. 25–37.

[26] Christel, L. A., et al., "Vibration Rectification in Silicon Micromachined Accelerometers," *Proc. IEEE Transducers '91*, San Francisco, CA, June 1991, pp. 89–92.

CHAPTER 5
Inertial Navigation

An inertial navigation system (INS), sometimes known as an inertial navigation unit (INU), is an example of a dead-reckoning navigation system. A position solution is maintained by integrating velocity, which, in turn, is maintained by integrating acceleration measurements obtained using an IMU. An attitude solution is also maintained by integrating the IMU's angular rate measurements. Following initialization, navigation can proceed without further information from the environment. Hence, inertial navigation systems are self-contained.

Inertial navigation has been used since the 1960s and 1970s for applications such as civil aviation, military aviation, submarines, military ships, and guided weapons. Some historical notes may be found in Section E.4 of Appendix E on the CD. These systems can typically operate either stand-alone or as part of an integrated navigation system. For newer applications, such as light aircraft, helicopters, unmanned air vehicles (UAVs), land vehicles, mobile mapping, and pedestrians, low-cost sensors are typically used and inertial navigation forms part of an INS/GNSS or multisensor integrated navigation system (Chapters 14 and 16).

As shown in Figure 5.1, an INS comprises an inertial measurement unit and a navigation processor. The IMU, described in the previous chapter, measures specific force and angular rate using a set of accelerometers and gyros. The discussion of IMU grades in Chapter 4 also applies to the INS as a whole. The navigation processor may be packaged with the IMU and the system sold as a complete INS. Alternatively, the navigation equations may be implemented on an integrated navigation processor or on the application's central processor. Marine, aviation, and intermediate grade inertial sensors tend to be sold as part of an INS, while tactical grade inertial sensors are usually sold as an IMU. In either case, the function is the same, so the term inertial navigation system is applied here to all architectures in which a three-dimensional navigation solution is obtained from inertial sensor measurements.

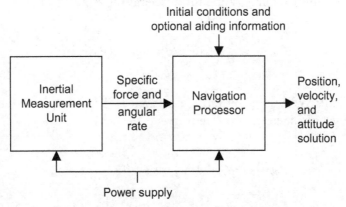

Figure 5.1 Basic schematic of an inertial navigation system.

This chapter focuses on the navigation processor. Section 5.1 introduces the main concepts of inertial navigation, illustrated by simple one- and two-dimensional examples. Three-dimensional navigation equations are then presented in Sections 5.2 to 5.5. A strapdown configuration, whereby the inertial sensors are fixed with respect to the vehicle body, is assumed throughout this chapter. The alternative platform configuration is described in Section E.5 of Appendix E on the CD.

Computation of an inertial navigation solution is an iterative process, making use of the solution from the previous iteration. Therefore, the navigation solution must be initialized before the INS can function. Section 5.6 describes the different methods of initializing the position, velocity, and attitude, including self-alignment and fine alignment processes.

Section 5.7 describes the error behavior of an INS. Errors can arise from the IMU, the initialization process, and the navigation equations. These then propagate through the navigation equations to give position, velocity, and attitude errors that vary with time. The short-term and long-term cases are examined. Finally, Section 5.8 discusses indexing, used to increase accuracy on ships and submarines, and Section 5.9 discusses inertial navigation using a partial IMU. Appendix E on the CD provides further information on a number of topics. A MATLAB inertial navigation simulation is also included on the accompanying CD.

Note that this chapter builds on the mathematical foundations introduced in Chapter 2.

5.1 Introduction to Inertial Navigation

An example of single-dimensional inertial navigation is considered first. A body, b, is constrained to move with respect to an Earth-fixed reference frame, p, in a straight line perpendicular to the direction of gravity. The body's axes are fixed with respect to frame p, so its motion has only one degree of freedom. Its Earth-referenced acceleration may be measured by a single accelerometer with its sensitive axis aligned along the direction of motion (neglecting the Coriolis force).

If the speed, v_{pb}, is known at an earlier time, t_0, it may be determined at a later time, t, simply by integrating the acceleration, a_{pb}:

$$v_{pb}(t) = v_{pb}(t_0) + \int_{t_0}^{t} a_{pb}(t')\, dt'. \tag{5.1}$$

Similarly, if the position, r_{pb}, at time t_0 is known, its value at time t may be obtained by integrating the velocity:

$$\begin{aligned} r_{pb}(t) &= r_{pb}(t_0) + \int_{t_0}^{t} v_{pb}(t')\, dt' \\ &= r_{pb}(t_0) + (t - t_0) v_{pb}(t_0) + \int_{t_0}^{t}\int_{t_0}^{t'} a_{pb}(t'')\, dt''\, dt' \end{aligned} \tag{5.2}$$

5.1 Introduction to Inertial Navigation

Extending the example to two dimensions, the body is now constrained to move within a horizontal plane defined by the x and y axes of the p frame. It may be oriented in any direction within this plane, but is constrained to remain level. It thus has one angular and two linear degrees of freedom. By analogy with the one-dimensional example, the position and velocity, resolved along the axes of the reference frame, p, are updated using

$$\begin{pmatrix} v^p_{pb,x}(t) \\ v^p_{pb,y}(t) \end{pmatrix} = \begin{pmatrix} v^p_{pb,x}(t_0) \\ v^p_{pb,y}(t_0) \end{pmatrix} + \int_{t_0}^{t} \begin{pmatrix} a^p_{pb,x}(t') \\ a^p_{pb,y}(t') \end{pmatrix} dt', \tag{5.3}$$

$$\begin{pmatrix} x^p_{pb}(t) \\ y^p_{pb}(t) \end{pmatrix} = \begin{pmatrix} x^p_{pb}(t_0) \\ y^p_{pb}(t_0) \end{pmatrix} + \int_{t_0}^{t} \begin{pmatrix} v^p_{pb,x}(t') \\ v^p_{pb,y}(t') \end{pmatrix} dt'. \tag{5.4}$$

Two accelerometers are required to measure the acceleration along two orthogonal axes. However, their sensitive axes will be aligned with those of the body, b. To determine the acceleration along the axes of frame p, the heading of frame b with respect to frame p, ψ_{pb}, is required. Figure 5.2 illustrates this. The rotation of the body with respect to the reference frame may be measured with a single gyro sensitive to rotation in the horizontal plane (neglecting Earth rotation). Thus, three inertial sensors are required to measure the three degrees of freedom of motion in two dimensions.

If the heading, ψ_{pb}, is known at the earlier time, t_0, it may be determined at the later time, t, by integrating the angular rate, $\omega^b_{pb,z}$:

$$\psi_{pb}(t) = \psi_{pb}(t_0) + \int_{t_0}^{t} \omega^b_{pb,z}(t') dt'. \tag{5.5}$$

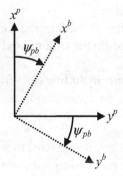

Figure 5.2 Orientation of body axes with respect to the resolving axes in a horizontal plane.

The accelerometer measurements may be transformed to the p-frame resolving axes using a 2×2 coordinate transformation matrix:

$$\begin{pmatrix} a_{pb,x}^{p}(t') \\ a_{pb,y}^{p}(t') \end{pmatrix} = \begin{pmatrix} \cos\psi_{pb}(t') & -\sin\psi_{pb}(t') \\ \sin\psi_{pb}(t') & \cos\psi_{pb}(t') \end{pmatrix} \begin{pmatrix} a_{pb,x}^{b}(t') \\ a_{pb,y}^{b}(t') \end{pmatrix}. \quad (5.6)$$

There is a clear dependency in processing the equations. The heading update must be computed before the accelerometer-output resolving frame transformation; the frame transformation must be computed before the velocity update; and the velocity update must be computed before the position update. Example 5.1 on the CD illustrates this over four measurement epochs and is editable using Microsoft Excel.

These one- and two-dimensional examples are presented only to aid understanding of the concepts of inertial navigation. For all practical applications, including ships, trains, and road vehicles, three-dimensional motion must be assumed. Although land and marine navigation is essentially a two-dimensional problem, strapdown inertial sensors will not remain in the horizontal plane due to terrain slopes or ship pitching and rolling. If the accelerometers are not in the horizontal plane, they will sense the reaction to gravity as well as the horizontal-plane acceleration. A platform tilt of just 10 mrad (0.57°) will produce an acceleration error of 0.1 m s^{-2}. If this is sustained for 100 seconds, the velocity error will be 10 m s^{-1} and the position error will be 500m. Tilts of 10 times this are quite normal for both cars and boats.

Motion in three dimensions generally has six degrees of freedom: three linear and three angular. Thus, six inertial sensors are required to measure that motion. A full strapdown IMU produces measurements of the specific force, $\mathbf{f}_{ib}^{b}$, and angular rate, $\boldsymbol{\omega}_{ib}^{b}$, of the IMU body frame with respect to inertial space in body-frame axes. Motion is *not* measured with respect to the Earth. Integrated specific force, $\mathbf{v}_{ib}^{b}$, and attitude increments, $\boldsymbol{\alpha}_{ib}^{b}$, may be output as an alternative. In general, none of the accelerometers may be assumed to be measuring pure acceleration. Therefore, a model of the gravitational acceleration must be used to determine inertially-referenced acceleration from the specific force measurements, while a gravity model must be used to obtain Earth-referenced acceleration (see Section 2.4.7).

Figure 5.3 shows a schematic of an inertial navigation processor. The IMU outputs are integrated to produce an updated position, velocity, and attitude solution in four steps:

1. The attitude update;
2. The transformation of the specific-force resolving axes from the IMU body frame to the coordinate frame used to resolve the position and velocity solutions;
3. The velocity update, including transformation of specific force into acceleration using a gravity or gravitation model;
4. The position update.

In an integrated navigation system, there may also be correction of the IMU outputs and the inertial navigation solution using estimates from the integration algorithm (see Section 14.1.1).

5.1 Introduction to Inertial Navigation

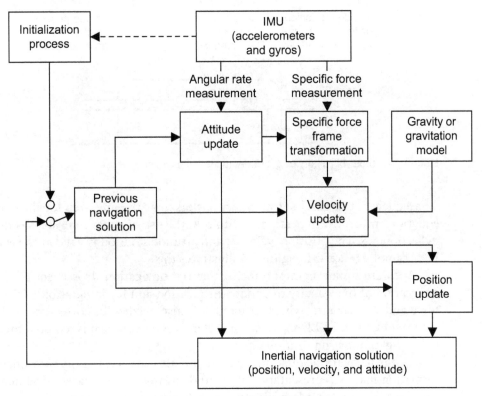

Figure 5.3 Schematic of an inertial navigation processor.

The form of the inertial navigation equations, also known as the strapdown computation, depends on the choice of reference frame and resolving axes (see Section 2.1). Section 5.2 describes the navigation equations for an Earth-centered inertial frame implementation, while Section 5.3 describes how they are modified for implementation in a rotating Earth-centered Earth-fixed frame. Section 5.4 presents an Earth-referenced local-navigation-frame implementation with curvilinear position. It also discusses the wander-azimuth-frame variant. In addition, Section E.7 of Appendix E on the CD presents a local tangent-plane implementation. Note that the navigation solution may be transformed to a different form for user output as described in Sections 2.4.3 and 2.5.

Continuous-time navigation equations, such as those presented for the one- and two-dimensional examples, physically describe a body's motion. Discrete-time navigation equations, also known as mechanization equations, provide a means of updating a navigation solution over a discrete time interval. They are an approximation of the continuous-time equations. Sections 5.2 to 5.4 present the continuous-time navigation equations together with the simplest practical mechanizations of the discrete-time equations, applying a number of first-order approximations and assuming that all stages are iterated at the IMU output rate. Section 5.5 then describes how the precision and/or efficiency of the inertial navigation equations may be improved and discusses which forms are appropriate for different applications.

As discussed in Section 4.3, the IMU provides outputs of specific force and angular rate averaged or integrated over a discrete sampling interval, τ_i. This provides a natural interval over which to compute each inertial navigation processing cycle.

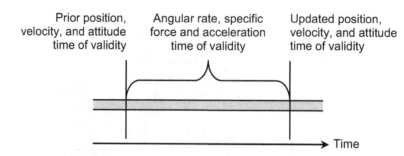

Figure 5.4 Times of validity of quantities in inertial navigation.

Consequently, this IMU output interval is taken as the time step for the navigation equations presented in Sections 5.2 to 5.4; the use of other time steps is discussed in Section 5.5. The position, velocity, and attitude solution is valid at the end of the IMU sampling interval. Figure 5.4 illustrates this.

Accurate timing is important for inertial navigation. It is needed for correct integration of the velocity to update the position and for integration of the specific force and angular rate, where required. It is also needed for correct transformation between ECI and ECEF resolving and references axes, which is required for all inertial navigation mechanization equations.

All the navigation equations presented in this chapter use the coordinate transformation matrix representation of attitude as this is the clearest. The quaternion form of the attitude update is described in Section E.6.3 of Appendix E on the CD. Navigation equations are also presented in [1, 2].

5.2 Inertial-Frame Navigation Equations

Figure 5.5 shows how the angular-rate and specific-force measurements made over the time interval t to $t + \tau_i$ are used to update the attitude, velocity, and position, expressed with respect to and resolved in the axes of an ECI coordinate frame. Each of the four steps is described in turn. The suffixes (–) and (+) are, respectively, used to denote values at the beginning of the navigation equations processing cycle, at time t, and at the end of the processing cycle, at time $t + \tau_i$. The inertial-frame navigation equations are the simplest of those presented here. However, a frame transformation must be applied to obtain an Earth-referenced solution for user output. Example 5.2 on the CD shows one processing cycle of the approximate navigation equations described in this section and is editable using Microsoft Excel.

5.2.1 Attitude Update

The attitude update step of the inertial navigation equations uses the angular-rate measurement from the IMU, ω_{ib}^b, to update the attitude solution, expressed as the body-to-inertial-frame coordinate transformation matrix, C_b^i.

From (2.56), the time derivative of the coordinate transformation matrix is

$$\dot{C}_b^i = C_b^i \Omega_{ib}^b \qquad (5.7)$$

5.2 Inertial-Frame Navigation Equations

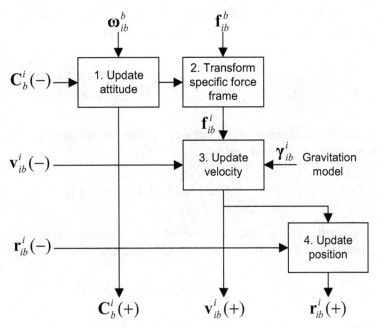

Figure 5.5 Block diagram of ECI-frame navigation equations.

recalling from Section 2.3.1 that $\boldsymbol{\Omega}_{ib}^b = [\boldsymbol{\omega}_{ib}^b \wedge]$, the skew-symmetric matrix of the angular rate. Integrating this gives

$$\mathbf{C}_b^i(t + \tau_i) = \mathbf{C}_b^i(t)\left[\lim_{n\to\infty}\prod_{i=1}^{n}\exp\left(\boldsymbol{\Omega}_{ib}^b\left(t + \frac{n-i}{n}\tau_i\right)\frac{\tau_i}{n}\right)\right], \quad (5.8)$$

noting (A.17) in Appendix A on the CD. If the angular rate is assumed to be constant over the attitude integration interval, this simplifies to

$$\begin{aligned}\mathbf{C}_b^i(t + \tau_i) &\approx \mathbf{C}_b^i(t)\exp(\boldsymbol{\Omega}_{ib}^b \tau_i) \\ &= \mathbf{C}_b^i(t)\exp([\boldsymbol{\omega}_{ib}^b \wedge]\tau_i)\end{aligned} \quad (5.9)$$

This assumption is often made where the attitude integration is performed at the IMU output rate. Applying (4.9), this may be expressed in terms of the attitude increment, $\boldsymbol{\alpha}_{ib}^b$:

$$\mathbf{C}_b^i(t + \tau_i) = \mathbf{C}_b^i(t)\exp([\boldsymbol{\alpha}_{ib}^b \wedge]). \quad (5.10)$$

Section 5.5.4 shows how a resultant attitude increment may be summed from successive increments in a way that correctly accounts for the noncommutivity of rotations. Therefore, (5.10) may be either exact or an approximation, depending on how $\boldsymbol{\alpha}_{ib}^b$ is calculated.

The exponent of a matrix is not the same as the matrix of the exponents of its components. Expressing (5.10) as a power series,

$$\mathbf{C}_b^i(t + \tau_i) = \mathbf{C}_b^i(t) \sum_{r=0}^{\infty} \frac{[\boldsymbol{\alpha}_{ib}^b \wedge]^r}{r!}. \tag{5.11}$$

The simplest form of the attitude update is obtained by truncating the power-series expansion to first order:

$$\mathbf{C}_b^i(+) \approx \mathbf{C}_b^i(-)\left(\mathbf{I}_3 + [\boldsymbol{\alpha}_{ib}^b \wedge]\right). \tag{5.12}$$

When the angular rate is assumed to be constant over the attitude integration interval, $\boldsymbol{\alpha}_{ib}^b \approx \boldsymbol{\omega}_{ib}^b \tau_i$. In this case, (5.12) becomes

$$\mathbf{C}_b^i(+) \approx \mathbf{C}_b^i(-)\left(\mathbf{I}_3 + \boldsymbol{\Omega}_{ib}^b \tau_i\right), \tag{5.13}$$

where

$$\mathbf{I}_3 + \boldsymbol{\Omega}_{ib}^b \tau_i = \begin{pmatrix} 1 & -\omega_{ib,z}^b \tau_i & \omega_{ib,y}^b \tau_i \\ \omega_{ib,z}^b \tau_i & 1 & -\omega_{ib,x}^b \tau_i \\ -\omega_{ib,y}^b \tau_i & \omega_{ib,x}^b \tau_i & 1 \end{pmatrix}. \tag{5.14}$$

This first-order approximation of (5.11) is a form of the small angle approximation, $\sin\theta \approx \theta$, $\cos\theta \approx 1$. The truncation of the power-series introduces errors in the attitude integration that will be larger at lower iteration rates (large τ_i) and higher angular rates. As discussed in Section 5.5.1, these errors are largest where the first-order approximation is used. In practice, the first-order approximation can be used for land vehicle applications where the dynamics are low, but not for high-dynamic applications, such as aviation. It is also unsuited to applications with regular periodic motion, such as pedestrian and boat navigation [3].

Precision may be improved by including higher-order terms in the power series, (5.11), breaking down the attitude update into smaller steps (see Section 5.5.5), or performing the exact attitude update, described in Section 5.5.1. All of these increase the complexity and processor load.

5.2.2 Specific-Force Frame Transformation

The IMU measures specific force along the body-frame resolving axes. However, for use in the velocity integration step of the navigation equations, it must be resolved about the same axes as the velocity—in this case, an ECI frame. The resolving axes are transformed simply by applying a coordinate transformation matrix:

$$\mathbf{f}_{ib}^i(t) = \mathbf{C}_b^i(t)\mathbf{f}_{ib}^b(t). \tag{5.15}$$

5.2 Inertial-Frame Navigation Equations

As the specific-force measurement is an average over time t to $t + \tau_i$, the coordinate transformation matrix should be similarly averaged. A simple implementation, assuming a constant angular rate, is

$$\mathbf{f}_{ib}^i \approx \tfrac{1}{2}\bigl(\mathbf{C}_b^i(-) + \mathbf{C}_b^i(+)\bigr)\mathbf{f}_{ib}^b. \tag{5.16}$$

However, the mean of two coordinate transformation matrices does not precisely produce the mean of the two attitudes. A more accurate form is presented in Section 5.5.2, while Section 5.5.4 shows how to account for variation in the angular rate over the update interval. The less the attitude varies over the time interval, the smaller the errors introduced by this approximation.

When the IMU outputs integrated specific force, this is transformed in the same way:

$$\begin{aligned}\boldsymbol{v}_{ib}^i &= \bar{\mathbf{C}}_b^i \boldsymbol{v}_{ib}^b \\ &\approx \tfrac{1}{2}\bigl(\mathbf{C}_b^i(-) + \mathbf{C}_b^i(+)\bigr)\boldsymbol{v}_{ib}^b\end{aligned}, \tag{5.17}$$

where $\bar{\mathbf{C}}_b^i$ is the average value of the coordinate transformation matrix over the interval from t to $t + \tau_i$.

5.2.3 Velocity Update

As given by (2.125), inertially referenced acceleration is obtained simply by adding the gravitational acceleration to the specific force:

$$\mathbf{a}_{ib}^i = \mathbf{f}_{ib}^i + \boldsymbol{\gamma}_{ib}^i(\mathbf{r}_{ib}^i), \tag{5.18}$$

where (2.141) models the gravitational acceleration, $\boldsymbol{\gamma}_{ib}^i$, as a function of Cartesian position in an ECI frame. Strictly, the position should be averaged over the interval t to $t + \tau_i$. However, this would require recursive navigation equations, and the gravitational field varies slowly with position, so it is generally sufficient to use[‡] $\mathbf{r}_{ib}^i(-)$.

When the reference frame and resolving axes are the same, the time derivative of velocity is simply acceleration, as shown by (2.77). Thus,

$$\dot{\mathbf{v}}_{ib}^i = \mathbf{a}_{ib}^i. \tag{5.19}$$

When variations in the acceleration over the velocity update interval are not known, as is the case when the velocity integration is iterated at the IMU output rate, the velocity update equation, obtained by integrating (5.19), is simply[‡]

$$\mathbf{v}_{ib}^i(+) = \mathbf{v}_{ib}^i(-) + \mathbf{a}_{ib}^i \tau_i. \tag{5.20}$$

[‡]This paragraph, up to this point, is based on material written by the author for QinetiQ, so comprises QinetiQ copyright material.

From (4.9), (5.18), and (5.19), the velocity update in terms of integrated specific force is

$$\mathbf{v}_{ib}^i(+) = \mathbf{v}_{ib}^i(-) + \mathbf{v}_{ib}^i + \gamma_{ib}^i \tau_i. \quad (5.21)$$

5.2.4 Position Update

In the inertial-frame implementation of the navigation equations, the time derivative of the Cartesian position is simply velocity as the reference frame and resolving axes are the same [see (2.68)]. Thus,

$$\dot{\mathbf{r}}_{ib}^i = \mathbf{v}_{ib}^i. \quad (5.22)$$

In the velocity update step where the variation in acceleration is unknown, $\mathbf{v}_{ib}^i$ is typically modeled as a linear function of time over the interval t to $t + \tau_i$. Integrating (5.22) thus leads to the position being modeled as a quadratic function of time. The velocity is known at the start and finish of the update interval, so the position is updated using

$$\begin{aligned}
\mathbf{r}_{ib}^i(+) &= \mathbf{r}_{ib}^i(-) + \left(\mathbf{v}_{ib}^i(-) + \mathbf{v}_{ib}^i(+)\right)\frac{\tau_i}{2} \\
&= \mathbf{r}_{ib}^i(-) + \mathbf{v}_{ib}^i(-)\tau_i + \mathbf{a}_{ib}^i \frac{\tau_i^2}{2} \\
&= \mathbf{r}_{ib}^i(-) + \mathbf{v}_{ib}^i(+)\tau_i - \mathbf{a}_{ib}^i \frac{\tau_i^2}{2}
\end{aligned} \quad (5.23)$$

where the three implementations are equally valid.[‡]

5.3 Earth-Frame Navigation Equations

An ECEF frame is commonly used as the reference frame and resolving axes for computation of satellite navigation solutions (Section 9.4), so, in an integrated system, there are benefits in using the same frame for computation of the inertial navigation solution. For some applications, such as airborne photogrammetry, the final navigation solution is more conveniently expressed in an ECEF frame [4]. A disadvantage of an ECEF-frame implementation, compared to an inertial-frame implementation, is that the rotation of the reference frame used for navigation solution computation with respect to an inertial reference, used for the inertial sensor measurements, introduces additional complexity. Figure 5.6 is a block diagram showing how the angular-rate and specific-force measurements are used to update the Earth-referenced attitude, velocity, and position. Each of the four steps is described in turn.

[‡]This paragraph, up to this point, is based on material written by the author for QinetiQ, so comprises QinetiQ copyright material.

5.3 Earth-Frame Navigation Equations

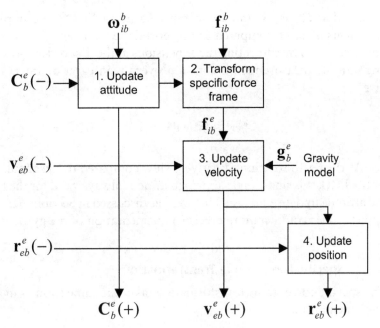

Figure 5.6 Block diagram of ECEF-frame navigation equations.

5.3.1 Attitude Update

The attitude update step of the ECEF-frame navigation equations uses the angular-rate measurement, $\boldsymbol{\omega}_{ib}^b$, to update the attitude solution, expressed as the body-to-Earth-frame coordinate transformation matrix, $\mathbf{C}_b^e$. From (2.56), (2.48), and (2.51), the time derivative is

$$\begin{aligned} \dot{\mathbf{C}}_b^e &= \mathbf{C}_b^e \boldsymbol{\Omega}_{eb}^b \\ &= \mathbf{C}_b^e \boldsymbol{\Omega}_{ib}^b - \boldsymbol{\Omega}_{ie}^e \mathbf{C}_b^e \end{aligned}, \tag{5.24}$$

where $\boldsymbol{\Omega}_{ib}^b$ is the skew-symmetric matrix of the IMU's angular-rate measurement, and $\boldsymbol{\Omega}_{ie}^e$ is the skew-symmetric matrix of the Earth-rotation vector. Thus, the rotation of the Earth must be accounted for in updating the attitude. From (2.122),

$$\boldsymbol{\Omega}_{ie}^e = \begin{pmatrix} 0 & -\omega_{ie} & 0 \\ \omega_{ie} & 0 & 0 \\ 0 & 0 & 0 \end{pmatrix}. \tag{5.25}$$

Integrating (5.24) gives

$$\begin{aligned} \mathbf{C}_b^e(t+\tau_i) &\approx \mathbf{C}_b^e(t)\exp(\boldsymbol{\Omega}_{eb}^b \tau_i) \\ &= \mathbf{C}_b^e(t)\exp\left[(\boldsymbol{\Omega}_{ib}^b - \boldsymbol{\Omega}_{ie}^e)\tau_i\right] \\ &= \mathbf{C}_b^e(t)\exp(\boldsymbol{\Omega}_{ib}^b \tau_i) - \mathbf{C}_b^e(t)\left[\exp(\boldsymbol{\Omega}_{ie}^e \tau_i) - \mathbf{I}_3\right] \\ &= \mathbf{C}_b^e(t)\exp\left([\boldsymbol{\alpha}_{ib}^b \wedge]\right) - \left[\exp(\boldsymbol{\Omega}_{ie}^e \tau_i) - \mathbf{I}_3\right]\mathbf{C}_b^e(t) \end{aligned}, \tag{5.26}$$

noting that $\mathbf{\Omega}_{ie}^e$ is assumed constant. As in an ECI-frame implementation, the exponents must be computed as power-series expansions. Applying the small angle approximation by truncating the expansions at the first order and assuming the IMU angular-rate measurement is constant over the integration interval (i.e., $\boldsymbol{\alpha}_{ib}^b \approx \boldsymbol{\omega}_{ib}^b \tau_i$) give

$$\mathbf{C}_b^e(+) \approx \mathbf{C}_b^e(-)\left(\mathbf{I}_3 + \mathbf{\Omega}_{ib}^b \tau_i\right) - \mathbf{\Omega}_{ie}^e \mathbf{C}_b^e(-)\tau_i. \tag{5.27}$$

As the Earth rotation rate is very slow compared to the angular rates measured by the IMU, this small angle approximation is always valid for the Earth rate term of the attitude update equation. However, as discussed in Sections 5.2.1 and 5.5.1, most applications require a more precise implementation of the gyro measurement term.

5.3.2 Specific-Force Frame Transformation

The specific-force frame transformation takes the same form as in an inertial-frame implementation:

$$\begin{aligned} \mathbf{f}_{ib}^e(t) &= \mathbf{C}_b^e(t)\mathbf{f}_{ib}^b(t) \\ \Rightarrow \mathbf{f}_{ib}^e &\approx \tfrac{1}{2}\left(\mathbf{C}_b^e(-) + \mathbf{C}_b^e(+)\right)\mathbf{f}_{ib}^b \end{aligned} \tag{5.28}$$

or

$$\begin{aligned} \mathbf{v}_{ib}^e &= \bar{\mathbf{C}}_b^e \mathbf{v}_{ib}^b \\ &\approx \tfrac{1}{2}\left(\mathbf{C}_b^e(-) + \mathbf{C}_b^e(+)\right)\mathbf{v}_{ib}^b \end{aligned} \tag{5.29}$$

5.3.3 Velocity Update

As in an inertial-frame implementation, the reference frame and resolving axes are the same so, from (2.76) and (2.77),

$$\dot{\mathbf{v}}_{eb}^e = \mathbf{a}_{eb}^e = \ddot{\mathbf{r}}_{eb}^e. \tag{5.30}$$

Now, applying (2.61), (2.65), and (2.66) in turn,

$$\begin{aligned} \mathbf{r}_{eb}^e &= \mathbf{r}_{ib}^e - \mathbf{r}_{ie}^e \\ &= \mathbf{r}_{ib}^e \end{aligned} \tag{5.31}$$

Substituting this into (5.30),

$$\dot{\mathbf{v}}_{eb}^e = \ddot{\mathbf{r}}_{ib}^e. \tag{5.32}$$

5.3 Earth-Frame Navigation Equations

Applying (2.81), noting that the Earth rate, $\boldsymbol{\omega}_{ie}^e$, is constant,

$$\dot{\mathbf{v}}_{eb}^e = -\boldsymbol{\Omega}_{ie}^e \boldsymbol{\Omega}_{ie}^e \mathbf{r}_{ib}^e - 2\boldsymbol{\Omega}_{ie}^e \dot{\mathbf{r}}_{eb}^e + \mathbf{a}_{ib}^e. \tag{5.33}$$

Thus, the rate of change of velocity resolved about the Earth-frame axes incorporates a centrifugal and a Coriolis term due to the rotation of the resolving axes as explained in Section 2.3.5. Applying (2.66) and (2.67),

$$\dot{\mathbf{v}}_{eb}^e = -\boldsymbol{\Omega}_{ie}^e \boldsymbol{\Omega}_{ie}^e \mathbf{r}_{eb}^e - 2\boldsymbol{\Omega}_{ie}^e \mathbf{v}_{eb}^e + \mathbf{a}_{ib}^e. \tag{5.34}$$

From (2.125), the applied acceleration, $\mathbf{a}_{ib}^e$, is the sum of the measured specific force, $\mathbf{f}_{ib}^e$, and the acceleration due to the gravitational force, $\boldsymbol{\gamma}_{ib}^e$. From (2.132), the acceleration due to gravity, $\mathbf{g}_b^e$, is the sum of the gravitational and centrifugal accelerations. Substituting these into (5.34),

$$\dot{\mathbf{v}}_{eb}^e = \mathbf{f}_{ib}^e + \mathbf{g}_b^e(\mathbf{r}_{eb}^e) - 2\boldsymbol{\Omega}_{ie}^e \mathbf{v}_{eb}^e. \tag{5.35}$$

An analytical solution is complex. However, as the Coriolis term will be much smaller than the specific-force and gravity terms, except for space applications, it is a reasonable approximation to neglect the variation of the Coriolis term over the integration interval. Thus,

$$\begin{aligned} \mathbf{v}_{eb}^e(+) &\approx \mathbf{v}_{eb}^e(-) + \left(\mathbf{f}_{ib}^e + \mathbf{g}_b^e(\mathbf{r}_{eb}^e(-)) - 2\boldsymbol{\Omega}_{ie}^e \mathbf{v}_{eb}^e(-)\right)\tau_i \\ &= \mathbf{v}_{eb}^e(-) + \boldsymbol{\upsilon}_{ib}^e + \left(\mathbf{g}_b^e(\mathbf{r}_{eb}^e(-)) - 2\boldsymbol{\Omega}_{ie}^e \mathbf{v}_{eb}^e(-)\right)\tau_i \end{aligned} \tag{5.36}$$

Most gravity models operate as a function of latitude and height, calculated from Cartesian ECEF position using (2.113). The gravity is converted from local navigation frame to ECEF resolving axes by premultiplying by $\mathbf{C}_n^e$, given by (2.150). Alternatively, a gravity model formulated in ECEF axes is presented in [4].

5.3.4 Position Update

In ECEF-frame navigation equations, the reference and resolving frames are the same, so, from (2.68),

$$\dot{\mathbf{r}}_{eb}^e = \mathbf{v}_{eb}^e. \tag{5.37}$$

Integrating this, assuming the velocity varies linearly over the integration interval,

$$\begin{aligned} \mathbf{r}_{eb}^e(+) &= \mathbf{r}_{eb}^e(-) + \left(\mathbf{v}_{eb}^e(-) + \mathbf{v}_{eb}^e(+)\right)\frac{\tau_i}{2} \\ &\approx \mathbf{r}_{eb}^e(-) + \mathbf{v}_{eb}^e(-)\tau_i + \left(\mathbf{f}_{ib}^e + \mathbf{g}_b^e(\mathbf{r}_{eb}^e(-)) - 2\boldsymbol{\Omega}_{ie}^e \mathbf{v}_{eb}^e(-)\right)\frac{\tau_i^2}{2} \end{aligned} \tag{5.38}$$

5.4 Local-Navigation-Frame Navigation Equations

In a local-navigation-frame implementation of the inertial navigation equations, an ECEF frame is used as the reference frame while a local navigation frame (north, east, down) comprises the resolving axes. Thus, attitude is expressed as the body-to-navigation-frame coordinate transformation matrix, C_b^n, and velocity is Earth-referenced in local navigation frame axes, v_{eb}^n. Position is expressed in the curvilinear form (i.e., as geodetic latitude, L_b, longitude, λ_b, and geodetic height, h_b) and is commonly integrated directly from the velocity rather than converted from its Cartesian form.

This form of navigation equations has the advantage of providing a navigation solution in a form readily suited for user output. However, additional complexity is introduced, compared to ECI- and ECEF-frame implementations, as the orientation of the resolving axes with respect to the reference frame depends on the position. Figure 5.7 is a block diagram showing how the angular-rate and specific-force measurements are used to update the attitude, velocity, and position in a local-navigation-frame implementation. Each of the four steps is described in turn. This is followed by a brief discussion of the related wander-azimuth implementation.

5.4.1 Attitude Update

The attitude update step of the local-navigation-frame navigation equations uses the position and velocity solution as well as the angular-rate measurement to update C_b^n. This is necessary because the orientation of the north, east, and down axes changes as the navigation system moves with respect to the Earth, as explained in Section 2.1.3. From (2.56), the time derivative of the coordinate transformation matrix is

$$\dot{C}_b^n = C_b^n \Omega_{nb}^b. \tag{5.39}$$

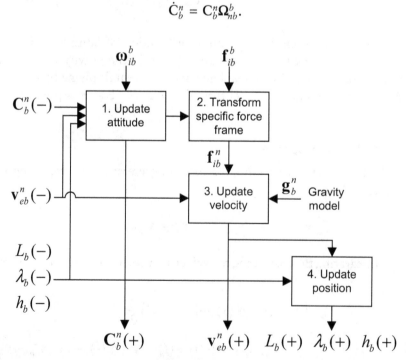

Figure 5.7 Block diagram of local-navigation-frame navigation equations.

5.4 Local-Navigation-Frame Navigation Equations

Using (2.48) and (2.51), this may be split into three terms:

$$\dot{C}_b^n = C_b^n \Omega_{ib}^b - (\Omega_{ie}^n + \Omega_{en}^n) C_b^n. \tag{5.40}$$

The first term is due to the inertially referenced angular rate, measured by the gyros, and the second is due to the rotation of the Earth with respect to an inertial frame. The third term, known as the *transport rate*, arises from the rotation of the local-navigation-frame axes as the frame center (i.e., the navigation system) moves with respect to the Earth. When the attitude of the body frame with respect to the local navigation frame remains constant, the gyros sense the Earth rotation and transport rate, which must be corrected for to keep the attitude unchanged.

The Earth-rotation vector in local navigation frame axes is given by (2.123), so the skew-symmetric matrix is

$$\Omega_{ie}^n = \omega_{ie} \begin{pmatrix} 0 & \sin L_b & 0 \\ -\sin L_b & 0 & -\cos L_b \\ 0 & \cos L_b & 0 \end{pmatrix}, \tag{5.41}$$

noting that this is a function of latitude.

From (2.56), the transport rate may be obtained by solving

$$\dot{C}_e^n = -\Omega_{en}^n C_e^n. \tag{5.42}$$

The ECEF-to-local-navigation-frame coordinate transformation matrix is given by (2.150). Taking the time derivative of this gives

$$\dot{C}_e^n = \left[\begin{pmatrix} -\dot{\lambda}_b \cos L_b \\ \dot{L}_b \\ \dot{\lambda}_b \sin L_b \end{pmatrix} \wedge \right] C_e^n. \tag{5.43}$$

Substituting this into (5.42), together with the derivatives of the latitude and longitude from (2.111) gives

$$\Omega_{en}^n = \begin{pmatrix} 0 & -\omega_{en,z}^n & \omega_{en,y}^n \\ \omega_{en,z}^n & 0 & -\omega_{en,x}^n \\ -\omega_{en,y}^n & \omega_{en,x}^n & 0 \end{pmatrix}$$

$$\omega_{en}^n = \begin{pmatrix} v_{eb,E}^n / (R_E(L_b) + h_b) \\ -v_{eb,N}^n / (R_N(L_b) + h_b) \\ -v_{eb,E}^n \tan L_b / (R_E(L_b) + h_b) \end{pmatrix}. \tag{5.44}$$

Integrating (5.39) gives

$$\begin{aligned}
\mathbf{C}_b^n(t+\tau_i) &\approx \mathbf{C}_b^n(t)\exp\left(\mathbf{\Omega}_{nb}^b\tau_i\right) \\
&= \mathbf{C}_b^n(t)\exp\left[\left(\mathbf{\Omega}_{ib}^b - \mathbf{\Omega}_{ie}^b - \mathbf{\Omega}_{en}^b\right)\tau_i\right] \\
&= \mathbf{C}_b^n(t)\exp\left(\mathbf{\Omega}_{ib}^b\tau_i\right) - \mathbf{C}_b^n(t)\left\{\exp\left[\left(\mathbf{\Omega}_{ie}^b + \mathbf{\Omega}_{en}^b\right)\tau_i\right] - \mathbf{I}_3\right\} \\
&= \mathbf{C}_b^n(t)\exp\left(\left[\boldsymbol{\alpha}_{ib}^b\wedge\right]\right) - \left\{\exp\left[\left(\mathbf{\Omega}_{ie}^n + \mathbf{\Omega}_{en}^n\right)\tau_i\right] - \mathbf{I}_3\right\}\mathbf{C}_b^n(t)
\end{aligned} \quad , \quad (5.45)$$

where the position and velocity, and hence $\mathbf{\Omega}_{ie}^n$ and $\mathbf{\Omega}_{en}^n$, are assumed constant over the attitude update interval. Accounting for their variation can require recursive navigation equations as discussed in Section 5.5. However, a reasonable approximation to (5.45) for most applications can be obtained by neglecting the position and velocity variation and truncating the power-series expansion of the Earth-rotation and transport rate terms to first order. Applying the first-order approximation to all terms gives

$$\mathbf{C}_b^n(+) \approx \mathbf{C}_b^n(-)\left(\mathbf{I}_3 + \mathbf{\Omega}_{ib}^b\tau_i\right) - \left(\mathbf{\Omega}_{ie}^n(-) + \mathbf{\Omega}_{en}^n(-)\right)\mathbf{C}_b^n(-)\tau_i, \quad (5.46)$$

where $\mathbf{\Omega}_{ie}^n(-)$ is calculated using $L_b(-)$ and $\mathbf{\Omega}_{en}^n(-)$ is calculated using $L_b(-)$, $h_b(-)$, and $\mathbf{v}_{eb}^n(-)$. As discussed in Section 5.2.1, a more precise implementation of the gyro measurement term is required for most applications. Higher precision solutions are discussed in Section 5.5.1.

5.4.2 Specific-Force Frame Transformation

The specific-force frame transformation is essentially the same as for the ECI and ECEF-frame implementations. Thus,

$$\begin{aligned}
\mathbf{f}_{ib}^n(t) &= \mathbf{C}_b^n(t)\mathbf{f}_{ib}^b(t) \\
\Rightarrow \mathbf{f}_{ib}^n &\approx \tfrac{1}{2}\left(\mathbf{C}_b^n(-) + \mathbf{C}_b^n(+)\right)\mathbf{f}_{ib}^b
\end{aligned} \quad , \quad (5.47)$$

or

$$\begin{aligned}
\boldsymbol{v}_{ib}^n &= \bar{\mathbf{C}}_b^n \boldsymbol{v}_{ib}^b \\
&\approx \tfrac{1}{2}\left(\mathbf{C}_b^n(-) + \mathbf{C}_b^n(+)\right)\boldsymbol{v}_{ib}^b
\end{aligned} \quad . \quad (5.48)$$

The accuracy of this approximation will be similar to that in an inertial frame as the gyro-sensed rotation will usually be much larger than the Earth rate and transport rate components.[‡]

[‡]This paragraph, up to this point, is based on material written by the author for QinetiQ, so comprises QinetiQ copyright material.

5.4.3 Velocity Update

In the local-navigation-frame navigation equations, the resolving axes of the velocity are not the same as its reference frame. From (2.73), the velocity is expressed in terms in terms of its counterpart in ECEF resolving axes by

$$\mathbf{v}_{eb}^n = \mathbf{C}_e^n \mathbf{v}_{eb}^e. \tag{5.49}$$

Differentiating this,

$$\dot{\mathbf{v}}_{eb}^n = \dot{\mathbf{C}}_e^n \mathbf{v}_{eb}^e + \mathbf{C}_e^n \dot{\mathbf{v}}_{eb}^e. \tag{5.50}$$

Thus, there is a transport-rate term in addition to the applied acceleration, centrifugal, and Coriolis terms found in the ECEF-frame velocity update described in Section 5.3.3. Applying (2.56) and (2.73) to the first term and substituting (5.34) for the second term,

$$\dot{\mathbf{v}}_{eb}^n = -\mathbf{\Omega}_{en}^n \mathbf{v}_{eb}^n + \mathbf{C}_e^n \left(-\mathbf{\Omega}_{ie}^e \mathbf{\Omega}_{ie}^e \mathbf{r}_{eb}^e - 2\mathbf{\Omega}_{ie}^e \mathbf{v}_{eb}^e + \mathbf{a}_{ib}^e \right). \tag{5.51}$$

Applying (2.51), (2.62), (2.73), and (2.83) to transform the resolving axes and rearranging give

$$\dot{\mathbf{v}}_{eb}^n = -\mathbf{\Omega}_{ie}^n \mathbf{\Omega}_{ie}^n \mathbf{r}_{eb}^n - \left(\mathbf{\Omega}_{en}^n + 2\mathbf{\Omega}_{ie}^n \right) \mathbf{v}_{eb}^n + \mathbf{a}_{ib}^n, \tag{5.52}$$

noting that the skew-symmetric matrices of the Earth rotation and transport rate are given by (5.41) and (5.44), respectively.

Expressing the acceleration in terms of the specific force, gravity, and centrifugal acceleration using (2.125) and (2.132) gives

$$\dot{\mathbf{v}}_{eb}^n = \mathbf{f}_{ib}^n + \mathbf{g}_b^n (L_b, h_b) - \left(\mathbf{\Omega}_{en}^n + 2\mathbf{\Omega}_{ie}^n \right) \mathbf{v}_{eb}^n, \tag{5.53}$$

where the acceleration due to gravity is modeled as a function of latitude and height. Again, obtaining a full analytical solution is complex. However, as the Coriolis and transport-rate terms will generally be the smallest, it is a reasonable approximation to neglect their variation over the integration interval. Again, the variation of the acceleration due to gravity over the integration interval can generally be neglected. Thus,

$$\begin{aligned}\mathbf{v}_{eb}^n(+) &\approx \mathbf{v}_{eb}^n(-) + \left[\mathbf{f}_{ib}^n + \mathbf{g}_b^n(L_b(-), h_b(-)) - \left(\mathbf{\Omega}_{en}^n(-) + 2\mathbf{\Omega}_{ie}^n(-) \right) \mathbf{v}_{eb}^n(-) \right] \tau_i \\ &= \mathbf{v}_{eb}^n(-) + \mathbf{\upsilon}_{ib}^n + \left[\mathbf{g}_b^n(L_b(-), h_b(-)) - \left(\mathbf{\Omega}_{en}^n(-) + 2\mathbf{\Omega}_{ie}^n(-) \right) \mathbf{v}_{eb}^n(-) \right] \tau_i\end{aligned} \tag{5.54}$$

5.4.4 Position Update

From (2.111), the derivatives of the latitude, longitude, and height are functions of the velocity, latitude, and height. Thus,*

---

*This and subsequent paragraphs are based on material written by the author for QinetiQ, so comprise QinetiQ copyright material.

$$L_b(+) = L_b(-) + \int_t^{t+\tau_i} \frac{v_{eb,N}^n(t')}{R_N(L_b(t')) + h_b(t')} dt'$$

$$\lambda_b(+) = \lambda_b(-) + \int_t^{t+\tau_i} \frac{v_{eb,E}^n(t')}{(R_E(L_b(t')) + h_b(t'))\cos L_b(t')} dt'. \quad (5.55)$$

$$h_b(+) = h_b(-) - \int_t^{t+\tau_i} v_{eb,D}^n(t') dt'$$

The variation of the meridian and transverse radii of curvature, R_N and R_E, with the geodetic latitude, L_b, is weak, so it is acceptable to neglect their variation with latitude over the integration interval. Assuming the velocity varies as a linear function of time over the integration interval, a suitable approximation for the position update is

$$\begin{aligned}
h_b(+) &= h_b(-) - \frac{\tau_i}{2}\left(v_{eb,D}^n(-) + v_{eb,D}^n(+)\right) \\
L_b(+) &\approx L_b(-) + \frac{\tau_i}{2}\left(\frac{v_{eb,N}^n(-)}{R_N(L_b(-)) + h_b(-)} + \frac{v_{eb,N}^n(+)}{R_N(L_b(-)) + h_b(+)}\right) \\
\lambda_b(+) &= \lambda_b(-) + \frac{\tau_i}{2}\left(\frac{v_{eb,E}^n(-)}{(R_E(L_b(-)) + h_b(-))\cos L_b(-)} + \frac{v_{eb,E}^n(+)}{(R_E(L_b(+)) + h_b(+))\cos L_b(+)}\right)
\end{aligned} \quad (5.56)$$

noting that the height, latitude, and longitude should be calculated in that order.[†] The longitude update does not work at the poles because $1/\cos L_b$ approaches infinity.

Alternatively, the position may be updated by solving

$$\dot{\mathbf{C}}_n^e = \mathbf{C}_n^e \mathbf{\Omega}_{en}^n, \quad (5.57)$$

where (2.150) and (2.151) provide the conversion between $\mathbf{C}_n^e$ and L_b and λ_b. A first-order solution is

$$\mathbf{C}_n^e(+) \approx \mathbf{C}_n^e(-)\left(\mathbf{I}_3 + \tfrac{1}{2}(\mathbf{\Omega}_{en}^n(-) + \mathbf{\Omega}_{en}^n(+))\tau_i\right), \quad (5.58)$$

where $\mathbf{\Omega}_{en}^n(-)$ and $\mathbf{\Omega}_{en}^n(+)$ are computed using (5.44) from $\mathbf{v}_{eb}^n(-)$ and $\mathbf{v}_{eb}^n(+)$, respectively. This approach also fails at the poles because $\omega_{en,z}^e$ approaches infinity.

5.4.5 Wander-Azimuth Implementation

Inertial navigation equations can be mechanized in the axes of a wander-azimuth frame to minimize the effects of the polar singularities that occur in a local navigation frame [5]. A wander-azimuth coordinate frame (see Section 2.1.6), denoted by w, is closely related to the corresponding local navigation frame. The z-axis is

---

[†]End of QinetiQ copyright material.

5.4 Local-Navigation-Frame Navigation Equations

coincidental, pointing down, but the *x*- and *y*-axes are rotated about the *z*-axis with respect to the local navigation frame by a wander angle that varies with position. The wander angle is simply the heading (or azimuthal) Euler angle from the local navigation frame to the wander azimuth frame, ψ_{nw}, although many authors use α. Thus, from (2.22) and (2.24),

$$\mathbf{C}_n^w = \begin{pmatrix} \cos\psi_{nw} & \sin\psi_{nw} & 0 \\ -\sin\psi_{nw} & \cos\psi_{nw} & 0 \\ 0 & 0 & 1 \end{pmatrix} \qquad \mathbf{C}_w^n = \begin{pmatrix} \cos\psi_{nw} & -\sin\psi_{nw} & 0 \\ \sin\psi_{nw} & \cos\psi_{nw} & 0 \\ 0 & 0 & 1 \end{pmatrix}. \quad (5.59)$$

The wander angle is generally initialized at zero at the start of navigation. Note that some authors use the wander angle with the opposing sign, ψ_{wn}, which may also be denoted as α.

Latitude and longitude in a wander-azimuth implementation are replaced by the ECEF frame to wander-azimuth frame coordinate transformation matrix, $\mathbf{C}_e^w$. From (2.15), (2.150), and (5.59), this may expressed in terms of the latitude, longitude, and wander angle using

$$\mathbf{C}_e^w = \begin{bmatrix} \begin{pmatrix} -\sin L_b \cos \lambda_b \cos\psi_{nw} \\ -\sin \lambda_b \sin\psi_{nw} \end{pmatrix} & \begin{pmatrix} -\sin L_b \sin \lambda_b \cos\psi_{nw} \\ +\cos \lambda_b \sin\psi_{nw} \end{pmatrix} & \cos L_b \cos\psi_{nw} \\ \begin{pmatrix} \sin L_b \cos \lambda_b \sin\psi_{nw} \\ -\sin \lambda_b \cos\psi_{nw} \end{pmatrix} & \begin{pmatrix} \sin L_b \sin \lambda_b \sin\psi_{nw} \\ +\cos \lambda_b \cos\psi_{nw} \end{pmatrix} & -\cos L_b \sin\psi_{nw} \\ -\cos L_b \cos \lambda_b & -\cos L_b \sin \lambda_b & -\sin L_b \end{bmatrix}. \quad (5.60)$$

Conversely,

$$\begin{aligned} L_b &= -\arcsin\left(C_{e,3,3}^w\right) \\ \lambda_b &= \arctan_2\left(-C_{e,3,2}^w, -C_{e,3,1}^w\right), \\ \psi_{nw} &= \arctan_2\left(-C_{e,2,3}^w, C_{e,1,3}^w\right) \end{aligned} \quad (5.61)$$

noting that the longitude and wander angle are undefined at the poles ($L_b = \pm 90°$) and may be subject to significant computational rounding errors near the poles.

The attitude, velocity, and height inertial navigation equations in a wander-azimuth frame are as those for a local navigation frame, presented earlier, with *w* substituted for *n*, except that the transport-rate term has no component about the vertical axis. Thus,

$$\boldsymbol{\omega}_{ew}^w = \mathbf{C}_n^w \begin{pmatrix} \omega_{en,N}^n \\ \omega_{en,E}^n \\ 0 \end{pmatrix}. \quad (5.62)$$

From (5.44), this may be obtained from the wander-azimuth-resolved velocity using

$$\omega_{ew}^w = C_n^w \begin{pmatrix} 0 & 1/(R_E(C_{e,3,3}^w) + h_b) & 0 \\ -1/(R_N(C_{e,3,3}^w) + h_b) & 0 & 0 \\ 0 & 0 & 0 \end{pmatrix} C_w^n v_{eb}^w$$

$$= \begin{pmatrix} \dfrac{\cos\psi_{nw} \sin\psi_{nw}}{R_E(C_{e,3,3}^w) + h_b} - \dfrac{\cos\psi_{nw} \sin\psi_{nw}}{R_N(C_{e,3,3}^w) + h_b} & \dfrac{\cos^2\psi_{nw}}{R_E(C_{e,3,3}^w) + h_b} + \dfrac{\sin^2\psi_{nw}}{R_N(C_{e,3,3}^w) + h_b} & 0 \\ -\dfrac{\cos^2\psi_{nw}}{R_N(C_{e,3,3}^w) + h_b} - \dfrac{\sin^2\psi_{nw}}{R_E(C_{e,3,3}^w) + h_b} & \dfrac{\cos\psi_{nw} \sin\psi_{nw}}{R_N(C_{e,3,3}^w) + h_b} - \dfrac{\cos\psi_{nw} \sin\psi_{nw}}{R_E(C_{e,3,3}^w) + h_b} & 0 \\ 0 & 0 & 0 \end{pmatrix} v_{eb}^w,$$

(5.63)

where from (2.105), (2.106), and (5.61), the meridian and transverse radii of curvature may be expressed directly in terms of $C_{e,3,3}^w$ using

$$R_N(C_{e,3,3}^w) = \frac{R_0(1-e^2)}{\left(1 - e^2 {C_{e,3,3}^w}^2\right)^{3/2}}, \qquad R_E(C_{e,3,3}^w) = \frac{R_0}{\sqrt{1 - e^2 {C_{e,3,3}^w}^2}}. \qquad (5.64)$$

At the poles, $R_N = R_E = R_0/\sqrt{1-e^2}$. Therefore, near the poles (e.g., where $|C_{e,3,3}^w| > 0.99995$), (5.63) may be replaced by

$$\omega_{ew}^w \approx \frac{1}{\dfrac{R_0}{\sqrt{1-e^2}} + h_b} \begin{pmatrix} 0 & 1 & 0 \\ -1 & 0 & 0 \\ 0 & 0 & 0 \end{pmatrix} v_{eb}^w, \qquad (5.65)$$

avoiding the need to compute the wander angle.

The Earth-rotation vector is, from (2.143),

$$\omega_{ie}^w = C_n^w \omega_{ie}^n. \qquad (5.66)$$

To the first order in time, the latitude and longitude may be updated using

$$C_e^w(+) \approx \left(I_3 - \tfrac{1}{2}(\Omega_{ew}^w(-) + \Omega_{ew}^w(+))\tau_i\right)C_e^w(-), \qquad (5.67)$$

where $\Omega_{ew}^w(-)$ and $\Omega_{ew}^w(+)$ are computed using (5.63) from $v_{eb}^w(-)$ and $v_{eb}^w(+)$, respectively. The height may be updated using (5.56), noting that $v_{eb,D}^n = v_{eb,z}^w$.

5.5 Navigation Equations Optimization

The inertial navigation equations presented in the preceding sections are approximate and exhibit errors that increase with the host vehicle dynamics, vibration level, and update interval. This section presents precision navigation equations that offer higher accuracy at the cost of greater complexity and processing load. This is followed by a discussion of the effects of the sensor sampling interval and vibration, including coning and sculling errors, and their mitigation. The section concludes with a discussion of the design tradeoffs that must be made in selecting suitable iteration rates and approximations for different inertial navigation applications. Factors to consider include performance requirements, operating environment, sensor quality, processing capacity, and available development time.

The MATLAB functions on the CD, Nav_equations_ECI, Nav_equations_ECEF, and Nav_equations_NED, respectively, implement the ECI-frame, ECEF-frame, and local-navigation-frame versions of the precision inertial navigation equations described in this section.

5.5.1 Precision Attitude Update

It is convenient to define the attitude update matrix as the coordinate transformation matrix from the body frame at the end of the attitude update step of the navigation equations to that at the beginning, C_{b+}^{b-} (some authors use $\mathbf{A}$). It may be used to define the attitude update step in an ECI frame; thus,

$$\begin{aligned} \mathbf{C}_b^i(+) &= \mathbf{C}_b^i(-)\mathbf{C}_{b+}^{b-} \\ \mathbf{C}_{b+}^{b-} &= \mathbf{C}_i^b(-)\mathbf{C}_b^i(+) \end{aligned} \quad (5.68)$$

Substituting (5.10) and (5.11) into (5.68) defines the attitude update matrix in terms of the attitude increment, α_{ib}^b:

$$\mathbf{C}_{b+}^{b-} = \mathbf{C}_{b(t+\tau_i)}^{b(t)} = \exp\left[\alpha_{ib}^b \wedge\right] = \sum_{r=0}^{\infty} \frac{\left[\alpha_{ib}^b \wedge\right]^r}{r!} \quad (5.69)$$

where a constant angular rate is assumed.

When the power-series expansion is truncated, errors arise depending on the step size of the attitude increment and the order at which the power series is truncated. Table 5.1 presents some examples [1]. Clearly, the third- and fourth-order algorithms perform significantly better than the first- and second-order algorithms. It should also be noted that the error varies as the square of the attitude increment for the first- and second-order algorithms, but as the fourth power for the third- and fourth-order variants. Thus, with the higher-order algorithms, increasing the iteration rate has more impact on the accuracy.

In practice, there are few applications where the host vehicle rotates continuously in the same direction, while errors arising from angular oscillation about

Table 5.1 Drift of First- to Fourth-Order Attitude Update Algorithms at an Update Rate of 100 Hz

Algorithm Order	Attitude Drift at 100-Hz update rate, rad s^{-1} (° hr^{-1})	
	$|\alpha| = 0.1$ rad step size	$|\alpha| = 0.05$ rad step size
1	0.033 (6,830)	8.3×10^{-3} (1,720)
2	0.017 (3,430)	4.2×10^{-3} (860)
3	3.4×10^{-5} (6.9)	2.5×10^{-6} (0.4)
4	8.3×10^{-6} (1.7)	5.2×10^{-7} (0.1)

a single axis cancel out over time. Problems generally occur when there is synchronized angular oscillation about two axes, known as coning, in which case using the first-order attitude update leads to an attitude drift about the mutually perpendicular axis that is generally proportional to the product of the amplitudes of the two oscillations and does not change sign. Thus, the ensuing attitude error increases with time. Similar errors occur in the presence of synchronized angular and linear oscillation, known as sculling. Coning and sculling are discussed further in Section 5.5.4.

The third and fourth powers of a skew-symmetric matrix have the following properties:

$$[\mathbf{x} \wedge]^3 = -|\mathbf{x}|^2 [\mathbf{x} \wedge] \\ [\mathbf{x} \wedge]^4 = -|\mathbf{x}|^2 [\mathbf{x} \wedge]^2. \qquad (5.70)$$

Substituting this into (5.69):

$$\mathbf{C}_{b+}^{b-} = \mathbf{I}_3 + \left(\sum_{r=0}^{\infty} (-1)^r \frac{|\boldsymbol{\alpha}_{ib}^b|^{2r}}{(2r+1)!}\right) [\boldsymbol{\alpha}_{ib}^b \wedge] + \left(\sum_{r=0}^{\infty} (-1)^r \frac{|\boldsymbol{\alpha}_{ib}^b|^{2r}}{(2r+2)!}\right) [\boldsymbol{\alpha}_{ib}^b \wedge]^2. \qquad (5.71)$$

The fourth-order approximation is then

$$\mathbf{C}_{b+}^{b-} \approx \mathbf{I}_3 + \left(1 - \frac{|\boldsymbol{\alpha}_{ib}^b|^2}{6}\right)[\boldsymbol{\alpha}_{ib}^b \wedge] + \left(\frac{1}{2} - \frac{|\boldsymbol{\alpha}_{ib}^b|^2}{24}\right)[\boldsymbol{\alpha}_{ib}^b \wedge]^2. \qquad (5.72)$$

However, the power-series expansions in (5.71) are closely related to those of the sine and cosine, so

$$\mathbf{C}_{b+}^{b-} = \mathbf{I}_3 + \frac{\sin|\boldsymbol{\alpha}_{ib}^b|}{|\boldsymbol{\alpha}_{ib}^b|}[\boldsymbol{\alpha}_{ib}^b \wedge] + \frac{1 - \cos|\boldsymbol{\alpha}_{ib}^b|}{|\boldsymbol{\alpha}_{ib}^b|^2}[\boldsymbol{\alpha}_{ib}^b \wedge]^2. \qquad (5.73)$$

This is known as Rodrigues' formula. To avoid division by zero, this should be replaced with the approximate version whenever $|\boldsymbol{\alpha}_{ib}^b|$ is very small.

The ECI-frame attitude update may thus be performed exactly. Note that the inverse of (5.73) gives the attitude increment vector in terms of the attitude update matrix:

$$\alpha_{ib}^b = \frac{\mu_{b+b-}}{2\sin\mu_{b+b-}} \begin{pmatrix} C_{b+3,2}^{b-} - C_{b+2,3}^{b-} \\ C_{b+1,3}^{b-} - C_{b+3,1}^{b-} \\ C_{b+2,1}^{b-} - C_{b+1,2}^{b-} \end{pmatrix}, \quad \mu_{b+b-} = \arccos\left[\frac{\mathrm{Tr}(C_{b+}^{b-}) - 1}{2}\right]. \quad (5.74)$$

A similar approach may be taken with the ECEF-frame attitude update. For precision, the first-order solution, (5.27), is replaced by

$$\begin{aligned} \mathbf{C}_b^e(+) &= \begin{pmatrix} \cos\omega_{ie}\tau_i & \sin\omega_{ie}\tau_i & 0 \\ -\sin\omega_{ie}\tau_i & \cos\omega_{ie}\tau_i & 0 \\ 0 & 0 & 1 \end{pmatrix} \mathbf{C}_b^e(-)\mathbf{C}_{b+}^{b-}, \\ &\approx \mathbf{C}_b^e(-)\mathbf{C}_{b+}^{b-} - \mathbf{\Omega}_{ie}^e \mathbf{C}_b^e(-)\tau_i \end{aligned} \quad (5.75)$$

where the attitude update matrix is given by (5.73) as before. Note that where the first-order approximation is retained for the Earth-rate term, it introduces an error of only 1.3×10^{-15} rad s^{-1} (7.4×10^{-14} ° hr^{-1}) at a 10-Hz update rate and 1.3×10^{-17} rad s^{-1} (7.4×10^{-16} ° hr^{-1}) at a 100-Hz update rate, which is much less that the bias of even the most accurate gyros. Thus, this is an exact solution for all practical purposes.

In the local-navigation-frame attitude update, there is also a transport-rate term, ω_{en}^n, given by (5.44). For velocities up to 467 m s^{-1} (Mach 1.4), this is less than the Earth-rotation rate, so for the vast majority of applications, it is valid to truncate the power-series expansion of $\exp(\mathbf{\Omega}_{en}^n \tau)$ to first order. Thus, for improved precision, the first-order solution, (5.46), is replaced by

$$\mathbf{C}_b^n(+) \approx \mathbf{C}_b^n(-)\mathbf{C}_{b+}^{b-} - \left(\mathbf{\Omega}_{ie}^e(-) + \mathbf{\Omega}_{en}^n(-)\right)\mathbf{C}_b^n(-)\tau_i. \quad (5.76)$$

However, for high-precision, high-dynamic applications, the variation of the transport rate over the update interval can be significant. When a high-precision specific-force frame transformation (Section 5.5.2) is implemented, the updated attitude is not required at that stage. This enables the attitude update step to be moved from the beginning to the end of the navigation equations processing cycle, enabling an averaged transport rate to be used for the attitude update:

$$\mathbf{C}_b^n(+) = \left[\mathbf{I}_3 - \left(\mathbf{\Omega}_{ie}^e(-) + \tfrac{1}{2}\mathbf{\Omega}_{en}^n(-) + \tfrac{1}{2}\mathbf{\Omega}_{en}^n(+)\right)\tau_i\right]\mathbf{C}_b^n(-)\mathbf{C}_{b+}^{b-}, \quad (5.77)$$

where $\mathbf{\Omega}_{en}^n(+)$ is calculated using $L_b(+)$, $h_b(+)$, and $\mathbf{v}_{eb}^n(+)$. Figure 5.8 shows the modified block diagram for the precision local-navigation-frame navigation equations.

Coordinate transformation matrices are orthonormal, (2.17), so the scalar product of any two rows or any two columns should be zero. Orthonormality is

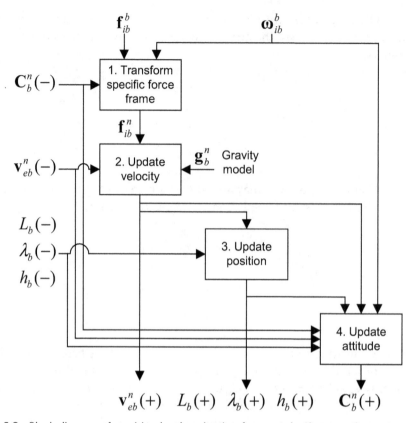

Figure 5.8 Block diagram of precision local-navigation-frame navigation equations.

maintained through exact navigation equations. However, the use of approximations and the presence of computational rounding errors can cause departures from this. Consequently, it can be useful to implement a reorthogonalization and renormalization algorithm at regular intervals.[*]

Breaking down the coordinate transformation matrix (frames omitted) into three rows,

$$\mathbf{C} = \begin{pmatrix} \mathbf{c}_1^\mathrm{T} \\ \mathbf{c}_2^\mathrm{T} \\ \mathbf{c}_3^\mathrm{T} \end{pmatrix}. \tag{5.78}$$

Orthogonalization is achieved by calculating $\Delta_{ij} = \mathbf{c}_i^\mathrm{T} \mathbf{c}_j$ for each pair of rows and apportioning a correction equally between them:[†]

---

[*]This and subsequent paragraphs are based on material written by the author for QinetiQ, so comprise QinetiQ copyright material.

[†]End of QinetiQ copyright material.

5.5 Navigation Equations Optimization

$$\begin{aligned}c_1(+) &\approx c_1(-) - \tfrac{1}{2}\Delta_{12}c_2(-) - \tfrac{1}{2}\Delta_{13}c_3(-) \\ c_2(+) &\approx c_2(-) - \tfrac{1}{2}\Delta_{12}c_1(-) - \tfrac{1}{2}\Delta_{23}c_3(-) \\ c_3(+) &\approx c_3(-) - \tfrac{1}{2}\Delta_{13}c_1(-) - \tfrac{1}{2}\Delta_{23}c_2(-) \end{aligned} \qquad (5.79)$$

Normalization is subsequently applied to each row by

$$\begin{aligned} c_i(+) &= \frac{1}{\sqrt{c_i^T(-)c_i(-)}} c_i(-) \\ &\approx \frac{2}{1 + c_i^T(-)c_i(-)} c_i(-) \end{aligned} \qquad (5.80)$$

The orthonormalization may also be performed column by column. Note that these corrections work best when the departure from orthonormality is small.

5.5.2 Precision Specific-Force Frame Transformation

The specific force in ECI-frame resolving axes is instantaneously related to that in the body-frame axes by [repeating (5.15)]:

$$\mathbf{f}_{ib}^i(t) = \mathbf{C}_b^i(t)\mathbf{f}_{ib}^b(t).$$

The IMU outputs the average specific force over the interval t to $t + \tau_i$ and the ECI-axes specific force is similarly averaged. The transformation is thus

$$\mathbf{f}_{ib}^i = \bar{\mathbf{C}}_b^i \mathbf{f}_{ib}^b, \qquad (5.81)$$

where the average coordinate transformation matrix over the time interval is

$$\bar{\mathbf{C}}_b^i = \frac{1}{\tau_i} \int_t^{t+\tau_i} \mathbf{C}_b^i(t')\,dt'. \qquad (5.82)$$

Substituting in (5.11), noting that the variation of the angular rate over the integration interval is unknown,

$$\begin{aligned} \bar{\mathbf{C}}_b^i &= \frac{1}{\tau_i}\mathbf{C}_b^i(-)\int_0^{t+\tau_i} \sum_{r=0}^{\infty} \frac{\{(t'/\tau_i)[\alpha_{ib}^b \wedge]\}^r}{r!} dt' \\ &= \mathbf{C}_b^i(-) \sum_{r=0}^{\infty} \frac{[\alpha_{ib}^b \wedge]^r}{(r+1)!} \end{aligned} \qquad (5.83)$$

Applying (5.70),

$$\bar{C}_b^i = C_b^i(-)C_{\bar{b}}^{b-}, \quad C_{\bar{b}}^{b-} = I_3 + \frac{1-\cos|\alpha_{ib}^b|}{|\alpha_{ib}^b|^2}[\alpha_{ib}^b \wedge] + \frac{1}{|\alpha_{ib}^b|^2}\left(1 - \frac{\sin|\alpha_{ib}^b|}{|\alpha_{ib}^b|}\right)[\alpha_{ib}^b \wedge]^2. \quad (5.84)$$

Again, this should be replaced with the approximate version whenever $|\alpha_{ib}^b|$ is very small to avoid division by zero.

Substituting this into (5.81) or (5.17), the specific force in ECI-frame resolving axes, f_{ib}^i, or the integrated specific force, v_{ib}^i, may be calculated exactly. Note that $\bar{C}_b^i$ is not an orthonormal matrix. Therefore, to reverse the transformation described by (5.81), $\bar{C}_b^i$ must be inverted.

Retaining the first-order approximation for the Earth-rate term, the precise transformation of the specific force to ECEF-frame axes is

$$f_{ib}^e = \bar{C}_b^e f_{ib}^b, \quad \bar{C}_b^e = C_b^e(-)C_{\bar{b}}^{b-} - \tfrac{1}{2}\Omega_{ie}^e C_b^e(-)\tau_i. \quad (5.85)$$

To transform the specific force to local-navigation-frame axes, the first-order approximation is also used for the transport-rate term as the velocity at time $t + \tau_i$ has yet to be computed:

$$f_{ib}^n = \bar{C}_b^n f_{ib}^b, \quad \bar{C}_b^n = C_b^n(-)C_{\bar{b}}^{b-} - \tfrac{1}{2}\left(\Omega_{ie}^n(-) + \Omega_{en}^n(-)\right)C_b^n(-)\tau_i. \quad (5.86)$$

To transform integrated specific force to ECEF and local-navigation-frame axes, $\bar{C}_b^e$ and $\bar{C}_b^n$ are substituted into (5.29) and (5.48), respectively.

The error arising from transforming the specific force as described in Sections 5.2.2, 5.3.2, and 5.4.2 varies approximately as the square of the attitude increment and is maximized where the rotation axis is perpendicular to the direction of the specific force. The maximum fractional error is 8.3×10^{-4} for $|\alpha_{ib}^b| = 0.1$ rad and 2.1×10^{-4} for $|\alpha_{ib}^b| = 0.05$ rad.

5.5.3 Precision Velocity and Position Updates

When the navigation equations are iterated at the IMU output rate and a constant acceleration may be assumed, the ECI-frame velocity and position update equations presented in Sections 5.2.3 and 5.2.4 are exact, except for the variation in gravitation over the update interval, which is small enough to be neglected. However, in the ECEF and local-navigation-frame implementations, exact evaluation of the Coriolis and transport-rate terms requires knowledge of the velocity at the end of the update interval, requiring a recursive solution. For most applications, the first-order approximation in (5.36) and (5.54) is sufficient. However, this may lead to significant errors for high-accuracy, high-dynamic applications. One solution is to predict forward the velocity using previous velocity solutions [2]. A better, but more processor-intensive, solution is a two-step recursive method, shown here for the local-navigation-frame implementation:

5.5 Navigation Equations Optimization

$$\mathbf{v}_{eb}^{n\prime} = \mathbf{v}_{eb}^n(-) + \left[\mathbf{f}_{ib}^n + \mathbf{g}_b^n(L_b(-), h_b(-)) - (\mathbf{\Omega}_{en}^n(-) + 2\mathbf{\Omega}_{ie}^n(-))\mathbf{v}_{eb}^n(-)\right]\tau_i$$

$$\mathbf{v}_{eb}^n(+) = \mathbf{v}_{eb}^n(-) + \begin{cases} \mathbf{f}_{ib}^n + \mathbf{g}_b^n(L_b(-), h_b(-)) - \tfrac{1}{2}[\mathbf{\Omega}_{en}^n(-) + 2\mathbf{\Omega}_{ie}^n(-)]\mathbf{v}_{eb}^n(-) \\ -\tfrac{1}{2}[\mathbf{\Omega}_{en}^n(L_b(-), h_b(-), \mathbf{v}_{eb}^{n\prime}) + 2\mathbf{\Omega}_{ie}^n(-)]\mathbf{v}_{eb}^{n\prime} \end{cases}\tau_i \quad (5.87)$$

Provided they are iterated at the same rate as the velocity update, the ECI- and ECEF-frame position updates introduce no further approximations beyond those made in the velocity update, while the effect of the meridian radius of curvature approximation in (5.56) is negligible.

When the latitude and longitude are updated using the coordinate transformation matrices from a local-navigation or wander-azimuth frame to an ECEF frame, greater accuracy may be obtained using Rodrigues' formula:

$$\begin{aligned} \mathbf{C}_n^e(+) &= \mathbf{C}_n^e(-)\left(\mathbf{I}_3 + \frac{\sin|\boldsymbol{\alpha}_{en}^n|}{|\boldsymbol{\alpha}_{en}^n|}[\boldsymbol{\alpha}_{en}^n \wedge] + \frac{1-\cos|\boldsymbol{\alpha}_{en}^n|}{|\boldsymbol{\alpha}_{en}^n|^2}[\boldsymbol{\alpha}_{en}^n \wedge]^2\right) \\ \mathbf{C}_w^e(+) &= \mathbf{C}_w^e(-)\left(\mathbf{I}_3 + \frac{\sin|\boldsymbol{\alpha}_{ew}^w|}{|\boldsymbol{\alpha}_{ew}^w|}[\boldsymbol{\alpha}_{ew}^w \wedge] + \frac{1-\cos|\boldsymbol{\alpha}_{ew}^w|}{|\boldsymbol{\alpha}_{ew}^w|^2}[\boldsymbol{\alpha}_{ew}^w \wedge]^2\right) \end{aligned} \quad (5.88)$$

where

$$\begin{aligned} \boldsymbol{\alpha}_{en}^n &= \int_t^{t+\tau_i} \boldsymbol{\omega}_{en}^n(t')\,dt' \approx \tfrac{1}{2}(\boldsymbol{\omega}_{en}^n(-) + \boldsymbol{\omega}_{en}^n(+))\tau_i \\ \boldsymbol{\alpha}_{ew}^w &= \int_t^{t+\tau_i} \boldsymbol{\omega}_{ew}^w(t')\,dt' \approx \tfrac{1}{2}(\boldsymbol{\omega}_{ew}^w(-) + \boldsymbol{\omega}_{ew}^w(+))\tau_i \end{aligned} \quad (5.89)$$

and approximate versions (see Section 5.5.1) should be used whenever $|\boldsymbol{\alpha}_{en}^n|$ or $|\boldsymbol{\alpha}_{ew}^w|$ is small to avoid division by zero.

Gravity model limitations can contribute several hundred meters to the position error over the course of an hour. Therefore, where precision inertial sensors and navigation equations are used, navigation accuracy can be significantly improved by using a precision gravity model (see Section 2.4.7) [6]. Alternatively, a gravity gradiometer (Section 13.4.1) can be used to measure gravitational variations in real time [7].

5.5.4 Effects of Sensor Sampling Interval and Vibration

The inertial sensor measurements enable the average specific force and angular rate over the sensor sampling interval, τ_i, to be determined. However, they do not give information on the variation in specific force and angular rate over that interval. Figure 5.9 shows different specific force or angular rate profiles that produce the same sensor output. Inertial navigation processing typically operates under the

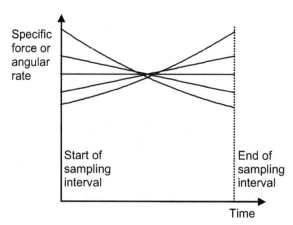

Figure 5.9 Different specific force or angular rate profiles producing the same sensor output.

assumption that the specific force and angular rate, as resolved in the body frame, are constant over the sampling interval. Figure 5.10 illustrates this.

If the direction of rotation remains constant over the gyro sampling interval, the same attitude update will be obtained from the average angular rate as from the true angular rate. However, if the direction of rotation changes, errors will occur because successive rotations about different directions do not commute (see Section 2.2).

Similarly, if the attitude of the IMU body remains constant over the accelerometer sampling interval, the same velocity update will be obtained from the average specific force as from the true specific force. However, if the body is rotating, any unknown variation in the specific force will result in an error in the transformation of the specific force into the resolving axes used for the velocity computation. A similar error will occur where the angular rate is changing even if the specific force is constant. Note also that assuming a constant acceleration in the presence of jerk (rate of change of acceleration) leads to an error in the position update.

Consider three examples, all assuming a 100-Hz IMU sampling rate. First, a 1 rad s^{-1} angular rate is combined with an angular acceleration of 1 rad s^{-2} about

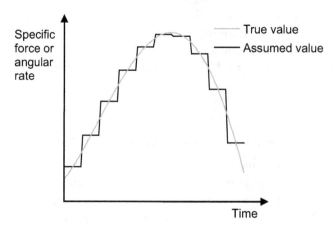

Figure 5.10 True and assumed sensor outputs.

a perpendicular axis. This leads to an angular rate error of 1.7×10^{-5} rad s^{-1} (3.5 ° hr^{-1}) about the mutually perpendicular axis. Second, a combination of a 1 rad s^{-2} angular acceleration with a specific force along a perpendicular axis of 10 m s^{-2} leads to a specific force frame transformation error of 8.3×10^{-5} m s^{-2} (8.5 μg) along the mutually perpendicular axis. Finally, the combination of a 100 m s^{-3} jerk with an angular rate of 1 rad s^{-1} about a perpendicular axis leads to a specific force frame transformation error along the mutually perpendicular axis of 8.3×10^{-4} m s^{-2} (985 μg). When such conditions result from dynamic maneuvers of the host vehicle, the duration over which the specific force and angular rate errors apply will typically be short, while successive maneuvers will often produce canceling errors. However, vibration-induced errors can have a more significant impact on inertial navigation performance.

The effects of vibration may be illustrated by the cases of coning and sculling motion. Coning motion is synchronized angular oscillation about two orthogonal axes as shown in Figure 5.11. Where there is a phase difference between the two oscillations, the resultant axis of rotation precesses, describing a cone-like surface. Note that mechanical dithering of an RLG triad (see Section 4.2.1.1) induces coning motion [8].

If the output of a triad of gyroscopes is integrated over a period, τ, in the presence of coning motion of angular frequency, ω_c, and angular amplitudes, θ_i and θ_j, with a phase difference, ϕ, between the two axes, it can be shown [1] that a false rotation, $\delta\omega_c$, is sensed about the axis orthogonal to θ_i and θ_j, where

$$\delta\omega_c = \omega_c \theta_i \wedge \theta_j \sin\phi \left(1 - \frac{\sin\omega_c\tau}{\omega_c\tau}\right). \tag{5.90}$$

This arises due to the difference between the actual and assumed order of rotation over the integration period. The coning error, $\delta\omega_c$, does not oscillate. Therefore, the attitude solution drifts under a constant coning motion. The higher the frequency of the coning motion and the longer the gyro outputs are integrated, the larger the drift will be. For example, if the coning amplitude is 1 mrad, the frequency is 100 rad s^{-1} (15.9 Hz), and the integration interval is 0.01 second, the maximum coning error is 1.59×10^{-5} rad s^{-1} (3.3 ° hr^{-1}). For a vibration frequency of 200 rad s^{-1}, the maximum error is 1.09×10^{-4} rad s^{-1} (22.5 ° hr^{-1}). These values assume the use of exact navigation equations. Much larger coning errors can occur where approximations are made, particularly in the attitude update step.

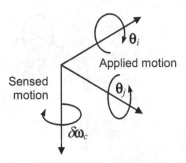

Figure 5.11 Coning motion. (*From:* [9]. ©2002 QinetiQ Ltd. Reprinted with permission.)

Sculling motion is synchronized angular oscillation about one axis and linear oscillation about an orthogonal axis as shown in Figure 5.12. This results in an error in the output of an accelerometer triad. If the angular frequency is ω_s and the acceleration amplitude is $\mathbf{a}_j$, a false acceleration, $\delta \mathbf{a}_s$, is sensed about the axis orthogonal to $\mathbf{\theta}_i$ and $\mathbf{a}_j$. From [1],

$$\delta \mathbf{a}_s = \tfrac{1}{2} \mathbf{\theta}_i \wedge \mathbf{a}_j \cos\phi \left(1 - \frac{\sin \omega_s \tau}{\omega_s \tau}\right). \tag{5.91}$$

Similarly, the sculling error, $\delta \mathbf{a}_s$, does not oscillate so the navigation solution drifts under constant sculling motion. Again, the resulting acceleration error is larger for longer integration times and higher sculling frequencies. For example, if the angular vibration amplitude is 1 mrad, the linear vibration amplitude is 1 mm, the frequency is 100 rad s^{-1} (15.9 Hz), and the integration interval is 0.01 second, the maximum sculling error is 7.9×10^{-4} m s^{-2} (79 μg). For a vibration frequency of 200 rad s^{-1}, the maximum error is 1.09×10^{-2} m s^{-1} (1.1 mg). Again, larger errors can occur when approximate navigation equations are used.

Although long periods of in-phase coning and sculling rarely occur in real systems, the navigation solution can still be significantly degraded by the effects of orthogonal vibration modes. Therefore, coning and sculling motion provides a useful test case for inertial navigation equations. The extent to which the navigation equations design must protect against the effects of vibration depends on both the accuracy requirements and the vibration environment. An example of a high-vibration environment is an aircraft wing pylon, where a guided weapon or sensor pod may be mounted.

The coning and sculling errors, together with the other errors that can arise from averaging specific force and angular rate, vary approximately as the square of the averaging interval. Consequently, when the navigation equations are iterated at a lower rate than the IMU output to reduce the processor load (see Section 5.5.5), successive IMU outputs should not be simply averaged.

The angular rate measurements should be combined using a method that minimizes coning errors. When the integration interval for the attitude update comprises n IMU output intervals, an exact attitude update matrix may be constructed by multiplying the attitude update matrices for each interval:

$$\mathbf{C}_{b+}^{b-} = \exp\left[\mathbf{\alpha}_{ib,1}^{b} \wedge\right] \exp\left[\mathbf{\alpha}_{ib,2}^{b} \wedge\right] \ldots \exp\left[\mathbf{\alpha}_{ib,n}^{b} \wedge\right], \tag{5.92}$$

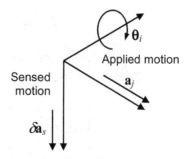

Figure 5.12 Sculling motion. (*From:* [9]. ©2002 QinetiQ Ltd. Reprinted with permission.)

5.5 Navigation Equations Optimization

where

$$\alpha_{ib,j}^b = \int_{t+(j-1)\tau_i/n}^{t+j\tau_i/n} \omega_{ib}^b(t')\, dt'. \tag{5.93}$$

Implementing (5.92) as it stands offers no computational saving over performing the attitude update at the IMU rate. From (2.43), the attitude update matrix may be expressed in terms of a rotation vector [10]:

$$C_{b+}^{b-} = \exp[\rho_{b-b+} \wedge]. \tag{5.94}$$

Note that the body-frame rotation vector, ρ_{b-b+}, is equal to the attitude increment of the body frame with respect to inertial space in body-frame axes, α_{ib}^b, over the same time interval. Thus,

$$\rho_{b-b+} = \int_t^{t+\tau_i} \omega_{ib}^b(t')\, dt'. \tag{5.95}$$

Note, however, that rotation vectors and attitude increments are not the same in general.

As the direction of rotation varies between successive measurements, the rotation vector is not simply the sum of the attitude increments. In physical terms, this is because the resolving axes vary between successive attitude increments. In mathematical terms, the skew-symmetric matrices of successive attitude increments do not commute.

From [10], the rate of change of the rotation vector varies with the angular rate as

$$\dot{\rho}_{b-b+} = \omega_{ib}^b + \frac{1}{2}\rho_{b-b+} \wedge \omega_{ib}^b + \frac{1}{|\rho_{b-b+}|^2}\left[1 - \frac{|\rho_{b-b+}|\sin|\rho_{b-b+}|}{2(1-\cos|\rho_{b-b+}|)}\right]\rho_{b-b+} \wedge \rho_{b-b+} \wedge \omega_{ib}^b. \tag{5.96}$$

From [2, 11], a second-order approximation incorporating only the first two terms of (5.96) gives the following solution:

$$\rho_{b-b+} \approx \sum_{j=1}^n \alpha_{ib,j}^b + \frac{1}{2}\sum_{j=1}^{n-1}\sum_{k=j+1}^n \alpha_{ib,j}^b \wedge \alpha_{ib,k}^b. \tag{5.97}$$

Note that, where sufficient processing capacity is available, it is both simpler and more accurate to iterate the attitude update at the IMU output rate.

Similarly, when the specific force in the resolving axes used for the velocity update is integrated over more than one IMU output interval, the specific-force transformation should account for the fact that each successive IMU specific-force measurement may be resolved about a different set of axes as the body-frame orientation changes. This minimizes the sculling error. A second-order transformation and summation of n successive IMU-specific force measurements into an ECI frame is, from [1, 2, 11],

$$v^i_{ib,\Sigma} \approx C^i_b(-)\left[\sum_{j=1}^{n} v^b_{ib,j} + \frac{1}{2}\sum_{j=1}^{n}\sum_{k=1}^{n} \alpha^b_{ib,j} \wedge v^b_{ib,k} + \frac{1}{2}\sum_{j=1}^{n-1}\sum_{k=j+1}^{n}\left(\alpha^b_{ib,j} \wedge v^b_{ib,k} - \alpha^b_{ib,k} \wedge v^b_{ib,j}\right)\right].$$

(5.98)

where $v^b_{ib,j}$ and $\alpha^b_{ib,j}$ are the jth integrated-specific-force and attitude-increment outputs from the IMU, and $v^i_{ib,\Sigma}$ is the summed integrated specific force in ECI resolving axes. Again, where there is sufficient processing capacity, it is simpler and more accurate to iterate the specific-force transformation at the IMU update rate.

The higher-order terms in (5.97) and (5.98) are sometimes known as coning and sculling corrections. When an IMU samples the gyros and accelerometers at a higher rate than it outputs angular rate and specific force, coning and sculling corrections may be applied by its processor prior to output.

When coning and sculling corrections are not applied within the IMU and the frequency of the vibration is less than half of the IMU output rate (i.e., the Nyquist rate), further reductions in the coning and sculling errors may be obtained by interpolating the IMU measurements to a higher rate. This makes use of earlier and, sometimes later, measurements to estimate the variation in specific force and angular rate over the sampling interval and may be performed in either the time domain or the frequency domain. Figure 5.13 illustrates this, noting that the average value of the interpolated measurement over each original measurement interval must equal the original measurement. The measurements may then be recombined using (5.93) to (5.98). Note that a processing lag is introduced if later measurements are used in the interpolation process. Also, the signal variation must exceed the IMU noise and quantization levels (see Section 4.4.3) for the interpolation to be useful; this is a particular issue for consumer-grade MEMS sensors.

When the position update interval is longer than the interval over which the IMU measurements are assumed to be constant, the assumption that the acceleration is constant over the update interval will introduce a correctable position error. This error will typically be small compared to the position accuracy requirement and/or other error sources. However, where necessary, it may be eliminated by applying a scrolling correction as described in [2, 11].

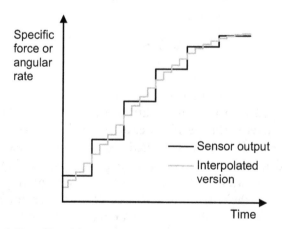

Figure 5.13 Interpolation of inertial sensor output.

5.5.5 Design Tradeoffs

The design of a set of inertial navigation equations is a tradeoff among accuracy, processing efficiency, and complexity. It is possible to optimize two of these, but not all three. In determining the accuracy requirements, it is important to consider the navigation system as a whole. For example, where the inertial sensors are relatively poor, an improvement in the accuracy of the navigation equations may have negligible impact on overall performance. Another consideration is the degree to which integration with other navigation sensors can correct the errors of the INS [12]. This can lead to a more demanding requirement for the attitude update accuracy than for the position and velocity as the latter are easier to correct using other sensors (see Section 14.2.1).

Traditionally, the accuracy requirements for inertial navigation have been high, as INS with high-quality inertial sensors have been used for sole-means navigation in the horizontal axes, or with infrequent position updates, for periods of hours. Until the 1990s, processing power was also at a premium. Hence considerable effort was expended developing highly accurate and highly efficient, but also highly complex, navigation algorithms (e.g., [2]). However, today, a faster processor can often be more cost effective than expending a large amount of effort designing, implementing, and debugging complex navigation equations.

The accuracy of the navigation equations is a function of three factors: the iteration rate, the nature of the approximations made, and the dynamic and vibration environment. The greater the level of dynamics or vibration, the greater the impact on navigation solution accuracy of an approximation in the navigation equations or a change in the iteration rate. At a given level of dynamics or vibration, the impact of an approximation is greater where the iteration rate is lower (i.e., the integration step is larger).

Different approximations in different stages of the navigation equations have differing impacts on the overall position, velocity, and attitude errors, depending on the magnitude and type of the dynamics and vibration. For example, in the ECEF-frame and local-navigation-frame implementations, the Earth-rate, transport-rate, and Coriolis terms tend to be much smaller than the terms derived from the accelerometer and gyro measurements. Consequently, approximating these terms and/or calculating them at a lower iteration rate will have less impact on overall navigation accuracy, providing an opportunity to improve the processing efficiency. A common approach is to combine successive IMU outputs using (5.94), (5.97), and (5.98) and then iterate precision inertial navigation equations (Sections 5.5.1 to 5.5.3) at a lower rate, for example, 50–200 Hz [2].

Section E.8 of Appendix E on the CD discusses a number of iteration rate issues, including using different iteration rates for different stages, using numerical integration, and iterating approximate forms of the navigation equations faster than the IMU output rate.

5.6 Initialization and Alignment

As Figure 5.3 shows, an INS calculates a navigation solution by integrating the inertial sensor measurements. Thus, each iteration of the navigation equations uses the previous navigation solution as its starting point. Therefore, before an INS can be used to provide a navigation solution, that navigation solution must be initialized.

Initial position and velocity must be provided from external information. Attitude may be initialized either from an external source or by sensing gravity and the Earth's rotation.[‡] The attitude initialization process is also known as alignment because, in a platform INS (Section E.5 of Appendix E on the CD), the inertial instruments are physically aligned with the axes of a local navigation frame.

The initialization is often followed by a period of calibration when stationary or against an external reference, typically lasting a few minutes. This is known as fine alignment, as its main role is to reduce the attitude initialization errors.

5.6.1 Position and Velocity Initialization

The INS position and velocity must be initialized using external information. When the host vehicle has not moved since the INS was last used, the last known position may be stored and used for initialization. However, an external position reference must be introduced at some point to prevent the navigation solution drift accumulating over successive periods of operation.

INS position may be initialized from another navigation system. This may be another INS, GNSS user equipment, or terrestrial radio navigation user equipment. Alternatively, the INS may be placed near a presurveyed point, or range and/or bearing measurements to known landmarks taken. In either case, the lever arm between the INS and the position reference must be measured. If this is only known in the body frame, the INS attitude will be required to transform the lever arm to the same coordinate frame as the position fix (see Section 2.5.5).

Velocity may be initialized simply by maintaining the INS stationary with respect to the Earth. Alternatively, another navigation system, such as GNSS, Doppler radar, or another INS, may be used as a reference. In that case, the lever arm and angular rate are required to calculate the lever arm velocity.

Further problems for velocity initialization are disturbance, vibration, and flexure. For example, when the INS is assumed to be stationary with respect to the Earth, the host vehicle could be disturbed by the wind or by human activity, such as refueling and loading. For ships and boats, water motion is also an issue. For in-motion initialization, the lever arm between the INS and the reference navigation system can be subject to flexure and vibration. The solution is to take initialization measurements over a few seconds and average them. Position can also be affected by flexure and vibration, but the magnitude is usually less than the accuracy required.

In the MATLAB INS/GNSS integration software on the CD, the inertial position and velocity solutions are initialized from the GNSS solution. For stand-alone inertial navigation, the MATLAB function, Initialize_NED, simply initializes the navigation solution to the truth offset by user-specified errors.

5.6.2 Attitude Initialization

When the INS is stationary, self-alignment can be used to initialize the roll and pitch with all but the poorest inertial sensors. However, accurate self-alignment of the

---

[‡]This paragraph, up to this point, is based on material written by the author for QinetiQ, so comprises QinetiQ copyright material.

5.6 Initialization and Alignment

heading requires aviation-grade gyros or better. Heading is often initialized using a magnetic compass, described in Section 6.1.1.

When the INS is initialized in motion, another navigation system must provide an attitude reference. For guided weapons, the host vehicle's INS is generally used. Multiple-antenna GNSS user equipment can also be used to measure attitude. However, this is very noisy unless long baselines and/or long averaging times are used, as described in Section 10.2.5. Another option for some applications is the star imager, described in Section 13.3.7. In all cases, the accuracy of the attitude initialization depends on how well the relative orientation of the initializing INS and the reference navigation system is known, as well as on the accuracy of the reference attitude. If there is significant flexure in the lever arm between the two systems, such as that which occurs for equipment mounted on an aircraft wing, the relative orientation may only be known to a few tens of milliradians (a degree or two).

For IMUs attached to most land vehicles, it can be assumed that the direction of travel defines the body x-axis except when the vehicle is turning (see Section 6.1.4). This enables a trajectory measured by a positioning system, such as GNSS, to be used to initialize the pitch and heading attitudes. When a portable IMU is used, there is no guarantee that the body x-axis will be aligned with the direction of travel. On a land vehicle, the normal direction of travel can be identified from the acceleration and deceleration that occurs when the vehicle starts and stops, which is normally accompanied by forward motion [13]. Once the IMU is aligned with the vehicle, its heading may be derived from the trajectory.

For aircraft and ships the direction of travel will only provide a rough attitude initialization as sideslip, due to wind or sea motion, results in an offset between the heading and the trajectory, while aircraft pitch is defined by the angle of attack needed to obtain lift and ship pitch oscillates due to the sea state. Trajectory-based heading alignment is thus context dependent.

Other alignment methods include memory, whereby the attitude is assumed to be the same as when the INS was last used; using a prealigned portable INS to transfer the attitude solution from a ready room; and aligning the host vehicle with a known landmark, such as a runway [14].

Self-alignment comprises two processes: a leveling process, which initializes the roll and pitch attitudes, and a gyrocompassing process, which initializes the heading. The leveling is normally performed first.

The principle behind *leveling* is that, when the INS is stationary (or traveling at constant velocity), the only specific force sensed by the accelerometers is the reaction to gravity, which is approximately in the negative down direction of a local navigation frame at the Earth's surface. Figure 5.14 illustrates this. Thus the attitude, $\mathbf{C}_b^n$, can be estimated by solving*

$$\mathbf{f}_{ib}^b = \mathbf{C}_n^b \mathbf{g}_b^n(L_b, h_b), \qquad (5.99)$$

given $\mathbf{a}_{eb}^\gamma = 0$. Taking the third column of $\mathbf{C}_n^b$, given by (2.22), (5.99) can be expressed in terms of the pitch, θ_{nb}, and roll, ϕ_{nb}, Euler angles:

---

*This and subsequent paragraphs are based on material written by the author for QinetiQ, so comprise QinetiQ copyright material.

$$\begin{pmatrix} f_{ib,x}^b \\ f_{ib,y}^b \\ f_{ib,z}^b \end{pmatrix} = \begin{pmatrix} \sin\theta_{nb} \\ -\cos\theta_{nb}\sin\phi_{nb} \\ -\cos\theta_{nb}\cos\phi_{nb} \end{pmatrix} g_{b,D}^n(L_b, h_b), \qquad (5.100)$$

where $g_{b,D}^n$ is the down component of the acceleration due to gravity. This solution is overdetermined. Therefore, pitch and roll may be determined without knowledge of gravity, and hence the need for position, using[†]

$$\theta_{nb} = \arctan\left(\frac{f_{ib,x}^b}{\sqrt{f_{ib,y}^{b\,2} + f_{ib,z}^{b\,2}}}\right), \qquad \phi_{nb} = \arctan_2\left(-f_{ib,y}^b, -f_{ib,z}^b\right), \qquad (5.101)$$

noting that a four-quadrant arctangent function must be used for roll.

When the INS is absolutely stationary, the attitude initialization accuracy is determined only by the accelerometer errors. For example, a 1-mrad roll and pitch accuracy is obtained from accelerometers accurate to 10^{-3} g. Disturbing motion, such as mechanical vibration, wind effects, and human activity, disrupts the leveling process. However, if the motion averages out over time, its effects on the leveling process may be mitigated simply by time-averaging the accelerometer measurements over a few seconds.

The pitch and roll initialization errors from leveling are then

$$\begin{aligned}
\delta\theta_{nb} &= \frac{\left(f_{ib,y}^{b\,2} + f_{ib,z}^{b\,2}\right)\delta f_{ib,x}^b - f_{ib,x}^b f_{ib,y}^b \delta f_{ib,y}^b - f_{ib,x}^b f_{ib,z}^b \delta f_{ib,z}^b}{\left(f_{ib,x}^{b\,2} + f_{ib,y}^{b\,2} + f_{ib,z}^{b\,2}\right)\sqrt{f_{ib,y}^{b\,2} + f_{ib,z}^{b\,2}}}, \\
\delta\phi_{nb} &= \frac{f_{ib,z}^b \delta f_{ib,y}^b - f_{ib,y}^b \delta f_{ib,z}^b}{f_{ib,y}^{b\,2} + f_{ib,z}^{b\,2}}
\end{aligned} \qquad (5.102)$$

where the accelerometer error model is described in Section 4.4.6.

The principle behind gyrocompassing is that, when the INS is stationary (or traveling in a straight line in an inertial frame), the only rotation it senses is that of the Earth, which is in the z direction of an ECEF frame. Measuring this rotation in the body frame enables the heading to be determined, except at or very near to the poles, where the rotation axis and gravity vector coincide. Figure 5.15 illustrates the concept. There are two types of gyrocompassing, direct and indirect.

Direct gyrocompassing measures the Earth rotation directly using the gyros. The attitude, C_b^n, may be obtained by solving

$$\omega_{ib}^b = C_n^b C_e^n(L_b, \lambda_b)\begin{pmatrix} 0 \\ 0 \\ \omega_{ie} \end{pmatrix}, \qquad (5.103)$$

[†]End of QinetiQ copyright material.

5.6 Initialization and Alignment

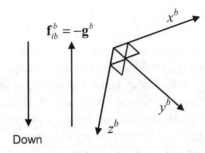

Figure 5.14 Principle of leveling. (*From:* [9]. © 2002 QinetiQ Ltd. Reprinted with permission.)

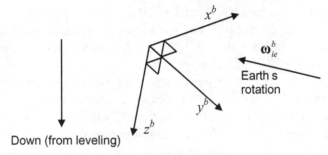

Figure 5.15 Principle of gyrocompassing. (*From:* [9]. © 2002 QinetiQ Ltd. Reprinted with permission.)

given that $\omega_{eb}^{\gamma} = 0$. Substituting in (2.150) and rearranging,

$$\begin{pmatrix} \cos L_b \omega_{ie} \\ 0 \\ -\sin L_b \omega_{ie} \end{pmatrix} = \mathbf{C}_b^n \boldsymbol{\omega}_{ib}^b, \qquad (5.104)$$

When the roll and pitch have already been obtained from leveling, the knowledge that the Earth's rotation vector has no east component in a local navigation frame can be used to remove the need for prior position knowledge. Thus, applying (2.24) to (5.104) and taking the second row give the heading Euler angle, ψ_{nb}, in terms of the roll, pitch, and gyro measurements:

$$\begin{aligned} \psi_{nb} &= \arctan_2\left(\sin\psi_{nb}, \cos\psi_{nb}\right) \\ \sin\psi_{nb} &= -\omega_{ib,y}^b \cos\phi_{nb} + \omega_{ib,z}^b \sin\phi_{nb} \\ \cos\psi_{nb} &= \omega_{ib,x}^b \cos\theta_{nb} + \omega_{ib,y}^b \sin\phi_{nb}\sin\theta_{nb} + \omega_{ib,z}^b \cos\phi_{nb}\sin\theta_{nb} \end{aligned} \qquad (5.105)$$

Again, a four-quadrant arctangent function must be used. Equations for performing leveling and direct gyrocompassing in one step are presented in a number of texts [1, 15, 16]. However, these require knowledge of the latitude.

Example 5.3 on the CD illustrates both leveling and direct gyrocompassing in the presence of accelerometer and gyro errors and may be edited using Microsoft Excel.

In the presence of angular disturbing motion, the gyro measurements used for direct gyrocompassing must be time averaged. However, even small levels of angular vibration will be much larger than the Earth-rotation rate. Therefore, if the INS is mounted on any kind of vehicle, an averaging time of many hours can be required. Thus, the application of direct gyrocompassing is limited.

Indirect gyrocompassing uses the gyros to compute a relative attitude solution, which is used to transform the specific-force measurements into inertial resolving axes. The direction of the Earth's rotation is then obtained from rotation about this axis of the inertially resolved gravity vector. Over a sidereal day, this vector forms a cone, while its time derivative rotates within the plane perpendicular to the Earth's rotation axis. Figure 5.16 illustrates this.

The process typically takes 2 to 10 minutes, depending on the amount of linear vibration and disturbance and the accuracy required. Indirect gyrocompassing is typically combined with fine alignment. A suitable quasi-stationary alignment algorithm is described in Section 15.2.

The accuracy of both gyrocompassing methods depends on gyro performance. Given that $\omega_{ie} \approx 7 \times 10^{-5}$ rad s^{-1}, to obtain a 1-mrad heading initialization at the equator, the gyros must be accurate to around 7×10^{-8} rad s^{-1} or about 0.01 ° hr^{-1}. Only aviation- and marine-grade gyros are this accurate. INSs with gyro biases exceeding about 5°/hr are not capable of gyrocompassing at all. Note that the accuracy of the roll and pitch initialization also affects the heading initialization.

The heading initialization error from gyrocompassing [17] is

$$\delta \psi_{nb} = -\frac{\delta f_{ib,y}^b}{g_{b,D}^n} \tan L_b + \frac{\delta \omega_{ib,y}^b}{\omega_{ie}} \sec L_b, \qquad (5.106)$$

where the accelerometer and gyro error models are presented in Section 4.4.6.

In principle, leveling and gyrocompassing techniques can be performed when the INS is not stationary if the acceleration, α_{eb}^b, and angular rate, ω_{eb}^b, with respect to the Earth are provided by an external sensor.‡ However, as the relative orientation of the external sensor must be known, this would be no more accurate than simply using the external sensor as an attitude reference.

5.6.3 Fine Alignment

Most inertial navigation applications require attitude to 1 mrad or better, if only to minimize position and velocity drift. Most attitude initialization techniques do not achieve this accuracy. It is therefore necessary to follow the initialization with a period of attitude calibration known as fine alignment.*

In fine alignment techniques, the residual attitude errors are sensed through the growth in the velocity errors. For example, a 1-mrad pitch or roll attitude error

---

‡This paragraph, up to this point, is based on material written by the author for QinetiQ, so comprises QinetiQ copyright material.

*This and subsequent paragraphs are based on material written by the author for QinetiQ, so comprise QinetiQ copyright material.

5.6 Initialization and Alignment

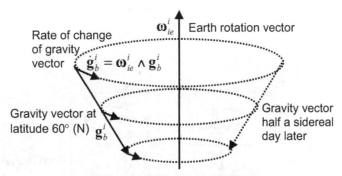

Figure 5.16 Earth rotation and gravity vectors resolved in ECI-frame axes.

will cause the horizontal velocity error to grow at a rate of ~10 mm s^{-2} due to false resolving of gravity.

There are three main fine alignment techniques, each providing a different reference to align against. *Quasi-stationary alignment* assumes that the position has been initialized and that the INS is stationary with respect to the Earth and uses zero velocity updates (ZVUs) or integrals thereof. *GNSS alignment*, or INS/GNSS integration, uses position and velocity derived from GNSS and can operate during the navigation phase as well as the alignment phase. Finally, *transfer alignment* uses position or velocity, and sometimes attitude, from another INS or INS/GNSS. It is generally used for aligning a guided-weapon INS between power-up and launch.[†] Alternatively, any other position-fixing or dead-reckoning technology, or combination thereof, that provides a 3-D position and velocity solution may be used as the reference for fine alignment. For foot-mounted inertial navigation, a ZVU can be performed during the stance phase of every step.

In all cases, measurements of the difference between the INS outputs and the reference are input to an estimation algorithm, such as a Kalman filter, which calibrates the velocity, attitude, and sometimes the position, depending on which measurements are used. Figure 5.17 illustrates this. Inertial instrument errors, such as accelerometer and gyro biases, are often estimated as well. However, when the INS is stationary, the effects of instrument errors cannot be fully separated from the attitude errors. For example, a 10 mm s^{-2} accelerometer bias can have the same effect on velocity as a 1-mrad attitude error. To separately observe these errors, maneuvers must be performed as discussed in Section 14.2.1. For example, if the INS is rotated, a given accelerometer error will have the same effect on velocity as a different attitude error. In quasi-stationary alignment, maneuvers are generally limited to heading changes, with the alignment process suspended during host vehicle maneuvers. For GNSS and transfer alignment, the maneuvers are limited only by the capabilities of the host vehicle. Even with maneuvers, there will still be some correlation between the residual INS errors following fine alignment.[‡]

[†]End of QinetiQ copyright material.

[‡]This paragraph, up to this point, is based on material written by the author for QinetiQ, so comprises QinetiQ copyright material.

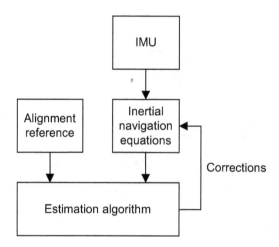

Figure 5.17 INS fine alignment architecture.

INS/GNSS integration algorithms are described in detail in Chapter 14, while quasi-stationary, transfer alignment, and ZVUs are described in Chapter 15. The use of other navigation systems to calibrate INS errors is described in Chapter 16, noting that an error-state integration architecture with the INS as the reference should be used.

The main differences between the techniques are the types of measurements used, although all three techniques can use velocity, and the characteristics of the noise on the measurements of differences between the aligning INS and the reference. In quasi-stationary alignment, where zero velocity and angular rate with respect to the Earth are assumed, the main noise source is buffeting of the host vehicle by wind or human activity, such as fuelling or loading. In GNSS alignment, the GNSS receiver measurements are noisy. In transfer alignment, noise arises from flexure and vibration of the lever arm between the host vehicle's INS and the aligning INS.*

Most fine alignment algorithms operate on the basis that position, velocity, and attitude are roughly known at the start of the process. This is important for determining how the system errors vary with time and may allow simplifications, such as the small angle approximation, to be made. For some applications, such as GNSS alignment of a tactical-grade INS, there may be no prior knowledge of heading. However, GNSS and transfer alignment algorithms may be adapted to handle this as discussed in Section 14.4.4 [18].

The type of fine alignment technique that is most suitable depends on the application. When the INS is stationary on the ground, a quasi-stationary alignment is usually best as the noise levels are lowest. Where there is a choice between transfer alignment and GNSS alignment for in-flight applications, the best option is transfer alignment using an INS/GNSS reference, as this combines the higher short-term accuracy and update rate of the INS with the high long-term accuracy of GNSS.†

---

*This and subsequent paragraphs are based on material written by the author for QinetiQ, so comprise QinetiQ copyright material.
†End of QinetiQ copyright material.

Other navigation technology should be considered where neither GNSS nor transfer alignment is available.

5.7 INS Error Propagation

The errors in an inertial navigation system's position, velocity, and attitude solution arise from three sources. These are errors in the accelerometer and gyro measurements, initialization errors, and processing approximations. The latter includes approximations in the discrete-time navigation equations, the effects of finite iteration rates, gravity modeling approximations, computational rounding errors, and timing errors.

The navigation equations integrate the accelerometer and gyro biases to produce position, velocity, and attitude errors that grow with time. Similarly, the velocity initialization error is integrated to produce a growing position error. Random accelerometer and gyro noise and navigation equations limitations have a cumulative effect on the navigation solution errors. In addition, the attitude errors contribute to the velocity and position errors and there is both positive and negative feedback of the position errors through the gravity model.

INS error propagation is also affected by the host vehicle trajectory. For example, the effect of scale factor and cross-coupling errors depends on the host vehicle dynamics, as does the coupling of the attitude errors, particularly heading, into velocity and position.

Full determination of INS error propagation is a complex problem and is invariably studied using simulation software. A number of inertial navigation demonstrations with different grades of IMU are included in the MATLAB software on the accompanying CD. Here, a number of simple examples are presented to illustrate the main principles. These are divided into the short-term and the medium- and long-term cases, followed by a discussion of the effects of maneuvers on error propagation. A more detailed treatment of INS error propagation may be found in a number of inertial navigation texts [1, 11, 17].

Generally, an INS error is simply the difference between an INS-indicated quantity, denoted by a "~", and the true value of that quantity. Thus, the Cartesian position, velocity and acceleration errors are

$$\begin{aligned} \delta r^{\gamma}_{\beta\alpha} &= \tilde{r}^{\gamma}_{\beta\alpha} - r^{\gamma}_{\beta\alpha} \\ \delta v^{\gamma}_{\beta\alpha} &= \tilde{v}^{\gamma}_{\beta\alpha} - v^{\gamma}_{\beta\alpha} \\ \delta a^{\gamma}_{\beta\alpha} &= \tilde{a}^{\gamma}_{\beta\alpha} - a^{\gamma}_{\beta\alpha} \end{aligned} \quad (5.107)$$

Similarly, the latitude, longitude, and height errors are

$$\begin{aligned} \delta L_b &= \tilde{L}_b - L_b \\ \delta \lambda_b &= \tilde{\lambda}_b - \lambda_b \\ \delta h_b &= \tilde{h}_b - h_b \end{aligned} \quad (5.108)$$

Coordinate transformation matrices should be used to calculate the attitude error. The coordinate transformation matrix form of the attitude error is defined by

$$\delta \mathbf{C}_\beta^\alpha = \tilde{\mathbf{C}}_\beta^\alpha \mathbf{C}_\alpha^\beta, \tag{5.109}$$

where the attitude error components are resolved about the axes of the α frame. This is because multiplying one coordinate transformation matrix by the transpose of another gives difference between the two attitudes that they represent. Note that

$$\begin{aligned}\delta \mathbf{C}_\alpha^\beta &= \tilde{\mathbf{C}}_\alpha^\beta \mathbf{C}_\beta^\alpha = \mathbf{C}_\alpha^\beta \left(\delta \mathbf{C}_\beta^\alpha\right)^{\mathrm{T}} \mathbf{C}_\beta^\alpha \\ \left(\delta \mathbf{C}_\beta^\alpha\right)^{\mathrm{T}} &= \mathbf{C}_\beta^\alpha \tilde{\mathbf{C}}_\alpha^\beta \end{aligned}, \tag{5.110}$$

where the components of $\delta \mathbf{C}_\alpha^\beta$ are resolved about the β frame axes.

Except under the small angle approximation, the attitude error in Euler angle form must be computed via coordinate transformation matrices (or quaternions or rotation vectors). When the small angle approximation applies, the attitude error may be expressed as a vector resolved about a chosen set of axes. $\delta \boldsymbol{\psi}_{\beta\alpha}^\gamma$ is the error in the INS indicated attitude of frame α with respect to frame β, resolved about the frame γ axes. From (2.26), the small angle attitude error may be expressed in terms of the coordinate transformation matrix form of the attitude error using

$$\left[\delta \boldsymbol{\psi}_{\beta\alpha}^\alpha \wedge\right] \approx \mathbf{I}_3 - \delta \mathbf{C}_\beta^\alpha, \qquad \left[\delta \boldsymbol{\psi}_{\beta\alpha}^\beta \wedge\right] \approx \delta \mathbf{C}_\alpha^\beta - \mathbf{I}_3. \tag{5.111}$$

Attitude errors are sometimes known as misalignments or misorientations. These terms are avoided here as they can be confused with the misalignments of the inertial-sensor sensitive axes with the body frame that produce cross-coupling errors (Section 4.4.2).

From Section 4.4.6, the accelerometer and gyro errors are [repeated from (4.18)]:

$$\begin{aligned}\delta \mathbf{f}_{ib}^b &= \tilde{\mathbf{f}}_{ib}^b - \mathbf{f}_{ib}^b \\ \delta \boldsymbol{\omega}_{ib}^b &= \tilde{\boldsymbol{\omega}}_{ib}^b - \boldsymbol{\omega}_{ib}^b\end{aligned}.$$

Simple models of gravity as a function only of latitude and height with few coefficients (see Section 2.4.7) are typically accurate to about 10^{-3} m s^{-2} (0.1 mg) in each direction [1, 17]. Consequently, they can be a significant source of error where higher precision inertial sensors are used.

The effect of timing errors is described in Section E.9 of Appendix E on the CD. Except for the highest precision applications, these errors are negligible compared to those arising from the inertial sensors.

5.7.1 Short-Term Straight-Line Error Propagation

The simplest INS error propagation scenario is short-term propagation when the host vehicle is traveling in a straight line at constant velocity and remains level. In considering only short-term error propagation, the effects of curvature and rotation of

5.7 INS Error Propagation

the Earth and gravity model feedback may be neglected, while there are no dynamics-induced errors where the host vehicle travels at constant velocity.

Figure 5.18 shows the position error growth with constant velocity, acceleration, attitude, and angular-rate errors. The position error is simply the integral of the velocity error, so with a constant velocity error,

$$\delta \mathbf{r}_{\beta b}^{\gamma}(t) = \delta \mathbf{v}_{\beta b}^{\gamma} t, \qquad (5.112)$$

where β is the reference frame and γ the resolving axes. There is no error propagation between axes. As Figure 5.18 illustrates, an 0.1 m s^{-1} initial velocity error produces a 30-m position error after 300 seconds (5 minutes).

The velocity error is the integral of the acceleration error, so the following velocity and position errors result from a constant accelerometer bias:

$$\delta \mathbf{v}_{\beta b}^{\gamma}(t) \approx \mathbf{C}_b^{\gamma} \mathbf{b}_a t, \qquad \delta \mathbf{r}_{\beta b}^{\gamma}(t) \approx \tfrac{1}{2} \mathbf{C}_b^{\gamma} \mathbf{b}_a t^2. \qquad (5.113)$$

There is no error propagation between axes where the attitude remains constant. As Figure 5.18 shows, an 0.01 m s^{-2} (~ 1 mg) accelerometer bias produces a 450-m position error after 300 seconds. Acceleration errors can also result from gravity modeling approximations, timing errors, and as a result of attitude errors.

Attitude errors produce errors in the transformation of the specific-force resolving axes from the body frame to an ECI, ECEF, or local-navigation frame, resulting

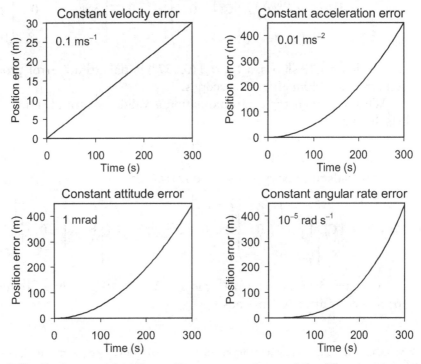

Figure 5.18 Short-term straight-line position error growth per axis for different error sources.

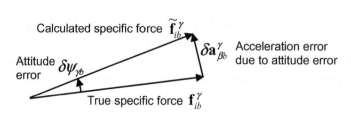

Figure 5.19 Acceleration error due to attitude error.

in errors in the acceleration resolved in that frame. Figure 5.19 illustrates this. When the attitude error may be expressed as a small angle, the resulting acceleration error is

$$\delta a^{\gamma}_{\beta b}(t) \approx \delta \psi^{\gamma}_{\gamma b} \wedge \left(C^{\gamma}_{b} \tilde{f}^{b}_{ib} \right) \\ = C^{\gamma}_{b} \left(\delta \psi^{b}_{\gamma b} \wedge \tilde{f}^{b}_{ib} \right)$$

(5.114)

In the constant-velocity and level example, the specific force comprises only the reaction to gravity. Thus, pitch (body-frame y-axis) attitude errors couple into along-track (body-frame x-axis) acceleration errors and roll (body-frame x-axis) attitude errors couple into across-track (body-frame y-axis) acceleration errors. These acceleration errors are integrated to produce the following velocity and position errors.

$$\delta v^{\gamma}_{\beta b}(t) \approx \delta \psi^{\gamma}_{\gamma b} \wedge \left[C^{\gamma}_{b} \begin{pmatrix} 0 \\ 0 \\ -g \end{pmatrix} \right] t = C^{\gamma}_{b} \left[\delta \psi^{b}_{\gamma b} \wedge \begin{pmatrix} 0 \\ 0 \\ -g \end{pmatrix} \right] t$$

$$\delta r^{\gamma}_{\beta b}(t) \approx \tfrac{1}{2} \delta \psi^{\gamma}_{\gamma b} \wedge \left[C^{\gamma}_{b} \begin{pmatrix} 0 \\ 0 \\ -g \end{pmatrix} \right] t^2 = \tfrac{1}{2} C^{\gamma}_{b} \left[\delta \psi^{b}_{\gamma b} \wedge \begin{pmatrix} 0 \\ 0 \\ -g \end{pmatrix} \right] t^2$$

(5.115)

As Figure 5.18 shows, a 1 mrad (0.057°) initial attitude error produces a position error of ~440m after 300 seconds.[*]

When the small angle approximation is valid, the attitude error due to a gyro bias, $\mathbf{b}_g$, is simply

$$\delta \psi^{b}_{ib} \approx \mathbf{b}_g t.$$

(5.116)

This leads to velocity and position errors of

$$\delta v^{\gamma}_{\beta b}(t) \approx \tfrac{1}{2} C^{\gamma}_{b} \left[\mathbf{b}_g \wedge \begin{pmatrix} 0 \\ 0 \\ -g \end{pmatrix} \right] t^2, \quad \delta r^{\gamma}_{\beta b}(t) \approx \tfrac{1}{6} C^{\gamma}_{b} \left[\mathbf{b}_g \wedge \begin{pmatrix} 0 \\ 0 \\ -g \end{pmatrix} \right] t^3.$$

(5.117)

As Figure 5.18 shows, a 10^{-5} rad s^{-1} (2.1 ° hr^{-1}) gyro bias produces a ~439m position error after 300 seconds.[†]

---

[*]This and subsequent paragraphs are based on material written by the author for QinetiQ, so comprise QinetiQ copyright material.

[†]End of QinetiQ copyright material.

5.7 INS Error Propagation

The other major source of error in this scenario is noise. In a well-designed system, the inertial sensor noise will be the largest noise source and may be considered white over timescales exceeding one second. If the single-sided accelerometer noise PSD is S_a, then, from (B.113) and (B.116) in Appendix B on the CD, the standard deviations of the ensuing velocity and position errors are

$$\begin{aligned} \sigma\left(\delta v^{\gamma}_{\beta b,i}\right) &= \sqrt{S_a t} \\ \sigma\left(\delta r^{\gamma}_{\beta b,i}\right) &= \sqrt{\tfrac{1}{3} S_a t^3} \end{aligned} \quad i \in x,y,z \qquad (5.118)$$

Similarly, if the gyro noise PSDs is S_g, then, from (B.113) and (B.116) in Appendix B on the CD, the standard deviations of the ensuing attitude errors and horizontal position and velocity errors are

$$\begin{aligned} \sigma\left(\delta\psi^{\gamma}_{\beta b,i}\right) &= \sqrt{S_g t} & i &\in x,y,z \\ \sigma\left(\delta v^{n}_{\beta b,j}\right) &= g\sqrt{\tfrac{1}{3} S_g t^3} & j &\in N,E . \\ \sigma\left(\delta r^{n}_{\beta b,j}\right) &= g\sqrt{\tfrac{1}{5} S_g t^5} \end{aligned} \qquad (5.119)$$

Figure 5.20 shows the growth in position error standard deviation due to sensor noise. If the accelerometer random noise PSD is 10^{-6} m² s⁻³ (corresponding to a root PSD of about 100 $\mu g\sqrt{Hz}$), the position error standard deviation after 300 seconds is 3m per axis. Similarly, if the gyro random noise PSD is 10^{-9} rad² s⁻¹ (a root PSD of ~0.1 °/√hr), the position error standard deviation after 300 seconds is ~22m per horizontal axis.

Figure 5.21 shows the horizontal position error standard deviation growth using tactical-grade and aviation-grade INSs with the characteristics listed in Table 5.2. The tactical-grade INS error is more than an order of magnitude bigger than that of the aviation-grade INS after 300 seconds. The difference in horizontal and vertical performance of the tactical-grade INS arises because the gyro bias dominates and, under constant velocity conditions, this only affects horizontal navigation. For the aviation-grade INS, the acceleration, roll, and pitch errors dominate. Note that the initial position error has little impact after the first minute. Example 5.4 on the CD shows the calculations and can be edited using Microsoft Excel.

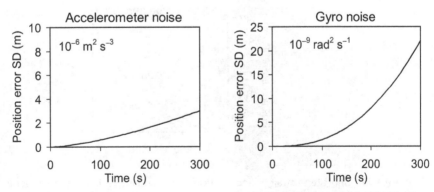

Figure 5.20 Short-term straight-line position error standard deviation growth per axis due to inertial sensor noise.

Table 5.2 Tactical-Grade and Aviation-Grade INS Characteristics

Sensor Grade	Tactical	Aviation
Initial position error standard deviation	10 m	10 m
Initial velocity error standard deviation	0.1 m s^{-1}	0.01 m s^{-1}
Initial (roll and pitch) attitude error standard deviation	1 mrad	0.1 mrad
Accelerometer bias standard deviation	0.01 m s^{-2} (1 mg)	0.001 m s^{-2} (0.1 mg)
Gyro bias standard deviation	5×10^{-5} rad s^{-1} (10 ° hr^{-1})	5×10^{-8} rad s^{-1} (0.01 ° hr^{-1})
Accelerometer noise PSD	10^{-6} m^2 s^{-3} (100 µg/√Hz)^2	10^{-7} m^2 s^{-3} (32 µg/√Hz)^2
Gyro noise PSD	10^{-9} rad^2 s^{-1} (0.1 °/√hr)^2	10^{-12} rad^2 s^{-1} (0.003 °/√hr)^2

The errors in Table 5.2 assume that no sensor calibration has been applied beyond that of the IMU manufacturer and that the roll and pitch have been initialized using a simple leveling procedure (see Section 5.6.2). Leveling correlates the roll and pitch errors with the accelerometer biases. Their effects on the velocity error largely cancel when the IMU orientation is the same as it was during leveling, reinforce when the IMU orientation is reversed within the horizontal plane, and are independent when the IMU is rotated by 90°. In Figure 5.21, the independent case is assumed.

Fine-alignment calibration (see Section 5.6.3 and Chapters 14 to 16) can significantly reduce the effective attitude errors and accelerometer and gyro biases. Figure 5.22 shows the horizontal position error standard deviation growth using a

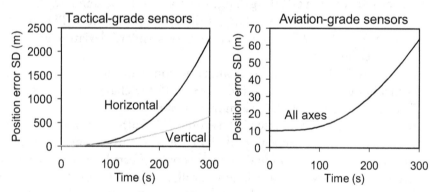

Figure 5.21 Short-term straight-line position error standard deviation growth per axis for tactical-grade and aviation-grade INSs.

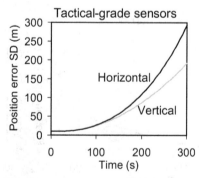

Figure 5.22 Short-term straight-line position error standard deviation growth per axis for a calibrated tactical-grade INS.

5.7 INS Error Propagation

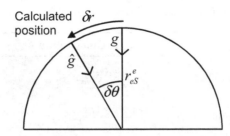

Figure 5.23 Gravity estimation from horizontal position error. (*From:* [9]. © 2002 QinetiQ Ltd. Reprinted with permission.)

calibrated tactical-grade INS where the residual roll and pitch errors are 0.3 mrad, the accelerometer biases 0.003 m s^{-2} (0.3 mg), and the gyro biases 5×10^{-6} rad s^{-1} (1 ° hr^{-1}). Comparing this with Figure 5.21, it can be seen that the calibration improves the position accuracy at 300 seconds by a factor of 8 horizontally and a factor of 3 vertically. This is also included in Example 5.4 on the CD.

5.7.2 Medium- and Long-Term Error Propagation

The gravity model within the inertial navigation equations, regardless of which coordinate frame they are mechanized in, acts to stabilize horizontal position errors and destabilize vertical channel errors.*

Consider a vehicle on the Earth's surface with a position error along that surface of δr_h. As a consequence, the gravity model assumes that gravity acts at an angle, $\delta\theta = \delta r / r^e_{eS}$, to its true direction, where r^e_{eS} is the geocentric radius. This is illustrated by Figure 5.23. Therefore, a false acceleration, $\delta \ddot{r}_h$, is sensed in the opposite direction to the position error. Thus, the horizontal position error is subject to negative feedback. Assuming the small angle approximation:

$$\delta \ddot{r}_h = -\frac{g}{r^e_{eS}} \delta r_h. \qquad (5.120)$$

This is the equation for simple harmonic motion with angular frequency $\sqrt{g/r^e_{eS}}$. This is known as the Schuler frequency and the process is known as the Schuler oscillation. A pendulum with its pivot at the center of the Earth and its bob at the INS is known as a Schuler pendulum.† More generally, the Schuler frequency for a navigation system at any location is $\omega_s = \sqrt{g_b/r^e_{eb}}$. The corresponding period of the Schuler oscillation is

$$\tau_s = \frac{2\pi}{\omega_s} = 2\pi \sqrt{\frac{r^e_{eb}}{g_b}}. \qquad (5.121)$$

As the strength of the gravity field and the distance from the INS to the center of the Earth varies with height and latitude, this period also varies. At the equator

---

*This and subsequent paragraphs are based on material written by the author for QinetiQ, so comprise QinetiQ copyright material.
†End of QinetiQ copyright material.

Table 5.3 Medium Term (Up to 4 Hours) Horizontal Position Error Growth from Selected Error Sources

Error Source	North Position Error, $\delta r^n_{eb,N}$	East Position Error, $\delta r^n_{eb,E}$
Initial velocity error, $\delta \mathbf{v}^n_{eb}$	$\dfrac{\sin \omega_s t}{\omega_s} \delta v^n_{eb,N}$	$\dfrac{\sin \omega_s t}{\omega_s} \delta v^n_{eb,E}$
Fixed accelerometer bias, $(\mathbf{C}^n_b \mathbf{b}_a)$	$\dfrac{1 - \cos \omega_s t}{\omega_s^2} (\mathbf{C}^n_b \mathbf{b}_a)_N$	$\dfrac{1 - \cos \omega_s t}{\omega_s^2} (\mathbf{C}^n_b \mathbf{b}_a)_E$
Initial attitude error, $\delta \boldsymbol{\psi}^n_{nb}$	$-(1 - \cos \omega_s t) r^e_{eS} \delta \psi^n_{nb,E}$	$(1 - \cos \omega_s t) r^e_{eS} \delta \psi^n_{nb,N}$
Fixed gyro bias, $(\mathbf{C}^n_b \mathbf{b}_g)$	$-\left(t - \dfrac{\sin \omega_s t}{\omega_s}\right) r^e_{eS} (\mathbf{C}^n_b \mathbf{b}_g)_E$	$\left(t - \dfrac{\sin \omega_s t}{\omega_s}\right) r^e_{eS} (\mathbf{C}^n_b \mathbf{b}_g)_N$

and at the Earth's surface, $\tau_s = 5{,}974$ seconds (84.6 minutes). Consequently, over periods of order an hour, position errors arising from an initial velocity error, an initial attitude error, or an accelerometer bias are bounded and position errors arising from a gyro bias grow linearly with time, as opposed to cubicly. Table 5.3 gives the horizontal position errors arising from different sources for periods of up to about 4 hours [1]. Note that, in practice, instrument biases are not fixed with respect to the north and east axes.[‡]

Figure 5.24 shows the position error magnitude over a 6,000-second (100-minute) period arising from a 0.1 m s^{-1} initial velocity error, a 0.01 m s^{-2} acceleration error, a 1-mrad initial attitude error, and a 10^{-5} rad s^{-1} angular rate error. Note that the position error due to the gyro bias is not bounded in the same way as that due to the other error sources. Because of this, much more effort has gone into precision gyro development than precision accelerometer development. Thus, there is much greater variation in gyro performance across different grades of INS and IMU. Figure 5.25 shows the overall position error standard deviation over the same period for the aviation-grade INS specified in Table 5.2, neglecting the effects of sensor noise.

In practice, the position error growth will be much more complex than Figures 5.24 and 5.25 show, in which constant velocity is effectively assumed. Whenever the host vehicle changes direction, the direction of the accelerometer and gyro biases with respect to the north and east axes will change. This will reset the Schuler cycles for these errors with the cumulative velocity and attitude errors at this point acting as the initial velocity and attitude errors for the new Schuler cycle. This effect is known as Schuler pumping. Further Schuler cycles, which are added to the existing Schuler oscillation, arise from dynamics-induced velocity and attitude errors (see Section 5.7.3). The inertial sensor noise and vibration-induced noise also triggers a tiny additive Schuler cycle each time the navigation solution is updated. The cumulative effect of these errors can often exceed those of the initialization errors.

In closed-loop integrated navigation systems in which the inertial navigation solution is constantly corrected (see Section 14.1.1), the Schuler oscillation is largely irrelevant.

[‡]This paragraph, up to this point, is based on material written by the author for QinetiQ, so comprises QinetiQ copyright material.

5.7 INS Error Propagation

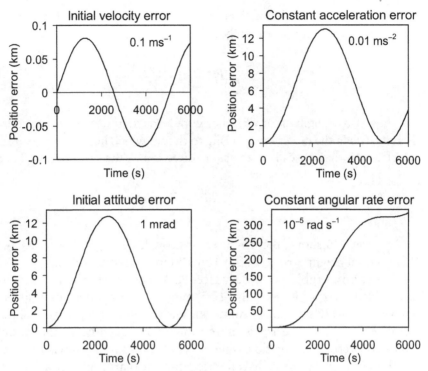

Figure 5.24 Horizontal position error growth per axis over a 6,000-second period for different error sources.

When the INS errors are resolved about the axes of an ECEF or local navigation frame, a further oscillation at the Earth rate, ω_{ie}, and amplitude modulation of the Schuler oscillation at angular frequency $\omega_{ie}\sin L_b$, known as the Foucault frequency, are seen. These are both due to feedback through the Coriolis force terms in the navigation equations. These oscillations are not observed in ECI-frame INS errors. However, they are present in an ECEF- or local-navigation-frame navigation solution converted from one computed in an ECI frame. Longer-term error propagation is discussed in more detail in [1, 15, 17].

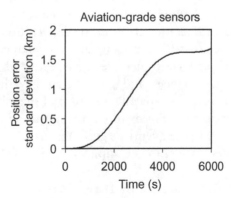

Figure 5.25 Horizontal position error standard deviation growth per axis over a 6,000-second period axis for an aviation-grade INS.

Considering now the vertical channel, as discussed in Section 2.4.7, the gravity varies with height approximately as[*]

$$g(h_b) \approx \left(1 - \frac{2h_b}{r_{eS}^e}\right)g_0. \qquad (5.122)$$

A positive height error, δh_b, therefore leads to gravity being underestimated. As gravity acts in the direction opposite to that in which height is measured, the virtual acceleration that arises is in the same direction as the height error. Thus,[†]

$$\delta \ddot{h}_b \approx \frac{2g}{r_{eS}^e} \delta h_b. \qquad (5.123)$$

Figure 5.26 shows the height error growth over 1,800 seconds arising from a 10-m initial height error and a 0.1 m s^{-1} initial vertical velocity error. The vertical position error is subject to positive feedback such that the height initialization error is doubled after ~750 seconds (12.5 minutes). Subsequent doublings occur after intervals of ~420 seconds (7 minutes). The height error growth due to the vertical velocity initialization error is more rapid. Consequently, an INS is only suited to long-term vertical navigation when it is aided by another navigation sensor.

For air applications, a barometric altimeter (baro) was always used for vertical aiding prior to the advent of GNSS and still forms a part of many integrated navigation systems. It measures the air pressure and then uses a standard atmospheric model to determine height. It exhibits errors that vary with the weather. A baro's operating principles and error sources are discussed in more detail in Section 6.2.1, while its integration with INS is described in Section 16.2.2.

For land and marine applications, it may be assumed that the average height above the terrain or sea surface is constant.

5.7.3 Maneuver-Dependent Errors

Much of the error propagation in inertial navigation depends on the maneuvers performed by the host vehicle. As discussed in Section 5.7.1, the effect of attitude errors on the velocity and position solutions depends on the specific force. At constant velocity, this is limited to the roll and pitch errors producing horizontal velocity errors. However, a linear acceleration or deceleration maneuver couples the heading error into the cross-track velocity and the pitch error into the vertical velocity. Similarly, a turn produces transverse acceleration, which couples the heading error into the along-track velocity and the roll error into the vertical velocity.

The heading error is typically an order of magnitude larger than the roll and pitch errors because heading is more difficult to align and calibrate (see Sections 5.6.2 and 14.2.1). Consequently, significant maneuvers can lead to rapid changes in velocity error. Consider the example of an aircraft flying north at 100 m s^{-1} with

---

[*]This and subsequent paragraphs are based on material written by the author for QinetiQ, so comprise QinetiQ copyright material.
[†]End of QinetiQ copyright material.

5.7 INS Error Propagation

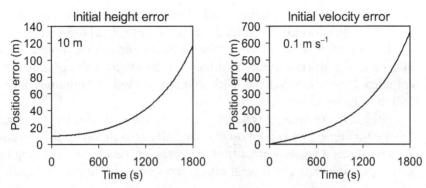

Figure 5.26 Vertical position error growth per axis over a 1,800-second period arising from height and velocity initialization errors.

north and east velocity errors of 0.05 m s^{-1} and 0.1 m s^{-1}, respectively, and a heading error of 1 mrad. The aircraft accelerates to 200 m s^{-1}, resulting in the east velocity error doubling to 0.2 m s^{-1}. It then undergoes a 90° turn to the west at constant speed; this maneuver increases the north velocity error to 0.25 m s^{-1} and drops the east velocity error to zero. Figure 5.27 illustrates this.

The effect of accelerometer and gyro scale factor and cross-coupling errors, gyro g-dependent errors, and higher-order inertial sensor errors (see Section 4.4) on navigation error growth also depends on the host vehicle maneuvers. In the previous example, a 500 ppm x-accelerometer scale factor error would produce an increase in north velocity error during the acceleration maneuver of 0.05 m s^{-1}, while a z-gyro scale factor error of –637 ppm would double the heading error to 2 mrad during the turn.

Velocity and direction changes often cancel out over successive maneuvers, so the effects of the scale factor and cross-coupling errors largely average out. An exception is circular and oval trajectories where the gyro scale factor and cross-coupling errors produce attitude errors that grow with time. The resulting velocity error will be oscillatory with the amplitude increasing with time, while the position error will be the sum of an oscillating term and a linear drift. Circling can occur when an aircraft

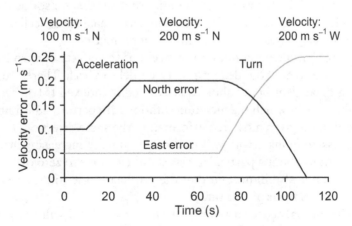

Figure 5.27 Illustration of the effect of maneuver on velocity error with a 1-mrad heading error.

is surveying an area or waiting in a holding pattern; it also occurs in motorsport. A similar problem occurs for guided weapons that spin about their roll axes.

Using tactical-grade gyros with scale factor and cross-coupling errors of around 300 ppm, the attitude errors will increase by about 0.1° per axis for each circuit completed by the host vehicle. With a circling period of 2 minutes, the position error will increase by about 400m per hour.

With a figure-of-eight trajectory, the attitude error due to gyro scale factor and cross-coupling errors will be oscillatory and correlated with the direction of travel. This produces a velocity error that increases with each circuit. Using tactical-grade gyros, position errors of several kilometers can build up over an hour.

5.8 Indexed IMU

In an *indexed* or *carouseling* IMU, the inertial sensor assembly is regularly rotated with respect to the casing, usually in increments of 90°. The rotation is typically performed about two axes or only about the vertical axis.

Indexing enables the cancellation over time of the position and velocity errors due to the accelerometer and gyro biases. The latter is particularly useful as gyro biases are the only major error source for which the horizontal position and velocity errors are not bounded over time by feedback through the gravity model (see Section 5.7.2). From (5.113) and (5.117), the growth in the position and velocity errors depends on the attitude of the IMU body frame with respect to the resolving axes of the navigation solution. Therefore, if the direction of an inertial sensor's sensitive axis is regularly reversed, its bias will lead to oscillatory position and velocity errors instead of continuously growing errors. To achieve this, it is rather more convenient to turn the inertial sensor assembly than to turn the entire host vehicle.

Single-axis indexing normally employs rotation of the inertial sensor assembly about the z-axis, generally the vertical. This enables cancellation of the effects of x- and y-axis accelerometer and gyro biases, but not the z-axis biases. The z-axis gyro bias has less impact on navigation accuracy as host-vehicle maneuvers are needed to couple the heading error into the position and velocity errors (see Section 5.7.3). The z-axis accelerometer bias mainly affects vertical positioning, which, as discussed in Section 5.7.2, always requires aiding from another sensor or a motion constraint, depending on the context. Dual-axis indexing enables cancellation of the effects of all six sensor biases on horizontal positioning [19].

Note that indexing does not cancel out the effects of gyro g-dependent biases, which are also not bounded by gravity-model feedback. Therefore, gyros that exhibit large g-dependent biases should be avoided in indexed IMUs.

The way in which the indexing rotations are performed is important. If all rotations about a given axis are performed in the same direction, the gyro scale-factor and cross-coupling errors will lead to continually increasing attitude errors. Thus, the rotations about a particular axis should average to zero over time. For dual-axis indexing, it is also important that the product of the rotations about any two axes also averages to zero over time.

The inertial sensor assembly of an indexed IMU will not generally be aligned with the host vehicle. The inertial navigation processor will compute the attitude of the inertial sensors with respect to the resolving frame. However, this is what it requires for positioning. The host vehicle attitude can be determined using the relative orientation of the sensor assembly, obtained from the indexing mechanism.

Indexed IMUs are typically deployed on submarines and military ships as these require the capability for stand-alone inertial navigation over many days and can handle the additional size, weight, and power consumption introduced by the indexing mechanism. However, rotation about the left-right axis may be used to limit the heading drift in foot-mounted inertial navigation [20].

5.9 Partial IMU

Normal land vehicle motion is subject to two motion constraints, also known as *nonholonomic constraints*. The velocity of the vehicle along the rotation axis of any of its wheels is zero. The velocity of the wheel rotation axes is also zero in the direction perpendicular to the road or rail surface. Note that, because of frame rotation, zero velocity does not necessarily imply zero acceleration. Consequently, under normal conditions (i.e., no wheel slip), the motion of a land vehicle has only four degrees of freedom, as opposed to six. By exploiting this context information, the vehicle motion can be measured using only four inertial sensors, known as a partial IMU. Vehicle motion in the presence of wheel slip is described in [21].

A common type of partial IMU has three accelerometers with mutually-perpendicular sensitive axes and a single gyro that measures the angular rate about the body z-axis, enabling changes in the vehicle heading to be measured. This is sometimes referred to as a 3A1G or 1G3A partial IMU. Partial IMU measurements are typically processed using conventional inertial navigation equations as described in Sections 5.2 to 5.5. The measurements from the missing inertial sensors are replaced with estimates, known as pseudo-measurements.

There are two main approaches to the generation of the pseudo-measurements and the application of the motion constraints. The first option is to simply assume the x- and y-axis angular rates are zero (i.e., $\omega_{ib,x}^b = \omega_{ib,y}^b = \alpha_{ib,x}^b = \alpha_{ib,y}^b = 0$). This will lead to errors in the position, velocity and attitude solution whenever the host vehicle pitches or rolls. These errors are then corrected by applying the motion constraints as Kalman filter measurement as described in Section 15.4.1 [22].

The second approach is to calculate the pseudo-measurements using the motion constraints and the remaining sensors. This is described in Section E.10 of Appendix E on the CD.

2A1G or 1G2A partial IMUs have also been proposed. These replace the z-axis accelerometer with a pseudo-measurement of $-g$, the reaction to gravity where that accelerometer is vertical [22]. However, even with the nonholonomic constraints, there is insufficient information to distinguish forward acceleration from changes in pitch. Consequently, sole-means navigation with such a sensor configuration requires a constant-gradient terrain along the direction of travel and external initialization of the pitch solution. In practice, 2A1G partial IMUs are only suitable for use in conjunction with other sensors, such as an odometer (Section 6.3), or GNSS (Chapters 8 to 10).

Problems and exercises for this chapter are on the accompanying CD.

References

[1] Titterton, D. H., and J. L. Weston, *Strapdown Inertial Navigation Technology*, 2nd ed., Stevenage, U.K.: IEE, 2004.

[2] Savage, P. G., "Strapdown Inertial Navigation Integration Algorithm Design Part 1: Attitude Algorithms," *Journal of Guidance Control and Dynamics*, Vol. 21, No. 1, 1998, pp. 19–28; "Strapdown Inertial Navigation Integration Algorithm Design Part 2: Velocity and Position Algorithms," *Journal of Guidance Control and Dynamics*, Vol. 21, No. 2, 1998, pp. 208–221.

[3] King, R., *The Effects of Approximations in the Processing of Strapdown Inertial Navigation Systems*, MSc Thesis, University College London, 2011.

[4] Wei, M., and K. P. Schwarz, "A Strapdown Inertial Algorithm Using an Earth-Fixed Cartesian Frame," *Navigation: JION*, Vol. 371, No.2, 1990, pp. 153–167.

[5] Jekeli, C., *Inertial Navigation Systems with Geodetic Applications*, Berlin, Germany: de Gruyter, 2000.

[6] Grejner-Brzezinska, D. A., et al., "Enhanced Gravity Compensation for Improved Inertial Navigation Accuracy," *Proc. ION GPS/GNSS 2003*, Portland, OR, September 2003. pp. 2897–2909.

[7] Jekeli, C., "Precision Free-Inertial Navigation with Gravity Compensation by an Onboard Gradiometer," *Journal of Guidance, Control, and Dynamics*, Vol. 29, No. 3, 2006, pp. 704–713.

[8] Johnson, D., "Frequency Domain Analysis for RLG System Design," *Navigation: JION*, Vol. 34, No. 3, 1987, pp. 178–189.

[9] Groves, P. D., "Principles of Integrated Navigation," Course Notes, QinetiQ Ltd., 2002.

[10] Bortz, J. E., "A New Mathematical Formulation for Strapdown Inertial Navigation," *IEEE Trans. on Aerospace and Electronic Systems*, Vol. AES-7, No. 1, 1971, pp. 61–66.

[11] Savage, P. G., *Strapdown Analytics, Parts 1 and 2*, Maple Plain, MN: Strapdown Associates, 2000.

[12] Farrell, J. L., "Strapdown at the Crossroads," *Navigation: JION*, Vol. 51, No. 4, 2004, pp. 249–257.

[13] Vinande, E., P. Axelrad, and D. Akos, "Mounting-Angle Estimation for Personal Navigation Devices," *IEEE Trans. on Vehicular Technology*, Vol. 59, No. 3, 2010, pp. 1129–1138.

[14] Tazartes, D. A., M. Kayton, and J. G. Mark, "Inertial Navigation," in *Avionics Navigation Systems*, 2nd ed., M. Kayton and W. R. Fried, (eds.), New York: Wiley, 1997, pp. 313–392,

[15] Farrell, J. A., *Aided Navigation: GPS with High Rate Sensors*, New York: McGraw-Hill, 2008.

[16] Rogers, R. M., *Applied Mathematics in Integrated Navigation Systems*, Reston, VA: AIAA, 2000.

[17] Britting, K. R., *Inertial Navigation Systems Analysis*, New York: Wiley, 1971 (Republished by Norwood, MA: Artech House, 2010).

[18] Rogers, R. M., "IMU In-Motion Alignment without Benefit of Attitude Initialization," *Navigation: JION*, Vol. 44, No.4, 1997, pp. 301–311.

[19] Levinson, E., and R. Majure, "Accuracy Enhancement Techniques Applied to the Marine Ring Laser Inertial Navigator (MARLIN)," *Navigation: JION*, Vol. 34, No. 1, 1987, pp. 64–86.

[20] Abdulrahim, K., *Heading Drift Mitigation for Low-Cost Inertial Pedestrian Navigation*, Ph.D. Thesis, University of Nottingham, 2012.

[21] Bevly, D. M., and S. Cobb, (eds.), *GNSS for Vehicle Control*, Norwood, MA: Artech House, 2010.

[22] El-Sheimy, N., "The Potential of Partial IMUs for Land Vehicle Navigation," *Inside GNSS*, Spring 2008, pp. 16–25.

CHAPTER 6
Dead Reckoning, Attitude, and Height Measurement

This chapter describes commonly used dead-reckoning techniques other than inertial navigation (Chapter 5), together with a number of techniques for measuring attitude, height, and depth. Although magnetic field and pressure measurements may also be classed as feature matching, they are described here as they are commonly used alongside dead-reckoning systems.

Dead reckoning measures the motion of the user with respect to the environment without the need for radio signals or extensive feature databases. Measurements are made in the sensor body frame. Consequently, a heading or full attitude solution (as appropriate) is required to resolve the motion in Earth-referenced coordinates. In addition, an initial position solution must be supplied, as described in Section 5.6.1 for inertial navigation.

Section 6.1 describes attitude measurement, including the magnetic compass, gyro-derived heading, and accelerometer leveling. Section 6.2 describes height and depth sensors. Section 6.3 describes odometry or wheel speed sensing, including combining odometry with a partial IMU. Section 6.4 describes pedestrian dead reckoning using step detection, and Section 6.5 describes Doppler radar and sonar. Finally, Section 6.6 discusses correlation-based velocity measurement, air data, and the ship's speed log. Additional dead-reckoning techniques based on environmental feature tracking are described in Chapter 13.

6.1 Attitude Measurement

This section describes a number of stand-alone attitude measurement techniques. Heading measurement techniques are described first, comprising the magnetic compass, marine gyrocompass, strapdown yaw-axis gyro, and trajectory-derived heading. This is followed by a discussion of multisensor integrated heading determination. Next, roll and pitch measurement using accelerometer leveling, tilt sensors, and horizon sensors is described. Finally, the attitude and heading reference system (AHRS) is introduced.

Attitude determination methods described elsewhere in the book include inertial navigation in Chapter 5, differential odometry in Section 6.3.2, multiple-antenna GNSS in Section 10.2.5, stellar imagery in Section 13.3.7, and the aircraft directional gyro in Section E.2.3 of Appendix E on the CD.

6.1.1 Magnetic Heading

The Earth's geomagnetic field points from the magnetic north pole to the magnetic south pole through the Earth, taking the opposite path through the upper atmosphere, as illustrated by Figure 6.1. The field is thus vertical at the magnetic poles and horizontal near the equator. The magnetic poles slowly move over time, with the north pole located on January 1, 2010, at latitude 80.08°, longitude −72.21° and the south pole at latitude −80.08°, longitude 107.79°, so the field is inclined at 9.98° to the Earth's axis of rotation [1].

A magnetic field is described by the magnetic flux density vector, such that the force per unit length due to magnetic inductance is the vector product of the flux density and current vectors. The SI unit of magnetic flux density is the Tesla (T), where $1\,\mathrm{T} = 1\,\mathrm{N\,A^{-1}m^{-1}}$. The standard notation for it is **B**. However, in the notation used here, this would be a matrix, while **b** clashes with the instrument biases, so **m** has been selected instead.

The flux density of the Earth's geomagnetic field, denoted by the subscript E, resolved about the axes of a local navigation frame, may be expressed using

$$\mathbf{m}_E^n(\mathbf{p}_b, t) = \begin{pmatrix} \cos\alpha_{nE}(\mathbf{p}_b, t)\cos\gamma_{nE}(\mathbf{p}_b, t) \\ \sin\alpha_{nE}(\mathbf{p}_b, t)\cos\gamma_{nE}(\mathbf{p}_b, t) \\ \sin\gamma_{nE}(\mathbf{p}_b, t) \end{pmatrix} B_E(\mathbf{p}_b, t), \qquad (6.1)$$

where B_E is the magnitude of the flux density, α_{nE} is the declination angle or magnetic variation, and γ_{nE} is the inclination or dip angle of the Earth's magnetic field. All three parameters vary as functions of position and time.

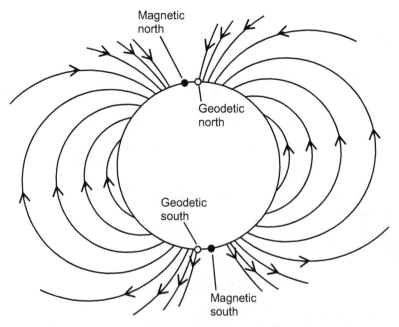

Figure 6.1 The Earth's geomagnetic field.

The flux density varies from about 30 μT at the equator to about 60 μT at the poles, while the dip is essentially the magnetic latitude so is within about 10° of the geodetic latitude, L_b. The declination angle gives the bearing of the magnetic field from true north and is the only one of the three parameters needed to determine a user's heading from magnetic field measurements. It may be calculated as a function of position and time using global models, such as the 275-coefficient International Geomagnetic Reference Field (IGRF) [2] or the 336-coefficient U.S./U.K. World Magnetic Model (WMM) [1]. Regional variations, correlated over a few kilometers, occur due to local geology. Global models are typically accurate to about 0.5°, but can exhibit errors of several degrees in places [3]. Higher-resolution national models are available for some countries. There is a diurnal (day-night) variation in the geomagnetic field of around 50 nT. Short-term temporal variations in the Earth's magnetic field also occur due to magnetic storms caused by solar activity. The effect on the declination angle varies from around 0.03° at the equator to more than 1° at latitudes over 80° [4].

Magnetometers measure the total magnetic flux density, denoted by the subscript m, resolved along the axes of their body frame. Assuming that the magnetometer sensitive axes are aligned with those of any inertial sensors used, the body frame is denoted b. The magnetometers thus measure

$$\mathbf{m}_m^b = \mathbf{C}_n^b \begin{pmatrix} \cos\alpha_{nm} \cos\gamma_{nm} \\ \sin\alpha_{nm} \cos\gamma_{nm} \\ \sin\gamma_{nm} \end{pmatrix} B_m, \qquad (6.2)$$

where B_m, α_{nm}, and γ_{nm} are, respectively, the magnitude, declination, and dip of the total magnetic flux density. Applying (2.22),

$$\mathbf{m}_m^b = \begin{pmatrix} \cos\theta_{nb} & 0 & -\sin\theta_{nb} \\ \sin\phi_{nb}\sin\theta_{nb} & -\cos\phi_{nb} & \sin\phi_{nb}\cos\theta_{nb} \\ \cos\phi_{nb}\sin\theta_{nb} & \sin\phi_{nb} & \cos\phi_{nb}\cos\theta_{nb} \end{pmatrix} \begin{pmatrix} \cos\psi_{mb} \cos\gamma_{nm} \\ \sin\psi_{mb} \cos\gamma_{nm} \\ \sin\gamma_{nm} \end{pmatrix} B_m, \qquad (6.3)$$

where ϕ_{nb} is the roll, θ_{nb} is the pitch, and the magnetic heading, ψ_{mb}, is given by

$$\psi_{mb} = \psi_{nb} - \alpha_{nm}. \qquad (6.4)$$

When the roll and pitch are zero, a magnetic heading measurement can be obtained from magnetometer measurements using

$$\tilde{\psi}_{mb} = \arctan_2\left(-\tilde{m}_{m,y}^b, \tilde{m}_{m,x}^b\right), \qquad (6.5)$$

whereas when they are nonzero but known, the magnetic heading measurement is

$$\tilde{\psi}_{mb} = \arctan_2\begin{pmatrix} -\tilde{m}_{m,y}^b \cos\hat{\phi}_{nb} + \tilde{m}_{m,z}^b \sin\hat{\phi}_{nb}, \\ \tilde{m}_{m,x}^b \cos\hat{\theta}_{nb} + \tilde{m}_{m,y}^b \sin\hat{\phi}_{nb}\sin\hat{\theta}_{nb} + \tilde{m}_{m,z}^b \cos\hat{\phi}_{nb}\sin\hat{\theta}_{nb} \end{pmatrix}, \qquad (6.6)$$

where a four-quadrant arctangent function should be used in both cases.

Floating-needle magnetic compasses have been used for centuries but do not provide an electronic readout. Electronic compasses use two or three orthogonally mounted magnetometers to measure the magnetic field and then calculate the magnetic heading using (6.5) or (6.6) as appropriate. The true heading is then given by

$$\tilde{\psi}_{nb} = \tilde{\psi}_{mb} + \alpha_{nE}. \qquad (6.7)$$

Types of magnetometer suitable for navigation systems are fluxgates, Hall-effect sensors, magnetoinductive sensors, and magnetoresistive sensors [3]. Magnetoinductive and magnetoresistive sensors are small and accurate to about 0.05 μT [5], which is good enough for most navigation applications, given the other error sources. Fluxgate sensors offer a better performance and can have dual sensitive axes, but are larger and more expensive, while the performance of Hall-effect sensors is much poorer.

A two-axis magnetic compass must be kept physically aligned with the horizontal plane to avoid errors in determining the heading. When the small angle approximation applies to the roll and pitch, the heading error is

$$\delta\psi_{mb} \approx (\theta_{nb}\sin\psi_{mb} - \phi_{nb}\cos\psi_{mb})\tan\gamma_{nm}. \qquad (6.8)$$

Thus, two-axis magnetic compasses are usually mounted in a set of gimbaled frames to keep them level, although, for road applications, changes in the magnitude of the magnetometer measurements may be used to estimate the pitch and correct the heading accordingly [6].

A three-axis, or strapdown, magnetic compass uses an accelerometer triad to measure the roll and pitch using leveling (Section 6.1.6) and is available for $50 (€40). However, the leveling is disrupted by acceleration, so the device is unsuited to high-dynamics applications. Acceleration-induced errors are also a problem for high-vibration applications, such as pedestrian navigation, but may be significantly reduced by smoothing measurements over the order of a second [7]. When roll and pitch from an INS or AHRS are available, these should always be used in preference to leveling measurements.

Gimbaled and floating-needle compasses are also disrupted by acceleration and mitigate this using mechanical damping [8]. However, as with INS, they have largely been superseded by their strapdown counterparts.

A major problem for land applications is that magnetic fields are produced by man-made objects, such as vehicles, buildings, bridges, lamp posts, and power lines [6, 7]. These can be significant several meters away and cannot easily be distinguished from the geomagnetic field. With a stand-alone magnetic compass, the only way of mitigating these local anomalies is to compare the magnitude of the magnetic flux density measurement, $|\mathbf{m}_{ml}^b|$, with an upper and a lower threshold to determine whether it is consistent with the Earth's magnetic field and reject magnetometer measurements that do not fall within the two thresholds. When the orientation of the magnetometer with respect to the vertical is known, the sensitivity of the anomaly detection may be improved by applying an additional test to the measured dip angle or separate tests to the horizontal and vertical magnetic flux density measurements [9]. However, this type of anomaly detection can still allow magnetic anomalies to produce undetected heading errors of several degrees, while forcing the navigation system to rely on an out-of-date heading measurement when an anomaly is detected.

When the magnetic compass is integrated with another heading sensor (see Sections 6.1.5, 6.1.8, and 16.2.1), the integration process naturally smooths out the effect of the local anomalies, while measurement innovation filtering (Section 17.3.1) can be used to reject the most corrupted magnetic compass measurements [7, 10]. Section 16.2.1.1 also discusses the use of magnetometer-derived angular rate.

The final obstacle to determining heading from magnetometer measurements is that, as well as the geomagnetic field and local anomalies, the magnetometers also measure the magnetic field of the navigation system itself, the host vehicle, and any equipment carried. This equipment magnetism is divided into hard-iron and soft-iron magnetism. Hard-iron magnetism is simply the magnetic fields produced by permanent magnets and electrical equipment. It is typically a few microTeslas, but can sometimes exceed the geomagnetic field. Soft-iron magnetism, however, is produced by materials that distort the underlying magnetic field. Soft-iron magnetism is relatively large in ships and can distort the magnetic field by the order of 10%, However, it is much smaller in most aircraft and road vehicles.

The total magnetic flux density measured by a set of magnetometers is thus

$$\mathbf{m}_m^b = \mathbf{b}_m + (\mathbf{I}_3 + \mathbf{M}_m)\mathbf{C}_n^b(\mathbf{m}_E^n + \mathbf{m}_A^n) + \mathbf{w}_m, \tag{6.9}$$

where $\mathbf{m}_E^n$ is the geomagnetic flux density as before, $\mathbf{m}_A^n$ is the flux density from local magnetic anomalies, $\mathbf{b}_m$ is the hard-iron flux density, resolved in the body frame, $\mathbf{M}_m$ is the soft-iron scale factor and cross-coupling matrix, and $\mathbf{w}_m$ is the magnetometer random noise. Hard-iron and soft-iron magnetism are thus analogous to the biases, scale-factor, and cross-coupling errors exhibited by inertial sensors (see Section 4.4). The magnetometers themselves also exhibit biases, scale-factor, and cross-coupling errors. However, these are typically much smaller than the errors due to hard-iron and soft-iron magnetism and are not distinguishable from them, so they do not need to be considered separately. A typical MEMS magnetometer exhibits measurement noise with a root PSD of 0.01 $\mu T/\sqrt{Hz}$.

Example 6.1 on the CD shows how the different sources of magnetic flux density contribute to the magnetometer measurements together with the calculation of the resulting magnetic and true heading measurements; it may be edited using Microsoft Excel.

The equipment and environmental magnetic flux densities may be distinguished by the fact that the equipment magnetism is referenced to the body frame, whereas the environmental magnetism is Earth-referenced. This enables $\mathbf{b}_m$ and $\mathbf{M}_m$ to be calibrated using a process known as *swinging*, whereby a series of measurements are taken with the magnetic compass at different orientations, with the roll and pitch varied as well as the heading. This is done at a fixed location, so the environmental magnetic flux density may be assumed constant. The calibration parameters and environmental flux density may then be estimated using a nonlinear estimation algorithm or an EKF [11–13], which is usually built into the magnetic compass. Following calibration, the magnetometer measurements are compensated using

$$\hat{\mathbf{m}}_m^b = \left(\mathbf{I}_3 + \hat{\mathbf{M}}_m\right)^{-1} \tilde{\mathbf{m}}_m^b - \hat{\mathbf{b}}_m, \qquad (6.10)$$

where $\hat{\mathbf{b}}_m$ and $\hat{\mathbf{M}}_m$ are the estimated hard- and soft-iron magnetism. When the magnetic compass is mounted in a large vehicle, a physical swinging process is not practical. Instead, some magnetic compasses perform electrical swinging using a self-generated magnetic field [8].

For applications where the magnetic compass is kept approximately level, a simpler four-coefficient calibration may be performed with the compass swung about the heading axis only. The correction is applied in the heading domain using [8]

$$\hat{\psi}_{mb} = \tilde{\psi}_{mb} + \hat{c}_{h1} \sin \tilde{\psi}_{mb} + \hat{c}_{h2} \cos \tilde{\psi}_{mb} + \hat{c}_{s1} \sin 2\tilde{\psi}_{mb} + \hat{c}_{s2} \cos 2\tilde{\psi}_{mb}, \quad (6.11)$$

where $\hat{c}_{h1}$ and $\hat{c}_{h2}$ are the hard-iron calibration coefficients and $\hat{c}_{s1}$ and $\hat{c}_{s2}$ the soft-iron coefficients. This calibration is only valid when the magnetic compass is level.

6.1.2 Marine Gyrocompass

The gyrocompass has been used for heading determination in ships from the early part of the twentieth century [14, 15]. It is related to the spinning-mass gyroscope used for angular rate measurement (Section E.2 of Appendix E on the CD). It consists of a single large spinning-mass gyro with its spin axis aligned along the north-south axis within the horizontal plane. The gyro assembly comprises a spinning disc with most of the mass around its edge, driven by a motor. The casing of the gyro assembly is either linked to the casing of the gyrocompass unit via a set of gimbals or floated in a bed of mercury. This isolates it from the motion of the host ship. The heading of the ship can thus be determined by reading off the angle between the gyro spin axis and the fore-aft axis of the ship.

Conservation of angular momentum keeps the gyro spin axis fixed with respect to inertial space (instrument errors excepted). Therefore as the Earth rotates and the ship moves, it will precess away from alignment with the north-south axis. With respect to a north, east, down frame, the gyro spin axis precesses both in azimuth and in pitch. Furthermore, if the spin axis is pointing to the east of due north, its pitch will increase as the Earth rotates. This relationship is exploited by the gravity control technique, whereby the rotor housing within the gimbals is imbalanced with the bottom heavier than the top (or vice versa). Consequently, when the spin axis is not horizontal, gravity exerts a torque on the rotor assembly. Due to conservation of angular momentum, the resulting rotation is about the axis perpendicular to the torque and the spin, so if the spin direction is set correctly, the spin axis will precess west when pitched upwards and east when pitched downwards. Over a complete cycle, the spin axis will describe an elliptical cone, centered about the north-south axis within the horizontal plane, a behavior known as north seeking.

To make the gyro spin axis home in on the north-south axis, damping is applied, typically in the form of an azimuthal torque proportional to the displacement of the spin axis from the horizontal plane. This damping also indirectly reduces the amplitude of the azimuthal precession. The damping time constant is typically

60–90 minutes. Due to the combined effect of the gravity control and the damping, the direction of the spin axis "spirals in" to alignment with north-south in the horizontal plane. A gyrocompass is thus self-aligning. This removes the need for an initial alignment process and ensures that the gyro sensor errors do not cause the gyrocompass heading error to grow with time. A gyrocompass typically requires a settling time of about an hour between activation and use.

For an aligned gyrocompass, the Earth's rotation will still precess the spin axis away from north-south at a rate proportional to the sine of the latitude. The gravity and damping control loops will counteract this. However, there will be a lag in applying the correction, resulting in a latitude-dependent heading bias. This is known as the latitude error and can be compensated by applying a latitude-dependent torque to the rotor assembly to compensate for the Earth rotation. Consequently, all gyrocompasses incorporate a latitude input.

A moving ship slowly rotates with respect to inertial space about its pitch axis (at the transport rate; see Section 5.4.1) to keep its xy-plane parallel to the sea surface (on average). When the ship is traveling north-south, its pitch axis coincides with that of the gyrocompass rotor, which does not rotate with the ship. Therefore, the spin axis slowly rotates with respect to the ship about its pitch axis. The gyrocompass gravity control loop responds to this, perturbing the heading solution by an amount proportional to $v^n_{eb,N}/\cos L_b$. At mid-latitudes, this equates to about 1° of displacement for a north-south speed of 20 m s^{-1}. This "steaming error" may be compensated either by applying a correction to the gyrocompass output or by applying a restoring torque to realign the gyro spin axis. Hence, gyrocompasses also have a north velocity input. When the ship is traveling east-west, its pitch axis coincides with the gyrocompass rotor's spin axis so the ship rotation does not impact the gyrocompass control loops.

At aircraft speeds, the steaming error cannot be effectively compensated, so a gyrocompass is not suitable for aircraft use. The steaming error also prevents gyrocompasses from working at the poles. However, this has not historically been a problem for shipping.

To avoid performance degradation, the cumulative instrument errors over the time constant of the gravity control and damping loops, which correct for them, must be small. Therefore, the instrument quality is comparable to that of an aviation-grade IMU (see Chapter 4). A typical gyrocompass has a heading accuracy of about 0.1° and costs around $15,000 (€12,000). The size is around (0.4–0.5m)^3.

6.1.3 Strapdown Yaw-Axis Gyro

The body xy-plane of a land vehicle is approximately parallel to the road surface. Therefore, when the road is level, a single yaw-axis gyro, sensitive to rotation about the vehicle body z-axis, will directly measure heading change. Heading measurement errors will then be introduced by sloping, banked, and uneven terrain; vehicle roll; gyro mounting misalignment; sensor errors; and Earth rotation, many of which may be compensated [16].

From (2.49), the angular rate measured by the yaw gyro is related to the angular rate of the vehicle body with respect to a local navigation frame by

$$\omega_{ib,z}^b = \mathbf{C}_{b3,1:3}^n(\boldsymbol{\omega}_{nb}^n + \boldsymbol{\omega}_{in}^n). \tag{6.12}$$

The heading solution may be updated approximately using

$$\hat{\psi}_{nb}(t + \tau_i) \approx \hat{\psi}_{nb}(t) + \omega_{ib,z}^b \tau_i, \tag{6.13}$$

where τ_i is the update interval.

Figure 6.2 illustrates the effect of a sloped road surface on yaw measurement. The rotation about the vehicle body z-axis, $\alpha_{ib,z}^b$, is less than the corresponding heading change, $\Delta\psi_{nb}$. Therefore, a yaw-axis gyro will underestimate heading changes on sloped terrain. In addition to ramps and hills, curves on fast roads are often banked to aid the stability of vehicles turning at speed. The heading change will be underestimated when following a banked curve and will be overestimated when turning on a banked slope. When the combined roll and pitch are within 0.14 rad (~8°), equivalent to a 10% slope, the angular rate scale factor error will be less than 1% [16].

The vehicle itself can also roll with respect to the road surface during turns, the amount depending on the vehicle design and the driving style; motorcycles undergo the largest rolls. However, the roll angle may be estimated as a function of the speed and yaw rate and used to compensate the gyro output [17].

Further errors occur on rough surfaces as a yaw gyro is sensitive to simultaneous rolling and pitching, resulting in a heading random walk error (assuming the pitch and roll motions are uncorrelated); however, this is only likely to impact off-road performance [16].

Finally, when the host vehicle is stationary, it may be assumed that the heading does not change. This not only prevents degradation of the heading solution, but can also be used to calibrate the gyro bias. Zero angular rate updates are described in Section 15.3.3.

In principle, a more sophisticated heading update algorithm could be implemented using pitch derived from the gradient of a terrain height database or from the horizontal and vertical velocity solution. However, it is difficult to determine the

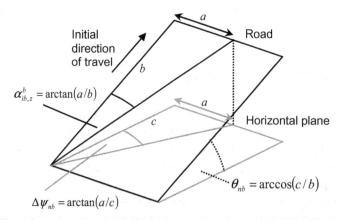

Figure 6.2 Effect of sloped surface on yaw measurement using a gyro with sensitive axis perpendicular to the road surface. (*After:* [16].)

6.1 Attitude Measurement

roll. Therefore, for applications requiring higher accuracy, a partial IMU (Section 5.9) or an AHRS (Section 6.1.8) is recommended.

6.1.4 Heading from Trajectory

When a land vehicle is traveling in a straight line, the direction of travel coincides with the vehicle-body-frame x-axis. This enables the heading to be determined from the Earth-referenced velocity solution using

$$\psi_{nb} = \arctan_2\left(v^n_{eb,E}, v^n_{eb,N}\right), \tag{6.14}$$

where a four-quadrant arctangent function must be used. To prevent positive feedback errors, the heading solution should not be computed from a velocity that is a function of the same heading solution. A velocity solution from a position-fixing system, such as GNSS, is suitable [16, 18]. The heading uncertainty is

$$\sigma_\psi = \frac{\sqrt{v^n_{eb,E}{}^2 \sigma^2_{v\_N} + v^n_{eb,N}{}^2 \sigma^2_{v\_E}}}{v^n_{eb,N}{}^2 + v^n_{eb,E}{}^2}, \tag{6.15}$$

where σ_{v_N} and σ_{v_E} are, respectively, the north and east velocity uncertainties. The accuracy of the velocity-derived heading solution thus depends on the speed. GNSS velocity should not be used for heading determination at speeds below 2 m s^{-1} [16]. An alternative is to estimate the trajectory from map matching (Section 13.1) [19].

When the vehicle is turning, the heading and trajectory no longer coincide. However, when the yaw angular rate is known, the velocity solution may be transformed from the body-frame origin to a point on the vertical plane containing the rear axle. Assuming the rear wheels are non-steering, this plane is constrained to move only along the body-frame x-axis (see Section 6.3.1). The heading thus becomes

$$\psi_{nb} = \arctan_2\left[\left(v^n_{eb,E} + \omega^n_{nb,z} l^b_{br,x} \cos\psi_{nb}\right), \left(v^n_{eb,N} - \omega^n_{nb,z} l^b_{br,x} \sin\psi_{nb}\right)\right], \tag{6.16}$$

where $l^b_{br,x}$ is the forwards distance from the b-frame origin to the rear axle. Note, however, that this may be degraded by wheel slip during turning. If the angular rate (or the lever arm) is not available, trajectory-derived heading measurements should be ignored whenever the heading is changing. More information on road vehicle motion may be found in [20].

For aircraft, ships, and boats, there is a divergence between the heading and the Earth-referenced trajectory due to wind or water currents. Consequently, trajectory-derived heading is only useful for providing a rough initialization for a heading alignment process and as a reference for fault detection (Chapter 17).

The pitch attitude of a land vehicle may be determined by the same method. Where the pitch is constant,

$$\theta_{nb} = \arctan\left(\frac{-v^n_{eb,D}}{\sqrt{v^n_{eb,N}{}^2 + v^n_{eb,E}{}^2}}\right). \tag{6.17}$$

6.1.5 Integrated Heading Determination

As discussed elsewhere, different heading determination methods exhibit different error characteristics. Magnetic heading measurements (Section 6.1.1) are subject to errors induced by accelerations and local magnetic anomalies. Multi-antenna GNSS-derived heading (Section 10.2.5) is noisy and vulnerable to signal interruption. Trajectory-derived heading (Section 6.1.4) relies on a velocity solution being available and is unreliable at low speeds. A gyroscope (Section 6.1.3) or differential odometer (Section 6.3.2) provides accurate measurements of short-term heading changes, but is subject to accumulation of errors due to sensor errors and the effects of vehicle tilts, slopes, and uneven roads.

A more stable and accurate heading solution can often be obtained by integrating an absolute heading measurement method, such as magnetic, GNSS-derived, or trajectory-derived heading, with a yaw-rate measurement method, such as a gyroscope or differential odometer. The yaw-rate sensor smooths out the noise on the absolute heading measurements, while the absolute heading measurements calibrate the gyro or odometer drift.

The sensors may be integrated with a fixed-gain smoothing filter. For example, a magnetic compass and gyro may be combined using

$$\hat{\psi}_{nb}(t) = W_m \hat{\psi}_{nb,m}(t) + (1 - W_m)\left[\hat{\psi}_{nb}(t - \tau) + \tilde{\omega}^b_{ib,z}\tau\right], \tag{6.18}$$

where $\hat{\psi}_{nb}$ is the integrated heading, $\hat{\psi}_{nb,m}$ is the magnetic compass indicated true heading, $\tilde{\omega}^b_{ib,z}$ is the gyro-measured angular rate, and W_m is the magnetic compass weighting. W_m may be set to zero whenever a magnetic anomaly is detected.

However, it is better to integrate heading measurements using a Kalman filter as described in Section 16.2.1.2. This enables the gyro or odometer bias to be calibrated, optimizes the sensor weighting, and provides more ways of filtering out anomalous absolute heading measurements.

6.1.6 Accelerometer Leveling and Tilt Sensors

As described in Section 5.6.2, the roll and pitch attitude components of an inertial navigation solution are commonly initialized using leveling. The accelerometer triad is used to detect the direction of the acceleration due to gravity, which, neglecting local variations, denotes the down axis of a local navigation frame. The pitch and roll are given by (5.101):

$$\theta_{nb} = \arctan\left(\frac{f^b_{ib,x}}{\sqrt{{f^b_{ib,y}}^2 + {f^b_{ib,z}}^2}}\right), \qquad \phi_{nb} = \arctan_2\left(-f^b_{ib,y}, -f^b_{ib,z}\right)$$

where $\mathbf{f}^b_{ib}$ is the specific force and a four-quadrant arctangent function is used for roll. The same principle is used to calibrate the roll and pitch errors from the rate of change of the velocity error in INS/GNSS integration, transfer alignment, quasi-stationary alignment, zero velocity updates, and other INS integration algorithms (see Chapters 14 to 16).

In a navigation system without gyros, which are more expensive than accelerometers, leveling may be used as the sole means of determining the roll and pitch. However, (5.101) makes the assumption that the accelerometers are stationary, so only the reaction to gravity is measured. Thus, any acceleration disrupts the leveling process. For example, a 1 m s^{-2} forward acceleration will lead to a pitch determination error of about 100 mrad (5.7°).

Tilt sensors, also known as inclinometers, determine the roll and pitch attitude by measuring the direction of the specific force, but not its magnitude. Thus, they also exhibit acceleration-induced errors. Some tilt sensors are accelerometer-based; other types include pendulums, liquid capacitive sensors, electrolytic sensors, and gas bubbles, commonly known as spirit levels. Sensors may be dual or single axis and prices are between $5 and $200 (€4 and €160).

Figure 6.3 depicts an electrolytic tilt sensor. This comprises a vial, partially filled with a conducting liquid that contains a number of parallel electrodes. The conductivity between any pair of electrodes is proportional to the length of electrode immersed in fluid. A single-axis sensor will typically have two or three electrodes and a dual-axis sensor four or five. Sensor noise is low, with a repeatability of about 0.01° for a wide-range sensor (better for a limited range sensor). However, systematic errors, which arise from electrode mounting misalignments, are significant with cross-coupling errors as large as 10%. Thus, for best performance, the sensors should be calibrated before use [21].

6.1.7 Horizon Sensing

Horizon sensing determines roll and pitch attitude by measuring the orientation of the Earth's horizon with respect to the sensor body frame. When the body's xy plane is parallel to the plane containing the horizon, the body is level. As horizon sensing does not involve the specific force, it is not affected by host vehicle acceleration. Simple horizon sensors assume a spherical Earth. However, for best accuracy, an ellipsoidal Earth must be assumed.

Horizon sensing has been used for orbital spacecraft attitude determination since the 1960s. Best results are obtained using the 5–15-μm region of the infrared

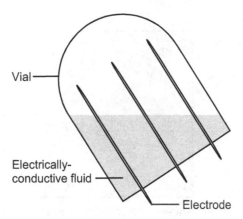

Figure 6.3 An electrolytic tilt sensor.

spectrum. Scanning-beam horizon sensors have now largely been replaced by imaging sensors.

In recent years, horizon sensing using a nose-mounted consumer video camera has been demonstrated on micro air vehicles (MAVs) [22, 23]. At low altitude, the horizon is approximately straight, enabling the roll angle to be determined from its gradient and the pitch from its vertical displacement within the image; Figure 6.4 illustrates this. The noise is of the order of a degree with a standard definition camera and outlier detection (see Chapter 17) should be used to filter out measurements when the image processing algorithm misidentifies the horizon.

Thermal horizon sensing operates on the principle that more heat is radiated from the ground than from the sky, enabling roll and pitch to be determined from measurements made by six thermocouples, facing in opposite directions along three mutually perpendicular axes. A set of sensors costs about $100 (€80) and is typically used for MAV applications.

6.1.8 Attitude and Heading Reference System

Figure 6.5 shows a basic schematic of an attitude and heading reference system, or heading and attitude reference system (HARS). This comprises a low-cost IMU with consumer- or tactical-grade sensors and a magnetic compass. It is typically used for low-cost aviation applications, such as private aircraft and UAVs, and provides a three-component inertial attitude solution without position and velocity. For marine applications, it is sometimes known as a strapdown gyrocompass. The attitude is computed by integrating the gyro measurements in the same way as in an INS (see Chapter 5), noting that the Earth-rotation and transport-rate terms in a local-navigation-frame implementation must be neglected when position and velocity are unknown.

The accelerometers measure the roll and pitch by leveling, as described in Section 6.1.6. This is used to correct the gyro-derived roll and pitch in a smoothing filter (see Section 6.1.5) with a low gain to minimize the corruption of the leveling measurements by host-vehicle maneuvers. The magnetic compass is used to correct the gyro-derived heading, again with a low-gain smoothing filter used to minimize the effects of short-term errors. The corrected gyro-indicated roll and pitch are used

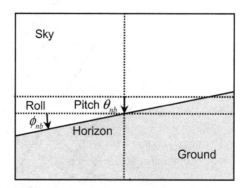

Figure 6.4 Roll and pitch attitude determination from low-altitude horizon detection.

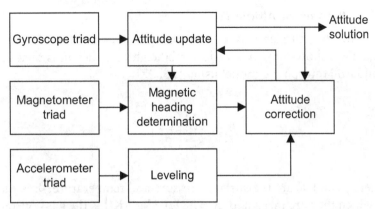

Figure 6.5 Basic schematic of an attitude and heading reference system.

to determine the magnetic heading from three-axis magnetometer measurements using (6.6). Many AHRS incorporate maneuver detection algorithms to filter out accelerometer measurements during high dynamics. This can be done by comparing the magnitude of the specific force with the acceleration due to gravity. Similarly, the magnetic heading measurements may be rejected when magnetic anomalies are detected. Using a Kalman filter to integrate data from the various sensors, as described in Section 16.2.1.4, enables the smoothing gains to be dynamically optimized and the gyro biases to be calibrated.

A typical AHRS provides roll and pitch to a 10-mrad (0.6°) accuracy and heading to a 20-mrad (1.2°) accuracy during straight and level flight, noting that performance depends on the quality of the inertial sensors and the type of processing used. Accuracy is typically degraded by a factor of 2 during high-dynamic maneuvers. More information on AHRS may be found in [8] while integration of AHRS with other navigation systems is discussed in Section 16.2.1.4.

Note that the term AHRS is sometimes used to describe a lower grade INS, rather than a device that only determines attitude.

6.2 Height and Depth Measurement

Height determination is essential for aircraft navigation, while depth determination is critical for a submarine, a diver, an underwater remotely-operated vehicle (ROV), or an autonomous underwater vehicle (AUV). Height is also needed for determining which floor of a building a pedestrian is on. Furthermore, for any vehicle, an independent height solution can be used to constrain GNSS positioning, improving robustness. This is also applicable to pseudolite, UWB, and acoustic positioning (Chapter 12). Land vehicle height may be determined as a function of latitude and longitude using a terrain height database [24, 25], while ship height may be determined using a geoid model and a tidal model.

This section describes independent height and depth measurement methods. The barometric altimeter, depth pressure sensor, and radar altimeter are covered.

6.2.1 Barometric Altimeter

A barometric altimeter uses a barometer to measure the ambient air pressure, p_b. Figure 6.6 shows how this varies with height. The height is then determined from a standard atmospheric model using [26, 27]

$$h_b = \frac{T_s}{k_T}\left[\left(\frac{p_b}{p_s}\right)^{-\left(\frac{Rk_T}{g_0}\right)} - 1\right] + h_s, \qquad (6.19)$$

where p_s and T_s are the surface pressure and temperature, h_s is the geodetic height at which they are measured, $R = 287.1$ J kg^{-1} K^{-1} is the gas constant, $k_T = 6.5 \times 10^{-3}$ K m^{-1} is the atmospheric temperature gradient, and $g_0 = 9.80665$ m s^{-2} is the average surface acceleration due to gravity. For differential barometry, the surface temperature and pressure are measured at a reference station and transmitted to the user. For stand-alone barometry, standard mean sea level values of $p_s = 101.325$ kPa and $T_s = 288.15$K are assumed, in which case, $h_b - h_s$ is the orthometric height, H_b, defined in Section 2.4.4. Note that (6.19) only applies at orthometric heights up to 10.769 km. Above this, a constant air temperature of 218.15K is assumed giving $h_b = 73{,}607 - 14{,}705 \log_{10} P_b$ m, where P_b is in Pa [28].

The baro measurement resolution is about 10 Pa [10], corresponding to 1m of height near the surface and about 3m at an altitude of 10 km. The pressure measurement can also exhibit significant lags during rapid climbs and dives and is disrupted by turbulence and sonic booms. For helicopter applications, the baro sensor must be carefully located and calibrated to prevent downwash from the rotors distorting the reading; further distortions can occur near the ground. However, the main source of error in barometric height measurement arises from differences between the true and modeled atmospheric temperature and pressure. For stand-alone barometry, height errors can be several hundred meters. For differential barometry, the error increases with the distance from the reference station and the age of the calibration data. Rapid changes in the barometric height error can occur when the navigation system passes through a weather front.

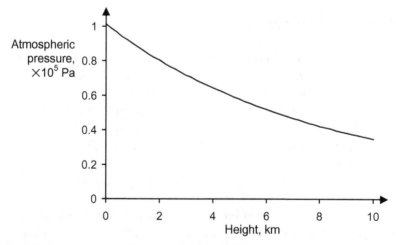

Figure 6.6 Variation of atmospheric pressure with height.

6.2 Height and Depth Measurement

Prior to the advent of GNSS, a baro was the only method of measuring absolute aircraft height at high altitudes, as the vertical channel of an INS is unstable (see Section 5.7.2). To maintain safe aircraft separation, it is more important for different aircraft to agree on a height measurement than for that height to be correct. Therefore, at heights above 5.486 km, all aircraft use the standard mean-sea-level values of p_s and T_s [28]. Furthermore, flight levels allocated by air traffic control are specified in terms of the barometric height, also known as pressure altitude, rather than the geodetic or orthometric height. Aircraft altimeters cost from $150 (€120) upwards.

Aircraft baros have traditionally been integrated with the vertical channel of the INS using a third-order control loop. Using variable gains, the baro data can calibrate the INS during straight and level flights, when the baro is more stable, without the baro scale factor errors contaminating the INS during climbs and dives [29, 30]. However, a baro-inertial loop presents a problem where the baro and INS are integrated with other navigation systems, such as GNSS, using a Kalman filter. This is because it is difficult to maintain an adequate model of the baro-inertial loop's behavior in the integration algorithm, particularly if the details are proprietary. Thus, the baro and INS should always be integrated as separate sensors. Baro integration is discussed in Section 16.2.2.

For land applications, barometric altimeters provide a compact and inexpensive way of measuring height changes. MEMS baros are available for around $30 (€25). To determine absolute height, they must be recalibrated at the beginning of each period of use to account for weather variations; this can be done using GNSS or a known starting point. In cooperative positioning, calibration parameters can also be shared between peers. When GNSS signal availability is poor, such as in an urban canyon, use of a calibrated baro enables a position solution to be obtained with signals from only three GNSS satellites [31]. Within a building, meter-level variations in barometric height from room to room can occur due to ventilation differences, particularly for stairwells, while opening a door will also perturb the reading. Within a car, variations of a few meters can occur when traveling through a tunnel, a window is opened, or the fan setting is changed [32].

6.2.2 Depth Pressure Sensor

A depth pressure sensor determines the depth, d_b, of a submarine, ROV, AUV, or diver from a measurement of the water pressure, p_b. Precise pressure to depth conversion is described in [33]. However, down to a few hundred meters, the depth may be modeled as a linear function of pressure using:

$$d_b = h_s - h_b \approx \frac{p_b - p_s}{\rho g}, \qquad (6.20)$$

where h_s and h_b are the geodetic heights of, respectively, the water surface and pressure sensor; p_s is the atmospheric pressure at the water surface; ρ is the water density; and g is the acceleration due to gravity. The water density varies as a function of the temperature and salinity, but is approximately 10^3 kg m^{-3} for fresh water and 1.03×10^3 kg m^{-3} for seawater. The pressure increases by about 1 atmosphere (10^5

Pa) for every 10m of depth. Pressure measurements are typically accurate to within 0.2%. Note that the surface height, h_s, may vary due to tidal motion.

6.2.3 Radar Altimeter

A radar altimeter (radalt) measures the height of an aircraft, missile, or UAV above the terrain by transmitting a radio signal downwards and measuring how long it takes the signal to return to the radalt after reflection off the ground below. Prices are typically in the $5,000–$10,000 (€4,000–€8,000) range. The height above terrain is normally used directly as a landing aid, for ground collision avoidance, or for performing terrain-following flight. However, it may be combined with a terrain height database to determine the geodetic or orthometric height of the host vehicle where the latitude and longitude are known. A radalt and terrain height database may also be used to perform terrain-referenced navigation as described in Section 13.2.

Radar altimeters generally transmit at 4.3 GHz, although some designs use 15.6 GHz. The range varies as the fourth root of the transmission power and is typically about 1,500m above the terrain. There are three main modulation techniques. A frequency-modulated continuous-wave (FMCW) radalt transmits a continuous signal at a varying frequency; the height above terrain is determined from the frequency difference between the transmitted and received signals. A pulsed radalt transmits a series of short pulses and determines the height from the time lag between the transmitted and received pulses. A spread-spectrum radalt operates in the same way as GNSS (see Sections 7.3.2 and 8.1.2). A PRN code is modulated on the transmitted signal, and the received signal is then correlated with a time-shifted replica of the same code. The time shift that produces the correlation peak determines the height above terrain. All three types of radalt use a tracking loop to smooth out the noise from successive measurements and filter out anomalous returns [34].

The measurement accuracy of radalt hardware is about 1m. However, the accuracy with which the height above the terrain directly below the aircraft can be determined is only 1%–3%. This is because the width of the transmitted radar beam is large, typically subtending ±60° in total with a full width at half maximum (FWHM) returned intensity of about 20° at 4.3 GHz. So, if the host vehicle is 1,000m above the terrain, the effective diameter of the radar footprint is about 350m. Thus, the terrain surrounding that directly below the aircraft has an effect on the height measurement. When the terrain is flat, the return path length for the center of the footprint is shorter than that for the edge, so the radalt processing is biased in favor of the earliest part of the return signal. This is also useful for obstacle avoidance. Height measurement errors are larger where the aircraft is higher, as this gives a larger footprint, and where there is more variation in terrain height within the footprint. Figure 6.7 illustrates this. The beam width may be reduced by using a larger antenna aperture or a higher frequency.

When the host vehicle is not level, the peak of the transmitted radar beam will not strike the terrain directly below. However, mitigating factors are that the terrain reflectivity is usually higher at normal incidence and the shortest return paths will still generally be from the terrain directly below. Thus, radalt measurements are valid for small pitch and roll angles, but should be rejected once those angles

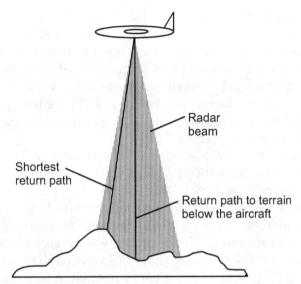

Figure 6.7 Effect of large footprint and terrain height variation on radar altimeter performance.

exceed a certain threshold. Wider beam radalts are more tolerant of pitch and roll than narrower beam designs.

A laser rangefinder has a much smaller footprint and can be scanned to find the shortest return. Hence, it offers a more accurate height above terrain measurement than a radalt. However, the range is both shorter and weather-dependent.

Ultrasonic altimeters offer high precision at low cost and weight, but their range is limited to a few meters. Hence, they are used for automated landing rather than navigation.

For ships, boats, submarines, and AUVs, acoustic echo sounding or sonar are similarly used to measure the depth of the sea bed or river bed below the vessel.

6.3 Odometry

Odometry is the determination of a land vehicle's speed and distance traveled by measuring the rotation of its wheels. Its earliest known use was in Roman chariots. The sensor, commonly known as an odometer, has traditionally been fitted to the transmission shaft. However, most new vehicles have a sensor on each wheel, known as a wheel speed sensor (WSS), which is used for the antilock braking system (ABS). Robots can also incorporate WSS.

By differentiating left and right WSS measurements, the yaw rate of the vehicle may be measured, a technique known as differential odometry. This was demonstrated by the Chinese in the third century CE with their *south-pointing chariot*.

To avoid mechanical wear, odometers use noncontact sensors, known as rotary encoders. In most devices, a toothed ferrous wheel is mounted on the transmission shaft or wheel axle. As each tooth passes through a sensor, the magnetic flux density varies. Measuring this produces a pulsed signal, with the number of pulses proportional to the distance traveled. These sensors are sometimes called wheel pulse or wheel tick sensors. Differentiating the pulsed signal gives the speed. Low-cost odometers and WSSs use passive sensors, based on variable reluctance. They exhibit poor signal-to-noise levels, so are vulnerable to vibration and interference. They also do not work at speeds below about 1 m s^{-1}, so are not recommended for navigation. Active sensors, often based on the Hall effect, give a strong signal at all speeds, but are more expensive [6, 35]. Optical sensors may also be used, but are vulnerable to dirt.

In road vehicles, WSS or odometer measurements can be accessed through the on-board diagnostics (OBD) interface. Different versions of OBD are mandated in different countries, including OBD-II in the United States, EODB in the European Union, and JODB in Japan [36]. A common protocol used for OBD is the controller area network (CAN). The speed measurements from the OBD interface often have large quantization errors. Higher-precision wheel speed measurements can usually be obtained by differentiating the wheel rotation data, normally expressed as a pulse count [37]. Note also that odometry-derived acceleration is noisy compared to acceleration obtained from an IMU.

Linear odometry is described first, followed by differential odometry, and finally the integrated odometer and partial IMU. Note that visual odometry, using a camera, is described in Section 13.3.5.

6.3.1 Linear Odometry

To describe navigation using odometry in a vehicle with front-wheel steering, it is useful to introduce three coordinate frames. The body frame, denoted as b, describes the point on the host vehicle for which a navigation solution is sought. The rear-wheel frame, denoted as r, is centered equidistant between the rear wheels along their axis of rotation and aligned with their direction of travel. This, in turn, is nominally aligned with the body frame, so $C_r^n \approx C_b^n$. The front-wheel frame, denoted as f, is centered equidistant between the front-wheel centers of rotation. It is aligned with the direction of travel of the front wheels, which is determined by the steering angle, ψ_{bf}. Thus,

$$C_f^n = C_b^n \begin{pmatrix} \cos\psi_{bf} & -\sin\psi_{bf} & 0 \\ \sin\psi_{bf} & \cos\psi_{bf} & 0 \\ 0 & 0 & 1 \end{pmatrix}. \qquad (6.21)$$

The lever arms from the body frame to the rear and front-wheel frames are l_{br}^b and l_{bf}^b, respectively. Transmission-shaft measurements give the speed of the rear or front wheel frame, v_{er} or v_{ef}, depending on which are the driving wheels. Wheel speed measurements give the speed of each wheel, v_{erL}, v_{erR}, v_{efL}, and v_{efR}. The rear and front-wheel-frame speeds are then

$$\begin{aligned} v_{er} &= \tfrac{1}{2}(v_{erL} + v_{erR}) \\ v_{ef} &= \tfrac{1}{2}(v_{efL} + v_{efR}) \end{aligned}. \qquad (6.22)$$

6.3 Odometry

When a road vehicle turns, each wheel travels at a different speed, while the front and rear wheels travel in different directions, moving along the forwards (x) axis of the rear or front wheel frame as appropriate. Thus,

$$\mathbf{v}_{erL}^r = \begin{pmatrix} v_{erL} \\ 0 \\ 0 \end{pmatrix}, \quad \mathbf{v}_{erR}^r = \begin{pmatrix} v_{erR} \\ 0 \\ 0 \end{pmatrix}, \quad \mathbf{v}_{efL}^f = \begin{pmatrix} v_{efL} \\ 0 \\ 0 \end{pmatrix}, \quad \mathbf{v}_{efR}^f = \begin{pmatrix} v_{efR} \\ 0 \\ 0 \end{pmatrix}. \qquad (6.23)$$

The velocity, in terms of both speed and direction, also varies across the vehicle body as Figure 6.8 illustrates [38]. The rear track width, T_r, is the distance between the centers of the rear wheels' contact surfaces with the road. It is also the perpendicular distance between the tracks of those wheels along the road. The front track width, T_f, is the distance between the centers of the front wheels' contact surfaces. However, the perpendicular distance between the tracks is $T_f \cos\psi_{bf}$.

From (2.165) and (6.23), the body-frame velocity may be obtained from the rear or front wheel measurements using

$$\mathbf{v}_{eb}^b = \mathbf{C}_r^b \begin{pmatrix} v_{er} \\ 0 \\ 0 \end{pmatrix} - \boldsymbol{\omega}_{eb}^b \wedge \mathbf{l}_{br}^b \approx \begin{pmatrix} v_{er} \\ 0 \\ 0 \end{pmatrix} - \boldsymbol{\omega}_{eb}^b \wedge \mathbf{l}_{br}^b \qquad (6.24)$$

or

$$\mathbf{v}_{eb}^b = \mathbf{C}_f^b \begin{pmatrix} v_{ef} \\ 0 \\ 0 \end{pmatrix} - \boldsymbol{\omega}_{eb}^b \wedge \mathbf{l}_{bf}^b = \begin{pmatrix} v_{ef} \cos\psi_{bf} \\ v_{ef} \sin\psi_{bf} \\ 0 \end{pmatrix} - \boldsymbol{\omega}_{eb}^b \wedge \mathbf{l}_{bf}^b, \qquad (6.25)$$

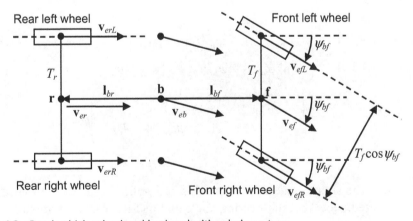

Figure 6.8 Road-vehicle wheel and body velocities during a turn.

where the yaw component of $\boldsymbol{\omega}_{eb}^b$ may be obtained from differential odometry (Section 6.3.2) or a yaw-axis gyro, correcting for the Earth rate. Neglecting the other components and the transport rate (Section 5.4.1), (6.24) and (6.25) simplify to

$$\begin{pmatrix} v_{eb,x}^b \\ v_{eb,y}^b \end{pmatrix} \approx \begin{pmatrix} v_{er} \\ 0 \end{pmatrix} + \begin{pmatrix} l_{br,y}^b \\ -l_{br,x}^b \end{pmatrix} \dot{\psi}_{nb} \qquad (6.26)$$

or

$$\begin{pmatrix} v_{eb,x}^b \\ v_{eb,y}^b \end{pmatrix} \approx \begin{pmatrix} \cos\psi_{bf} \\ \sin\psi_{bf} \end{pmatrix} v_{ef} + \begin{pmatrix} l_{bf,y}^b \\ -l_{bf,x}^b \end{pmatrix} \dot{\psi}_{nb}. \qquad (6.27)$$

Neglecting vehicle roll and pitch, the change in position from time t to time $t + \tau_o$ is

$$\begin{pmatrix} \Delta r_{eb,N}^n(t, t+\tau_o) \\ \Delta r_{eb,E}^n(t, t+\tau_o) \end{pmatrix} \approx \int_t^{t+\tau_o} \begin{pmatrix} \cos\psi_{nb}(t') & -\sin\psi_{nb}(t') \\ \sin\psi_{nb}(t') & \cos\psi_{nb}(t') \end{pmatrix} \begin{pmatrix} v_{eb,x}^b(t') \\ v_{eb,y}^b(t') \end{pmatrix} dt'. \qquad (6.28)$$

When the sensor(s) measure the average velocity from time t to $t + \tau_o$ and the heading rate and steering angle are both known, substituting (6.26) or (6.27) into (6.28) and integrating gives

$$\begin{pmatrix} \Delta r_{eb,N}^n(t, t+\tau_o) \\ \Delta r_{eb,E}^n(t, t+\tau_o) \end{pmatrix} \approx \begin{pmatrix} \cos\psi_{nb}(t) - \tfrac{1}{2}\dot{\psi}_{nb}\tau_o \sin\psi_{nb}(t) \\ \sin\psi_{nb}(t) + \tfrac{1}{2}\dot{\psi}_{nb}\tau_o \cos\psi_{nb}(t) \end{pmatrix} v_{er}\tau_o$$

$$+ \begin{pmatrix} \cos\psi_{nb}(t) & -\sin\psi_{nb}(t) \\ \sin\psi_{nb}(t) & \cos\psi_{nb}(t) \end{pmatrix} \begin{pmatrix} l_{br,y}^b \\ -l_{br,x}^b \end{pmatrix} \dot{\psi}_{nb}\tau_o \qquad (6.29)$$

or

$$\begin{pmatrix} \Delta r_{eb,N}^n(t, t+\tau_o) \\ \Delta r_{eb,E}^n(t, t+\tau_o) \end{pmatrix} \approx \begin{pmatrix} \cos[\psi_{nb}(t) + \psi_{bf}(t)] - \tfrac{1}{2}(\dot{\psi}_{nb} + \dot{\psi}_{bf})\tau_o \sin[\psi_{nb}(t) + \psi_{bf}(t)] \\ \sin[\psi_{nb}(t) + \psi_{bf}(t)] + \tfrac{1}{2}(\dot{\psi}_{nb} + \dot{\psi}_{bf})\tau_o \cos[\psi_{nb}(t) + \psi_{bf}(t)] \end{pmatrix} v_{ef}\tau_o$$

$$+ \begin{pmatrix} \cos\psi_{nb}(t) & -\sin\psi_{nb}(t) \\ \sin\psi_{nb}(t) & \cos\psi_{nb}(t) \end{pmatrix} \begin{pmatrix} l_{bf,y}^b \\ -l_{bf,x}^b \end{pmatrix} \dot{\psi}_{nb}\tau_o ,$$

$$(6.30)$$

where the small angle approximation is applied to $\dot{\psi}_{nb}\tau_o$ and $\dot{\psi}_{bf}\tau_o$, and $\dot{\psi}_{nb}^2$ is neglected. Note that the steering angle and its rate of change are needed to navigate

6.3 Odometry

using front-wheel speed sensors. In either case, the latitude and longitude solutions are updated using

$$L_b(t + \tau_o) = L_b(t) + \frac{\Delta r^n_{eb,N}(t, t + \tau_o)}{R_N(L_b(t)) + h_b(t)}$$
$$\lambda_b(t + \tau_o) = \lambda_b(t) + \frac{\Delta r^n_{eb,E}(t, t + \tau_o)}{[R_E(L_b(t)) + h_b(t)] \cos L_b(t)}, \qquad (6.31)$$

where R_N and R_E are given by (2.105) and (2.106).

Odometers or WSSs measure the distance traveled over ground, not the distance traveled in the horizontal plane. Thus, if the host vehicle is traveling on a slope, the horizontal distance will be overestimated as shown in Figure 6.9. Vehicle roll and road banking do not affect speed and distance measurement. For slopes of up to 140 mrad (8°), the error will be less than 1%. If the pitch is known, the slope effects may be corrected by multiplying the speed, velocity, and/or distance traveled by $\cos\theta_{nb}$. The pitch may be estimated using an IMU, GNSS velocity measurements, or a terrain height database. It may also be determined from the rate of change of barometric height with distance traveled using

$$\hat{\theta}_{nb} = \arctan\left(\frac{\Delta h_b}{\sqrt{\Delta r^n_{eb,N}{}^2 + \Delta r^n_{eb,E}{}^2}}\right). \qquad (6.32)$$

The dominant error source in linear odometry is the scale factor error due to uncertainty in the wheel radii. Tire wear reduces the radius by up to 3% over the lifetime of a tire, while variations of order 1% can occur due to changes in pressure, temperature, load, and speed [39–41]. Thus, it is standard practice to calibrate the scale factor error using other navigation sensors, such as GNSS, as described in Section 16.2.3.

Example 6.2 presents an example of rear-wheel odometry in the presence of scale-factor errors and is editable using Microsoft Excel.

Quantization resulting from the ferrous wheel teeth can be a significant source of short-term velocity errors. However, as quantization errors are always corrected by subsequent measurements, the long-term position error is negligible [38]. Random errors also arise from road surface unevenness.

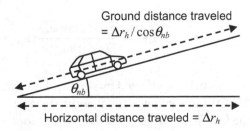

Figure 6.9 Effect of slope on measurement of distance traveled.

Odometers and wheel speed sensors will produce false measurements of vehicle velocity where a wheel slips or skids due to rapid acceleration or braking on a slippery road [20]. These can often by detected and filtered out using integrity monitoring techniques as described in Chapter 17. Vehicle ABS and traction control systems detect wheel slips and skids by comparing the WSS measurements with automotive-grade accelerometer measurements. The driving wheels are subject to more slippage so odometry using the nondriving wheels is more reliable [41]. Odometry is a context-dependent navigation technique, so will also give misleading information when the vehicle is on a ferry, train, or trailer.

Odometry is unreliable for rail applications because of high levels of wheel slip. Driving wheel sensors typically overestimate the train speed while trailing wheel sensors underestimate it, particularly during acceleration. During braking, all sensors will typically underestimate the train speed. Furthermore, to enable the train to follow the track without the need for steering, train wheels are conical and mounted on a solid axle. Consequently, variation in the rail spacing (typically ±1 cm) leads to variation in the scale factor errors.

6.3.2 Differential Odometry

Differential odometry may be implemented where individual wheel speed measurements are available. The yaw rate is

$$\dot{\psi}_{nb} = \frac{v_{erL} - v_{erR}}{T_r} \tag{6.33}$$

from the rear wheels or

$$\dot{\psi}_{nb} = \frac{v_{efL} - v_{efR}}{T_f \cos\psi_{bf}} - \dot{\psi}_{bf} \tag{6.34}$$

from the front wheels. When odometry measurements are made over the interval t to $t + \tau_o$, the heading is updated using

$$\psi_{nb}(t + \tau_o) = \psi_{nb}(t) + \frac{1}{T_r}(v_{erL} - v_{erR})\tau_o \tag{6.35}$$

or

$$\psi_{nb}(t + \tau_o) = \psi_{nb}(t) + \frac{2 + \dot{\psi}_{bf}\tau_o \tan\psi_{bf}}{2T_f \cos\psi_{bf}}(v_{efL} - v_{efR})\tau_o - \dot{\psi}_{bf}\tau_o. \tag{6.36}$$

Differential odometry is affected by sloping, banked, and uneven terrain (but not by vehicle roll) in the same way as yaw-axis gyro measurements as described in Section 6.1.3. When the heading change is a small angle, slope effects may be compensated by dividing the yaw rate or heading change by $\cos\theta_{nb}$. However, compensation for banked terrain is more complex. When the roll and pitch are unknown, the heading error must be corrected through integration with other navigation sensors.

Differential odometry is very sensitive to scale factor errors. For example, a 1% difference in scale factor error between the left and right wheels leads to a yaw-rate error of about 3° s^{-1} at a speed of 10 m s^{-1}. Scale factor errors for yaw-rate measurement are affected by errors in the track widths, T_r and T_f, as well as errors in the assumed tire radii. The track width can change when tires are replaced, particularly in countries where separate winter and summer tires are used [42]. Thus, the yaw-rate scale factor error should be calibrated using measurements from other navigation sensors. Differential odometry is included in Example 6.2 on the CD.

Road surface unevenness is a major source of random errors for differential odometry, much more than for velocity measurement. For example, a pot hole or bump that affects only one side of the vehicle might produce a 1° heading error, but only a 1.5-cm error in distance traveled. Furthermore, the camber of the road surface (which is designed to aid water run-off) can introduce both positive and negative scale factor errors during turns, while changes in camber on straight roads produce false turn readings [36]. A curved road camber can also bias differential odometry by changing the effective tire radii; this may be compensated using vehicle roll measurements from accelerometer leveling (Section 6.1.6) [37].

Differential odometry is also affected by wheel slips and skids, while sensor quantization errors affect short-term angular-rate measurement, but are negligible in measuring longer term yaw changes.

6.3.3 Integrated Odometry and Partial IMU

As explained in Section 5.9, a 2A1G partial IMU has insufficient degrees of freedom to measure land vehicle motion, even with the application of nonholonomic motion constraints. However, the addition of linear odometry provides the necessary additional information. The odometry measurements may be integrated as a separate sensor as described in Section 16.2.3. However, a combined navigation solution may also be determined directly from the sensor measurements [43].

The velocity resolved about body-frame axes, $\mathbf{v}_{eb}^b$, is obtained from the odometer or WSSs as described in Section 6.3.1; this already incorporates the nonholonomic constraints of land vehicle motion, noting that $v_{eb,z}^b = 0$. From (2.17), (2.56), (2.67) and (2.77),

$$\mathbf{a}_{eb}^b = \dot{\mathbf{v}}_{eb}^b + \mathbf{\Omega}_{eb}^b \mathbf{v}_{eb}^b. \tag{6.37}$$

If the Earth rotation and the roll- and pitch-axis components of the angular rate are neglected, this may be approximated as

$$\mathbf{a}_{eb}^b \approx \begin{pmatrix} \dot{v}_{eb,x}^b - \omega_{ib,z}^b v_{eb,y}^b \\ \dot{v}_{eb,y}^b + \omega_{ib,z}^b v_{eb,x}^b \\ 0 \end{pmatrix}. \tag{6.38}$$

Otherwise, $\omega_{ib,x}^b$ and $\omega_{ib,y}^b$ must be determined recursively from the current and previous attitude solutions.

An estimate of the acceleration due to gravity resolved about the body-frame axes can then be obtained from the accelerometer measurements using

$$\mathbf{g}_b^b \approx \begin{pmatrix} -f_{ib,x}^b + a_{eb,x}^b \\ -f_{ib,y}^b + a_{eb,y}^b \\ \cos\phi_{nb}\cos\theta_{nb} g_{b,D}^n(L_b, h_b) \end{pmatrix}, \qquad (6.39)$$

where $g_{b,D}^n(L_b, h_b)$ is obtained from a gravity model (see Section 2.4.7). Using leveling (see Sections 5.6.2), an estimate of the roll and pitch is obtained using

$$\theta_{nb} = \arctan\left(\frac{-g_{b,x}^b}{\sqrt{{g_{b,y}^b}^2 + {g_{b,z}^b}^2}}\right), \qquad \phi_{nb} = \arctan_2\left(g_{b,y}^b, g_{b,z}^b\right) \qquad (6.40)$$

where (6.39) and (6.40) are iterated until convergence, using the previous roll and pitch solution in the first iteration of (6.39).

Neglecting the Earth rotation and transport rate, the heading may then be updated using

$$\psi_{nb}(t + \tau_o) \approx \psi_{nb}(t) + \frac{\cos\phi_{nb}}{\cos\theta_{nb}}\omega_{ib,z}^b \tau_o, \qquad (6.41)$$

where the pitch-axis angular rate may be neglected (otherwise an additional term, $(\sin\phi_{nb}/\cos\theta_{nb})\omega_{ib,y}^b \tau_o$ is required).

Having obtained a full attitude solution, the odometry-measured velocity is resolved about local-navigation-frame axes using

$$\mathbf{v}_{eb}^n = \mathbf{C}_b^n \mathbf{v}_{eb}^b \qquad (6.42)$$

and the position solution updated as described in Section 5.4.4.

6.4 Pedestrian Dead Reckoning Using Step Detection

Pedestrian navigation is one of the most challenging applications of navigation technology. A pedestrian navigation system must work in urban areas, under tree cover, and even indoors, where coverage of GNSS and most other radio navigation systems is poor. Inertial sensors can be used to measure forward motion by dead reckoning. However, for pedestrian use, they must be small, light, consume minimal power, and, for most applications, be low-cost. Thus, MEMS sensors must be used. However, these provide very poor inertial navigation performance stand alone, while the combination of low dynamics and high vibration limits the calibration available from GNSS or other positioning systems. One solution is to use a shoe-mounted

6.4 Pedestrian Dead Reckoning Using Step Detection

IMU and combine conventional inertial navigation (Chapter 5) with zero velocity updates (Section 15.3) performed on every step. However, shoe-mounted sensors are not practical for every application.

This section describes the other solution, which is to use the inertial sensors for step counting. Note that a step is the movement of one foot with the other remaining stationary, while a stride is the successive movement of both feet. This approach is known as pedestrian dead reckoning (PDR). Note that some authors use this term to describe shoe-mounted inertial navigation with ZVUs. Here, *pedestrian dead reckoning* means the step detection method. For sensors mounted on the user's body or in a handheld device, PDR using step detection gives significantly better performance than conventional inertial navigation, even when tactical-grade sensors are used [7].

Most PDR implementations use only accelerometers to sense motion; some also use the gyroscopes. PDR can use a single accelerometer, mounted vertically on the body or along the forward axis of a shoe. However, using an accelerometer triad or full IMU allows the sensors to be placed almost anywhere on the body or in a handheld device, and enables PDR to operate independently of the user's orientation [44]. It also aids motion classification. An accelerometer triad or a full IMU, using consumer-grade sensors, is a common feature on a smart phone.

PDR has also been demonstrated using other sensors. Impact sensors may be mounted in the soles of the user's shoes to detect footfalls [45]. A downward-pointing camera may be used to measure foot motion [46]. Electromyography (EMG) can be used to detect motion by measuring the electric field from the leg muscles; sensors must be strapped to one or both legs [47]. Use of an IMU or accelerometers is assumed in the following discussion. However, much of it is applicable to the other sensors.

A pedestrian dead-reckoning algorithm comprises three phases: step detection, step length estimation, and navigation-solution update. This is illustrated by Figure 6.10. The step-detection phase identifies that a step has taken place. For shoe-mounted accelerometers, the measured specific force is constant when the foot is on the ground and variable when the foot is swinging, enabling steps to be easily identified [48]. For body-mounted or device-mounted sensors, the vertical or root sum of squares (RSS) accelerometer signals exhibit a double-peaked oscillatory pattern during walking

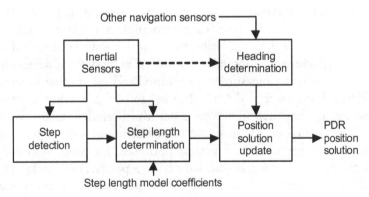

Figure 6.10 Pedestrian dead-reckoning processing.

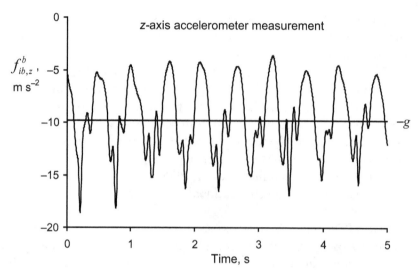

Figure 6.11 Vertical accelerometer signal during walking motion. (Data courtesy of QinetiQ Ltd.)

as Figure 6.11 shows. Steps can be detected from the "acceleration zero crossings" where the specific force rises above or drops below the acceleration due to gravity, with a recognition window used to limit false detections [44]. Alternatively, steps may be detected from the peaks in the accelerometer signals [49].

The length of a step varies not only between individuals, but also according to the slope and texture of the terrain, whether there are obstacles to be negotiated, whether an individual is tired, whether they are carrying things, and whether they are walking alone or with others. Thus, PDR implementations that assume a fixed step length for each user are only accurate to about 10% of distance traveled [50]. However, the step length varies approximately as a linear function of the step frequency [49]. It is also correlated with the variance of the accelerometer measurements [51] and the slope of the terrain [10] or vertical velocity [52].

The PDR-estimated step length, Δr_P, may thus be modeled as follows [53]:

$$\Delta r_P = c_{P0} + \frac{c_{P1}}{\tau_P} + c_{P2}\sigma_f^2 + c_{P3}\hat{\theta}_{nb}, \tag{6.43}$$

where τ_P is the interval between successive steps, σ_f^2 is the variance of the specific force measurements, $\hat{\theta}_{nb}$ is the estimated angle of the slope, and c_{P0}, c_{P1}, c_{P2}, and c_{P3} are the model coefficients. Using this approach, an accuracy of about 3% of distance traveled may be obtained [51, 52]. The model coefficients for each user may be estimated using measurements from GNSS or another positioning system. An EKF-based approach is discussed in Section 16.2.4, while calibration using fuzzy logic and artificial neural networks has also been demonstrated [45].

How inertial sensors respond to pedestrian motion depends on their location. Thus, step-length model coefficients optimized for waist-mounted sensors may not give the best results for sensors located in a pocket, in a backpack, in a device held by the user, or mounted on a shoe. This a particular issue for sensors located within

6.4 Pedestrian Dead Reckoning Using Step Detection

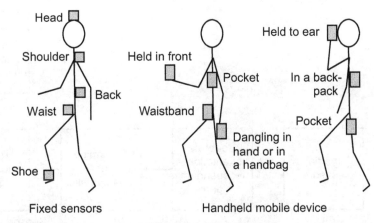

Figure 6.12 Possible inertial sensor locations on the human body.

a mobile device, such as a phone, which will be moved between different locations around the body. Figure 6.12 shows possible locations for both body-fixed and mobile-device-mounted sensors.

A basic PDR algorithm makes the assumption that all steps detected are forward walking, so backward and sideways steps lead to false measurements. Furthermore, step-length model coefficients optimized for walking will not give good results for running, turning, and climbing stairs or steps. Military and emergency service personnel may also crawl, roll, and jump, while many different kinds of motion are possible during sports. Figure 6.13 summarizes the different classes of motion.

PDR is thus doubly context-dependent. The appropriate configuration of the algorithms depends on both the sensor location and the activity. A robust implementation of PDR should thus incorporate a real-time classification system that detects both the motion type and sensor location and tunes both the step-detection and step-length-estimation algorithms accordingly [54–58]. Figure 6.14 illustrates a typical approach. The first step is to generate orientation-independent signals from the sensor outputs. Motion classification requires at least a full accelerometer triad,

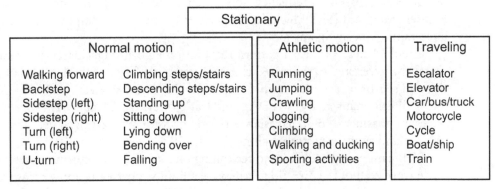

Figure 6.13 Human motion categories.

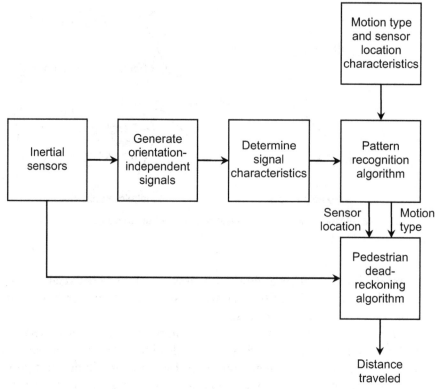

Figure 6.14 Motion classification and sensor location determination algorithm.

while sensor location determination needs a full IMU. Suitable signals include the magnitudes of the accelerometer and gyro triads, $|\mathbf{f}_{ib}^b|$ and $|\boldsymbol{\omega}_{ib}^b|$, and the dynamic acceleration, $|\mathbf{f}_{ib}^b| - g_b$. For applications in which the sensor location and orientation are known, separate horizontal and vertical signals may be used [56].

The second step is to determine the characteristics of each signal using a few seconds of data. Suitable time-domain characteristics include the mean, standard deviation, root mean squared (RMS), interquartile range, mean absolute deviation, maximum–minimum, maximum magnitude, number of zero crossings, and number of mean crossings. Frequency-domain characteristics, determined from the fast Fourier transform (FFT), include the peak frequency, peak amplitude, and energy in certain frequency bands [55–58].

The final step is to use a pattern recognition algorithm to match the measured signal characteristics to the stored characteristics of different combinations of activities and sensor locations. Possible algorithms include k-nearest-neighbors (KNN), linear discriminant analysis (LDA), quadratic discriminant analysis (QDA), naïve Bayesian classifier (NBC), Bayesian network, decision tree, artificial neural network (ANN), and support vector machine (SVM) [58].

PDR cannot be used for dead reckoning on its own, as it only measures the distance traveled, not the direction. It may be combined with a heading measurement (Section 6.1.5), noting that PDR may share the accelerometer triad of an AHRS (Section 6.1.8), in which case the position solution is updated using

$$L_b(+) = L_b(-) + \frac{\Delta r_P \cos(\psi_{nb} + \psi_{bh})}{R_N(L_b(-)) + h_b(-)}$$

$$\lambda_b(+) = \lambda_b(-) + \frac{\Delta r_P \sin(\psi_{nb} + \psi_{bh})}{[R_E(L_b(-)) + h_b(-)]\cos L_b(-)}, \quad (6.44)$$

where ψ_{bh} is the boresight angle, and the suffixes (–) and (+) denote before and after the update, respectively. The boresight angle is the angle in the horizontal plane between the forward axis of the sensor pack used for heading determination and the direction of motion. It is zero where the sensors are aligned with the direction of motion. Otherwise, it may be calibrated alongside the step-length-estimation coefficients [44].

Alternatively, PDR measurements may be used to calibrate the drift of an INS, sharing its accelerometers, as described in Section 16.2.4 [7, 59]. When a tactical-grade IMU is used, this smooths step-length estimation errors. However, there is little benefit in computing an inertial position solution using uncalibrated consumer-grade sensors.

Step detection may also be used to vary the system noise according to the level of motion in a total-state navigation filter inputting measurements from GNSS and/or other position-fixing systems (see Sections 9.4.2, 10.5, 16.1.7, and 16.2.4).

6.5 Doppler Radar and Sonar

When a radio or sound wave is transmitted to a receiver that is moving with respect to the transmitter, the receiver moves towards or away from the signal, causing the wavefronts to arrive at the receiver at a faster or slower rate than that at which they are transmitted. Thus, the frequency of the received signal is shifted, to the first order, by

$$\Delta f_{tr} \approx -\frac{f_t}{c} \mathbf{u}_{tr}^{\gamma\,\mathrm{T}} \mathbf{v}_{tr}^{\gamma}, \quad (6.45)$$

where f_t is the transmitted frequency, c is the speed of light or sound, $\mathbf{u}_{tr}^{\gamma}$ is the line of sight unit vector from transmitter to receiver, and $\mathbf{v}_{tr}^{\gamma}$ is the velocity of the receiver with respect to the transmitter. This is the Doppler effect. When the transmitter and receiver are coincident on a body, b, but the signal is reflected off a surface, s, the Doppler shifts in each direction add, so

$$\Delta f_{tr} \approx -\frac{2f_t}{c} \mathbf{u}_{bs}^{b\,\mathrm{T}} \mathbf{v}_{bs}^{b}. \quad (6.46)$$

By reflecting three or more noncoplanar radio or sound beams off a surface and measuring the Doppler shifts, the velocity of the body with respect to that surface can be obtained. This is the principle of Doppler radar and sonar navigation. Most systems use a four-beam Janus configuration as shown in Figure 6.15. The direction of each beam, indexed by s, with respect to the unit's body frame is given by a (negative) elevation angle, θ_{bs}, and an azimuth, ψ_{bs}, giving a line of sight vector of

$$\mathbf{u}_{bs}^b = \begin{pmatrix} \cos\psi_{bs} \cos\theta_{bs} \\ \sin\psi_{bs} \cos\theta_{bs} \\ -\sin\theta_{bs} \end{pmatrix}. \tag{6.47}$$

The elevation is typically −60° for sonar and −65° to −80° for radar and is nominally the same for each beam [14, 34]. Nominal azimuths are either 30–45°, 135–150°, 210–225°, and 315–330°, as in Figure 6.15, or 0°, 90°, 180°, and 270°. The actual elevations and azimuths will vary due to manufacturing tolerances and may be calibrated and programmed into the unit's software. The beam width is around 3°–4° [14, 34].

The return signal to the Doppler unit comes from scattering of radar or sonar by objects in the beam footprint, not specular reflection off the surface. Thus, the Doppler shift is a function of the relative velocity of the scatterers with respect to the host vehicle, not the range rate of the beams. When the scatterers are fixed to the Earth's surface, $\mathbf{v}_{bs}^b = -\mathbf{v}_{eb}^b$, so the Doppler unit measures velocity with respect to the Earth. The measurement model is thus

$$\begin{pmatrix} \Delta\tilde{f}_{tr1} \\ \Delta\tilde{f}_{tr2} \\ \Delta\tilde{f}_{tr3} \\ \Delta\tilde{f}_{tr4} \end{pmatrix} = \mathbf{H}_D^b \mathbf{v}_{eb}^b + \begin{pmatrix} w_{m1} \\ w_{m2} \\ w_{m3} \\ w_{m4} \end{pmatrix}, \tag{6.48}$$

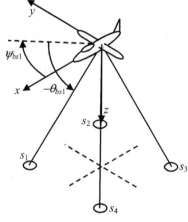

Figure 6.15 Typical four-beam Janus Doppler radar configuration.

where w_{mi} is the measurement noise of the ith beam and the measurement matrix, $\mathbf{H}_D^b$, is

$$\mathbf{H}_D^b = \frac{2f_t}{c} \begin{pmatrix} \mathbf{u}_{b1}^{b\,\mathrm{T}} \\ \mathbf{u}_{b2}^{b\,\mathrm{T}} \\ \mathbf{u}_{b3}^{b\,\mathrm{T}} \\ \mathbf{u}_{b4}^{b\,\mathrm{T}} \end{pmatrix}. \tag{6.49}$$

The Earth-referenced velocity may be obtained by least-squares estimation (see Section 7.3.3), noting that iteration is not necessary as the relationship between Doppler shift and velocity is linear. Thus,

$$\hat{\mathbf{v}}_{eb}^b = \left(\mathbf{H}_D^{b\,\mathrm{T}} \mathbf{H}_D^b\right)^{-1} \mathbf{H}_D^{b\,\mathrm{T}} \begin{pmatrix} \Delta \tilde{f}_{tr1} \\ \Delta \tilde{f}_{tr2} \\ \Delta \tilde{f}_{tr3} \\ \Delta \tilde{f}_{tr4} \end{pmatrix}, \tag{6.50}$$

noting that the velocity is overdetermined when four (or more) radar or sonar beams are used. This enables consistency checks (see Section 17.4) to be performed to identify faulty measurements, which may occur when a moving animal or vehicle interrupts one of the beams.

To maintain a position solution, the attitude, $\mathbf{C}_b^n$, is required. Doppler radar and sonar do not measure attitude, so an AHRS, INS, or other attitude sensor must be used. The latitude, longitude, and height are then updated using

$$\begin{pmatrix} L_b(+) \\ \lambda_b(+) \\ h^b(+) \end{pmatrix} = \begin{pmatrix} L_b(-) \\ \lambda_b(-) \\ h^b(-) \end{pmatrix}$$
$$+ \begin{pmatrix} 1/[R_N(L_b(-)) + h_b(-)] & 0 & 0 \\ 0 & 1/\{[R_E(L_b(-)) + h_b(-)]\cos L_b(-)\} & 0 \\ 0 & 0 & -1 \end{pmatrix} \hat{\mathbf{C}}_b^n \hat{\mathbf{v}}_{eb}^b$$
$$\tag{6.51}$$

Noise arises because the Doppler shift varies across the footprint of each beam while the scatterers are distributed randomly. The noise standard deviation varies as the square root of the velocity [34]. It also increases with the height above terrain or seabed as the footprint increases and the returned signal strength decreases, while different types of terrain have different scattering properties. Dynamic response lags, typically 0.1 second for radar, arise due to the use of frequency tracking loops in the receiver to smooth the noise. Velocity cross-coupling errors also arise due to residual beam misalignment and misalignment of the body frames of the Doppler unit and the attitude sensor. In addition, host vehicle maneuvers involving extreme roll or pitch angles can cause interruptions in the Doppler velocity measurement as this requires at least three beams to be reflected off the ground within the system's range.

Modern Doppler radars operate at 13.325 GHz and are usually frequency-modulated. The technology was developed in the mid-1940s, becoming established for military applications in the 1950s and for civil applications in the 1960s [60]. The typical accuracy of body-frame-resolved velocity over land, where noise is the dominant error source, is 0.06 m s^{-1} $\pm 0.2\%$, although high-performance designs are about a factor of 2 better. Long-term position accuracy is about 1% of the distance traveled with AHRS attitude and 0.15% with INS attitude [34]. The maximum altitude is at least 3,000m above terrain [61].

Performance is poorer over water due to a large variation in the scattering coefficient with the angle of incidence. This causes the velocity to be underestimated by 1%–5%, with the larger errors occurring over smooth water. Older Doppler radar systems provided the user with a land/sea calibration switch, reducing the residual errors to 0.3%–0.6% (1σ). Newer designs typically use a modified beam shape, reducing the velocity errors to within 0.2%, while high-performance units measure the variation in scattering coefficient using additional or steerable beams [34]. In addition, the velocity is measured with respect to the water surface, not the Earth. Correction for this requires real-time calibration data or integration with a positioning system, such as GNSS.

Doppler radar is typically used for helicopter applications, where the slower speeds (below 100 m s^{-1}), compared to fixed-wing aircraft, lead to smaller velocity errors, while aircraft-grade INS are usually too expensive. Doppler radar units for aircraft typically have a size around 400×400×50 mm, a mass of around 5 kg, and cost around $10,000 (€8,000).

Two-beam Doppler radar, omitting cross-track velocity, is sometimes used for rail applications. This avoids wheel slip-induced errors but can be affected by debris on the track that is disturbed by the passage of the train. Doppler radar units are now available that are sufficiently compact for road vehicle or even pedestrian use. For pedestrians, Doppler radar must be integrated with a gyro triad to track the constant changes in sensor orientation that arise from human motion [62].

A Doppler sonar system, also known as a Doppler velocity log (DVL), is used underwater to measure the velocity with respect to the bottom; it is applicable to the navigation of ships, submarines, ROVs, AUVs, and divers. DVLs typically cost around $10,000 (€8,000) and have a mass of 3–20 kg, a diameter of 120–250 mm, and a length of around 200 mm. Sonar transducers typically both transmit and receive. Operating frequencies vary between 100 kHz and 1 MHz; pulsed signals are typically used [14]. The range is a few hundred meters, with lower frequencies having a longer range, but producing noisier measurements. The speed of sound in water is about 1,500 m s^{-1}, but varies with temperature, depth, and salinity by a few percent. To get the best performance out of sonar, this must be correctly modeled.

Sonar is also subject to the effects of acoustic noise, while, in murky water, scattering of the sound by particles in the water above the bottom can introduce water-current-dependent errors. Large errors can also occur in the presence of cavitation (water bubbles), while the sensor must be regularly cleared of barnacles. A well-calibrated and aligned Doppler sonar navigation system is accurate to 0.2%–0.5% of distance traveled [14, 63, 64].

6.6 Other Dead-Reckoning Techniques

This section briefly reviews a number of other techniques, each designed for a specific context, that may be used to measure or calibrate velocity, resolved about the body frame. Correlation-based velocity measurement, air data, and the ship's speed log are discussed. In each case, the velocity must be combined with an attitude measurement to update the position solution.

6.6.1 Correlation-Based Velocity Measurement

For marine applications, a correlation velocity log (CVL), also known as an acoustic correlation log, transmits a wide beam of sonar pulses straight down through water. The sonar is scattered by the bottom such that an interference pattern is produced. This is then measured by an array of receiving transducers on the host vessel. By correlating the interference patterns received from successive sonar pulses, an estimate of host vessel velocity, resolved in the body frame, is obtained [65, 66]. A single-dimensional receiving array gives only forward velocity, while a two-dimensional array gives both horizontal components.

With a much wider beam, a CVL can operate at a much lower frequency than a Doppler velocity log with the same transducer size, giving a longer range. Frequencies of 20–40 kHz are typical [67]. A CVL can operate at least 3,500m above the sea bed. Its velocity measurements are noisier than those of a DVL, but are not affected by variations in the speed of sound. The long-term accuracy is similar at 0.1%–0.5% of distance traveled.

For land vehicles, accelerometers may be used to sense bumps in the road. By measuring the time interval between the front and rear wheels hitting a bump, the forward speed may be determined [68]. This method will not provide continuous speed measurements, but could be used to calibrate other dead-reckoning sensors.

The velocity of low-flying aircraft may be determined by comparing fore and aft laser scanners as discussed in Section 13.2.4.

6.6.2 Air Data

Air speed is the forward component of an aircraft's velocity with respect to the air, as opposed to the ground. It is measured by differencing the pressure measured in a forward-pointing tube, known as a pitot, with that measured from a static port on the side of the aircraft [28]. It is accurate to about 2 m s^{-1} at speeds above 50 m s^{-1}, but can be less accurate below this. Air speed is essential for flight control as the aircraft flies with respect to the air. However, it is a poor indicator of speed with respect to the ground, so is not generally used for navigation. Another navigation sensor, such as GNSS, can be used to calibrate the wind speed to within about 1 m s^{-1} [69].

6.6.3 Ship's Speed Log

A ship's speed log measures the speed of a ship with respect to the water. Impellers or turbines are typically used on yachts due to their simplicity and low cost. The principle of operation is simple: the rate of rotation of the turbine is proportional to its speed through water. Typically, a pulse is emitted for each rotation, from which the speed and distance traveled may be computed. Thus, an impeller speed log is the marine equivalent of the odometer (Section 6.3). It may be either mounted on the hull or dragged behind the vessel by a length of cable. Regular maintenance is required as the impeller becomes clogged with debris; the cables of trailed impellers can also become entangled.

For large vessels, an electromagnetic (EM) speed log or sonar is used to measure water speed on ships. An EM speed log induces an electromagnetic field in the water close to the vessel's hull and in a direction perpendicular to the direction of travel. Because salt water is an electrical conductor, a potential difference is induced that is proportional to the vessel's water speed and in a direction perpendicular to both the magnetic field and the vessel's motion. This potential difference is measured using a pair of electrodes. EM speed logs are accurate to around 0.03 m s^{-1} and can also be used to measure the transverse speed [14, 15]. However, they cannot operate in fresh water.

Both DVLs and CVLs may be used to determine the velocity with respect to the water by measuring the sound scattered by particles suspended in water. These signals may be distinguished from those returned from the bottom by timing, though the water speed measurements are often only used when bottom reflections cannot be received.

EM speed logs and hull-mounted impellers must be calibrated for the effects of water flow around the hull in order to obtain the best available accuracy. This is because the water in the immediate vicinity of the hull can move relative to the main water mass. This effect depends on the location of the speed log sensor(s) on the hull.

In principle, the water current can be calibrated through integration of the speed log with other navigation sensors, enabling the water speed measurements to be converted to Earth-referenced speed.

Some historical speed logs are described in Section K.7.5 of Appendix K on the CD.

Problems and exercises for this chapter are on the accompanying CD.

References

[1] Maus, S., et al., *The US/UK World Magnetic Model for 2010–2015*, Technical Report NESDIS/NGDC, Washington, D.C.: National Oceanic and Atmosphere Administration, and Edinburgh, U.K.: British Geological Survey, 2010.

[2] Finlay, C. C., et al., "International Geomagnetic Reference Field: The Eleventh Generation," *Geophysical Journal International*, Vol. 183, No. 3, 2010, pp. 1216–1230.

[3] Langley, R. B., "Getting Your Bearings: The Magnetic Compass and GPS," *GPS World*, September 2003, pp. 70–81.

[4] Goldenberg, F., "Magnetic Heading, Achievements and Prospective," *Proc. ION NTM*, San Diego, CA, January 2007, pp. 743–755.

6.6 Other Dead-Reckoning Techniques

[5]   Caruso, M. J., "Applications of Magnetic Sensors for Low Cost Compass Systems," *Proc. IEEE PLANS 2000*, San Diego, CA, March 2000, pp. 177–184.

[6]   Zhao, Y., *Vehicle Location and Navigation Systems*, Norwood, MA: Artech House, 1997.

[7]   Mather, C. J., P. D. Groves, and M. R. Carter, "A Man Motion Navigation System Using High Sensitivity GPS, MEMS IMU and Auxiliary Sensors," *Proc. ION GNSS 2006*, Fort Worth, TX, September 2006, pp. 2704–2714.

[8]   Kayton, M., and W. G. Wing, "Attitude and Heading References," in *Avionics Navigation Systems*, 2nd ed., M. Kayton and W. R. Fried, (eds.), New York: Wiley, 1997, pp. 426–448.

[9]   Afzal, M. H., V. Renaudin, and G. Lachapelle, "Magnetic Field Based Heading Estimation for Pedestrian Navigation Environments," *Proc. Indoor Positioning and Indoor Navigation*, Guimarães, Portugal, September 2011.

[10] Ladetto, Q., et al., "Digital Magnetic Compass and Gyroscope for Dismounted Soldier Position & Navigation," *Proc. NATO RTO Symposium on Emerging Military Capabilities Enabled by Advances in Navigation Sensors*, Istanbul, Turkey, October 2002.

[11] Gebre-Egziabher, D., et al., "Calibration of Strapdown Magnetometers in Magnetic Field Domain," *Journal of Aerospace Engineering*, Vol. 19, No. 2, 2006, pp. 87–102.

[12] Siddharth, S., et al., "A Game-Theoretic Approach for Calibration of Low-Cost Magnetometers Under Noise Uncertainty," *Measurement Science and Technology*, Vol. 23, No. 2, 2012, paper 025003.

[13] Guo, P., et al., "The Soft Iron and Hard Iron Calibration Method Using Extended Kalman Filter for Attitude and Heading Reference System," *Proc. IEEE/ION PLANS*, Monterey, CA, May 2008, pp. 1167–1174.

[14] Tetley, L., and D. Calcutt, *Electronic Aids to Navigation*, London, U.K.: Edward Arnold, 1986.

[15] Appleyard, S. F., R. S. Linford, and P. J. Yarwood, *Marine Electronic Navigation*. 2nd ed., London, U.K.: Routledge & Kegan Paul, 1988.

[16] Groves, P. D., R. J. Handley, and S. T. Parker, "Vehicle Heading Determination Using Only Single-Antenna GPS and a Single Gyro," *Proc. ION GNSS 2009*, Savannah, GA, September 2009, pp. 1775–1784.

[17] Coaplen, J. P., et al., "On Navigation Systems for Motorcycles: The Influence and Estimation of Roll Angle," *Journal of Navigation*, Vol. 58, No. 3, 2005, pp. 375–388.

[18] Gu, D., and N. El-Sheimy, "Heading Accuracy Improvement of MEMS IMU/DGPS Integrated Navigation System for Land Vehicle," *Proc. IEEE/ION PLANS*, Monterey, CA, May 2008, pp. 1292–1296.

[19] Fouque, C., P. Bonnifait, and D. Bétaille, "Enhancement of Global Vehicle Localization Using Navigable Road Maps and Dead-Reckoning," *Proc. IEEE/ION PLANS*, Monterey, CA, May 2008, pp. 1286–1291.

[20] Brown, L., and D. Edwards, "Vehicle Modeling," in *GNSS for Vehicle Control*, D. M. Bevly and S. Cobb, (eds.), Norwood, MA: Artech House, 2010, pp. 61–89.

[21] Pheifer, D., and W. B. Powell, "The Electrolytic Tilt Sensor," *Sensors*, May 2000.

[22] Winkler, S., et al., "Improving Low-Cost GPS/MEMS-Based INS Integration for Autonomous MAV Navigation by Visual Aiding," *Proc. ION GNSS 2004*, Long Beach, CA, September 2004, pp. 1069–1075.

[23] Ettinger, S. M., et al., "Vision-Guided Flight Stability and Control for Micro Air Vehicles," *Proc. IEEE International Conference on Intelligent Robots and Systems*, Lausanne, Switzerland, October 2002, pp. 2134–2140.

[24] Amt, J. H. R., and J. F. Raquet, "Positioning for Range-Based Land Navigation Systems Using Surface Topography," *Proc. ION GNSS 2006*, Fort Worth, TX, September 2006, pp. 1494–1505.

[25] Zheng, Y., and M. Quddus, "Accuracy Performances of Low Cost Tightly Coupled GPS, DR Sensors and DEM Integration System for ITS Applications," *Proc. ION GNSS 2009*, Savannah, GA, September 2009, pp. 2195–2204.

[26] *Manual of ICAO Standard Atmosphere*, Document 7488/2, Montreal, Canada: International Civil Aviation Organization, 1964.

[27] Kubrak, D., C. Macabiau, and M. Monnerat, "Performance Analysis of MEMS Based Pedestrian Navigation Systems," *Proc. ION GNSS 2005*, Long Beach, CA, September 2005, pp. 2976–2986.

[28] Osder, S. S., "Air-Data Systems," in *Avionics Navigation Systems*, 2nd ed., M. Kayton and W. R. Fried, (eds.), New York: Wiley, 1997, pp. 393–425.

[29] Ausman, J. S., "Baro-Inertial Loop for the USAF Standard RLG INU," *Navigation: JION*, Vol. 38, No. 2, 1991, pp. 205–220.

[30] Bekir, E., *Introduction to Modern Navigation Systems*, Singapore: World Scientific, 2007.

[31] Käppi, J., and K. Alanen, "Pressure Altitude Enhanced AGNSS Hybrid Receiver for a Mobile Terminal," *Proc. ION GNSS 2005*, Long Beach, CA, September 2005, pp. 1991–1997.

[32] Parviainen, J., J. Kantola, and J. Collin, "Differential Barometry in Personal Navigation," *Proc. IEEE/ION PLANS*, Monterey, CA, May 2008, pp. 148–152.

[33] Fofonoff, N. P., and R. C. Millard, Jr., *Algorithms for Computation of Fundamental Properties of Seawater*, Unesco Technical Papers in Marine Science 44, Paris, France: Unesco, 1983.

[34] Fried, W. R., H. Buell, and J. R. Hager, "Doppler and Altimeter Radars," in *Avionics Navigation Systems*, 2nd ed., M. Kayton and W. R. Fried, (eds.), New York: Wiley, 1997, pp. 449–502.

[35] Hay, C., "Turn, Turn, Turn: Wheel-Speed Dead Reckoning for Vehicle Navigation," *GPS World*, October 2005, pp. 37–42.

[36] Wilson, J. L., "Low-Cost PND Dead Reckoning Using Automotive Diagnostic Links," *Proc. ION GNSS 2007*, Fort Worth, TX, September 2007, pp. 2066–2074.

[37] Wilson, J. L., and M. J. Slade, "Accelerometer Compensated Differential Wheel Pulse Based Dead Reckoning," *Proc. ION GNSS 2009*, Savannah, GA, September 2009, pp. 3087–3095.

[38] Carlson, C. R., J. C. Gerdes, and J. D. Powell, "Error Sources When Land Vehicle Dead Reckoning with Differential Wheelspeeds," *Navigation: JION*, Vol. 51, No. 1, 2004, pp. 13–27.

[39] Bullock, J. B., et al., "Integration of GPS with Other Sensors and Network Assistance," in *Understanding GPS Principles and Applications*, 2nd ed., E. D. Kaplan and C. J. Hegarty, (eds.), Norwood, MA: Artech House, 2006, pp. 459–558.

[40] French, R. L., "Land Vehicle Navigation and Tracking," in *Global Positioning System: Theory and Applications Volume II*, B. W. Parkinson and J. J. Spilker, Jr., (eds.), Washington, D.C.: AIAA, 1996, pp. 275–301.

[41] Stephen, J., and G. Lachapelle, "Development and Testing of a GPS-Augmented Multi-Sensor Vehicle Navigation System," *Journal of Navigation*, Vol. 54, No. 2, 2001, pp. 297–319.

[42] Hollenstein, C., et al., "Performance of a Low-Cost Real-Time Navigation System Using Single-Frequency GNSS Measurements Combined with Wheel-Tick Data," *Proc. ION GNSS 2008*, Savannah, GA, September 2008, pp. 1610–1618.

[43] Georgy, J., et al., "Low-Cost Three-Dimensional Navigation Solution for RISS/GPS Integration Using Mixture Particle Filter," *IEEE Trans. on Vehicular Technology*, Vol. 59, No. 2, 2010, pp. 599–615.

[44] Käppi, J., J. Syrjärinne, and J. Saarinen, "MEMS-IMU Based Pedestrian Navigator for Handheld Devices," *Proc. ION GPS 2001*, Salt Lake City, UT, September 2001, pp. 1369–1373.

[45] Moafipoor, S., D. A. Grejner-Brzezinska, and C. K. Toth, "Multi-Sensor Personal Navigator Supported by Adaptive Knowledge Based System: Performance Assessment," *Proc. IEEE/ION PLANS*, Monterey, CA, May 2008, pp. 129–140.

[46] Aubeck, F., C. Isert, and D. Gusenbauer, "Camera Based Step Detection on Mobile Phones," *Proc. Indoor Positioning and Indoor Navigation*, Guimarães, Portugal, September 2011.

[47] Chen, W., et al., "Comparison of EMG-Based and Accelerometer-Based Speed Estimation Methods in Pedestrian Dead Reckoning," *Journal of Navigation*, Vol. 64, No. 2, 2011, pp. 265–280.

[48] Cho, S. Y., et al., "A Personal Navigation System Using Low-Cost MEMS/GPS/Fluxgate," *Proc. ION 59th AM*, Albuquerque, NM, June 2003, pp. 122–127.

[49] Judd, T., "A Personal Dead Reckoning Module," *Proc. ION GPS-97*, Kansas, MO, September 1997, pp. 47–51.

[50] Collin, J., O. Mezentsev, and G. Lachapelle, "Indoor Positioning System Using Accelerometry and High Accuracy Heading Sensors," *Proc. ION GPS/GNSS 2003*, Portland, OR, September 2003, pp. 1164–1170.

[51] Ladetto, Q., "On Foot Navigation: Continuous Step Calibration Using Both Complementary Recursive Prediction and Adaptive Kalman Filtering," *Proc. ION GPS 2000*, Salt Lake City, UT, September 2000, pp. 1735–1740.

[52] Leppäkoski, H., et al., "Error Analysis of Step Length Estimation in Pedestrian Dead Reckoning," *Proc. ION GPS 2002*, Portland, OR, September 2002, pp. 1136–1142.

[53] Groves, P. D., et al., "Inertial Navigation Versus Pedestrian Dead Reckoning: Optimizing the Integration," *Proc. ION GNSS 2007*, Fort Worth, TX, September 2007, pp. 2043–2055.

[54] Park, C. G., et al., "Adaptive Step Length Estimation with Awareness of Sensor Equipped Location for PNS," *Proc. ION GNSS 2007*, Fort Worth, TX, September 2007, pp. 1845–1850.

[55] Kantola, J., et al., "Context Awareness for GPS-Enabled Phones," *Proc. ION ITM*, San Diego, CA, January 2010, pp. 117–124.

[56] Frank, K., et al., "Reliable Real-Time Recognition of Motion Related Human Activities Using MEMS Inertial Sensors," *Proc. ION GNSS 2010*, Portland, OR, September 2010, pp. 2919–2932.

[57] Saeedi, S., et al., "Context Aware Mobile Personal Navigation Using Multi-Level Sensor Fusion," *Proc. ION GNSS 2011*, Portland, OR, September 2011, pp. 1394–1403.

[58] Pei, L., et al., "Using Motion-Awareness for the 3D Indoor Personal Navigation on a Smartphone," *Proc. ION GNSS 2011*, Portland, OR, September 2011, pp. 2906–2913.

[59] Soehren, W., and W. Hawkinson, "A Prototype Personal Navigation System," *Proc. IEEE/ION PLANS*, San Diego, CA, April 2006, pp. 539–546.

[60] Tull, W. J., "The Early History of Airborne Doppler Systems," *Navigation: JION*, Vol. 43, No. 1, 1996, pp. 9–24.

[61] Buell, H., "Doppler Radar Systems for Helicopters," *Navigation: JION*, Vol. 27, No. 2, 1980, pp. 124–131.

[62] McCroskey, R., et al., "GLANSER—An Emergency Responder Locator System for Indoor and GPS-Denied Applications," *Proc. ION GNSS 2010*, Portland, OR, September 2010, pp. 2901–2909.

[63] Jourdan, D. W., "Doppler Sonar Navigator Error Propagation and Correction," *Navigation: JION*, Vol. 32, No. 1, 1985, pp. 29–56.

[64] Butler, B., and R. Verrall, "Precision Hybrid Inertial/Acoustic Navigation System for a Long-Range Autonomous Underwater Vehicle," *Navigation: JION*, Vol. 48, No. 1, 2001, pp. 1–12.

[65] Grose, B. L., "The Application of the Correlation Sonar to Autonomous Underwater Vehicle Navigation," *Proc. IEEE Symposium on Autonomous Underwater Vehicle Technology*, Washington, D.C., June 1992, pp. 298–303.

[66] Boltryk, P., et al., "Improvement of Velocity Estimate Resolution for a Correlation Velocity Log Using Surface Fitting Methods," *Proc. MTS/IEEE Oceans '02*, October 2002, pp. 1840–1848.

[67] Griffiths, G., and S. E. Bradley, "A Correlation Speed Log for Deep Waters," *Sea Technology*, Vol. 39, No. 3, 1998, pp. 29–35.

[68] Shih, P., and H. Weinberg, "A Useful Role for the ADXL202 Dual-Axis Accelerometer in Speedometer-Independent Car-Navigation Systems," *Analog Dialogue*, Vol. 35, No. 4, 2001, pp. 1–3.

[69] An, D., J. A. Rios, and D. Liccardo, "A UKF Based GPS/DR Positioning System for General Aviation," *Proc. ION GNSS 2005*, Long Beach, CA, September 2005, pp. 989–998.

CHAPTER 7
Principles of Radio Positioning

This chapter explains the physical principles of radio positioning and discusses the characteristics that are common across the different technologies, both space-based and terrestrial. Section 7.1 compares different positioning configurations and describes the different methods. Section 7.2 discusses the properties of positioning signals. Section 7.3 describes the main features of radio navigation user equipment, including the calculation of a two-dimensional position solution from ranging measurements. Finally, Section 7.4 discusses how various error sources and the signal geometry determine the positioning accuracy.

7.1 Radio Positioning Configurations and Methods

There are a number of different configurations and methods for obtaining position information from radio signals. Each has advantages and disadvantages that determine its suitability for different applications. This section begins by comparing different positioning configurations, such as whether signals are transmitted from known to unknown or unknown to known locations, and whether dedicated or existing signals are used. Relative positioning is then discussed. This is followed by descriptions of the five main classes of positioning method: proximity, ranging, angular positioning, pattern matching, and Doppler positioning [1, 2]. Most classes of positioning method can potentially be used with any radio signal. Furthermore, multiple positioning methods may be used simultaneously with the same set of signals. Different positioning methods may also be applied to the same signals depending on the context, such as ranging in open environments, pattern matching indoors, and a mixture of the two in urban areas.

7.1.1 Self-Positioning and Remote Positioning

Radio positioning systems may be classed as either remote positioning or self-positioning. In a *remote*, *network-based*, or *multilateral* positioning system, a transmitter is located at the object whose position is to be determined, while multiple receivers are placed at known locations. Measurements from the receivers are forwarded to a master station, where the position of the transmitter is calculated. In a *self-*, *mobile-based*, or *unilateral* positioning system, multiple transmitters operate from known locations, while a receiver is located at the object whose position is to be determined. Position is calculated at the receiver. Figure 7.1 depicts both configurations. In both cases, an object whose position is to be determined is often referred to as a mobile station, while a receiver or transmitter at a known location is referred to as a base station, particularly when its position is fixed [1].

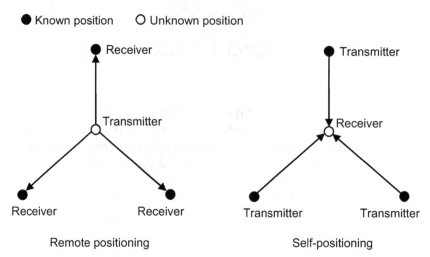

Figure 7.1 Remote- and self-positioning configurations (one-way transmission).

Remote positioning is best suited to tracking applications, for which the positioning information is required at the master station rather than at the object itself, sometimes known as the target. However, it can be used for navigation if the position solution is transmitted to the target. In these cases, the target's transmissions will incorporate identification information. A drawback of remote positioning is that there is a limit to the number of targets that may be tracked at any one time. To a certain extent, this may be increased by reducing the solution update rate. The limiting factors are the available radio frequency (RF) spectrum, the number of signals each receiver can process simultaneously, and the master station processing capacity.

Self-positioning is more suited to navigation applications, for which the positioning information is required at the object whose position is being determined (i.e., the navigation system user). Most self-positioning systems have the advantage that there is no practical limit to the number of users that can be supported.

Self-positioning systems must incorporate a method for conveying transmitter position information to the user equipment, except for pattern matching and some proximity methods that require a signal reception database instead. In the oldest terrestrial radio navigation systems, such as DME/VOR/TACAN (Section 11.1) and older versions of Loran (Section 11.2), the transmitter positions are simply prestored in the user equipment. Typically, a transmitter database is preloaded by the manufacturer, while details of additional transmitters are input manually. In newer systems, such as GNSS (Chapters 8 to 10), the position, and, where appropriate, the velocity, of each transmitter is modulated onto its signals. This is essential for systems with moving transmitters as their trajectories are not entirely predictable. A further option is to convey the transmitter position information by a separate data link. Assisted GNSS (Section 10.5.2) does this to improve on the download speed and robustness of stand-alone GNSS.

Some radio positioning systems transmit signals in both directions. However, they may still be classified as remote or self-positioning, depending on whether the position solutions are calculated by the user equipment or a master station. Note that self-positioning with bidirectional transmission has the same limitation on the number of users as remote positioning.

Both remote- and self-positioning techniques may use signals of opportunity, which are signals designed for purposes other than positioning, such as broadcasting or communications. In the self-positioning case, this dramatically reduces the infrastructure cost. When a suitable receiver is already present (e.g., for communication purposes), it can also reduce the user equipment cost. However, positioning performance using SOOP may not be as good as that obtained using signals designed specifically for positioning. Furthermore, the signals may not contain transmitter location information, in which case a database or separate data link is required.

The vast majority of radio positioning systems used for navigation are self-positioning. Therefore, in this book, self-positioning is assumed unless stated otherwise.

7.1.2 Relative Positioning

In *relative positioning*, signals are transmitted between participants to determine their relative positions. Thus, each participant must both transmit and receive. Relative positioning may be combined with either self-positioning or remote positioning to determine the absolute position of the participants. Relative positioning is a component of cooperative positioning (see Section 1.4.5), also known as collaborative and peer-to-peer positioning.

There are two configurations of relative positioning: a chain and a network. In a self-positioning chain, illustrated by Figure 7.2, each participant broadcasts a signal that includes its own position and sometimes the uncertainty thereof. The participants at the beginning of the chain are at known locations. The others obtain their position using the signals from participants earlier in the chain. This can be any positioning method or a combination of methods. Thus, the further down the chain a participant is, the less accurate his or her position solution will be. To prevent positive feedback, a strict hierarchy must be maintained. However, this will change as the participants move. Examples include the Joint Tactical Information Distribution System (JTIDS) and Multi-functional Information Distribution System (MIDS) (Section 11.1.4), and some UWB positioning systems (Section 12.2). A

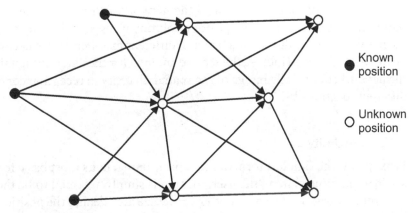

Figure 7.2 Relative- and self-positioning chain.

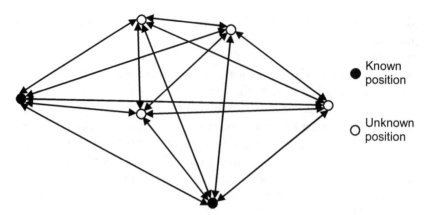

Figure 7.3 Relative navigation network.

remote-positioning chain operates on the same principle, but with the transmission and reception roles in each link of the chain reversed.

In a relative positioning network, illustrated by Figure 7.3, positioning signals are exchanged between all participants within range of each other typically using short-range communications systems (Section 12.3) or UWB. In a remote-positioning network, all of the signal measurements are then relayed to a master station that calculates the relative position of all participants. In a self-positioning network, the position determination processing is distributed between the participants. Certain participants may act as master nodes, determining the position of themselves and their neighbors [3–5]. A large network may be divided into clusters, each with a master node, while individual participants may move from one cluster to another as they move around.

Using ranging, at least four participants are needed for a two-dimensional relative position solution and at least six for a three-dimensional solution. To obtain absolute positions for the network without orientation information, at least two participants must be at known locations for a 2-D solution and three for a 3-D solution. In addition, the approximate position of another participant is required to break mirror symmetry. When both the position and the orientation of one of the participants are known, angular positioning measurements may be used to obtain the absolute positions of the rest of the network. Alternatively, the measurements required to obtain an absolute position solution using signals or features external to the network may be distributed between different members of the network. This may be thought of as distributing the receive antenna for the external signals throughout the network, taking advantage of the spatial diversity in reception conditions across different locations [3].

7.1.3 Proximity

Proximity is the simplest form of radio positioning. In its most basic form, assuming a self-positioning system, the user position is simply assumed to be the same as the transmitter position. The transmitter's coverage area defines the position uncertainty. When the transmitter is not at the center of the coverage area (e.g., where it uses a directional antenna or there is an obstruction), the user position may be taken to be the center of the coverage area if this is known.

When short-range transmitters are used, proximity is sufficient to meet the accuracy requirements for many applications. For example, radio frequency identification (RFID), wireless personal area network (WPAN), and wireless local area network (WLAN) technology, all described in Section 12.3, can provide a position accuracy of a few meters or tens of meters using proximity.

Proximity positioning using mobile phones (Section 11.3.1) is known as cell identification and can result in errors of up to 1 km in urban areas and 35 km in rural areas [6]. Errors using public broadcasting signals can be larger. However, such an approximate position solution is useful for aiding the signal acquisition process in a long-range positioning system, such as GNSS (see Section 10.5.1).

The accuracy of proximity positioning can be improved by using multiple transmitters. The simplest approach is to set the position solution to the average of the positions of the transmitters received. However, lower-power transmitters will typically be nearer, as will transmitters received with a higher signal strength. Therefore, a weighted average will usually be more accurate.

A more sophisticated approach is containment intersection. The coverage area of each transmitter received may be considered as a containment area within which the user may be found. Therefore, if multiple transmitters are received, the user will be located within the intersection of their containment areas as illustrated by Figure 7.4. This also enables the uncertainty of the position solution to be determined. In practice, the containment areas are not simple circles as obstructions and variations in terrain height, and the transmit antenna gain pattern will cause the coverage radius to vary with direction. There can also be gaps in a transmitter's coverage area due to shadowing by buildings and other obstacles.

A further complication is that the boundaries of coverage areas are not sharply defined. Fading (see Section 7.1.4.6) and multipath interference (Section 7.4.2) cause localized variations in signal strength. Receiver sensitivity varies and receive antennas can have directional gain patterns, which may also be frequency-dependent. People

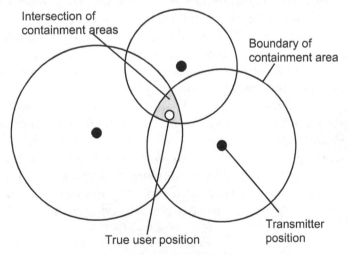

Figure 7.4 Proximity positioning by intersection of containment areas.

and vehicles can also cause temporary shadowing. One solution to this problem is to set inner and outer containment areas, corresponding to the inner and outer limits of the coverage area, respectively. If a signal is received, the user is assumed within the outer containment area, whereas if a signal is not received, the user is assumed to be outside the inner containment area. Received signal strength (RSS) measurements may be used to enhance containment intersection, enabling the containment area to be divided into a series of overlapping zones based on RSS.

7.1.4 Ranging

Positioning by ranging is the determination of an object's position by measuring the range to a number of objects at known locations. It is also known as lateration and rho-rho positioning with trilateration denoting ranging using three signals and multilateration ranging using more than three signals. Ranging is the most common method used in radio navigation. For self-positioning, ranges are measured from transmitters at known locations to a receiver at an unknown location. Range is usually obtained by measuring the signal's time of flight (TOF), but may also be estimated from the received signal strength. Signal timing measurement is discussed in Section 7.3.2.

In this section, it is generally assumed that the antennas of all transmitters and receivers are located within a plane and that only a two-dimensional (2-D) position solution within that plane is required. Three-dimensional (3-D) positioning by ranging is described in Sections 8.1.3, 9.4 and 12.2.3, while horizontal positioning accounting for antenna height differences is described in Section 11.1.1.2.

When a self-positioning ranging measurement from a single transmitter is used, the position of the user's receiving antenna within a plane containing both the transmitting and receiving antennas can be anywhere on a circle centered on the transmitter's antenna. The radius of the circle is equal to the distance between the two antennas, known as the geometric range. This circle is an example of a line of position. More generally, a LOP is a locus of candidate positions and its shape depends on the positioning method.

When a second transmitter is introduced with its antenna in the same plane, the locus of the user's position is limited to the intersection of two LOPs, comprising circles of radii r_1 and r_2, centered at the antennas of transmitters 1 and 2, respectively. Figure 7.5 illustrates this. The two circles intersect at two points. Therefore, the 2-D position solution obtained only from two ranging measurements is ambiguous. This ambiguity may be resolved by introducing a ranging measurement from a third transmitter, also shown in Figure 7.5. However, the ambiguity can sometimes be resolved using prior information. For example, the combination of a previous position solution with knowledge of the maximum distance travelable during the intervening period can be used to constrain the current position solution. Three-dimensional positioning requires one more range measurement than 2-D positioning.

In the 2-D case, each geometric range, r_{at}, may be expressed in terms of the user position, (x_{pa}^p, y_{pa}^p), and transmitter position, (x_{pt}^p, y_{pt}^p), by

$$r_{at} = \sqrt{\left(x_{pt}^p - x_{pa}^p\right)^2 + \left(y_{pt}^p - y_{pa}^p\right)^2}, \qquad (7.1)$$

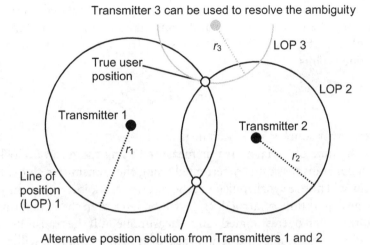

Figure 7.5 Lines of position from single, dual, and triple ranging measurements in two dimensions.

where a is the user antenna body frame, t is the transmit antenna body frame, and p is a planar coordinate frame, defined such that the user antenna and all of the transmit antennas lie within its xy plane. Note that geometric range is independent of direction, so $r_{ta} = r_{at}$. Timing, signal propagation, and frame rotation effects, together with measurement errors, are neglected for the moment.

An estimate of the user position may be obtained from a set of range measurements by solving a set of simultaneous equations of the form given by (7.1). These equations are nonlinear and must generally be solved iteratively as described in Section 7.3.3.

There are five types of TOF-based ranging measurement from which a position solution may be determined. These are:

- Passive ranging or time of arrival (TOA);
- Time difference of arrival (TDOA) across transmitters or hyperbolic ranging;
- Differential ranging or TDOA across receivers;
- Double-differenced ranging across transmitters and receivers;
- Two-way ranging.

The terms TOA and TDOA are typically used to describe short-range positioning systems, while passive, hyperbolic, and differential ranging are typically used to describe long-range systems. Each type of measurement is described in turn, followed by a discussion of RSS-based ranging. The calculation of position from ranging measurements is described in Section 7.3.3.

7.1.4.1 Positioning from Passive Ranging or Time of Arrival

In passive ranging or TOA measurement, the receiver measures the time of arrival, $t^t_{sa,a}$, at receive antenna a of a particular feature of the signal that was transmitted at a known time, $t^t_{st,a}$, from transmit antenna t. The transmission time may be a predetermined feature of the system or may be modulated onto the signal. By differencing the times of arrival and transmission and then multiplying by the speed of light, c, which is 299,792,458 m s^{-1} in free space, a range measurement may be obtained. Thus,

$$r_{at} = \left(t^t_{sa,a} - t^t_{st,a}\right) c, \tag{7.2}$$

where error sources have been neglected.

The time of signal arrival is measured using the receiver clock, while the time of signal transmission is determined using the transmitter clock. In practice, these clocks will not be synchronized. If the receiver clock is running ahead of system time, the measured time of arrival, $\tilde{t}^t_{sa,a}$, will be later than the actual time of arrival, $t^t_{st,a}$, resulting in an overestimated range measurement. If the transmitter clock is running ahead of system time, the actual time of transmission, $t^t_{st,a}$, will be earlier than the intended time of transmission, $\tilde{t}^t_{st,a}$, which is that deduced by the user equipment from the signal modulation. This will result in an underestimated range measurement. If the receiver clock is ahead by δt^a_c and the transmitter clock ahead by δt^t_c, the range measurement, neglecting other error sources, is

$$\begin{aligned} \rho^t_a &= \left(\tilde{t}^t_{sa,a} - \tilde{t}^t_{st,a}\right)c \\ &= \left(t^t_{sa,a} + \delta t^a_c - t^t_{st,a} - \delta t^t_c\right)c, \\ &= r_{at} + \left(\delta t^a_c - \delta t^t_c\right)c \end{aligned} \tag{7.3}$$

where ρ^t_a is known as the *pseudo-range* to distinguish it from the range measured in the absence of clock errors. Note that the superscript refers to the transmitter and the subscript to the receiver. Figure 7.6 illustrates this.

Unlike geometric range, pseudo-range depends on the direction of transmission. The pseudo-range for a signal transmitted from a to t is thus

$$\begin{aligned} \rho^a_t &= r_{ta} + \left(\delta t^t_c - \delta t^a_c\right)c \\ &= r_{at} + \left(\delta t^t_c - \delta t^a_c\right)c \\ &= \rho^t_a - 2\left(\delta t^a_c - \delta t^t_c\right)c \end{aligned} \tag{7.4}$$

In practice, pseudo-range and range measurements will also be subject to propagation errors as discussed in Sections 7.4.1 and 7.4.2; this *raw* measurement is denoted by the subscript R. There will also be receiver measurement errors as discussed in Section 7.4.3; their presence is denoted by ~. Thus, a measurement of the pseudo-range from transmit antenna t to user antenna a is denoted as $\tilde{\rho}^t_{a,R}$. Corrections applied to account for propagation and/or timing errors are denoted by the subscript C, giving $\tilde{\rho}^t_{a,C}$.

For self-positioning using passive ranging to work, the transmitter clocks must be synchronized with each other. Otherwise, differential positioning (Section 7.1.4.3) must be used. There are three ways in which transmitters may be synchronized. The first option is to synchronize all transmitters to a common timebase, such as

7.1 Radio Positioning Configurations and Methods

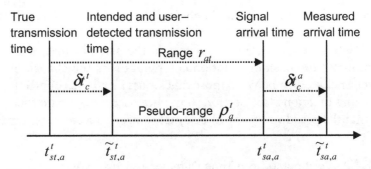

Figure 7.6 Effect of unsynchronized transmitter and receiver clocks on range measurement.

Coordinated Universal Time (UTC). This is commonly implemented where different transmitters use the same frequency at different times. The second option is to measure the time offset between each transmitter and a common timebase and modulate this information onto the signals. The user equipment then corrects the pseudo-range measurements for the transmitter clock offset, δt_c^t. This method is used by GNSS (Chapters 8 to 10). The final option is chain synchronization, whereby each station transmits at a fixed interval after receipt of the transmission from the preceding station in the chain. Transmitter chains may be looped or there may be a master station that transmits first.

With the transmitter clocks synchronized, only the receiver clock offset with respect to the transmission timebase remains. This is treated as an additional unknown in the position solution. Thus, in the 2-D case, each pseudo-range measurement, corrected for any transmitter clock error, $\tilde{\rho}_{a,C}^t$, may be expressed as

$$\tilde{\rho}_{a,C}^t = \sqrt{\left[x_{pt}^p(t_{st,a}^t) - x_{pa}^p(t_{sa,a}^t)\right]^2 + \left[y_{pt}^p(t_{st,a}^t) - y_{pa}^p(t_{sa,a}^t)\right]^2} + \delta\rho_c^a(t_{sa,a}^t), \quad (7.5)$$

where $\delta\rho_c^a = \delta t_c^a c$ and other errors are neglected. Note that the distance between transmitter and receiver may change during the time it takes for the signal to travel from one to the other. Therefore, in computing the position solution, the receiver position must be considered at the time of signal arrival and the transmitter position at the time of signal transmission. This is particularly important for GNSS where the transmission distances are long and the transmitters are moving with respect to the Earth.

The need to solve for the receiver clock offset increases the number of measurements required. Thus, passive ranging requires at least three measurements for a 2-D position solution and four measurements for a 3-D solution. When there is insufficient prior information to resolve the LOP intersection ambiguity shown in Figure 7.5, four measurements are required for a 2-D solution and five for a 3-D solution. If a highly stable receiver clock is used (see Section 9.1.2), the clock offset can be assumed constant for tens of minutes, enabling positioning to be achieved using one less passive ranging measurement once the clock has been calibrated.

If different groups of transmitters are synchronized to different timebases and the differences between those timebases are unknown, they can be estimated as part of the position solution. One additional ranging measurement per timescale difference is required. However, if the timebase differences are stable, the additional ranging measurements are only required for the initial calibration.

Note that significant differences in the arrival times of different signals can complicate the position solution computation as the change in user position and receiver clock drift between those times must then be accounted for. Generally, signal tracking functions (see Section 7.3.2) are used to estimate the pseudo-range rates so that the pseudo-ranges can be synchronized to a common time of signal arrival.

7.1.4.2 Positioning from Time Difference of Arrival Across Transmitters

In positioning by TDOA across transmitters, ranging measurements are differenced across transmitters to eliminate the receiver clock offset (assuming self-positioning). Transmitters must be synchronized as for passive ranging and the same number of signals is required. Range difference, or delta-range, measurements may be obtained simply by differencing corrected pseudo-range measurements obtained from passive ranging. Thus, for transmitters s and t,

$$\Delta \tilde{\rho}_{a,C}^{st} = \tilde{\rho}_{a,C}^{t} - \tilde{\rho}_{a,C}^{s}. \tag{7.6}$$

Alternatively, the TDOA of two signals, $\Delta \tilde{t}_{TD,a}^{st} = \tilde{t}_{sa,a}^{t} - \tilde{t}_{sa,a}^{s}$, may be measured directly. This is sometimes called a time difference (TD) measurement. The range difference is then

$$\Delta \tilde{\rho}_{a,C}^{st} = \left(\Delta \tilde{t}_{TD,a}^{st} - \Delta \tilde{t}_{NED,a}^{st}\right) c. \tag{7.7}$$

where $\Delta \tilde{t}_{NED,a}^{st} = \tilde{t}_{st,a}^{t} - \tilde{t}_{st,a}^{s}$ is the difference between the nominal times of transmission of the two signals, sometimes known as the nominal emission delay (NED). Direct TDOA measurements are commonly made in systems with chain-synchronized transmitters. Measurements should not be made across signals from different chains that are not synchronized.

Historically, positioning by TDOA across transmitters was called hyperbolic positioning because a line of position obtained from a range difference in 2-D positioning [substituting (7.5) into (7.6)] is a hyperbola. Figure 7.7 shows some example LOPs. In the 3-D case, the position locus obtained from a range difference is a hyperboloid.

7.1.4.3 Differential Positioning

In differential self-positioning, ranging measurements from the same transmitter are differenced across receivers, usually a user receiver and a reference receiver. A separate data link conveys measurements from the reference receiver to the user. Figure 7.8 illustrates this. This technique cancels out the transmitter clock offset in the differencing process, which is essential in systems where the transmitters are unsynchronized. However, the difference in clock offset between the two receivers is left to be determined as part of the position solution. When the receivers are relatively close together, differential positioning can also be used to cancel out transmitter position

7.1 Radio Positioning Configurations and Methods

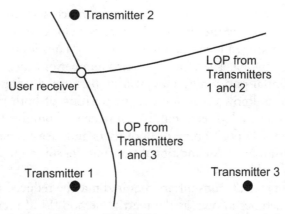

Figure 7.7 Hyperbolic lines of position from TDOA measurements in two dimensions using three transmitters (alternate intersection not shown).

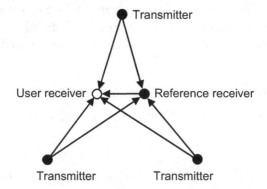

Figure 7.8 Differential positioning.

and signal propagation errors (Section 7.4). This method is sometimes known as positioning by TDOA across receivers.

Range difference measurements may be obtained by differencing raw pseudorange measurements, denoted by the subscript R. Thus

$$\nabla \tilde{\rho}^t_{ra,R} = \tilde{\rho}^t_{a,R} - \tilde{\rho}^t_{r,R}, \tag{7.8}$$

where subscript a denotes the user receiver antenna as before and subscript r denotes the reference receiver antenna. A range difference may also be obtained directly by comparing the signal received by the user with that received by the reference and then transmitted to the user [7]. However, this typically requires much greater datalink bandwidth and short-term storage of the received signals by the user.

When the reference receiver position is known, the user position, in the 2-D case, is obtained by solving simultaneous equations of the form

$$\nabla \tilde{\rho}^t_{ra,R} = \sqrt{\left[x^p_{pt}(t^t_{st,a}) - x^p_{pa}(t^t_{sa,a})\right]^2 + \left[y^p_{pt}(t^t_{st,a}) - y^p_{pa}(t^t_{sa,a})\right]^2} - r_{rt} + \nabla \rho^{ra}_c(t^t_{sa,a}), \tag{7.9}$$

where r_{rt} is the true range from the reference station to the transmitter and $\nabla \rho_c^{ra} = \delta \rho_c^a - \delta \rho_c^r$ is the differential receiver clock offset. Thus, with a reference receiver at a known location, the number of quantities to determine is the same as for passive ranging. Therefore, the same number of transmitters is required.

Differential positioning may also be performed using a pair of receivers that are both at unknown locations. If the position of both receivers is required, the data link must be bidirectional. Measurements from five transmitting stations are required for 2-D positioning of both users and measurements from seven stations for 3-D positioning. An additional transmitting station may be required where the solution is ambiguous.

The number of transmitters required may be reduced by using more receivers. This is sometimes known as the matrix method [8]. If there are n receivers and m transmitters, there are mn undifferenced pseudo-range measurements, $m + n - 1$ unknown relative clock offsets, and either $2n$ or $3n$ position components, depending on whether positioning is 2-D or 3-D. Thus, for a d-dimensional position solution to be obtainable, the condition $mn \geq m + (d + 1)n - 1$ must hold. The minimum number of transmitting stations required is four for 2-D positioning and five for 3-D positioning. Again, a further station may be required for ambiguity resolution.

7.1.4.4 Positioning from Double-Differenced Ranging

Differential and TDOA positioning may be combined to produce double-differenced measurements. Thus,

$$\begin{aligned} \nabla \Delta \tilde{\rho}_{ra,R}^{st} &= \tilde{\rho}_{a,R}^{t} - \tilde{\rho}_{a,R}^{s} - \tilde{\rho}_{r,R}^{t} + \tilde{\rho}_{r,R}^{s} \\ &= \Delta \tilde{\rho}_{a,R}^{st} - \Delta \tilde{\rho}_{r,R}^{st} \\ &= \nabla \tilde{\rho}_{ra,R}^{t} - \nabla \tilde{\rho}_{ra,R}^{s} \end{aligned} \quad (7.10)$$

These are commonly used in GNSS carrier-phase positioning (Section 10.2).

Note that triple-differenced measurements may be formed by differencing double-differenced measurements across time; these measure changes in position.

7.1.4.5 Positioning from Two-Way Ranging

In a self-positioning two-way ranging system, a mobile user transmits to the base stations, either together or in turn. The base stations then transmit back to the user after a fixed interval. The user measures the round-trip time (RTT), $\Delta t_{rt,a}^{t}$, where the subscript denotes the transmitter of the initial signal and receiver of the response signal and the superscript denotes the receiver of the initial signal and the transmitter of the response signal. From this, an average range may be estimated using

$$\tilde{r}_{at} = \tfrac{1}{2}\bigl(\Delta \tilde{t}_{rt,a}^{t} - \tau_r^t\bigr)c, \quad (7.11)$$

where $\sim$ denotes a measurement and τ_r^t is the base station response time interval, which will either be fixed or included in the signal from the base station.

7.1 Radio Positioning Configurations and Methods

No time synchronization of the base stations is required as their transmissions are triggered by the incoming signals. The effect of the mobile user's clock offset largely cancels between the outgoing and incoming transmissions. Thus, the timing error in the RTT measurement is

$$\Delta \tilde{t}^t_{rt,a} - \Delta t^t_{rt,a} = \delta t^a_c(t^t_{sa,a}) - \delta t^a_c(t^a_{st,t}), \quad (7.12)$$

where δt^a_c is the user clock offset, $t^a_{st,t}$ is the time of transmission of the signal from user to base station, and $t^t_{sa,a}$ is the time of arrival of the signal from base station to user. Figure 7.9 illustrates this. For a constant clock drift, the RTT error will increase with distance from the base station. However, even with a low-cost oscillator, the ranging error will be no more than one part in 10^5 (the relative frequency error of the oscillator). A further error arises due to the base station's response timing. However, this should be noise-like provided the response time is kept short as the timing resolution contribution should far exceed the contribution from the base station clock drift.

When the user is moving with respect to the base station, (7.1) will give the average range over the round-trip measurement time. If a signal tracking function (see Section 7.3.2) is used to estimate the rate of change of the RTT, $\Delta \dot{t}^t_{rt,a}$, the range at a time, t, may be estimated using

$$\tilde{r}_{at}(t) = \tfrac{1}{2}\left[\Delta \tilde{t}^t_{rt,a} - \tau^t_r + \left(t - \tfrac{1}{2}\tilde{t}^t_{sa,a} - \tfrac{1}{2}\tilde{t}^a_{st,t}\right)\Delta \dot{\tilde{t}}^t_{rt,a}\right]c. \quad (7.13)$$

This is also useful for synchronizing measurements from different base stations.

When the timing errors are negligible, a 2-D position solution can be obtained by substituting (7.13) into (7.1). If prior information is available to resolve the ambiguity, only two base stations are required, while three are required for 3-D positioning. Thus, two-way ranging has the advantage over the other methods that one less base station is required. However, if the user clock drift-induced errors are too large, the

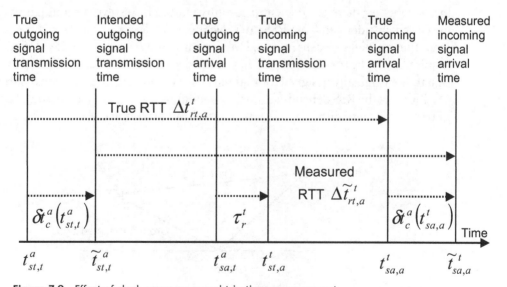

Figure 7.9 Effect of clock errors on round-trip time measurement.

clock drift must be determined alongside the position solution, requiring an extra base station. Consequently, two-way ranging is better suited to short- and medium-range positioning than to long-range.

Some two-way ranging systems implement a symmetric double-sided ranging protocol [9], whereby A transmits to B, then B transmits to A, and finally, A transmits to B a second time. This provides both A and B with an RTT measurement from which they can calculate the range between them. This is particularly useful for relative positioning (Section 7.1.2).

7.1.4.6 Ranging Using Received Signal Strength

In free space more than two wavelengths away from the transmitter, the received signal strength varies inversely as the square of the distance between the transmitter and receiver. However, in a terrestrial environment, the relationship between RSS and distance is more complicated. At frequencies above about 30 MHz, the ground acts as a near-perfect reflector, reversing the phase of the signal. The line-of-sight and ground-reflected signals interfere at the receiver as shown in Figure 7.10. This interference may be either constructive or destructive, depending on the distance between the transmit and receive antennas and the height above ground of both antennas. Thus, the RSS tends to oscillate as the receiver moves with respect to the transmitter, a phenomenon known as fading [10]. Further interference, known as multipath (Section 7.4.2) can arise from reflected or diffracted signals, while the direct line-of-sight signal is sometimes blocked or attenuated by an obstacle. An additional complication is that RSS measurements may be affected by directional variation of the receiving antenna gain and shielding by the host vehicle or user's body.

Horizontal range may be inferred from an RSS measurement using a semi-empirical model appropriate to the frequency, reception environment and base station antenna height. Some examples may be found in [1, 11]. RSS-derived range measurements are not affected by time synchronization errors so a 2-D position solution may be obtained by solving (7.1) using two or three base stations, depending on whether there is sufficient prior information to resolve the ambiguity. Range measurements derived from RSS are typically much less accurate that those derived from time-of-flight measurements. However, no knowledge of the signal structure is required, only the transmitted power, and the derivation of ranging information from signals not originally designed for that purpose does not require additional hardware.

Indoors, the RSS depends as much on building layout as distance from the transmitter so it is very difficult to derive a useful range from it.

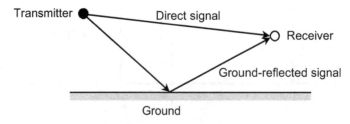

Figure 7.10 Line-of-sight and ground-reflected signals.

7.1.5 Angular Positioning

In angular positioning, also known as angulation or the angle of arrival (AOA) method, position is determined from the directions of the lines of sight from the user to two or more known locations. Each line of sight (LOS) forms a line of position. The user position is then at the intersection of the two lines. In contrast to ranging, there is no ambiguity (except if signals travel all of the way around the Earth). In self-positioning, transmitters are situated at the known locations and a receiver at the user location.

The angle, ψ_{nu}^{at}, between true north at the user and the projection within the horizontal plane of the line-of-sight vector, $\mathbf{u}_{at}^{n}$ (see Section 8.5.3), from user a to transmitter t is known as the bearing with respect to true north or the *azimuth*. It is the same as the heading (see Section 2.2.1) of a body at the user that has been turned to face the transmitter. Note that the azimuth measured at the transmitter will be slightly different due to the curvature of the Earth. By measuring azimuths to two transmitters, the user position in two dimensions may be determined, as Figure 7.11 shows.

To facilitate a Cartesian approach, it is convenient to present the positioning equations in a local-tangent-plane frame (see Section 2.1.4), denoted l, with its x-, y-, and z-axes, respectively, aligned with north, east, and down at the user. Thus, each azimuth may be expressed in terms of the user antenna position, (x_{la}^l, y_{la}^l), and the transmitter position, (x_{lt}^l, y_{lt}^l), by

$$\tan \psi_{nu}^{at} = \frac{y_{lt}^l(t_{st,a}^t) - y_{la}^l(t_{sa,a}^t)}{x_{lt}^l(t_{st,a}^t) - x_{la}^l(t_{sa,a}^t)}. \tag{7.14}$$

The horizontal position solution using two transmitters is then

$$x_{la}^l(t_{sa,a}^{1,2}) = \frac{x_{l1}^l(t_{st,a}^1)\tan\psi_{nu}^{a1} - x_{l2}^l(t_{st,a}^2)\tan\psi_{nu}^{a2} - y_{l1}^l(t_{st,a}^1) + y_{l2}^l(t_{st,a}^2)}{\tan\psi_{nu}^{a1} - \tan\psi_{nu}^{a2}}$$

$$y_{la}^l(t_{sa,a}^{1,2}) = \frac{\left(x_{l1}^l(t_{st,a}^1) - x_{l2}^l(t_{st,a}^2)\right)\tan\psi_{nu}^{a1}\tan\psi_{nu}^{a2} - y_{l1}^l(t_{st,a}^1)\tan\psi_{nu}^{a2} + y_{l2}^l(t_{st,a}^2)\tan\psi_{nu}^{a1}}{\tan\psi_{nu}^{a1} - \tan\psi_{nu}^{a2}}, \tag{7.15}$$

where the body frames of the two transmit antennas are denoted 1 and 2 and a common time of signal arrival is assumed.

An iterated least-squares method for obtaining position from an overdetermined set of azimuth measurements (more than two) is presented in Section F.1 of Appendix F on the CD. If the bearings of the lines of sight are known with respect to an arbitrary reference, but the absolute azimuths (i.e., with respect to north) are unknown, the horizontal user position may still be determined. However, measurements from at least three transmitters are required. An iterated least-squares (ILS) method for this is also shown in Section F.1 of Appendix F on the CD.

The angle at the user between the horizontal plane and the line of sight is known as the elevation, θ_{nu}^{at}. It is the same as the elevation (see Section 2.2.1) of a body at the user that has been turned to face the transmitter. Again, the elevation measured at the transmitter will be slightly different. The user height may be determined from

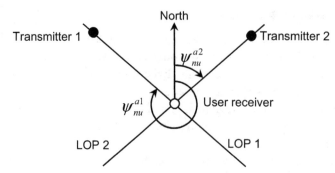

Figure 7.11 Angular positioning in the horizontal plane with absolute azimuth measurements.

an elevation measurement to a single transmitter, provided the range to that transmitter is already known. Figure 7.12 illustrates this.

Using the local-tangent-plane frame and assuming straight line propagation (see Section 7.4.1), the elevation may be expressed in terms of the user and transmitter positions by

$$\tan\theta_{nu}^{at} = -\frac{z_{lt}^{l}(t_{st,a}^{t}) - z_{la}^{l}(t_{sa,a}^{t})}{\sqrt{\left(x_{lt}^{l}(t_{st,a}^{t}) - x_{la}^{l}(t_{sa,a}^{t})\right)^2 + \left(y_{lt}^{l}(t_{st,a}^{t}) - y_{la}^{l}(t_{sa,a}^{t})\right)^2}}. \tag{7.16}$$

If the tangent plane is defined such that its xy plane intersects the ellipsoid (see Section 2.4) at the normal from the ellipsoid to the user, the height of the user antenna is given by

$$\begin{aligned}h_a(t_{sa,a}^t) &= -z_{la}^l(t_{sa,a}^t) \\ &= -\left[z_{lt}^l(t_{st,a}^t) + \tan\theta_{nu}^{at}\sqrt{\left(x_{lt}^l(t_{st,a}^t) - x_{la}^l(t_{sa,a}^t)\right)^2 + \left(y_{lt}^l(t_{st,a}^t) - y_{la}^l(t_{sa,a}^t)\right)^2}\right]\end{aligned}. \tag{7.17}$$

An ILS method for obtaining 3-D position from an overdetermined set of azimuth and elevation measurements is presented in Section F.1 of Appendix F on the CD.

The angle of arrival may be determined by two methods: direction finding and nonisotropic transmission. In a direction-finding system, the user antenna is used to determine the AOA. The simplest form of direction finder is a directional antenna. In most cases, this is rotated to minimize the received signal strength, as most antennas have sharper minima than maxima in their gain patterns. A consumer amplitude modulation (AM) radio antenna or loop television antenna is suitable for this. To avoid physically rotating the antenna, two orthogonally-mounted directional antennas whose signals are combined with a goniometer may be used. By varying the gain and phase of the signal combination, the sensitive direction of the antenna system may be varied [12]. Using multiple goniometers, the direction of multiple transmitters may be determined simultaneously.

A rotating antenna or a goniometer system can measure the AOA of a signal to an accuracy of about a degree, noting that the effective positioning accuracy can sometimes be poorer due to signal propagation effects. This method tends to be used

7.1 Radio Positioning Configurations and Methods

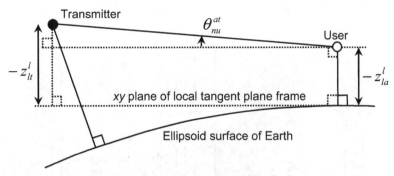

Figure 7.12 Determination of user height from elevation.

only for azimuth determination. There is a 180° ambiguity; however, this does not affect position determination.

Direction finding may also be performed using a controlled reception pattern antenna (CRPA) system or smart antenna [11]. This comprises an array of nondirectional antennas, typically mounted half a wavelength apart, and a control unit. By combining the signals from each antenna element with varying gains and phases, a varying directional reception pattern may be generated for the array as a whole. The greater the number of antennas in the array, the higher the direction-finding accuracy. Size constraints limit practical antenna arrays to higher-frequency signals. They are also expensive.

Self-positioning using direction finding can operate with any signal, provided the transmitter location is known. No knowledge of the signal structure or timing is required. However, the size of the equipment tends to limit it to aircraft and shipping use. For land applications, remote positioning by direction finding, where the direction-finding antennas are located at the base stations, may be used.

Note that where the user position is known, direction finding can be used to determine the attitude of the user antenna and hence that of the host vehicle.

A nonisotropic transmission comprises a signal whose modulation varies with direction. In self-positioning, users may determine their azimuth and/or elevation from the transmitter simply by demodulating the signal. Users do not require directional antenna systems or an attitude solution. In theory, position determination is complicated by the fact that the azimuth and/or elevation are measured at the transmitter, not the receiver. However, the range and accuracy of nonisotropic transmissions in practice are not sufficient for this to be a major issue. An example is VOR (Section 11.1.2).

Angular positioning may be combined with ranging to determine a full two- or three-dimensional position solution using only one base station. However, in cases where there is a 180° ambiguity in the azimuth, there will be two candidate position solutions that must be distinguished using prior information.

7.1.6 Pattern Matching

Pattern matching determines position by comparing properties of the received signal that vary as a function of position with a database of their values at different

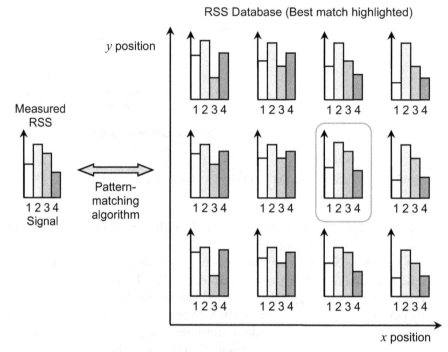

Figure 7.13 Pattern matching using received signal strength.

locations. The simplest form of pattern matching just compares the measured and predicted signal availability. However, RSS pattern matching, also known as fingerprinting or RSS mapping, is most common [1, 2, 13–15]. Multiple signals are used to reduce the ambiguity as the availability or RSS of a given signal will be similar at multiple locations. Thus, the availability or RSS of a set of signals at a given position forms the location signature of that position. Figure 7.13 illustrates this. A heterogeneous combination of different types of signals, such as WLAN, mobile-phone, and FM broadcasting signals, may be used. Pattern matching may also be applied to environmental features as discussed in Chapter 13.

All pattern-matching techniques require an approximate position solution to identify the region of the database to search. For short-range radio signals, proximity may be used. Thus, the area where there is an approximate match between the signals received and those in the database comprises the search region. An exact correspondence may not be achieved due to signal shadowing and changes since the database was compiled.

A test statistic must then be computed for each location within the search region. The Euclidean distance is widely used. This is the root mean square of the difference between the measured and database-indicated logarithmic signal strengths. The test statistic for RSS fingerprinting at candidate position (x_{pa}^p, y_{pa}^p) is thus

$$d_S\left(x_{pa}^p, y_{pa}^p\right) = \sqrt{\frac{1}{m} \sum_{j=1}^{m} \left[\tilde{S}_j - S_{D,j}\left(x_{pa}^p, y_{pa}^p\right)\right]^2}, \qquad (7.18)$$

where $\tilde{S}_j$ is the measured signal strength of the jth signal in decibels, $S_{D,j}(x_{pa}^p, y_{pa}^p)$ is the database-indicated signal strength at the candidate position, and m is the number of signals that are both measured and appear in the database. Some form of calibration is required to account for differences in the RSS measurements of the same signal by different designs of user equipment.

An alternative test statistic is the variance of the difference between measured and database-indicated logarithmic signal strengths:

$$\sigma_S^2\left(x_{pa}^p, y_{pa}^p\right) = \frac{1}{m}\sum_{j=1}^{m}\left[\tilde{S}_j - S_{D,j}\left(x_{pa}^p, y_{pa}^p\right)\right]^2 - \frac{1}{m^2}\left(\sum_{j=1}^{m}\left[\tilde{S}_j - S_{D,j}\left(x_{pa}^p, y_{pa}^p\right)\right]\right)^2. \quad (7.19)$$

This helps to account for differences in receiver sensitivity and antenna gain between different types of equipment. Alternatively, pattern matching may be performed using the rank order of the RSS measurements instead of their values, in which case test statistics are formed using the rankings instead of the signal-strength measurements [16].

The most likely position is that at which the test statistic is smallest. The simplest approach, known as nearest neighbor (NN), simply takes the candidate position with the smallest test statistic as the position solution. However, there may be multiple candidates with similar scores. The k-nearest-neighbors method sets the position solution to the simple average of the k candidate positions with the lowest test statistics, where k is typically 3 or 4. For a more statistically rigorous approach, a likelihood surface may be constructed using

$$\Lambda_S\left(x_{pa}^p, y_{pa}^p\right) = \exp\left[-\frac{d_S^2\left(x_{pa}^p, y_{pa}^p\right)}{2d_{S,R}^2}\right], \quad (7.20)$$

or

$$\Lambda_S\left(x_{pa}^p, y_{pa}^p\right) = \exp\left[-\frac{\sigma_S^2\left(x_{pa}^p, y_{pa}^p\right)}{2\sigma_{S,R}^2}\right], \quad (7.21)$$

where $d_{S,R}$ is the expected Euclidean distance and $\sigma_{S,R}^2$ is the expected variance at the true position. If there is a single peak, the position may be obtained by fitting a bivariate Gaussian distribution to the surrounding points in the likelihood function. If there are multiple competing peaks, the position solution is ambiguous.

One method of resolving an ambiguous fix is to determine candidate positions from successive sets of RSS or signal availability measurements as the user moves around. Candidate position solutions from successive measurement sets that are close together are more likely to be correct. When the user is equipped with dead-reckoning technology, the distance traveled between successive measurements is known, enabling the RSS or availability measurements to be combined into a transect and matched with the database together. As a transect-based location signature contains more information, a unique match is more likely. More information on obtaining position fixes from pattern-matching likelihood surfaces may be found in Section 13.2.2.

In self-positioning, the RSS or signal availability database must be either preinstalled or downloaded by the user as they enter each operational area. However, RSS fingerprinting is often implemented in remote positioning with the position solution relayed from a server to the user where required. In remote positioning, the RSS of transmissions from the mobile user to a set of base stations may also be used.

Signal availability and RSS databases may be generated either by collecting large amounts of measurement data or by using a 3-D model of the environment to predict the signal propagation characteristics. Combinations of the two methods may also be used. Measurement campaigns are complicated by the need to determine the true positions of the survey points. One solution is to use simultaneous localization and mapping (see the introduction to Chapter 13 and Section 16.3.6) to determine the signal characteristics and position information simultaneously. A pedestrian or robot equipped with a suitable receiver and dead-reckoning equipment simply patrols the area to be surveyed until enough measurements have been gathered to build the database [17, 18].

Building a GNSS signal availability or RSS database is not practical because the transmitters are constantly moving. Instead, signal shadowing must be predicted when required using a model of the environment stored in the user equipment or downloaded as required (see Section 10.6).

RSS measurements and signal availability at a given location can vary with time due to the opening and closing of doors and the movement of vehicles, people, furniture, and equipment. The position accuracy from pattern matching can therefore be improved by using a number of reference stations at known locations to measure these effects in real time [19].

Compared to other RSS-based positioning methods, such as ranging and enhanced proximity, pattern matching gives by far the best performance in areas, such as indoor environments, where the relationship between RSS and position is highly nonlinear. However, it is expensive to implement as a database must be generated. Also, the positioning algorithms can be processor intensive. For example, multiple-hypothesis filtering (Section 3.4.5) or particle filtering (Section 3.5) may be used to handle ambiguous measurement-database matches.

7.1.7 Doppler Positioning

When there is significant relative motion between a transmitter and receiver, position information may be derived from the Doppler shift of the signal. In practice, Doppler positioning is used where either the transmitter or receiver is onboard a satellite. It is used for Iridium satellite positioning (Section 11.4.1) and may also be used to compute an approximate GNSS position solution [20].

Neglecting relativistic effects, the range rate, $\dot{r}_{at}$, is obtained from the Doppler shift, $\Delta f_{ca,a}^t$, using

$$\dot{r}_{at} \approx -\left(\frac{\Delta f_{ca,a}^t}{f_{ca}} + \delta \dot{t}_c^a - \delta \dot{t}_c^t\right)c, \qquad (7.22)$$

where f_{ca} is the carrier frequency, $\delta \dot{t}_c^a$ is the receiver clock drift, and $\delta \dot{t}_c^t$ is the transmitter clock drift. The range rate may be expressed in terms of the 3-D inertially referenced positions, $\mathbf{r}_{ia}^i$ and $\mathbf{r}_{it}^i$, and velocities, $\mathbf{v}_{ia}^i$ and $\mathbf{v}_{it}^i$, of the receive and transmit antennas, respectively, using

$$\dot{r}_{at} = \frac{\left(\mathbf{r}_{it}^i(t_{st,a}^t) - \mathbf{r}_{ia}^i(t_{sa,a}^t)\right)^{\mathrm{T}} \left(\mathbf{v}_{it}^i(t_{st,a}^t) - \mathbf{v}_{ia}^i(t_{sa,a}^t)\right)}{\left|\mathbf{r}_{it}^i(t_{st,a}^t) - \mathbf{r}_{ia}^i(t_{sa,a}^t)\right|}. \tag{7.23}$$

Assuming self-positioning, when the transmitter position and velocity, user antenna velocity with respect to inertial space, and clock drifts are known, a single Doppler shift measurement defines a conical surface of position on which the user is located. The point of the cone is at the transmitter and its axis of symmetry is the line intersecting the transmitter in the direction of the relative velocity of the receive and transmit antennas.

In practice, the inertially-referenced velocity is usually unknown if the user position is unknown. However, the Earth-referenced velocity, $\mathbf{v}_{ea}^e$, may be known, particularly when the user is stationary, enabling $\mathbf{v}_{ia}^i$ to be expressed in terms $\mathbf{v}_{ea}^e$ and the Earth-referenced position, $\mathbf{r}_{ea}^e$. In this case, the SOP is a distorted cone. Substituting (2.146), (2.147), and (7.22) into (7.23),

$$-\frac{\Delta f_{ca,a}^t c}{f_{ca}} =$$

$$\frac{\left(\mathbf{r}_{it}^i(t_{st,a}^t) - \mathbf{C}_e^i(t_{sa,a}^t)\mathbf{r}_{ea}^e(t_{sa,a}^t)\right)^{\mathrm{T}} \left[\mathbf{v}_{it}^i(t_{st,a}^t) - \mathbf{C}_e^i(t_{sa,a}^t)\left(\mathbf{v}_{ea}^e(t_{sa,a}^t) + \mathbf{\Omega}_{ie}^e \mathbf{r}_{ea}^e(t_{sa,a}^t)\right)\right]}{\left|\mathbf{r}_{it}^i(t_{st,a}^t) - \mathbf{C}_e^i(t_{sa,a}^t)\mathbf{r}_{ea}^e(t_{sa,a}^t)\right|}$$

$$+ \left(\delta \dot{t}_c^a - \delta \dot{t}_c^t\right)c. \tag{7.24}$$

Using (7.24), the user position, $\mathbf{r}_{ea}^e$, and receiver clock drift, $\delta \dot{t}_c^a$, may be obtained from a minimum of four Doppler-shift measurements. As three conical surfaces (and the equivalent in four dimensions) can intersect at up to eight points, additional measurements may be required to resolve ambiguity, depending on the signal geometry and any constraints on the position solution. When the user velocity is unknown, it can be determined as part of the navigation solution if at least three additional Doppler-shift measurements are available.

The signal geometry changes as the transmitter and/or receiver move. Therefore, if insufficient signals are available to determine a single-epoch position solution, a position may be determined over multiple epochs provided that the clock drift is stable and the user motion is known. The following substitution may be made in (7.24):

$$\mathbf{r}_{ea}^e(t_{sa,a}^t) = \mathbf{r}_{ea}^e(t_0) + \Delta \mathbf{r}_{ea}^e(t_0, t_{sa,a}^t), \tag{7.25}$$

where $\Delta \mathbf{r}_{ea}^e(t_0, t_{sa,a}^t)$ is the displacement of the user position between times t_0 and $t_{sa,a}^t$, which is assumed to be known, and $\mathbf{r}_{ea}^e(t_0)$ is the position at time t_0, which is to be determined using Doppler positioning. This way, a Doppler position solution may be obtained over time from a single satellite.

7.2 Positioning Signals

The simplest form of radio signal is an unmodulated carrier. This carries no information and is identified only by its frequency. However, it may easily be used for positioning by proximity, AOA, and pattern matching. For TOF-based ranging, the carrier phase of the signal must be measured. This leads to a one-wavelength ambiguity in the resulting range or pseudo-range measured. A pseudo-range measurement may be derived from a carrier phase measurement, $\tilde{\phi}_a^t$ using

$$\tilde{\rho}_a^t = -\left(\frac{\tilde{\phi}_a^t}{2\pi} + N_a^t\right)\lambda_{ca}, \tag{7.26}$$

where λ_{ca} is the wavelength and N_a^t is an unknown integer, which is often negative. The sign change occurs because $\tilde{\phi}_a^t$ is a measure of the time of transmission with respect to the receiver time. For a long-wavelength (low-frequency) signal, the ambiguity may be resolvable using prior information. Another option is to use multiple frequencies. If the phases can be measured with sufficient precision, the ambiguity distance is increased to a multiple of the lowest common multiple (LCM) of the wavelength. If the LCM wavelength is greater than the maximum range of the transmitter plus the clock uncertainties (expressed as ranges), the ambiguity is removed.

When information is modulated onto the carrier, transmitter identification, position, and timing data may be conveyed to the user. Furthermore, if the time of transmission of certain features of the modulation is known or comparisons with a reference station are made, the modulation may be used for ranging measurements. Modulation-based ranging may or may not be subject to an ambiguity, depending on whether the repetition interval of the signal features used exceeds the time taken for the signal to propagate over the maximum range plus the clock uncertainties. However, the ambiguity distance from a modulation measurement will always be much greater than that from a carrier phase measurement. The rest of this section discusses modulation types and the radio spectrum.

7.2.1 Modulation Types

There are three ways of modulating an analog signal onto a carrier. Amplitude modulation (AM) varies the amplitude of the carrier as a function of the modulating signal, frequency modulation (FM) varies the frequency of the carrier, and phase modulation (PM) varies the phase. These modulation types may be combined. Quadrature amplitude modulation (QAM) comprises two carriers on the same frequency, 90° out of phase, each amplitude modulated with a different signal. The two signals are referred to as being in phase quadrature. They can be separated without interference because the product of a sine and cosine of the same argument averages to zero.

The simplest form of digital modulation is on-off keying (OOK), whereby the carrier is simply switched on and off to convey information. The digital equivalents of AM, FM, and PM are, respectively, amplitude shift keying (ASK), frequency shift keying (FSK), and phase shift keying (PSK). A digital signal comprises a sequence of symbols, each of which may be in one of an integer number of states. If the number of states is 2^k, then each symbol conveys k bits of information. However, the number of states need not be a power of 2.

A digital modulation is usually denoted by the number of states followed by the modulation type. So, for example, an 8-FSK modulation comprises a carrier that hops between eight different frequencies. A two-state system is denoted "bi" [e.g., biphase shift keying (BPSK)], while a four-state system is denoted "quadrature" [e.g., quadrature-amplitude shift keying (QASK)]. Otherwise, the numeral is used.

An n-QAM digital modulation comprises two $\sqrt{n}$-state signals in phase quadrature that are amplitude and binary-phase shift keyed (i.e., they have positive and negative values). A 4-QAM modulation is thus the same as quadrature-phase shift keying (QPSK). For systems with more than eight states, QAM is more efficient than PSK.

An orthogonal frequency division multiplex (OFDM) comprises a series of closely-spaced carriers, each with a low-symbol-rate PSK or QAM modulation incorporating a guard interval between symbols. OFDM enables multiple transmitters to broadcast identical signals on the same frequency without destructive interference. The data transmission is also highly resistant to multipath interference (Section 7.4.2).

Digital signals may also comprise a sequence of pulses, obtained by modulating the amplitude. Information may be conveyed by a number of methods. A pulse may be present or absent at a particular time. The timing of a pulse may be varied, known as pulse position modulation (PPM). The phase of the carrier with respect to the pulse envelope may be varied. Finally, the pulse shape itself may be varied, known as chirping.

The minimum double-sided bandwidth required to transmit (and receive) a digital radio signal is twice the symbol rate. Further bandwidth may be occupied by harmonics. Higher-bandwidth signals provide higher timing resolution (see Section 7.4.3), which is desirable for ranging. The power required to transmit a signal a given distance at a given frequency, subject to a given level of interference, is proportional to the symbol rate and the number of states per symbol. However, this only applies to symbols that are not already known to the receiver.

Spread spectrum techniques obtain the resolution benefits of a higher bandwidth signal without increased transmission power by further modulating the signal with a spreading code that is known to the receiver. Direct-sequence spread spectrum (DSSS), described in Section 8.1.2, applies the spreading code using BPSK, while frequency-hopping spread spectrum (FHSS) uses FSK, and time-hopping spread spectrum (THSS) uses PPM. Wideband pulsing can also be used, which is sometimes known as chirp spread spectrum (CSS). Other benefits of spread spectrum are resistance to narrowband interference, an inability to decode the signal without knowing the spreading code, and the ability to share spectrum with minimal interference.

7.2.2 Radio Spectrum

Figure 7.14 depicts the spectrum used for navigation, broadcasting and telecommunications. It also depicts the terms used by the International Telecommunications Union (ITU) and the Institute of Electrical and Electronic Engineers (IEEE) to

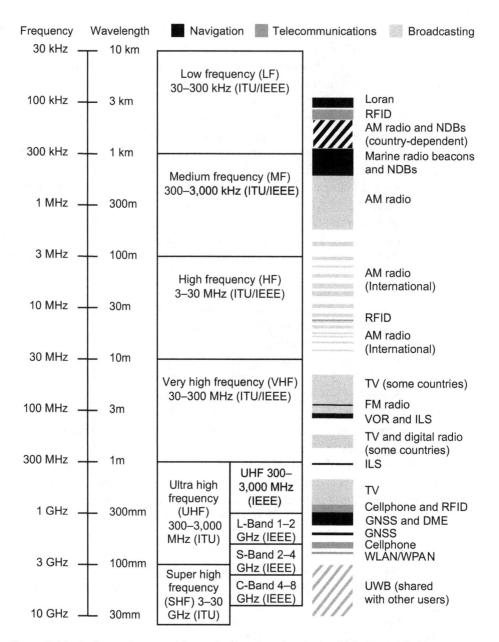

Figure 7.14 Radio spectrum used for navigation, broadcasting and telecommunications.

describe various regions of the spectrum. As a general rule, lower-frequency signals propagate further from terrestrial transmitters, while higher-frequency allocations allow higher bandwidth.

The optimal size for a radio antenna is a quarter or half of the signal wavelength. Consequently, lower-frequency radio systems use very large antennas at base stations, combined with high transmission powers to overcome the limitations of the small inefficient antennas that must be used in mobile equipment [2]. Given that it is impractical to transmit at high power from mobile equipment, VLF, LF, and MF

7.3 User Equipment

spectrum is much more suited to one-way self-positioning than to remote positioning or two-way approaches.

There are three main ways in which a region of radio spectrum may be shared between different transmitters. In a frequency division multiple access (FDMA) system, each transmission receivable at a given location uses a separate frequency. In a time division multiple access (TDMA) system, each transmission uses a separate time slot. In a code division multiple access (CDMA) system, different transmissions share the same frequency and timeslot and are distinguished by different DSSS or FHSS spreading codes. FDMA, TDMA, and CDMA may be combined.

7.3 User Equipment

This section provides an overview of radio positioning user equipment and processing, focusing on features that are common to different systems. Discussions of user equipment architecture and signal timing measurement are followed by a basic description of position determination from ranging. A detailed description of GNSS signal processing may be found in Chapter 9, while more information on other radio navigation systems is in Chapters 11 and 12.

7.3.1 Architecture

Figure 7.15 illustrates the architecture of receiving-only user equipment for a radio self-positioning system. A receiving antenna converts an electromagnetic signal into an electrical signal so that it may be processed by a radio receiver. A transmitting antenna performs the reverse operation. The gain of an antenna varies with frequency. Therefore, an antenna for radio navigation must be sensitive across the frequency band used by the relevant positioning signals. For proximity positioning, ranging, and pattern matching, a nondirectional antenna gives the best performance under good reception conditions.

A receiver front end processes the signals from the antenna in the analog domain. This is known as *signal conditioning* and comprises band-limiting of the signals to remove out-of-band interference and, usually, downconversion from the radio

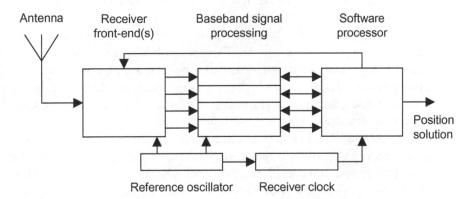

Figure 7.15 User equipment for a radio self-positioning system (receiving only).

frequency (RF) to an intermediate frequency (IF). The IF signals are then sampled by an analog-to-digital converter (ADC). Typically, there is one front end for each frequency band on which signals are simultaneously received. In an FDMA system, the user equipment software will determine on which frequency each front end operates. In low-cost, low-performance user equipment, a single front end may cycle between frequencies. The combined strength of the received signal and in-band noise is determined at the front end.

Each baseband signal processor demodulates one of the signals, outputting the data modulated onto that signal. It may also output measurements from which the signal timing and/or the signal-to-noise ratio may be determined. This stage is usually implemented in digital hardware, but may also be implemented in software.

Every radio receiver requires a reference oscillator to ensure correct frequency selection and signal demodulation. In ranging systems, the oscillator is used to drive a receiver clock, enabling signal timing measurements. The discussion of GNSS receiver oscillators in Section 9.1.2 is largely applicable to radio navigation systems in general.

The software processor performs several functions. It determines which signals to use, operates signal acquisition and tracking (Section 7.3.2), decodes the data modulated onto the signals, and calculates the position solution.

User equipment that receives signals from different radio positioning systems usually has a separate antenna, front end(s), and signal processors for each system, but a common oscillator and software processor.

7.3.2 Signal Timing Measurement

The simplest method of timing a signal in a ranging-based positioning system is to log the time of arrival of a known feature of the signal modulation. This is sometimes called energy detection. The feature could be the rise, fall, or peak of a pulse or it could be the end of a particular symbol sequence. Accuracy is limited by the clock resolution, while RF noise can introduce errors in the determination of the sampling point (see Section 7.4.3) [1]. Therefore, ranges are usually determined from the average of several successive timing measurements. The averaging time is limited by the rate at which the range changes unless its rate of change, the range rate, may be determined from the signal.

Most modern user equipment determines the signal timing by correlating the known features of the incoming signal modulation with an internally-generated replica (or the signal from another receiver). This is sometimes called a matched filter. The correlation process multiplies the two signals together at each sampling point and sums the result over a time interval. This accumulated correlator output is maximized when the replica is exactly aligned with the incoming signal. Figure 7.16 illustrates this. Signal correlation gives a higher resolution than simple feature timing at a given signal-to-noise level and enables a useful timing measurement to be obtained at lower signal-to-noise levels. This is because it effectively times all of the known features of the incoming signal rather than just selected features. The main drawback is that greater processing power is required, particularly for high-bandwidth signals.

Signal correlation comprises two processes: acquisition and tracking. In acquisition, the signal timing is unknown or partially known, so each possible replica signal alignment must be correlated with the incoming signal to determine which

7.3 User Equipment

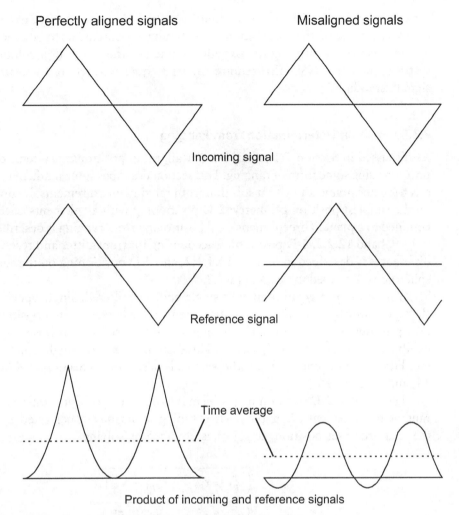

Figure 7.16 Correlation of perfectly-aligned and misaligned signals.

is correct. This may be done in series, in parallel, or in a combination of both. In some systems, the acquisition process must also determine the Doppler shift of the incoming signal (plus the oscillator frequency bias).

In tracking, an accurate prediction of the signal timing is available from previous measurements. Typically, a pair of replica signals that are, respectively, early and late, compared to the predicted signal timing, are correlated with the incoming signal and the results compared to generate a correction to the predicted timing. If the output of the early signal correlation is greater than that of the late, the predicted time of arrival is too late and vice versa. Because acquisition requires many more signal alignments to be tested than tracking does, it requires a greater processing capacity and/or a higher signal-to-noise ratio.

Obtaining absolute ranging information from the signal's carrier alone requires signals transmitted at multiple carrier frequencies to remove the ambiguity (see Section 7.2). However, the change in range over time or the range rate can be determined much more accurately from a single-frequency carrier than from its modulation. This difference in accuracy is proportional to the modulation bandwidth divided by the carrier frequency. Therefore, making use of measurements from carrier tracking enables the noise on the modulation-based timing measurements to be averaged over a longer period, improving the overall accuracy of the navigation solution. Some systems, such as GNSS, also require carrier frequency tracking to ensure correct signal demodulation.

7.3.3 Position Determination from Ranging

As discussed in Section 7.1, most, but not all, radio positioning systems determine position using some form of ranging. This section describes some mathematical methods for determining a position solution from ranging measurements. It concentrates on the simplest position geometry: 2-D positioning with all elements of the system confined to a plane. Three-dimensional positioning from ranging is described in Sections 9.4 and 12.2.3, 2-D positioning accounting for transmitter and receiver height differences is described in Section 11.1.1.2, and 2-D positioning on the surface of a spheroid is described in Section 11.2.2.

A positioning algorithm may be single-epoch or filtered. Single-epoch, or snapshot, positioning uses only current measurements, whereas filtered positioning also uses previous measurement data. A filtered position solution is less noisy but can exhibit a lag in response to dynamics. This section focuses on single-epoch positioning. Filtered positioning using radio signals is described in Sections 9.4.2, Chapter 14, and Chapter 16.

The simplest 2-D positioning problem is to obtain a position from two two-way ranging measurements, $\tilde{r}_{a1}$ and $\tilde{r}_{a2}$, where timing errors may be neglected. From (7.1), the measured user position, $(\tilde{x}_{pa}^p, \tilde{y}_{pa}^p)$, is obtained by solving

$$\begin{aligned} \tilde{r}_{a1} &= \sqrt{\left(x_{p1}^p - \tilde{x}_{pa}^p\right)^2 + \left(y_{p1}^p - \tilde{y}_{pa}^p\right)^2} \\ \tilde{r}_{a2} &= \sqrt{\left(x_{p2}^p - \tilde{x}_{pa}^p\right)^2 + \left(y_{p2}^p - \tilde{y}_{pa}^p\right)^2} \end{aligned}, \quad (7.27)$$

where the transmitter positions, (x_{p1}^p, y_{p1}^p) and (x_{p2}^p, y_{p2}^p), are known, noting that the transmit antenna body frames are numbered 1 and 2. This is typically solved by first generating a predicted user position, $(\hat{x}_{pa}^{p-}, \hat{y}_{pa}^{p-})$, from which predicted ranges, $\hat{r}_{a1}^-$ and $\hat{r}_{a2}^-$, are calculated using

$$\hat{r}_{aj}^- = \sqrt{\left(x_{pj}^p - \hat{x}_{pa}^{p-}\right)^2 + \left(y_{pj}^p - \hat{y}_{pa}^{p-}\right)^2} \quad j \in 1,2. \quad (7.28)$$

The predicted position is typically the previous position solution. Subtracting the predicted ranges from the measured ranges and applying a first-order Taylor expansion about the predicted user position gives

$$\begin{pmatrix} \tilde{r}_{a1} - \hat{r}_{a1}^- \\ \tilde{r}_{a2} - \hat{r}_{a2}^- \end{pmatrix} = \mathbf{H}_R^p \begin{pmatrix} \tilde{x}_{pa}^p - \hat{x}_{pa}^{p-} \\ \tilde{y}_{pa}^p - \hat{y}_{pa}^{p-} \end{pmatrix} + \begin{pmatrix} \delta r_{L1} \\ \delta r_{L2} \end{pmatrix}, \quad (7.29)$$

where δr_{L1} and δr_{L2} are the linearization errors and $\mathbf{H}_R^p$ is the measurement matrix, also known as the geometry matrix, observation matrix, or design matrix. $\mathbf{H}_R^p$ is given by

$$\mathbf{H}_R^p = \begin{pmatrix} \dfrac{\partial r_{a1}}{\partial x_{pa}^p} & \dfrac{\partial r_{a1}}{\partial y_{pa}^p} \\ \dfrac{\partial r_{a2}}{\partial x_{pa}^p} & \dfrac{\partial r_{a2}}{\partial y_{pa}^p} \end{pmatrix}\Bigg|_{(x_{pa}^p,y_{pa}^p)=(\hat{x}_{pa}^{p-},\hat{y}_{pa}^{p-})} = \begin{pmatrix} -\dfrac{x_{p1}^p - \hat{x}_{pa}^{p-}}{\hat{r}_{a1}^-} & -\dfrac{y_{p1}^p - \hat{y}_{pa}^{p-}}{\hat{r}_{a1}^-} \\ -\dfrac{x_{p2}^p - \hat{x}_{pa}^{p-}}{\hat{r}_{a2}^-} & -\dfrac{y_{p2}^p - \hat{y}_{pa}^{p-}}{\hat{r}_{a2}^-} \end{pmatrix}, \quad (7.30)$$

Rearranging (7.29), a user position estimate may be obtained using

$$\begin{pmatrix} \hat{x}_{pa}^{p+} \\ \hat{y}_{pa}^{p+} \end{pmatrix} = \begin{pmatrix} \hat{x}_{pa}^{p-} \\ \hat{y}_{pa}^{p-} \end{pmatrix} + \mathbf{H}_R^{p\,-1}\begin{pmatrix} \tilde{r}_{a1} - \hat{r}_{a1}^- \\ \tilde{r}_{a2} - \hat{r}_{a2}^- \end{pmatrix}, \quad (7.31)$$

where, from Section A.4 of Appendix A on the CD, a 2×2 matrix may be inverted using

$$\mathbf{H}^{-1} = \dfrac{1}{H_{11}H_{22} - H_{12}H_{21}}\begin{pmatrix} H_{22} & -H_{12} \\ -H_{21} & H_{11} \end{pmatrix}. \quad (7.32)$$

From (7.29) and (7.31), the position solution has an error due to the linearization process of

$$\begin{pmatrix} \hat{x}_{pa}^{p+} - \tilde{x}_{pa}^p \\ \hat{y}_{pa}^{p+} - \tilde{y}_{pa}^p \end{pmatrix} = \mathbf{H}_R^{p\,-1}\begin{pmatrix} \delta r_{L1} \\ \delta r_{L2} \end{pmatrix}. \quad (7.33)$$

A more accurate solution may then be obtained by resetting $(\hat{x}_{pa}^{p-}, \hat{y}_{pa}^{p-})$ to $(\hat{x}_{pa}^{p+}, \hat{y}_{pa}^{p+})$ and repeating the preceding process. Iteration should continue until the difference between successive position solutions is within the required precision. Figure 7.17 summarizes the process. Example 7.1 on the CD, which is editable using Microsoft Excel, illustrates this. If the required precision is not achieved within a certain number of iterations, there is a convergence problem and the calculation should be aborted.

As Figure 7.5 shows, (7.27) actually has two solutions, which are reflected about the line joining the two transmitters. The solution obtained by this process will be whichever is closer to the initial predicted user position. An exception is where the predicted solution lies on the line joining the transmitters, in which case $\mathbf{H}_R^p$ will be singular and have no inverse.

When there are three or more range measurements, the ambiguity is removed. However, due to range measurement errors, the lines of position from each range measurement will not intersect at a single point. Figure 7.18 illustrates this. It is therefore necessary to introduce the measurement residual, $\delta r_{aj,\varepsilon}^+$, which is not the same as the measurement error. For the jth signal, the residual is defined as the

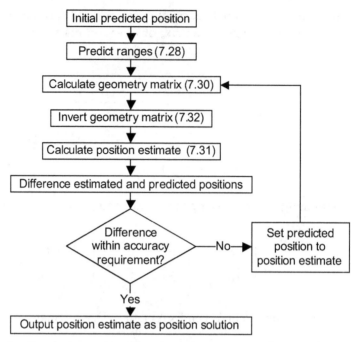

Figure 7.17 Iterative single-epoch position determination process (not overdetermined).

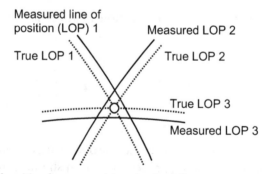

Figure 7.18 Lines of position for an overdetermined solution.

difference between the measured range and that predicted from the position solution, $\hat{r}_{aj}^+$ Beware that some authors use the opposite sign convention. Thus,

$$\delta r_{aj,\varepsilon}^+ = \tilde{r}_{aj} - \hat{r}_{aj}^+ \tag{7.34}$$

and

$$\tilde{r}_{aj} = \sqrt{\left(x_{pj}^p - \hat{x}_{pa}^{p+}\right)^2 + \left(y_{pj}^p - \hat{y}_{pa}^{p+}\right)^2} + \delta r_{aj,\varepsilon}^+, \tag{7.35}$$

where $(\hat{x}_{pa}^{p+}, \hat{y}_{pa}^{p+})$ is the estimated position solution used to predict $\hat{r}_{aj}^+$.

The best position solution is that which is most consistent with the measurements. The iterative position determination algorithm described previously is thus modified to find the position solution that minimizes the sum of the squares of the measurement residuals. This is known as an iterated least-squares algorithm and is derived in Section D.1 of Appendix D on the CD. When the measurement has more components than the solution, the solution is overdetermined. Consequently, the measurement matrix, H, is not square so has no inverse. The pseudo-inverse (see Section A.4 of Appendix A on the CD) is used instead. Thus, in the ILS algorithm, (7.31) is replaced by

$$\begin{pmatrix} \hat{x}_{pa}^{p+} \\ \hat{y}_{pa}^{p+} \end{pmatrix} = \begin{pmatrix} \hat{x}_{pa}^{p-} \\ \hat{y}_{pa}^{p-} \end{pmatrix} + \left(\mathbf{H}_R^{p\mathrm{T}}\mathbf{H}_R^p\right)^{-1}\mathbf{H}_R^{p\mathrm{T}} \begin{pmatrix} \tilde{r}_{a1} - \hat{r}_{a1}^- \\ \tilde{r}_{a2} - \hat{r}_{a2}^- \\ \vdots \\ \tilde{r}_{am} - \hat{r}_{am}^- \end{pmatrix}. \quad (7.36)$$

where m is the number of measurements, the measurement matrix is given by

$$\mathbf{H}_R^p = \begin{pmatrix} -\left(x_{p1}^p - \hat{x}_{pa}^{p-}\right)/\hat{r}_{a1}^- & -\left(y_{p1}^p - \hat{y}_{pa}^{p-}\right)/\hat{r}_{a1}^- \\ -\left(x_{p2}^p - \hat{x}_{pa}^{p-}\right)/\hat{r}_{a2}^- & -\left(y_{p2}^p - \hat{y}_{pa}^{p-}\right)/\hat{r}_{a2}^- \\ \vdots & \vdots \\ -\left(x_{pm}^p - \hat{x}_{pa}^{p-}\right)/\hat{r}_{am}^- & -\left(y_{pm}^p - \hat{y}_{pa}^{p-}\right)/\hat{r}_{am}^- \end{pmatrix}, \quad (7.37)$$

and the predicted measurements are given by (7.28) as before. Example 7.2 on the CD illustrates this.

When passive ranging or TOA positioning is used, the range measurements are replaced by pseudo-range measurements and the receiver clock offset, $\delta\rho_c^a$, expressed here as a range, must be solved as part of the position solution. This is obtained by solving

$$\tilde{\rho}_{a,C}^j = \sqrt{\left(x_{pj}^p - \hat{x}_{pa}^{p+}\right)^2 + \left(y_{pj}^p - \hat{y}_{pa}^{p+}\right)^2} + \delta\hat{\rho}_c^{a+} + \delta\rho_{a,\varepsilon}^{j+} \quad j \in 1,2,\ldots,m, \quad (7.38)$$

where $\delta\rho_{a,\varepsilon}^{j+}$ is the jth measurement residual and the pseudo-range measurements have been corrected for any transmitter clock offsets.

Using an ILS algorithm, the position may be obtained by iterating

$$\begin{pmatrix} \hat{x}_{pa}^{p+} \\ \hat{y}_{pa}^{p+} \\ \delta\hat{\rho}_c^{a+} \end{pmatrix} = \begin{pmatrix} \hat{x}_{pa}^{p-} \\ \hat{y}_{pa}^{p-} \\ \delta\hat{\rho}_c^{a-} \end{pmatrix} + \left(\mathbf{H}_R^{p\mathrm{T}}\mathbf{H}_R^p\right)^{-1}\mathbf{H}_R^{p\mathrm{T}} \begin{pmatrix} \tilde{\rho}_{a,C}^1 - \hat{\rho}_{a,C}^{1-} \\ \tilde{\rho}_{a,C}^2 - \hat{\rho}_{a,C}^{2-} \\ \vdots \\ \tilde{\rho}_{a,C}^m - \hat{\rho}_{a,C}^{m-} \end{pmatrix}, \quad (7.39)$$

where the measurement matrix is

$$\mathbf{H}_R^p = \begin{pmatrix} -\dfrac{x_{p1}^p - \hat{x}_{pa}^{p-}}{\hat{\rho}_{a,C}^{1-} - \delta\hat{\rho}_c^{a-}} & -\dfrac{y_{p1}^p - \hat{y}_{pa}^{p-}}{\hat{\rho}_{a,C}^{1-} - \delta\hat{\rho}_c^{a-}} & 1 \\ -\dfrac{x_{p2}^p - \hat{x}_{pa}^{p-}}{\hat{\rho}_{a,C}^{2-} - \delta\hat{\rho}_c^{a-}} & -\dfrac{y_{p2}^p - \hat{y}_{pa}^{p-}}{\hat{\rho}_{a,C}^{2-} - \delta\hat{\rho}_c^{a-}} & 1 \\ \vdots & \vdots & \vdots \\ -\dfrac{x_{pm}^p - \hat{x}_{pa}^{p-}}{\hat{\rho}_{a,C}^{m-} - \delta\hat{\rho}_c^{a-}} & -\dfrac{y_{pm}^p - \hat{y}_{pa}^{p-}}{\hat{\rho}_{a,C}^{m-} - \delta\hat{\rho}_c^{a-}} & 1 \end{pmatrix}, \qquad (7.40)$$

and the predicted measurements are given by

$$\hat{\rho}_{a,C}^{j-} = \sqrt{\left(x_{pj}^p - \hat{x}_{pa}^{p-}\right)^2 + \left(y_{pj}^p - \hat{y}_{pa}^{p-}\right)^2} + \delta\hat{\rho}_c^{a-} \quad j \in 1,2,\ldots,m. \qquad (7.41)$$

When TDOA positioning with one receiver is used, measurements are of the form, repeating (7.6),

$$\Delta\tilde{\rho}_{a,C}^{st} = \tilde{\rho}_{a,C}^t - \tilde{\rho}_{a,C}^s.$$

Therefore, using ILS, the position solution is obtained by iterating

$$\begin{pmatrix} \hat{x}_{pa}^{p+} \\ \hat{y}_{pa}^{p+} \end{pmatrix} = \begin{pmatrix} \hat{x}_{pa}^{p-} \\ \hat{y}_{pa}^{p-} \end{pmatrix} + \left(\mathbf{H}_R^{pT}\mathbf{H}_R^p\right)^{-1}\mathbf{H}_R^{pT} \begin{pmatrix} \Delta\tilde{\rho}_{a,C}^{s1} - \hat{\rho}_{a,C}^{1-} + \hat{\rho}_{a,C}^{s-} \\ \Delta\tilde{\rho}_{a,C}^{s2} - \hat{\rho}_{a,C}^{2-} + \hat{\rho}_{a,C}^{s-} \\ \vdots \\ \Delta\tilde{\rho}_{a,C}^{sm} - \hat{\rho}_{a,C}^{m-} + \hat{\rho}_{a,C}^{s-} \end{pmatrix}, \qquad (7.42)$$

where the measurement matrix is given by

$$\mathbf{H}_R^p = \begin{pmatrix} -\dfrac{x_{p1}^p - \hat{x}_{pa}^{p-}}{\hat{\rho}_{a,C}^{1-}} + \dfrac{x_{ps}^p - \hat{x}_{pa}^{p-}}{\hat{\rho}_{a,C}^{s-}} & -\dfrac{y_{p1}^p - \hat{y}_{pa}^{p-}}{\hat{\rho}_{a,C}^{1-}} + \dfrac{y_{ps}^p - \hat{y}_{pa}^{p-}}{\hat{\rho}_{a,C}^{s-}} \\ -\dfrac{x_{p2}^p - \hat{x}_{pa}^{p-}}{\hat{\rho}_{a,C}^{2-}} + \dfrac{x_{ps}^p - \hat{x}_{pa}^{p-}}{\hat{\rho}_{a,C}^{s-}} & -\dfrac{y_{p2}^p - \hat{y}_{pa}^{p-}}{\hat{\rho}_{a,C}^{2-}} + \dfrac{y_{ps}^p - \hat{y}_{pa}^{p-}}{\hat{\rho}_{a,C}^{s-}} \\ \vdots & \vdots \\ -\dfrac{x_{pm}^p - \hat{x}_{pa}^{p-}}{\hat{\rho}_{a,C}^{m-}} + \dfrac{x_{ps}^p - \hat{x}_{pa}^{p-}}{\hat{\rho}_{a,C}^{s-}} & -\dfrac{y_{pm}^p - \hat{y}_{pa}^{p-}}{\hat{\rho}_{a,C}^{m-}} + \dfrac{y_{ps}^p - \hat{y}_{pa}^{p-}}{\hat{\rho}_{a,C}^{s-}} \end{pmatrix}, \qquad (7.43)$$

and the predicted measurements given by (7.28).

Three-dimensional position and velocity solutions obtained using a weighted ILS algorithm are described in Section 9.4.1, while angular positioning using ILS is described in Section F.1 of Appendix F on the CD.

7.4 Propagation, Error Sources, and Positioning Accuracy

Radio positioning accuracy is affected by a wide range of phenomena: ionosphere, troposphere, and surface propagation effects; attenuation, reflection, multipath, and diffraction; resolution, noise, and tracking errors; and transmitter location and timing errors. Each of these is discussed in turn, followed by a discussion of how signal geometry determines the impact of measurement errors on the position solution. Self-positioning is assumed throughout this section.

7.4.1 Ionosphere, Troposphere, and Surface Propagation Effects

The ionosphere extends from about 50 to 1,000 km above the Earth's surface and comprises gases that may be ionized to a plasma of ions and free electrons. As the ionization is caused by solar radiation, the characteristics of the ionosphere vary with time of day, latitude, and season, and are also affected by solar storm activity. The effects of the ionosphere on signal propagation are frequency dependent.

LF and MF signals are absorbed by the lowest layer of the ionosphere, known as the D layer, during the day. However, this layer reduces at night with the result that lower elevation signals are reflected by the E and F layers of the ionosphere. These reflected signals are known as sky waves and are a source of co-channel interference. Consequently, the useful coverage area of an LF or MF transmitter is generally smaller at night.

At frequencies above about 30 MHz, ionospheric reflection is negligible. However, signals passing through the ionosphere are affected by frequency-dependent refraction, causing a modulation delay and phase advance. Ionospheric effects on GNSS are described in Section 9.3.2.

The troposphere is the lowest region of the Earth's atmosphere and extends to about 12 km above the surface. Refraction by the tropospheric gases causes a frequency-independent delay in both modulation and phase, compared to a signal propagating through free space. At the Earth's surface, propagation is slowed by an average factor of 1.000292. The refractive index decreases with height because of the decreasing troposphere density. Consequently, propagation paths are curved towards the Earth. This has a number of effects. The first effect is that radio signals will propagate beyond the optical horizon. Over flat terrain, the radius of the radio horizon (i.e., the maximum range of the signal), approximated from Pythagoras' theorem, is

$$r_H \approx \sqrt{\tfrac{8}{3} R_0 (h_t - h_T)}, \tag{7.44}$$

where h_t is the transmit antenna height, h_T is the terrain height (with respect to the same datum), and the Earth radius, R_0, is multiplied by 4/3 to account for the curved propagation. The distance over which reception is possible, r_R, also depends on the receive antenna height, h_a:

$$r_R \approx \sqrt{\tfrac{8}{3} R_0 (h_t - h_T)} + \sqrt{\tfrac{8}{3} R_0 (h_a - h_T)}. \tag{7.45}$$

The second effect is that, for terrestrial systems, the variation in troposphere-induced signal propagation delay with the distance between transmitter and receiver will have a nonlinear component, complicating range determination. The third effect is that the elevation angle of the received signal will diverge from that of the transmitter-receiver line of sight with increasing distance, affecting the determination of user height from signal elevation (see Section 7.1.5).

The refractive index of the troposphere varies with the weather, particularly the water vapor content. This can result in propagation delays varying with a standard deviation of about 10%. Troposphere effects on GNSS are also described in Section 9.3.2.

Three further tropospheric effects are worth noting. First, spatial variation in the refractive index can result in a signal reaching the receiver by multiple paths, causing multipath interference as discussed in Section 7.4.2. For ground-based receivers and transmitters, this limits the maximum reliable range to about 80 km for the 300–1,000-MHz band, longer for lower frequencies and shorter for higher frequencies [10]. Second, at VHF and UHF frequencies, ducting can occasionally occur, causing signals to travel much further than normal, causing cochannel interference. Finally, at frequencies above about 10 GHz, rain can significantly attenuate the signals.

At the Earth's surface, currents are induced that slow signal propagation. This curves the signal path, causing it to follow the surface and propagate beyond the radio horizon. Signals propagated in this way are known as ground waves and are subject to attenuation proportional to the frequency and the distance traveled. Consequently, ground wave propagation is only suited to LF and MF signals. The attenuation also depends on the terrain and is much less over sea water than over land [10]. Ground waves cannot be used for height determination.

7.4.2 Attenuation, Reflection, Multipath, and Diffraction

Radio signals are attenuated by objects within the path between transmitter and receiver. The degree of attenuation depends on the object's material, structure, and thickness, together with the frequency of the signal. If the attenuation is sufficient to reduce the signal strength below that required for demodulation, that signal is blocked or obstructed.

LF and MF signals are difficult to block as they will diffract around objects smaller than their wavelength, though this does affect angular positioning and carrier-phase-based ranging. Also, ground-wave propagation enables the signal to pass over most terrain obstructions. However, steel-framed buildings, steel-reinforced concrete, and metal vehicle bodies can act as Faraday cages, blocking reception completely.

VHF and UHF (including L-band) signals can be attenuated by terrain, buildings, vehicles, foliage, and people. Metal walls block radio signals completely. Windows typically attenuate signals much less than walls, although metalized windows attenuate more than nonmetalized. So, within a building, signals might be receivable from one side only. Deep inside a large building or within a basement, there may be no useful reception at all.

When RF energy encounters any surface, some of it is absorbed or transmitted through and some of it is reflected. When the surface is smooth compared to the wavelength, *specular reflection* occurs; otherwise, *scattering*, also known as diffuse reflection, occurs [12]. At optical frequencies, liquids, metals, and glass often exhibit specular reflection, while most other surfaces scatter light.

With specular reflection, energy incident to the surface from a particular direction is reflected in a particular direction. The components of a vector describing the signal's direction of travel that are parallel to the reflected surface are the same for the incident and reflected signals, whereas the component perpendicular to the surface has the opposite sign. Figure 7.19 illustrates this. Components of the signal polarized within the plane containing the incident and reflected paths undergo a phase reversal (or 180° phase shift) for angles of incidence less than Brewster's angle, while components polarized perpendicular to this plane are unaffected. This is also shown in Figure 7.19. A large proportion of the incident signal energy can undergo specular reflection.

Scattering comprises a very large number of small reflections. Consequently, RF energy is emitted from the surface in many directions, but relatively little energy is emitted in a particular direction. Scattered energy received at a particular point will have been reflected at multiple points on the surface so the phase shift of the resultant is effectively random.

A reflected signal will travel a longer path from transmitter to receiver than its line-of-sight counterpart, its angle of arrival will be different, and its signal strength will be reduced. Consequently, all forms of radio positioning are affected. When the reflection is specular and the location of the reflecting surface is known, corrections can be applied in principle, although this is not easy in practice.

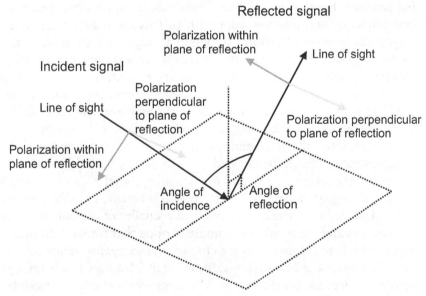

Figure 7.19 Specular reflection of a signal.

Often, user equipment will receive a reflected signal in addition to the line-of-sight signal and/or multiple reflections of the same signal will be received. In these cases, multipath interference will occur. Interference can be constructive or destructive, depending on whether the signals are in phase or out of phase. This is sometimes known as fading as the RSS varies with the relative phase of the interference. Thus, position determination from RSS is directly affected by multipath. Multipath interference also impacts direction finding as the RSS minima are no longer perpendicular to the direction of the directly-received signal. However, with suitable processing, signal separation should be possible within the resolution limits of the antenna system.

The impact of multipath on range measurement is complex, depending on the relative amplitudes. Ranges may be underestimated as well as overestimated. In general, multipath components may be separated by receiver processing where the path difference exceeds c/B_{PC}, where B_{PC} is the double-sided modulation bandwidth. Multipath may also be mitigated using advanced antenna designs, by mapping the reflections, and within the navigation processor. The effects of multipath on GNSS and their mitigation are described in Sections 9.3.4 and 10.4, respectively.

Diffraction occurs at the edges of an obstacle blocking the signal, bending its path. This results in attenuated reception of the signal in areas where the direct line of sight is blocked. Diffraction effects at VHF and UHF frequencies are normally highly local. However, diffraction can affect the propagation of LF and MF signals over wide areas, with interference between signals received via different diffracted paths being common.

7.4.3 Resolution, Noise, and Tracking Errors

As described in Section 7.3.2, range is determined by measuring the time of arrival of one or more known features of a signal transmitted at a known time. When only a single feature is measured, the clock resolution can have a major impact on ranging accuracy. For instance, if the reference oscillator produces pulses at 3 MHz, the one-way ranging resolution will be 100m. This resolving error may be reduced by timing multiple signal features to obtain a range measurement. If the features are evenly spaced, the resolving error is inversely proportional to the number of features, subject to a minimum value. The best timing resolution obtainable depends on the relationship between the signal feature and reference oscillator repetition intervals, which may vary due to the Doppler shift and oscillator frequency drift, respectively. Consequently, the oscillator frequency must be chosen carefully. Simple relationships, such as one interval being a multiple of the other or the two intervals having a large common factor, should be avoided.

When the receiver correlates the incoming signal with an internally-generated replica, timing of multiple signal features is inherent. The effective number of features measured over a given interval is the smaller of the number of times the signal changes and the number of samples, noting that timing information cannot be extracted where a signal does not change over successive samples.

The time taken to measure sufficient signal features to obtain a given ranging resolution depends on the rate at which suitable features are modulated onto the signal. For a pulsed signal, this will be the pulse repetition rate. For a continuous signal, it will typically be a function of the modulation bandwidth and the proportion of the duty cycle used for ranging measurement.

RF noise and thermal noise within the receiver distort the incoming signal, causing errors in the timing of signal features. Figure 7.20 shows an example of this. For a given signal-to-noise ratio, noise-induced timing errors are smaller for a higher-bandwidth signal because the signal changes more rapidly. Noise-induced ranging errors are also reduced by timing multiple signal features. For band-limited white noise, the error standard deviation is inversely proportional to the square root of the number of measurements (see Section B.4.2 of Appendix B on the CD). In most practical navigation systems, noise has a much greater impact on positioning accuracy than clock resolution.

Time averaging minimizes the effects of RF noise and clock resolution on ranging accuracy. The problem is that a range or pseudo-range can change over the averaging interval due to receiver motion, transmitter motion, and/or clock drift. Consequently, tracking filters are normally used instead of simple averaging. A first-order range tracking filter will, however, exhibit a lag in responding to changes in the range. This can be mitigated using a second-order filter, which also tracks the range rate. A second-order filter will exhibit a lag in responding to changes in range rate, which can be mitigated using a third-order filter and so forth. However, the useful order of a tracking filter is limited by noise. Consequently, there will always be tracking errors resulting from the lag in responding to dynamics.

The lower the bandwidth (and thus the longer the time constant) of the tracking filter, the lower the noise-induced tracking errors and the higher the dynamics-induced tracking errors will be. This is discussed in more detail for GNSS in Section 9.3.3. Consequently, the tracking filter tuning parameters that minimize the overall range-tracking error will depend on both the signal-to-noise ratio and the dynamics. When these vary, an adaptive tracking filter, which varies its bandwidth according to the conditions, may be used.

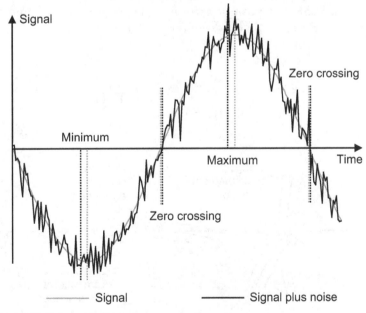

Figure 7.20 Noise-induced timing errors.

7.4.4 Transmitter Location and Timing Errors

How errors in specifying the transmitter locations affect the user position error depends on the positioning method. For proximity, the transmitter position error is added directly to the other sources of position error. For ranging, the component of the transmitter position error along the line of sight from the transmitter to the user dominates. Errors perpendicular to the line of sight have negligible impact where they are much smaller than the distance from the transmitter to the user's receiver. For angular positioning, it is the components of the transmitter position error perpendicular to the line of sight that have the most impact. For both ranging and angular positioning, the impact of an error on the user position solution also depends on the signal geometry as discussed in Section 7.4.5.

Transmitter location errors do not affect positioning by pattern matching. This is affected by database errors instead.

Transmitter timing errors only directly affect positioning by ranging, where they have the same effect as an error in the transmitter position along the line of sight. When a transmitter is moving, timing errors can also cause the user to compute an erroneous transmitter position.

7.4.5 Effect of Signal Geometry

The accuracy of a position solution obtained from ranging measurements depends not only on the accuracy of the measurements, but also on the signal geometry. Figure 7.21 illustrates this for the simplest case of a 2-D position solution from two two-way ranging measurements. The arcs show the mean and error bounds for each ranging measurement, while the shaded areas show the uncertainty bounds for the position solution and the arrows show the line-of-sight vectors from the user to the transmitters. The overall position error for a given ranging accuracy is minimized where the line-of-sight vectors are perpendicular.

From (7.36), the position error in a planar coordinate frame may be expressed in terms of the errors in two-way range measurements using

$$\begin{pmatrix} \delta x_{pa}^p \\ \delta y_{pa}^p \end{pmatrix} = \begin{pmatrix} \hat{x}_{pa}^p - x_{pa}^p \\ \hat{y}_{pa}^p - y_{pa}^p \end{pmatrix} = \left(\mathbf{H}_R^{p\mathrm{T}} \mathbf{H}_R^p \right)^{-1} \mathbf{H}_R^{p\mathrm{T}} \begin{pmatrix} \delta r_{a1} \\ \delta r_{a2} \\ \vdots \\ \delta r_{am} \end{pmatrix}, \qquad (7.46)$$

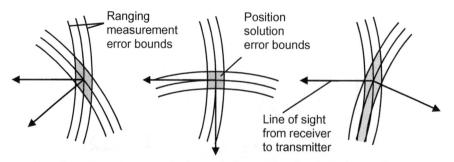

Figure 7.21 Effect of signal geometry on the position accuracy from 2-D ranging. (*After:* [21].)

7.4 Propagation, Error Sources, and Positioning Accuracy

where $\delta r_{aj} = \tilde{r}_{aj} - r_{aj}$ and the measurement or geometry matrix, $\mathbf{H}_R^p$, is as given by (7.37). Squaring and taking expectations, the error covariance matrix of the position solution (see Section B.2.1 of Appendix B on the CD) is

$$\mathbf{P} = \begin{pmatrix} \sigma_x^2 & P_{x,y} \\ P_{y,x} & \sigma_y^2 \end{pmatrix} = \left(\mathbf{H}_R^{pT}\mathbf{H}_R^p\right)^{-1}\mathbf{H}_R^{pT} \begin{pmatrix} \sigma_{r1}^2 & 0 & \cdots & 0 \\ 0 & \sigma_{r2}^2 & \cdots & 0 \\ \vdots & \vdots & \ddots & \vdots \\ 0 & 0 & \cdots & \sigma_{rm}^2 \end{pmatrix} \mathbf{H}_R^p \left(\mathbf{H}_R^{pT}\mathbf{H}_R^p\right)^{-1}, \quad (7.47)$$

where σ_x^2 and σ_y^2 are, respectively, the variances of the x- and y-axis position errors, $P_{x,y} = P_{y,x}$ is their covariance, σ_{rj}^2 is the variance of the jth range measurement error, and the errors on each range measurement are assumed to be independent. When all measurement errors have the same variance, σ_r^2, this simplifies to

$$\mathbf{P} = \left(\mathbf{H}_R^{pT}\mathbf{H}_R^p\right)^{-1}\sigma_r^2. \quad (7.48)$$

Figures 7.22 to 7.25 show the line-of-sight vectors and corresponding position error ellipses obtained with different geometries of two-way ranging measurements with equal error standard deviation. An error ellipse links the error standard deviations in each direction. A general rule is that the position information along a given axis obtainable from a given ranging signal is maximized when the angle between that axis and the signal line of sight is minimized.

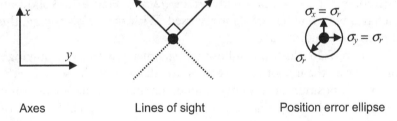

Figure 7.22 Two two-way ranging measurements with optimal geometry in two dimensions.

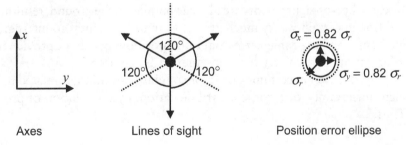

Figure 7.23 Three two-way ranging measurements with optimal geometry in two dimensions.

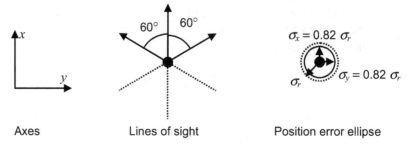

Figure 7.24 Three two-way ranging measurements with one line of sight reversed from the optimal geometry in two dimensions.

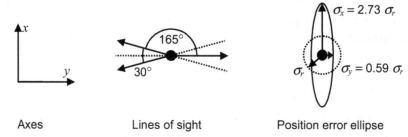

Figure 7.25 Three two-way ranging measurements with suboptimal geometry in two dimensions.

In Figure 7.22, there are two measurements with orthogonal lines of sight. This is the optimal geometry and leads to a circular position error distribution with radius equal to the range measurement error standard deviation. Figure 7.23 illustrates the optimum geometry with three ranging measurements. This leads to a circular position error distribution with radius $\sqrt{2/3}\sigma_r$. Figure 7.24 shows the Figure 7.23 scenario with one line of sight reversed. With two-way ranging, this has no effect on the position accuracy.

Figure 7.25 illustrates suboptimal geometry with three ranging measurements. In this case, the line-of-sight vectors are all close to the y axis with the result that the x-axis position uncertainty is much greater than its y-axis counterpart. This is representative of an urban canyon, where buildings on either side of a street block many of the radio signals.

A more extreme version of the geometry shown in Figure 7.25 can be used to represent a vertical section of positioning using terrestrial radio transmitters. In this case, the line-of-sight vectors are all near-parallel to the ground, resulting in a vertical position uncertainty very much greater than its horizontal counterpart. This is why medium- and long-range terrestrial radio positioning cannot provide useful vertical information, particularly for land- and sea-based users.

The position uncertainty, or error standard deviation, is related to the measurement uncertainty, or error standard deviation, by the dilution of precision (DOP). Thus,

$$\sigma_x = D_x \sigma_r, \quad \sigma_y = D_y \sigma_r, \tag{7.49}$$

7.4 Propagation, Error Sources, and Positioning Accuracy

where D_x and D_y are, respectively, the x-axis and y-axis DOPs. The DOPs are defined in terms of the measurement matrix by the cofactor matrix. This is

$$\Pi^p = \begin{pmatrix} D_x^2 & \cdot \\ \cdot & D_y^2 \end{pmatrix} = \left(\mathbf{H}_R^{p\,\mathrm{T}} \mathbf{H}_R^p \right)^{-1}. \tag{7.50}$$

When 2-D positioning is performed in the xy plane of a local navigation (n) frame or a local-tangent-plane (l) frame aligned with the local navigation frame at the user, the north DOP, D_N, and east DOP, D_E, are given by

$$\begin{pmatrix} D_N^2 & \cdot \\ \cdot & D_E^2 \end{pmatrix} = \left(\mathbf{H}_R^{nC\,\mathrm{T}} \mathbf{H}_R^{nC} \right)^{-1} = \left(\mathbf{H}_R^{l\,\mathrm{T}} \mathbf{H}_R^l \right)^{-1}. \tag{7.51}$$

As well as using (7.37) (with l substituted for p), the measurement matrix, $\mathbf{H}_R^{nC}$ or $\mathbf{H}_R^l$, may be calculated using the azimuths [given by (7.14)]. Thus,

$$\mathbf{H}_R^{nC} = \mathbf{H}_R^l = \begin{pmatrix} -\cos\psi_{nu}^{a1} & -\sin\psi_{nu}^{a1} \\ -\cos\psi_{nu}^{a2} & -\sin\psi_{nu}^{a2} \\ \vdots & \vdots \\ -\cos\psi_{nu}^{am} & -\sin\psi_{nu}^{am} \end{pmatrix}. \tag{7.52}$$

The horizontal dilution of precision (HDOP) is defined as

$$D_H = \sqrt{D_N^2 + D_E^2}. \tag{7.53}$$

For 2-D positioning with one-way ranging measurements, the position error and residual receiver clock error, $\delta\delta\rho_c^a$, are expressed in terms of the pseudo-range measurement errors using

$$\begin{pmatrix} \delta x_{pa}^p \\ \delta y_{pa}^p \\ \delta\delta\rho_c^a \end{pmatrix} = \left(\mathbf{H}_R^{p\,\mathrm{T}} \mathbf{H}_R^p \right)^{-1} \mathbf{H}_R^{p\,\mathrm{T}} \begin{pmatrix} \delta\rho_{a,C}^1 \\ \delta\rho_{a,C}^2 \\ \vdots \\ \delta\rho_{a,C}^m \end{pmatrix}, \tag{7.54}$$

where $\mathbf{H}_R^p$ is given by (7.40) and $\delta\rho_{a,C}^j = \tilde{\rho}_{a,C}^j - \rho_{a,C}^j$.

When all pseudo-range measurement errors have the same variance, σ_p^2, the error covariance matrix of the position and time solution is

$$\mathbf{P} = \begin{pmatrix} \sigma_x^2 & P_{x,y} & P_{x,c} \\ P_{y,x} & \sigma_y^2 & P_{y,c} \\ P_{c,x} & P_{c,y} & \sigma_c^2 \end{pmatrix} = \left(\mathbf{H}_R^{p\,\mathrm{T}} \mathbf{H}_R^p \right)^{-1} \sigma_p^2, \tag{7.55}$$

where σ_c^2 is the variance of the residual clock error (expressed as a range) and $P_{x,c} = P_{c,x}$ and $P_{y,c} = P_{c,y}$ are covariances. Note that this can be used to compute the accuracy of TDOA positioning as well as passive ranging.

With one-way ranging, the DOP is defined as

$$\begin{pmatrix} D_x^2 & \cdot & \cdot \\ \cdot & D_y^2 & \cdot \\ \cdot & \cdot & D_T^2 \end{pmatrix} = \left(\mathbf{H}_R^{p\,\mathrm{T}} \mathbf{H}_R^p \right)^{-1}. \quad (7.56)$$

or

$$\begin{pmatrix} D_N^2 & \cdot & \cdot \\ \cdot & D_E^2 & \cdot \\ \cdot & \cdot & D_T^2 \end{pmatrix} = \left(\mathbf{H}_R^{nC\,\mathrm{T}} \mathbf{H}_R^{nC} \right)^{-1} = \left(\mathbf{H}_R^{l\,\mathrm{T}} \mathbf{H}_R^{l} \right)^{-1}, \quad (7.57)$$

where D_T is the time dilution of precision (TDOP) and the measurement matrix may be calculated in terms of the azimuths using

$$\mathbf{H}_R^{nC} = \mathbf{H}_R^l = \begin{pmatrix} -\cos\psi_{nu}^{a1} & -\sin\psi_{nu}^{a1} & 1 \\ -\cos\psi_{nu}^{a2} & -\sin\psi_{nu}^{a2} & 1 \\ \vdots & \vdots & \vdots \\ -\cos\psi_{nu}^{am} & -\sin\psi_{nu}^{am} & 1 \end{pmatrix}. \quad (7.58)$$

Figures 7.26 and 7.27 show the position error ellipses obtained using three one-way ranging measurements with the same geometry as the two-way ranging examples shown in Figures 7.23 and 7.24, respectively. In Figure 7.26, the lines of sight are equally spaced. This is the optimum geometry and gives the same dilution of precision as the same geometry for two-way ranging (Figure 7.23).

If the direction of one of the signals is reversed, as shown in Figure 7.27, the accuracy in that direction is degraded (by a factor of 3 in this example). This is because signals in opposing directions are needed to fully decorrelate the receiver clock offset and position solutions.

To illustrate this, consider one-dimensional positioning using two passive ranging measurements. The signals may come from either the same or opposing directions. A change in receiver clock offset will always have the same impact on both pseudo-range measurements. When the signals come from the same direction, a change in user position will also have the same impact on both measurements, whereas, when the signals come from opposing directions, a change in position will result in opposing changes to the two pseudo-ranges. Consequently, separate position and clock offset solutions can only be obtained where the signals are in opposing directions.

Returning to 2-D positioning, accuracy with the minimum three signals is significantly degraded when two of the lines of sight are close together and severely degraded where all three lines of sight are close. For example, with three lines of

7.4 Propagation, Error Sources, and Positioning Accuracy

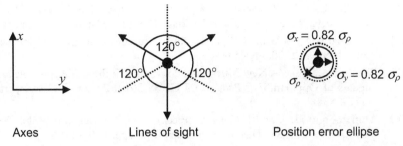

Figure 7.26 Three one-way ranging measurements with optimal geometry in two dimensions.

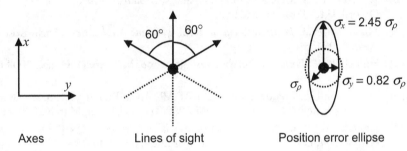

Figure 7.27 Three one-way ranging measurements with one line of sight reversed from the optimal geometry in two dimensions.

sight spanning 10° in azimuth, the position accuracy is 400 times poorer than with optimal geometry.

The effect of signal geometry on passive ranging in three dimensions is described in Section 9.4.3.

Signal geometry also affects angular positioning. The best accuracy is obtained with the same geometries as for two-way ranging. However, DOP is not a useful concept as the accuracy of positioning information obtained from each signal will be different. This is because, for a given attitude measurement, the corresponding position error is proportional to the distance between transmitter and receiver.

Problems and exercises for this chapter are on the accompanying CD.

References

[1] Bensky, A., *Wireless Positioning Technologies and Applications*, Norwood, MA: Artech House, 2008.

[2] Fuller, R., "Combining GNSS with RF Systems," in *GNSS Applications and Methods*, S. Gleason and D. Gebre-Egziabher, (eds.), Norwood, MA: Artech House, 2009, pp. 211–244.

[3] Grejner-Brzezinska, D. A., et al., "Challenged Positions: Dynamic Sensor Network, Distributed GPS Aperture, and Inter-Nodal Ranging Signals," *GPS World*, September 2010, pp. 35–42, 56.

[4] Figueiras, J., and S. Frattasi, "Cooperative Mobile Positioning," in *Mobile Positioning and Tracking from Conventional to Cooperative Techniques*, J. Figueiras and S. Frattasi, (eds.), New York: Wiley, 2010, pp. 213–250.

[5]   Garello, R., et al., "Peer-to-Peer Cooperative Positioning Part II: Hybrid Devices with GNSS & Terrestrial Ranging Capability," *Inside GNSS*, July/August 2012, pp. 56–64.

[6]   Kitching, T. D., "GPS and Cellular Radio Measurement Integration," *Journal of Navigation*, Vol. 53, No. 3, 2000, pp. 451–463.

[7]   Webb, T. A., et al., "A New Differential Positioning Technique Applicable to Generic FDMA Signals of Opportunity," *Proc. ION GNSS 2011*, Portland, OR, September 2011, pp. 3527–3538.

[8]   Duffett-Smith, P. J. and P. Hansen, "Precise Time Transfer in a Mobile Radio Terminal," *Proc. ION NTM*, San Diego, CA, January 2005, pp. 1101–1106.

[9]   Sahinoglu, Z., S. Gezici, and I. Guvenc, *Ultra-Wideband Positioning Systems: Theoretical Limits, Ranging Algorithms, and Protocols*, New York: Cambridge University Press, 2008.

[10] Wharton, W., S. Metcalfe, and G. C. Platts, *Broadcast Transmission Engineering Practice*, Oxford, U.K.: Focal Press, 1992.

[11] Munoz, D., et al., *Position Location Techniques and Applications*, Burlington, MA: Academic Press, 2009.

[12] Forssell, B. *Radionavigation Systems*, Norwood, MA: Artech House, 2008 (first published 1991).

[13] Bahl, P., and V. N. Padmanabhan, "RADAR: An In-Building RF-Based User Location and Tracking System," *Proc. INFOCOM 2000*, Tel Aviv, Israel, March 2000, pp. 775–784.

[14] Eissfeller, B., et al., "Indoor Positioning Using Wireless LAN Radio Signals," *Proc. ION GNSS 2004*, Long Beach, CA, September 2004, pp. 1936–1947.

[15] Hatami, A., and K. Pahlavan, "A Comparative Performance Evaluation of RSS-Based Positioning Algorithms Used in WLAN Networks," *Proc. IEEE Wireless Communications and Networking Conference*, March 2005, pp. 2331–2337.

[16] Machaj, J., P. Brida, and R. Piché, "Rank Based Fingerprinting Algorithm for Indoor Positioning," *Proc. Indoor Positioning and Indoor Navigation*, Guimarães, Portugal, September 2011.

[17] Bruno, L., and P. Robertson, "WiSLAM: Improving FootSLAM with WiFi," *Proc. Indoor Positioning and Indoor Navigation*, Guimarães, Portugal, September 2011.

[18] Faragher, R. M., C. Sarno, and M. Newman, "Opportunistic Radio SLAM for Indoor Navigation Using Smartphone Sensors," *Proc. IEEE/ION PLANS*, Myrtle Beach, SC, April 2012, pp. 120–128.

[19] Chey, J., et al., "Indoor Positioning Using Wireless LAN Signal Propagation Model and Reference Points," *Proc. ION NTM*, San Diego, CA, January 2005, pp. 1107–1112.

[20] Van Diggelen, F., *A-GPS: Assisted GPS, GNSS, and SBAS*, Norwood, MA: Artech House, 2009.

[21] Misra, P., and P. Enge, *Global Positioning System Signals, Measurements, and Performance*, 2nd ed., Lincoln, MA: Ganga-Jamuna Press, 2006.

CHAPTER 8
GNSS: Fundamentals, Signals, and Satellites

Global navigation satellite systems is the collective term for those navigation systems that provide the user with a 3-D positioning solution by passive ranging using radio signals transmitted by orbiting satellites. GNSS is thus a self-positioning system; the position solution is calculated by the user equipment, which does not transmit any signals for positioning purposes.

A number of systems aim to provide global coverage. The most well-known is the Navigation by Satellite Ranging and Timing (NAVSTAR) Global Positioning System, owned and operated by the United States government and usually known simply as GPS. The Russian GLONASS is also fully operational. At the time of this writing, the European Galileo system was being deployed, while a global version of the Chinese Beidou system was under development. In addition, a number of regional satellite navigation systems enhance and complement the global systems.

Some authors use the term GPS to describe satellite navigation in general, while the term GNSS is sometimes reserved for positioning using signals from more than one satellite navigation system. Here, the term GPS is reserved explicitly for the NAVSTAR system, while the term GNSS is used to describe features common to all of the systems. Similarly, the terms GLONASS, Galileo, Beidou, and so forth are used to describe features specific to those systems.

This chapter provides an introduction to satellite navigation and describes the satellite signals and orbits. Section 8.1 describes the basic principles of GNSS, including system architecture, signal properties, ranging, positioning method, and error sources. Section 8.2 summarizes the main features of the different GNSS systems. Section 8.3 describes the signals and Section 8.4 discusses the navigation data messages. Finally, Section 8.5 describes the satellite orbits and geometry, including determination of satellite position, velocity, range, range rate, line of sight, azimuth, and elevation.

A detailed description of GNSS user equipment processing is provided in Chapter 9. This follows the signal path from the antenna and receiver hardware, through signal acquisition and tracking, to the generation of the navigation solution and includes a description of the error sources. Chapter 10 describes how basic GNSS technology may be enhanced to provide greater accuracy and improved robustness in difficult environments. Additional information on a number of GNSS topics is provided in Appendix G on the CD. A basic GNSS simulation using MATLAB is also included on the CD.

8.1 Fundamentals of Satellite Navigation

Before presenting the details of the various satellite navigation systems and services, the fundamental concepts must be introduced. First, the architecture of GNSS, in terms of the space, control, and user segments and their functions, is described. Then the structure of the GNSS signals and how this is used to obtain ranging measurements is described. Finally, the determination of the user position and velocity from ranging measurements is explained and the error sources and performance limitations are summarized. This follows the generic description of radio positioning principles in Chapter 7, noting that most of the examples presented in Chapter 7 are for 2-D positioning, whereas GNSS provides 3-D positioning.

8.1.1 GNSS Architecture

Figure 8.1 shows the architecture of a satellite navigation system, which consists of three components: the space segment, the control or ground segment, and the user segment, which, in turn, comprises multiple pieces of user equipment [1–3]. Each GNSS has its own independent space and control segments, whereas user equipment may use signals from one, two, or multiple GNSS.

The space segment comprises the satellites, collectively known as a *constellation*, which broadcast signals to both the control segment and the users. Some authors use the term space vehicle (SV) instead of satellite. A typical GNSS satellite has a mass of around 1,000 kg and is about 5m across, including solar panels. Each fully operational constellation comprises at least 24 satellites. By 2020, there could be more than 100 GNSS satellites in orbit.

GPS, GLONASS, Galileo, and Beidou satellites are distributed among a number of medium Earth orbits (MEOs). GPS satellites orbit at a radius of 26,580 km, perform two orbits per sidereal day (see Section 2.4.6), and move at about 3,800 m s^{-1}. To provide suitable signal geometry for 3-D positioning (see Section 9.4.3), the

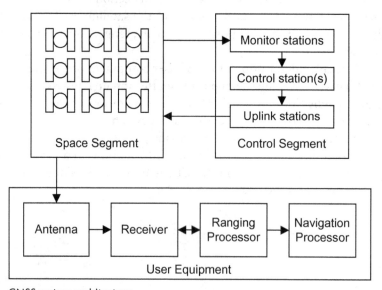

Figure 8.1 GNSS system architecture.

8.1 Fundamentals of Satellite Navigation

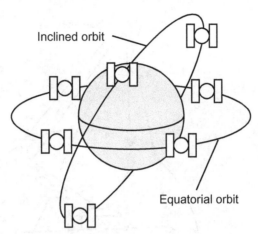

Figure 8.2 Equatorial and inclined orbits.

satellites in each constellation must be distributed across several nonparallel orbital planes. Therefore, in contrast to the equatorial orbits of geostationary satellites, GNSS orbital planes are inclined with respect to the equator (at 55° for GPS). This also provides better coverage in polar regions. Figure 8.2 illustrates this. With a clear line of sight, between 5 and 14 transmitting GPS satellites are visible at most times. More satellites are visible in equatorial and polar regions than at mid-latitudes [4]. The orbits of all the satellite constellations are described in Section 8.5, while more information on the satellite hardware may be found in Section G.1 of Appendix G on the CD.

GNSS satellites broadcast multiple signals on several frequencies. These can incorporate both ranging codes and navigation data messages. The ranging codes enable the user equipment to determine the time at which the received signals were transmitted, while a data message includes timing parameters and information about the satellite orbits, enabling the satellite positions to be determined. A number of atomic clocks aboard each satellite maintain a stable time reference. The signals and navigation data are described in Sections 8.3 and 8.4, respectively.

The control segment, or ground segment, consists of a network of monitor stations, one or more control stations, and a number of uplink stations, as shown in Figure 8.3. The GPS control segment comprises 16 monitor stations, 12 uplink stations, and two control stations. The monitor stations obtain ranging measurements from the satellites and send these to the control station(s). The monitor stations are at precisely surveyed locations and have synchronized clocks, enabling their ranging measurements to be used to determine the satellite orbits and calibrate the satellite clocks. Radar and laser tracking measurements may also be used.

The control stations calculate the navigation data message for each satellite and determine whether any maneuvers must be performed. This information is then transmitted to the space segment by the uplink stations. Most satellite maneuvers are small infrequent corrections, known as station keeping, which are used to maintain the satellites in their correct orbits. However, major relocations are performed in the event of satellite failure, with the failed satellite moved to a different orbit and a new satellite moved to take its place. Satellites are not moved from one orbital

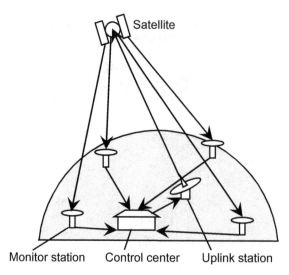

Figure 8.3 Control segment operation.

plane to another. Details of each system's control segment may be found in Section G.1 of Appendix G on the CD.

Nearly all GNSS user equipment receives either GPS signals alone or GPS together with one or more of the other systems. GNSS user equipment is commonly described as GPS, GLONASS, Galileo, Beidou, and GNSS receivers, as appropriate. However, as Figure 8.1 shows, the receiver forms only part of each set of user equipment. The antenna converts the incoming GNSS radio signals to electrical signals. These are input to the receiver, which demodulates the signals using a clock to provide a time reference. The ranging processor uses acquisition and tracking algorithms to determine the range from the antenna to each of the satellites used from the receiver outputs. It also controls the receiver and decodes the navigation messages. Lastly, the navigation processor uses the ranging measurements to compute a position, velocity, and time (PVT) solution.

There are many different types of GNSS user equipment, designed for different applications. User equipment may be supplied as a complete unit, including the power supply and user interface, with either an integrated or external antenna. The receiver and navigation processor may also be supplied as a module, either boxed or on a card. This is often called an original equipment manufacturer (OEM) receiver. An OEM receiver requires a direct current (DC) power supply and an external antenna. It communicates with other modules via a data link and may form part of a multisensor integrated navigation system. Finally, a GNSS receiver may also be supplied simply as a chipset, in which case some of the processing, such as the position calculation, may be performed on the host system's processor. GNSS chipsets are typically found in smartphones and other mobile devices.

Consumer-grade GNSS user equipment is designed to minimize cost and power consumption. It typically operates on one frequency only and its accuracy can be relatively poor. Chipsets can be produced for about $1 while integrated devices for car navigation, yachting, or walking typically cost around $100 (€80). Professional-grade user equipment is designed for high performance in terms of both accuracy and reliability. It typically operates on two or more frequencies and is optimized for a particular application such as aviation, shipping, or surveying. Top-of-the-range equipment costs more than $10,000 (€8,000). Finally, military user equipment is designed for maximum robustness and typically uses separate signals where available.

Note that it is wrong to describe user equipment as "a GNSS" or "a GPS," as these terms apply to the whole system, not just the user segment. Chapter 9 describes the user equipment in detail.

8.1.2 Signals and Range Measurement

Most GNSS signals are broadcast within the 1–2-GHz L-band region of the electromagnetic spectrum. Each satellite transmits on several frequencies, usually with multiple signals on each frequency. Right-handed circular polarization (RHCP) is always used.

A GNSS signal is the combination of a carrier with a spreading or ranging code and, in many cases, a navigation data message. In the majority of signals, the code and navigation data are applied to the carrier using biphase shift key (BPSK) modulation (see Section 7.2.1). This shifts the carrier phase by either 0 or 180°, which is equivalent to multiplying the carrier by a sequence of plus ones and minus ones (as opposed to the ones and zeroes of a basic binary sequence). The spreading code and navigation data are simply multiplied together. Figure 8.4 shows how a carrier is modulated with a BPSK code.

The amplitude of each BPSK-modulated GNSS signal, s, is given by

$$s(t) = \sqrt{2P}C(t)D(t)\cos(2\pi f_{ca}t + \phi_0), \tag{8.1}$$

where P is the signal power, C is the spreading code, D is the navigation data, f_{ca} is the carrier frequency, t is time, and ϕ_0 is a phase offset [3, 5]. Both C and D take values of ±1, varying with time. The most commonly used GNSS signal is the GPS coarse/acquisition (C/A) code. This has a data-message rate, f_d, of 50 symbol s^{-1} and a spreading-code rate, f_{co}, of 1.023 Mchip s^{-1}. Details of the different types of GNSS signal are presented in Section 8.3. Note that it is a matter of convention to describe the data message in terms of symbols and the spreading code in terms of chips; mathematically, the two terms are interchangeable. The term chip (as opposed

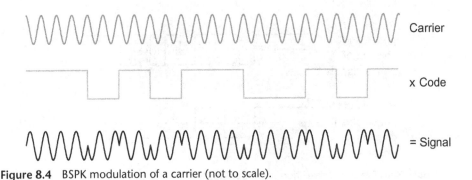

Figure 8.4 BSPK modulation of a carrier (not to scale).

to bit or symbol) is used for the spreading code sequence because it does not carry any information.

The spreading code consists of a pseudo-random noise (PRN) sequence, which is known to the receiver. It is known as a spreading code because multiplying the carrier and navigation data by the code increases the double-sided bandwidth of the signal's main spectral lobe to twice the spreading-code chipping rate while proportionately decreasing the power spectral density. The signal bandwidth is then much larger than the minimum required to transmit the data (where applicable). This technique is known as direct-sequence spread spectrum. Other types of spread spectrum are listed in Section 7.2.1.

In the receiver, the incoming spread-spectrum signal is multiplied by a replica of the spreading code, a process known as correlation or despreading. If the phase of the receiver-generated spreading code matches that of the incoming signal, the product of the two codes is maximized and the original carrier and navigation data may be recovered. If the two codes are out of phase, their product varies in sign and averages to a low value over time, so the carrier and navigation data are not recovered. Figure 8.5 illustrates this.

By adjusting the phase of the receiver-generated PRN code until the correlation peak is found (i.e., the carrier and navigation data are recoverable), the phase of the incoming PRN code is measured. From this, the signal transmission time (from satellite s), $\tilde{t}^s_{st,a}$, may be deduced. The time of signal arrival, $\tilde{t}^s_{sa,a}$, is determined from the receiver clock. A raw GNSS pseudo-range measurement, $\tilde{\rho}^s_{a,R}$, is obtained by

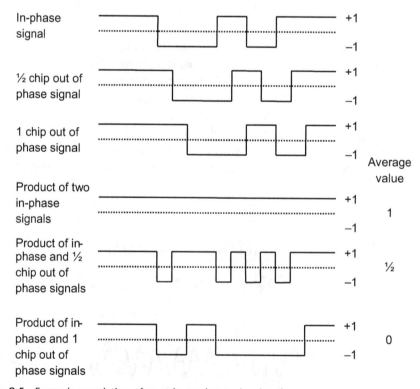

Figure 8.5 Example correlation of pseudo-random noise signals.

8.1 Fundamentals of Satellite Navigation

differencing measurements of the times of arrival and transmission and multiplying by the speed of light, c. Thus,

$$\tilde{\rho}_{a,R}^s = \left(\tilde{t}_{sa,a}^s - \tilde{t}_{st,a}^s\right)c, \tag{8.2}$$

where error sources have been neglected. Hence, the PRN code is also known as a ranging code. As explained in Section 7.1.4.1, pseudo-range differs from range due to synchronization errors between the transmitter and receiver clocks.

The receiver-generated PRN code also spreads interference over the code bandwidth. Following the correlation process, the receiver bandwidth may be reduced to that required to decode the navigation data message, rejecting most of the interference. Consequently, GNSS signals can be broadcast at a substantially lower power per unit bandwidth (after spreading) than thermal noise. Figure 8.6 illustrates the spread-spectrum modulation and demodulation process.

When the signal and receiver-generated spreading codes are different, the correlation between them is much less than if they are the same and aligned. Consequently, a number of different signals may be broadcast simultaneously on the same carrier frequency, provided they each use a different spreading code. The receiver then selects the spreading code for the desired signal. This is an example of code-division multiple access (see Section 7.2.2) and is used for all GPS, Galileo, Beidou, and regional system signals, and for some GLONASS signals.

If the incoming signal is correlated only with the PRN code, as shown in Figure 8.7, the result is a sinusoid due to the carrier component of the signal; this averages to zero. Therefore, to detect a GNSS signal, it must also be multiplied by a replica of the carrier. The result is a sinusoid of twice the frequency of the incoming and reference signals. As Figure 8.8 shows, if the incoming and reference signals are in phase, the product is always positive, whereas if the signals are 90° out of phase, the product averages to zero, leaving the signal undetected. However, if the incoming signal is correlated with two reference signals, 90° apart in phase, then the sum of squares of the two products is always positive, regardless of the phase of the reference and incoming signals. Thus carrier phase alignment is not required (though it can be used to improve precision). For best signal-to-noise performance (see Section 9.1.4), the two correlation products should be summed separately over at least a millisecond before squaring and combining them. Consequently, large variations in the phase difference between the incoming and reference signals over the summation period must be avoided. Otherwise, as Figure 8.9 shows, the average correlation product

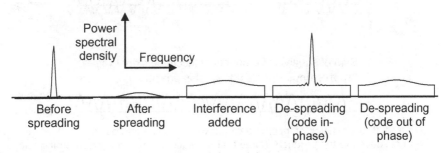

Figure 8.6 Spread-spectrum modulation and demodulation.

may be close to zero. Therefore, the carrier frequency of the reference signals must be aligned with that of the corresponding incoming signal.

When the pseudo-range is unknown, all phases of the receiver-generated spreading code must be searched until the correlation peak is found, a process known as *signal acquisition*. However, where a pseudo-range prediction from a previous measurement is available, it is only necessary to vary the receiver-generated code phase slightly; this is *signal tracking*. Once the acquisition of a GNSS signal is complete,

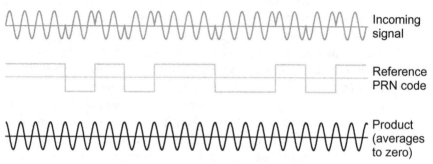

Figure 8.7 Correlation of incoming signal with reference code only.

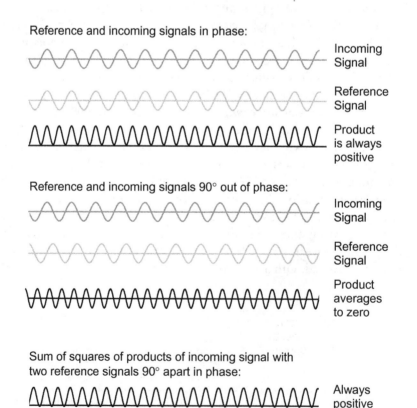

Figure 8.8 Correlation of incoming carrier with in-phase and out-of-phase references and the sum of squares of the products. (Note that, in practice, squaring and summing happen after correlation.)

8.1 Fundamentals of Satellite Navigation

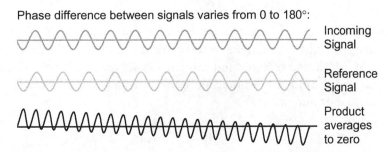

Figure 8.9 Correlation of incoming and reference carrier with variable phase difference.

the user equipment switches to tracking mode for that signal. In most receivers, tracking can operate in much poorer signal-to-noise environments than acquisition.

The carrier frequency of the received GNSS signals varies due to the Doppler effect. The perceived carrier frequency also varies due to the receiver clock drift. Therefore, to ensure effective demodulation of the signals, the carrier frequency must also be tracked by GNSS receivers (sometimes with the carrier phase). Carrier tracking may also be used both as an aid to code tracking and to provide low-noise pseudo-range rate measurements. Pseudo-range rate is simply the rate of change of the pseudo-range. Consequently, when the pseudo-range rate is unknown, the acquisition process must also search for the signal's Doppler shift. Signal correlation, acquisition, and tracking are described in Sections 9.1.4 and 9.2.

8.1.3 Positioning

A GNSS position solution is determined by passive ranging (see Section 7.1.4.1) in three spatial dimensions [6]. Using a range measurement from a single satellite signal, the user position can be anywhere on the surface of a sphere of radius r centered on that satellite. This is a surface of position (SOP). When ranges to two satellites are used, the locus of the user position is the circle of intersection of the surfaces of two spheres of radii r_{a1} and r_{a2}. Adding a third range measurement limits the user position to two points on that circle as illustrated by Figure 8.10. For most applications,

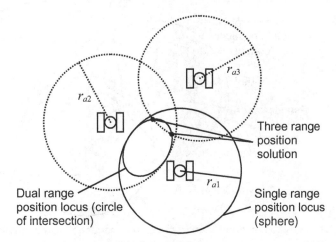

Figure 8.10 Position loci from single, dual, and triple range measurements in three dimensions.

only one position solution will be viable in practice; the other may be in space, inside the Earth, or simply outside the user's area of operation. When both solutions are viable, an additional ranging measurement can be used to resolve the ambiguity.

In GNSS, the receiver and satellite clocks are not synchronized. The measurements made are pseudo-range, not range. From (7.3), the pseudo-range from satellite s to user antenna a is

$$\rho_a^s = r_{as} + \left(\delta t_c^a - \delta t_c^s\right)c, \tag{8.3}$$

where r_{as} is the corresponding range, δt_c^a, is the receiver clock offset from system time, and δt_c^s is the satellite clock offset. The satellite clock offsets are measured by the control segment and transmitted in the navigation data message. Therefore, the navigation processor is able to correct for them. The receiver clock offset is unknown, but is common to all simultaneous pseudo-range measurements made using a given receiver. Therefore, it is determined as part of the navigation solution alongside the user position. Thus, unless constraints are applied, a GNSS navigation solution is four dimensional with three position dimensions and one time dimension.

Determination of a four-dimensional navigation solution requires signals from at least four different GNSS satellites to be measured. Figure 8.11 illustrates this geometrically. If a sphere of radius equal to the pseudo-range is placed around each of four satellites, there is normally no point at which all four spheres intersect. However, if the range error due to the receiver clock offset, $\delta\rho_c^a = \delta t_c^a c$, is subtracted from each pseudo-range, this will leave the corresponding range. Spheres of radii equal to the ranges will then intersect at the user location. Thus, $\delta\rho_c^a$ could be determined by adjusting the radii of the four spheres by equal amounts until they intersect. In practice, the position and clock offset are solved simultaneously. Note that there are two solutions, only one of which normally gives a viable position solution.

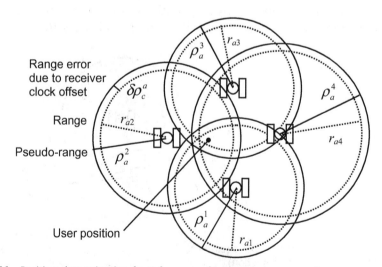

Figure 8.11 Position determination from four pseudo-range measurements.

8.1 Fundamentals of Satellite Navigation

Each pseudo-range measurement, corrected for the satellite clock error (and other known errors), $\tilde{\rho}_{a,C}^s$, may be expressed in terms of the satellite position, $\mathbf{r}_{is}^i$, at the time of signal transmission, $t_{st,a}^s$, the user antenna position, $\mathbf{r}_{ia}^i$, at the time of signal arrival, $t_{sa,a}^s$, and the receiver clock offset by

$$\tilde{\rho}_{a,C}^s = \sqrt{\left(\mathbf{r}_{is}^i(t_{st,a}^s) - \mathbf{r}_{ia}^i(t_{sa,a}^s)\right)^T \left(\mathbf{r}_{is}^i(t_{st,a}^s) - \mathbf{r}_{ia}^i(t_{sa,a}^s)\right)} + \delta\rho_c^a(t_{sa,a}^s), \qquad (8.4)$$

where measurement noise is neglected. Figure 8.12 illustrates this. The satellite position is obtained from the set of parameters broadcast in the navigation data message describing the satellite orbit, known as the *ephemeris* (see Section 8.5.2), together with the corrected measurement of the time of signal transmission. The four unknowns, the antenna position and receiver clock error, are common to the pseudo-range equations for each of the satellites, assuming a common time of signal arrival. Therefore, they may be obtained by solving simultaneous equations for four pseudo-range measurements. Similarly, the velocity of the user antenna, together with the receiver clock drift rate, may be obtained from a set of pseudo-range rate measurements. Calculation of the GNSS navigation solution is described in detail in Section 9.4.

As well as navigation and positioning, GNSS may also be used as a timing service to synchronize a network of clocks. More information is available in [3, 7].

8.1.4 Error Sources and Performance Limitations

A number of GNSS error sources are illustrated by Figure 8.13. Errors in the GNSS navigation solution calculation arise from differences between the true and broadcast ephemeris and satellite clock errors. Signal propagation delays arise from refraction in the ionosphere and troposphere, resulting in measured pseudo-ranges which are too

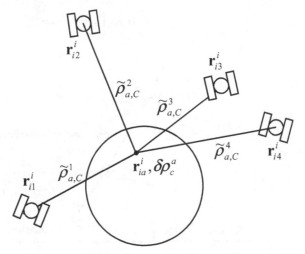

Figure 8.12 Determination of a position solution using four satellite navigation signals.

large. These may be partially calibrated using models; however, if ranging measurements from a given satellite are made on more than one frequency, the ionosphere propagation delay may be determined from the difference. Receiver measurement errors arise due to delays in responding to dynamics, receiver noise, radio frequency (RF) interference, and multipath interference. Multipath occurs where the signal is received by more than one path as GNSS signals can be reflected by buildings and the ground.

The elevation of a satellite is the angle between the horizontal plane and the line of sight from the user to the satellite (see Section 8.5.4). Low-elevation signals exhibit much larger ionosphere and troposphere propagation delays and are also more vulnerable to multipath interference. Most GNSS receivers therefore ignore signals from below a certain elevation, known as the masking angle. This is typically set at between 5° and 15°. Error sources are described in detail in Section 9.3.

Under good reception conditions, GNSS position solutions are typically accurate to a few meters, with performance depending on which signals are used, as discussed in Section 9.4.4. Accuracy may be improved to meter level by making use of calibration information from one or more reference stations at known locations. This is known as differential GNSS (DGNSS) and is described in Section 10.1. Centimeter-level positioning may be obtained in benign environments using carrier-phase-based differential techniques as discussed in Section 10.2. Reference stations may also be used to detect faults in the GNSS signals, a process known as integrity monitoring and discussed in Chapter 17.

GNSS performance is degraded in challenging environments. Signals may be blocked by buildings, mountainous terrain, and parts of the user equipment's host vehicle. They may also be received via reflected paths only, known as nonline-of-sight (NLOS) reception, which introduces large positive range errors. Figure 8.14 illustrates this. Low-elevation signals are most affected. Signal blockage and multipath interference is a particular problem in streets surrounded by tall buildings, known as *urban canyons*, where the ratio of the building height to their separation

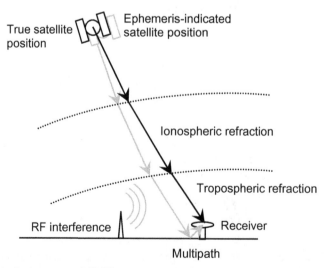

Figure 8.13 Principal sources of GNSS error.

determines how many signals get through. The number of GNSS constellations that are received has a large impact on the performance achievable in urban areas [8]. Signal blockage is also a problem in mountainous areas, where lower elevation signals will not be receivable in valleys.

Consequently, it is not always possible to receive the four signals necessary to compute a position solution, particularly where only one satellite constellation is used. A solution can sometimes be obtained for a limited period with fewer satellites by predicting forward the receiver clock errors or assuming a constant user height. Even where four or more signals can be received, the geometry may be poor, leading to much poorer accuracy in one direction than another (see Section 9.4.3). Multipath and NLOS mitigation is discussed in Section 10.4 while Section 10.6 describes shadow matching, a new positioning technique that uses a 3-D city model to enhance GNSS in urban canyons.

GNSS performance is also degraded where the signal is attenuated, such as indoors or under trees. RF interference from communications signals in neighboring frequency bands, poorly filtered harmonics of strong signals on any frequency, and deliberate jamming of the GNSS signals themselves causes similar degradation. GNSS signals are particularly vulnerable to attenuation and interference because they are very weak compared to other types of radio signal. Techniques for improving performance in poor signal-to-noise environments are discussed in Sections 10.3 and 10.5.1.

The final limitation of GNSS discussed here is the time to first fix (TTFF). It can take over a minute to obtain a position solution after GNSS user equipment is first switched on. First, the code phase and Doppler shift of the first four signals acquired must be determined by trial and error. Then, for each signal, the ephemeris data must be downloaded from the satellite. This can take up to 30 seconds using the GPS C/A-code navigation data message. Reacquisition of a signal after an interruption is much quicker as the user equipment then knows the satellite positions and velocities and the approximate user position and receiver clock parameters. If this information can be supplied to the user equipment in advance of initial acquisition, the TTFF can be much reduced; this is the principle of assisted GNSS (AGNSS), described in Section 10.5.2.

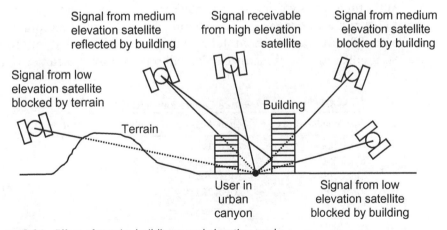

Figure 8.14 Effect of terrain, buildings, and elevation angle.

8.2 The Systems

This section introduces the systems that comprise GNSS. The global satellite navigation systems, GPS, GLONASS, Galileo, and Beidou, are described in turn. Each system operates both open and restricted services using different signals. The open services are available free of charge to all users with suitable equipment, whereas the restricted services are only available to authorized users. The regional navigation systems, the Quasi-Zenith Satellite System (QZSS) and the Indian Regional Navigation System (IRNSS), are then described, followed by the space-based and ground-based augmentation systems. The section concludes by discussing the compatibility of the different systems. For each system, both the current status and future plans are described. Note, however, that GNSS development and deployment programs often fall behind schedule. In addition, Section G.2 of Appendix G on the CD provides some historical notes on GNSS development.

8.2.1 Global Positioning System

NAVSTAR GPS was developed by the United States government as a military navigation system. Its controlling body is the GPS Directorate, which operates under the auspices of the Department of Defense (DOD). Development started in 1973 with the merger of several earlier programs. The first operational prototype satellite was launched in 1978, initial operational capability (IOC) was declared at the end of 1993, and full operational capability (FOC) was attained at the end of 1994 [9].

GPS offers two navigation services: an open or civil service, known as the Standard Positioning Service (SPS), and a restricted or military service known as the Precise Positioning Service (PPS). The PPS is only available to users licensed by the U.S. government, including U.S. and NATO military forces and their suppliers, and its signals are encrypted.

Table 8.1 lists the present and planned generations of GPS satellite. Since the late 1990s, GPS has been undergoing a modernization process, improving the control segment, improving the satellite design, and introducing additional signals for both SPS and PPS users as described in Section 8.3.2. The Block III satellites will introduce further new signals and broadcast some signals at a higher power.

At the time of this writing, the basic SPS provides a horizontal accuracy of about 3.8m (1σ) and a vertical accuracy of 6.2m (1σ) under good reception conditions, while the PPS accuracy is about 1.2m (1σ) horizontally and 1.9m (1σ) vertically (see Section 9.4.4). The modernized SPS will offer a similar accuracy to the PPS.

Table 8.1 Present and Future Generations of GPS Satellites

GPS Satellite Block	Launch Dates	Number of Satellites
Block IIA	1990–1997	19
Block IIR	1997–2004	12[a]
Block IIR-M	2005–2009	7[b]
Block IIF	2010–2015	12
Block III	2015–2024	24 (planned)

[a]Excludes failed launches. [b]Excludes faulty satellite.

8.2.2 GLONASS

GLONASS (Global'naya Navigatsionnaya Sputnikovaya Sistema) was developed as a military navigation system in the mid-1970s, in parallel to GPS, initially by the USSR and then by Russia. Like GPS, it was designed to offer both a civil and a military positioning service. The first satellite was launched in 1982. A full satellite constellation was briefly achieved in 1995, but then decayed, reaching a nadir of six satellites in 2001. A modernization program was then instigated, rebuilding the constellation, introducing new signals, and updating the control segment. IOC with 18 satellites was achieved again in 2010, with a full 24-satellite constellation following in 2011.

Table 8.2 lists the present and planned generations of GLONASS satellite. GLONASS-M satellites only broadcast frequency-division multiple access signals, with each satellite broadcasting the same PRN codes on different carrier frequencies. GLONASS-K satellites broadcast additional CDMA signals and add a search-and-rescue (SAR) service. When fully modernized, GLONASS is intending to offer similar positioning accuracy to GPS.

8.2.3 Galileo

Development of the Galileo satellite navigation system was initiated in 1999 by the European Union (EU) and European Space Agency (ESA). The first test satellite was launched in 2005. Initial operational capability (with 18 satellites) has been planned for 2015 with full operational capability (with 26 satellites) in 2016. Unlike GPS and GLONASS, Galileo has been developed essentially as a civil navigation system. It is managed by the European GNSS Agency, sometimes referred to as the GSA. Development is funded mainly by the EU. A number of non-European countries are also participating, but are not involved in critical aspects of the program.

Galileo is initially offering two navigation services, an open service (OS) and a public regulated service (PRS), together with a search-and-rescue service [10, 11]. The open service provides signals in two frequency bands to all users. From FOC, it will offer a similar performance level to the modernized SPS GPS service, with a horizontal accuracy of the order of 2m (1σ) and a vertical accuracy of the order of 4m (1σ).

The PRS is a restricted service, intended to provide high integrity, continuity, and some interference resistance to trusted subscribers in EU member states, such as emergency services, security services, and the military. However, the accuracy will be slightly poorer than that obtained from the open service at 3m (1σ) horizontally

Table 8.2 Present and Future Generations of GLONASS Satellites

Satellite Block[a]	Launch Dates	Number of Satellites
GLONASS-M	2003–2015	50[a]
GLONASS-K1	2011–2013	2
GLONASS-K2	2015–	25 (planned)
GLONASS-KM	To be determined	To be determined

[a]Excludes failed launches.

and 6m (1σ) vertically. It uses encrypted signals in two frequency bands and only limited information is available publicly for security reasons.

Further navigation services, a safety-of-life (SOL) service and commercial services (CS), may be implemented in the future [10, 11]. The safety-of-life service would use the same signals as the open service, but would add signal integrity and authentication data, which validates that the Galileo signal is genuine, protecting against spoofing (transmission of fake signals). The commercial service was conceived as a restricted service, offering higher performance to paying subscribers.

8.2.4 Beidou

Beidou, commonly known as "Compass" between 2007 and 2012, is being developed in three distinct phases. The experimental phase 1 system, which used two-way ranging, is described in Section F.2 of Appendix F on the CD. The first test satellite using GNSS technology was launched in 2007. Phase 2, completed in 2012, provides a regional GNSS-based service to China and surrounding countries using 12 satellites. Phase 3, currently under development, was intended to provide a global GNSS by 2020. Phase 2 and 3 satellites broadcast both open and restricted signals.

8.2.5 Regional Systems

QZSS will provide positioning services primarily to in-car and personal receivers in Japan, although the signals will be receivable across much of East Asia and Oceania. The positioning service is intended to supplement GPS by increasing the number of high-elevation satellites visible in urban canyons and mountainous regions. It will also provide high-resolution GPS differential corrections [12]. The first QZSS satellite launched in 2010 and FOC was expected in 2013.

IRNSS is intended to provide a fully independent GNSS service for India. It will be under the sole control of that country and was planned to be operational by 2014 to 2015, with the first satellite launch in 2013. The service area will be from longitudes 40° to 140° and the accuracy within India will be about 10m (1σ) horizontally and vertically [13].

8.2.6 Augmentation Systems

There are two definitions of a GNSS augmentation system: a broad definition and a narrow definition. The broad concept encompasses any system that supplements GNSS by providing differential corrections to improve accuracy (see Sections 10.1 and 10.2), assistance data to speed up signal acquisition (see Section 10.5.2), and/or integrity alerts to protect users from the effects of erroneous GNSS signals (see Section 17.5). Note that, at the time of this writing, the GPS control segment only monitored the health of the PPS signals, and it could take over 2 hours to alert users of problems. Some authors consider augmentation systems to be the fourth segment of a GNSS (see Section 8.1.1).

The narrow definition of an augmentation system is one that supplies both differential corrections and integrity alerts that meet the needs of safety-critical applications, such as civil aviation. There are two main types of augmentation system meeting this definition. Space-based augmentation systems (SBAS) are designed to serve a large country or small continent and broadcast to their users via geostationary satellites. Ground-based augmentation systems (GBAS) serve a local area, such as an airfield, providing a higher precision service than SBAS and broadcasting to users via ground-based transmitters.

There are six SBAS systems, at varying stages of development at the time of this writing, as summarized in Table 8.3. Each uses a network of several tens of reference stations across its coverage area to monitor the GNSS signals. The differential corrections are only valid within the region spanned by the reference stations. However, the satellite signal failure alerts can be used throughout the coverage area of each geostationary satellite, which typically spans latitudes from $-70°$ to $+70°$ and longitudes within $70°$ of the satellite. The full-service coverage area of an SBAS system may be expanded within the signal footprint by adding additional reference stations [14–16].

SBAS differential corrections comprise individual-satellite clock and ephemeris data, together with ionosphere model coefficients that may be used with any GNSS signal. At the time of this writing, WAAS, EGNOS, and MSAS only provided integrity alerts and clock and ephemeris corrections for the GPS satellites. Proposals to add Galileo data to EGNOS, known as the Multi-Constellation Regional System (MRS), are under consideration. SDCM will transmit both GPS and GLONASS corrections, but only for satellites in view of the coverage area.

Some SBAS signals may also be used for ranging, increasing coverage, and assisting user-equipment-based integrity monitoring (see Chapter 17). WAAS offers SBAS ranging but EGNOS does not.

GBAS differential corrections and integrity data are broadcast in the 108–118-MHz VHF band. GBAS is being deployed at airports in many different countries to enable GPS to be safely used for category I landing. Research and development to meet the more demanding category II and III landing requirements are ongoing. Within the United States, GBAS is sometimes known as the Local Area Augmentation System (LAAS) [17], while the military Joint Precision Approach and Landing

Table 8.3 Space-Based Augmentation Systems

SBAS	Full Service Coverage Area	Status
Wide Area Augmentation System (WAAS)	North America	Operational
European Geostationary Navigation Overlay System (EGNOS)	Europe and surrounding countries	Operational
Multi-function Transport Satellite (MTSat) Satellite Augmentation System (MSAS)	Japan	Operational
GPS/GLONASS and GEO Augmented Navigation (GAGAN)	India	Testing
System of Differential Corrections and Monitoring (SDCM)	Russia	Deployment and testing
Satellite Navigation Augmentation System (SNAS)	China	Under development

System (JPALS) is based on GBAS. The GBAS concept could be extended to incorporate additional ranging signals provided by ground-based GPS-like transmitters, known as pseudolites (see Section 12.1).

Note also that, in civil aviation, the term aircraft-based augmentation system (ABAS) is used to describe the integration of GNSS with inertial navigation (Chapter 14), other sensors, such as a barometric altimeter (Chapter 16), and aircraft-based integrity monitoring (Chapter 17).

8.2.7 System Compatibility

GPS and GLONASS were originally developed as military systems during the Cold War. Compatibility between them was not a requirement. Physical collision between the satellites was avoided by using different orbits, while signal interference was not an issue as different frequencies were used.

By the time Galileo development started at the end of the 1990s, the vast majority of GNSS users were civil. It was recognized that existing GPS open-service users would be more likely to use Galileo alongside GPS than instead of it. The Galileo open services were therefore designed to minimize the cost of dual-standard user equipment by using the same frequencies as some of the GPS signals. This raised concerns about interference to GPS, resulting in bilateral negotiations between the United States and the European Union to agree compatible signal formats for Galileo and the modernized GPS.

China and Russia subsequently announced plans to use the same frequencies for similar reasons. This resulted in the establishment of the International Committee on GNSS (ICG) at the end of 2005 under the auspices of the United Nations to provide a forum for multilateral negotiations between the GNSS service providers.

At the time of this writing, there were plans for all GNSS satellites (except IRNSS) to transmit open-service signals with common frequencies and modulations by about 2020. This is beneficial to users in challenging environments, such as dense urban area and indoors. However, in open areas, there is concern that the increased intersatellite interference will actually degrade overall performance [18].

A further requirement for compatibility is alignment of the different reference datums and timescales used by the different systems. GPS uses the WGS 84 datum, whereas Galileo uses the GTRF datum, both based on the ITRF. WGS84, GTRF, and ITRF differ by only a few centimeters, so this is only an issue for high-precision users. GLONASS uses the PZ-90.02 frame. This is aligned with the ITRF, but the origin is offset by about 0.4m. The CGCS 2000 datum used by Beidou is also nominally aligned with the ITRF. However, all four systems use different time bases. This is discussed in Section 8.4.5.

There is a clear demand for multiconstellation GNSS user equipment to boost position solution availability in challenging environments, improve accuracy through averaging out of noise and error sources across more measurements (see Section 9.4.3), and enhance the robustness of consistency-based integrity checks (Section 17.4). Between 2010 and 2012, a large number of manufacturers introduced combined GPS/GLONASS user equipment despite the increased cost and complexity arising from using FDMA GLONASS signals. Russia has imposed import restrictions on GNSS user equipment that does not receive GLONASS signals.

The performance benefits of using two GNSS constellations instead of one outweigh those of using one constellation in preference to another. However, once more than two full constellations are available, many equipment manufacturers will select a subset of the systems to limit hardware costs and power consumption. Criteria may include signal design, quality of satellite clock and ephemeris data, and provision of timely integrity alerts.

Military users require their signals to use separate spectrum so that they can jam the signals used by their opponents while minimizing the impact on their own GNSS use. If another GNSS then uses the same spectrum, this presents a problem.

8.3 GNSS Signals

Once modernization is complete, GNSS satellites will typically transmit around 10 signals each, spread over three or four frequencies. Different signals are needed for the open and restricted services, while frequency diversity enables calibration of the ionosphere propagation delay, reduces the impact of interference on one frequency, and aids carrier-phase positioning (see Section 10.2.4). Furthermore, different types of signal are suited to different applications and operating conditions. Some receivers use a single signal type, such as GPS C/A code, while others use multiple signals from each satellite. Table 8.4 and Figure 8.15 show the frequency bands used by the main GNSSs.

This section first compares the different types of GNSS signals and discusses their advantages and drawbacks. It then describes the signals of GPS, GLONASS, Galileo, and Beidou, in turn, concluding with the regional and augmentation systems. In addition, Section G.3 of Appendix G on the CD discusses signal multiplexing and ranging code design.

The nominal signal powers listed in this section are minimum values. Satellites initially transmit at higher powers, but the power drops as the satellite ages. Some details of the restricted signals are not publicly available, so they are marked as restricted to authorized users (RAU) in the tables within this section. Details of

Table 8.4 GPS, GLONASS, Galileo, and Proposed Compass Phase 3 Frequency Bands

Band Name	Lower Limit (MHz)	Carrier Frequency (MHz)	Upper Limit (MHz)	Bandwidth (MHz)
Galileo E5 and Beidou B2	1,145.76	1,191.795	1,237.83	92.07
Galileo E5a and Compass B2a	1,145.76	1,176.45	1,191.795	46.04
GPS and GLONASS L5	1,161.105	1,176.45	1,191.795	30.69
Galileo E5b and Compass B2b	1,191.795	1,207.14	1,237.83	46.04
GLONASS L3	1,192.002	1,202.025	1,212.258	20.46
GPS L2	1,212.255	1,227.60	1,242.945	30.69
GLONASS L2	1,237.8275	Varying	1,258.29	20.46
Galileo E6 and Compass B3	1,258.29	1,278.75	1,299.21	40.92
Galileo E1 and Compass B1	1,554.96	1,575.42	1,595.88	40.92
GPS L1 and GLONASS L1OCM	1,560.075	1,575.42	1,590.765	30.69
GLONASS L1 (main)	1,590.765	Varying	1,611.225	20.46

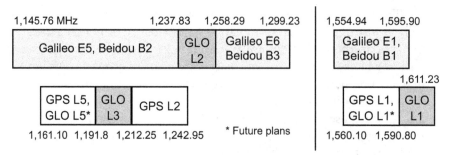

Figure 8.15 GPS, GLONASS, Galileo, and proposed Compass Phase 3 frequency bands.

open signals which were not available at the time of writing are marked as to be determined (TBD).

8.3.1 Signal Types

GNSS signals differ in three main respects: DSSS modulation, code repetition length, and navigation data modulation. Each is discussed in turn.

Higher spreading-code chipping rates offer better resistance against narrowband interference. They can also offer high-precision ranging (see Section 9.3.3) and resistance to multipath errors (Section 9.3.4) simultaneously. However, receivers require greater computational capacity to process them. Consequently, the newest GNSS satellites offer a range of open-access signals with different chipping rates.

Many of the newer GNSS signals use binary offset carrier (BOC) modulation instead of BPSK. This adds an extra component, the subcarrier, S, giving a total signal amplitude of

$$s(t) = \sqrt{2P} S(t) C(t) D(t) \cos(2\pi f_{ca} t + \phi_{ca}). \tag{8.5}$$

The subcarrier function repeats at a rate, f_s, which spreads the signal into two sidebands or sidelobes, centered at $f_{ca} \pm f_s$. To separate the main lobes of these sidebands, f_s must be at least the spreading-code chipping rate, f_{co}. BOC modulation can be used to minimize interference with BPSK signals sharing the same carrier frequency. It can also give better code tracking performance and multipath resistance than a BPSK signal with the same spreading-code chipping rate [19, 20]. However, BOC signals require a more complex receiver design (see Sections 9.1.4 and 9.2).

For a basic BOC modulation, the subcarrier function is simply a square wave with chipping rate $2f_s$. This may be sine-phased, in which case the subcarrier function transitions are in-phase with the spreading code transitions, or cosine-phased, in which the transitions are a quarter of a subcarrier function period out of phase [20]. Figure 8.16 illustrates this. More complex subcarrier functions may also be used, such as the alternate BOC used for the Galileo E5 signal (see Section 8.3.4). BOC modulation is described using the shorthand $BOC_s(f_s, f_{co})$ for sine-phased and $BOC_c(f_s, f_{co})$ for cosine-phased subcarrier functions, where f_s and f_{co} are usually expressed as multiples of 1.023×10^6 chip s^{-1}. The terms $BOC_{\sin}(f_s, f_{co})$ and $BOC_{\cos}(f_s, f_{co})$ are also used. Figure 8.17 shows the power spectral density of BPSK, sine-phased BOC, and cosine-phased BOC-modulated signals.

8.3 GNSS Signals

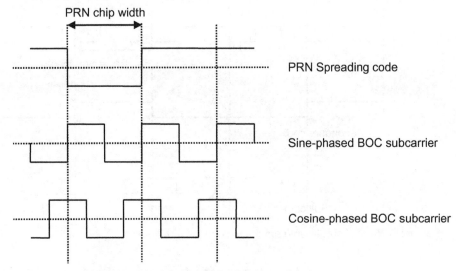

Figure 8.16 Sine-phased and cosine-phased BOC(1,1) modulation.

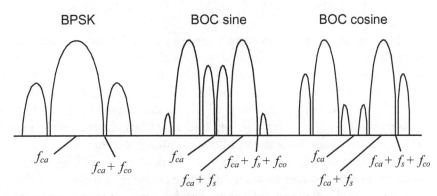

Figure 8.17 Power spectral density of BPSK and BOC-modulated signals (log scale).

Spreading codes with short repetition lengths, in terms of the number of chips, require fewer options to be searched in the acquisition process. However, there are false correlation peaks, both between unsynchronized copies of the same code and between the different spreading codes used for cochannel signals in CDMA systems. The shorter the code repetition length, the larger these false peaks are. This conflict may be resolved using layered, or tiered, codes. A short-repetition-length primary code is multiplied by a secondary code, with chip size equal to the repetition length of the primary code, giving a longer repetition length for the full code. The receiver may then correlate the incoming signal with either the primary code or the full code. Figure 8.18 illustrates this. The codes used for the restricted signals all have very long repetition lengths to prevent unauthorized users determining them by analyzing the signals.

A faster data message rate enables more information to be broadcast or a given amount of information to be downloaded more quickly. However, faster data rates require a higher postcorrelation bandwidth, reducing interference rejection (see Sections 9.1.4 and 10.3). The best interference resistance is obtained by omitting

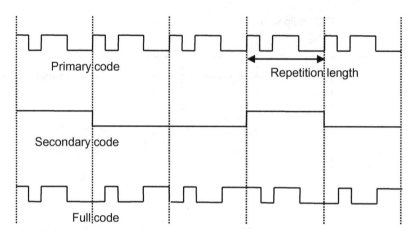

Figure 8.18 A simple layered spreading code.

the data altogether [i.e., omitting D in (8.1)]. This also gives better carrier-tracking performance as discussed in Section 9.2.3. These data-free signals are also known as pilot signals.

Most of the new GNSS signals are transmitted either as data-modulated and pilot pairs, or with data modulated only on alternate spreading code chips. The latter approach is known as a time-division multiplex (TDM) or time-division data multiplex (TDDM) and the data-modulated and pilot components may either share a PRN code or use separate codes, which are interleaved by applying them to alternate chips [21–23]. For both paired and TDM signals, the receiver may track the data-modulated component, the pilot component, or both.

Four open signals are common to all modernized GNSS (except IRNSS). These are data-modulated and pilot 10.23 Mchip s^{-1} BPSK signals at a carrier frequency of 1,176.45 MHz (known as L5, E5a, and B2a), and data-modulated and pilot multiplexed BOC (MBOC) or BOC$_s$(1,1) signals at 1,575.42 MHz (known as L1, E1, and B1). The MBOC signal is a combination of a higher-powered BOC$_s$(1,1) signal and a lower-powered BOC$_s$(6,1) signal. Receivers have the option of ignoring the BOC$_s$(6,1) component to reduce the processing load.

8.3.2 Global Positioning System

There are 10 different GPS navigation signals, broadcast across three bands, known as link 1 (L1), link 2 (L2), and link 5 (L5). The C/A and precise (encrypted precise) (P(Y))-code signals are known as the legacy GPS signals as they predate the modernization program. The other signals are being introduced as part of GPS modernization and are not broadcast by all satellites. The signals are summarized in Table 8.5 and their PSDs are illustrated by Figure 8.19 [24–27]. The time-multiplexed BOC (TMBOC) signal comprises a BOC$_s$(1,1) modulation for 29/33 of the time and a BOC$_s$(6,1) modulation for 4/33 of the time. See Section G.3.1 of Appendix G on the CD for more information. Note that the Block III satellites will transmit some signals at higher power with equal power in the L1 and L2 bands.

8.3 GNSS Signals

Table 8.5 GPS Signal Properties

Signal	Band and Carrier Frequency (MHz)	Service	Modulation and Chipping Rate ($\times 1.023$ Mchip s^{-1})	Navigation Message Rate (symbol s^{-1})	Minimum Received Signal Power (dBW)	Satellite Blocks
C/A	L1, 1,575.42	SPS/PPS	BPSK 1	50	−158.5	All
P(Y)	L1, 1,575.42	PPS	BPSK 10	50	−161.5	All
M code	L1, 1,575.42	PPS	BOC$_s$(10,5)	TDM, RAU	RAU	From IIR-M
L1C-d	L1, 1,575.42	PPS	BOC$_s$(1,1)	100	−163	From III
L1C-p	L1, 1,575.42	PPS	TMBOC	None	−158.3	From III
L2C	L2, 1,227.60	SPS	BPSK 1	TDM, 50	−160	From IIR-M
P(Y)	L2, 1,227.60	PPS	BPSK 10	50	−164.5	All
M code	L2, 1,227.60	PPS	BOC$_s$(10,5)	TDM, RAU	RAU	From IIR-M
L5I	L5, 1,176.45	SPS	BPSK 10	100	−158	From IIF
L5Q	L5, 1,176.45	SPS	BPSK 10	None	−158	From IIF

Table 8.6 gives the code lengths of the open signals, many of which are layered [24, 26, 27]. The coarse/acquisition code is so named because it was intended to provide less accurate positioning than the P(Y) code and most PPS user equipment acquires the C/A-code signal before the P(Y)-code signals (see Section 9.2.1). Because the C/A code repeats every millisecond (i.e., at 1 kHz), it is relatively easy to acquire. However, the correlation properties are poor with cross-correlation peaks only 21–24 dB below the main autocorrelation peak.

The link 2 civil (L2C) signal is a time-division multiplex of two components, one carrying the civil-moderate (CM) code with a navigation data message and the other carrying the civil-long (CL) code data free. As the total L2C chipping rate is 1.023 Mchip s^{-1}, the rate of each code is 511.5 kchip s^{-1} [24]. The CM code can be acquired more quickly or with less processing power than the CL code, while the data-free CL code gives more accurate carrier tracking and better performance in poor signal-to-noise environments. The CM code is more difficult to acquire than the C/A code, but offers an increased margin of 45 dB between the main autocorrelation peak and cross-correlation peaks [21, 22]. The L2C signal is not suited to safety-of-life applications because of the amount of interference in the L2 band.

The encrypted precise (Y) code comprises the publicly known precise (P) code multiplied by an encryption code, which is only available to licensed PPS users. This encryption acts as an antispoofing (AS) measure because it makes it difficult for hostile forces to deceive GPS user equipment by broadcasting replica signals, which is known as spoofing. All GPS satellites normally broadcast Y code, but can be switched to broadcast P code. P code is also used by GPS signal simulators to test PPS user equipment without the need for encryption data. The notation P(Y) code is commonly used to refer to the P and Y codes collectively [5]. The power of the P(Y)-code signals may be reduced and the phasing of the signals changed after 2020.

The military (M)-code signals were the first GNSS signals to use BOC modulation. This is done to provide spectral separation from the SPS signals, enabling use of jamming to prevent hostile use of GPS and allowing higher-power PPS signals to be broadcast without disrupting the civil GPS service [28]. The M code is intended

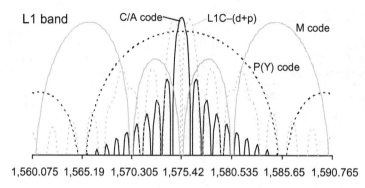
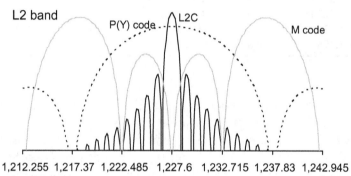
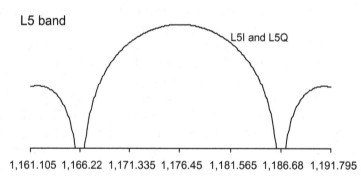

Figure 8.19 GPS signal multiplex power spectral densities (log scale).

Table 8.6 GPS Open-Signal Code Lengths

	Primary Code		Secondary Code	Full Code	
Signal	Length	Repetition Interval	Length	Length	Repetition Interval
C/A	1,023	1 ms	—	1,023	1 ms
L1C-d	10,230	10 ms	—	10,230	10 ms
L1C-p	10,230	10 ms	1,800	184,140,000	18s
CM (L2C)	10,230	20 ms	—	10,230	20 ms
CL (L2C)	767,250	1.5 seconds	—	767,250	1.5s
L5I	10,230	1 ms	10	102,300	10 ms
L5Q	10,230	1 ms	20	204,600	20 ms

8.3.3 GLONASS

The GLONASS FDMA signals are summarized in Table 8.7 [29]. FDMA offers better rejection of intersatellite interference between signals using short ranging codes than CDMA. However, more complex receivers are required that are more expensive to produce. The C/A code has a 511-chip length and 1-ms repetition period. The P code, which may be encrypted, is truncated to a 5,110,000-chip length, giving a 1-second repetition period. Each satellite is allocated a channel number, k, and broadcasts on $1{,}602 + 0.5625k$ MHz in the L1 band and $1{,}246 + 0.4375k$ MHz in the L2 band. Channel numbers from -7 to $+6$ are used with satellites in opposite slots in the same orbital plane sharing the same channels. This only causes interference to space-based users [30].

The GLONASS CDMA signals are summarized in Table 8.8 [31]. These are only broadcast by the newest satellites, with different signals to be introduced with different generations of satellite. It has been planned to transmit CDMA signals from all satellites by 2020. Open signals are denoted by an "O" in the name and restricted signals by an "S." Table 8.9 gives the code lengths of the L3OC signals, which are layered.

Table 8.7 Properties of GLONASS FDMA Signals (Transmitted by All Satellites)

Signal	Carrier Frequency Range (MHz)	Modulation and Chipping Rate ($\times 1.022$ Mchip s^{-1})	Navigation Message Rate (symbol s^{-1})	Minimum Received Signal Power (dBW)
L1OF (C/A)	1,598.0625–1,605.375	BPSK 0.5	50	−161
L1SF (P)	1,598.0625–1,605.375	BPSK 5	50	−161
L2OF (C/A)	1,242.9375–1,248.625	BPSK 0.5	50	−167
L2SF (P)	1,242.9375–1,248.625	BPSK 5	50	−167

Table 8.8 GLONASS CDMA Signal Properties (All Signals are Subject to Change)

Signal	Carrier Frequency (MHz)	Modulation and Chipping Rate ($\times 1.023$ Mchip s^{-1})	Navigation Message Rate (symbol s^{-1})	Minimum Received Signal Power (dBW)	Satellite Blocks
L1OC-d	1,600.995[a]	BSPK 1	TBD	TBD	From K2
L1OC-p	1,600.995[a]	BOC$_s$(1,1)	None	TBD	From K2
L1SC	1,600.995	BOC$_s$(5,2.5)	RAU	RAU	From K2
L1OCM-d	1,575.42	BOC$_s$(1,1)	TBD	TBD	From KM
L1OCM-p	1,575.42	BOC$_s$(1,1)	TBD	TBD	From KM
L2OC	1,248.06[a]	BOC$_s$(1,1)	None	TBD	From K2 or KM
L2SC-a	1,248.06[a]	BSPK 1	RAU	RAU	From K2
L2SC-b	1,248.06	BOC$_s$(5,2.5)	RAU	RAU	From K2
L3OC-d	1,202.025	BPSK 10	200	−158	From K1
L3OC-p	1,202.025	BPSK 10	None	−158	From K1
L5OCM-d	1,176.45	BPSK 10	TBD	TBD	From KM
L5OCM-p	1,176.45	BPSK 10	TBD	TBD	From KM

[a]A time division multiplex of alternating chips is proposed for the two L1OC signals and for L2OC and L2SC-a.

Table 8.9 GLONASS L3OC-Signal Code Lengths

Signal	Primary Code		Secondary Code	Full Code	
	Length	Repetition Interval	Length	Length	Repetition Interval
L3OC-d	10,230	1 ms	5	512,150	5 ms
L3OC-p	10,230	1 ms	10	102,300	10 ms

8.3.4 Galileo

Galileo broadcasts 10 different navigation signals across three frequency bands, E5, E6, and E1 [11, 20, 32]. The E1 band and signals are sometimes referred to as L1, in common with GPS terminology. The interface control document (ICD) uses both terms. Table 8.10 summarizes the signals, while Figure 8.20 illustrates their PSDs. The E1-A and E6 signals use encrypted ranging codes. The composite BOC (CBOC) signal is a type of MBOC comprising the sum of a 10/11 power $BOC_s(1,1)$ modulation and a 1/11 power $BOC_s(6,1)$ modulation. See Section G.3.1 of Appendix G on the CD for more information.

Table 8.11 gives the code lengths and repetition intervals for the Galileo OS, SOL, and CS ranging codes [32]. Layered codes are generally used. The total code length for the navigation-data-message-modulated signals is set to the data symbol length. For the data-free, or pilot, signals, a 100-ms code repetition period is used to ensure that the code length is not less than the satellite-to-user distance. Different primary codes are used for the data and pilot signals.

The Galileo satellites broadcast the E5a and E5b signals coherently, providing the users with the option of tracking a single wideband signal, instead of separate signals, to obtain more accurate pseudo-range measurements. However, because the E5a-I and E5b-I signals carry different navigation data messages, standard BOC modulation cannot be used. Instead, an alternate-binary-offset-carrier (AltBOC) modulation scheme has been developed with a 15.345-MHz subcarrier frequency and 10.23-Mchip s^{-1} spreading code chipping rate. The wideband AltBOC signal, centered at 1,191.795 MHz, has 8PSK modulation (see Section 7.2.1). This permits

Table 8.10 Galileo Signal Properties

Signal	Band and Carrier Frequency (MHz)	Services	Modulation and Chipping Rate ($\times$ 1.023 Mchip s^{-1})	Navigation Message Rate (symbol s^{-1})	Minimum Received Signal Power (dBW)
E1-A	E1, 1,575.42	PRS	$BOC_c(15,2.5)$	TDM, RAU	−157
E1-B	E1, 1,575.42	OS, SOL, CS	CBOC	250	−160
E1-C	E1, 1,575.42	OS, SOL, CS	CBOC	None	−160
E5a-I	E5a, 1,176.45	OS, CS	BPSK 10	50	−158
E5a-Q	E5a, 1,176.45	OS, CS	BPSK 10	None	−158
E5b-I	E5b, 1,207.14	OS, SOL, CS	BPSK 10	250	−158
E5b-Q	E5b, 1,207.14	OS, SOL, CS	BPSK 10	None	−158
E6-A	E6, 1,278.75	PRS	$BOC_c(10,5)$	TDM, RAU	−155
E6-B	E6, 1,278.75	CS	BPSK 5	1,000	−158
E6-C	E6, 1,278.75	CS	BPSK 5	None	−158

8.3 GNSS Signals

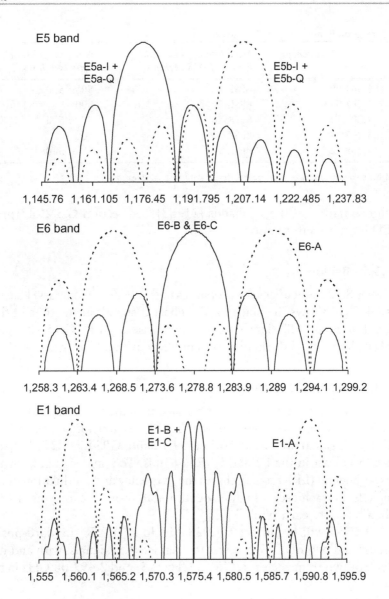

Figure 8.20 Galileo signal multiplex power spectral densities (log scale).

Table 8.11 Galileo OS, SOL, and CS Code Lengths

Signal	Primary Code		Secondary Code	Full Code	
	Length	Repetition Interval	Length	Length	Repetition Interval
E5a-I	10,230	1 ms	20	204,600	20 ms
E5b-I	10,230	1 ms	4	40,920	4 ms
E5a-Q, E5b-Q	10,230	1 ms	100	1,023,000	100 ms
E6-B	5,115	1 ms	—	5,115	1 ms
E6-C	5,115	1 ms	100	511,500	100 ms
E1-B	4,092	4 ms	—	4,092	4 ms
E1-C	4,092	4 ms	25	102,300	100 ms

Table 8.12 Beidou Phase 2 Signal Properties

Name	Carrier Frequency (MHz)	Modulation and Chipping Rate (× 1.023 Mchip s^{-1})	Navigation Message Rate (symbol s^{-1})	Service
B1I	1,561.098	BPSK 2	50 or 500	Open
B1Q	1,561.098	BPSK 2	500	Authorized
B2I	1,207.14	BPSK 10	50 or 500	Open
B2Q	1,207.14	BPSK 10	500	Authorized
B3	1,268.52	BPSK 10	50 or 500	Authorized

Note: The higher navigation message rates apply to the geostationary satellites.

differentiation of the sidebands [11, 33]. See Section G.3.2 of Appendix G on the CD for more information.

8.3.5 Beidou

Tables 8.12 summarizes the properties of the Beidou (Compass) Phase 2 signals, while Table 8.13 shows the plans for the Phase 3 signals at the time of this writing [34]. The four Phase 3 B2 signals together comprise an AltBOC(15,10) signal at 1,191.795 MHz. Very little information was publicly available at the time of this writing.

8.3.6 Regional Systems

QZSS transmits navigation signals in four bands. A standard version of the GPS C/A code, a high-data-rate (500 symbol s^{-1}) version, known as L1-SAIF, and the new GPS L1C signal are broadcast in the L1 band. GPS-like L2C, L5I, and L5Q signals are broadcast in the L2 and L5/E5a bands. The final signal, known as the L-band experimental (LEX) signal, shares the frequency and modulation of the Galileo E6-B signals, but achieves a high navigation data rate at 250 symbol s^{-1} by encoding 8 data bits onto each symbol [35].

IRNSS will broadcast a 1.023 Mchip s^{-1} BPSK standard-positioning-service signal and two BOC$_s$(5,2) restricted-service (RS) signals, with and without data, on each of two frequencies, 1,176.45 GHz (L5) and 2,492.08 GHz in the S-band [13].

Table 8.13 Proposed Beidou Phase 3 Signal Properties

Name	Carrier Frequency (MHz)	Modulation and Chipping Rate (×1.023 Mchip s^{-1})	Navigation Message Rate (symbol s^{-1})	Service
B1-CD	1,575.42	BOC$_s$(1,1) or MBOC[a]	100	Open
B1-CP	1,575.42	BOC$_s$(1,1) or MBOC[a]	No	Open
B1D	1,575.42	BOC$_s$(14,2)	100	Authorized
B1P	1,575.42	BOC$_s$(14,2)	No	Authorized
B2aD	1,176.45	BPSK 10	50	Open
B2aP	1,176.45	BPSK 10	No	Open
B2bD	1,207.14	BPSK 10	100	Open
B2bP	1,207.14	BPSK 10	No	Open
B3	1,268.52	QPSK 10	500	Authorized
B3-AD	1,268.52	BOC$_c$(15,2.5)	100	Authorized
B3-AP	1,268.52	BOC$_c$(15,2.5)	No	Authorized

[a]This MBOC signal comprises a 10/11 power BOC$_s$(1,1) modulation and a 1/11 power BOC$_s$(6,1) modulation.

8.3.7 Augmentation Systems

All SBAS systems broadcast a common signal format, originally developed for WAAS, enabling the same receivers to be used. A signal is broadcast on the GPS L1 carrier frequency with the same chipping rate and code length as GPS C/A code, but different PRN codes and a different navigation data message at a rate of 500 symbol s^{-1} [36–38]. A second signal, based on the GPS L5I signal, is being added to WAAS in 2012/13. This may also be added to EGNOS after 2018.

8.4 Navigation Data Messages

The navigation data message serves two main purposes. It enables the complete time of transmission to be deduced, removing the ambiguity introduced by the repetition of the PRN codes. It also enables satellite positions and clock offsets to be determined for which two types of data are provided. Precision ephemeris data is repeated relatively quickly but only describes the satellite that is transmitting it. Almanac data, comprising approximate ephemeris parameters, clock calibration, signal health, and navigation data health information, is provided for the whole constellation. However, the precision is relatively poor and the repetition rate relatively slow. It is intended to aid the user equipment in selecting which satellites to use and acquiring the signals.

Most satellites broadcast several different types of navigation message and both the data rates and the format vary. Message formats may be fixed frame or variable frame. In a fixed-frame format, the data is always transmitted in the same order with the same repetition intervals. This makes it easy for receivers to combine information from successive transmissions when reception is poor. In a variable-frame format, a series of fixed-length messages may be transmitted in any order. This enables integrity alerts to be broadcast quickly when a fault occurs, provides greater flexibility to transmit different information at different rates, and allows new message types to be added.

The newer GNSS navigation data messages incorporate forward error correction (FEC). This introduces redundancy into the data, allowing correction of decoding errors, which enables the data to be successfully decoded in a poorer signal-to-noise environment. However, assuming binary modulation, the rate at which the signal must be modulated, known as the symbol rate, must be higher than the rate at which information is conveyed, known as the data rate. GNSS uses 1/2-rate FEC, so if the symbol rate is 100 symbol s^{-1}, the data rate is 50 bit s^{-1}. The navigation message rates given in Section 8.3 are symbol rates. For a binary message with no FEC, the data and symbol rates are the same.

In this section, the main features of the navigation data messages broadcast by GPS, GLONASS, Galileo, and, SBAS are summarized in turn. More information may be found in the ICDs [24–27, 29, 32, 35–38]. Information about the Beidou navigation messages was not available at the time of this writing. The section concludes by discussing time base synchronization.

8.4.1 GPS

There are four different GPS navigation data messages. The legacy navigation message is broadcast simultaneously on the C/A and both P(Y)-code signals, MNAV messages are broadcast on the M-code signals, and CNAV messages are due to be broadcast on the L2C signal (CM component) and L5I signal. A further, C2NAV, message will be introduced with the L1C signals.

The legacy navigation message is broadcast in a fixed-frame format with no FEC at a data rate of 50 bit s^{-1}. It is divided into 30-bit words of a 0.6-second duration, each incorporating a parity check, while the full message lasts 12.5 minutes [24, 39].

The satellite clock calibration data (see Section 9.3.1) and ephemeris information, expressed as a set of 16 Keplerian orbital parameters (see Section 8.5.2), for the transmitting satellite are broadcast every 30 seconds. Issue of Data Ephemeris (IODE) and Issue of Data Clock (IODC) integers are incremented each time this navigation data is updated, currently every 2 hours. The handover word (HOW), which aids the transition from C/A code to P(Y) code tracking by indicating the number of 1.5-second P(Y)-code periods that have occurred thus far in the week, is transmitted every 6 seconds.

Almanac data for up to 32 satellites is only broadcast every 12.5 minutes. It is valid for longer than the precise ephemeris data, giving satellite positions to an accuracy of 900m up to a day from transmission, 1,200m for one week, and 3,600m for up to two weeks. Also broadcast every 12.5 minutes are the eight coefficients of the Klobuchar ionosphere propagation delay correction model for single-frequency users (see Section 9.3.2) and GPS-UTC time conversion data.

The MNAV and CNAV messages have a variable-frame format with FEC applied. MNAV subframes are 400 bits long and CNAV subframes 300 bits long. The CNAV message data rate is 25 bit s^{-1} on L2C and 50 bit s^{-1} on L5I. These messages incorporate higher-precision ephemeris and satellite clock parameters than the legacy message. The C2NAV message has a hybrid format with a mixture of fixed and variable features.

8.4.2 GLONASS

GLONASS broadcasts different navigation data messages on the C/A-code and P-code signals. Both messages employ a fixed-frame format with no FEC and a data rate of 50 bit s^{-1}. The messages are divided into lines of 100 bits, lasting 2 seconds, each with a parity check. The full C/A-code message repeats every 2.5 minutes, while the P-code message repeats every 12 minutes. The ephemeris and satellite clock information for the transmitting satellite is broadcast every 30 seconds for C/A code and 10 seconds for P code, while the almanac is repeated at the full message rate. GLONASS does not broadcast ionosphere model parameters. The GLONASS navigation message is multiplied by a 100 chip s^{-1} "meander sequence" of length 30 chips [29].

The ephemeris for the transmitting satellite is expressed simply as an ECEF-frame position and velocity, together with the lunisolar acceleration and a reference time, rather than as Keplerian parameters. The user equipment then determines the current position and velocity using a force model. These parameters are quicker to transmit but must be updated every 30 minutes at 15 and 45 minutes past the hour.

The L3 message has a fixed-frame format and a data rate of 100 bit s^{-1}. It comprises eight or ten 15-second frames, repeating every 2 or 2.5 minutes. The ephemeris is repeated every frame and the almanac is repeated at the full message rate [40].

8.4.3 Galileo

Galileo has four different data messages. The Freely-accessible, FNAV, message is carried on the E5a-I signal; the Integrity, INAV, message is carried on the E5b-I and E1-B signals, the Commercial, CNAV, message is centered on the E6-B signal, and the Government-access, GNAV, message is carried on both PRS signals. All messages have a variable-frame structure and use FEC. The data rates are 25 bit s^{-1} for FNAV, 125 bit s^{-1} for INAV and 500 bit s^{-1} for CNAV. The INAV messages on the E5b-I and E1-b signals are staggered, enabling users tracking both signals to download the data more quickly.

Ephemeris and almanac data are similar to that in the GPS legacy message, while the satellite clock parameters are at a higher resolution for both the transmitting satellite and the constellation. A common Issue of Data Navigation (IODNav) integer is incremented when the ephemeris and clock data are updated, every 3 hours. There is also an Issue of Data Almanac (IODA) integer. Three coefficients for the NeQuick ionosphere model are transmitted instead of the Klobuchar model parameters.

Integrity data, including three levels of integrity alert and authentification data, are proposed for transmission on both the INAV and GNAV messages.

8.4.4 SBAS

The SBAS navigation message on the L1 frequency is broadcast in a variable-frame format with FEC at a data rate of 250 bit s^{-1}. Messages are 250 bits long and take 1 second to transmit. The data includes differential corrections for the GPS signals, ionosphere model parameters, data which can be used to estimate the accuracy of the SBAS-corrected pseudo-range measurements, and, for some systems, SBAS satellite position and velocity. Fast corrections messages, normally transmitted every 10 seconds, allow the differential corrections, accuracy, and satellite health data to be updated rapidly. In the event of a rapid satellite signal failure, the fast corrections message can be brought forward to provide an integrity alert to the user within 6 seconds of detection.

8.4.5 Time Base Synchronization

Each GNSS uses a slightly different time base. GPS time is synchronized with Universal Coordinated Time (UTC) as maintained by the U.S. Naval Observatory (USNO). However, it is not subject to leap seconds and is expressed in terms of a week number and the number of seconds from the start of that week (midnight Saturday/Sunday). The week number "rolls over" every 1,024 weeks (19 years and 227/228 days). At the time of this writing, GPS time exhibited meter-order jumps with respect to UTC at each day boundary (i.e., 00:00 UTC). This can make it look as though the receiver clock offset is suddenly changing and must be accounted for by user equipment designers.

GLONASS system time is synchronized with the Russian version of UTC with a 3-hour offset corresponding to Moscow local time. Unlike GPS, leap seconds are applied.

Galileo System Time (GST) is maintained within 50 ns of International Atomic Time, not UTC. Like GPS time, GST is expressed in weeks and seconds, but with a "roll over" after 4096 weeks (about 78 years). Beidou uses Beidou time (BDT), which is also nominally aligned to UTC.

Although all of these time bases are nominally synchronized, the differences are significant in GNSS-ranging terms. Time base conversion data is therefore needed to use signals from different constellations in the same position solution computation. The Galileo data messages include both GST-UTC and Galileo-GPS time conversion data, while GLONASS-GPS time conversion data is included within the GLONASS almanac. The GPS CNAV and C2NAV messages have a flexible data format that can broadcast both Galileo-GPS and GLONASS-GPS time conversion data and can be upgraded to support Beidou and any future systems.

When time conversion data is not available, the interconstellation timing bias must be treated as an additional unknown in the navigation solution. The GPS-GLONASS timescale offset is of the order of 100m in range terms.

8.5 Satellite Orbits and Geometry

This section describes the satellite orbits and signal geometry. It begins by summarizing the orbits of each satellite constellation. The calculation of the satellite positions and velocities from the information in the navigation data message is then described. This is followed by a discussion of range and range-rate computation, including the effect of Earth rotation and the impact of different errors at different processing stages. Finally, the direction of the satellite signal from the user antenna is defined in terms of line-of-sight vector, elevation, and azimuth.

8.5.1 Satellite Orbits

GPS operates with a nominal constellation of 24 satellites. However, spare satellites provide a full service so up to 36 satellites may be transmitting at once. GLONASS and Galileo are designed to operate with 24 and 27 satellites, respectively. Their spare satellites do not transmit to users, but are instead kept on standby until an older satellite fails. All GPS, GLONASS, and Galileo satellites are in mid-Earth orbits. Beidou phase 2 comprises four satellites in geostationary orbit, five satellites in inclined geosynchronous orbit (IGSO), and three MEO satellites. Phase 3 will comprise three geostationary satellites, three IGSO satellites, and 27 MEO satellites.

The properties of the mid-Earth orbits for all four constellations are listed in Table 8.14 [4, 15, 41]. The orbital planes are evenly spaced in longitude and are

Table 8.14 Properties of GNSS Mid-Earth Orbits

Constellation	Number of Planes	Radius (km)	Height (km)	Period	Orbits per Sidereal Day	Ground-Track Repeat Period (Sidereal Days)	Inclination Angle
GPS	6	26,580	20,180	11 hr, 58 min	2	1	55°
GLONASS	3	25,500	19,100	11 hr, 15 min	2.125	8	64.8°
Galileo	3	29,620	23,220	14 hr, 5 min	1.7	10	56°
Beidou	3	27,840	21,440	12 hr, 52 min	1.857	7	55°

depicted in Figure 8.21. The orbital periods listed are with respect to inertial space. The ground track of a satellite is the locus of points directly below the satellite on the surface of the Earth. The interval over which it repeats is the lowest common multiple of the Earth rotation period and satellite orbit period. As GPS ground tracks nearly repeat every sidereal day, the constellation precesses with respect to the Earth's surface by roughly 4 minutes per solar day. The inclination angle is the angle between the orbital and equatorial planes.

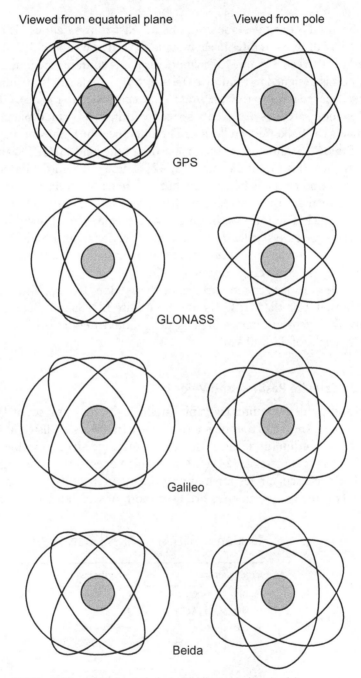

Figure 8.21 GNSS satellite orbits (to scale but not aligned in longitude).

Table 8.15 Properties of GNSS Inclined Geosynchronous Orbits

Constellation	Number of Satellites	Inclination Angle	Longitude(s) of Equatorial Crossing(s)
Beidou	5 (phase 2)	55°	118°
QZSS	3	45°	126° and 149°
IRNSS	2	55°	55°
IRNSS	2	55°	112°

Each GPS orbital plane contains at least four satellites. These are not evenly spaced, with two satellites in each plane separated by about 30° and the others separated by between 92° and 137° where the plane contains the minimum four satellites [4]. This is designed to minimize the effect of a single satellite outage. New satellites may be placed close to the satellites they are intended to replace. GLONASS, Galileo, and Beidou satellites are evenly spaced within their orbital planes, a configuration known as a Walker constellation. The tolerance for Galileo is ±2°.

Table 8.15 summarizes the properties of the Beidou, QZSS, and IRNSS inclined geosynchronous satellite orbits [15, 42]. These have an orbital and ground-track-repeat period of one sidereal day, but unlike geostationary orbits, they move with respect to the Earth's surface with varying latitude, producing a figure-of-eight ground track The QZSS ground track is asymmetric about the equator, ensuring that there is always at least one satellite over Japan at a high elevation angle. This is the origin of the term *quasi-zenith*. IGSO satellites have also been proposed for extending SBAS coverage to polar regions.

Table 8.16 lists the longitudes of the Beidou, IRNSS, and SBAS geostationary satellites, noting that GAGAN and SNAS use IRNSS and Beidou satellites, respectively. Both geostationary and geosynchronous satellites orbit at a radius of 42,200 km (a height of 35,800 km).

8.5.2 Satellite Position and Velocity

GPS and Galileo satellites transmit satellite orbit data as a set of 16 quasi-Keplerian parameters, known as the ephemeris. These parameters are listed in Table 8.17, including the resolution (in terms of the least significant bit) applicable to the legacy GPS navigation data message [24] and the Galileo open-service messages [32]. Note that the ICDs use semicircles for many of the angular terms, whereas radians are used here. Two further parameters are used, both of which are considered constant: the

Table 8.16 Summary of GNSS and SBAS Geostationary Satellites

Constellation	Longitudes
Beidou and SNAS	58.8°, 84°, 139.9°, 160°
IRNSS and GAGAN	34°, 83°, 132°
WAAS	−133°, −107.3°, −98°
EGNOS	−15.5°, 21.5°, 25°
MSAS	140°, 145°
SDCM	−16°, 95°, 167°

8.5 Satellite Orbits and Geometry

Table 8.17 GPS and Galileo Satellite Orbit Ephemeris Parameters

Symbol	Description	Resolution (LSB)
t_{oe}	Reference time of the ephemeris	$2^4 = 16$ seconds
M_0	Mean anomaly at the reference time	$2^{-31}\pi = 1.462918079 \times 10^{-9}$ rad
e_o	Eccentricity of the orbit	$2^{-33} = 1.164153218 \times 10^{-10}$
$a^{1/2}$	Square root of the semi-major axis	$2^{-19} = 1.907348633 \times 10^{-6}$ m$^{-0.5}$
Ω_0	Right ascension of ascending node of orbital plane at the weekly epoch	$2^{-31}\pi = 1.462918079 \times 10^{-9}$ rad
i_0	Inclination angle at the reference time	$2^{-31}\pi = 1.462918079 \times 10^{-9}$ rad
ω	Argument of perigee	$2^{-31}\pi = 1.462918079 \times 10^{-9}$ rad
Δn	Mean motion difference from computed value	$2^{-43}\pi = 3.57158 \times 10^{-13}$ rad s^{-1}
$\dot{\Omega}_d$	Rate of change of longitude of the ascending node at the reference time	$2^{-43}\pi = 3.5715773 \times 10^{-13}$ rad s^{-1}
$\dot{i}_d$	Rate of inclination	$2^{-43}\pi = 3.5716 \times 10^{-13}$ rad s^{-1}
C_{uc}	Amplitude of the cosine harmonic correction term to the argument of latitude	$2^{-29} = 1.86265 \times 10^{-9}$ rad
C_{us}	Amplitude of the sine harmonic correction term to the argument of latitude	$2^{-29} = 1.86265 \times 10^{-9}$ rad
C_{rc}	Amplitude of the cosine harmonic correction term to the orbit radius	$2^{-5} = 0.03125$m
C_{rs}	Amplitude of the sine harmonic correction term to the orbit radius	$2^{-5} = 0.03125$m
C_{ic}	Amplitude of the cosine harmonic correction term to the angle of inclination	$2^{-29} = 1.86265 \times 10^{-9}$ rad
C_{is}	Amplitude of the sine harmonic correction term to the angle of inclination	$2^{-29} = 1.86265 \times 10^{-9}$ rad

Earth-rotation rate, ω_{ie} (see Section 2.4.6), and the Earth's gravitational constant, μ (see Section 2.4.7).

Although most GNSS satellite orbits are nominally circular, the eccentricity of the orbit must be accounted for in order to accurately determine the satellite position. A two-body Keplerian model is used as the baseline for the satellite motion. This assumes the satellite moves in an ellipse, subject to the gravitational force of a point source at one focus of the ellipse [3, 6]. Seven parameters are used to describe a pure Keplerian orbit: a reference time, t_{oe}, three parameters describing the satellite orbit within the orbital plane, and three parameters describing the orientation of that orbit with respect to the Earth.

Figure 8.22 illustrates the satellite motion within the orbital plane. The size of the orbit is defined by the length of the semi-major axis, a. This is simply the radius of the orbit at its largest point. The shape of the orbit is defined by the eccentricity, e_o, where the subscript o has been added to distinguish it from the eccentricity of the Earth's surface. The two foci are each located at a distance $e_o a$ along the semi-major axis from the center of the ellipse. The center of the Earth is at one focus. The *perigee* is defined as the point of the orbit that approaches closest to the center of the Earth and is located along the semi-major axis. The direction of perigee points from the center of the ellipse to the perigee, via the center of the Earth. Finally, the location of the satellite within the orbit at the reference time is defined by the true anomaly, v, which is the angle in the counterclockwise direction from the direction of perigee to the line of sight from the center of the Earth to the satellite. The true anomaly does not vary at a constant rate over the orbit, so GNSS satellites broadcast

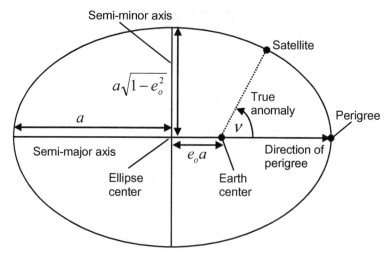

Figure 8.22 Satellite motion within the orbital plane.

the mean anomaly, ν, which does vary at a constant rate and from which the true anomaly can be calculated.

Figure 8.23 illustrates the orientation of the orbital plane with respect to the Earth. The inclination angle, i, is the angle subtended by the normal to the orbital plane and the polar axis of the Earth and takes values between 0° and 90°. The *ascending node* is the point where the orbit crosses the Earth's equatorial plane while the satellite is moving in the positive z-direction of an ECI or ECEF frame (i.e., south to north). The *descending node* is where the orbit crosses the equatorial plane in the opposite direction. The ascending and descending nodes are nominally fixed in an ECI frame, but move within an ECEF frame as the Earth rotates. Therefore, the longitude of the ascending node, Ω, also known as the right ascension, is defined at the reference time.

The final term determining the orbit is the orientation of the direction of perigree within the orbital plane. This is defined using the argument of perigree, ω, which is the angle in the counterclockwise direction from the direction of the ascending node from the center of the Earth to the direction of perigree. Figure 8.24 illustrates this, together with the axes of an orbital coordinate frame, which is denoted by the symbol o and centered at the Earth's center of mass, like ECI and ECEF frames. The x-axis of an orbital frame defines the direction of the ascending node and lies in the Earth's equatorial plane. The z-axis defines the normal to the equatorial plane in the Earth's northern hemisphere, as shown in Figure 8.23, and the y-axis completes the right-handed orthogonal set.

GNSS satellites depart from pure Keplerian motion due to a combination of nonuniformity of the Earth's gravitational field, the gravitational fields of the Sun and Moon, solar radiation pressure, and other effects. These are approximated by the remaining ephemeris parameters: the mean motion correction, rates of change of the inclination and longitude of the ascending node, and the six harmonic correction terms.

Calculation of the satellite position comprises two steps: determination of the position within an orbital coordinate frame and transformation of this to an ECEF

8.5 Satellite Orbits and Geometry

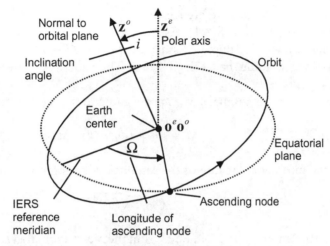

Figure 8.23 Orientation of an orbital plane with respect to the equatorial plane.

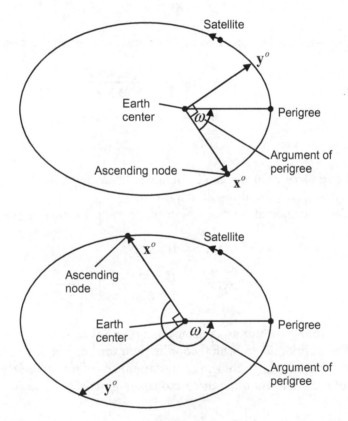

Figure 8.24 The argument of perigree and orbital coordinate frame axes.

or ECI frame as required. However, the time of signal transmission relative to the ephemeris reference time must first be determined:

$$\Delta t = t_{st,a}^s - t_{oe}. \tag{8.6}$$

The GPS ephemeris reference time is transmitted relative to the start of the GPS week (see Section 8.4.1). Assuming the same for the time of signal transmission, $t^s_{st,a}$, it is sometimes necessary to apply a ±604,800-second correction when the two times straddle the week crossover. This should be done where $|\Delta t| > 302{,}400$ seconds.

Next, the mean anomaly, M, is propagated to the signal transmission time using

$$M = M_0 + (\bar{\omega}_{is} + \Delta n)\Delta t, \tag{8.7}$$

where the mean angular rate of the satellite's orbital motion, $\bar{\omega}_{is}$, is given by

$$\bar{\omega}_{is} = \sqrt{\mu/a^3}. \tag{8.8}$$

The true anomaly, n, is obtained from the mean anomaly via the eccentric anomaly, E. The eccentric anomaly, explained in [3, 6], is obtained using Kepler's equation:

$$M = E - e_o \sin E, \tag{8.9}$$

which must be solved iteratively. A common numerical solution is

$$\begin{aligned} E_0 &= M + \frac{e_o \sin M}{1 - \sin(M + e_o) + \sin M} \\ E_i &= M + e_o \sin E_{i-1} \quad i = 1, 2 \ldots n \\ E &= E_n \end{aligned} \tag{8.10}$$

Centimetric accuracy can be obtained from 20 iterations (i.e., $n = 20$), with millimetric accuracy requiring 22 iterations.

The true anomaly is then obtained from the eccentric anomaly using

$$\begin{aligned} v &= \arctan_2(\sin v, \cos v) \\ &= \arctan_2\left[\left(\frac{\sqrt{1 - e_o^2}\sin E}{1 - e_o \cos E}\right), \left(\frac{\cos E - e_o}{1 - e_o \cos E}\right)\right], \end{aligned} \tag{8.11}$$

where a four-quadrant arctangent function must be used.

The position in an orbital coordinate frame may be expressed in polar coordinates comprising the radius, r^o_{os}, and argument of latitude, Φ, which is simply the sum of the argument of perigee and true anomaly, so

$$\Phi = \omega + v. \tag{8.12}$$

The orbital radius varies as a function of the eccentric anomaly, while harmonic perturbations are applied to both terms, giving

$$\begin{aligned} r^o_{os} &= a(1 - e_o \cos E) + C_{rs}\sin 2\Phi + C_{rc}\cos 2\Phi \\ u^o_{os} &= \Phi + C_{us}\sin 2\Phi + C_{uc}\cos 2\Phi \end{aligned}, \tag{8.13}$$

8.5 Satellite Orbits and Geometry

where u_{os}^o is the corrected argument of latitude.

The satellite position in an orbital frame is then

$$x_{os}^o = r_{os}^o \cos u_{os}^o, \qquad y_{os}^o = r_{os}^o \sin u_{os}^o, \qquad z_{os}^o = 0. \qquad (8.14)$$

The position in an ECEF or ECI frame is obtained by applying a coordinate transformation matrix as orbital frames have the same origin. Thus,

$$\mathbf{r}_{es}^e = \mathbf{C}_o^e \mathbf{r}_{os}^o, \qquad \mathbf{r}_{is}^i = \mathbf{C}_o^i \mathbf{r}_{os}^o. \qquad (8.15)$$

The Euler rotation from an ECEF to an orbital frame comprises a yaw rotation through the longitude of the ascending node, Ω, followed by a roll rotation through the inclination angle, i. For GPS, the longitude of the ascending node is transmitted at the week epoch, rather than the reference time, so its value at the time of signal transmission is

$$\Omega = \Omega_0 - \omega_{ie}(\Delta t + t_{oe}) + \dot{\Omega}_d \Delta t \qquad (8.16)$$

while the inclination angle is corrected using

$$i = i_0 + \dot{i}_d \Delta t + C_{is} \sin 2\Phi + C_{ic} \cos 2\Phi. \qquad (8.17)$$

Applying (2.24) with $\psi_{eo} = \Omega$, $\phi_{eo} = i$, and $\theta_{eo} = 0$ gives

$$\mathbf{C}_o^e = \begin{pmatrix} \cos\Omega & -\cos i \sin\Omega & \sin i \sin\Omega \\ \sin\Omega & \cos i \cos\Omega & -\sin i \cos\Omega \\ 0 & \sin i & \cos i \end{pmatrix}. \qquad (8.18)$$

Thus, from (8.15), the ECEF-frame satellite position is

$$\mathbf{r}_{es}^e = \begin{pmatrix} x_{os}^o \cos\Omega - y_{os}^o \cos i \sin\Omega \\ x_{os}^o \sin\Omega + y_{os}^o \cos i \cos\Omega \\ y_{os}^o \sin i \end{pmatrix} \qquad (8.19)$$

and, applying (2.145) and (2.146), the ECI-frame satellite position is

$$\mathbf{r}_{is}^i = \begin{Bmatrix} x_{os}^o \cos[\Omega + \omega_{ie}(t_{st,a}^s - t_0)] - y_{os}^o \cos i \sin[\Omega + \omega_{ie}(t_{st,a}^s - t_0)] \\ x_{os}^o \sin[\Omega + \omega_{ie}(t_{st,a}^s - t_0)] + y_{os}^o \cos i \cos[\Omega + \omega_{ie}(t_{st,a}^s - t_0)] \\ y_{os}^o \sin i \end{Bmatrix}, \qquad (8.20)$$

where t_0 is the time of coincidence of the ECI and ECEF-frame axes.

From (2.67), the satellite velocity is obtained simply by differentiating the position with respect to $t^s_{st,a}$. Differentiating (8.9) to (8.14) gives the satellite velocity in an orbital frame:

$$\dot{E} = \frac{\bar{\omega}_{is} + \Delta_n}{1 - e_o \cos E}, \tag{8.21}$$

$$\dot{\Phi} = \frac{\sin v}{\sin E} \dot{E}, \tag{8.22}$$

$$\begin{aligned} \dot{r}^o_{os} &= (ae_o \sin E)\dot{E} + 2(C_{rs} \cos 2\Phi - C_{rc} \sin 2\Phi)\dot{\Phi} \\ \dot{u}^o_{os} &= (1 + 2C_{us} \cos 2\Phi - 2C_{uc} \sin 2\Phi)\dot{\Phi} \end{aligned}, \tag{8.23}$$

$$\begin{aligned} \dot{x}^o_{os} &= \dot{r}^o_{os} \cos u^o_{os} - r^o_{os} \dot{u}^o_{os} \sin u^o_{os} \\ \dot{y}^o_{os} &= \dot{r}^o_{os} \sin u^o_{os} + r^o_{os} \dot{u}^o_{os} \cos u^o_{os}. \\ \dot{z}^o_{os} &= 0 \end{aligned} \tag{8.24}$$

Differentiating (8.16) and (8.17) gives

$$\dot{\Omega} = \dot{\Omega}_d - \omega_{ie}, \tag{8.25}$$

$$\dot{i} = \dot{i}_d + 2(C_{is} \cos 2\Phi - C_{ic} \sin 2\Phi)\dot{\Phi}. \tag{8.26}$$

Differentiating (8.19) and (8.20) then gives the ECEF-frame and ECI-frame satellite velocities:

$$\mathbf{v}^e_{es} = \begin{pmatrix} \dot{x}^o_{os} \cos\Omega - \dot{y}^o_{os} \cos i \sin\Omega + \dot{i} y^o_{os} \sin i \sin\Omega \\ \dot{x}^o_{os} \sin\Omega + \dot{y}^o_{os} \cos i \cos\Omega - \dot{i} y^o_{os} \sin i \cos\Omega \\ \dot{y}^o_{os} \sin i + \dot{i} y^o_{os} \cos i \end{pmatrix} + (\omega_{ie} - \dot{\Omega}_d) \begin{pmatrix} x^o_{os} \sin\Omega + y^o_{os} \cos i \cos\Omega \\ -x^o_{os} \cos\Omega + y^o_{os} \cos i \sin\Omega \\ 0 \end{pmatrix}, \tag{8.27}$$

$$\mathbf{v}^i_{is} = \begin{Bmatrix} \dot{x}^o_{os} \cos[\Omega + \omega_{ie}(t^s_{st,a} - t_0)] - \dot{y}^o_{os} \cos i \sin[\Omega + \omega_{ie}(t^s_{st,a} - t_0)] + \dot{i} y^o_{os} \sin i \sin[\Omega + \omega_{ie}(t^s_{st,a} - t_0)] \\ \dot{x}^o_{os} \sin[\Omega + \omega_{ie}(t^s_{st,a} - t_0)] + \dot{y}^o_{os} \cos i \cos[\Omega + \omega_{ie}(t^s_{st,a} - t_0)] - \dot{i} y^o_{os} \sin i \cos[\Omega + \omega_{ie}(t^s_{st,a} - t_0)] \\ \dot{y}^o_{os} \sin i + \dot{i} y^o_{os} \cos i \end{Bmatrix} \\ -\dot{\Omega}_d \begin{Bmatrix} x^o_{os} \sin[\Omega + \omega_{ie}(t^s_{st,a} - t_0)] + y^o_{os} \cos i \cos[\Omega + \omega_{ie}(t^s_{st,a} - t_0)] \\ -x^o_{os} \cos[\Omega + \omega_{ie}(t^s_{st,a} - t_0)] + y^o_{os} \cos i \sin[\Omega + \omega_{ie}(t^s_{st,a} - t_0)] \\ 0 \end{Bmatrix}. \tag{8.28}$$

8.5 Satellite Orbits and Geometry

When the satellite position has been calculated at an approximate time of transmission, $\tilde{t}^s_{st,a}$, it may be corrected using

$$\mathbf{r}^e_{es}(t^s_{st,a}) \approx \mathbf{r}^e_{es}(\tilde{t}^s_{st,a}) + \left(t^s_{st,a} - \tilde{t}^s_{st,a}\right)\mathbf{v}^e_{es}(\tilde{t}^s_{st,a})$$
$$\mathbf{r}^i_{is}(t^s_{st,a}) \approx \mathbf{r}^i_{is}(\tilde{t}^s_{st,a}) + \left(t^s_{st,a} - \tilde{t}^s_{st,a}\right)\mathbf{v}^i_{is}(\tilde{t}^s_{st,a})$$
(8.29)

provided the time correction does not exceed 1 second.

Section G.4 of Appendix G on the CD presents the corresponding accelerations and a description of how to use a force model to determine the satellite position, velocity, and acceleration from the ECEF-frame position, velocity, and lunisolar acceleration broadcast in the GLONASS FDMA navigation messages.

Table 8.18 presents the mean orbital radius, inertially referenced speed, and angular rate for the GPS, GLONASS, Galileo, and Beidou constellations.

8.5.3 Range, Range Rate, and Line of Sight

The true range, r_{as}, is the distance between the satellite, s, at the time of signal transmission, $t^s_{st,a}$, and the user's antenna, a, at the time of signal arrival, $t^s_{sa,a}$. As Figure 8.25 shows, it is important to account for the signal transit time as the satellite-user distance generally changes over this interval, even where the user is stationary with respect to the Earth. The user equipment obtains pseudo-range measurements by multiplying its transit-time measurements by the speed of light. The speed of light in free space is only constant in an inertial frame, where $c = 299{,}792{,}458$ m s^{-1}. In a rotating frame, such as ECEF, the speed of light varies. Consequently, the true range calculation is simplest in an ECI frame. Thus,

$$r_{as} = \left|\mathbf{r}^i_{is}(t^s_{st,a}) - \mathbf{r}^i_{ia}(t^s_{sa,a})\right| = \sqrt{\left(\mathbf{r}^i_{is}(t^s_{st,a}) - \mathbf{r}^i_{ia}(t^s_{sa,a})\right)^{\mathrm{T}}\left(\mathbf{r}^i_{is}(t^s_{st,a}) - \mathbf{r}^i_{ia}(t^s_{sa,a})\right)}. \quad (8.30)$$

As the user position is computed with respect to the Earth and the GPS interface standard [24] gives formulas for computing the satellite position in ECEF coordinates, it is convenient to compute the range in an ECEF frame. However, this neglects the rotation of the Earth during the signal transit time, causing the range to be overestimated or underestimated as Figure 8.25 illustrates. At the equator, the range error can be up to 41m [43]. To compensate for this, a correction, $\delta\rho^s_{ie,a}$, known as the Sagnac or Earth-rotation correction, must be applied. Thus,

$$r_{as} = \left|\mathbf{r}^e_{es}(t^s_{st,a}) - \mathbf{r}^e_{ea}(t^s_{sa,a})\right| + \delta\rho^s_{ie,a}. \quad (8.31)$$

Table 8.18 GNSS Satellite Orbital Radii, Speeds, and Angular Rates

Constellation	GPS	GLONASS	Galileo	Beidou
Mean orbital radius, $\bar{r}_{es}$ (km)	26,580	25,500	29,620	27,840
Mean satellite speed, $\bar{v}_{is}$ (m s^{-1})	3,870	3,950	3,670	3,780
Mean orbital angular rate, $\bar{\omega}_{is}$ (rad s^{-1})	1.46×10^{-4}	1.55×10^{-4}	1.24×10^{-4}	1.36×10^{-4}

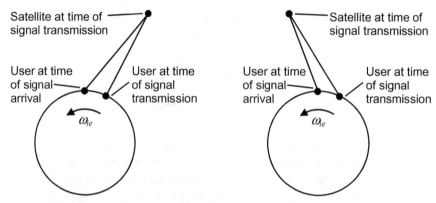

Figure 8.25 Effect of Earth rotation on range calculation (inertial frame perspective, user stationary with respect to the Earth).

However, computing the Sagnac correction exactly requires calculation of ECI-frame satellite and user positions, so an approximation is generally used:

$$\delta\rho_{ie,a}^s \approx \frac{\omega_{ie}}{c}\left[y_{es}^e(t_{st,a}^s)x_{ea}^e(t_{sa,a}^s) - x_{es}^e(t_{st,a}^s)y_{ea}^e(t_{sa,a}^s)\right]. \tag{8.32}$$

Example 8.1 on the CD illustrates this and is editable using Microsoft Excel.

The convenience of an ECEF-frame calculation can be combined with the accuracy of an ECI-frame calculation by aligning the ECI-frame axes with the ECEF-frame axes at the time of signal arrival or transmission [43]. From (2.145) and (2.146),

$$\mathbf{r}_{Ia}^I(t_{sa,a}^s) = \mathbf{r}_{ea}^e(t_{sa,a}^s) \qquad \mathbf{r}_{Is}^I(t_{st,a}^s) = \mathbf{C}_e^I(t_{st,a}^s)\mathbf{r}_{es}^e(t_{st,a}^s), \tag{8.33}$$

where I denotes an ECI frame synchronized with a corresponding ECEF frame at the time of signal arrival and

$$\mathbf{C}_e^I(t) = \begin{pmatrix} \cos\omega_{ie}(t - t_{sa,a}^s) & -\sin\omega_{ie}(t - t_{sa,a}^s) & 0 \\ \sin\omega_{ie}(t - t_{sa,a}^s) & \cos\omega_{ie}(t - t_{sa,a}^s) & 0 \\ 0 & 0 & 1 \end{pmatrix}. \tag{8.34}$$

The range is then given by

$$r_{as} = \left|\mathbf{C}_e^I(t_{st,a}^s)\mathbf{r}_{es}^e(t_{st,a}^s) - \mathbf{r}_{ea}^e(t_{sa,a}^s)\right|. \tag{8.35}$$

The small angle approximation may be applied to the rotation of the Earth during the signal transit time. Therefore,

$$\mathbf{C}_e^I(t_{st,a}^s) \approx \begin{pmatrix} 1 & \omega_{ie}(t_{sa,a}^s - t_{st,a}^s) & 0 \\ -\omega_{ie}(t_{sa,a}^s - t_{st,a}^s) & 1 & 0 \\ 0 & 0 & 1 \end{pmatrix} = \begin{pmatrix} 1 & \omega_{ie}r_{as}/c & 0 \\ -\omega_{ie}r_{as}/c & 1 & 0 \\ 0 & 0 & 1 \end{pmatrix}, \tag{8.36}$$

8.5 Satellite Orbits and Geometry

as

$$r_{as} = \left(t^s_{sa,a} - t^s_{st,a}\right) c, \qquad (8.37)$$

noting that the range, not the pseudo-range, must be used.

The direction from which a satellite signal arrives at the user antenna may be described by a unit vector. The unit vector describing the direction of the origin of frame α with respect to the origin of frame β, resolved about the axes of frame γ, is denoted $\mathbf{u}^\gamma_{\beta\alpha}$ (some authors use l or e). Unit vectors have the property

$$\mathbf{u}^{\gamma\,\mathrm{T}}_{\beta\alpha}\mathbf{u}^\gamma_{\beta\alpha} \equiv \mathbf{u}^\gamma_{\beta\alpha} \cdot \mathbf{u}^\gamma_{\beta\alpha} = 1, \qquad (8.38)$$

and the resolving axes are transformed using a coordinate transformation matrix:

$$\mathbf{u}^\delta_{\beta\alpha} = \mathbf{C}^\delta_\gamma \mathbf{u}^\gamma_{\beta\alpha}. \qquad (8.39)$$

The line-of-sight unit vector from the user antenna, a, to satellite, s, resolved about ECI-frame axes, is

$$\mathbf{u}^i_{as} = \frac{\mathbf{r}^i_{is}(t^s_{st,a}) - \mathbf{r}^i_{ia}(t^s_{sa,a})}{\left|\mathbf{r}^i_{is}(t^s_{st,a}) - \mathbf{r}^i_{ia}(t^s_{sa,a})\right|} = \frac{\mathbf{r}^i_{is}(t^s_{st,a}) - \mathbf{r}^i_{ia}(t^s_{sa,a})}{r_{as}}, \qquad (8.40)$$

The corresponding ECEF-frame line-of-sight vector is

$$\mathbf{u}^e_{as} = \mathbf{C}^e_i(t^s_{sa,a})\mathbf{u}^i_{as} \approx \frac{\mathbf{r}^e_{es}(t^s_{st,a}) - \mathbf{r}^e_{ea}(t^s_{sa,a})}{\left|\mathbf{r}^e_{es}(t^s_{st,a}) - \mathbf{r}^e_{ea}(t^s_{sa,a})\right|}. \qquad (8.41)$$

The range rate is the rate of change of the range. Differentiating (8.30),

$$\dot{r}_{as} = \frac{\left(\mathbf{r}^i_{is}(t^s_{st,a}) - \mathbf{r}^i_{ia}(t^s_{sa,a})\right)^{\mathrm{T}}\left(\dot{\mathbf{r}}^i_{is}(t^s_{st,a}) - \dot{\mathbf{r}}^i_{ia}(t^s_{sa,a})\right)}{r_{as}}. \qquad (8.42)$$

Thus, applying (8.40), the range rate is obtained by resolving the satellite–antenna velocity difference along the line-of-sight unit vector:

$$\dot{r}_{as} = \mathbf{u}^{i\,\mathrm{T}}_{as}\left(\mathbf{v}^i_{is}(t^s_{st,a}) - \mathbf{v}^i_{ia}(t^s_{sa,a})\right). \qquad (8.43)$$

The maximum range rate for a user that is stationary with respect to the Earth is 1,200 m s^{-1}. The largest range rates occur at the equator where the inertially referenced velocity due to Earth rotation is maximized.

Applying (2.147), the range rate may be obtained from ECEF-frame velocities using

$$\dot{r}_{as} = \mathbf{u}^{e\,\mathrm{T}}_{as}\left[\mathbf{C}^I_e(t^s_{st,a})\left(\mathbf{v}^e_{es}(t^s_{st,a}) + \mathbf{\Omega}^e_{ie}\mathbf{r}^e_{es}(t^s_{st,a})\right) - \left(\mathbf{v}^e_{ea}(t^s_{sa,a}) + \mathbf{\Omega}^e_{ie}\mathbf{r}^e_{ea}(t^s_{sa,a})\right)\right]. \qquad (8.44)$$

or, from (8.31),

$$\dot{r}_{as} = \mathbf{u}_{as}^{e\,\mathrm{T}}\left(\mathbf{v}_{es}^{e}(t_{st,a}^{s}) - \mathbf{v}_{ea}^{e}(t_{sa,a}^{s})\right) + \delta\dot{\rho}_{ie,a}^{s}, \tag{8.45}$$

where the range-rate Sagnac correction is approximately

$$\delta\dot{\rho}_{ie,a}^{s} \approx \frac{\omega_{ie}}{c}\begin{pmatrix} v_{es,y}^{e}(t_{st,a}^{s})x_{ea}^{e}(t_{sa,a}^{s}) + y_{es}^{e}(t_{st,a}^{s})v_{ea,x}^{e}(t_{sa,a}^{s}) \\ -v_{es,x}^{e}(t_{st,a}^{s})y_{ea}^{e}(t_{sa,a}^{s}) - x_{es}^{e}(t_{st,a}^{s})v_{ea,y}^{e}(t_{sa,a}^{s}) \end{pmatrix}, \tag{8.46}$$

Applying (8.45) without the Sagnac correction leads to a range-rate error of up to 2 mm s^{-1}. Line-of-sight unit vector and range rate calculations are also included in Example 8.1 on the CD.

Section G.4 of Appendix G on the CD describes how range acceleration may be calculated.

The true range and range rate are only of academic interest. A number of different ranges, pseudo-ranges, range rates, and pseudo-range rates apply at different stages of the GNSS processing chain. The effective range that would be measured if the receiver and satellite clocks were perfectly synchronized is longer than the true range due to the refraction of the signal by the ionosphere and troposphere. Furthermore, the receiver actually measures the pseudo-range, which is also perturbed by the satellite and receiver clock errors as described in Section 8.1.3. The pseudo-range and pseudo-range rate measured by the user equipment for satellite s signal l are given by

$$\begin{aligned} \rho_{a,R}^{s,l} &= r_{as} + \delta\rho_{I,a}^{s,l} + \delta\rho_{T,a}^{s} - \delta\rho_{c}^{s,l} + \delta\rho_{c}^{a} \\ \dot{\rho}_{a,R}^{s,l} &= \dot{r}_{as} + \delta\dot{\Phi}_{I,a}^{s,l} + \delta\dot{\rho}_{T,a}^{s} - \delta\dot{\rho}_{c}^{s} + \delta\dot{\rho}_{c}^{a} \end{aligned}, \tag{8.47}$$

where $\delta\rho_{I,a}^{s,l}$, $\delta\Phi_{I,a}^{s,l}$, and $\delta\rho_{T,a}^{s}$ are, respectively, the modulation ionosphere, carrier ionosphere, and troposphere propagation errors (see Section 9.3.2), $\delta\rho_{c}^{s,l}$ is the range error due to the satellite clock (see Section 9.3.1), $\delta\rho_{c}^{a}$ the range error due to the receiver clock (see Section 9.1.2), and $\delta\dot{\rho}_{I,a}^{s,l}$, $\delta\dot{\Phi}_{I,a}^{s,l}$, $\delta\dot{\rho}_{T,a}^{s}$, $\delta\dot{\rho}_{c}^{s}$, and $\delta\dot{\rho}_{c}^{a}$ are their range-rate counterparts.

The raw pseudo-range and pseudo-range-rate measurements made by the receiver incorporate additional errors:

$$\begin{aligned} \tilde{\rho}_{a,R}^{s,l} &= \rho_{a,R}^{s,l} + \delta\rho_{M,a}^{s,l} + w_{\rho,a}^{s,l} \\ \tilde{\dot{\rho}}_{a,R}^{s,l} &= \dot{\rho}_{a,R}^{s,l} + \delta\dot{\rho}_{M,a}^{s,l} + w_{r,a}^{s,l} \end{aligned}, \tag{8.48}$$

where $w_{\rho,a}^{s,l}$ and $w_{r,a}^{s,l}$ are the tracking errors (Section 9.3.3) and $\delta\rho_{M,a}^{s,l}$ and $\delta\dot{\rho}_{M,a}^{s,l}$ are the errors due to multipath and/or NLOS reception (Section 9.3.4).

The navigation processor uses pseudo-range and pseudo-range-rate measurements with corrections applied. These are

8.5 Satellite Orbits and Geometry

$$\begin{aligned}\tilde{\rho}_{a,C}^{s,l} &= \tilde{\rho}_{a,R}^{s,l} - \delta\hat{\rho}_{I,a}^{s,l} - \delta\hat{\rho}_{T,a}^{s} + \delta\hat{\rho}_{c}^{s,l} \\ \tilde{\rho}_{a,C}^{s,l} &= \tilde{\rho}_{a,R}^{s,l} + \delta\hat{\rho}_{c}^{s}\end{aligned} \qquad (8.49)$$

where $\delta\hat{\rho}_{I,a}^{s,l}$ and $\delta\hat{\rho}_{T,a}^{s}$ are, respectively, the estimated modulation ionosphere and troposphere propagation errors (see Section 9.3.2), and $\delta\hat{\rho}_{c}^{s,l}$ and $\delta\hat{\rho}_{c}^{s}$ are the estimated satellite clock offset and drift (see Section 9.3.1).

Finally, most navigation processors make use of an estimated pseudo-range and pseudo-range rate given by

$$\begin{aligned}\hat{\rho}_{a,C}^{s} &= \left|\hat{\mathbf{r}}_{is}^{i}(\hat{t}_{st,a}^{s}) - \hat{\mathbf{r}}_{ia}^{i}(\hat{t}_{sa,a}^{s})\right| + \delta\hat{\rho}_{c}^{a}(\hat{t}_{sa,a}^{s}) \\ &= \left|\mathbf{C}_{e}^{I}(\hat{t}_{st,a}^{s})\hat{\mathbf{r}}_{es}^{e}(\hat{t}_{st,a}^{s}) - \hat{\mathbf{r}}_{ea}^{e}(\hat{t}_{sa,a}^{s})\right| + \delta\hat{\rho}_{c}^{a}(\hat{t}_{sa,a}^{s})\end{aligned} \qquad (8.50)$$

and

$$\begin{aligned}\dot{\hat{\rho}}_{a,C}^{s} &= \hat{\mathbf{u}}_{as}^{i}{}^{T}\left(\hat{\mathbf{v}}_{is}^{i}(\hat{t}_{st,a}^{s}) - \hat{\mathbf{v}}_{ia}^{i}(\hat{t}_{sa,a}^{s})\right) + \delta\dot{\hat{\rho}}_{c}^{a}(\hat{t}_{sa,a}^{s}) \\ &= \hat{\mathbf{u}}_{as}^{e}{}^{T}\left[\mathbf{C}_{e}^{I}(\hat{t}_{st,a}^{s})\left(\hat{\mathbf{v}}_{es}^{e}(\hat{t}_{st,a}^{s}) + \mathbf{\Omega}_{ie}^{e}\hat{\mathbf{r}}_{es}^{e}(\hat{t}_{st,a}^{s})\right) - \left(\hat{\mathbf{v}}_{ea}^{e}(\hat{t}_{sa,a}^{s}) + \mathbf{\Omega}_{ie}^{e}\hat{\mathbf{r}}_{ea}^{e}(\hat{t}_{sa,a}^{s})\right)\right] + \delta\dot{\hat{\rho}}_{c}^{a}(\hat{t}_{sa,a}^{s})\end{aligned} \qquad (8.51)$$

where $\hat{\mathbf{r}}_{is}^{i}$ or $\hat{\mathbf{r}}_{es}^{e}$ and $\hat{\mathbf{v}}_{is}^{i}$ or $\hat{\mathbf{v}}_{es}^{e}$ are the estimated satellite position and velocity, obtained from the navigation data message, $\hat{\mathbf{r}}_{ia}^{i}$ or $\hat{\mathbf{r}}_{ea}^{e}$ and $\hat{\mathbf{v}}_{ia}^{i}$ or $\hat{\mathbf{v}}_{ea}^{e}$ are the navigation processor's estimates of the user antenna position and velocity, $\delta\hat{\rho}_{c}^{a}$ and $\delta\dot{\hat{\rho}}_{c}^{a}$ are the estimates of the receiver clock offset and drift, $\hat{\mathbf{u}}_{as}^{i}$ and $\hat{\mathbf{u}}_{as}^{e}$ is the line-of-sight vector obtained from the estimated satellite and user positions, and $\hat{t}_{st,a}^{s}$ and $\hat{t}_{sa,a}^{s}$ are the estimated times of signal transmission and arrival. These are determined by

$$\begin{aligned}\hat{t}_{sa,a}^{s} &= \tilde{t}_{sa,a}^{s} - \delta\hat{t}_{c}^{a} & \hat{t}_{st,a}^{s} &= \hat{t}_{sa,a}^{s} - \hat{r}_{as}/c \\ &= \tilde{t}_{sa,a}^{s} - \delta\hat{\rho}_{c}^{a}/c & &= \tilde{t}_{sa,a}^{s} - \hat{\rho}_{c,C}^{a}/c\end{aligned} \qquad (8.52)$$

The estimated range and range rate are

$$\begin{aligned}\hat{r}_{as} &= \hat{\rho}_{a,C}^{s} - \delta\hat{\rho}_{c}^{a}(\hat{t}_{sa,a}^{s}) \\ &= \left|\hat{\mathbf{r}}_{is}^{i}(\hat{t}_{st,a}^{s}) - \hat{\mathbf{r}}_{ia}^{i}(\hat{t}_{sa,a}^{s})\right| \\ &= \left|\mathbf{C}_{e}^{I}(\hat{t}_{st,a}^{s})\hat{\mathbf{r}}_{es}^{e}(\hat{t}_{st,a}^{s}) - \hat{\mathbf{r}}_{ea}^{e}(\hat{t}_{sa,a}^{s})\right|\end{aligned} \qquad (8.53)$$

and

$$\begin{aligned}\dot{\hat{r}}_{as} &= \dot{\hat{\rho}}_{a,C}^{s} - \delta\dot{\hat{\rho}}_{c}^{a}(\hat{t}_{sa,a}^{s}) \\ &= \hat{\mathbf{u}}_{as}^{i}{}^{T}\left(\hat{\mathbf{v}}_{is}^{i}(\hat{t}_{st,a}^{s}) - \hat{\mathbf{v}}_{ia}^{i}(\hat{t}_{sa,a}^{s})\right) \\ &= \hat{\mathbf{u}}_{as}^{e}{}^{T}\left[\mathbf{C}_{e}^{I}(\hat{t}_{st,a}^{s})\left(\hat{\mathbf{v}}_{es}^{e}(\hat{t}_{st,a}^{s}) + \mathbf{\Omega}_{ie}^{e}\hat{\mathbf{r}}_{es}^{e}(\hat{t}_{st,a}^{s})\right) - \left(\hat{\mathbf{v}}_{ea}^{e}(\hat{t}_{sa,a}^{s}) + \mathbf{\Omega}_{ie}^{e}\hat{\mathbf{r}}_{ea}^{e}(\hat{t}_{sa,a}^{s})\right)\right]\end{aligned} \qquad (8.54)$$

The errors in the estimated range and range rate are

$$\begin{aligned}\hat{r}_{as} - r_{as} &= \delta\rho_e^s - \mathbf{u}_{as}^{i\,T}\delta\mathbf{r}_{ia}^i(\hat{t}_{sa,a}^s) \\ &= \delta\rho_e^s - \mathbf{u}_{as}^{e\,T}\delta\mathbf{r}_{ea}^e(\hat{t}_{sa,a}^s)\end{aligned}, \quad (8.55)$$

and

$$\begin{aligned}\hat{\dot{r}}_{as} - \dot{r}_{as} &= \delta\dot{\rho}_e^s - \mathbf{u}_{as}^{i\,T}\delta\mathbf{v}_{ia}^i(\hat{t}_{sa,a}^s) \\ &= \delta\dot{\rho}_e^s - \mathbf{u}_{as}^{e\,T}\delta\mathbf{v}_{ea}^e(\hat{t}_{sa,a}^s)\end{aligned}, \quad (8.56)$$

where $\delta\rho_e^s$ and $\delta\dot{\rho}_e^s$ are the range and range-rate errors due to the ephemeris data in the navigation message (see Section 9.3.1), while $\delta\mathbf{r}_{ia}^i$ or $\delta\mathbf{r}_{ia}^i$ and $\delta\mathbf{v}_{ia}^i$ or $\delta\mathbf{v}_{ea}^e$ are the errors in the user position and velocity solution.

8.5.4 Elevation and Azimuth

The direction of a GNSS satellite from the user antenna is commonly described by an elevation, θ_{nu}^{as}, and azimuth, ψ_{nu}^{as}. These angles define the orientation of the line-of-sight vector with respect to the north, east, and down axes of a local navigation frame, as shown in Figure 8.26, and correspond to the elevation and azimuth angles used to describe the attitude of a body (see Section 2.2.1). They may be obtained from the line-of-sight vector in the local navigation frame, $\mathbf{u}_{as}^n = (u_{as,N}^n, u_{as,E}^n, u_{as,D}^n)$, using

$$\theta_{nu}^{as} = -\arcsin(u_{as,D}^n), \quad \psi_{nu}^{as} = \arctan_2(u_{as,E}^n, u_{as,N}^n), \quad (8.57)$$

where a four-quadrant arctangent function must be used. Example 8.1 on the CD illustrates this. The reverse transformation is

$$\mathbf{u}_{as}^n = \begin{pmatrix} \cos\theta_{nu}^{as}\cos\psi_{nu}^{as} \\ \cos\theta_{nu}^{as}\sin\psi_{nu}^{as} \\ -\sin\theta_{nu}^{as} \end{pmatrix}. \quad (8.58)$$

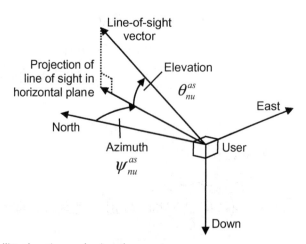

Figure 8.26 Satellite elevation and azimuth.

The local navigation frame line-of-sight vector is transformed to and from its ECEF and ECI-frame counterparts using (8.39) and (2.150) or (2.154).

Problems and exercises for this chapter are on the accompanying CD.

References

[1] Dorsey, A. J., et al., "GPS System Segments," in *Understanding GPS Principles and Applications*, 2nd ed., E. D. Kaplan and C. J. Hegarty, (eds.), Norwood, MA: Artech House, 2006, pp. 67–112.

[2] Spilker, J. J., Jr., and B. W. Parkinson, "Overview of GPS Operation and Design," in *Global Positioning System: Theory and Applications Volume I*, B. W. Parkinson and J. J. Spilker, Jr., (eds.), Washington, D.C.: AIAA, 1996, pp. 29–55.

[3] Misra, P., and P. Enge, *Global Positioning System Signals, Measurements, and Performance*, 2nd ed., Lincoln, MA: Ganga-Jamuna Press, 2006.

[4] Spilker, J. J., Jr., "Satellite Constellation and Geometric Dilution of Precision," in *Global Positioning System: Theory and Applications, Volume I*, B. W. Parkinson and J. J. Spilker, Jr., (eds.), Washington, D.C.: AIAA, 1996, pp. 177–208.

[5] Spilker, J. J., Jr., "Signal Structure and Theoretical Performance," in *Global Positioning System: Theory and Applications, Volume I*, B. W. Parkinson and J. J. Spilker, Jr., (eds.), Washington, D.C.: AIAA, 1996, pp. 57–119.

[6] Kaplan, E. D., et al., "Fundamentals of Satellite Navigation," in *Understanding GPS Principles and Applications*, 2nd ed., E. D. Kaplan and C. J. Hegarty, (eds.), Norwood, MA: Artech House, 2006, pp. 21–65.

[7] Klepczynski, W. J., "GPS for Precise Time and Time Interval Measurement," in *Global Positioning System: Theory and Applications, Volume II*, B. W. Parkinson and J. J. Spilker, Jr., (eds.), Washington, D.C.: AIAA, 1996, pp. 483–500.

[8] Wang, L., P. D. Groves, and M. K. Ziebart, "Multi-Constellation GNSS Performance Evaluation for Urban Canyons Using Large Virtual Reality City Models," *Journal of Navigation*, Vol. 65, No. 3, 2012, pp. 459–476.

[9] Parkinson, B. W., "Introduction and Heritage of NAVSTAR," in *Global Positioning System: Theory and Applications, Volume I*, B. W. Parkinson and J. J. Spilker, Jr., (eds.), Washington, D.C.: AIAA, 1996, pp. 3–28.

[10] Ruiz, L., R. Crescinberi, and E. Breeuwer, "Galileo Services Definition and Navigation Performance," *Proc. ENC-GNSS 2004*, Rotterdam, the Netherlands, May 2004.

[11] Falcone, M., P. Erhard, and G. W Hein, "Galileo," in *Understanding GPS Principles and Applications*, 2nd ed., E. D. Kaplan and C. J. Hegarty, (eds.), Norwood, MA: Artech House, 2006, pp. 559–594.

[12] Petrovsky, I. G., "QZSS: Japan's New Integrated Communication and Positioning Service for Mobile Users," *GPS World*, June 2003, pp. 24–29.

[13] Rao, V. J., G. Lachapelle, and V. Kumar, "Analysis of IRNSS over Indian Subcontinent," *Proc. ION ITM*, San Diego, CA, January 2011, pp. 1150–1162.

[14] Habereder, H., I. Schempp, and M. Bailey, "Performance Enhancements for the Next Phase of WAAS," *Proc. ION GNSS 2004*, Long Beach, CA, September 2004, pp. 1350–1358.

[15] Hein, G. W., et al., "Envisioning a Future GNSS System of Systems, Part 1," *Inside GNSS*, January/February 2007, pp. 58–67.

[16] Enge, P. K., and A. J., Van Dierendonck, "Wide Area Augmentation System," in *Global Positioning System: Theory and Applications, Volume II*, B. W. Parkinson and J. J. Spilker, Jr., (eds.), Washington, D.C.: AIAA, 1996, pp. 117–142.

[17] Braff, R., "Description of the FAA's Local Area Augmentation System (LAAS)," *Navigation: JION*, Vol. 44, No. 4, 1997, pp. 411–423.

[18] Gibbons, G., "GNSS Interoperability: Not So Easy, After All," *Inside GNSS*, January/February 2011, pp. 28–31.

[19] Betz, J. W., "Binary Offset Carrier Modulation for Radionavigation," *Navigation: JION*, Vol. 48, No. 4, 2001, pp. 227–246.

[20] Pratt, A. R., "New Navigation Signals and Future Systems in Evolution," in *GNSS Applications and Methods*, S. Gleason and D. Gebre-Egziabher, (eds.), Norwood, MA: Artech House, 2009, pp. 437–483.

[21] Fontana, R. D., et al., "The New L2 Civil Signal," *Proc. ION GPS 2001*, Salt Lake City, UT, September 2001, pp. 617–631.

[22] Tran, M., "Performance Evaluation of the New GPS L5 and L2 Civil (L2C) Signals," *Navigation: JION*, Vol. 51, No. 3, 2004, pp. 199–212.

[23] Dafesh, P., et al., "Description and Analysis of Time-Multiplexed M-Code Data," *Proc. ION 58th AM*, Albuquerque, NM, June 2002, pp. 598–611.

[24] *Navstar GPS Space Segment/Navigation User Interfaces*, IS-GPS-200, Revision F, GPS Directorate, September 2011.

[25] *Navstar GPS Military-Unique Space Segment/ User Segment Interfaces*, ICD-GPS-700, Revision A, ARINC, September 2004.

[26] *Navstar GPS Space Segment/User Segment L5 Interfaces*, IS-GPS-705, Revision B, GPS Directorate, September 2011.

[27] *Navstar GPS Space Segment/User Segment L1C Interface*, IS-GPS-800, Revision B, GPS Directorate, September 2011.

[28] Barker, B. C., et al., "Overview of the GPS M Code Signal," *Proc. ION NTM*, Anaheim, CA, January 2000, pp. 542–549.

[29] *Global Navigation Satellite System GLONASS Interface Control Document*, Edition 5.1, Russian Institute of Space Device Engineering, 2008.

[30] Feairheller, S., and R., Clark, "Other Satellite Navigation Systems," in *Understanding GPS Principles and Applications*, 2nd ed., E. D. Kaplan and C. J. Hegarty, (eds.), Norwood, MA: Artech House, 2006, pp. 595–634.

[31] Stupak, G., "GLONASS Signals Development," *Proc. International Summit on Satellite Navigation*, Munich, Germany, March 2011.

[32] *European GNSS (Galileo) Open Service Signal in Space Interface Control Document*, Issue 1 Revision 1, GNSS Supervisory Authority, September 2010.

[33] Issler, J. -L., et al., "Spectral Measurements of GNSS Satellite Signals Need for Wide Transmitted Bands," *Proc. ION GPS/GNSS 2003*, Portland, OR, September 2003, pp. 445–460.

[34] Chengqi, R., "Beidou Navigation Satellite System," *Proc. 5th Meeting of International Committee on GNSS*, Turin, Italy, October 2010.

[35] *Quasi-Zenith Satellite System: Interface Specification for QZSS*, Draft V1.2, Japan Aerospace Exploration Agency, March 2010.

[36] *Minimum Operational Performance Standards for Global Positioning System/Wide Area Augmentation System Airborne Equipment*, RTCA/DO229C, November 2001.

[37] *U.S. Department of Transportation Federal Aviation Administration Specification for the Wide Area Augmentation System*, DTFA01-96-C-00025 Modification No. 0111, August 2001.

[38] *User Guide for EGNOS Application Developers*, Edition 1.1., CNES, European Commission, and ESA, July 2009.

[39] Spilker, J. J., Jr., "GPS Navigation Data," in *Global Positioning System: Theory and Applications Volume I*, B. W. Parkinson and J. J. Spilker, Jr., (eds.), Washington, D.C.: AIAA, 1996, pp. 121–176.

[40] Urlichich, Y., et al., "GLONASS Developing Strategy," *Proc. ION GNSS 2010*, Portland, OR, September 2010, pp. 1566–1571.

[41] Dinwiddy, S. E., E. Breeuwer, and J. H. Hahn, "The Galileo System," *Proc. ENC-GNSS 2004*, Rotterdam, the Netherlands, May 2004.

[42] Maeda, H., "QZSS Overview and Interoperability," *Proc. ION GNSS 2005*, Long Beach, CA, September 2005.

[43] Di Esposti, R., "Time-Dependency and Coordinate System Issues in GPS Measurement Models," *Proc. ION GPS 2000*, Salt Lake City, UT, September 2000, pp. 1925–1929.

CHAPTER 9
GNSS: User Equipment Processing and Errors

This chapter describes how GNSS user equipment processes the signals from the satellites to obtain ranging measurements and then a navigation solution. It also reviews the error sources and describes the effect of the geometry of the navigation signals. It follows on from the fundamentals of satellite navigation described in Section 8.1.

Different authors describe GNSS user equipment architecture in different ways [1–4]. Here, it is divided into four functional blocks, as shown in Figure 9.1: the antenna, receiver hardware, ranging processor, and navigation processor. This approach splits up the signal processing, ranging, and navigation functions, matching the different INS/GNSS integration architectures described in Chapter 14.

Section 9.1 describes the antenna and receiver hardware, with an emphasis on signal processing. Section 9.2 describes the ranging processor, including acquisition, code and carrier tracking, lock detection, navigation message demodulation, signal-to-noise measurement, and generation of the pseudo-range, pseudo-range rate, and carrier-phase measurements. Section 9.3 discusses the error sources leading to ranging errors, including ephemeris and satellite clock errors; ionosphere and troposphere propagation errors; tracking errors; and multipath interference, NLOS

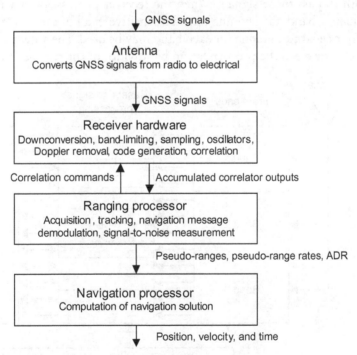

Figure 9.1 GNSS user equipment functional diagram.

349

reception, and diffraction. The satellite clock, ionosphere, and troposphere errors are partially corrected by the user equipment, either within the ranging processor or the navigation processor. Finally, Section 9.4 describes the navigation processor, covering both single-epoch and filtered navigation solutions. The effect of navigation solution geometry on positioning accuracy is also described and the section concludes with a discussion of navigation error budgets.

9.1 Receiver Hardware and Antenna

This section describes the hardware components of GNSS user equipment, shown in Figure 9.2. A brief discussion of antennas and the reference oscillator is followed by a description of the processing performed by the receiver hardware in the front end and then the baseband signal processor. In the front end, the GNSS signals are amplified, filtered, downconverted, and sampled. In the baseband signal processor, the signals are correlated with internally-generated code and carrier and summed to produce the accumulated correlator outputs, which are provided to the ranging processor. Notation is simplified in this section, omitting the transmitting satellite and receiver designations.

9.1.1 Antennas

GNSS user equipment must incorporate an antenna that has peak sensitivity near to the carrier frequency of the signals processed by the receiver and sufficient bandwidth to pass those signals. When the receiver processes signals in more than one frequency band, the antenna must be sensitive in all of the bands required. Alternatively, a separate antenna for each band may be used. The antenna bandwidth should match or exceed the precorrelation bandwidth of the receiver (see Section 9.1.3).

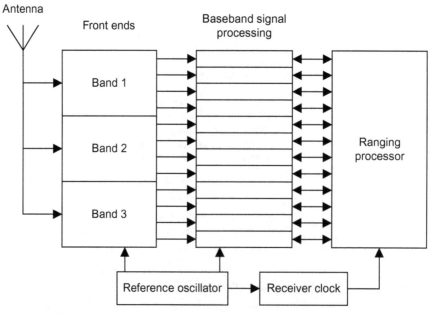

Figure 9.2 GNSS receiver hardware architecture.

A GNSS antenna should generally be sensitive to signals from all directions. A typical GNSS antenna has a gain of 2 to 4 dB for signals at normal incidence (i.e., at the antenna zenith). This drops as the angle of incidence increases and is generally negative (in decibel terms) for angles of incidence greater than 75°. For a horizontally mounted antenna, a 75° incidence angle corresponds to a satellite signal at a 15° elevation angle. Some typical GPS antenna gain patterns are shown in [1, 2].

GNSS signals are transmitted with right-handed circular polarization (RHCP). On surface reflection at an angle of incidence less than Brewster's angle, this is reversed to left-handed circular polarization (LHCP). Elliptical polarization, which mixes RHCP and LHCP, can also arise when the reflecting surface is not flat. Therefore, to minimize multipath problems (Section 9.3.4), the antenna should be sensitive only to RHCP signals.

Although signals are received throughout the antenna, GNSS user equipment will determine a navigation solution for a particular point in space, known as the electrical phase center. This does not necessarily coincide with the physical center of the antenna and may even be outside the antenna casing. For a given antenna, the phase center can vary with the elevation, azimuth, and frequency of the incoming signals by around a centimeter (less for high-grade antennas). Phase center calibration is therefore important for high-precision applications.

Basic GNSS antennas come in a number of shapes and sizes. Patch, or microstrip, antennas have the advantage of being low-cost, flat, and rugged, but their polarization varies with the angle of incidence. Better performance can be obtained from a dome or helical (volute) antenna. More advanced antenna technology may be used to limit the effects of radio frequency (RF) interference sources and/or multipath. This is discussed in Sections 10.3.2 and 10.4.1.

Antennas for tracking devices, smartphones, and other mobile devices are designed to minimize size and cost at the expense of performance. They are often linearly polarized. This reduces the gain, by 3 dB typically, but by up to 20 dB where the polarization axis of the antenna coincides with the line of sight to the satellite. Linear polarization also increases sensitivity to multipath interference. The effective gain is sometimes reduced further by mounting the GNSS antenna at the bottom of the device due to space constraints.

For handheld applications, the antenna is usually encased with the rest of the user equipment, whereas for vehicles, a separate antenna is generally mounted on the vehicle body. The cable between the antenna and the receiver imposes a common-mode lag on the incoming signal. However, the effects of antenna cable lag and receiver clock offset are indistinguishable, so the navigation processor simply accounts for the lag as part of its clock offset estimate. Antenna cables attenuate the signal by 0.3 dB or more per meter; this may be mitigated by including an amplifier in the antenna.

More detailed descriptions of GNSS antenna technology may be found in [5–7].

9.1.2 Reference Oscillator

The timing in a GNSS receiver is controlled by the reference oscillator. This provides a frequency standard which drives both the receiver clock, which provides a time reference for the ranging and navigation processors, and the various oscillators used in the receiver front end and baseband processor. Long-term errors and drift in the receiver's frequency standard are compensated in the navigation processor, so do not present a problem, provided the frequency error is not large enough to disrupt the front end. It can be reset from GNSS system time once the user equipment has decoded the navigation data message. However, short-term variation in the oscillator frequency over the correlator coherent integration interval (see Section 9.1.4) and the time constant of the carrier tracking loop (see Section 9.2.3) can present a problem, particularly when the user equipment is optimized for poor signal-to-noise environments (see Section 10.3).

A basic GNSS receiver may use a quartz crystal oscillator (XO) as the frequency standard. The dominant source of error is variation in frequency with temperature and the oscillator frequency can vary by one part in 10^5 or 10^6 over typical operating temperature ranges [4]. A temperature-compensated crystal oscillator (TCXO) is more common. This typically costs a few dollars or Euros. It uses a temperature sensor to vary the oscillator control voltage, stabilizing the frequency variation to within one part in 10^8 over a 1-second interval, although the overall error is still a few parts in 10^6. This corresponds to a range-rate bias of the order of 1,000 m s^{-1} or a clock drift of a few hundred milliseconds per day. The frequency normally varies continuously, subject to quantization in the control process, but can experience sudden changes, known as microjumps [8].

An oven-controlled crystal oscillator (OCXO) uses an oven to maintain the oscillator at a fixed temperature. This achieves a frequency variation of about one part in 10^{11} over a second, with a frequency bias of one part in 10^8, corresponding to a range-rate bias of the order of 3 m s^{-1}. However, an OCXO is relatively large, consuming significant power and costing over a $100 (€80), so its use is restricted to specialized applications, such as survey receivers [3, 8].

Quartz oscillators also exhibit frequency errors proportional to the applied specific force. The coefficient of this g-dependent error is different for each of the three axes of the oscillator body. Frequency errors vary between one part in 10^{12} and one part in 10^8 per m s^{-2} of acceleration. Large g-dependent errors can disrupt carrier tracking in high-dynamics and high-vibration environments [9, 10].

Reference stations used for wide area differential GNSS, signal monitoring, and the control of the GNSS systems themselves need to provide accurate measurements of the range errors. To do this, they require a precise time reference. Therefore, they use a cesium or rubidium atomic clock instead of a crystal oscillator, giving a short-term stability of one part in 10^{11} and a long-term stability of one part in 10^{12} to 10^{13} [3].

Conventional atomic clocks are large and have relatively high power consumption. A chip-scale atomic clock (CSAC) has a mass of about 35g and a power consumption of 100 mW, making it practical for navigation. It is stable to one part in 10^{10} over 1 second and one part in 10^{11} over 1 hour, corresponding to a 10-m pseudo-range drift [11, 12]. At the time of this writing, CSACs were available for $1,500 (€1,200), but the technology was new.

9.1.3 Receiver Front End

The receiver front end processes the GNSS signals from the antenna in the analog domain, known as *signal conditioning*, and then digitizes the signal for output to the baseband signal processor [3, 13]. All signals in the same frequency band are processed together. Multiband receivers normally incorporate one front end for each frequency band, although multiband GNSS front ends have been developed [14]. Front ends for neighboring bands, such as GPS L1 and GLONASS FDMA L1, may also share some components [15]. GLONASS FDMA typically implements a wideband front end, covering all FDMA channels in a given band.

Figure 9.3 shows a typical front-end architecture, comprising an RF processing stage, followed by two intermediate frequency (IF) downconversion stages, and then the analog-to-digital converter [4]. Some receivers employ a single downconversion stage, while others use more than two. An alternative, direct digitization or direct conversion, approach is discussed in [16, 17].

The carrier frequency is downconverted from the original L-band radio frequency to a lower IF to enable a lower sampling rate to be used, reducing the baseband processing load. At each downconversion stage, the incoming signal at a carrier frequency f_i is multiplied with a receiver-generated sinusoid of frequency f_o. This produces two signals at carrier frequencies $|f_i - f_o|$ and $f_i + f_o$, each with the same modulation as the incoming signal. The higher-frequency signal is normally eliminated using a bandpass filter (BPF), which also limits out-of-band interference [3]. The signal must be amplified by about seven orders of magnitude between the antenna and the ADC. To prevent feedback problems, the amplification is distributed between the RF and IF processing stages [4].

The bandwidth of the conditioned signals entering the ADC is known as the *precorrelation bandwidth*. The minimum double-sided bandwidth required is about twice the chipping rate for a BPSK signal and $2(f_s + f_{co})$ for a BOC(f_s, f_{co}) signal (see Section 8.3.1). Table 9.1 lists the minimum precorrelation bandwidths for the main GNSS signals. However, a wider precorrelation bandwidth sharpens the code correlation function (see Section 9.1.4.1), which can improve performance, particularly multipath mitigation (Section 10.4.2). The maximum useful precorrelation bandwidth is the transmission bandwidth of the GNSS satellite, listed in Table 8.4.

To process a single sidelobe of a BOC signal as a BPSK-like signal, to reduce the processing load, the other sidelobe must be filtered out, which is difficult to achieve when $f_s \leq 2f_{co}$ [e.g., BOC$_s$(1,1)]. This filtering may be conducted within the front end, in which case dual front ends must be used if separate processing of both sidelobes is required. There are also post-ADC filtering techniques.

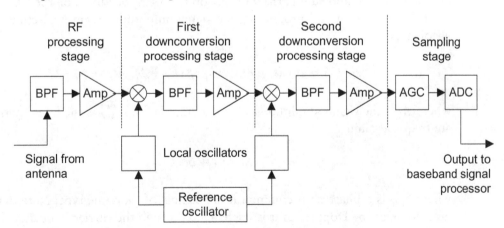

Figure 9.3 A typical GNSS receiver front-end architecture.

Table 9.1 Minimum Receiver Precorrelation Bandwidths for Selected GNSS Signals

Signals	Minimum Precorrelation Bandwidth (MHz)
GLONASS C/A (single channel)	1.022
GPS C/A and L2C	2.046
GPS L1C and Galileo E1-B/C ($BOC_s(1,1)$ component)	4.092
GLONASS C/A (all L2 channels)	6.7095
GLONASS C/A (all L1 channels)	8.3345
GLONASS P (single channel)	10.22
Galileo E6-B/C	10.23
GPS L1C-p and Galileo E1-B/C (full MBOC signal)	14.322
GLONASS P (all L2 channels)	15.9075
GLONASS P (all L1 channels)	17.5325
GPS P(Y) & L5, GLONASS L3 & L5, and Galileo E5a & E5b	20.46
GPS M code and Galileo E6-A	30.69
Galileo E1-A	35.805
Galileo E5 Alt BOC	51.15

Assuming a BPSK signal for simplicity, the amplitude of a satellite signal received at the antenna phase center, neglecting band-limiting effects, is [3]

$$s_a(t_{sa}) = \sqrt{2P} C(t_{st}) D(t_{st}) \cos\left[2\pi(f_{ca} + \Delta f_{ca})t_{sa} + \phi_0\right], \quad (9.1)$$

where P is the signal carrier power, C is the spreading code, D is the navigation data message, and ϕ_0 is the phase offset, as defined in Section 8.1.2. Here, f_{ca} is the transmitted carrier frequency, while Δf_{ca} is the Doppler shift due to the relative motion of the satellite and user-equipment antennas, t_{sa} is the time of signal arrival at the antenna, and t_{st} is the time of signal transmission. The Doppler shift is given by

$$\Delta f_{ca} = -\frac{f_{ca}}{c}\frac{\partial \rho_R}{\partial t_{sa}} \approx -\frac{f_{ca}}{c}\dot{\rho}_R, \quad (9.2)$$

noting that there are additional frequency shifts due to relativistic time dilation as described in [18] and summarized in Section G.5 of Appendix G on the CD.

Following front-end processing, the signal amplitude, again neglecting band-limiting, is

$$s_{IF}(t_{sa}) = A_a C(t_{st}) D(t_{st}) \cos\left[2\pi(f_{IF} + \Delta f_{ca})t_{sa} + \phi_{IF}\right], \quad (9.3)$$

where A_a is the signal amplitude following amplification, f_{IF} is the final intermediate frequency, and

$$\phi_{IF} = \phi_0 + \delta\phi_{IF}, \quad (9.4)$$

where $\delta\phi_{IF}$ is a phase shift common to all signals of the same type. Note that the magnitude of the Doppler shift is unchanged through the carrier-frequency downconversion process.

To prevent aliasing effects from disrupting carrier tracking, the ADC sampling rate must be at least twice the IF [3], while to prevent spectral foldover distorting the signal, the IF must exceed the single-sided precorrelation bandwidth. The sampling rate should be asynchronous with both the IF and the code-chipping rate. This ensures that the samples vary in code and carrier phase, collectively encompassing the whole signal waveform.

Low-cost receivers use single-bit sampling. However, this reduces the effective signal to noise, known as an implementation loss. Better performance is obtained using a quantization level of 2 bits or more, together with an automatic gain control (AGC). The AGC varies the amplification of the input to the ADC to keep it matched to the dynamic range of the quantization process. As the noise dominates the signal prior to correlation, an AGC ensures a roughly constant noise standard deviation within the baseband signal processor and the measurements used by the ranging processor, while the signal level can vary. With a fast response rate, the AGC can be used to suppress pulsed interference [1].

The sampling rate and quantization level of the ADC determines the processing power needed for the baseband signal processor. Consequently, there is a tradeoff between receiver performance and cost. It is important to match the IF to the precorrelation bandwidth to avoid a sampling rate higher than necessary.

In multifrequency receivers, timing biases arise between the different frequency bands and even between different GLONASS FDMA channels within the same band or between different BOC lobes. These are known as interfrequency biases and are typically calibrated by the receiver manufacturer; however, residual effects remain [19].

9.1.4 Baseband Signal Processor

The baseband signal processor demodulates the sampled and conditioned GNSS signals from the receiver front end by correlating them with internally generated replicas of the ranging (or spreading) code and carrier. The correlated samples are then summed and sent to the ranging processor, which controls the internally generated code and carrier.

Many authors class the ranging processor as part of the baseband processor as the control loops used to acquire and track the signals span both the ranging-processor software and baseband signal-processor hardware. Here the two are treated separately because the interface to the ranging processor traditionally marks the boundary between the hardware and software parts of current GNSS user equipment. In addition, advanced user equipment can combine the functions of the ranging and navigation processors (see Section 10.3.7).

The baseband signal processor is split into a series of parallel channels, one for each signal processed. A basic C/A-code-only GPS receiver typically has 10 or 12 channels, while a multiconstellation, multifrequency GNSS receiver can have over 100 channels. Figure 9.4 shows the architecture of a typical GNSS baseband signal processor channel. Advanced designs may implement more than six correlators to speed up acquisition (Section 9.2.1), mitigate multipath (Section 10.4.2), acquire and track BOC signals (Sections 9.2.1 and 9.2.2), and mitigate poor signal-to-noise environments (Section 10.3).

In contemporary GNSS receivers, the baseband signal processing is generally implemented digitally in hardware using an application-specific integrated circuit (ASIC). A number of experimental receivers, known as software receivers or software-defined receivers (or radios) (SDRs), implement the baseband signal processing in software on a general-purpose processor and/or digital signal processor (DSP) [16, 20–22]. This enables the signal processor to be reconfigured to adapt to different contexts, such as high dynamics, low signal to noise, and high multipath. It also allows the signal samples to be stored, making it easier to resolve synchronization errors. However, it is less efficient than an ASIC implementation in terms of processing power per unit cost and power consumption. Receiver implementations based on programmable logic devices (PLDs) or field-programmable gate arrays (FPGAs) offer a compromise between the hardware and software approaches.

The baseband signal processing order varies according to the receiver design, but the outputs to the ranging processor are the same. Here a typical approach is described.

The first stage in each channel is in-phase (I) and quadraphase (Q) sampling, also known as carrier wipeoff or Doppler wipeoff. Successive samples from the receiver front end have a different carrier phase, so a summation of these will tend to zero, regardless of whether the signal and receiver-generated codes are aligned. The in-phase and quadraphase sampling, also known as phase quadrature sampling, transforms the precorrelation signal samples into two streams, I_0 and Q_0, each

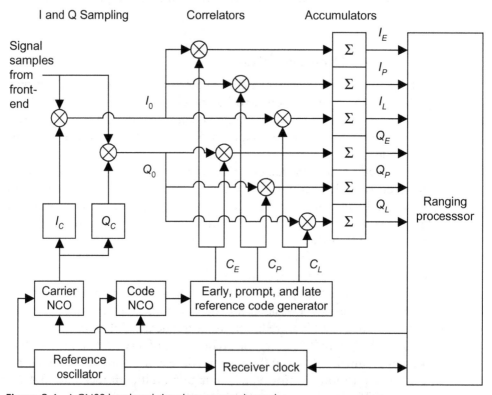

Figure 9.4 A GNSS baseband signal processor channel.

with a nominally constant carrier phase. There is a 90° difference in carrier phase between the in-phase and quadraphase streams. Taking the sum of squares of the I and Q streams enables noncoherent processing, which is independent of the carrier phase (see Section 8.1.2), while the carrier may be measured by comparing the two streams. Carrier wipeoff is performed separately for each channel because the Doppler-shifted carrier frequency is different for each signal.

The I and Q samples are generated by multiplying the samples from the ADC, which comprise the wanted signal, s_{IF}, the unwanted signals on the same frequency, and noise, by in-phase and quadraphase samples of the receiver-generated carrier, I_C and Q_C, given by [3]

$$I_C(t_{sa}) = \cos\left[2\pi\left(f_{IF} + \Delta\tilde{f}_{ca}\right)t_{sa} + \tilde{\phi}_{IF}\right]$$
$$Q_C(t_{sa}) = \sin\left[2\pi\left(f_{IF} + \Delta\tilde{f}_{ca}\right)t_{sa} + \tilde{\phi}_{IF}\right], \quad (9.5)$$

where $\Delta\tilde{f}_{ca}$ is the ranging processor's measurement of the Doppler shift and $\tilde{\phi}_{IF}$ is its measurement of the carrier phase offset after front-end processing. Approximate sine and cosine functions reduce the processor load [13]. The frequency $f_{IF} + \Delta\tilde{f}_{ca}$ is generated by the carrier numerically controlled oscillator (NCO), also known as a digitally controlled oscillator (DCO), which is driven by the reference oscillator and controlled by the ranging processor. For GLONASS FDMA signals processed with a common front end, f_{IF} is different for each individual signal. Therefore, the appropriate value must be used by each channel's NCO.

Applying (9.3) and (9.5) and neglecting the component at frequency $2f_{IF}$ and other harmonics, which are filtered out, the in-phase and quadraphase signal samples are then

$$I_0(t_{sa}) = A_0 C(t_{st}) D(t_{st}) \cos\left[2\pi\left(\Delta f_{ca} - \Delta\tilde{f}_{ca}\right)t_{sa} + \phi_{IF} - \tilde{\phi}_{IF}\right] + w_{I0}(t_{sa})$$
$$Q_0(t_{sa}) = A_0 C(t_{st}) D(t_{st}) \sin\left[2\pi\left(\Delta f_{ca} - \Delta\tilde{f}_{ca}\right)t_{sa} + \phi_{IF} - \tilde{\phi}_{IF}\right] + w_{Q0}(t_{sa}), \quad (9.6)$$

where A_0 is the signal amplitude following the AGC and ADC, and w_{I0} and w_{Q0} represent the noise from the receiver, RF interference, and the other satellite signals. When the ranging processor's carrier phase and frequency estimates are correct, all of the signal power is in the in-phase samples. If the phase estimate is out by 90°, the signal power is in the quadraphase samples. When there is an error in the frequency estimate, the signal power oscillates between the in-phase and quadraphase samples; the larger the frequency error, the shorter the period of oscillation.

Some receivers, including all direct digitization receivers, perform phase quadrature sampling at the ADC. However, they still require a carrier wipeoff process to apply the Doppler shift for each signal to the common carrier frequency used in the front end [1].

The next stage of baseband signal processing is the code correlation, introduced in Section 8.1.2. For BPSK signals, this comprises the multiplication of the precorrelation signal samples, I_0 and Q_0, with the early, prompt, and late reference codes, given by

$$C_E(t_{sa}) = C(\tilde{t}_{st} + d/2f_{co})$$
$$C_P(t_{sa}) = C(\tilde{t}_{st}) \qquad , \qquad (9.7)$$
$$C_L(t_{sa}) = C(\tilde{t}_{st} - d/2f_{co})$$

where $\tilde{t}_{st}$ is the ranging processor's measurement of the time of signal transmission and d is the code-phase offset in chips between the early and late reference signals. The prompt reference signal phase is halfway between the other two. The early–late correlator spacing varies between 0.05 and 1 chip, depending on the type of signal and the receiver design. The code correlation is also known as code wipeoff. The phase of the reference code generator is the integral of the code NCO output. This is driven by the reference oscillator and controlled by the ranging processor. Signals with layered, or tiered, codes, may be correlated with either the full code or just the primary code.

To reduce the processing load, some receivers implement early-minus-late correlators, instead of separate early and late correlators, in which case I_0 and Q_0 are multiplied by

$$\begin{aligned} C_{E-L}(t_{sa}) &= C_E(t_{sa}) - C_L(t_{sa}) \\ &= C(\tilde{t}_{st} + d/2f_{co}) - C(\tilde{t}_{st} - d/2f_{co}) \end{aligned} . \qquad (9.8)$$

The correlator outputs are accumulated over an interval, τ_a, of at least 1 ms and then sent to the ranging processor. Although the accumulation is strictly a summation, there are sufficient samples to treat it as an integration for analytical purposes, and the accumulation is often known as *integrate and dump*. The early, prompt, and late in-phase and quadraphase accumulated correlator outputs are thus given by

$$I_E(t_{sa}) = f_a \int_{t_{sa}-\tau_a}^{t_{sa}} I_0(t)C_E(t)\,dt, \qquad Q_E(t_{sa}) = f_a \int_{t_{sa}-\tau_a}^{t_{sa}} Q_0(t)C_E(t)\,dt$$

$$I_P(t_{sa}) = f_a \int_{t_{sa}-\tau_a}^{t_{sa}} I_0(t)C_P(t)\,dt, \qquad Q_P(t_{sa}) = f_a \int_{t_{sa}-\tau_a}^{t_{sa}} Q_0(t)C_P(t)\,dt , \qquad (9.9)$$

$$I_L(t_{sa}) = f_a \int_{t_{sa}-\tau_a}^{t_{sa}} I_0(t)C_L(t)\,dt, \qquad Q_L(t_{sa}) = f_a \int_{t_{sa}-\tau_a}^{t_{sa}} Q_0(t)C_L(t)\,dt$$

where f_a is the ADC sampling frequency and noting that the time tag is applied to the end of the correlation interval here. These are commonly known simply as Is and Qs.

Substituting (9.6) and (9.7) into (9.9) and assuming that an AGC is used, it may be shown that the accumulated correlator outputs are [1]

$$I_E(t_{sa}) = \sigma_{IQ}\left[\sqrt{2(c/n_0)\tau_a}R(x - d/2)D(t_{st})\text{sinc}(\pi\delta f_{ca}\tau_a)\cos(\delta\phi_{ca}) + w_{IE}(t_{sa})\right]$$
$$I_P(t_{sa}) = \sigma_{IQ}\left[\sqrt{2(c/n_0)\tau_a}R(x)D(t_{st})\text{sinc}(\pi\delta f_{ca}\tau_a)\cos(\delta\phi_{ca}) + w_{IP}(t_{sa})\right]$$
$$I_L(t_{sa}) = \sigma_{IQ}\left[\sqrt{2(c/n_0)\tau_a}R(x + d/2)D(t_{st})\text{sinc}(\pi\delta f_{ca}\tau_a)\cos(\delta\phi_{ca}) + w_{IL}(t_{sa})\right]$$
$$Q_E(t_{sa}) = \sigma_{IQ}\left[\sqrt{2(c/n_0)\tau_a}R(x - d/2)D(t_{st})\text{sinc}(\pi\delta f_{ca}\tau_a)\sin(\delta\phi_{ca}) + w_{QE}(t_{sa})\right], \quad (9.10)$$
$$Q_P(t_{sa}) = \sigma_{IQ}\left[\sqrt{2(c/n_0)\tau_a}R(x)D(t_{st})\text{sinc}(\pi\delta f_{ca}\tau_a)\sin(\delta\phi_{ca}) + w_{QP}(t_{sa})\right]$$
$$Q_L(t_{sa}) = \sigma_{IQ}\left[\sqrt{2(c/n_0)\tau_a}R(x + d/2)D(t_{st})\text{sinc}(\pi\delta f_{ca}\tau_a)\sin(\delta\phi_{ca}) + w_{QL}(t_{sa})\right]$$

where σ_{IQ} is the noise standard deviation, c/n_0 is the carrier power to noise density, R is the code correlation function, x is the code tracking error in chips, δf_{ca} is the carrier frequency tracking error, $\text{sinc}(x) = \sin(x)/x$, $\delta\phi_{ca}$ is the carrier phase tracking error, and $w_{IE}, w_{IP}, w_{IL}, w_{QE}, w_{QP},$ and w_{QL} are the normalized I and Q noise terms. For the data-free, or pilot, GNSS signals, D is omitted.

The tracking errors are defined as follows:

$$x = (t_{st} - \tilde{t}_{st})f_{co}$$
$$\delta f_{ca} = \Delta f_{ca} - \Delta\tilde{f}_{ca} \quad (9.11)$$
$$\delta\phi_{ca} = \phi_{IF} - \tilde{\phi}_{IF} + (2\pi t_{sa} - \pi\tau_a)\delta f_{ca}$$

From (8.2), (8.47), (8.48), and (9.2), the tracking errors may be expressed in terms of the pseudo-range and pseudo-range rate measurement errors by

$$x = (\tilde{\rho}_R - \rho_R)f_{co}/c$$
$$\delta f_{ca} = (\dot{\tilde{\rho}}_R - \dot{\rho}_R)f_{ca}/c \quad (9.12)$$

The BPSK code correlation function, BOC correlation, carrier power to noise density, and noise properties are each described below, followed by discussions of the accumulation interval, signal multiplex processing and semicodeless correlation.

9.1.4.1 BPSK Code Correlation Function

The code correlation function provides a measurement of the alignment between the signal and reference codes. Neglecting band-limiting effects, from (9.6) to (9.11), the correlation function for BPSK signals is

$$R_{BPSK}(x) = \frac{1}{\tau_a}\int_{t_{st}-\tau_a}^{t_{st}} C(t)C(t - x/f_{co})\,dt. \quad (9.13)$$

As stated in Section 8.1.2, the ranging code, C, takes values of ± 1. Therefore, when the signal and reference codes are aligned, the correlation function $R(0)$ is unity.

BPSK ranging codes are pseudo-random, so for tracking errors in excess of one code chip, there is a near-equal probability of the signal and reference code product being +1 or −1 at a given snapshot in time. Integrating this over the accumulation interval gives a nominal correlation function of zero. When the tracking error is half a code chip, the signal and reference codes will match for half the accumulation interval and their product will average to near zero for the other half, giving a correlation function of $R(0.5) = 0.5$. Figure 8.5 illustrates this. More generally, the smaller the tracking error, the more the signal and reference will match. Thus, the correlation function is approximately

$$R_{BPSK}(x) \approx \begin{matrix} 1 - |x| & |x| \leq 1 \\ 0 & |x| \geq 1 \end{matrix}. \qquad (9.14)$$

This is illustrated by Figure 9.5. In practice, the precorrelation band-limiting imposed at the satellite transmitter and the receiver front end rounds the transitions of the signal code chips as illustrated by Figure 9.6. This occurs because a rectangular waveform is comprised of an infinite Fourier series of sinusoids, which is truncated by the band-limiting. The smoothing of the code chips results in a smoothing of the correlation function as Figure 9.5 also shows.

One method of approximating the band-limited signal code is to replace the rectangular waveform with a trapezoidal waveform of rise time $\Delta x \approx 0.88 f_{co}/B_{PC}$, where B_{PC} is the double-sided precorrelation bandwidth [23]. This is shown in Figure 9.6. The correlation function under the trapezium approximation is [24]

$$R(x, \Delta x) \approx \begin{matrix} 1 - \dfrac{\Delta x}{4} - \dfrac{x^2}{\Delta x} & 0 \leq |x| \leq \Delta x/2 \\ 1 - |x| & \Delta x/2 \leq |x| \leq 1 - \Delta x/2 \\ \dfrac{1}{2\Delta x} - \left(\dfrac{|x|}{\Delta x} - \dfrac{1}{2}\right)\left(1 - \dfrac{|x|}{2} + \dfrac{\Delta x}{4}\right) & 1 - \Delta x/2 \leq |x| \leq 1 + \Delta x/2 \\ 0 & 1 + \Delta x/2 \leq |x| \end{matrix}. \qquad (9.15)$$

The band-limited correlation function may be represented more precisely using Fourier analysis.

For code tracking errors greater than one chip, the auto-correlation function of a PRN code sequence is not exactly zero. Instead, it has noise-like behavior with a standard deviation of $1/\sqrt{n}$, where n is the number of code chips over the code repetition length or the accumulation interval, whichever is fewer [3]. The cross-correlation function between the reference code and a different PRN code of the same length also has a standard deviation of $1/\sqrt{n}$. The ranging codes used for GNSS signals are not randomly selected. For example, the GPS C/A codes are selected to limit the cross-correlation and minor autocorrelation peaks to +0.064 and −0.062.

9.1 Receiver Hardware and Antenna

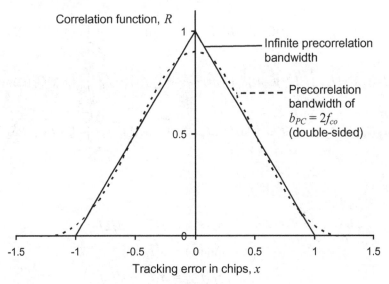

Figure 9.5 BPSK code correlation function for unlimited and band-limited signals.

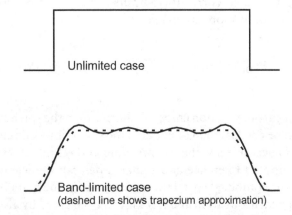

Figure 9.6 Comparison of unlimited and band-limited code chips.

9.1.4.2 BOC Correlation

For BOC signals, the samples following carrier wipeoff, I_0 and Q_0, are given by

$$I_0(t_{sa}) = A_0 C(t_{st}) S(t_{st}) D(t_{st}) \cos\left[2\pi\left(\Delta f_{ca} - \Delta \tilde{f}_{ca}\right) t_{sa} + \phi_{IF} - \tilde{\phi}_{IF}\right] + w_{I0}(t_{sa})$$
$$Q_0(t_{sa}) = A_0 C(t_{st}) S(t_{st}) D(t_{st}) \sin\left[2\pi\left(\Delta f_{ca} - \Delta \tilde{f}_{ca}\right) t_{sa} + \phi_{IF} - \tilde{\phi}_{IF}\right] + w_{Q0}(t_{sa})$$
(9.16)

To demodulate them, they must also be multiplied by a reference subcarrier function. This may be an early, prompt, or late subcarrier, aligned with the reference code. Thus,

$$I_E(t_{sa}) = f_a \int_{t_{sa}-\tau_a}^{t_{sa}} I_0(t)C_E(t)S_E(t)\,dt, \qquad Q_E(t_{sa}) = f_a \int_{t_{sa}-\tau_a}^{t_{sa}} Q_0(t)C_E(t)S_E(t)\,dt$$

$$I_P(t_{sa}) = f_a \int_{t_{sa}-\tau_a}^{t_{sa}} I_0(t)C_P(t)S_P(t)\,dt, \qquad Q_P(t_{sa}) = f_a \int_{t_{sa}-\tau_a}^{t_{sa}} Q_0(t)C_P(t)S_P(t)\,dt, \qquad (9.17)$$

$$I_L(t_{sa}) = f_a \int_{t_{sa}-\tau_a}^{t_{sa}} I_0(t)C_L(t)S_L(t)\,dt, \qquad Q_L(t_{sa}) = f_a \int_{t_{sa}-\tau_a}^{t_{sa}} Q_0(t)C_L(t)S_L(t)\,dt$$

where

$$\begin{aligned} S_E(t_{sa}) &= S(\tilde{t}_{st} + d/2f_{co}) \\ S_P(t_{sa}) &= S(\tilde{t}_{st}) \\ S_L(t_{sa}) &= S(\tilde{t}_{st} - d/2f_{co}) \end{aligned} \qquad (9.18)$$

The accumulated correlator outputs are as given by (9.10). However, the correlation function is more complex:

$$R_{BOC}(x) = \frac{1}{\tau_a} \int_{t_{st}-\tau_a}^{t_{st}} C(t)C(t - x/f_{co})S(t)S(t - x/f_{co})\,dt. \qquad (9.19)$$

The subcarrier function chips are shorter than the spreading-code chips and have a repetition period less than or equal to the spreading-code chip size. Therefore, if the code tracking error is less than a spreading-code chip, but greater than a subcarrier-function chip, the reference and signal codes can be negatively correlated. Figure 9.7 shows the combined correlation functions for the main BOC GNSS signals [25].

BOC signal acquisition and tracking may be aided by also correlating the signals with alternative functions of the subcarrier. An example is the differenced subcarrier:

$$I_{P\Delta}(t_{sa}) = f_a \int_{t_{sa}-\tau_a}^{t_{sa}} I_0(t)C_P(t)S_\Delta(t)\,dt, \qquad Q_{P\Delta}(t_{sa}) = f_a \int_{t_{sa}-\tau_a}^{t_{sa}} Q_0(t)C_P(t)S_\Delta(t)\,dt, \qquad (9.20)$$

where

$$S_\Delta(t_{sa}) = S(\tilde{t}_{st} + e/2f_s) - S(\tilde{t}_{st} - e/2f_s) \qquad (9.21)$$

and e is the offset in subcarrier periods between the two versions of the subcarrier.

9.1.4.3 Carrier Power to Noise Density and Noise Properties

The carrier power to noise density, c/n_0, is the ratio of the received signal power to the single-sided noise PSD, weighted by the GNSS signal spectrum. It is the primary

9.1 Receiver Hardware and Antenna

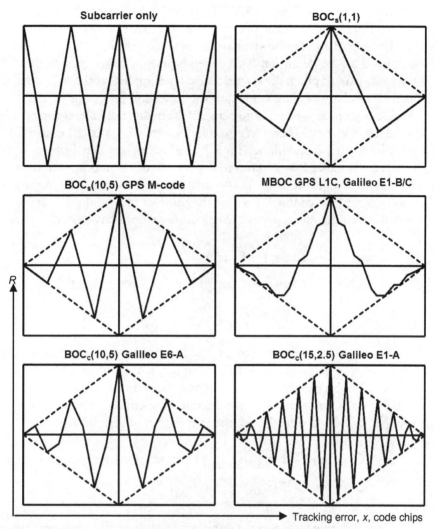

Figure 9.7 BOC combined spreading and subcarrier code correlation functions (neglecting band-limiting; dashed lines show equivalent BPSK correlation functions).

measure of the signal-to-noise environment used to determine GNSS performance. It is commonly expressed in decibel form where, to avoid confusion, the upper case equivalent, C/N_0, is used (some authors use C/No):

$$C/N_0 = 10\log_{10}(c/n_0), \qquad c/n_0 = 10^{\frac{C/N_0}{10}}. \qquad (9.22)$$

Determination of the carrier power to noise density as a function of the signal strength, interference levels, receiver, and antenna design is discussed in [3, 26]. For a strong GPS C/A-code signal at normal incidence to a good antenna in the absence of interference, C/N_0 should exceed 45 dB-Hz. Measurement of c/n_0 by the user equipment is discussed in Section 9.2.6, while the effect of c/n_0 on range measurement errors is discussed in Section 9.3.3. Section G.6.1 of Appendix G on the CD describes the relationship between signal to noise and carrier power to noise density.

From (9.10), the noise standard deviation on the I and Q accumulated correlator outputs is σ_{IQ} by definition. This depends on four factors: the noise standard deviation prior to sampling, the quantization level applied at the ADC, the accumulation interval, and any scaling applied to the baseband processor's I and Q outputs. The AGC maintains a constant ratio between the quantization level and presampling noise standard deviation. Therefore, for all receivers with a continuous AGC, σ_{IQ} is constant for a given accumulation interval, but varies between different receiver designs.

The normalized noise terms in the correlator outputs have unit variance by definition and zero mean. There is also no correlation between the noise in the in-phase and quadraphase channels because the product of I_C and Q_C averages close to zero over the accumulation interval. Thus, the noise terms have the following expectations

$$\begin{aligned} \mathrm{E}(w_{I\alpha}^2) = \mathrm{E}(w_{Q\alpha}^2) &= 1 \\ \mathrm{E}(w_{I\alpha}) = \mathrm{E}(w_{Q\alpha}) &= 0 \quad, \qquad \alpha, \beta \in E, P, L. \\ \mathrm{E}(w_{I\alpha} w_{Q\beta}) &= 0 \end{aligned} \qquad (9.23)$$

The noise on the early, prompt, and late correlator outputs is correlated because the same noise sequences, $w_{I0}(t_{sa})$ and $w_{Q0}(t_{sa})$, are multiplied by the same reference codes, offset by less than one chip. Thus, the noise is the same over the proportion of the correlation interval where the reference codes are aligned. When precorrelation band-limiting is neglected, the correlation properties are

$$\begin{aligned} \mathrm{E}(w_{IE} w_{IP}) = \mathrm{E}(w_{QE} w_{QP}) = \mathrm{E}(w_{IP} w_{IL}) = \mathrm{E}(w_{QP} w_{QL}) &= 1 - d/2 \\ \mathrm{E}(w_{IE} w_{IL}) = \mathrm{E}(w_{QE} w_{QL}) &= 1 - d \end{aligned} \qquad (9.24)$$

Band-limiting increases the correlation between the early, prompt, and late correlator outputs because it introduces time correlation to the input noise sequences, $w_{I0}(t_{sa})$ and $w_{Q0}(t_{sa})$ [27].

9.1.4.4 Accumulation Time

The choice of the time interval, τ_a, over which to accumulate the correlator outputs is a tradeoff between four factors: signal to noise, signal coherence, navigation-data-bit handling, and ranging-processor bandwidth. As shown in Section G.6.1 of Appendix G on the CD, the signal-to-noise ratio of the baseband signal processor's I and Q outputs is optimized by maximizing τ_a. However, the other factors are optimized by short accumulation times.

If the residual carrier phase error after carrier wipeoff varies over the accumulation interval, the summed samples will interfere with each other, as shown in Figure 8.9. This is accounted for by the $\mathrm{sinc}(\pi \delta f_{ca} \tau_a)$ term in (9.10). Total cancellation occurs where the phase error changes by an integer number of cycles over the accumulation interval. To maximize the I and Q signal to noise, a constant phase error must be maintained over the accumulation time. This is known as maintaining signal coherence, while linear summation of the Is and Qs is known as coherent integration. To limit the signal power loss due to carrier phase interference to a factor of 2, the following conditions must be met

$$\operatorname{sinc}(\pi \delta f_{ca}\tau_a) < 1/\sqrt{2}$$
$$|\delta f_{ca}| < \frac{0.443}{\tau_a} \qquad (9.25)$$
$$|\delta \dot{\rho}| < 0.443 \frac{c}{f_{ca}\tau_a}$$

Thus, for $\tau_a = 20$ ms and a signal in the L1 band, the pseudo-range-rate error must be less than 4.2 m s^{-1}.

Squaring and adding the Is and Qs eliminates carrier phase interference over periods longer than the accumulation interval, so summation of $I^2 + Q^2$ is known as noncoherent integration. However, the power signal to noise for noncoherent integration varies as the square root of the integration time, as opposed to linearly for coherent integration. Therefore, signal to noise is optimized by performing coherent summation up to the point when carrier phase interference starts to be a problem and then performing noncoherent summation beyond that.

All of the legacy GNSS signals and about half of the new signals incorporate a navigation data message. If coherent summation is performed over two message data bits, there is an equal probability of those bits having the same or opposite signs. When the bits are different, the signal component changes sign halfway through the summation and the accumulated signal power is cancelled out. Accumulating correlator outputs over more than one data bit also prevents navigation message demodulation. Therefore, the data-bit length acts as the effective limit to the accumulation time for data-carrying signals, varying from 1 ms for the Galileo E6-B to 20 ms for the legacy signals, GPS L2C, and Galileo E5a-I (see Chapter 8). Specialist techniques for circumventing this limit are discussed in Section 10.3.5.

For the newer navigation-data-modulated GNSS signals, the code repetition interval is greater than or equal to the data bit length, so the data bit edges are determined from the ranging code. However, the GPS and GLONASS C/A codes repeat 20 times per data bit, requiring the data bit edges to be found initially using a search process (Section 9.2.5). When the data bit edges are unknown, the only way of preventing summation across data bit boundaries is to limit the accumulation time to the 1-ms code length. However, for accumulation times below 20 ms, less than a quarter of the signal power is lost through summation across data bit boundaries. For GLONASS FDMA signals, the 100 chip s^{-1} meander sequence (see Section 8.4.2) adds further complications.

When a signal with a layered code is correlated only with the primary code (e.g., during acquisition), coherent summation across successive secondary code boundaries should be similarly avoided, noting that these boundaries are always known.

The final factor in determining the accumulation time is the Nyquist criteria for the ranging processor. In tracking mode, the sampling rate of the tracking function must be at least twice the tracking loop bandwidth, which is larger for carrier tracking. The tracking function sampling rate is generally the inverse of the accumulation time, τ_a, and is known as the postcorrelation or predetection bandwidth. In acquisition mode, longer accumulation times result in either a longer acquisition time or the need to employ more correlators (see Section 9.2.1).

The optimum accumulation time depends on whether the ranging processor is in acquisition or tracking mode and can also depend on the signal-to-noise environment. The accumulation time can be varied within the baseband processor. However, it is simpler to fix it at the minimum required, typically 1 ms, and perform additional summation in the ranging processor.

9.1.4.5 Signal Multiplex Processing

The GPS L5 and L1C signals, some GLONASS CDMA signals, and most Galileo and Beidou phase 3 signals are broadcast as in-phase and quadraphase pairs with a navigation data message on one component only. The GPS L2C, GPS M-code, Galileo PRS, and some GLONASS CDMA signals are broadcast as time-division multiplexes with their codes alternating between navigation-message-modulated and unmodulated bits. Many receivers will process both signals and many of the baseband processor functions can be shared between them. As the code and carrier of these multiplexed signal pairs are always in phase, the code and carrier NCOs can be shared.

A TDM signal pair may be treated as a single signal up until the code correlator outputs, with samples corresponding to alternate reference-code bits sent to separate data and pilot-channel accumulators. Alternatively, separate code correlators may be used for each component, with the reference code taking values of +1, –1, and 0, such that the reference code for one component is zero when the code for the other component is nonzero [28, 29].

Separate code correlators must be used for the in-phase and quadraphase signal multiplexes as both components are transmitted simultaneously. However, the in-phase and quadraphase sampling or carrier wipeoff phase may be shared as the in-phase samples for one signal are the quadraphase samples for the other.

9.1.4.6 Semi-Codeless Correlation

In principle, the GPS P(Y) code, when encrypted to Y code, is only available to authorized users. However, other users can take advantage of the layered property of this code to track it. Y code comprises the publicly-known 10.23-Mbit s^{-1} P code multiplied by a 0.5115-Mbit s^{-1} encryption code. Therefore, the Y code in the L2 band can be acquired and tracked by correlating it with the P code, provided that the coherent integration interval of the correlator outputs is limited to the 20-bit length of each encryption code chip. Beyond this, noncoherent accumulation must be used, summing $I^2 + Q^2$ or the root thereof. This technique is known as semi-codeless tracking [30] and brings a signal-to-noise penalty of about 18 dB over correlation with the full Y code. Alternative, codeless, techniques are also described in [30].

9.2 Ranging Processor

The GNSS ranging processor uses the accumulated correlator outputs from the receiver to determine the pseudo-range, pseudo-range rate, and carrier phase and to control the receiver's generation of the reference code and carrier. This section describes acquisition of GNSS signals and tracking of the code and carrier, followed by a discussion of tracking lock detection, navigation-message demodulation, signal-to-noise measurement, and generation of the measurements output to the navigation processor. Simplified notation, omitting the transmitter and receiver designations, is used in Sections 9.2.1–9.2.6.

9.2.1 Acquisition

When GNSS user equipment is switched on or a new satellite signal comes into view, the code phase of that signal is unknown. To determine this and obtain the time of signal transmission, the reference code phase must be varied until it matches that of the signal. When one of the reference codes is within one chip of the signal code, the despread signal is observed in the receiver's accumulated correlator outputs (see Section 9.1.4). However, the Doppler-shifted carrier frequency of the signal must also be known to sufficient accuracy to maintain signal coherence over the accumulation interval. Otherwise, the reference Doppler shift must also be varied. This searching process is known as *acquisition* [1, 3, 13]. Acquisition algorithm design is a tradeoff among speed, sensitivity, reliability, and processing load.

Each code phase and Doppler shift searched is known as a *bin* while each combination of the two is known as a *cell*. The time spent correlating the signal for each cell is known as the *dwell time* and may comprise coherent and noncoherent integration. The code-phase bins are usually set half a chip apart. Except for long dwell times, the spacing of the Doppler bins is dictated by the coherent integration interval, τ_a, and is around $1/2\tau_a$. For each cell, a test statistic combining the in-phase and quadraphase channels, $I^2 + Q^2$, is compared against a threshold. If the threshold is exceeded, the signal is deemed to be found.

Traditional acquisition algorithms, such as the Tong method, start at the center of the Doppler search window and move outwards, alternating from side to side. Each code phase at a given Doppler shift is searched before moving onto the next Doppler bin. Code phase is searched from early to late so that directly received signals are usually found before reflected signals. As each baseband processor channel typically has three I and Q correlator pairs, three code phases may be searched simultaneously. Narrow correlator spacing should not be used for acquisition. Parallel baseband processor channels are used either to increase the number of parallel cells searched or for acquisition of other signals. When the acquisition threshold is exceeded, the test for that cell is repeated with further samples and the search is stopped if signal acquisition is confirmed. This confirmation process enables the false detection and missed detection probabilities to be minimized without having to implement a long dwell time for every bin searched. Figure 9.8 depicts the acquisition process.

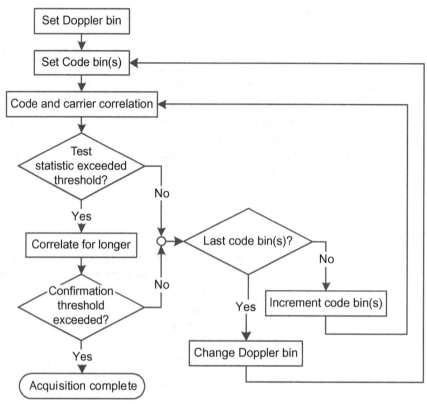

Figure 9.8 Time-domain signal acquisition process.

Acquisition searches can find more than one peak for a given signal. Smaller auto-correlation and cross-correlation peaks arise in a code-phase search due to the limitations in the code correlation function (see Section 9.1.4.1); the longer the code repetition length, the smaller these peaks are. Smaller peaks arise in a Doppler search due to the minor peaks in the sinc function of (9.10), as illustrated by Figure 9.9. These are the same for all GNSS signals and are larger than the code-phase minor peaks. To prevent the acquisition algorithm finding a minor peak first, the threshold may be set higher than the minor peaks in a strong signal-to-noise environment. The threshold is only reduced if no signal is found on the first search, noting that the signal-to-noise level cannot be measured until a signal has been acquired.

A related problem occurs when acquiring weak signals with short code repetition lengths, such as GPS C/A-code. Cross-correlation peaks between the reference code and stronger C/A-code signals can be mistaken for the signal being acquired. A well-designed acquisition algorithm will acquire the strongest signals first. The on-frequency cross-correlation peaks will then be known and algorithms can be designed to eliminate them from the search. Cross-correlation peaks can also be found at 1-kHz frequency offsets and must be identified through a mismatch between carrier cycles and code chips [31].

When an AGC is used, the noise level is constant. Otherwise, the detection threshold is varied with the noise level to maintain a constant probability of false acquisition. The probability of missed detection thus increases as the signal to noise

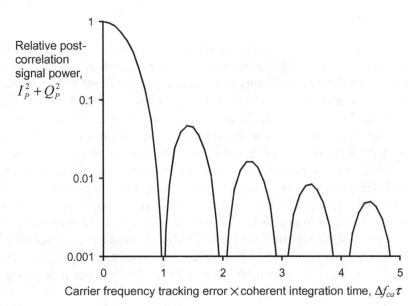

Figure 9.9 Relative postcorrelation signal power as a function of frequency tracking error.

decreases. Therefore, GNSS acquisition in poor signal-to-noise environments requires long dwell times for each cell in the search so that the signal can be identified above the noise and/or interference. For coherent integration, the required dwell time varies as $(c/n_0)^{-1}$, whereas, for noncoherent integration, it varies as $(c/n_0)^{-2}$. Thus, more benefit is obtained from increasing the coherent integration time, τ_a, although this does require a reduced Doppler bin spacing, increasing the number of Doppler bins to search [32].

Long noncoherent integrations also require more closely spaced Doppler bins because the change in code phase over the total integration interval takes over from the change in carrier phase over the coherent integration interval as the limiting factor. The threshold is about 20 seconds for GPS C/A code and 2 seconds for P(Y) code (assuming $\tau_a = 20$ ms). The number of Doppler bins required is then proportional to the chipping rate multiplied by the dwell time.

Implementing long dwell times when the host vehicle is maneuvering can be problematic due to the Doppler shift changing. This may be compensated using external aiding (see Section 10.5.1).

In most cases, known as *warm starts*, the user equipment has the approximate time, user position, and satellite almanac data from when it was last used or through integration with other navigation systems. Situations where this information is not available prior to acquisition are known as *cold starts*. Situations in which the current ephemeris data is available and the time is known to within a millisecond are sometimes called *hot starts*. The size of the code search window is thus determined by the time and position uncertainty and can be set to the 3σ bounds. For short codes, such as the GPS and GLONASS C/A codes, the search window is limited to the code repetition length, significantly reducing the number of cells to search when prior information is poor. Conversely, very long codes, such as the GPS P(Y) and M codes and the Galileo PRS, cannot practically be acquired without prior knowledge of time.

For a given search window, the number of code bins is directly proportional to the chipping rate. The size of the Doppler search window depends on the satellite and user velocity uncertainty and again is generally set to the 3σ bounds up to a maximum determined by the maximum user velocity and maximum satellite velocity along the line of sight ($\sim$1,200 m s^{-1}). However, when a low-cost reference oscillator (Section 9.1.2) is used, a wider Doppler search window, typically $\pm$2,500 m s^{-1}, is needed for the first signal acquired to determine the receiver clock drift.

When four or more satellite signals have been acquired, enabling calculation of a navigation solution, and current almanac data is available, the code search window for acquiring signals from further satellites or reacquiring lost signals (known as *reacquisition*) is small, while a Doppler search may not be required at all. The search window is similarly small where other signals from the same satellite have already been acquired. So, in most PPS GPS user equipment, the P(Y) code is acquired after the C/A code. The same method may be used to acquire other codes with longer lengths and higher chipping rates.

For a cold-start acquisition, the number of GPS C/A-code bins is 2,046. In a strong signal-to-noise environment, an adequate dwell time is 1 ms, giving about 50 Doppler bins with a stationary receiver using a TCXO. Searching three cells at a time, acquisition can take place using a single channel within about 15 seconds. However, acquisition of low-C/N_0 signals using longer dwell times, acquiring higher-chipping-rate codes, and acquiring codes with longer repetition lengths where prior information is poor all takes much longer using traditional techniques. Note that higher-chipping-rate codes offer greater resistance against narrowband interference, while there is a requirement to acquire the GPS M code independently of other signals.

Solving these more challenging acquisition tasks, or simply speeding up the acquisition process, requires more processing hardware. Receivers with massively parallel correlator arrays can search thousands of code bins in parallel, as opposed to 30–36 cells per frequency band for basic GPS receivers [33, 34]. However, acquisition techniques exploiting the fast Fourier transform are much more efficient.

FFT-based acquisition algorithms are based on the principle that multiplication in the frequency domain is the same as convolution in the time domain. They are applied to the in-phase and quadraphase signal samples following carrier wipeoff, I_0 and Q_0. An FFT is used to transform a series of signal samples over the correlation accumulation interval, τ_a, and at a given Doppler shift, to the frequency domain. The reference code is similarly transformed to the frequency domain where it is multiplied by the corresponding signal samples. An inverse FFT then produces test statistics for all of the code bins simultaneously [16, 22, 35, 36]. Figure 9.10 depicts the FFT-based acquisition process. Note that an FFT may only be applied to 2^n samples, where n is any integer. Zero padding of the reference code may be used to avoid correlation across navigation-data-bit and secondary-code-chip boundaries (see Section 9.1.4.4) at the cost of reduced processing efficiency.

Both massively parallel correlation and FFT-based acquisition require substantially more processing power than tracking. Consequently, they are often implemented using a dedicated processor, known as an *acquisition engine*, that is only powered up when required.

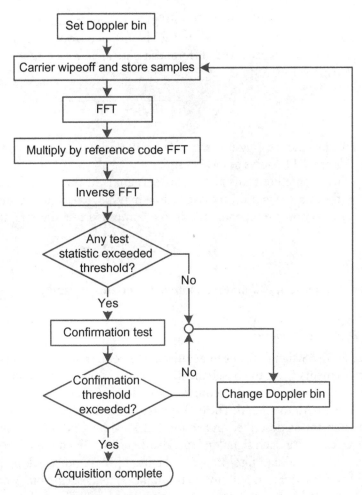

Figure 9.10 FFT-based signal acquisition process.

The GPS L5 codes and most of the Galileo codes are layered. This can be used to speed up acquisition. By initially limiting the coherent integration interval to the primary-code repetition interval, the number of code bins in the acquisition search is limited to the primary-code length. Once the primary-code phase has been acquired, the full code is acquired. A longer coherent integration interval and dwell time must be used to determine the correct secondary alignment, but the number of code bins required is no greater than the secondary-code length.

Acquisition of a full BOC signal requires a code bin separation of a quarter of a subcarrier-function repetition interval because of the narrowing of the correlation function peak (see Section 9.1.4.2). If acquisition is performed using one sidelobe only or by noncoherently combining separately correlated sidelobes, the code bin separation need only be half a spreading-code chip. This reduces the number of cells to search by a factor of 2 for the GPS L1C and Galileo E1-B/C signals (assuming only the $BOC_s(1,1)$ component is used), a factor of 4 for GPS M code and Galileo E6-A, and a factor of 12 for Galileo E1-A. However, single-sidelobe operation halves the effective signal-to-noise ratio, while separate correlation of the sidelobes requires additional hardware.

Subcarrier cancellation (SCC) offers the same performance and code bin size as separate sidelobe correlation, but with only one front end. Signals are correlated with both synchronized and differenced subcarriers (see Section 9.1.4.2) and the test statistic is [37, 38]

$$s_{SCC} = \sqrt{I_P^2 + Q_P^2} + \sqrt{I_{P\Delta}^2 + Q_{P\Delta}^2}. \qquad (9.26)$$

This produces a ziggurat-shaped correlation function for sine-phased BOC signals; Figure 9.11 shows some examples.

When a satellite transmits a pair of signals in the same frequency band, with and without a navigation message, they may be acquired together, combining the correlator outputs noncoherently into a combined test statistic, such as

$$s_{d+p} = I_d^2 + Q_d^2 + I_p^2 + Q_p^2, \qquad (9.27)$$

where the subscripts d and p denote data-modulated and pilot, respectively [39].

9.2.2 Code Tracking

Once a GNSS signal has been acquired, the code tracking process uses the I and Q measurements from the baseband signal processor to refine its measurement of the code phase, which is used to control the code NCO, maintaining the reference code's alignment with the signal. The code phase is also used to calculate the pseudo-range measurement as described in Section 9.2.7. Most GNSS user equipment performs code tracking for each signal independently using a fixed-gain delay lock loop (DLL) [23], while a Kalman filter may also be used [32]. Code tracking may also be combined with navigation-solution determination as described in Section 10.3.7.

Figure 9.12 shows a typical code tracking loop. The early, prompt, and late in-phase and quadraphase accumulated correlator outputs from the receiver are input to a discriminator function, which calculates a measurement of the code tracking error. This is used to correct the tracking loop's code-phase estimate, which is then predicted forward in time and used to generate a code NCO command, which is sent to the receiver. The prediction phase is usually aided with range-rate information

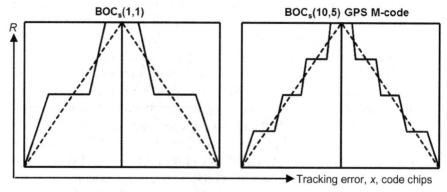

Figure 9.11 Effective BOC correlation functions using subcarrier cancellation (neglecting band-limiting; dashed lines show equivalent BPSK correlation functions).

9.2 Ranging Processor

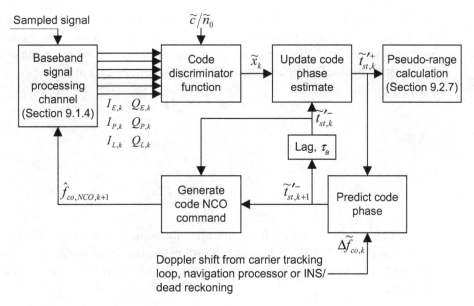

Figure 9.12 Code tracking loop.

from the carrier tracking function, the navigation processor, or an INS or other dead-reckoning system (see Section 10.5.1). The carrier tracking function of another signal from the same satellite may also be used to provide range-rate aiding [40]. Each step is now described.

The discriminator function may coherently integrate the Is and Qs over a navigation-message bit. However, there is no benefit in performing noncoherent integration as the tracking loop does this inherently. The Is and Qs accumulated over a total time, τ_a, are then used to produce a discriminator function, D, which is proportional to the code tracking error, x. The most common discriminators are the dot-product power (DPP), early-minus-late power (ELP), and early-minus-late envelope (ELE) noncoherent discriminators [1, 13]:

$$D_{DPP} = (I_E - I_L)I_P + (Q_E - Q_L)Q_P$$
$$D_{ELP} = (I_E^2 + Q_E^2) - (I_L^2 + Q_L^2) \qquad (9.28)$$
$$D_{ELE} = \sqrt{I_E^2 + Q_E^2} - \sqrt{I_L^2 + Q_L^2}$$

Note that only the dot-product discriminator uses the prompt correlator outputs. Its formulation enables early-minus-late correlators to be implemented (see Section 7.1.4).

A coherent discriminator is less noisy, but requires carrier phase tracking to be maintained to keep the signal power in the in-phase channel. However, as carrier phase tracking is much less robust than code tracking, only noncoherent discriminators may be used in poor signal-to-noise environments. An example of a coherent discriminator is the decision-directed discriminator

$$D_{Coh} = (I_E - I_L)\operatorname{sign}(I_P). \qquad (9.29)$$

These discriminators all work on the principle that the signal power in the early and late correlation channels is equal when the prompt reference code is synchronized with the signal. To obtain a measurement of the code tracking error, the discriminator must be multiplied by a normalization function, N_D. Thus,

$$\tilde{x}_k = N_D D, \tag{9.30}$$

where

$$N_D = \frac{x}{\lim_{x \to 0} \mathrm{E}[D(x)]}, \tag{9.31}$$

noting that the discriminator functions are only linear functions of x where x is small.

From (9.10), (9.14), (9.23), and (9.24), neglecting precorrelation band-limiting and assuming $|x| < 1 - d/2$ and $\delta f_{ca} \approx 0$, the expectations of the discriminator functions are

$$\begin{aligned}
\mathrm{E}(D_{DPP}) &\approx 2\sigma_{IQ}^2 (c/n_0) \tau_a (1 - |x|)(|x + d/2| - |x - d/2|) \\
\mathrm{E}(D_{ELP}) &\approx 2\sigma_{IQ}^2 (c/n_0) \tau_a (2 - |x + d/2| - |x - d/2|)(|x + d/2| - |x - d/2|) \\
\mathrm{E}(D_{ELE}) &\approx \sigma_{IQ} \sqrt{2(c/n_0)\tau_a} (|x + d/2| - |x - d/2|) \\
\mathrm{E}(D_{Coh}) &\approx \sigma_{IQ} \sqrt{2(c/n_0)\tau_a} (|x + d/2| - |x - d/2|)\cos(\delta\phi_{ca})
\end{aligned} \tag{9.32}$$

From (9.31), the normalization functions are thus

$$\begin{aligned}
N_{DPP} &= \frac{1}{4\sigma_{IQ}^2 (\tilde{c}/\tilde{n}_0) \tau_a} \\
N_{ELP} &= \frac{1}{4(2-d)\sigma_{IQ}^2 (\tilde{c}/\tilde{n}_0) \tau_a} \\
N_{ELE} &= \frac{1}{2\sigma_{IQ} \sqrt{2(\tilde{c}/\tilde{n}_0) \tau_a}} \\
N_{Coh} &= \frac{1}{2\sigma_{IQ} \sqrt{2(\tilde{c}/\tilde{n}_0) \tau_a}}
\end{aligned} \tag{9.33}$$

noting that the measured carrier power to noise density has been substituted for its true counterpart. In some receivers, the normalization is performed by dividing by $I^2 + Q^2$ or its root (as appropriate). However, $I^2 + Q^2$ is only proportional to $(c/n_0)\tau_a$ in strong signal-to-noise environments.

Figure 9.13 shows the discriminator input-output curves, neglecting noise and band-limiting, for early-late correlator spacings of 0.1 and 1 chips. With the larger correlator spacing, the discriminator can respond to larger tracking errors. It has a larger pull-in range and a longer linear region. However, as shown in Section 9.3.3, the tracking noise is larger.

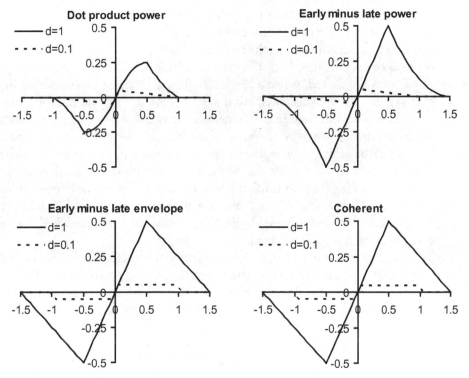

Figure 9.13 Code discriminator input-output curves (units: chips).

The code-phase estimate of the tracking function is denoted here by $\tilde{t}'_{st}$ as it is offset from the time of signal transmission, t_{st}, by an integer number of code repetition periods. This is updated using the discriminator output:

$$\tilde{t}'^{+}_{st,k} = \tilde{t}'^{-}_{st,k} + K_{co}\tilde{x}_k/f_{co}, \tag{9.34}$$

where K_{co} is the code loop gain and, in analogy with Kalman filter notation, the subscript k denotes the iteration, while the superscripts – and + denote before and after the update, respectively. The loop gain is set at less than unity to smooth out the noise on the correlator outputs. The double-sided noise bandwidth of the code tracking loop, B_{L_CO}, is given by [1, 13]

$$B_{L\_CO} = K_{co}/4\tau_a. \tag{9.35}$$

Conversely,

$$K_{co} = 4B_{L\_CO}\tau_a. \tag{9.36}$$

The code tracking bandwidth typically takes values between 0.05 and 1 Hz. The narrower the bandwidth, the greater the noise resistance, but the longer it takes to respond to dynamics (see Section 9.3.3). Thus, its selection is a tradeoff, dependent on the operating context. The code tracking loop does not need to track the absolute code phase, only the error in the code phase obtained by integrating the range-rate aiding. The most accurate source of aiding is from the corresponding carrier phase tracking loop, in which case, the code tracking loop need only track the code-carrier divergence due to ionospheric dispersion (Section 9.3.2). However, carrier phase cannot always be tracked. When a carrier frequency tracking loop, another signal from that satellite, or another navigation system provides the aiding, the code tracking loop must track the error in the range-rate aiding. When two or more aiding sources are available, they may be weighted according to their respective uncertainties.

When the aiding is provided by the GNSS navigation processor, it may only be able to supply the range rate due to the satellite motion and Earth rotation, leaving the code tracking loop to track the user dynamics. This is only possible where the user velocity is less than $4B_{L\_CO}$ times the maximum recoverable tracking error (see Section 9.3.3). With a 1-Hz code tracking bandwidth, land vehicle dynamics can be tracked using most GNSS signals.

Another issue is the navigation-solution update rate required. The interval between statistically independent code-phase measurements is approximately $1/4B_{L\_CO}$. Consequently, the lowest code tracking bandwidths tend to be used for static applications.

The code-phase estimate is predicted forward to the next iteration using

$$\tilde{t}'^{-}_{st,k+1} = \tilde{t}'^{+}_{st,k} + \frac{f_{co} + \Delta\tilde{f}_{co,k}}{f_{co}}\tau_a, \qquad (9.37)$$

where f_{co} is the transmitted code chipping rate and $\Delta\tilde{f}_{co,k}$ is its Doppler shift, obtained from the aiding source. This may be estimated from the carrier Doppler shift or pseudo-range rate using

$$\Delta\tilde{f}_{co} \approx \frac{f_{co}}{f_{ca}}\Delta\tilde{f}_{ca} \approx -\frac{f_{co}}{c}\tilde{\dot{\rho}}_R, \qquad (9.38)$$

noting that there will be a small discrepancy due to code-carrier divergence that will be corrected by the code tracking loop.

Most GNSS receivers do not allow step changes to the reference code, so code-phase corrections are made by running the code NCO faster or slower than the Doppler-shifted code chipping rate. This can be done by setting the code NCO frequency to the following:

$$\hat{f}_{co,NCO,k+1} = \frac{\tilde{t}'^{-}_{st,k+1} - \tilde{t}'^{-}_{st,k}}{\tau_a}f_{co}. \qquad (9.39)$$

When step changes are permitted, the reference code is shifted by $\tilde{t}'^{+}_{st,k} - \tilde{t}'^{-}_{st,k}$ and the code NCO frequency is set to $f_{co} + \Delta\tilde{f}_{co,k}$.

Except in a software receiver, the processing of the code tracking function and the signal correlation occur simultaneously. Therefore, there is a lag of one correlation period, τ_a, in applying the NCO control corrections to the receiver. Figure 9.14 illustrates this. However, this is not a problem as the lag is much less than the time constant of the tracking loop.

9.2 Ranging Processor

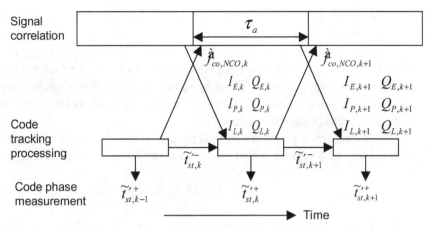

Figure 9.14 Timing of signal correlation and code tracking processing.

For tracking of BOC signals, the correlator spacing must be carefully selected so that the early, prompt, and late correlators all lie on the central peak of the correlation function (see Figure 9.7) to ensure that the discriminator function has the correct sign for small code tracking errors [41]. The discriminator function will still exhibit sign errors for large tracking errors [42], limiting the tracking errors that can be recovered from or locking onto a minor peak of the correlation function. Furthermore, the wrong peak may be locked onto following acquisition.

A number of techniques may be used to track BOC signals unambiguously. The *double estimator* tracks the subcarrier and PRN code separately, implementing separate discriminators, tracking loops, and NCOs [38, 43]. The subcarrier tracking provides a precise but ambiguous measurement of the pseudo-range and the code tracking is used to resolve the ambiguity. *Bump jumping* uses additional very early and very late correlators to measure the neighboring peaks of the correlation function. If one of these is found to be larger than the peak that is currently tracked, a half-subcarrier-period correction is applied to the code-phase estimate [44]. More details of both methods are presented in Section G.6.2 of Appendix G on the CD, together with a summary of some other BOC-tracking techniques.

When a pair of signals from the same satellite in the same frequency band, one with and one without a navigation message (see Section 9.1.4.5), are both tracked, they may share a common tracking function, with the discriminator outputs from the two signals averaged. When a longer coherent integration time is used for the pilot signal, its discriminator should be given higher weighting [28]. Alternatively, measurements from the pilot signal may be used to maintain tracking of both signals.

9.2.3 Carrier Tracking

The primary purpose of carrier tracking in GNSS user equipment is to maintain a measurement of the Doppler-shifted carrier frequency. This is used to maintain signal coherence over the correlator accumulation interval. It is also used to aid the code tracking loop and to provide a less noisy measurement of the pseudo-range rate to the navigation processor. Either the carrier phase or the carrier frequency may be tracked, noting that a carrier phase tracking function also tracks the frequency.

Carrier phase tracking enables the navigation data message to be demodulated more easily and allows precision carrier-phase positioning techniques (Section 10.2) to be used. Carrier frequency tracking is more robust in poor signal-to-noise and high-dynamics environments because tracking lock may be maintained with larger errors. Consequently, many GNSS user equipment designs implement frequency tracking as a reversionary mode to phase tracking and as an intermediate step between acquisition and phase tracking.

A GNSS carrier phase tracking function tracks the phase of the received signal with respect to a reference signal at the carrier frequency, f_{ca}. This is defined as

$$\phi_{ca} = 2\pi(\Delta f_{ca} t_{sa} - N_0) + \phi_{IF}, \qquad (9.40)$$

where N_0 is an arbitrary but constant integer. It is typically set at carrier tracking initialization either to minimize $|\phi_{ca}|$ or such that $\phi_{ca} \approx 2\pi\rho_R/\lambda_{ca}$, where λ_{ca} is the carrier wavelength and ρ_R is the pseudo-range obtained from code tracking. Note that ϕ_{ca} is not limited to a $-\pi$ to π or 0 to 2π range, enabling changes in pseudo-range to be measured by time differencing it. However, in some user equipment designs, the integer cycle count and the phase within the current cycle are stored separately [13].

Most GNSS user equipment performs carrier tracking independently for each signal, using a fixed-gain phase lock loop (PLL) for phase tracking and a frequency lock loop (FLL) for frequency tracking. Figures 9.15 and 9.16 show typical carrier phase and carrier frequency tracking loops. These are similar to the code tracking loop. The main differences are that only the prompt correlator outputs from the baseband signal processor are used; there is usually no external aiding information; and the loop estimates three quantities for phase tracking and two for frequency tracking. The carrier frequency estimate aids maintenance of the carrier phase estimate and the rate of frequency change estimate aids maintenance of the frequency estimate. A combined PLL and FLL using both types of discriminator may also be implemented [45]. Carrier tracking of BOC signals is no different from that of BPSK signals.

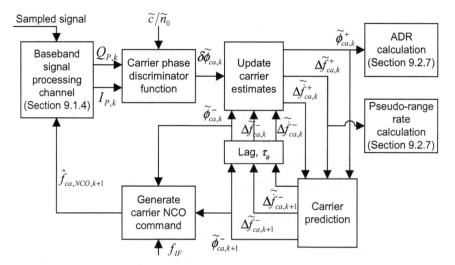

Figure 9.15 Carrier phase-tracking loop.

9.2 Ranging Processor

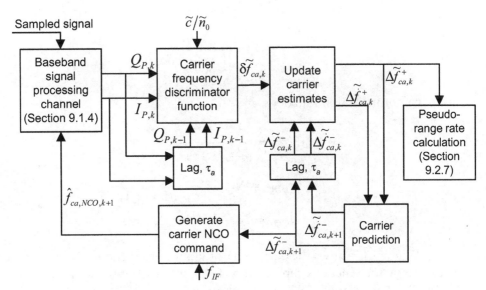

Figure 9.16 Carrier frequency-tracking loop.

For navigation message-modulated signals, the I and Q samples have a common-mode sign ambiguity due to the navigation data bit. To prevent this from disrupting carrier tracking, Costas discriminators may be used, which give the same result regardless of the data-bit sign. Examples include the IQ-product (IQP), decision-directed-Q (DDQ), Q-over-I (QOI) and two-quadrant arctangent (ATAN) discriminators [1, 13]:

$$P_{IQP} = Q_P I_P \qquad P_{DDQ} = Q_P \text{sign}(I_P)$$
$$P_{QOI} = Q_P / I_P \qquad P_{ATAN} = \arctan(Q_P / I_P) \qquad (9.41)$$

The Costas discriminators are only sensitive to carrier phase tracking errors in the range $-90° < \delta\phi_{ca} < 90°$, exhibiting a sign error outside this range. The Q-over-I discriminator exhibits singularities at $\pm 90°$, so upper and lower limits must be applied to prevent tracking instability.

Costas discriminators may also be used for the pilot signals. However, it is better to use PLL discriminators, which are sensitive to the full range of tracking errors. Examples include the quadraphase-channel (QC) and four-quadrant arctangent (ATAN2) discriminators:

$$P_{QC} = Q_P \qquad P_{ATAN2} = \arctan_2(Q_P, I_P). \qquad (9.42)$$

To obtain a measurement of the carrier phase error, the discriminator is normalized:

$$\delta\tilde{\phi}_{ca} = N_P P, \qquad (9.43)$$

where

$$N_P = \frac{\delta\phi_{ca}}{\underset{\delta\phi\to 0}{\text{Lim}} E[P(\delta\phi_{ca})]}, \qquad (9.44)$$

giving normalization functions of

$$N_{IQP} = \frac{1}{2\sigma_{IQ}^2(\tilde{c}/\tilde{n}_0)\tau_a}$$

$$N_{DDQ} = N_{QC} = \frac{1}{\sigma_{IQ}\sqrt{2(\tilde{c}/\tilde{n}_0)\tau_a}} \cdot \qquad (9.45)$$

$$N_{QOI} = N_{ATAN} = N_{ATAN2} = 1$$

Figure 9.17 shows the discriminator input-output curves. Note that the Costas discriminator functions repeat every 180°, so a carrier-tracking loop using one is equally likely to track 180° out-of-phase as in-phase.

When a common carrier tracking function is used for a pair of signals, with and without a navigation message, in the same frequency band, it is better to use the pilot signal for the carrier discriminator, though a weighted average may also be used [28].

Carrier frequency discriminators use the current and previous correlator outputs. The decision-directed cross-product (DDC), crossover-dot product (COD), and ATAN discriminators are Costas frequency discriminators and may be used across data-bit transitions:

$$\begin{aligned}
F_{DDC} &= \left(I_{P,k-1}Q_{P,k} - I_{P,k}Q_{P,k-1}\right)\text{sign}\left(I_{P,k-1}I_{P,k} + Q_{P,k-1}Q_{P,k}\right) \\
F_{COD} &= \frac{I_{P,k-1}Q_{P,k} - I_{P,k}Q_{P,k-1}}{I_{P,k-1}I_{P,k} + Q_{P,k-1}Q_{P,k}} \\
F_{ATAN} &= \arctan\left(\frac{I_{P,k-1}Q_{P,k} - I_{P,k}Q_{P,k-1}}{I_{P,k-1}I_{P,k} + Q_{P,k-1}Q_{P,k}}\right) = \arctan\left(\frac{Q_{P,k}}{I_{P,k}}\right) - \arctan\left(\frac{Q_{P,k-1}}{I_{P,k-1}}\right)
\end{aligned} \qquad (9.46)$$

where upper and lower limits must be applied to the COD discriminator to prevent tracking instability in the event of singularities.

The cross-product (CP) and ATAN2 discriminators are FLL discriminators and cannot be used across data-bit transitions:

$$\begin{aligned}
F_{CP} &= I_{P,k-1}Q_{P,k} - I_{P,k}Q_{P,k-1} \\
F_{ATAN2} &= \arctan_2\left[\left(I_{P,k-1}Q_{P,k} - I_{P,k}Q_{P,k-1}\right),\left(I_{P,k-1}I_{P,k} + Q_{P,k-1}Q_{P,k}\right)\right] \\
&= \arctan_2\left(Q_{P,k},I_{P,k}\right) - \arctan_2\left(Q_{P,k-1},I_{P,k-1}\right)
\end{aligned} \qquad (9.47)$$

To obtain a measurement of the carrier frequency error, the discriminator is normalized:

$$\delta\tilde{f}_{ca} = N_F F, \qquad (9.48)$$

9.2 Ranging Processor

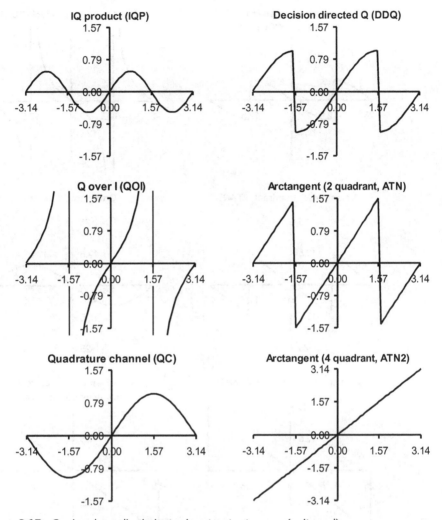

Figure 9.17 Carrier phase discriminator input-output curves (units: rad).

where

$$N_F = \frac{\delta f_{ca}}{\underset{\delta\phi\to 0}{\text{Lim}}\,\text{E}\big[F(\delta f_{ca})\big]}, \tag{9.49}$$

giving normalization functions of

$$N_{DDC} = N_{CP} = \frac{1}{4\pi\sigma_{IQ}^2(\tilde{c}/\tilde{n}_0)\tau_a^2}$$

$$N_{COD} = N_{ATAN} = N_{ATAN2} = \frac{1}{2\pi\tau_a}, \tag{9.50}$$

where it is assumed that τ_a is the interval between I and Q samples as well as the accumulation period. Figure 9.18 shows the discriminator input-output curves. Note

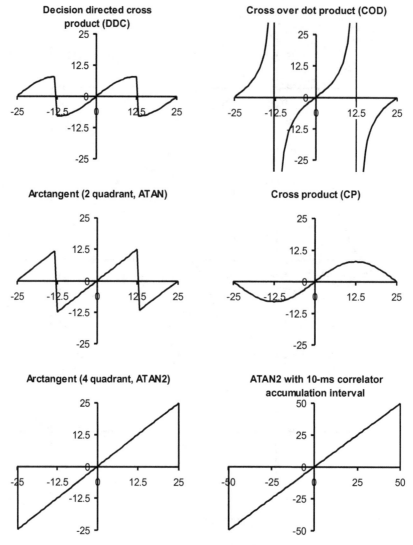

Figure 9.18 Carrier frequency discriminator input-output curves (units: Hz; $\tau_a = 20$ ms correlator accumulation interval, except where indicated otherwise).

that the maximum frequency error for all discriminators is inversely proportional to the accumulation interval [13].

In a carrier phase tracking function, the PLL is typically third order and the estimates of the carrier phase, $\tilde{\phi}_{ca}$, Doppler frequency shift, $\Delta\tilde{f}_{ca}$, and rate of change of Doppler, $\Delta\tilde{\dot{f}}_{ca}$, are updated using

$$\tilde{\phi}^+_{ca,k} = \tilde{\phi}^-_{ca,k} + K_{ca1}\delta\tilde{\phi}_{ca,k}$$
$$\Delta\tilde{f}^+_{ca,k} = \Delta\tilde{f}^-_{ca,k} + \frac{K_{ca2}}{2\pi\tau_a}\delta\tilde{\phi}_{ca,k}, \qquad (9.51)$$
$$\Delta\tilde{\dot{f}}^+_{ca,k} = \Delta\tilde{\dot{f}}^-_{ca,k} + \frac{K_{ca3}}{2\pi\tau_a^2}\delta\tilde{\phi}_{ca,k}$$

where K_{ca1}, K_{ca2}, and K_{ca3} are the tracking loop gains and k, $-$, and $+$ are as defined for code tracking. The carrier phase tracking bandwidth is then [13]

$$B_{L\_CA} = \frac{K_{ca1}^2 K_{ca2} + K_{ca2}^2 - K_{ca1}K_{ca3}}{4(K_{ca1}K_{ca2} - K_{ca3})\tau_a}. \quad (9.52)$$

A commonly used set of gains is [1]

$$K_{ca1} = 2.4 B_{L\_CA}\tau_a, \quad K_{ca2} = 2.88(B_{L\_CA}\tau_a)^2, \quad K_{ca3} = 1.728(B_{L\_CA}\tau_a)^3. \quad (9.53)$$

As for code tracking, narrower bandwidths give more noise smoothing, while wider bandwidths give better dynamics response. For applications where the user antenna is stationary, the signal dynamics change very slowly. However, the receiver's oscillator noise must be tracked to maintain carrier phase lock. A tracking bandwidth of 5 Hz is sufficient for this. When the antenna is subject to high dynamics or vibration, a higher bandwidth of 15–18 Hz is needed to track changes in line-of-sight acceleration or jerk.

The update phase of a carrier frequency tracking function using a second-order FLL is

$$\begin{aligned}\Delta \tilde{f}_{ca,k}^+ &= \Delta \tilde{f}_{ca,k}^- + K_{cf1}\delta \tilde{f}_{ca,k} \\ \Delta \tilde{\dot{f}}_{ca,k}^+ &= \Delta \tilde{\dot{f}}_{ca,k}^- + \frac{K_{cf2}}{\tau_a}\delta \tilde{f}_{ca,k}\end{aligned}, \quad (9.54)$$

and the carrier frequency tracking bandwidth is

$$B_{L\_CF} = \frac{K_{cf1}^2 + K_{cf2}}{4K_{cf1}\tau_a}, \quad (9.55)$$

with a typical value of 2 Hz [13]. A suitable pair of gains is

$$K_{cf1} = 3.4 B_{L\_CA}\tau_a, \quad K_{cf2} = 2.04(B_{L\_CA}\tau_a)^2. \quad (9.56)$$

Note that the frequency tracking bandwidth of a carrier phase tracking loop is

$$B_{L\_CF} = \frac{K_{ca2}^2 + K_{ca3}}{4K_{ca2}\tau_a}. \quad (9.57)$$

The carrier phase tracking loop's estimates are predicted forward to the next iteration using

$$\begin{aligned}\tilde{\phi}_{ca,k+1}^- &= \tilde{\phi}_{ca,k}^+ + 2\pi \Delta \tilde{f}_{ca,k}^+ \tau_a + \pi \Delta \tilde{\dot{f}}_{ca,k}^+ \tau_a^2 \\ \Delta \tilde{f}_{ca,k+1}^- &= \Delta \tilde{f}_{ca,k}^+ + \Delta \tilde{\dot{f}}_{ca,k}^+ \tau_a \\ \Delta \tilde{\dot{f}}_{ca,k+1}^- &= \Delta \tilde{\dot{f}}_{ca,k}^+\end{aligned}, \quad (9.58)$$

while the estimates of the frequency tracking loop are predicted forward using

$$\Delta \tilde{f}_{ca,k+1}^- = \Delta \tilde{f}_{ca,k}^+ + \Delta \dot{\tilde{f}}_{ca,k}^+ \tau_a$$
$$\Delta \dot{\tilde{f}}_{ca,k+1}^- = \Delta \dot{\tilde{f}}_{ca,k}^+$$
(9.59)

The reference signal carrier phase in the receiver is advanced and retarded by running the carrier NCO faster or slower than the Doppler-shifted carrier frequency. Thus, in user equipment implementing carrier phase tracking, the carrier NCO frequency is set to

$$\begin{aligned}\hat{f}_{ca,NCO,k+1} &= f_{IF} + \Delta \tilde{f}_{ca,k+1}^- + \frac{\tilde{\phi}_{ca,k}^+ - \tilde{\phi}_{ca,k}^-}{2\pi\tau_a} \\ &= f_{IF} + \Delta \tilde{f}_{ca,k+1}^- + \frac{K_{ca1}\delta\tilde{\phi}_{ca,k}}{2\pi\tau_a}\end{aligned}$$
(9.60)

When carrier frequency tracking is used, the carrier NCO frequency is simply set to the ranging processor's best estimate:

$$\hat{f}_{ca,NCO,k+1} = f_{IF} + \Delta \tilde{f}_{ca,k+1}^-.$$
(9.61)

Block-diagram treatments of the code and carrier loop filters may be found in other texts [1, 13].

9.2.4 Tracking Lock Detection

GNSS user equipment must detect when it is no longer tracking the code from a given signal so that contamination of the navigation processor with incorrect pseudo-range data is avoided and the acquisition mode reinstigated to try and recover the signal.

Code can no longer be tracked when the tracking error exceeds the pull-in range of the discriminator. This is the region within which the discriminator output in the absence of noise has the same sign as the tracking error. As Figure 9.13 shows, the pull-in range depends on the correlator spacing and discriminator type.

Tracking lock is lost when the carrier power to noise density, C/N_0, is too low and/or the signal dynamics is too high. Whether there is sufficient signal to noise to maintain code tracking is determined by measuring C/N_0 and comparing it with a minimum value. The threshold should match the code discriminator pull-in range to about three times the code tracking noise standard deviation (see Section 9.3.3) and allow a margin for C/N_0 measurement error. A threshold of around 19 dB-Hz is suitable with a 1-Hz code-tracking bandwidth, with a lower threshold suitable for a smaller tracking bandwidth. The same test can be used to detect loss of code lock due to dynamics, as this causes the measured C/N_0 to be underestimated [24].

Loss of carrier phase tracking lock must be detected to enable the ranging processor to transition to carrier frequency tracking and prevent erroneous Doppler measurements from disrupting code tracking and the navigation processor. As with code tracking, a C/N_0 measurement threshold can be used to determine whether there is sufficient signal to noise to track carrier phase. A suitable threshold is typically 24–30 dB-Hz, depending on the tracking bandwidth, discriminator type, and requirements of the application.

A C/N_0-based lock detector will not detect dynamics-induced loss of lock. However, because the carrier discriminator function repeats every 180° (Costas) or 360° (PLL), carrier phase lock can be spontaneously recovered. During the interval between loss and recovery, the carrier phase estimate can advance or retard by a multiple of 180° or 360° with respect to truth. This is known as a cycle slip and affects the accumulated delta range measurements (see Section 9.2.7) and navigation-message demodulation (see Section 9.2.5). For applications in which cycle-slip detection is required, a phase lock detector based on carrier-phase discriminator statistics should be employed [1, 13]. Alternatively, a parallel FLL may be used as a cycle slip detector [46].

Carrier frequency lock is essential for maintaining code tracking as it ensures signal coherence over the correlator accumulation interval (see Section 9.1.4.4), while the C/N_0 level needed to maintain carrier frequency tracking is only slightly higher than that needed for code tracking. However, as Figure 9.18 shows, carrier frequency discriminators repeat a number of times within the main peak of the signal power versus tracking error curve (Figure 9.9). Consequently, the FLL can undergo false lock at an offset of $n/2\tau_a$ from the true carrier frequency, where n is an integer (assuming a Costas discriminator). This produces a pseudo-range-rate measurement error of a few m s^{-1} and disrupts navigation-message demodulation. A PLL can also exhibit false frequency lock following a cycle slip. To prevent this, a false-frequency-lock detector must be implemented. This simply compares the Doppler shift from the FLL or PLL with that obtained from the code tracking loop.

9.2.5 Navigation-Message Demodulation

In stand-alone GNSS user equipment, the navigation data message must be demodulated to obtain the satellite positions and velocities and resolve any ambiguities in the time of transmission. When carrier phase tracking is in lock, the data bit is given simply by

$$D(t) = \text{sign}(I_P(t)). \tag{9.62}$$

When carrier frequency tracking is used, the data-bit transitions are detected by observing the 180° changes in $\arctan_2(I_P, Q_P)$. This gives noisier data demodulation than phase tracking. In both cases, there is a sign ambiguity in the demodulated data-bit stream. In frequency tracking, this occurs because the sign of the initial bit is unknown, whereas in phase tracking, it occurs because it is unknown whether the tracking loop is locked in-phase or 180° out-of-phase. The ambiguity is resolved using the parity check information broadcast in the message itself. This must be checked continuously as phase tracking is vulnerable to cycle slips and frequency tracking to missed detection of the bit transitions [1].

The signal to noise on the data demodulation is optimized by matching the correlator accumulation interval (Section 9.1.4.4) to the length of the data bit. Accumulating over data-bit transitions should be avoided. For the newer GNSS signals, the timing of the bit transitions is indicated by the ranging code. However, for GPS and GLONASS C/A code, there is an ambiguity as there are 20 code repetition intervals per data bit, so the ranging processor has to search for the bit transitions. This is done by forming 20 test statistics, each summed coherently over 20 ms with a different offset and then noncoherently over n data bits. The test statistics are

$$T_r = \sum_{i=1}^{n} \left(\sum_{j=1}^{20} I_{P,(20i+j+r)} \right)^2, \quad (9.63)$$

for carrier phase tracking and

$$T_r = \sum_{i=1}^{n} \left[\left(\sum_{j=1}^{20} I_{P,(20i+j+r)} \right)^2 + \left(\sum_{j=1}^{20} Q_{P,(20i+j+r)} \right)^2 \right], \quad (9.64)$$

for frequency tracking, where r takes values from 1 to 20 and the accumulation interval for the Is and Qs is 1 ms. The largest test statistic corresponds to the correct bit synchronization [13, 32].

Reliable demodulation of the legacy navigation messages requires a C/N_0 of about 30 dB-Hz. Newer messages, incorporating FEC, may be demodulated at lower signal-to-noise levels. When there is insufficient C/N_0 for reliable demodulation or frequent interruptions to the signal (e.g., in urban areas), the data may be reconstructed by combining information from successive message transmission cycles [32, 47]. This is more straightforward for fixed-frame message formats (see Section 8.4).

9.2.6 Carrier-Power-to-Noise-Density Measurement

Measurements of the carrier power to noise density, c/n_0, defined in Section 9.1.4.3, are needed for tracking lock detection. They may also be used to determine the weighting of measurements in the navigation processor and for adapting the tracking loops to the signal-to-noise environment.

To correctly determine receiver performance, c/n_0 must be measured after the signal is correlated with the reference code. A suitable method is narrow-to-wide power-ratio measurement [1, 24]. This computes the coherently summed narrow-band power, P_N, and noncoherently summed wideband power, P_W, over an interval τ_{aN}, generally the data-bit interval:

$$P_N = \left(\sum_{i=1}^{M} I_{P,i} \right)^2 + \left(\sum_{i=1}^{M} Q_{P,i} \right)^2 \qquad P_W = \sum_{i=1}^{M} \left(I_{P,i}^2 + Q_{P,i}^2 \right), \quad (9.65)$$

where $I_{P,i}$ and $Q_{P,i}$ are accumulated over time $\tau_{aW} = \tau_{aN}/M$, typically 1 ms. The power ratio is then computed and averaged over n iterations to reduce noise:

$$\bar{P}_{N/W} = \frac{1}{n}\sum_{r=1}^{n}\frac{P_{N,r}}{P_{W,r}}. \qquad (9.66)$$

A typical averaging time is 1 second. Taking expectations,

$$E(\bar{P}_{N/W}) \approx \frac{M[(c/n_o)\tau_{aN}+1]}{M+(c/n_o)\tau_{aN}}. \qquad (9.67)$$

The carrier-power-to-noise-density measurement is then

$$\tilde{c}/\tilde{n}_o = \frac{M(\bar{P}_{N/W}-1)}{\tau_{aN}(M-\bar{P}_{N/W})}. \qquad (9.68)$$

C/N_0 is then obtained using (9.22). Some other methods are described in Section G.6.3 of Appendix G on the CD. All methods are very noisy at low c/n_0, requiring a longer averaging time to produce useful measurements. The averaging time may be varied to optimize the tradeoff between noise and response time [24].

9.2.7 Pseudo-Range, Pseudo-Range-Rate, and Carrier-Phase Measurements

GNSS ranging processors can output four types of measurement: pseudo-range, pseudo-range rate or Doppler shift, delta range, and accumulated delta range (ADR), often known as carrier phase. The pseudo-range measurement is obtained from code tracking and the others from carrier tracking. Note that some authors, particularly within the surveying and geodesy community, use the term *observation* instead of measurement and the term *observable* to denote a parameter that may be measured.

The raw measured pseudo-range for signal l from satellite s to user antenna a is given by

$$\tilde{\rho}_{a,R}^{s,l} = \left(\tilde{t}_{sa,a}^{s,l} - \tilde{t}_{st,a}^{s,l}\right)c, \qquad (9.69)$$

where $\tilde{t}_{sa,a}^{s,l}$ is the time of signal arrival, measured by the receiver clock, and $\tilde{t}_{st,a}^{s,l}$ is the measured time of signal transmission. To obtain the transmission time from the code phase, $\tilde{t}_{st}'$, measured by the code tracking loop (Section 9.2.2), an integer number of code repetition periods, determined from the navigation message, must usually be added.

For the GPS and GLONASS C/A codes, the additional step of determining the data-bit transitions must also be performed (see Section 9.2.5). Bit-synchronization errors produce errors in $\tilde{t}_{st,a}^{s,l}$ of multiples of 1 ms, leading to pseudo-range errors of multiples of 300 km. The navigation processor should check for these errors.

The Doppler-shift measurement, $\Delta \tilde{f}_{ca,a}^{s,l}$, is obtained directly from the carrier tracking loop (Section 9.2.3). This may be transformed to a pseudo-range-rate measurement using

$$\tilde{\rho}_{a,R}^{s,l} \approx -\frac{c}{f_{ca}^l}\Delta\tilde{f}_{ca,a}^{s,l}. \tag{9.70}$$

For users with a high velocity (with respect to the Earth), a correction for the effects of residual relativistic time dilation should be applied as described in [18] and summarized in Section G.5 of Appendix G on the DVD.

The delta range is the integral of the pseudo-range rate over the interval since the last measurement:

$$\Delta\tilde{\rho}_{a,R}^{s,l}(\tilde{t}_{sa,a,k}^{s,l}) = \int_{\tilde{t}_{sa,a,k-1}^{s,l}}^{\tilde{t}_{sa,a,k}^{s,l}} \dot{\tilde{\rho}}_{a,R}^{s,l}(t)\, dt, \tag{9.71}$$

The ADR, $\tilde{\Phi}_{a,R}^{s,l}$, is simply the corresponding carrier phase measurement, $\tilde{\phi}_{ca,a}^{s,l}$, converted to the range domain. Thus,

$$\tilde{\Phi}_{a,R}^{s,l} = -\frac{c}{2\pi f_{ca}^l}\tilde{\phi}_{ca,a}^{s,l} = -\frac{\lambda_{ca}^l}{2\pi}\tilde{\phi}_{ca,a}^{s,l}. \tag{9.72}$$

Note that there is a sign change because an advance in the phase of the incoming signal with respect to the reference oscillator denotes a decrease in the pseudo-range. The ADR comprises the sum of the phase within the current carrier cycle, a count of the integer change in carrier cycles since carrier tracking initialization, and, in some cases, an offset (also an integer number of carrier cycles). Delta range may also be determined by time-differencing the ADR:

$$\Delta\tilde{\rho}_{a,R}^{s,l}(\tilde{t}_{sa,a,k}^{s,l}) = \tilde{\Phi}_{a,R}^{s,l}(\tilde{t}_{sa,a,k}^{s,l}) - \tilde{\Phi}_{a,R}^{s,l}(\tilde{t}_{sa,a,k-1}^{s,l}). \tag{9.73}$$

The navigation processor will only use one carrier-derived measurement as they all convey the same information. Similarly, user equipment often only outputs one type of carrier-derived measurement. The ADR and delta-range measurements have the advantage of smoothing out the carrier tracking noise where the navigation-processor update rate is less than the carrier-tracking bandwidth.

The duty cycles of the tracking loops are commonly aligned with the navigation-data-bit transitions and/or the code repetition period. Consequently, the tracking loops for the different signals are not synchronized, producing measurements corresponding to different times of arrival. However, navigation-solution computation is much simpler if a common time of signal arrival can be assumed. Consequently, the measurements are typically predicted forward to a common time of arrival as described in [13].

The code and carrier measurements can be combined to produce a smoothed pseudo-range, $\tilde{\rho}_S$:

$$\tilde{\rho}_{a,S}^{s,l}(t) = W_{co}\tilde{\rho}_{a,R}^{s,l}(t) + (1 - W_{co})\left[\tilde{\rho}_{a,S}^{s,l}(t-\tau) + \tilde{\Phi}_{a,R}^{s,l}(t) - \tilde{\Phi}_{a,R}^{s,l}(t-\tau)\right], \tag{9.74}$$

where W_{co} is the code weighting factor and τ is the update interval. This is known as a Hatch filter. It improves the accuracy of a single-epoch position solution or receiver autonomous integrity monitoring (RAIM) algorithm (see Section 17.4.1), but does not benefit filtered positioning or integrity monitoring, where smoothing is implicit. The time constant, τ/W_{co}, is typically set at 100 seconds, limiting the effects of code–carrier ionosphere divergence (see Section 9.3.2) and cycle slips. This smooths the code tracking noise by about an order of magnitude [48].

Pseudo-range rate measurements may also be used to smooth the pseudo-range:

$$\tilde{\rho}_{a,S}^{s,l}(t) = W_{co}\tilde{\rho}_{a,R}^{s,l}(t) + (1 - W_{co})\left[\tilde{\rho}_{a,S}^{s,l}(t - \tau) + \tau\dot{\tilde{\rho}}_{a,R}^{s,l}(t)\right]. \quad (9.75)$$

They are less accurate than ADR, but more robust [49].

GNSS user equipment may output raw pseudo-ranges and pseudo-range rates or it may apply corrections for the satellite clock, ionosphere propagation, and troposphere propagation errors as described in the next section. Some user equipment also subtracts the Sagnac correction, $\delta\rho_{ie}$, forcing the navigation processor to use an ECEF coordinate frame; this is not always documented.

Some user equipment corrects its pseudo-range outputs with its current receiver clock offset estimates, a process known as *clock steering*. The carrier-derived measurements may or may not be similarly corrected. More common is the application of periodic millisecond corrections to the pseudo-ranges to keep the receiver clock within a millisecond of GPS time. Note that when these clock jumps occur, there may also be a 1-ms discrepancy between the difference in time tags of successive measurements and the difference in the actual measurement times.

9.3 Range Error Sources

The pseudo-range, pseudo-range rate, and ADR measurements made by GNSS user equipment are subject to two main types of error: time-correlated and noise-like. The satellite clock errors and the ionosphere and troposphere propagation errors are correlated over the order of an hour and are partially corrected for by the user equipment using (8.49). The errors prior to the application of corrections are known as raw errors and those remaining after the correction process are known as residual errors.

Tracking errors are correlated over less than a second and cannot be corrected, only smoothed. Errors due to multipath interference and NLOS reception are typically correlated over a few seconds for most navigation applications and can be mitigated using a number of techniques as discussed in Section 10.4.

Building on the general discussion in Section 7.4, this section describes each of these error sources in turn, together with the ephemeris prediction error, which affects the navigation solution through the computation of the satellite position and velocity. Further information on range error corrections is provided in Section G.7 of Appendix G on the CD, while Section G.5 presents a correction for the relativistic frequency shift.

9.3.1 Ephemeris Prediction and Satellite Clock Errors

The ephemeris prediction error is simply the error in the control segment's prediction of the satellite position. Its components, expressed in orbital-frame cylindrical coordinates, δr^o_{os}, δu^o_{os}, and δz^o_{os}, as shown in Figure 9.19, are correlated over the order of an hour and change each time the ephemeris data in the navigation message is updated. The range error due to the ephemeris error is

$$\delta \rho^s_e = \frac{\mathbf{r}^\beta_{\beta s}}{|\mathbf{r}^\beta_{\beta s}|} \cdot \mathbf{u}^\beta_{as} \delta r^o_{os} + \frac{\mathbf{v}^\beta_{\beta s}}{|\mathbf{v}^\beta_{\beta s}|} \cdot \mathbf{u}^\beta_{as} r^o_{os} \delta u^o_{os} + \frac{\mathbf{r}^\beta_{\beta s} \wedge \mathbf{v}^\beta_{\beta s}}{|\mathbf{r}^\beta_{\beta s}||\mathbf{v}^\beta_{\beta s}|} \cdot \mathbf{u}^\beta_{as} \delta z^o_{os}, \quad \beta \in i,e,I. \quad (9.76)$$

This varies with the signal geometry, so is different for users at different locations, but is dominated by the radial component, δr^o_{os}.

The satellite clock error arises due to the cumulative effect of oscillator noise. It is mostly corrected for using three calibration coefficients, a^s_{f0}, a^s_{f1}, and a^s_{f2}, and a reference time, t^s_{oc}, transmitted in the navigation data message and common to all signals from that satellite. An additional term, $\Delta a^{s,l}_{is}$, is added to account for intersignal timing biases within the satellite. Furthermore, a relativistic correction is applied to account for the variation in satellite clock speed with the velocity and gravitational potential over the satellite's elliptical orbit [18, 50]. The total satellite clock correction for satellite s signal l is

$$\delta \hat{\rho}^{s,l}_c = \left[a^s_{f0} + a^s_{f1}\left(t^{s,l}_{st,a} - t^s_{oc}\right) + a^s_{f2}\left(t^{s,l}_{st,a} - t^s_{oc}\right)^2 + \Delta a^{s,l}_{is} \right] c - 2\frac{\mathbf{r}^e_{es} \cdot \mathbf{v}^e_{es}}{c}, \quad (9.77)$$

where a ±604,800-second correction is applied to t_{oc} where $\left|t^{s,l}_{st,a} - t^s_{oc}\right| > 302{,}400$ s to account for week crossovers. Calculation of $\Delta a^{s,l}_{is}$ from the navigation data is described in Section G.7.1 of Appendix G on the CD. For multiconstellation operation, a correction for the appropriate interconstellation timing bias (see Section 8.4.5) must be added to (9.77).

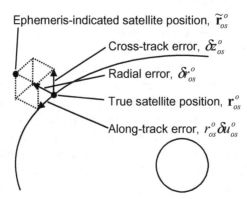

Figure 9.19 Components of the ephemeris prediction error.

9.3 Range Error Sources

The range-rate satellite clock correction is

$$\delta\hat{\dot{\rho}}_c^s = \left[a_{f1}^s + \frac{a_{f2}^s}{2}\left(t_{st,a}^{s,l} - t_{oc}^s\right)\right]c, \qquad (9.78)$$

noting that the relativistic term is either neglected or accounted for separately as described in Section G.5 of Appendix G on the CD.

The residual satellite clock and ephemeris errors depend on the quality of the control-segment orbit and clock modeling, the quantization and latency of the broadcast navigation data, the stability of the satellite and control-segment clocks, and the size of the control segment's monitor network. A larger monitor station network enables better separation of the three ephemeris error components and the satellite clock error through improved observation geometry.

The GNSS operators typically quote the signal-in-space (SIS) error, which is the combined range error standard deviation due to ephemeris and satellite clock errors. The average SIS error for the GPS constellation was 0.9m in 2011 [51]. However, this varies between satellite designs due to advances in the clock technology. For Block IIR and IIR-M, the SIS error is about 0.5m, while for Block IIF satellites it is 0.3m. A lower SIS error is also obtained using the newer CNAV, MNAV, and C2NAV messages. The GLONASS SIS error was 1.6m in 2012 [52].

9.3.2 Ionosphere and Troposphere Propagation Errors

As discussed in Section 7.4.1, GNSS signals are refracted in both the ionosphere and troposphere regions of the Earth's atmosphere. Signals from low-elevation satellites experience much more refraction than signals from high-elevation satellites as they pass through more atmosphere as Figure 9.20 shows.

The ionosphere propagation delay varies with the elevation angle approximately as [3, 50]

$$\delta\rho_{I,a}^{s,l} \propto \left[1 - \left(\frac{R\cos\theta_{nu}^{as}}{R + h_i}\right)^2\right]^{-1/2}, \qquad (9.79)$$

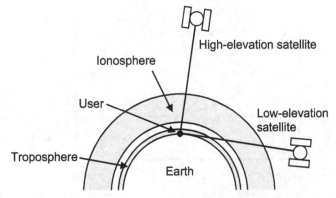

Figure 9.20 Ionosphere and troposphere propagation for high- and low-elevation satellites (not to scale).

where R is the average Earth radius and h_i is the mean ionosphere height, about 350 km. The troposphere propagation delay varies approximately as [3, 50]

$$\delta\rho^s_{T,a} \propto \left[1 - \left(\frac{\cos\theta^{as}_{nu}}{1.001}\right)^2\right]^{-1/2}. \quad (9.80)$$

These are known as obliquity factors or mapping functions and are shown in Figure 9.21. They are unity for satellites at zenith or normal incidence, which is 90° elevation. Most GNSS user equipment implements a minimum elevation threshold, known as the mask angle, of between 5° and 15°, below which signals are excluded from the navigation solution.

The ionosphere is a dispersive medium, meaning that the propagation velocity varies with the frequency. As with nondispersive refraction, the signal modulation (PRN code and navigation data) is delayed. However, the carrier phase is advanced by approximately the same amount [50]. Thus, the modulation and carrier ionosphere propagation errors, respectively, $\delta\rho^{s,l}_{I,a}$ and $\delta\Phi^{s,l}_{I,a}$, are related by

$$\delta\rho^{s,l}_{I,a} \approx -\delta\Phi^{s,l}_{I,a}. \quad (9.81)$$

The time variations of the code-based and carrier-based ranging errors due to the ionosphere are thus opposite. This is known as code-carrier divergence. As the ionization of the ionosphere gases is caused by solar radiation, there is more refraction during the day than at night. The signal modulation delay for a satellite at zenith varies from 1–3m around 02:00 to 5–15m around 14:00 local time. More than 99% of the propagation delay/advance varies as f^{-2}_{ca} [53]. The higher-order effects only account for a few centimeters of ranging error [54].

The troposphere is a nondispersive medium, so all GNSS signals are delayed equally and there is no code–carrier divergence. On average, about 90% of the delay is attributable to the dry gases in the atmosphere and is relatively stable. The remaining delay is due to water vapor and varies considerably. The total tropospheric delay at zenith is about 2.5m and varies by about ±10% with the climate and weather.

When a GNSS receiver tracks signals on more than one frequency, they may be combined to eliminate most of the ionosphere propagation delay. For older GPS

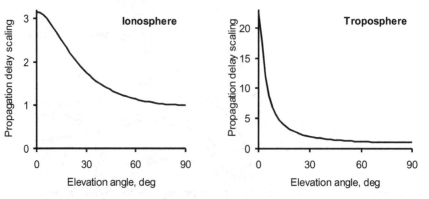

Figure 9.21 Ionosphere and troposphere delay mapping functions.

satellites with open signals on the L1 frequency only, unauthorized users may access the P(Y) code signals on L2 using semi-codeless correlation (Section 9.1.4.6). The ionosphere-corrected pseudo-range is then

$$\tilde{\rho}_{a,R}^{s,IC} = \frac{(f_{ca}^\alpha)^2 \tilde{\rho}_{a,R}^{s,\alpha} - (f_{ca}^\beta)^2 \tilde{\rho}_{a,R}^{s,\beta}}{(f_{ca}^\alpha)^2 - (f_{ca}^\beta)^2}, \qquad (9.82)$$

where the superscripts α and β denote the signals on the two different frequencies. A similarly weighted intersignal timing bias must be used for the satellite clock correction as described in Section G.7.1 of Appendix G on the CD. However, this combination brings a penalty in the form of increased tracking noise. The tracking error standard deviation of the corrected pseudo-range is [55]

$$\sigma_{pw}^{IC} = \frac{\sqrt{(f_{ca}^\alpha)^4 (\sigma_{pt}^\alpha)^2 + (f_{ca}^\beta)^4 (\sigma_{pt}^\beta)^2}}{\left|(f_{ca}^\alpha)^2 - (f_{ca}^\beta)^2\right|}, \qquad (9.83)$$

where σ_{pw}^α and σ_{pw}^β are the code tracking error standard deviations for the two signals. The closer together the two frequencies, the more the tracking error is scaled up. For users of GPS L1 and L2 signals, $\sigma_{pw}^{IC}/\sigma_{pw}^{L1} \approx 3.36$, noting that $\sigma_{pw}^{L2}/\sigma_{pw}^{L1} \approx \sqrt{2}$ due to the different transmission powers. The ratio is higher where semi-codeless correlation is used. Multipath errors are increased by a similar ratio.

As the ionosphere propagation delay varies slowly, with correlation times of about half an hour [56], the ionosphere correction may be smoothed over time to reduce the tracking error. Applying the smoothing over m iterations and using k to denote the current iteration, the corrected pseudo-range is then

$$\tilde{\rho}_{a,R,k}^{s,IC} = \tilde{\rho}_{a,R,k}^{s,\alpha} - \delta\hat{\rho}_{I,a,k}^{s,\alpha}, \qquad (9.84)$$

where

$$\delta\hat{\rho}_{I,a,k}^{s,\alpha} = \frac{(f_{ca}^\beta)^2}{m\left[(f_{ca}^\alpha)^2 - (f_{ca}^\beta)^2\right]} \sum_{r=k+1-m}^{k} \left(\tilde{\rho}_{a,R,r}^{s,\beta} - \tilde{\rho}_{a,R,r}^{s,\alpha}\right). \qquad (9.85)$$

Alternatively, if the measurements on the two frequencies are weighted to minimize the tracking noise,

$$\tilde{\rho}_{a,R,k}^{s,IC} = W_\alpha \tilde{\rho}_{a,R,k}^{s,\alpha} + (1 - W_\alpha) \tilde{\rho}_{a,R,k}^{s,\beta} - \delta\hat{\rho}_{I,a,k}^{s,W}, \qquad (9.86)$$

where

$$\delta\hat{\rho}_{I,a,k}^{s,W} = \frac{(1 - W_\alpha)(f_{ca}^\alpha)^2 + W_a(f_{ca}^\beta)^2}{m\left[(f_{ca}^\alpha)^2 - (f_{ca}^\beta)^2\right]} \sum_{r=k+1-m}^{k} \left(\tilde{\rho}_{a,R,r}^{s,\beta} - \tilde{\rho}_{a,R,r}^{s,\alpha}\right) \qquad (9.87)$$

and

$$W_\alpha = \frac{\left(\sigma_{pw}^\beta\right)^2}{\left(\sigma_{pw}^\alpha\right)^2 + \left(\sigma_{pw}^\beta\right)^2}. \tag{9.88}$$

The residual ionosphere propagation error following smoothed dual-frequency correction is of the order of 0.1m [50].

The carrier-smoothed pseudo-range may be corrected for the ionosphere propagation delay using

$$\tilde{\rho}_{a,S}^{s,IC}(t) = \frac{W_{co}\left(\left(f_{ca}^\alpha\right)^2 \tilde{\rho}_{a,R}^{s,\alpha}(t) - \left(f_{ca}^\beta\right)^2 \tilde{\rho}_{a,R}^{s,\beta}(t)\right)}{\left(f_{ca}^\alpha\right)^2 - \left(f_{ca}^\beta\right)^2}$$
$$+ (1 - W_{co})\left[\tilde{\rho}_{a,S}^{s,IC}(t - \tau) + \frac{\left(f_{ca}^\alpha\right)^2 \left(\tilde{\Phi}_{a,R}^{s,\alpha}(t) - \tilde{\Phi}_{a,R}^{s,\alpha}(t - \tau)\right) - \left(f_{ca}^\beta\right)^2 \left(\tilde{\Phi}_{a,R}^{s,\beta}(t) - \tilde{\Phi}_{a,R}^{s,\beta}(t - \tau)\right)}{\left(f_{ca}^\alpha\right)^2 - \left(f_{ca}^\beta\right)^2}\right],$$
(9.89)

As this also corrects for code-carrier ionosphere divergence, it allows a longer smoothing time constant to be used [48].

For wideband GNSS signals, such as Galileo E5 AltBOC, the variation in ionosphere propagation error across the bandwidth of the signal can be significant. This leads to errors of a few centimeters in the basic dual-frequency ionosphere correction and also distorts the correlation function, resulting in increased tracking noise that varies with the ionosphere [57].

Single-frequency users use a model to estimate the ionosphere propagation delay as a function of time, the user latitude and longitude, and the elevation and azimuth of each satellite line of sight. One ionosphere model may be used to correct measurements from all GNSS constellations. The Klobuchar model, described in Section G.7.2 of Appendix G on the CD, is the most widely used and corrects about 50% of the propagation delay. This incorporates eight parameters, common to all satellites, which are broadcast in the GPS navigation data message [53, 58]. Galileo satellites broadcast three parameters for the more sophisticated NeQuick ionosphere model [59, 60], while GLONASS does not broadcast any ionosphere data. The SBAS ionosphere model, described in Section G.7.3 of Appendix G on the CD, provides the most accurate corrections, but is only valid within the relevant full-service coverage area (see Section 8.2.6).

Models are also used to correct the troposphere propagation delay. The NATO Standardization Agreement (STANAG) model simply represents the propagation delay as a function of elevation angle and orthometric user height. The zenith delay is given by [61]

$$\delta\hat{\rho}_{TZ} = \begin{array}{ll} \left[2.464 - 3.248 \times 10^{-4} H_a + 2.2395 \times 10^{-8} H_a^2\right]\text{m} & H_a \leq 1{,}000\text{m} \\ \left[2.284\exp\left(-0.1226\left\{10^{-3}H_a - 1\right\}\right) - 0.122\right]\text{m} & 1{,}000\text{m} \leq H_a \leq 9{,}000\text{m}. \\ 0.7374\exp\left(1.2816 - 1.424 \times 10^{-4} H_a\right)\text{m} & 9{,}000\text{m} \leq H_a \end{array} \tag{9.90}$$

The estimated troposphere delay for an individual signal is then

$$\delta\hat{\rho}^s_{T,a} = \frac{\delta\hat{\rho}_{TZ}}{\sin\theta^{as}_{nu} + \dfrac{0.00143}{\tan\theta^{as}_{nu} + 0.0455}}. \quad (9.91)$$

Residual errors using this model are of the order of 0.6m. The initial WAAS and University of New Brunswick 3 (UNB3) models, described respectively in Sections G.7.4 and G.7.5 of Appendix G on the CD, also account for variations in latitude and season. Residual errors of around 0.2m are achieved using the UNB3 model [50]. Further models are described in [3, 4, 62, 63]. Best performance is obtained using current temperature, pressure, and humidity data. However, incorporation of meteorological sensors is not practical for most navigation applications. One option is to transmit weather forecast data to users requiring high precision [64, 65].

For high-precision applications, the residual troposphere propagation errors may be calibrated as part of the navigation solution. This exploits the high degree of correlation between the errors on signals from different satellites and improves positioning accuracy by a few centimeters.

During periods of high solar storm activity, the ionosphere refractive index can fluctuate on a localized basis from second to second, a process known as scintillation. This is most prevalent in two bands centered at geomagnetic latitudes of ±20° between sunset and midnight, but also occurs in polar regions where auroras are seen [26, 66, 67]. Scintillation normally affects only a portion of the sky, but it can occasionally impact all GNSS signals received at a given location.

Ionospheric scintillation has two effects on GNSS user equipment. First, the rapid fluctuation of the refractive index, known as phase scintillation, can result in low-bandwidth carrier-phase tracking loops struggling to maintain lock (see Section 9.3.3). Second, the spatial variation in the refractive index results in signals arriving at the user antenna via multiple paths. The resulting multipath interference (see Section 9.3.4) introduces additional ranging errors. As the amplitudes of the different signal paths can be similar and the interference can be constructive or destructive, the amplitude of the resultant signal varies rapidly and can dip below the tracking threshold, a phenomenon called amplitude scintillation. When possible, scintillation-affected signals should be excluded from the navigation solution.

9.3.3 Tracking Errors

The code and carrier discriminator functions, described in Sections 9.2.2 and 9.2.3, exhibit random errors due to receiver thermal noise, RF interference, and other GNSS signals on the same frequency. Interference from other GNSS signals exceeds thermal noise where more than two constellations broadcast a similar signal on the same frequency [68]. When the signal is attenuated, a given amount of noise will produce larger tracking errors. Sources of interference and attenuation are discussed in Section 10.3.1.

Neglecting precorrelation band-limiting, it may be shown [69] that the code discriminator noise variances for a BPSK signal are

$$\sigma^2(N_{DPP}D_{DPP}) \approx \frac{d}{4(c/n_0)\tau_a}\left[1 + \frac{1}{(c/n_0)\tau_a}\right]$$

$$\sigma^2(N_{ELP}D_{ELP}) \approx \frac{d}{4(c/n_0)\tau_a}\left[1 + \frac{2}{(2-d)(c/n_0)\tau_a}\right],\quad (9.92)$$

$$\sigma^2(N_{ELE}D_{ELE}) \approx \frac{d}{4(c/n_0)\tau_a},\quad (c/n_0)\tau_a \gg 1$$

$$\sigma^2(N_{Coh}D_{Coh}) \approx \frac{d}{4(c/n_0)\tau_a}$$

where the infinite precorrelation bandwidth approximation is valid for $d \geq \pi f_{co}/B_{PC}$ (otherwise, $\pi f_{co}/B_{PC}$ should be substituted for d). The variances of D_{DPP} and D_{ELP} under the trapezium approximation are given in [24], while [27] discusses the general case.

The tracking loop smooths out the discriminator noise, but also introduces a lag in responding to dynamics. The code discriminator output may be written as

$$\tilde{x}_k = k_{ND}(x_k)x_k + w_{ND}, \quad (9.93)$$

where w_{ND} is the discriminator noise and k_{ND} is the discriminator gain, which may be obtained from the slopes of Figure 9.13. The gain is unity for small tracking errors by definition, but drops as the tracking error approaches the pull-in limits of the discriminator. From (9.11), (9.34), and (9.93), the code tracking error is propagated as

$$\begin{aligned}x_k^+ &= x_k^- - K_{co}(k_{ND}x_k^- + w_{ND})\\ &= (1 - K_{co}k_{ND})x_k^- - K_{co}w_{ND}\end{aligned}. \quad (9.94)$$

The code tracking error has zero mean and standard deviation σ_x, while the discriminator noise has zero mean and standard deviation σ_{ND} as given by (9.92). Squaring (9.94) and applying the expectation operator,

$$\sigma_x^2 = (1 - K_{co}\bar{k}_{ND})^2 \sigma_x^2 + K_{co}^2 \sigma_{ND}^2. \quad (9.95)$$

where $\bar{k}_{ND}$ is the average discriminator gain across the tracking error distribution. Assuming $K_{co} \ll 1$ and $\bar{k}_{ND} \approx 1$, which is valid except on the verge of tracking loss,

$$\sigma_x^2 \approx \tfrac{1}{2} K_{co} \sigma_{ND}^2. \quad (9.96)$$

Substituting in (9.36),

$$\sigma_x^2 \approx 2B_{L\_CO}\tau_a \sigma_{ND}^2. \quad (9.97)$$

9.3 Range Error Sources

The code tracking noise standard deviation in chips, is thus

$$\sigma_x \approx \begin{cases} \sqrt{\dfrac{B_{L\_co}d}{2(c/n_0)}\left[1+\dfrac{1}{(c/n_0)\tau_a}\right]} & D = D_{DPP} \\[1em] \sqrt{\dfrac{B_{L\_co}d}{2(c/n_0)}\left[1+\dfrac{2}{(2-d)(c/n_0)\tau_a}\right]} & D = D_{ELP} \\[1em] \sqrt{\dfrac{B_{L\_co}d}{2(c/n_0)}} & D = D_{ELE}, (c/n_0)\tau_a \gg 1 \text{ or } D = D_{Coh} \end{cases} \quad , \quad (9.98)$$

noting that precorrelation band-limiting is neglected. When $d < \pi f_{co}/B_{PC}$, an approximation may be obtained by substituting $\pi f_{co}/B_{PC}$ for d. A more precise model for an early-minus-late power discriminator, accounting for precorrelation band-limiting, is presented in [13, 27]. Figure 9.22 depicts the code tracking noise standard deviation as a function of C/N_0 for different tracking bandwidths.

The pseudo-range error standard deviation due to tracking noise is

$$\sigma_{pw} = \frac{c}{f_{co}}\sigma_x. \quad (9.99)$$

For a BOC(f_s, f_{co}) signal where precorrelation band-limiting can be neglected, the code tracking noise standard deviation is that of a f_{co} chipping-rate BPSK signal multiplied by $\frac{1}{2}\sqrt{f_{co}/f_s}$ [70]. In practice, this only applies to a BOC$_s$(1,1) signal as the other GNSS BOC signals require narrow correlator spacings, so the effect of precorrelation band-limiting is significant. The tracking noise standard deviation for the GPS M code is given in [13, 41].

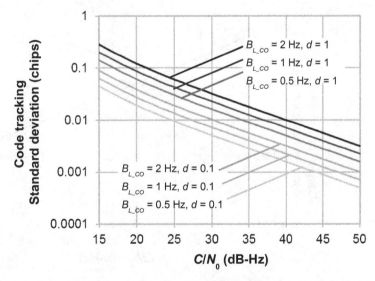

Figure 9.22 Code tracking noise standard deviation with a dot product power discriminator and $\tau_a = 20$ ms.

The carrier phase discriminator noise variance is [1, 3]

$$\sigma^2(N_P P) \approx \frac{1}{2(c/n_0)\tau_a}\left[1 + \frac{1}{2(c/n_0)\tau_a}\right] \quad (9.100)$$

with a Costas discriminator and

$$\sigma^2(N_P P) \approx \frac{1}{2(c/n_0)\tau_a} \quad (9.101)$$

with a PLL discriminator. The carrier tracking noise standard deviation is then

$$\sigma_{\delta\phi} \approx \begin{array}{l} \sqrt{\dfrac{B_{L\_CA}}{(c/n_0)}\left[1 + \dfrac{1}{2(c/n_0)\tau_a}\right]} \quad \text{Costas} \\ \\ \sqrt{\dfrac{B_{L\_CA}}{(c/n_0)}} \quad \text{PLL} \end{array} \quad (9.102)$$

Figure 9.23 shows the carrier phase tracking noise standard deviation as a function of C/N_0 for different tracking bandwidths. Without cycle slips, the ADR standard deviation due to tracking noise is

$$\sigma(\widetilde{\Phi}) = \frac{c}{2\pi f_{ca}} \sigma_{\delta\phi}. \quad (9.103)$$

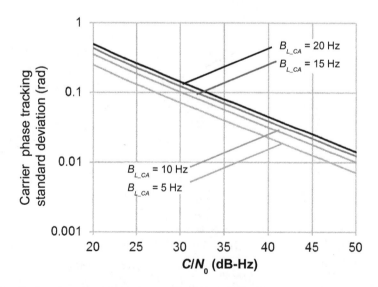

Figure 9.23 Carrier phase tracking noise standard deviation with a Costas discriminator and τ_a = 20 ms.

9.3 Range Error Sources

The carrier frequency tracking noise standard deviation with carrier phase tracking is

$$\sigma_{\delta f} \approx \sqrt{\frac{0.72 B_{L\_CA}}{\tau_a} \frac{\sigma_{\delta\phi}}{2\pi}}. \qquad (9.104)$$

The noise variance of a carrier frequency discriminator is

$$\sigma^2(N_F F) \approx \begin{matrix} \dfrac{1}{2\pi^2 (c/n_0)\tau_a^3}\left[1+\dfrac{1}{(c/n_0)\tau_a}\right] & \text{Costas} \\[1em] \dfrac{1}{2\pi^2 (c/n_0)\tau_a^3} & \text{FLL} \end{matrix}, \qquad (9.105)$$

giving a frequency tracking noise standard deviation of [13]

$$\sigma_{\delta f} \approx \begin{matrix} \dfrac{1}{\pi\tau_a}\sqrt{\dfrac{B_{L\_CF}}{(c/n_0)}}\left[1+\dfrac{1}{(c/n_0)\tau_a}\right] & \text{Costas} \\[1em] \dfrac{1}{\pi\tau_a}\sqrt{\dfrac{B_{L\_CF}}{(c/n_0)}} & \text{FLL} \end{matrix}. \qquad (9.106)$$

Figures 9.24 and 9.25 show the carrier frequency tracking noise standard deviation as a function of C/N_0 for different tracking bandwidths using a PLL and an FLL, respectively. The pseudo-range-rate error standard deviation due to tracking noise is

$$\sigma_{rw} = \frac{c}{f_{ca}}\sigma_{\delta f}. \qquad (9.107)$$

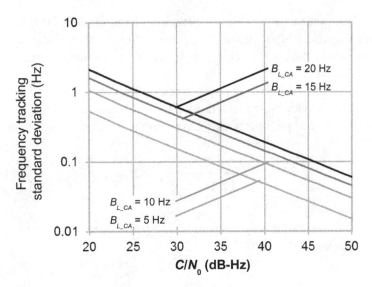

Figure 9.24 Carrier frequency tracking noise standard deviation with a Costas discriminator, and τ_a = 20 ms using a PLL.

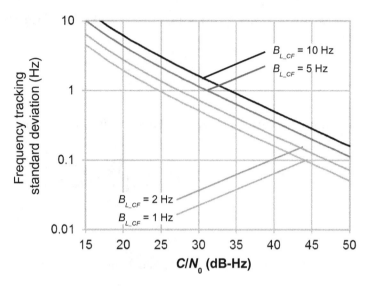

Figure 9.25 Carrier frequency tracking noise standard deviation with a Costas discriminator, and $\tau_a = 20$ ms using a FLL.

The code tracking error due to the lag in responding to dynamics depends on the tracking loop bandwidth and error in the range-rate aiding, $\tilde{\dot{\rho}}_{a,R}^{s,l} - \dot{\rho}_{a,R}^{s}$. The steady-state tracking error due to a constant range-rate error is

$$\delta\rho_{a,lag}^{s,l} = \frac{\left(\tilde{\dot{\rho}}_{a,R}^{s,l} - \dot{\rho}_{a,R}^{s}\right)\tau_a}{K_{co}} = \frac{\tilde{\dot{\rho}}_{a,R}^{s,l} - \dot{\rho}_{a,R}^{s}}{4B_{L\_CA}}$$

$$x_{a,lag}^{s,l} = \frac{f_{co}^{l}}{4B_{L\_CA}c}\left(\tilde{\dot{\rho}}_{a,R}^{s,l} - \dot{\rho}_{a,R}^{s}\right)$$

(9.108)

Note that with a 1-Hz code tracking bandwidth, a 20-ms coherent integration interval, and a BPSK signal, range-rate aiding errors will cause loss of signal coherence before the code tracking error is pushed outside the pull-in range of the code discriminator.

A third-order carrier phase tracking loop does not exhibit tracking errors in response to velocity or acceleration, but is susceptible to line-of-sight jerk. The steady state ADR and phase tracking errors due to a constant line-of-sight jerk are [1]

$$\delta\Phi_{a,lag}^{s} = -\frac{\dddot{\rho}_{a,R}^{s}}{\left(1.2B_{L\_CA}\right)^{3}}, \quad \delta\phi_{ca,a,lag}^{s} = \frac{2\pi f_{ca}}{c}\frac{\dddot{\rho}_{a,R}^{s}}{\left(1.2B_{L\_CA}\right)^{3}}$$

(9.109)

where the tracking loop gains in (9.53) are assumed. To prevent cycle slips, the jerk must thus be limited to

$$\left|\dddot{\rho}_{a,R}^{s}\right| < \frac{\left(1.2B_{L\_CA}\right)^{3}c}{4f_{ca}},$$

(9.110)

with a Costas discriminator and twice this with a PLL discriminator, noting that the threshold applies to the average jerk over the time constant of the carrier tracking loop, $1/4B_{L\_CA}$. Thus, higher jerks may be tolerated for very short periods. In practice, a lower threshold should be assumed to prevent cycle slips due to a mixture of jerk and noise. The steady-state range-rate and Doppler errors are

$$\delta \dot{\rho}^s_{a,lag} = -\frac{\ddot{\rho}^s_{a,R}}{(1.2 B_{L\_CA})^2}, \quad \delta \Delta f^s_{ca,a,lag} = \frac{f_{ca}}{c} \frac{\ddot{\rho}^s_{a,R}}{(1.2 B_{L\_CA})^2}. \tag{9.111}$$

A second-order carrier frequency tracking loop exhibits the following steady-state errors due to a constant jerk [1, 13]

$$\delta \dot{\rho}^s_{a,lag} = -\frac{\ddot{\rho}^s_{a,R}}{(1.885 B_{L\_CF})^2}, \quad \delta \Delta f^s_{ca,a,lag} = \frac{f_{ca}}{c} \frac{\ddot{\rho}^s_{a,R}}{(1.885 B_{L\_CF})^2}. \tag{9.112}$$

To prevent false lock, the line-of-sight jerk must then be limited to

$$\left|\ddot{\rho}^s_{a,R}\right| < \frac{(1.885 B_{L\_CF})^2 c}{2 f_{ca} \tau_a}, \tag{9.113}$$

with a Costas discriminator and twice this otherwise. Again, a lower threshold should be set in practice due to noise. Table 9.2 lists jerk tolerances for selected PLL and FLL tracking loop bandwidths.

9.3.4 Multipath, Nonline-of-Sight, and Diffraction

GNSS user equipment may receive reflected signals from a given satellite in addition to or instead of the direct signals. For land applications, most signal reflections occur within the surrounding environment, such as the ground, buildings, vehicles, or trees. For air, sea, and space applications, reflections off the host-vehicle body are more common. Water, glass, and metal can produce particularly strong specular reflections (see Section 7.4.3) with the reflected signal attenuated by as little as 2–3 dB. Rainwater also enhances the reflectivity of other surfaces, such as roads, foliage, and buildings. Low-elevation-angle signals are more likely than high-elevation-angle signals to be received via reflections by vertical surfaces.

Table 9.2 Jerk Tolerances for Selected PLL and FLL Tracking Loop Bandwidths

PLL Tracking Loop Bandwidth	PLL Jerk Tolerance	FLL Tracking Loop Bandwidth	FLL Jerk Tolerance
5 Hz	10.6 m s^{-3}	1 Hz	4.37 m s^{-3}
10 Hz	85.0 m s^{-3}	2 Hz	17.5 m s^{-3}
15 Hz	287 m s^{-3}	5 Hz	109 m s^{-3}
20 Hz	680 m s^{-3}	10 Hz	427 m s^{-3}

Reflected signals are always delayed with respect to direct signals and have a lower amplitude unless the direct signals are attenuated (e.g., by a building or foliage). When a signal is received via a reflected path only, known as nonline-of-sight (NLOS) reception, the pseudo-range measurement errors are potentially unbounded and always positive. Although NLOS measurement errors are normally within a few hundred meters, errors of several kilometers occasionally occur when a signal is reflected by a distant tall building. The range-rate errors can result in the user's apparent direction of motion being reflected in the object reflecting the signal. Consequently, reflectors perpendicular to the direction of travel can produce much larger navigation errors than those parallel to the trajectory, particularly with a filtered navigation solution (Section 9.4.2). Also, for a reflector close to a moving user antenna, the pseudo-range error may be small, but the range-rate error large.

When a signal is partially blocked by an obstacle, diffraction can occur, bending the path of the signal. The attenuation increases with the diffraction angle with useable GNSS signals receivable at deflections of up to 5° [71]. Diffracted signals are also delayed, but typically only by decimeters. A diffracted signal is normally received instead of the direct signal, but may occasionally be received in addition.

The signal path between satellite and user is not a simple ray, but is instead determined by Fresnel zones. Consequently, the radius of the effective signal footprint at a reflecting or diffracting object is $\sqrt{r\lambda_{ca}}$, where r is the distance of the object from the user antenna. Irregularities in the object on this scale will therefore affect the properties of the reflected or diffracted signal.

When a reflected (or diffracted) signal is received in addition to the direct signal and/or multiple reflected or diffracted signals are received, multipath interference occurs. Figure 9.26 illustrates an example.

Within the receiver (i.e., after the user antenna), each reflected or diffracted signal may be described by an amplitude, α_i, range lag, Δ_i, and carrier phase offset, φ_i, with respect to the direct signal (or the strongest signal if no direct signal is received). There is also a carrier frequency offset, δf_{mi}, which is larger where the user is moving with respect to the reflecting surface [72]. The relative amplitude is given by

$$\alpha_i = \sqrt{\frac{G_i R_i}{G_0 R_0}}, \qquad (9.114)$$

where G_i and G_0 are the antenna gains for the ith and strongest signals, respectively, and R_i and R_0 are the reflection coefficients. When the strongest signal is the direct

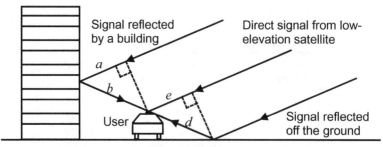

Figure 9.26 Example of a multipath interference scenario.

9.3 Range Error Sources

signal, $R_0 = 1$. For the building-reflected signal in Figure 9.26, the range lag is $\Delta = a + b$, while for the ground-reflected signal it is $\Delta = d - e$. The phase offset is given by

$$\varphi_i = \left(\frac{2\pi\Delta_i}{\lambda_{ca}} + \varphi_{Ri}\right) \text{MOD} 2\pi, \qquad (9.115)$$

where the MOD operator gives the remainder from integer division and φ_{Ri} is the phase shift on reflection, which is π radians for a totally flat specular reflector at an angle of incidence less than Brewster's angle. The frequency offset is

$$\delta f_{M,i} = \frac{\partial}{\partial t}\left(\frac{\Delta_i}{\lambda_{ca}} + \frac{\varphi_{Ri}}{2\pi}\right). \qquad (9.116)$$

By analogy with (9.1), the total received signal is

$$s_a(t_{sa}) = \sqrt{2P} \sum_{i=0}^{n} \left\{ \begin{array}{l} \alpha_i C(t_{st} - \Delta_i/c) D(t_{st} - \Delta_i/c) \\ \times \cos\left[2\pi\left(f_{ca} + \Delta f_{ca} + \delta f_{M,i}\right) t_{sa} + \phi_{ca} + \varphi_i\right] \end{array} \right\}, \qquad (9.117)$$

where n is the number of reflected or diffracted signals and, by definition, $\alpha_0 = 1$ and $\Delta_0 = \varphi_0 = \delta f_{m0} = 0$. The accumulated correlator outputs, given by (9.10), then become

$$I_E(t_{sa}) = \sigma_{IQ}\left\{\sqrt{2(c/n_0)\tau_a} D(t_{st}) \sum_{i=0}^{n} \begin{bmatrix} \alpha_i R(x - \delta_i - d/2)\text{sinc}\left(\pi(\delta f_{ca} + \delta f_{M,i})\tau_a\right) \\ \times \cos(\delta\phi_{ca} + \varphi_i) \end{bmatrix} + w_{IE}(t_{sa})\right\}$$

$$I_P(t_{sa}) = \sigma_{IQ}\left\{\sqrt{2(c/n_0)\tau_a} D(t_{st}) \sum_{i=0}^{n} \begin{bmatrix} \alpha_i R(x - \delta_i)\text{sinc}\left(\pi(\delta f_{ca} + \delta f_{M,i})\tau_a\right) \\ \times \cos(\delta\phi_{ca} + \varphi_i) \end{bmatrix} + w_{IP}(t_{sa})\right\}$$

$$I_L(t_{sa}) = \sigma_{IQ}\left\{\sqrt{2(c/n_0)\tau_a} D(t_{st}) \sum_{i=0}^{n} \begin{bmatrix} \alpha_i R(x - \delta_i + d/2)\text{sinc}\left(\pi(\delta f_{ca} + \delta f_{M,i})\tau_a\right) \\ \times \cos(\delta\phi_{ca} + \varphi_i) \end{bmatrix} + w_{IL}(t_{sa})\right\}$$

$$Q_E(t_{sa}) = \sigma_{IQ}\left\{\sqrt{2(c/n_0)\tau_a} D(t_{st}) \sum_{i=0}^{n} \begin{bmatrix} \alpha_i R(x - \delta_i - d/2)\text{sinc}\left(\pi(\delta f_{ca} + \delta f_{M,i})\tau_a\right) \\ \times \sin(\delta\phi_{ca} + \varphi_i) \end{bmatrix} + w_{QE}(t_{sa})\right\}$$

$$Q_P(t_{sa}) = \sigma_{IQ}\left\{\sqrt{2(c/n_0)\tau_a} D(t_{st}) \sum_{i=0}^{n} \begin{bmatrix} \alpha_i R(x - \delta_i)\text{sinc}\left(\pi(\delta f_{ca} + \delta f_{M,i})\tau_a\right) \\ \times \sin(\delta\phi_{ca} + \varphi_i) \end{bmatrix} + w_{QP}(t_{sa})\right\}$$

$$Q_L(t_{sa}) = \sigma_{IQ}\left\{\sqrt{2(c/n_0)\tau_a} D(t_{st}) \sum_{i=0}^{n} \begin{bmatrix} \alpha_i R(x - \delta_i + d/2)\text{sinc}\left(\pi(\delta f_{ca} + \delta f_{M,i})\tau_a\right) \\ \times \sin(\delta\phi_{ca} + \varphi_i) \end{bmatrix} + w_{QL}(t_{sa})\right\}$$

$$(9.118)$$

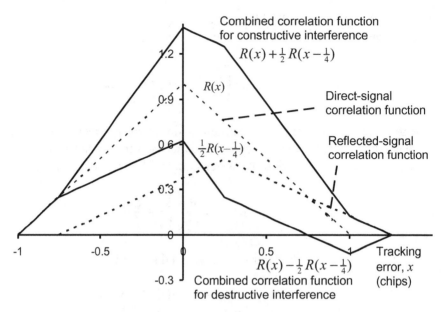

Figure 9.27 Direct-, reflected-, and combined-signal correlation functions for $\delta = 1/4$, $\alpha = 1/2$, $\varphi_i = 0$ and $180°$ (neglecting precorrelation band-limiting).

where the effect of multipath on navigation data reception is neglected and $\delta_i = \Delta_i f_{co}/c$ is the lag in code chips.

Code tracking errors due to multipath are maximized when the carrier phase offset is 0° or 180°. The multipath interference is constructive where $-90° < \varphi_i < 90°$ and destructive otherwise. Figure 9.27 shows the direct-signal, reflected-signal, and combined correlation functions for a single interfering signal with $\delta = 1/4$, $\alpha = 1/2$, and $\varphi_i = 0$ and $180°$; precorrelation band-limiting is neglected. The shape of the correlation function is thus distorted by the multipath interference. Note that there is no interference from the main correlation peak where $\delta > 1 + d/2$. Consequently, higher chipping-rate signals are less susceptible to multipath interference, as the range lag, Δ, must be smaller for the reflected signal to affect the main correlation peak.

The code tracking loop acts to equate the signal powers in the early and late correlation channels, so the tracking error in the presence of multipath is obtained by solving

$$I_E^2 + Q_E^2 - I_L^2 - Q_L^2 = 0. \tag{9.119}$$

As Figure 9.28 shows, the tracking error depends on the early-late correlator spacing. Multipath has less impact on the peak of the correlation function, so a narrower correlator spacing often leads to a smaller tracking error [69]. However, when precorrelation band-limiting is significant, the correlation function is rounded, reducing the benefit of narrowing the correlator spacing as Figure 9.29 illustrates.

An analytical solution to (9.119) is possible where there is a single delayed signal (i.e., specular reflection from a single object), the lag is small, the frequency offset is negligible, and precorrelation band-limiting may be neglected:

9.3 Range Error Sources

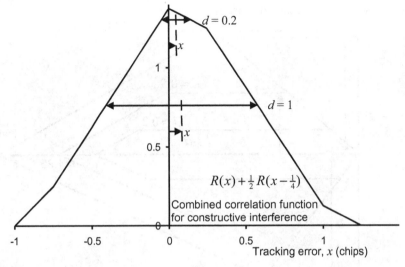

Figure 9.28 Effect of early-late correlator spacing on multipath error (neglecting precorrelation band-limiting).

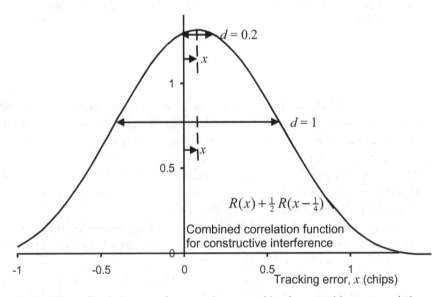

Figure 9.29 Effect of early-late correlator spacing on multipath error with a precorrelation bandwidth of $B_{PC} = 2f_{co}$.

$$x = \frac{\alpha^2 + \alpha\cos\varphi}{\alpha^2 + 2\alpha\cos\varphi + 1}\delta$$

$$\delta\rho_M = \frac{\alpha^2 + \alpha\cos\varphi}{\alpha^2 + 2\alpha\cos\varphi + 1}\Delta \quad |x - \delta| < d/2. \quad (9.120)$$

Otherwise, numerical methods must be used.

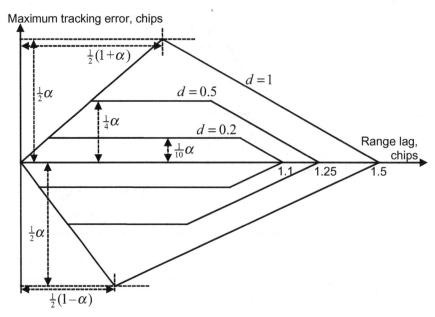

Figure 9.30 Limits of code tracking error due to multipath (neglecting precorrelation band-limiting).

Figure 9.30 shows the limits of the code tracking error for different correlator spacings, assuming a BPSK signal [73]. The actual tracking error oscillates as the carrier phase offset changes. Note that the mean tracking error, averaged over the carrier phase offset, is nonzero. For BOC signals, the code tracking error exhibits $2f_s/f_{co} - 1$ nodes, evenly distributed in range error [42, 74]. For the Galileo $BOC_c(15,2.5)$ signals, multipath interference can distort the code correlation function sufficiently to prevent identification of the correct peak [75].

When the range lag is several chips, tracking errors of up to a meter can be caused by interference from one of the minor peaks of the GPS or GLONASS C/A code correlation function [76].

Multipath interference also produces carrier phase tracking errors. For a single delayed signal, the carrier phase error is

$$\delta\phi_M = \arctan_2\left(\frac{\alpha R(x-\delta)\sin\varphi}{1+\alpha R(x-\delta)\cos\varphi}\right). \quad (9.121)$$

For $\alpha < 1$, this does not exceed 90°, corresponding to 4.8 cm in the L1 band.

When the user moves with respect to the reflecting surface, the pseudo-range-rate errors are significant. These are given by

$$\delta\dot{\rho}_M = \frac{\alpha\sin\varphi c\dfrac{\partial R(x-\delta)}{\partial \delta}\dfrac{\partial \delta}{\partial t} - \left[1+\alpha R(x-\delta)\cos\varphi + \alpha^2 R^2(x-\delta)\right]c\dfrac{\partial \phi}{\partial t}}{2\pi\left[1+2\alpha R(x-\delta)\cos\varphi + \alpha^2 R^2(x-\delta)\right]f_{ca}}. \quad (9.122)$$

Multipath, NLOS propagation, and diffraction-induced errors all vary with time as the satellites and user move, with the latter dominating. If the user moves perpendicularly to the reflector, the sign of the pseudo-range and carrier phase multipath errors will fluctuate rapidly as the carrier phase offset of the reflection varies, enabling much of the error to be averaged out.

Multipath and NLOS mitigation techniques suitable for navigation applications are discussed in Section 10.4. Note that many authors classify NLOS reception and diffraction as multipath, which is incorrect. These are different phenomena that can occur both separately and in combination.

9.4 Navigation Processor

This section describes how a GNSS navigation processor calculates the user position and velocity and calibrates the receiver clock errors using pseudo-range and pseudo-range rate (or equivalent) measurements. It builds on the generic description of position determination from ranging in Section 7.3.3 and the introduction to GNSS positioning in Section 8.1.3. Note that positioning using the carrier-phase-derived ADR measurements is discussed in Section 10.2.

The ranging processor outputs measurements for all the satellites tracked at a common time of signal arrival. When it does not apply corrections for the satellite clock errors and the ionosphere and troposphere propagation delays, these corrections (see Section 9.3) should be applied by the navigation processor using [repeating (8.49)]

$$\tilde{\rho}_{a,C}^{s,l} = \tilde{\rho}_{a,R}^{s,l} - \delta\hat{\rho}_{I,a}^{s,l} - \delta\hat{\rho}_{T,a}^{s} + \delta\hat{\rho}_{c}^{s,l}$$
$$\tilde{\rho}_{a,C}^{s,l} = \tilde{\rho}_{a,R}^{s,l} + \delta\hat{\rho}_{c}^{s}.$$

The navigation processor estimates the user position, $\hat{r}_{ia}^i = (\hat{x}_{ia}^i, \hat{y}_{ia}^i, \hat{z}_{ia}^i)$, and receiver clock offset, $\delta\hat{\rho}_c^a$, at the time of signal arrival, $t_{sa,a}^{s,l}$. Each corrected pseudo-range measurement, $\tilde{\rho}_{a,C}^{s,l}$, may be expressed in terms of these estimates by

$$\begin{aligned}
\tilde{\rho}_{a,C}^{s,l} &= \sqrt{\left[\hat{r}_{is}^i(\tilde{t}_{st,a}^{s,l}) - \hat{r}_{ia}^i(t_{sa,a}^{s,l})\right]^T \left[\hat{r}_{is}^i(\tilde{t}_{st,a}^{s,l}) - \hat{r}_{ia}^i(t_{sa,a}^{s,l})\right]} + \delta\hat{\rho}_c^a(t_{sa,a}^{s,l}) + \delta\rho_{a,\varepsilon}^{s,l+} \\
&= \sqrt{\left[\hat{x}_{is}^i(\tilde{t}_{st,a}^{s,l}) - \hat{x}_{ia}^i(t_{sa,a}^{s,l})\right]^2 + \left[\hat{y}_{is}^i(\tilde{t}_{st,a}^{s,l}) - \hat{y}_{ia}^i(t_{sa,a}^{s,l})\right]^2 + \left[\hat{z}_{is}^i(\tilde{t}_{st,a}^{s,l}) - \hat{z}_{ia}^i(t_{sa,a}^{s,l})\right]^2}, \\
&\quad + \delta\hat{\rho}_c^a(t_{sa,a}^{s,l}) + \delta\rho_{a,\varepsilon}^{s,l+}
\end{aligned}$$

(9.123)

where $\hat{r}_{is}^i$ is the satellite position obtained from the navigation data message as described in Section 8.5.2, $\tilde{t}_{st,a}^{s,l}$ is the measured transmission time, and $\delta\rho_{a,\varepsilon}^{s,l+}$ is the measurement residual, given by

$$\delta\rho_{a,\varepsilon}^{s,l+} = \tilde{\rho}_{a,C}^{s,l} - \hat{\rho}_{a,C}^{s,l+}, \qquad (9.124)$$

where $\hat{\rho}_{a,C}^{s,l+}$ is the pseudo-range predicted from the navigation solution. Note that the residuals are only nonzero in an overdetermined or filtered navigation solution.

The transmission time may be output by the ranging processor. Otherwise, it is given by

$$\tilde{t}_{st,a}^{s,l} = \tilde{t}_{sa,a}^{s,l} - \left(\tilde{\rho}_{a,R}^{s,l} + \delta\hat{\rho}_c^{s,l}\right)/c, \quad (9.125)$$

noting that the receiver clock errors in the pseudo-range and time of signal arrival cancel.

When measurements from more than one GNSS constellation are used and the interconstellation timing biases (Section 8.4.5) are not included in the satellite clock corrections, they must be estimated as part of the navigation solution (see Section G.8.1 of Appendix G on the CD).

Each pseudo-range-rate measurement, $\tilde{\dot{\rho}}_{a,C}^{s,l}$, is expressed in terms of the navigation processor's user velocity, $\hat{v}_{ia}^i$, and receiver clock drift, $\delta\hat{\rho}_c^a$, estimates by

$$\tilde{\dot{\rho}}_{a,C}^{s,l} = \mathbf{u}_{as}^{i\,T}\left[\hat{\mathbf{v}}_{is}^i\left(\tilde{t}_{st,a}^{s,l}\right) - \hat{\mathbf{v}}_{ia}^i\left(t_{sa,a}^{s,l}\right)\right] + \delta\hat{\dot{\rho}}_c^a\left(t_{sa,a}^{s,l}\right) + \delta\dot{\rho}_{a,\varepsilon}^{s,l+}, \quad (9.126)$$

where calculation of the satellite velocity, $\hat{\mathbf{v}}_{is}^i$, and the line-of-sight vector, $\mathbf{u}_{as}^i$, are described in Section 8.5.2 and the measurement residual is

$$\delta\dot{\rho}_{a,\varepsilon}^{s,l+} = \tilde{\dot{\rho}}_{a,C}^{s,l} - \hat{\dot{\rho}}_{a,C}^{s,l+}. \quad (9.127)$$

When the user position and velocity are estimated in an ECEF frame, (9.123) and (9.126) are replaced by

$$\begin{aligned}\tilde{\rho}_{a,C}^{s,l} - \delta\rho_{ie,a}^s &= \sqrt{\left[\hat{\mathbf{r}}_{es}^e\left(\tilde{t}_{st,a}^{s,l}\right) - \hat{\mathbf{r}}_{ea}^e\left(t_{sa,a}^{s,l}\right)\right]^T\left[\hat{\mathbf{r}}_{es}^e\left(\tilde{t}_{st,a}^{s,l}\right) - \hat{\mathbf{r}}_{ea}^e\left(t_{sa,a}^{s,l}\right)\right]} + \delta\hat{\rho}_c^a\left(t_{sa,a}^{s,l}\right) + \delta\rho_{a,\varepsilon}^{s,l+}\\ \tilde{\dot{\rho}}_{a,C}^{s,l} - \delta\dot{\rho}_{ie,a}^s &= \mathbf{u}_{as}^{e\,T}\left[\hat{\mathbf{v}}_{es}^e\left(\tilde{t}_{st,a}^{s,l}\right) - \hat{\mathbf{v}}_{ea}^e\left(t_{sa,a}^{s,l}\right)\right] + \delta\hat{\dot{\rho}}_c^a\left(t_{sa,a}^{s,l}\right) + \delta\dot{\rho}_{a,\varepsilon}^{s,l+}\end{aligned}, \quad (9.128)$$

where the Sagnac corrections are given by (8.32) and (8.46), or by

$$\begin{aligned}\tilde{\rho}_{a,C}^{s,l} &= \sqrt{\left[\mathbf{C}_e^I\left(\tilde{t}_{st,a}^{s,l}\right)\hat{\mathbf{r}}_{es}^e\left(\tilde{t}_{st,a}^{s,l}\right) - \hat{\mathbf{r}}_{ea}^e\left(t_{sa,a}^{s,l}\right)\right]^T\left[\mathbf{C}_e^I\left(\tilde{t}_{st,a}^{s,l}\right)\hat{\mathbf{r}}_{es}^e\left(\tilde{t}_{st,a}^{s,l}\right) - \hat{\mathbf{r}}_{ea}^e\left(t_{sa,a}^{s,l}\right)\right]} + \delta\hat{\rho}_c^a\left(t_{sa,a}^{s,l}\right) + \delta\rho_{a,\varepsilon}^{s,l+}\\ \tilde{\dot{\rho}}_{a,C}^{s,l} &= \mathbf{u}_{as,j}^{e\,T}\left[\mathbf{C}_e^I\left(\tilde{t}_{st,a}^{s,l}\right)\left(\hat{\mathbf{v}}_{es}^e\left(\tilde{t}_{st,a}^{s,l}\right) + \mathbf{\Omega}_{ie}^e\hat{\mathbf{r}}_{es}^e\left(\tilde{t}_{st,a}^{s,l}\right)\right) - \left(\hat{\mathbf{v}}_{ea}^e\left(t_{sa,a}^{s,l}\right) + \mathbf{\Omega}_{ie}^e\hat{\mathbf{r}}_{ea}^e\left(t_{sa,a}^{s,l}\right)\right)\right] + \delta\hat{\dot{\rho}}_c^a\left(t_{sa,a}^{s,l}\right) + \delta\dot{\rho}_{a,\varepsilon}^{s,l+}\end{aligned}$$

(9.129)

Note that the I frame is synchronized with the ECEF frame at the true time of signal arrival, not the measured time of arrival. Therefore, $\mathbf{C}_e^I$ is calculated using (8.36) as a function of the range between the user antenna and satellite, not the pseudo-range.

9.4 Navigation Processor

The navigation solution is obtained by solving the above equations with pseudo-range and pseudo-range-rate measurements from at least four satellites. A *single-epoch* or *snapshot* navigation solution only uses the current set of ranging processor measurements and is described first. A filtered navigation solution, described next, also makes use of previous measurement data. The filtered navigation solution is much less noisy, but can exhibit dynamic-response lags, while successive single-epoch solutions are independent, so they can highlight erroneous measurement data more quickly. Thus, the single-epoch solution is useful for integrity monitoring (see Chapter 17). The accuracy of the single-epoch solution can be improved by using carrier-smoothed pseudo-range measurements, but the noise on successive solutions is then not independent. A single-epoch solution is also needed to initialize the filtered solution.

The section concludes with discussions of the effect of signal geometry on navigation solution accuracy and position error budgets. Section G.8 of Appendix G on the CD describes interconstellation timing bias estimation, solutions based on TDOA across satellites, solutions using delta range measurements, and signal geometry with a chip-scale atomic clock. The MATLAB functions on the CD, GNSS_Least_Squares and GNSS_Kalman_Filter, simulate GNSS with single-epoch and filtered navigation solutions, respectively.

9.4.1 Single-Epoch Navigation Solution

A position solution cannot easily be obtained analytically from a set of pseudo-range measurements using (9.123). Therefore, the equations are linearized by performing a Taylor expansion about a predicted user position, $\hat{r}_{ia}^{i-}$, and clock offset, $\delta\hat{\rho}_c^{a-}$, in analogy with the linearized Kalman filter (Section 3.4.1). The predicted user position and clock offset is generally the solution from the previous set of pseudo-range measurements. At initialization, the solution may have to be iterated two or three times to minimize the linearization errors. Thus, (9.123) is replaced by

$$\begin{pmatrix} \tilde{\rho}_{a,C}^1 - \hat{\rho}_{a,C}^{1-} \\ \tilde{\rho}_{a,C}^2 - \hat{\rho}_{a,C}^{2-} \\ \vdots \\ \tilde{\rho}_{a,C}^m - \hat{\rho}_{a,C}^{m-} \end{pmatrix} = \mathbf{H}_G^i \begin{pmatrix} \hat{x}_{ia}^{i+} - \hat{x}_{ia}^{i-} \\ \hat{y}_{ia}^{i+} - \hat{y}_{ia}^{i-} \\ \hat{z}_{ia}^{i+} - \hat{z}_{ia}^{i-} \\ \delta\hat{\rho}_c^{a+} - \delta\hat{\rho}_c^{a-} \end{pmatrix} + \begin{pmatrix} \delta\rho_{a,\varepsilon}^{1+} \\ \delta\rho_{a,\varepsilon}^{2+} \\ \vdots \\ \delta\rho_{a,\varepsilon}^{m+} \end{pmatrix}, \qquad (9.130)$$

where the number of measurements, m, is at least four. Note that the linearization errors are included in the residuals. Using j to denote the combination of a satellite, s, and signal, l, from that satellite, the predicted pseudo-range for the jth measurement is

$$\hat{\rho}_{a,C}^{j-} = \sqrt{\left[\hat{r}_{ij}^i\left(\tilde{t}_{st,a}^j\right) - \hat{r}_{ia}^{i-}\left(\tilde{t}_{sa,a}^j\right)\right]^{\mathrm{T}}\left[\hat{r}_{ij}^i\left(\tilde{t}_{st,a}^j\right) - \hat{r}_{ia}^{i-}\left(\tilde{t}_{sa,a}^j\right)\right]} + \delta\hat{\rho}_c^{a-}, \qquad (9.131)$$

and the measurement or geometry matrix, $\mathbf{H}_G^i$, is

$$\mathbf{H}_G^i = \begin{pmatrix} \frac{\partial \rho_a^1}{\partial x_{ia}^i} & \frac{\partial \rho_a^1}{\partial y_{ia}^i} & \frac{\partial \rho_a^1}{\partial z_{ia}^i} & \frac{\partial \rho_a^1}{\partial \rho_c^a} \\ \frac{\partial \rho_a^2}{\partial x_{ia}^i} & \frac{\partial \rho_a^2}{\partial y_{ia}^i} & \frac{\partial \rho_a^2}{\partial z_{ia}^i} & \frac{\partial \rho_a^2}{\partial \rho_c^a} \\ \vdots & \vdots & \vdots & \vdots \\ \frac{\partial \rho_a^m}{\partial x_{ia}^i} & \frac{\partial \rho_a^m}{\partial y_{ia}^i} & \frac{\partial \rho_a^m}{\partial z_{ia}^i} & \frac{\partial \rho_a^m}{\partial \rho_c^a} \end{pmatrix}_{\mathbf{r}_{ia}^i = \hat{\mathbf{r}}_{ia}^{i-}}. \qquad (9.132)$$

Differentiating (9.123) with respect to the user position and clock offset gives

$$\mathbf{H}_G^i = \begin{pmatrix} -u_{a1,x}^i & -u_{a1,y}^i & -u_{a1,z}^i & 1 \\ -u_{a2,x}^i & -u_{a2,y}^i & -u_{a2,z}^i & 1 \\ \vdots & \vdots & \vdots & \vdots \\ -u_{am,x}^i & -u_{am,y}^i & -u_{am,z}^i & 1 \end{pmatrix}_{\mathbf{r}_{ia}^i = \hat{\mathbf{r}}_{ia}^{i-}}, \qquad (9.133)$$

where the line-of-sight unit vectors are obtained from (8.40) using the predicted user position. When there are four pseudo-range measurements (i.e., $m = 4$), the number of measurements matches the number of unknowns, so the measurement residuals are zero. The position and clock solution is then

$$\begin{pmatrix} \hat{\mathbf{r}}_{ia}^{i+} \\ \delta \hat{\rho}_c^{a+} \end{pmatrix} = \begin{pmatrix} \hat{\mathbf{r}}_{ia}^{i-} \\ \delta \hat{\rho}_c^{a-} \end{pmatrix} + \mathbf{H}_G^{i\,-1} \begin{pmatrix} \tilde{\rho}_{a,C}^1 - \hat{\rho}_{a,C}^{1-} \\ \tilde{\rho}_{a,C}^2 - \hat{\rho}_{a,C}^{2-} \\ \tilde{\rho}_{a,C}^3 - \hat{\rho}_{a,C}^{3-} \\ \tilde{\rho}_{a,C}^4 - \hat{\rho}_{a,C}^{4-} \end{pmatrix}. \qquad (9.134)$$

When there are more than four pseudo-range measurements, the solution is overdetermined and, without the measurement residual terms, the set of measurements would not produce a consistent navigation solution. However, the extra measurements provide the opportunity to smooth out some of the measurement noise. As discussed in Section 7.3.3, an iterated least-squares algorithm (see Section D.1 of Appendix D on the CD) is used. Applying this to (9.130), the position and clock offset solution is [77]:

$$\begin{pmatrix} \hat{\mathbf{r}}_{ia}^{i+} \\ \delta \hat{\rho}_c^{a+} \end{pmatrix} = \begin{pmatrix} \hat{\mathbf{r}}_{ia}^{i-} \\ \delta \hat{\rho}_c^{a-} \end{pmatrix} + \left(\mathbf{H}_G^{i\,\mathrm{T}} \mathbf{H}_G^i \right)^{-1} \mathbf{H}_G^{i\,\mathrm{T}} \begin{pmatrix} \tilde{\rho}_{a,C}^1 - \hat{\rho}_{a,C}^{1-} \\ \tilde{\rho}_{a,C}^2 - \hat{\rho}_{a,C}^{2-} \\ \vdots \\ \tilde{\rho}_{a,C}^m - \hat{\rho}_{a,C}^{m-} \end{pmatrix}. \qquad (9.135)$$

9.4 Navigation Processor

Example 9.1 on the CD illustrates this with a five-satellite ECI-frame position solution and is editable using Microsoft Excel.

When the accuracy of the pseudo-range measurements is known to differ, for example, due to variation in c/n_0 or the residual ionosphere and troposphere propagation errors, which depend on the elevation angle, a weighted least-squares estimate can be computed [77]:

$$\begin{pmatrix} \hat{\mathbf{r}}_{ia}^{i+} \\ \delta\hat{\rho}_c^{a+} \end{pmatrix} = \begin{pmatrix} \hat{\mathbf{r}}_{ia}^{i-} \\ \delta\hat{\rho}_c^{a-} \end{pmatrix} + \left(\mathbf{H}_G^{iT}\mathbf{C}_\rho^{-1}\mathbf{H}_G^i\right)^{-1}\mathbf{H}_G^{iT}\mathbf{C}_\rho^{-1} \begin{pmatrix} \tilde{\rho}_{a,C}^1 - \hat{\rho}_{a,C}^{1-} \\ \tilde{\rho}_{a,C}^2 - \hat{\rho}_{a,C}^{2-} \\ \vdots \\ \tilde{\rho}_{a,C}^m - \hat{\rho}_{a,C}^{m-} \end{pmatrix}, \quad (9.136)$$

where the diagonal elements of the measurement error covariance matrix, $\mathbf{C}_\rho$, are the predicted variances of each pseudo-range error and the off-diagonal terms account for any correlations between the pseudo-range errors. Note that $\mathbf{C}_\rho$ includes time-correlated errors (i.e., biases), while the Kalman filter measurement noise covariance matrix, $\mathbf{R}$, does not. A commonly used elevation-dependent model is

$$\mathbf{C}_\rho = \begin{pmatrix} \sigma_\rho^2(\theta_{nu}^{a1}) & 0 & \cdots & 0 \\ 0 & \sigma_\rho^2(\theta_{nu}^{a2}) & \cdots & 0 \\ \vdots & \vdots & \ddots & \vdots \\ 0 & 0 & \cdots & \sigma_\rho^2(\theta_{nu}^{am}) \end{pmatrix}, \quad \sigma_\rho(\theta_{nu}^{aj}) = \frac{\sigma_{pz}}{\sin(\theta_{nu}^{aj})}, \quad (9.137)$$

where σ_{pz} is the zenith pseudo-range error standard deviation, noting that an arbitrary value may be used as it does not affect the ILS solution.

The weighted least-squares velocity and receiver clock drift solution is

$$\begin{pmatrix} \hat{\mathbf{v}}_{ia}^{i+} \\ \delta\hat{\rho}_c^{a+} \end{pmatrix} = \begin{pmatrix} \hat{\mathbf{v}}_{ia}^{i-} \\ \delta\hat{\rho}_c^{a-} \end{pmatrix} + \left(\mathbf{H}_G^{iT}\mathbf{C}_r^{-1}\mathbf{H}_G^i\right)^{-1}\mathbf{H}_G^{iT}\mathbf{C}_r^{-1} \begin{pmatrix} \tilde{\dot{\rho}}_{a,C}^1 - \hat{\dot{\rho}}_{a,C}^{1-} \\ \tilde{\dot{\rho}}_{a,C}^2 - \hat{\dot{\rho}}_{a,C}^{2-} \\ \vdots \\ \tilde{\dot{\rho}}_{a,C}^m - \hat{\dot{\rho}}_{a,C}^{m-} \end{pmatrix}, \quad (9.138)$$

where

$$\hat{\dot{\rho}}_{a,C}^{j-} = \hat{\mathbf{u}}_{aj}^{i-T}\left[\hat{\mathbf{v}}_{ij}^i(\tilde{t}_{st,a}^j) - \hat{\mathbf{v}}_{ia}^{i-}(\tilde{t}_{sa,a}^j)\right] + \delta\hat{\dot{\rho}}_c^{a-}, \quad (9.139)$$

and noting that the measurement matrix is the same. A similar model for the measurement error covariance may be used:

$$\mathbf{C}_r = \begin{pmatrix} \sigma_r^2(\theta_{nu}^{a1}) & 0 & \cdots & 0 \\ 0 & \sigma_r^2(\theta_{nu}^{a2}) & \cdots & 0 \\ \vdots & \vdots & \ddots & \vdots \\ 0 & 0 & \cdots & \sigma_r^2(\theta_{nu}^{am}) \end{pmatrix}, \quad \sigma_r(\theta_{nu}^{aj}) = \frac{\sigma_{rz}}{\sin(\theta_{nu}^{aj})}, \quad (9.140)$$

where σ_{rz} is the zenith pseudo-range rate error standard deviation.

When an ECEF frame is used, the weighted least-squares solution is

$$\begin{pmatrix} \hat{\mathbf{r}}_{ea}^{e+} \\ \delta\hat{\rho}_c^{a+} \end{pmatrix} = \begin{pmatrix} \hat{\mathbf{r}}_{ea}^{e-} \\ \delta\hat{\rho}_c^{a-} \end{pmatrix} + \left(\mathbf{H}_G^{e\mathrm{T}} \mathbf{C}_\rho^{-1} \mathbf{H}_G^e\right)^{-1} \mathbf{H}_G^{e\mathrm{T}} \mathbf{C}_\rho^{-1} \begin{pmatrix} \tilde{\rho}_{a,C}^1 - \hat{\rho}_{a,C}^{1-} \\ \tilde{\rho}_{a,C}^2 - \hat{\rho}_{a,C}^{2-} \\ \vdots \\ \tilde{\rho}_{a,C}^m - \hat{\rho}_{a,C}^{m-} \end{pmatrix},$$

$$\begin{pmatrix} \hat{\mathbf{v}}_{ea}^{e+} \\ \delta\hat{\dot{\rho}}_c^{a+} \end{pmatrix} = \begin{pmatrix} \hat{\mathbf{v}}_{ea}^{e-} \\ \delta\hat{\dot{\rho}}_c^{a-} \end{pmatrix} + \left(\mathbf{H}_G^{e\mathrm{T}} \mathbf{C}_r^{-1} \mathbf{H}_G^e\right)^{-1} \mathbf{H}_G^{e\mathrm{T}} \mathbf{C}_r^{-1} \begin{pmatrix} \tilde{\dot{\rho}}_{a,C}^1 - \hat{\dot{\rho}}_{a,C}^{1-} \\ \tilde{\dot{\rho}}_{a,C}^2 - \hat{\dot{\rho}}_{a,C}^{2-} \\ \vdots \\ \tilde{\dot{\rho}}_{a,C}^m - \hat{\dot{\rho}}_{a,C}^{m-} \end{pmatrix}, \quad (9.141)$$

where

$$\begin{aligned} \hat{\rho}_{a,C}^{j-} &= \sqrt{\left[\hat{\mathbf{r}}_{ej}^e(\tilde{t}_{st,a}^j) - \hat{\mathbf{r}}_{ea}^{e-}(\tilde{t}_{sa,a}^j)\right]^\mathrm{T} \left[\hat{\mathbf{r}}_{ej}^e(\tilde{t}_{st,a}^j) - \hat{\mathbf{r}}_{ea}^{e-}(\tilde{t}_{sa,a}^j)\right]} + \delta\hat{\rho}_c^{a-} + \delta\hat{\rho}_{ie,a}^{j-} \\ \hat{\dot{\rho}}_{a,C}^{j-} &= \hat{\mathbf{u}}_{aj}^{e-\mathrm{T}} \left[\hat{\mathbf{v}}_{ej}^e(\tilde{t}_{st,a}^j) - \hat{\mathbf{v}}_{ea}^{e-}(\tilde{t}_{sa,a}^j)\right] + \delta\hat{\dot{\rho}}_c^{a-} + \delta\hat{\dot{\rho}}_{ie,a}^{j-} \end{aligned} \quad (9.142)$$

or

$$\begin{aligned} \hat{\rho}_{a,C}^{j-} &= \sqrt{\left[\mathbf{C}_e^I(\tilde{t}_{st,a}^j)\hat{\mathbf{r}}_{ej}^e(\tilde{t}_{st,a}^j) - \hat{\mathbf{r}}_{ea}^{e-}(\tilde{t}_{sa,a}^j)\right]^\mathrm{T} \left[\mathbf{C}_e^I(\tilde{t}_{st,a}^j)\hat{\mathbf{r}}_{ej}^e(\tilde{t}_{st,a}^j) - \hat{\mathbf{r}}_{ea}^{e-}(\tilde{t}_{sa,a}^j)\right]} + \delta\hat{\rho}_c^{a-} \\ \hat{\dot{\rho}}_{a,C}^{j-} &= \hat{\mathbf{u}}_{aj}^{e-\mathrm{T}} \left[\mathbf{C}_e^I(\tilde{t}_{st,a}^j)\left(\hat{\mathbf{v}}_{ej}^e(\tilde{t}_{st,a}^j) + \mathbf{\Omega}_{ie}^e \hat{\mathbf{r}}_{ej}^e(\tilde{t}_{st,a}^j)\right) - \left(\hat{\mathbf{v}}_{ea}^{e-}(\tilde{t}_{sa,a}^j) + \mathbf{\Omega}_{ie}^e \hat{\mathbf{r}}_{ea}^{e-}(\tilde{t}_{sa,a}^j)\right)\right] + \delta\hat{\dot{\rho}}_c^{a-} \end{aligned}, \quad (9.143)$$

and

$$\mathbf{H}_G^e = \begin{pmatrix} -u_{a1,x}^e & -u_{a1,y}^e & -u_{a1,z}^e & 1 \\ -u_{a2,x}^e & -u_{a2,y}^e & -u_{a2,z}^e & 1 \\ \vdots & \vdots & \vdots & \vdots \\ -u_{am,x}^e & -u_{am,y}^e & -u_{am,z}^e & 1 \end{pmatrix}_{\mathbf{r}_{ea}^e = \hat{\mathbf{r}}_{ea}^{e-}}. \quad (9.144)$$

It is easier to obtain the curvilinear position, (L_a, λ_a, h_a), and the velocity in local navigation frame axes, $\mathbf{v}_{ea}^n$, from $\mathbf{r}_{ea}^e$ and $\mathbf{v}_{ea}^e$ using (2.113) and (2.73), where $\mathbf{C}_e^n$ is given by (2.150), than to calculate them directly.

The estimated receiver clock offset may be fed back to the ranging processor to correct the clock itself, either on every iteration or when it exceeds a certain threshold, such as 1 ms. The clock drift may also be fed back.

The MATLAB function, GNSS_LS_position_velocity, on the CD implements unweighted ECEF-frame position and velocity solutions using iterated least squares.

9.4.2 Filtered Navigation Solution

Most GNSS user equipment designed for real-time applications implements a filtered navigation solution. Unlike a single-epoch solution, a filtered solution makes use of information derived from previous measurements. Thus, the prior clock offset and drift solution is used to predict the current clock offset and drift while the prior position and velocity solution is used to predict the current position and velocity. The current pseudo-range and pseudo-range rate measurements are then used to correct the predicted navigation solution. A Kalman-filter-based estimation algorithm, described in Chapter 3, is used to maintain optimum weighting of the current set of pseudo-range and pseudo-range-rate measurements against the estimates obtained from previous measurements.

The filtered navigation solution has a number of advantages. The carrier-derived pseudo-range-rate measurements smooth out the code tracking noise on the position solution. This also reduces the impact of pseudo-range multipath errors on the navigation solution, particularly when moving. A navigation solution can be maintained for a limited period with only three satellites where the clock errors are well calibrated. This is known as *clock coasting*. Furthermore, a rough navigation solution can be maintained for a few seconds when all GNSS signals are blocked, such as in tunnels.

The choice of states to estimate is now discussed, followed by descriptions of the system and measurement models, and discussions of the measurement noise covariance and the handling of range biases, constellation changes, and ephemeris updates. The underlying extended Kalman filter algorithm is described in Sections 3.2.2 and 3.4.1. Commercial user equipment manufacturers implement sophisticated navigation filters, often with adaptive system noise and measurement noise models, the details of which are kept confidential.

9.4.2.1 State Selection

The GNSS navigation solution comprises the Kalman filter state vector, an example of total-state estimation. The appropriate states to estimate depend on the application [78]. The position must always be estimated, and whenever the user equipment is moving, the velocity must also be estimated. For most land and marine applications, when the dynamics are low, the acceleration may be modeled as system noise. When the dynamics are high, such as for fighter aircraft, guided weapons, motorsport, and space launch vehicles, the acceleration should be estimated as Kalman filter states.

However, in practice, an INS/GNSS integrated navigation system (Chapter 14) is typically used for these applications.

Any coordinate frame may be used for the navigation states. An ECI-frame implementation has the simplest system and measurement models. Estimating latitude, longitude, and height with Earth-referenced velocity in local navigation frame axes avoids the need to convert the navigation solution for output. A Cartesian ECEF-frame implementation is a common compromise.

The receiver clock offset and drift must always be estimated. For very-high-dynamic applications, the clock g-dependent error may also be modeled as described in Section 14.2.7. When multiple GNSS constellations are used, it is necessary to estimate the interconstellation timing biases if they cannot be determined from the navigation data messages. This is described in Section G.8.1 of Appendix G on the CD.

Strictly, the correlated range errors due to ephemeris errors and the residual satellite clock, ionosphere and troposphere errors (see Section 9.3) should also be estimated to ensure that the error covariance matrix, $\mathbf{P}$, is representative of the true system. These range biases typically have standard deviations of a few meters and correlations times of around 30 minutes. They are unobservable when signals from only four satellites are used and partially observable otherwise. Range bias estimation imposes a significantly higher processor load for only a small performance benefit. Therefore, it is rarely used and the range biases are typically modeled in an ad hoc manner as discussed in Section 9.4.2.5. Range biases for single-frequency users may be partially accounted for by estimating corrections to some of the ionosphere model coefficients.

When a dual-frequency receiver is used, pseudo-range measurements on each frequency may be input separately and the ionosphere propagation delays estimated as Kalman filter states, in which case smoothing of the ionosphere corrections is implicit. However, using combined-frequency ionosphere-corrected pseudo-ranges (Section 9.3.2) is more computationally efficient.

When carrier-smoothed pseudo-ranges (see Section 9.2.7) are used, they exhibit significant time-correlated tracking errors, which may be estimated as additional Kalman filter states. This enables the pseudo-range rate measurements to be omitted altogether as the range rate information can be inferred from the changes in the carrier-smoothed pseudo-ranges from epoch to epoch. This typically reduces the processor load where a vector measurement update is used and increases it where a sequential update is implemented.

For the navigation filter system and measurement models described here, eight states are estimated: the user antenna position and velocity and the receiver clock offset and drift. The ECI-frame implementation is described first, followed by discussions of the ECEF and local-navigation-frame variants. The state vectors are

$$\mathbf{x}^i = \begin{pmatrix} \mathbf{r}^i_{ia} \\ \mathbf{v}^i_{ia} \\ \delta\rho^a_c \\ \delta\dot{\rho}^a_c \end{pmatrix}, \quad \mathbf{x}^e = \begin{pmatrix} \mathbf{r}^e_{ea} \\ \mathbf{v}^e_{ea} \\ \delta\rho^a_c \\ \delta\dot{\rho}^a_c \end{pmatrix}, \quad \mathbf{x}^n = \begin{pmatrix} L_a \\ \lambda_a \\ h_a \\ \mathbf{v}^e_{ea} \\ \delta\rho^a_c \\ \delta\dot{\rho}^a_c \end{pmatrix}, \quad (9.145)$$

9.4.2.2 System Model

The Kalman filter system model (Section 3.2.3) describes how the states and their uncertainties are propagated forward in time to account for the user motion and receiver clock dynamics between successive measurements from the GNSS ranging processor. It also maintains a rough navigation solution during signal outages.

The system model for GNSS navigation in an ECI frame is simple. From (2.67), the time derivative of the position is the velocity, while the time derivative of the clock offset is the clock drift. The velocity and clock drift are not functions of any of the Kalman filter states, so the expectations of their time derivatives are zero. The state dynamics are thus

$$\dot{\mathbf{r}}_{ia}^i = \mathbf{v}_{ia}^i, \qquad \frac{\partial}{\partial t}\delta\rho_c^a = \delta\dot{\rho}_c^a$$
$$\mathrm{E}(\mathbf{v}_{ia}^i) = 0, \qquad \mathrm{E}\left(\frac{\partial}{\partial t}\delta\dot{\rho}_c^a\right) = 0 \qquad (9.146)$$

Substituting this into (3.26) gives the system matrix:

$$\mathbf{F}^i = \begin{pmatrix} 0 & 0 & 0 & 1 & 0 & 0 & 0 & 0 \\ 0 & 0 & 0 & 0 & 1 & 0 & 0 & 0 \\ 0 & 0 & 0 & 0 & 0 & 1 & 0 & 0 \\ 0 & 0 & 0 & 0 & 0 & 0 & 0 & 0 \\ 0 & 0 & 0 & 0 & 0 & 0 & 0 & 0 \\ 0 & 0 & 0 & 0 & 0 & 0 & 0 & 0 \\ 0 & 0 & 0 & 0 & 0 & 0 & 0 & 1 \\ 0 & 0 & 0 & 0 & 0 & 0 & 0 & 0 \end{pmatrix}. \qquad (9.147)$$

To save space, the Kalman filter matrices may be expressed in terms of submatrices corresponding to the vector subcomponents of the state vector (i.e., position, velocity, clock offset, and clock drift). Thus,

$$\mathbf{F}^i = \begin{pmatrix} \mathbf{0}_3 & \mathbf{I}_3 & \mathbf{0}_{3,1} & \mathbf{0}_{3,1} \\ \mathbf{0}_3 & \mathbf{0}_3 & \mathbf{0}_{3,1} & \mathbf{0}_{3,1} \\ \mathbf{0}_{1,3} & \mathbf{0}_{1,3} & 0 & 1 \\ \mathbf{0}_{1,3} & \mathbf{0}_{1,3} & 0 & 0 \end{pmatrix}, \qquad (9.148)$$

where $\mathbf{I}_n$ is the $n \times n$ identity matrix, $\mathbf{0}_n$ is the $n \times n$ null matrix, and $\mathbf{0}_{n,m}$ is the $n \times m$ null matrix.

The state dynamics for an ECEF-frame implementation are the same, so $\mathbf{F}^e = \mathbf{F}^i$. When a local-navigation-frame implementation with curvilinear position is used, the time derivative of the position is given by (2.111). Strictly, this violates the linearity assumption of the Kalman filter system model. However, the denominators may be treated as constant over the state propagation interval, so

$$\mathbf{F}^n \approx \begin{pmatrix} \mathbf{0}_3 & \mathbf{F}^n_{12} & \mathbf{0}_{3,1} & \mathbf{0}_{3,1} \\ \mathbf{0}_3 & \mathbf{0}_3 & \mathbf{0}_{3,1} & \mathbf{0}_{3,1} \\ \mathbf{0}_{1,3} & \mathbf{0}_{1,3} & 0 & 1 \\ \mathbf{0}_{1,3} & \mathbf{0}_{1,3} & 0 & 0 \end{pmatrix},$$

$$\mathbf{F}^n_{12} = \begin{pmatrix} 1/[R_N(\hat{L}_a) + \hat{h}_a] & 0 & 0 \\ 0 & 1/[(R_E(\hat{L}_a) + \hat{h}_a)\cos\hat{L}_a] & 0 \\ 0 & 0 & -1 \end{pmatrix}$$

(9.149)

where R_N and R_E are given by (2.105) and (2.106).

The higher order terms in (3.34) are zero for all three implementations, so the transition matrix is simply

$$\Phi_{k-1} = \mathbf{I}_8 + \mathbf{F}_{k-1}\tau_s, \qquad (9.150)$$

where τ_s is the state propagation interval.

The main sources of increased uncertainty of the state estimates are changes in velocity due to user motion and the random walk of the receiver clock drift. There is also some additional phase noise on the clock offset. From (3.43), the system noise covariance is obtained by integrating the power spectral densities of these noise sources over the state propagation interval, accounting for the deterministic system model. Thus,

$$\begin{aligned}\mathbf{Q}^\gamma_{k-1} &= \int_0^{\tau_s} \exp(\mathbf{F}^\gamma_{k-1}t') \begin{pmatrix} \mathbf{0}_3 & \mathbf{0}_3 & \mathbf{0}_{3,1} & \mathbf{0}_{3,1} \\ \mathbf{0}_3 & \mathbf{S}^\gamma_a & \mathbf{0}_{3,1} & \mathbf{0}_{3,1} \\ \mathbf{0}_{1,3} & \mathbf{0}_{1,3} & S^a_{c\phi} & 0 \\ \mathbf{0}_{1,3} & \mathbf{0}_{1,3} & 0 & S^a_{cf} \end{pmatrix} \exp(\mathbf{F}^{\gamma\,\mathrm{T}}_{k-1}t')\,dt' \\ &= \int_0^{\tau_s} (\mathbf{I}_8 + \mathbf{F}^\gamma_{k-1}t') \begin{pmatrix} \mathbf{0}_3 & \mathbf{0}_3 & \mathbf{0}_{3,1} & \mathbf{0}_{3,1} \\ \mathbf{0}_3 & \mathbf{S}^\gamma_a & \mathbf{0}_{3,1} & \mathbf{0}_{3,1} \\ \mathbf{0}_{1,3} & \mathbf{0}_{1,3} & S^a_{c\phi} & 0 \\ \mathbf{0}_{1,3} & \mathbf{0}_{1,3} & 0 & S^a_{cf} \end{pmatrix} (\mathbf{I}_8 + \mathbf{F}^{\gamma\,\mathrm{T}}_{k-1}t')\,dt'\end{aligned} \quad \gamma \in i,e,n, \quad (9.151)$$

where S_a^γ is the acceleration PSD matrix resolved about the axes of frame γ, S_{cf}^a is the receiver clock frequency-drift PSD, and $S_{c\phi}^a$ is the receiver clock phase-drift PSD. Assuming the PSDs are constant and substituting in (9.148) and (9.149),

$$Q_{k-1}^\gamma = \int_0^{\tau_s} \begin{pmatrix} I_3 & I_3 t' & 0_{3,1} & 0_{3,1} \\ 0_3 & I_3 & 0_{3,1} & 0_{3,1} \\ 0_{1,3} & 0_{1,3} & 1 & t' \\ 0_{1,3} & 0_{1,3} & 0 & 1 \end{pmatrix} \begin{pmatrix} 0_3 & 0_3 & 0_{3,1} & 0_{3,1} \\ 0_3 & S_a^\gamma & 0_{3,1} & 0_{3,1} \\ 0_{1,3} & 0_{1,3} & S_{c\phi}^a & 0 \\ 0_{1,3} & 0_{1,3} & 0 & S_{cf}^a \end{pmatrix} \begin{pmatrix} I_3 & 0_3 & 0_{3,1} & 0_{3,1} \\ I_3 t' & I_3 & 0_{3,1} & 0_{3,1} \\ 0_{1,3} & 0_{1,3} & 1 & 0 \\ 0_{1,3} & 0_{1,3} & t' & 1 \end{pmatrix} dt',$$

$$= \begin{pmatrix} \tfrac{1}{3} S_a^\gamma \tau_s^3 & \tfrac{1}{2} S_a^\gamma \tau_s^2 & 0_{3,1} & 0_{3,1} \\ \tfrac{1}{2} S_a^\gamma \tau_s^2 & S_a^\gamma \tau_s & 0_{3,1} & 0_{3,1} \\ 0_{1,3} & 0_{1,3} & S_{c\phi}^a \tau_s + \tfrac{1}{3} S_{cf}^a \tau_s^3 & \tfrac{1}{2} S_{cf}^a \tau_s^2 \\ 0_{1,3} & 0_{1,3} & \tfrac{1}{2} S_{cf}^a \tau_s^2 & S_{cf}^a \tau_s \end{pmatrix} \qquad \gamma \in i,e$$

(9.152)

and

$$Q_{k-1}^n = \int_0^{\tau_s} \begin{pmatrix} I_3 & F_{12}^n t' & 0_{3,1} & 0_{3,1} \\ 0_3 & I_3 & 0_{3,1} & 0_{3,1} \\ 0_{1,3} & 0_{1,3} & 1 & t' \\ 0_{1,3} & 0_{1,3} & 0 & 1 \end{pmatrix} \begin{pmatrix} 0_3 & 0_3 & 0_{3,1} & 0_{3,1} \\ 0_3 & S_a^n & 0_{3,1} & 0_{3,1} \\ 0_{1,3} & 0_{1,3} & S_{c\phi}^a & 0 \\ 0_{1,3} & 0_{1,3} & 0 & S_{cf}^a \end{pmatrix} \begin{pmatrix} I_3 & 0_3 & 0_{3,1} & 0_{3,1} \\ F_{12}^{nT} t' & I_3 & 0_{3,1} & 0_{3,1} \\ 0_{1,3} & 0_{1,3} & 1 & 0 \\ 0_{1,3} & 0_{1,3} & t' & 1 \end{pmatrix} dt'$$

$$= \begin{pmatrix} \tfrac{1}{3} F_{12}^n S_a^n F_{12}^{nT} \tau_s^3 & \tfrac{1}{2} F_{12}^n S_a^n \tau_s^2 & 0_{3,1} & 0_{3,1} \\ \tfrac{1}{2} S_a^n F_{12}^{nT} \tau_s^2 & S_a^n \tau_s & 0_{3,1} & 0_{3,1} \\ 0_{1,3} & 0_{1,3} & S_{c\phi}^a \tau_s + \tfrac{1}{3} S_{cf}^a \tau_s^3 & \tfrac{1}{2} S_{cf}^a \tau_s^2 \\ 0_{1,3} & 0_{1,3} & \tfrac{1}{2} S_{cf}^a \tau_s^2 & S_{cf}^a \tau_s \end{pmatrix}.$$

(9.153)

For small propagation intervals, depending on the dynamics, this may be approximated to

$$Q_{k-1}^\gamma \approx Q_{k-1}'^\gamma = \begin{pmatrix} 0_{3,3} & 0_{3,3} & 0_{3,1} & 0_{3,1} \\ 0_{3,3} & S_a^\gamma \tau_s & 0_{3,1} & 0_{3,1} \\ 0_{1,3} & 0_{1,3} & S_{c\phi}^a \tau_s & 0 \\ 0_{1,3} & 0_{1,3} & 0 & S_{cf}^a \tau_s \end{pmatrix} \qquad \gamma \in i,e,n, \qquad (9.154)$$

which may be used in conjunction with (3.46). However, for propagation intervals of 1 second and longer, the exact version, given by (9.152) or (9.153), is recommended.

In a Kalman filter, the system noise sources are assumed to be white. However, the real velocity and clock behavior is much more complex, so the system noise covariance model must overbound the true behavior to maintain a stable filter.

The acceleration PSD matrix may be expressed as

$$\mathbf{S}_a^i = \mathbf{C}_n^i \begin{pmatrix} S_{aH} & 0 & 0 \\ 0 & S_{aH} & 0 \\ 0 & 0 & S_{aV} \end{pmatrix} \mathbf{C}_i^n, \quad \mathbf{S}_a^e = \mathbf{C}_n^e \begin{pmatrix} S_{aH} & 0 & 0 \\ 0 & S_{aH} & 0 \\ 0 & 0 & S_{aV} \end{pmatrix} \mathbf{C}_e^n, \quad \mathbf{S}_a^n = \begin{pmatrix} S_{aH} & 0 & 0 \\ 0 & S_{aH} & 0 \\ 0 & 0 & S_{aV} \end{pmatrix}, \quad (9.155)$$

where $\mathbf{C}_n^i$ and $\mathbf{C}_n^e$ are, respectively, given by (2.154) and (2.150) and S_{aH} and S_{aV} are, respectively, the horizontal and vertical acceleration PSDs, modeled as

$$S_{aH} = \frac{\sigma^2 \left(v_{eb,N}^n(t+\tau_s) - v_{eb,N}^n(t) \right)}{\tau_s} = \frac{\sigma^2 \left(v_{eb,E}^n(t+\tau_s) - v_{eb,E}^n(t) \right)}{\tau_s}$$

$$S_{aV} = \frac{\sigma^2 \left(v_{eb,D}^n(t+\tau_s) - v_{eb,D}^n(t) \right)}{\tau_s}. \quad (9.156)$$

These depend on the dynamics of the application. Thus, the system noise is inherently context-dependent. Suitable values for S_{aH} are around 1 m² s⁻³ for a pedestrian or ship, 10 m² s⁻³ for a car, and 100 m² s⁻³ for a military aircraft. The vertical acceleration PSD is usually smaller. More sophisticated models may vary the PSDs as a function of speed and assume separate along-track and across-track values.

The clock PSDs are similarly modeled as

$$S_{cf}^a = \frac{\sigma^2 \left(\delta \dot{\rho}_c^a(t+\tau_s) - \delta \dot{\rho}_c^a(t) \right)}{\tau_s} \qquad S_{c\phi}^a = \frac{\sigma^2 \left(\delta \rho_c^a(t+\tau_s) - \delta \rho_c^a(t) - \delta \dot{\rho}_c^a(t) \tau_s \right)}{\tau_s}. \quad (9.157)$$

Typical values for a TCXO are $S_{cf}^a \approx 0.04$ m² s⁻³ and $S_{c\phi}^a \approx 0.01$ m² s⁻¹ [79].

For applications where velocity is not estimated, system noise based on the velocity PSD must be modeled on the position states. Even where the user is stationary, a small system noise should be modeled to keep the Kalman filter receptive to new measurements.

As GNSS navigation is a total-state implementation of the Kalman filter, a nonzero initialization of the state estimates is required. The position and clock offset may be initialized using a single-epoch navigation solution (Section 9.4.1). The same approach may be used for the velocity and clock drift. However, for many applications, the velocity may be initialized to that of the Earth at the initial position and the clock drift estimate initialized at zero.

The initial values of the error covariance matrix, **P**, must reflect the precision of the initialization process. Thus, if the clock drift state is initialized at zero, its initial uncertainty must match the standard deviation of the actual receiver clock drift.

The MATLAB function, GNSS_KF_Epoch, on the CD implements a single EKF cycle including the ECEF-frame version of the system model described in this section.

9.4.2.3 Measurement Model

The measurement model (Section 3.2.4) of a GNSS navigation filter updates the navigation solution using the measurements from the ranging processor and is analogous to the single-epoch solution described in Section 9.4.1. The measurement vector comprises the pseudo-ranges and pseudo-range rates output by the navigation processor, Thus, for m satellites tracked,

$$\mathbf{z}_G = \left\{ \tilde{\rho}_{a,C}^1, \ \tilde{\rho}_{a,C}^2, \ \cdots \ \tilde{\rho}_{a,C}^m, \ \middle| \ \tilde{\dot{\rho}}_{a,C}^1, \ \tilde{\dot{\rho}}_{a,C}^2, \ \cdots \ \tilde{\dot{\rho}}_{a,C}^m \right\}, \quad (9.158)$$

where the subscript G denotes a GNSS measurement. Note that using carrier-smoothed pseudo-ranges does not bring significant benefits over using unsmoothed pseudo-ranges together with pseudo-range rates because the Kalman filter performs the same smoothing.

The pseudo-ranges and pseudo-range rates, modeled by (9.123) to (9.129), are not linear functions of the states estimated. Therefore, an extended Kalman filter measurement model (Section 3.4.1) must be used. The measurement innovation vector is given by

$$\delta \mathbf{z}_{G,k}^- = \mathbf{z}_{G,k} - \mathbf{h}_G(\hat{\mathbf{x}}_k^-), \quad (9.159)$$

where

$$\mathbf{h}_G(\hat{\mathbf{x}}_k^-) = \left(\hat{\rho}_{a,C}^{1-}, \ \hat{\rho}_{a,C}^{2-}, \ \cdots \ \hat{\rho}_{a,C}^{m-}, \ \middle| \ \hat{\dot{\rho}}_{a,C}^{1-}, \ \hat{\dot{\rho}}_{a,C}^{2-}, \ \cdots \ \hat{\dot{\rho}}_{a,C}^{m-} \right)_k. \quad (9.160)$$

The predicted pseudo-ranges and pseudo-range rates are the same as in the single-epoch solution except that the predicted user position and velocity and receiver clock offset and drift are replaced by the Kalman filter estimates, propagated forward using the system model. Thus, in an ECI-frame implementation,

$$\begin{aligned}
\hat{\rho}_{a,C,k}^{j-} &= \sqrt{\left[\hat{\mathbf{r}}_{ij}^i(\tilde{t}_{st,a,k}^j) - \hat{\mathbf{r}}_{ia,k}^{i-}\right]^T \left[\hat{\mathbf{r}}_{ij}^i(\tilde{t}_{st,a,k}^j) - \hat{\mathbf{r}}_{ia,k}^{i-}\right]} + \delta\hat{\rho}_{c,k}^{a-} \\
\hat{\dot{\rho}}_{a,C,k}^{j-} &= \hat{\mathbf{u}}_{aj,k}^{i-\,T}\left[\hat{\mathbf{v}}_{ij}^i(\tilde{t}_{st,a,k}^j) - \hat{\mathbf{v}}_{ia,k}^{i-}\right] + \delta\hat{\dot{\rho}}_{c,k}^{a-}
\end{aligned}, \quad (9.161)$$

where j denotes the combination of a satellite, s, and signal, l, from that satellite, and the line-of-sight unit vector is obtained from (8.40) using $\hat{\mathbf{r}}_{ia,k}^{i-}$. From (3.90), the measurement matrix is

$$\mathbf{H}_{G,k}^{i} = \left(\begin{array}{ccccccccc} \frac{\partial \rho_a^1}{\partial x_{ia}^i} & \frac{\partial \rho_a^1}{\partial y_{ia}^i} & \frac{\partial \rho_a^1}{\partial z_{ia}^i} & 0 & 0 & 0 & \frac{\partial \rho_a^1}{\partial \rho_c^a} & 0 \\ \frac{\partial \rho_a^2}{\partial x_{ia}^i} & \frac{\partial \rho_a^2}{\partial y_{ia}^i} & \frac{\partial \rho_a^2}{\partial z_{ia}^i} & 0 & 0 & 0 & \frac{\partial \rho_a^2}{\partial \rho_c^a} & 0 \\ \vdots & \vdots & \vdots & \vdots & \vdots & \vdots & \vdots & \vdots \\ \frac{\partial \rho_a^m}{\partial x_{ia}^i} & \frac{\partial \rho_a^m}{\partial y_{ia}^i} & \frac{\partial \rho_a^m}{\partial z_{ia}^i} & 0 & 0 & 0 & \frac{\partial \rho_a^m}{\partial \rho_c^a} & 0 \\ \hline \frac{\partial \dot{\rho}_a^1}{\partial x_{ia}^i} & \frac{\partial \dot{\rho}_a^1}{\partial y_{ia}^i} & \frac{\partial \dot{\rho}_a^1}{\partial z_{ia}^i} & \frac{\partial \dot{\rho}_a^1}{\partial v_{ia,x}^i} & \frac{\partial \dot{\rho}_a^1}{\partial v_{ia,y}^i} & \frac{\partial \dot{\rho}_a^1}{\partial v_{ia,z}^i} & 0 & \frac{\partial \dot{\rho}_a^1}{\partial \dot{\rho}_c^a} \\ \frac{\partial \dot{\rho}_a^2}{\partial x_{ia}^i} & \frac{\partial \dot{\rho}_a^2}{\partial y_{ia}^i} & \frac{\partial \dot{\rho}_a^2}{\partial y_{ia}^i} & \frac{\partial \dot{\rho}_a^2}{\partial v_{ia,x}^i} & \frac{\partial \dot{\rho}_a^2}{\partial v_{ia,y}^i} & \frac{\partial \dot{\rho}_a^2}{\partial v_{ia,z}^i} & 0 & \frac{\partial \dot{\rho}_a^2}{\partial \dot{\rho}_c^a} \\ \vdots & \vdots & \vdots & \vdots & \vdots & \vdots & \vdots & \vdots \\ \frac{\partial \dot{\rho}_a^m}{\partial x_{ia}^i} & \frac{\partial \dot{\rho}_a^m}{\partial y_{ia}^i} & \frac{\partial \dot{\rho}_a^m}{\partial y_{ia}^i} & \frac{\partial \dot{\rho}_a^m}{\partial v_{ia,x}^i} & \frac{\partial \dot{\rho}_a^m}{\partial v_{ia,y}^i} & \frac{\partial \dot{\rho}_a^m}{\partial v_{ia,z}^i} & 0 & \frac{\partial \dot{\rho}_a^m}{\partial \dot{\rho}_c^a} \end{array} \right)_{\mathbf{x} = \hat{\mathbf{x}}_k^-}, \quad (9.162)$$

noting that the pseudo-ranges are not functions of the user velocity or clock drift, while the pseudo-range rates are not functions of the clock offset. The dependence of the pseudo-range rates on position is weak with a 1-m position error having a similar impact to a ~5×10⁻⁵ m s⁻¹ velocity error, so the $\partial \dot{\rho}/\partial \mathbf{r}$ terms are commonly neglected. Thus, from (9.123) and (9.126)

$$\mathbf{H}_{G,k}^{i} \approx \left(\begin{array}{cccccccc} -u_{a1,x}^i & -u_{a1,y}^i & -u_{a1,z}^i & 0 & 0 & 0 & 1 & 0 \\ -u_{a2,x}^i & -u_{a2,y}^i & -u_{a2,z}^i & 0 & 0 & 0 & 1 & 0 \\ \vdots & \vdots & \vdots & \vdots & \vdots & \vdots & \vdots & \vdots \\ -u_{am,x}^i & -u_{am,y}^i & -u_{am,z}^i & 0 & 0 & 0 & 1 & 0 \\ \hline 0 & 0 & 0 & -u_{a1,x}^i & -u_{a1,y}^i & -u_{a1,z}^i & 0 & 1 \\ 0 & 0 & 0 & -u_{a2,x}^i & -u_{a2,y}^i & -u_{a2,z}^i & 0 & 1 \\ \vdots & \vdots & \vdots & \vdots & \vdots & \vdots & \vdots & \vdots \\ 0 & 0 & 0 & -u_{am,x}^i & -u_{am,y}^i & -u_{am,z}^i & 0 & 1 \end{array} \right)_{\mathbf{x} = \hat{\mathbf{x}}_k^-}, \quad (9.163)$$

noting that the components are the same as those of the measurement matrix for the single-epoch least-squares solution.

For a Cartesian ECEF-frame implementation, the predicted pseudo-ranges and pseudo-range rates are

$$\begin{aligned} \hat{\rho}_{a,C,k}^{j-} &= \sqrt{\left[\hat{\mathbf{r}}_{ej}^e\left(\tilde{t}_{st,a,k}^j\right) - \hat{\mathbf{r}}_{ea,k}^{e-} \right]^T \left[\hat{\mathbf{r}}_{ej}^e\left(\tilde{t}_{st,a,k}^j\right) - \hat{\mathbf{r}}_{ea,k}^{e-} \right]} + \delta \hat{\rho}_{c,k}^{a-} + \delta \rho_{ie}^j \\ \hat{\dot{\rho}}_{a,C,k}^{j-} &= \hat{\mathbf{u}}_{aj,k}^{e-\,T} \left[\hat{\mathbf{v}}_{ej}^e\left(\tilde{t}_{st,a,k}^j\right) - \hat{\mathbf{v}}_{ea,k}^{e-} \right] + \delta \hat{\dot{\rho}}_{c,k}^{a-} + \delta \rho_{ie}^j \end{aligned} \quad (9.164)$$

9.4 Navigation Processor

or

$$\hat{\rho}_{a,C,k}^{j-} = \sqrt{\left[\mathbf{C}_e^I\left(\tilde{t}_{st,a,k}^j\right)\hat{\mathbf{r}}_{ej}^e\left(\tilde{t}_{st,a,k}^j\right) - \hat{\mathbf{r}}_{ea,k}^{e-}\right]^T \left[\mathbf{C}_e^I\left(\tilde{t}_{st,a,k}^j\right)\hat{\mathbf{r}}_{ej}^e\left(\tilde{t}_{st,a,k}^j\right) - \hat{\mathbf{r}}_{ea,k}^{e-}\right]} + \delta\hat{\rho}_{c,k}^{a-},$$

$$\hat{\dot{\rho}}_{a,C,k}^{j-} = \hat{\mathbf{u}}_{as,j}^{e-T}\left[\mathbf{C}_e^I\left(\tilde{t}_{st,a,k}^j\right)\left(\hat{\mathbf{v}}_{ej}^e\left(\tilde{t}_{st,a,k}^j\right) + \mathbf{\Omega}_{ie}^e\hat{\mathbf{r}}_{ej}^e\left(\tilde{t}_{st,a,k}^j\right)\right) - \left(\hat{\mathbf{v}}_{ea,k}^{e-} + \mathbf{\Omega}_{ie}^e\hat{\mathbf{r}}_{ea,k}^{e-}\right)\right] + \delta\hat{\dot{\rho}}_{c,k}^{a-}$$

(9.165)

while the measurement matrix, $\mathbf{H}_{G,k}^e$, is as $\mathbf{H}_{G,k}^i$ with $\mathbf{u}_{aj}^e$ substituted for $\mathbf{u}_{aj}^i$. This is implemented within the MATLAB function, GNSS_KF_Epoch, on the CD.

For a local-navigation-frame implementation, it is easiest to compute the predicted pseudo-ranges and pseudo-range rates as above using the Cartesian position, calculated using (2.112), and ECEF velocity, calculated using (2.152). The measurement matrix is

$$\mathbf{H}_{G,k}^n \approx \left(\begin{array}{cccccccc} h_L u_{a1,N}^n & h_\lambda u_{a1,E}^n & u_{a1,D}^n & 0 & 0 & 0 & 1 & 0 \\ h_L u_{a2,N}^n & h_\lambda u_{a2,E}^n & u_{a2,D}^n & 0 & 0 & 0 & 1 & 0 \\ \vdots & \vdots & \vdots & \vdots & \vdots & \vdots & \vdots & \vdots \\ h_L u_{am,N}^n & h_\lambda u_{am,E}^n & u_{am,D}^n & 0 & 0 & 0 & 1 & 0 \\ \hline 0 & 0 & 0 & -u_{a1,N}^n & -u_{a1,E}^n & -u_{a1,D}^n & 0 & 1 \\ 0 & 0 & 0 & -u_{a2,N}^n & -u_{a2,E}^n & -u_{a2,D}^n & 0 & 1 \\ \vdots & \vdots & \vdots & \vdots & \vdots & \vdots & \vdots & \vdots \\ 0 & 0 & 0 & -u_{am,N}^n & -u_{am,E}^n & -u_{am,D}^n & 0 & 1 \end{array}\right)_{\mathbf{x}=\hat{\mathbf{x}}_k^-},$$

(9.166)

where

$$h_L = -\left[R_N\left(\hat{L}_a\right) + \hat{h}_a\right], \qquad h_\lambda = -\left[R_E\left(\hat{L}_a\right) + \hat{h}_a\right]\cos\hat{L}_a,$$

(9.167)

If closed-loop correction of the receiver clock offset is implemented, the state estimate must be zeroed within the Kalman filter after each time it is fed back to the ranging processor. This may occur every iteration, periodically, or when a certain threshold, such as 1 ms, is exceeded. The same applies to clock drift feedback.

If the navigation filter is implemented outside the GNSS user equipment and 1-ms corrections are applied within, it will be necessary to detect and respond to the discontinuities in the pseudo-range measurements at the start of the measurement update process using the algorithm shown in Figure 9.31.

9.4.2.4 Measurement Noise Covariance

The measurement noise covariance matrix, $\mathbf{R}_G$, models the noise-like errors on the pseudo-range and pseudo-range-rate measurements, such as tracking errors, multipath variations, and satellite clock noise. In many GNSS navigation filters, $\mathbf{R}_G$ is

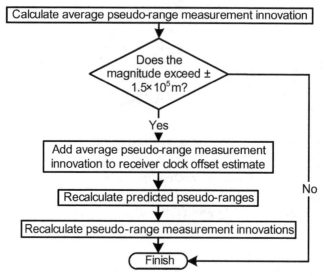

Figure 9.31 Clock jump detection and correction algorithm.

modeled as diagonal and constant, but there can be benefits in varying it as a function of $\tilde{c}/\tilde{n}_0$ and/or the level of dynamics, where known.

The noise-like errors on the pseudo-range and pseudo-range-rate measurements are generally uncorrelated with each other as the former are derived from code tracking and the latter from carrier tracking. An exception is if carrier-smoothed pseudo-range measurements, (9.74) or (9.75), are used, in which case correlation must be modeled.

In theory, the measurement noise covariance should not account for the bias-like errors due to the ionosphere, troposphere, and satellite clock. In principle, these errors should be estimated as states, but this is not usually practical (see Section 9.4.2.1). Therefore, to enable those measurements with lower range bias standard deviations to receive a stronger weighting in the navigation solution, $\mathbf{R}_G$ may also incorporate elevation-angle-dependent terms.

A measurement noise covariance model that accounts for elevation, c/n_0, and range acceleration is

$$\mathbf{R}_G = \begin{pmatrix} \sigma_{\rho 1}^2 & 0 & \cdots & 0 & 0 & 0 & \cdots & 0 \\ 0 & \sigma_{\rho 2}^2 & \cdots & 0 & 0 & 0 & \cdots & 0 \\ \vdots & \vdots & \ddots & \vdots & \vdots & \vdots & \ddots & \vdots \\ 0 & 0 & \cdots & \sigma_{\rho m}^2 & 0 & 0 & \cdots & 0 \\ \hline 0 & 0 & \cdots & 0 & \sigma_{r1}^2 & 0 & \cdots & 0 \\ 0 & 0 & \cdots & 0 & 0 & \sigma_{r2}^2 & \cdots & 0 \\ \vdots & \vdots & \ddots & \vdots & \vdots & \vdots & \ddots & \vdots \\ 0 & 0 & \cdots & 0 & 0 & 0 & \cdots & \sigma_{rm}^2 \end{pmatrix},$$

$$\sigma_{\rho j}^2 = \frac{1}{\sin^2\left(\theta_{nu}^{aj}\right)} \left(\sigma_{\rho Z}^2 + \frac{\sigma_{\rho c}^2}{(c/n_0)_j} + \sigma_{\rho a}^2 \ddot{r}_{aj}^2 \right),$$

$$\sigma_{rj}^2 = \frac{1}{\sin^2\left(\theta_{nu}^{aj}\right)} \left(\sigma_{rZ}^2 + \frac{\sigma_{rc}^2}{(c/n_0)_j} + \sigma_{ra}^2 \ddot{r}_{aj}^2 \right),$$

(9.168)

where the coefficients σ_{pZ}, σ_{pc}, σ_{pa}, σ_{rZ}, σ_{rc}, and σ_{ra} should be determined empirically, and Section G.4.1 of Appendix G on the CD shows how to calculate the range acceleration, $\ddot{r}_{aj}$.

An assumption of the EKF is that the measurement noise is white. However, in practice, it is not. When measurement updates are performed at a faster rate than about 2 Hz for the pseudo-ranges or 10 Hz for the pseudo-range rates, it may be necessary to account for the time correlation of the tracking noise, depending on the design of the tracking loops. The correlation time of the multipath and NLOS errors, which is often longer than that of the tracking noise, must also be considered. Time-correlated measurement noise may be accounted for by increasing $\mathbf{R}_G$ (see Section 3.4.3). A suitable value for a component of the measurement noise covariance is thus the variance of the pseudo-range or pseudo-range rate error multiplied by the ratio of the error correlation time to the measurement update interval. Typical values for a 1-Hz update interval are $(1–5\text{m})^2$ for pseudo-range and $(0.1–1 \text{ m s}^{-1})^2$ for pseudo-range rate, with the larger values used under poorer GNSS reception conditions. Some experimentation will be required to determine the optimum values of the model coefficients for a particular application.

Carrier-smoothed pseudo-range measurements (see Section 9.2.7) exhibit much less noise than unsmoothed pseudo-ranges. However, that noise is time correlated over the smoothing interval. Thus, the weighting of these measurements should be the same as for their unsmoothed counterparts, requiring a similar measurement noise variance to be used. The exception is where the time-correlated tracking errors are estimated as additional Kalman filter states, in which case a much smaller measurement noise covariance should be used.

9.4.2.5 Range Biases, Constellation Changes, and Ephemeris Updates

When the correlated range errors due to residual ionosphere, troposphere, satellite clock, and ephemeris errors are not estimated by the Kalman filter, they will bias the position and clock offset estimates away from their true values. To account for this, an extra term should be added to the state uncertainty modeled by the Kalman filter. The corrected position and timing offset uncertainties are then

$$\begin{pmatrix} \sigma_x \\ \sigma_y \\ \sigma_z \\ \sigma_T \end{pmatrix} = \begin{pmatrix} \sqrt{P_{1,1} + \Delta\sigma_x^2} \\ \sqrt{P_{2,2} + \Delta\sigma_y^2} \\ \sqrt{P_{3,3} + \Delta\sigma_z^2} \\ \sqrt{P_{7,7} + \Delta\sigma_T^2} \end{pmatrix} \quad (9.169)$$

for ECI and Cartesian ECEF frame position, where $\Delta\sigma_x$, $\Delta\sigma_y$, and $\Delta\sigma_z$ are the position error standard deviations due to the correlated range errors resolved along the x, y, and z axes of an ECI or ECEF frame, and $\Delta\sigma_T$ is the corresponding clock offset standard deviation, expressed as a range. Approximate values may be obtained by multiplying the correlated range error standard deviation by the appropriate DOP (see Section 9.4.3). The off-diagonal elements of the position and clock error covariance may be similarly corrected using the off-diagonal elements of the cofactor matrix, defined in Section 9.4.3.

The corrected curvilinear position uncertainties are given by

$$\begin{pmatrix} \sigma_L \\ \sigma_\lambda \\ \sigma_h \\ \sigma_T \end{pmatrix} = \begin{matrix} \sqrt{P_{1,1} + \Delta\sigma_L^2} \\ \sqrt{P_{2,2} + \Delta\sigma_\lambda^2} \\ \sqrt{P_{3,3} + \Delta\sigma_D^2} \\ \sqrt{P_{7,7} + \Delta\sigma_T^2} \end{matrix}, \quad (9.170)$$

where

$$\Delta\sigma_L = \frac{\Delta\sigma_N}{R_N(L_a) + h_a}, \quad \Delta\sigma_\lambda = \frac{\Delta\sigma_E}{[R_E(L_a) + h_a]\cos L_a} \quad (9.171)$$

and $\Delta\sigma_N$, $\Delta\sigma_E$, and $\Delta\sigma_D$ are, respectively, the north, east, and vertical position error standard deviations due to the correlated range errors. The radial distance RMS error (see Section B.2.3 of Appendix B on the CD) is

$$r_D = \sqrt{\sigma_N^2 + \sigma_E^2} = \sqrt{[R_N(L_a) + h_a]^2 P_{1,1} + [R_E(L_a) + h_a]^2 \cos^2 L_a P_{2,2} + \Delta\sigma_N^2 + \Delta\sigma_E^2}. \quad (9.172)$$

When there is a change in the satellites tracked by the receiver, known as a constellation change, or there is an ephemeris update, the error in the navigation solution due to the correlated range errors will change. The Kalman filter will respond more quickly to this change if the position and clock-offset state uncertainties are boosted. When the range biases are estimated as states, the relevant state should instead be reset on a constellation change or ephemeris update.

For the ephemeris update at the day boundary (i.e., around 00:00 UTC), there is currently a discontinuity in GPS system time that corresponds to a meter-order range jump. This should be modeled by increasing the uncertainty of the clock offset state.

9.4.3 Signal Geometry and Navigation Solution Accuracy

The accuracy of a GNSS navigation solution depends not only on the accuracy of the ranging measurements, but also on the signal geometry. Signal geometry in two dimensions is discussed in Section 7.4.5; here this is extended to 3-D positioning.

The effect of signal geometry on the navigation solution is quantified using the dilution of precision (DOP) concept [50]. The uncertainty of each pseudo-range measurement, known as the user-equivalent range error (UERE), is σ_ρ. The DOP is then used to relate the uncertainty of various parts of the navigation solution to the pseudo-range uncertainty using

$$\begin{matrix} \sigma_N = D_N \sigma_\rho & \sigma_E = D_E \sigma_\rho & \sigma_D = D_V \sigma_\rho & \sigma_H = D_H \sigma_\rho \\ \sigma_x = D_x \sigma_\rho & \sigma_y = D_y \sigma_\rho & \sigma_z = D_z \sigma_\rho & \sigma_P = D_P \sigma_\rho \\ \sigma_T = D_T \sigma_\rho & \sigma_G = D_G \sigma_\rho & & \end{matrix}, \quad (9.173)$$

9.4 Navigation Processor

Table 9.3 Uncertainties and Corresponding Dilutions of Precision

Uncertainty	Dilution of Precision
σ_N, north position	D_N, north dilution of precision
σ_E, east position	D_E, east dilution of precision
σ_D, vertical position	D_V, vertical dilution of precision (VDOP)
σ_H, horizontal position	D_H, horizontal dilution of precision (HDOP)
σ_x, x-axis position	D_x, x-axis dilution of precision
σ_y, y-axis position	D_y, y-axis dilution of precision
σ_z, z-axis position	D_z, z-axis dilution of precision
σ_P, overall position	D_P, position dilution of precision (PDOP)
σ_T, receiver clock offset	D_T, time dilution of precision (TDOP)
σ_G, total position and clock	D_G, geometric dilution of precision (GDOP)

where the various uncertainties and their DOPs are defined in Table 9.3.

Consider a GNSS receiver tracking signals from m satellites, each with a pseudo-range error $\delta\rho_a^s$. Using the line-of-sight unit vectors, the vector of errors in the pseudo-ranges estimated from the navigation solution, $\delta\rho$, may be expressed in terms of the navigation solution position error, δr_{ea}^n, and residual receiver clock error, $\delta\delta\rho_c^a$:

$$\delta\rho = \begin{pmatrix} \delta\rho_a^1 \\ \delta\rho_a^2 \\ \vdots \\ \delta\rho_a^m \end{pmatrix} = \begin{pmatrix} -u_{a1,N}^n & -u_{a1,E}^n & -u_{a1,D}^n & 1 \\ -u_{a2,N}^n & -u_{a2,E}^n & -u_{a2,D}^n & 1 \\ \vdots & \vdots & \vdots & \vdots \\ -u_{am,N}^n & -u_{am,E}^n & -u_{am,D}^n & 1 \end{pmatrix} \begin{pmatrix} \delta r_{ea,N}^n \\ \delta r_{ea,E}^n \\ \delta r_{ea,D}^n \\ \delta\delta\rho_c^a \end{pmatrix} = \mathbf{H}_G^{nC}\left(\delta\mathbf{r}_{ea}^n, \delta\delta\rho_c^a\right), \quad (9.174)$$

where $\mathbf{H}_G^{nC}$ is the local-navigation-frame Cartesian measurement or geometry matrix for the single-epoch solution. Similarly,

$$\delta\rho = \begin{pmatrix} \delta\rho_a^1 \\ \delta\rho_a^2 \\ \vdots \\ \delta\rho_a^m \end{pmatrix} = \begin{pmatrix} -u_{a1,x}^\beta & -u_{a1,y}^\beta & -u_{a1,z}^\beta & 1 \\ -u_{a2,x}^\beta & -u_{a2,y}^\beta & -u_{a2,x}^\beta & 1 \\ \vdots & \vdots & \vdots & \vdots \\ -u_{am,x}^\beta & -u_{am,y}^\beta & -u_{am,z}^\beta & 1 \end{pmatrix} \begin{pmatrix} \delta r_{\beta a,x}^\beta \\ \delta r_{\beta a,y}^\beta \\ \delta r_{\beta a,z}^\beta \\ \delta\delta\rho_c^a \end{pmatrix} = \mathbf{H}_G^\beta\left(\delta\mathbf{r}_{\beta a}^\beta, \delta\delta\rho_c^a\right), \quad \beta \in i,e,$$

(9.175)

where $\mathbf{H}_G^i$ and $\mathbf{H}_G^e$ are the ECI-frame and ECEF-frame geometry matrices.

Squaring both sides of (9.174) and taking expectations,

$$E\left(\delta\rho\,\delta\rho^T\right) = \mathbf{H}_G^{nC} E\left[\left(\delta\mathbf{r}_{ea}^n, \delta\delta\rho_c^a\right)\left(\delta\mathbf{r}_{ea}^n, \delta\delta\rho_c^a\right)^T\right] \mathbf{H}_G^{nC\,T}. \quad (9.176)$$

The error covariance matrix of the Cartesian local-navigation-frame navigation solution is

$$\mathbf{P} = \mathrm{E}\left[\left(\delta\mathbf{r}_{ea}^n,\delta\delta\rho_c^a\right)\left(\delta\mathbf{r}_{ea}^n,\delta\delta\rho_c^a\right)^{\mathrm{T}}\right] = \begin{pmatrix} \sigma_N^2 & P_{N,E} & P_{N,D} & P_{N,T} \\ P_{E,N} & \sigma_E^2 & P_{E,D} & P_{E,T} \\ P_{D,N} & P_{D,E} & \sigma_D^2 & P_{D,T} \\ P_{T,N} & P_{T,E} & P_{T,D} & \sigma_T^2 \end{pmatrix}. \quad (9.177)$$

Assuming that the pseudo-range errors are independent and have the same uncertainties gives

$$\mathrm{E}\left(\delta\rho\delta\rho^{\mathrm{T}}\right) = \mathbf{I}_n \sigma_\rho^2, \quad (9.178)$$

noting that, in reality, this does not apply to the ionosphere and troposphere propagation errors. If the measurements are weighted within the navigation solution to account for the variation in pseudo-range error uncertainty, DOP provides a better estimate of positioning performance than if they are unweighted.

Substituting (9.177) and (9.178) into (9.176) and rearranging gives

$$\mathbf{P} = \mathbf{H}_G^{nC-1}\left(\mathbf{H}_G^{nC\mathrm{T}}\right)^{-1} \sigma_\rho^2 = \left(\mathbf{H}_G^{nC\mathrm{T}} \mathbf{H}_G^{nC}\right)^{-1} \sigma_\rho^2. \quad (9.179)$$

From (9.173) and (9.177), the DOPs are then defined in terms of the measurement matrix by

$$\mathbf{\Pi}^n = \begin{pmatrix} D_N^2 & \cdot & \cdot & \cdot \\ \cdot & D_E^2 & \cdot & \cdot \\ \cdot & \cdot & D_V^2 & \cdot \\ \cdot & \cdot & \cdot & D_T^2 \end{pmatrix} = \left(\mathbf{H}_G^{nC\mathrm{T}} \mathbf{H}_G^{nC}\right)^{-1}, \quad (9.180)$$

where $\mathbf{\Pi}^n$ is the local-navigation-frame cofactor matrix and

$$\begin{aligned} D_H &= \sqrt{D_N^2 + D_E^2} \\ D_P &= \sqrt{D_N^2 + D_E^2 + D_V^2} \\ D_G &= \sqrt{D_N^2 + D_E^2 + D_V^2 + D_T^2} = \sqrt{\mathrm{tr}\left[\left(\mathbf{H}_G^{nC\mathrm{T}} \mathbf{H}_G^{nC}\right)^{-1}\right]} \end{aligned}. \quad (9.181)$$

Similarly,

$$\mathbf{\Pi}^\gamma = \begin{pmatrix} D_x^2 & \cdot & \cdot & \cdot \\ \cdot & D_y^2 & \cdot & \cdot \\ \cdot & \cdot & D_z^2 & \cdot \\ \cdot & \cdot & \cdot & D_T^2 \end{pmatrix} = \left(\mathbf{H}_G^{\gamma\mathrm{T}} \mathbf{H}_G^\gamma\right)^{-1} \quad \gamma \in i,e, \quad (9.182)$$

9.4 Navigation Processor

where $\mathbf{\Pi}^i$ and $\mathbf{\Pi}^e$ are the ECI-frame and ECEF-frame cofactor matrices and

$$D_P = \sqrt{D_x^2 + D_y^2 + D_z^2}$$
$$D_G = \sqrt{D_x^2 + D_y^2 + D_z^2 + D_T^2} = \sqrt{\mathrm{tr}\left[\left(\mathbf{H}_G^{i\,\mathrm{T}}\mathbf{H}_G^i\right)^{-1}\right]} = \sqrt{\mathrm{tr}\left[\left(\mathbf{H}_G^{e\,\mathrm{T}}\mathbf{H}_G^e\right)^{-1}\right]}. \qquad (9.183)$$

From (9.174),

$$\mathbf{H}_G^{n\mathrm{C}\,\mathrm{T}}\mathbf{H}_G^{n\mathrm{C}} = \begin{pmatrix} g_{NN} & g_{NE} & g_{ND} & g_{NT} \\ g_{NE} & g_{EE} & g_{ED} & g_{ET} \\ g_{ND} & g_{ED} & g_{DD} & g_{DT} \\ g_{NT} & g_{ET} & g_{DT} & n \end{pmatrix}, \qquad (9.184)$$

where

$$g_{NN} = \sum_{j=1}^{m} u_{aj,N}^{n\,2} \quad g_{NT} = -\sum_{j=1}^{m} u_{aj,N}^{n} \quad g_{NE} = \sum_{j=1}^{m} u_{aj,N}^{n} u_{aj,E}^{n}$$

$$g_{EE} = \sum_{j=1}^{m} u_{aj,E}^{n\,2} \quad g_{ET} = -\sum_{j=1}^{m} u_{aj,E}^{n} \quad g_{ND} = \sum_{j=1}^{m} u_{aj,N}^{n} u_{aj,D}^{n}. \qquad (9.185)$$

$$g_{DD} = \sum_{j=1}^{m} u_{aj,D}^{n\,2} \quad g_{DT} = -\sum_{j=1}^{m} u_{aj,D}^{n} \quad g_{ED} = \sum_{j=1}^{m} u_{aj,E}^{n} u_{aj,D}^{n}$$

As $\mathbf{H}_G^{n\,\mathrm{T}}\mathbf{H}_G^n$ is symmetric about the diagonal, the matrix inversion is simplified. Matrix inversion techniques are discussed in Section A.4 of Appendix A on the CD. The other cofactor matrices are calculated in the same way, substituting x, y, and z for N, E, and D. The cofactor matrices transform as

$$\mathbf{\Pi}^{\beta} = \begin{pmatrix} \mathbf{C}_\alpha^\beta & \mathbf{0}_{3,1} \\ \mathbf{0}_{1,3} & 1 \end{pmatrix} \mathbf{\Pi}^\alpha \begin{pmatrix} \mathbf{C}_\beta^\alpha & \mathbf{0}_{3,1} \\ \mathbf{0}_{1,3} & 1 \end{pmatrix} \qquad (9.186)$$

so DOP information calculated in one frame may easily be transformed to another.

As discussed in Section 7.4.5, the position information along a given axis obtainable from a given ranging signal is maximized when the angle between that axis and the signal line of sight is minimized. However, as GNSS uses passive ranging, signals from opposite directions are required to separate position and timing information. Therefore, the horizontal GNSS positioning accuracy is optimized where signals from low-elevation satellites are available and the line-of-sight vectors are evenly distributed in azimuth. Vertical accuracy is optimized when signals from a range of different elevations, including high elevations, are available. However, because signals from negative-elevation satellites are normally blocked by the Earth, vertical accuracy is normally poorer than horizontal accuracy.

Figure 9.32 illustrates the optimal geometry for four GNSS satellites, while Figures 9.33 to 9.35 illustrate a number of poor-geometry cases. In Figure 9.33, the azimuths vary by only 60°, resulting in poor accuracy both along the direction the signals come from and vertically because it is difficult to separate position from time. This can occur where two perpendicular walls block most of the signals. In Figure 9.34, all of the signals are from high-elevation satellites, resulting in poor vertical accuracy because the height is difficult to separate from the time. This can occur in dense urban and mountainous areas. Finally, in Figure 9.35, signals are received from two opposing directions, resulting in poor horizontal accuracy perpendicular to this, but good separation of position and time. This geometry can occur in urban streets with tall buildings on either side and in deep valleys. Example 9.2 on the CD shows all of these DOP calculations and is editable using Microsoft Excel.

Very poor geometry can also occur coincidentally. Three LOS vectors will always define the surface of a cone. Any additional LOS vector that fits the surface of the cone can be expressed as a linear combination of the other three. Consequently, the rows of the measurement matrix, $\mathbf{H}$, are not linearly independent. To obtain a position and clock offset solution, at least four rows of $\mathbf{H}$ must be linearly independent. Otherwise, $\mathbf{H}^T\mathbf{H}$ has no inverse and some or all of the DOPs are infinite. These DOP

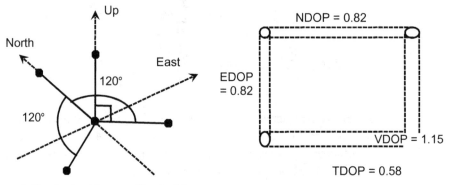

Figure 9.32 Optimal four-satellite GNSS geometry.

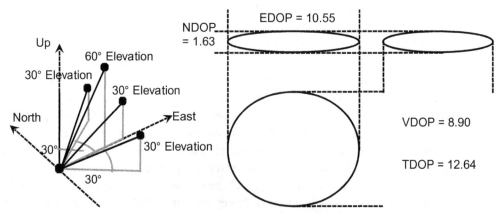

Figure 9.33 Poor GNSS geometry due to lack of azimuth variation.

9.4 Navigation Processor

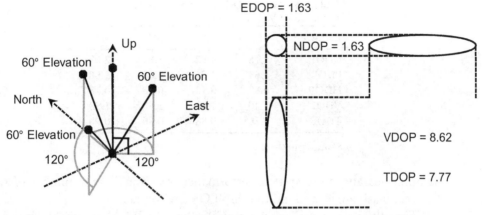

Figure 9.34 Poor GNSS geometry due to high elevations.

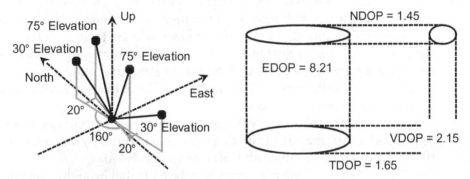

Figure 9.35 Poor GNSS geometry due to signal reception from opposing directions only.

singularities are common where signals from only four satellites are tracked, but are much rarer with more signals.

When a subset of GNSS signals is selected for position computation or signal tracking, it is important to select a combination with a good geometry to give good positioning performance.

Table 9.4 gives average DOPs of a nominal 24-satellite GPS constellation at a range of latitudes, assuming signals are tracked from all satellites in view. Note that the VDOP is much larger than the HDOP, particularly in polar regions, even though the HDOP accounts for two axes. Overall performance is best in equatorial regions, though the difference is not great.

Sections G.8.1 and G.8.4 of Appendix G on the CD, respectively, discuss the impact of interconstellation timing bias estimation and use of a chip-scale atomic clock on dilution of precision.

9.4.4 Position Error Budget

The error in a GNSS position solution is determined by the range errors and signal geometry. Every published GNSS error budget is different, making varying

Table 9.4 Average DOPs for a Nominal GPS Constellation and an All-in-View Receiver

Latitude	0°	30°	60°	90°
GDOP	1.78	1.92	1.84	2.09
PDOP	1.61	1.71	1.65	1.88
HDOP	0.80	0.93	0.88	0.75
VDOP	1.40	1.43	1.40	1.73
TDOP	0.76	0.88	0.80	0.90

Source: QinetiQ Ltd.

assumptions about the system performance, receiver design, number of satellites tracked, mask angle, and multipath/NLOS environment.

The values presented here assume GPS Block IIR/IIR-M satellite clock and ephemeris errors [51]. BPSK(1), $BOC_s(1,1)$, BPSK(10), and $BOC_s(10,5)$ signal modulations are considered. Tracking noise is calculated assuming $C/N_0 = 40$ dB-Hz, $\tau_a = 20$ ms, a dot-product power code discriminator is used, $B_{L\_CO} = 1$ Hz, and the receiver precorrelation bandwidth matches the transmission bandwidth. Three multipath environments are considered, each with a factor of 5 attenuation of the delayed signal and a uniform delay distribution of 0–2m for the short-range model, 0–20m for the medium-range model, and 1–200m for the long-range model. NLOS reception is neglected. Table 9.5 lists the standard deviations of the range error components, averaged across elevation angle.

Note that for positioning with respect to the Earth's surface, as opposed to the reference ellipsoid, there is an additional position uncertainty of 0.3m vertically and 0.05m horizontally due to Earth tides (see Section 2.4.4).

The average positioning accuracy may be estimated by multiplying the average range error standard deviation by the average DOP. Taking a weighted average of the DOP values in Table 9.4, assuming a 24-satellite GPS constellation with all satellites

Table 9.5 Contributions to the Average Range Error Standard Deviation

Source	Range Error Standard Deviation (m)
Residual satellite clock and ephemeris errors	0.5
Residual ionosphere error (single-frequency user)	4.0
Residual ionosphere error (dual-frequency user)	0.1
Residual troposphere error (assuming latitude- and season-dependent model)	0.2
Tracking noise for:	
BPSK(1) signal, $d = 0.1$	0.67
$BOC_s(1,1)$ signal, $d = 0.1$	0.39
BPSK(10) signal, $d = 1$	0.21
$BOC_s(10,5)$ signal	0.06
Short-range multipath error	0.1
Medium-range multipath error	0.94
Long-range multipath error for:	
BPSK(1) signal, $d = 0.1$	1.44
$BOC_s(1,1)$ signal, $d = 0.1$	1.33
BPSK(10) signal, $d = 1$	0.12
$BOC_s(10,5)$ signal [13]	0.23

Table 9.6 Single-Constellation Position Error Budget

			\multicolumn{3}{c}{Position Error Standard Deviation (m)}		
Frequencies	Multipath	Signal	Total	Horizontal (radial)	Vertical
Single	Short-range	BPSK(1)	6.8	3.6	5.8
		$BOC_s(1,1)$	6.8	3.6	5.8
		BPSK(10)	6.8	3.6	5.8
		$BOC_s(10,5)$	6.8	3.6	5.8
	Medium-range	BPSK(1)	7.0	3.7	6.0
		$BOC_s(1,1)$	6.9	3.7	5.9
		BPSK(10)	6.9	3.7	5.9
		$BOC_s(10,5)$	6.9	3.7	5.9
	Long-range	BPSK(1)	7.2	3.8	6.2
		$BOC_s(1,1)$	7.1	3.8	6.1
		BPSK(10)	6.8	3.6	5.7
		$BOC_s(10,5)$	6.8	3.6	5.7
Dual	Short-range	BPSK(1)	1.5	0.8	1.2
		$BOC_s(1,1)$	1.1	0.6	1.0
		BPSK(10)	1.0	0.5	0.8
		$BOC_s(10,5)$	1.0	0.5	0.8
	Medium-range	BPSK(1)	2.1	1.1	1.8
		$BOC_s(1,1)$	1.9	1.0	1.6
		BPSK(10)	1.8	1.0	1.6
		$BOC_s(10,5)$	1.8	1.0	1.6
	Long-range	BPSK(1)	2.8	1.5	2.4
		$BOC_s(1,1)$	2.5	1.3	2.1
		BPSK(10)	1.0	0.5	0.9
		$BOC_s(10,5)$	1.0	0.5	0.9

in view tracked, the PDOP is 1.67, HDOP is 0.88, and VDOP is 1.42. However, caution should be exercised in using this approach for individual positioning scenarios due to the great variation in all of the error sources.

Table 9.6 presents the position error budget for a single constellation. Divide these values by $\sqrt{2}$ to obtain approximate values for two constellations, by $\sqrt{3}$ for three constellations and by 2 for four constellations. These multiconstellation values will be overoptimistic because the impact of correlations between the ranging errors of different satellites becomes more significant as the number of satellites used increases.

Problems and exercises for this chapter are on the accompanying CD.

References

[1] Van Dierendonck, A. J., "GPS Receivers," in *Global Positioning System: Theory and Applications, Volume I*, B. W. Parkinson and J. J. Spilker, Jr., (eds.), Washington, D.C.: AIAA, 1996, pp. 329–407.

[2] Dorsey, A. J., et al., "GPS System Segments" in *Understanding GPS Principles and Applications*, 2nd ed., E. D. Kaplan and C. J. Hegarty, (eds.), Norwood, MA: Artech House, 2006, pp. 67–112.

[3] Misra, P., and P. Enge, *Global Positioning System Signals, Measurements, and Performance*, 2nd ed., Lincoln, MA: Ganga-Jamuna Press, 2006.

[4] Grewal, M. S., L. R. Weill, and A. P. Andrews, *Global Positioning Systems, Inertial Navigation, and Integration*, 2nd ed., New York: Wiley, 2007.
[5] Chen, X., et al., *Antennas for Global Navigation Satellite Systems*, New York: Wiley, 2012.
[6] Rama Rao, B., W. Kunysz, and K. McDonald, *GNSS Antennas*, Norwood, MA: Artech House, 2012.
[7] Modernaut, G. J. K., and D. Orban, "GNSS Antennas: An Introduction to Bandwidth, Gain Pattern, Polarization, and All That," *GPS World*, February 2009, pp. 42–48.
[8] Vittorini, L. D., and B. Robinson, "Receiver Frequency Standards: Optimizing Indoor GPS Performance," *GPS World*, November 2003, pp. 40–48.
[9] Pratt, A. R., "g-Effects on Oscillator Performance in GPS Receivers," *Navigation: JION*, Vol. 36, No. 1, 1989, pp. 63–75.
[10] Chiou, T.-Y., et al., "Model Analysis on the Performance for an Inertial Aided FLL-Assisted-PLL Carrier-Tracking Loop in the Presence of Ionospheric Scintillation," *Proc. ION NTM*, San Deigo, CA, January 2007, pp. 1276–1295.
[11] Kitching, J., "Time for a Better Receiver: Chip-Scale Atomic Frequency References," *GPS World*, November 2007, pp. 52–57.
[12] DeNatale, J. F., et al., "Compact, Low-Power Chip-Scale Atomic Clock," *Proc. IEEE/ION PLANS*, Monterey, CA, May 2008, pp. 67–70.
[13] Ward, P. W., J. W. Betz, and C. J. Hegarty, "Satellite Signal Acquisition, Tracking and Data Demodulation," in *Understanding GPS Principles and Applications*, 2nd ed., E. D. Kaplan and C. J. Hegarty, (eds.), Norwood, MA: Artech House, 2006, pp. 153–241.
[14] Adane, Y., A. Ucar, and I. Kale, "Dual-Tracking Multi-Constellation GNSS Front-End for High-Performance Receiver Applications," *Proc. ION GNSS 2011*, Portland, OR, September 2011, pp. 803–807.
[15] Mattos, P. G., "Adding GLONASS to the GPS/Galileo Consumer Receiver, with Hooks for Compass," *Proc. ION GNSS 2010*, Portland, OR, September 2010, pp. 2835–2839.
[16] Bao-Yen Tsui, J., *Fundamentals of Global Positioning System Receivers: A Software Approach*, 2nd ed., New York: Wiley, 2004.
[17] Weiler, R., et al., "L1/E5 Receiver: Pulling in Wideband," *GPS World*, June 2009, pp. 12–29.
[18] Ashby, N., and J. J. Spilker, Jr., "Introduction to Relativistic Effects on the Global Positioning System," in *Global Positioning System: Theory and Applications, Volume I*, B. W. Parkinson and J. J. Spilker, Jr., (eds.), Washington, D.C.: AIAA, 1996, pp. 623–697.
[19] Yamada, H., et al., "Evaluation and Calibration of Receiver Inter-Channel Biases for RTK-GPS/GLONASS," *Proc. ION GNSS 2010*, Portland, OR, September 2010, pp. 1580–1587.
[20] Akos, D. M., et al., "Real-Time GPS Software Radio Receiver," *Proc. ION NTM*, Long Beach, CA, January 2001, pp. 809–816.
[21] Borre, K., et al., *A Software-Defined GPS and Galileo Receiver: A Single Frequency Approach*, Boston: MA, Birkhäuser, 2007.
[22] Pany, T., *Navigation Signal Processing for GNSS Software Receivers*, Norwood, MA: Artech House, 2010.
[23] Spilker, J. J., Jr., "Fundamentals of Signal Tracking Theory," in *Global Positioning System: Theory and Applications Volume I*, B. W. Parkinson and J. J. Spilker, Jr., (eds.), Washington, D.C.: AIAA, 1996, pp. 245–327.
[24] Groves, P. D., "GPS Signal to Noise Measurement in Weak Signal and High Interference Environments," *Navigation: JION*, Vol. 52, No. 2, 2005, pp. 83–92.
[25] Hein, G. W., et al., "Performance of Galileo L1 Signal Candidates," *Proc. ENC GNSS 2004*, Rotterdam, the Netherlands, May 2004.
[26] Ward, P. W., J. W. Betz, and C. J. Hegarty, "Interference, Multipath and Scintillation," in *Understanding GPS Principles and Applications*, 2nd ed., E. D. Kaplan and C. J. Hegarty, (eds.), Norwood, MA: Artech House, 2006, pp. 243–299.

[27] Betz, J. W., and K. R. Kolodziejski, "Extended Theory of Early-Late Code Tracking for a Bandlimited GPS Receiver," *Navigation: JION*, Vol. 47, No. 3, 2000, pp. 211–226.

[28] Tran, M., and C. Hegarty, "Receiver Algorithms for the New Civil GPS Signals," *Proc. ION NTM*, San Diego, CA, January 2002, pp. 778–789.

[29] Dafesh, P., et al., "Description and Analysis of Time-Multiplexed M-Code Data," *Proc. ION 58th AM*, Albuquerque, NM, June 2002, pp. 598–611.

[30] Woo, K. T., "Optimum Semicodeless Carrier-Phase Tracking on L2," *Navigation: JION*, Vol. 47, No. 2, 2000, pp. 82–99.

[31] Mattos, P. G., "High Sensitivity GNSS Techniques to Allow Indoor Navigation with GPS and with Galileo," *Proc. GNSS 2003, ENC*, Graz, Austria, April 2003.

[32] Ziedan, N. I., *GNSS Receivers for Weak Signals*, Norwood, MA: Artech House, 2006.

[33] Harrison, D., et al., "A Fast Low-Energy Acquisition Technology for GPS Receivers," *Proc. ION 55th AM*, Cambridge, MA, June 1999, pp. 433–441.

[34] Lee, W. C., et al., "Fast, Low Energy GPS Navigation with Massively Parallel Correlator Array Technology," *Proc. ION 55th AM*, Cambridge, MA, June 1999, pp. 443–450.

[35] Scott, L., A. Jovancevic, and S. Ganguly, "Rapid Signal Acquisition Techniques for Civilian & Military User Equipment Using DSP Based FFT Processing," *Proc. ION GPS 2001*, Salt Lake City, UT, September 2001, pp. 2418–2427.

[36] Lin, D. M., and J. B. Y. Tsui, "An Efficient Weak Signal Acquisition Algorithm for a Software GPS Receiver," *Proc. ION GPS 2001*, Salt Lake City, UT, September 2001, pp. 115–119.

[37] Ward, P. W., "A Design Technique to Remove the Correlation Ambiguity in Binary Offset Carrier (BOC) Spread Spectrum Signals," *Proc. ION 59th AM*, Albuquerque, NM, June 2003, pp. 146–155.

[38] Blunt, P. D., "GNSS Signal Acquisition and Tracking," in *GNSS Applications and Methods*, S. Gleason and D. Gebre-Egziabher, (eds.), Norwood, MA: Artech House, 2009, pp. 23–54.

[39] Borio, D., C. O'Driscoll, and G. Lachapelle, "Coherent, Noncoherent, and Differentially Coherent Combining Techniques for Acquisition of New Composite GNSS Signals," *IEEE Trans. on Aerospace and Electronic Systems*, Vol. 45, No. 3, 2009, pp. 1227–1240.

[40] Qaisar, S. U., "Performance Analysis of Doppler Aided Tracking Loops in Modernized GPS Receivers," *Proc. ION GNSS 2009*, Savannah, GA, September 2009, pp. 209–218.

[41] Betz, J. W., "Design and Performance of Code Tracking for the GPS M Code Signal," *Proc. ION GPS 2000*, Salt Lake City, UT, September 2000, pp. 2140–2150.

[42] Betz, J. W., "Binary Offset Carrier Modulation for Radionavigation," *Navigation: JION*, Vol. 48, No. 4, 2001, pp. 227–246.

[43] Hodgart, M. S., and P. D. Blunt, "Dual Estimate Receiver of Binary Offset Carrier Modulated Signals for Global Navigation Satellite Systems," *Electronics Letters*, Vol. 43, No. 16, 2007, pp. 877–878.

[44] Fine, P., and W. Wilson, "Tracking Algorithm for GPS Offset Carrier Signals," *Proc. ION NTM*, San Diego, CA, January 1999, pp. 671–676.

[45] Ward, P., "Performance Comparisons Between FLL, PLL and a Novel FLL-Assisted PLL Carrier Tracking Loop Under RF Interference Conditions," *Proc. ION GPS-98*, Nashville, TN, September 1998, pp. 783–795.

[46] So, H., et al., "On-Line Detection of Tracking Loss in Aviation GPS Receivers Using Frequency-Lock Loops," *Journal of Navigation*, Vol. 62, No. 2, 2009, pp. 263–281.

[47] Duffett-Smith, P. J., and A. R. Pratt, "Reconstruction of the Satellite Ephemeris from Time-Spaced Snippets," *Proc. ION GNSS 2007*, Fort Worth, TX, September 2007, pp. 1867–1875.

[48] Hwang, P. Y., G. A. McGraw, and J. R. Bader, "Enhanced Differential GPS Carrier-Smoothed Code Processing Using Dual-Frequency Measurements," *Navigation: JION*, Vol. 46, No. 2, 1999, pp. 127–137.

[49] Bah'rami, M., "Getting Back on the Sidewalk: Doppler-Aided Autonomous Positioning with Single-Frequency Mass Market Receivers in Urban Areas," *Proc. ION GNSS 2009*, Savannah, GA, September 2009, pp. 1716–1725.

[50] Conley, R., et al., "Performance of Stand-Alone GPS," in *Understanding GPS Principles and Applications*, 2nd ed., E. D. Kaplan and C. J. Hegarty, (eds.), Norwood, MA: Artech House, 2006, pp. 301–378.

[51] Gruber, B., "GPS Program Update," *Civil GPS Service Interface Committee (CGSIC) meeting*, Nashville, TN, September 2012.

[52] Revnivykh, S., "GLONASS Status and Modernization," *Proc. ION GNSS 2012*, Nashville, TN, September 2012.

[53] Klobuchar, J. A., "Ionosphere Effects on GPS," in *Global Positioning System: Theory and Applications, Volume I*, B. W. Parkinson and J. J. Spilker, Jr., (eds.), Washington, D.C.: AIAA, 1996, pp. 485–515.

[54] Morton, Y. T., et al., "Assessment of the Higher Order Ionosphere Error on Position Solutions," *Navigation: JION*, Vol. 56, No. 3, 2009, pp. 185–193.

[55] Groves, P. D., and S. J. Harding, "Ionosphere Propagation Error Correction for Galileo," *Journal of Navigation*, Vol. 56, No. 1, 2003, pp. 45–50.

[56] Olynik, M., et al., "Temporal Variability of GPS Error Sources and Their Effect on Relative Position Accuracy," *Proc. ION NTM*, San Diego, CA, January 2002, pp. 877–888.

[57] Gao, G. X., et al., "Ionosphere Effects for Wideband GNSS Signals," *Proc. ION 63rd AM*, Cambridge, MA, April 2007, pp. 147–155.

[58] *Navstar GPS Space Segment/Navigation User Interfaces*, IS-GPS-200, Revision F, GPS Directorate, September 2011.

[59] Radicella, S. M., and R. Leitinger, "The Evolution of the DGR Approach to Model Electron Density Profiles," *Advances in Space Research*, Vol. 27, No. 1, 2001, pp. 35–40.

[60] *European GNSS (Galileo) Open Service Signal in Space Interface Control Document*, Issue 1 Revision 1, GNSS Supervisory Authority, September 2010.

[61] Collins, J. P., *Assessment and Development of a Tropospheric Delay Model for Aircraft Users of the Global Positioning System*, Technical Report No. 203, University of New Brunswick, September 1999.

[62] Spilker, J. J., Jr., "Tropospheric Effects on GPS," in *Global Positioning System: Theory and Applications, Volume I*, B. W. Parkinson and J. J. Spilker, Jr., (eds.), Washington, D.C.: AIAA, 1996, pp. 517–546.

[63] Mendes, V. B., and R. B. Langley, "Tropospheric Zenith Delay Prediction Accuracy for High-Precision GPS Positioning and Navigation," *Navigation: JION*, Vol. 46, No. 1, 1999, pp. 25–34.

[64] Powe, M., J. Butcher, and J. Owen, "Tropospheric Delay Modelling and Correction Dissemination Using Numerical Weather Prediction Fields," *Proc. GNSS 2003, ENC*, Graz, Austria, April 2003.

[65] Jupp, A., et al., "Use of Numerical Weather Prediction Fields for the Improvement of Tropospheric Corrections in Global Positioning Applications," *Proc. ION GPS/GNSS 2003*, Portland, OR, September 2003, pp. 377–389.

[66] Conker, R. S., et al., "Modeling the Effects of Ionospheric Scintillation on GPS/Satellite-Based Augmentation System Availability," *Radio Science*, Vol. 38, No. 1, 1001, 2003.

[67] Nichols, J., et al., "High-Latitude Measurements of Ionospheric Scintillation Using the NSTB," *Navigation: JION*, Vol. 47, No. 2, 2000, pp. 112–120.

[68] Gibbons, G., "GNSS Interoperability: Not So Easy, After All," *Inside GNSS*, January/February 2011, pp. 28–31.

[69] Van Dierendonck, A. J., P. Fenton, and T. Ford, "Theory and Performance of a Narrow Correlator Spacing in a GPS Receiver," *Navigation: JION*, Vol. 39, No. 3, 1992, pp. 265–283.

[70] Ries, L., et al., "Tracking and Multipath Performance Assessments of BOC Signals Using a Bit-Level Signal Processing Simulator," *Proc. ION GPS/GNSS 2003*, Portland, OR, September 2003, pp. 1996–2010.

[71] Bradbury, J., "Prediction of Urban GNSS Availability and Signal Degradation Using Virtual Reality City Models," *Proc. ION GNSS 2007*, Fort Worth, TX, September 2007, pp. 2696–2706.

[72] Braasch, M. S., "Multipath Effects," in *Global Positioning System: Theory and Applications, Volume I*, B. W. Parkinson and J. J. Spilker, Jr., (eds.), Washington, D.C.: AIAA, 1996, pp. 547–568.

[73] Van Nee, R. D. J., "GPS Multipath and Satellite Interference," *Proc. ION 48th AM*, Washington, D.C., June 1992, pp. 167–177.

[74] Irsigler, M., and B. Eissfeller, "Comparison of Multipath Mitigation Techniques with Consideration of Future Signal Structures," *Proc. ION GPS/GNSS 2003*, Portland, OR, September 2003, pp. 2584–2592.

[75] Hodgart, M. S., "Galileo's Problem with PRS or What's in a Phase?" *International Journal of Navigation and Observation*, 2011, Article ID 247360.

[76] Braasch, M. S., "Autocorrelation Sidelobe Considerations in the Characterization of Multipath Errors," *IEEE Trans. on Aerospace and Electronic Systems*, Vol. 33, No. 1, 1997, pp. 290–295.

[77] Hegarty, C. J., "Least-Squares and Weighted Least-Squares Estimates," in *Understanding GPS Principles and Applications*, 2nd ed., E. D. Kaplan and C. J. Hegarty, (eds.), Norwood, MA: Artech House, 2006, pp. 663–669.

[78] Axelrad, P., and R. G. Brown, "Navigation Algorithms," in *Global Positioning System: Theory and Applications, Volume I*, B. W. Parkinson and J. J. Spilker, Jr., (eds.), Washington, D.C.: AIAA, 1996, pp. 409–493.

[79] Brown, R. G., and P. Y. C. Hwang, *Introduction to Random Signals and Applied Kalman Filtering*, 3rd ed., New York: Wiley, 1997.

CHAPTER 10
GNSS: Advanced Techniques

The preceding chapters described the satellite navigation systems and their user equipment. This chapter reviews a number of techniques that enhance the accuracy, robustness, and reliability of GNSS.

Section 10.1 discusses how additional infrastructure may be used to improve GNSS positioning accuracy using differential techniques, while Section 10.2 describes how carrier phase techniques may be used to obtain high-precision position and attitude measurements under good conditions. Sections 10.3 and 10.4, respectively, review techniques for improving GNSS robustness in poor signal-to-noise environments and for mitigating the effects of multipath and NLOS reception. Section 10.5 discusses aiding, assistance, and orbit prediction, while Section 10.6 describes shadow matching, a positioning technique for urban canyons based on pattern matching.

A context-adaptive or cognitive receiver can reconfigure itself in real time to respond to changes in the environment, such as varying signal-to-noise levels and multipath, and to variations in the user or host vehicle dynamics [1, 2]. It can also trade different performance requirements, such as accuracy, sensitivity, and TTFF against power consumption. The number and configuration of the correlators, the acquisition and tracking algorithms, and the navigation processor can all be varied. Context detection is discussed further in Section 16.1.10.

10.1 Differential GNSS

The correlated range errors due to ephemeris prediction errors and residual satellite clock, ionosphere, and troposphere errors vary slowly with time and user location. Therefore, by comparing pseudo-range measurements with those made by equipment at a presurveyed location, known as a *reference station* or base station, the correlated range errors may be calibrated out. This improves the navigation solution accuracy, leaving just the signal tracking and multipath errors, and is the principle behind differential GNSS (DGNSS). Figure 10.1 illustrates the concept. An additional benefit is that Earth tide effects (see Section 2.4.4) will largely cancel between the user and reference station.

This section describes some different implementations of DGNSS, covering a local area with a single reference station or a regional or wide area with multiple reference stations. Before this, the spatial and temporal correlation properties of the various GNSS error sources are discussed, while the section concludes with a description of relative GNSS.

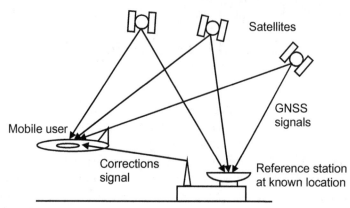

Figure 10.1 Schematic of differential GNSS.

10.1.1 Spatial and Temporal Correlation of GNSS Errors

Table 10.1 gives typical values for the variation of correlated GNSS error sources with time and space [3, 4]. This gives an indication of how the accuracy of the DGNSS navigation solution varies with the separation of the user from the reference station and the latency of the calibration data. The divergence in correlated range errors as the user moves away from a reference station is known as *spatial decorrelation*, while the divergence due to differences in measurement time is known as *time decorrelation*.

The satellite clock errors are the same for all observers, while the spatial variation of the ephemeris errors is very small. The temporal variation of these errors is also small. The variation in ionosphere and troposphere propagation errors is much greater and depends on the elevation angle, time of day, and weather. The spatial variation of the troposphere error is largest where there is a weather front between receivers.

The tracking, multipath, and NLOS errors are uncorrelated between users at different locations, so they cannot be corrected using DGNSS. Therefore, these errors must be minimized in the reference station to prevent them from disrupting the mobile user's navigation solution. A narrow early-late correlator spacing and narrow tracking-loop bandwidths (see Section 9.3.3), combined with carrier-smoothing of the pseudo-range measurements (Section 9.2.7) and use of a high-performance reference oscillator (Section 9.1.2), minimize the tracking errors. The narrow correlator spacing also reduces the impact of multipath interference (Section 9.3.4), while further multipath mitigation techniques are discussed in Section 10.4.

Table 10.1 Typical Variation of Correlated GNSS Error Sources over Time and Space

Error Source	Variation over 100 Seconds	Variation over 100 km Horizontal Separation	Variation over 1 km Vertical Separation
Residual satellite clock	~0.1m	None	None
Ephemeris	~0.01m	~0.002m	Negligible
Ionosphere (uncorrected)	0.1–0.4m	0.2–0.5m	Negligible
Troposphere (uncorrected)	0.1–1.5m	0.1–1.5m	1–2m*

*Ground reference station

10.1.2 Local and Regional Area DGNSS

In a local area DGNSS (LADGNSS) system, corrections are transmitted from a single reference station to mobile users, sometimes known as rovers, within the range of its transmitter. The closer the user is to the reference station, the more accurate the navigation solution is. Users within a 150-km horizontal radius typically achieve an accuracy of about 1m.

Transmitting corrections to the position solution requires all users and the reference station to use the same set of satellites so that the correlated errors affecting each satellite are cancelled out. This is not practical as satellite signals are intermittently blocked by buildings, terrain, and sometimes the host vehicle body. Instead, range corrections are transmitted, allowing the user to select any combination of the satellites tracked by the reference station. The corrections may be subject to reference-station receiver clock errors. However, this does not present a problem as the user's navigation processor simply solves for the relative clock offset and drift between the user and reference instead of the user receiver clock errors.

To obtain differentially corrected pseudo-range measurements, $\tilde{\rho}_{a,DC}^{s,l}$, differential corrections, $\nabla\rho_{dc}^{s,l}$, may be applied either in place of the satellite clock, ionosphere, and troposphere corrections:

$$\tilde{\rho}_{a,DC}^{s,l} = \tilde{\rho}_{a,R}^{s,l} + \nabla\rho_{dc}^{s,l}, \qquad (10.1)$$

or in addition to these corrections:

$$\tilde{\rho}_{a,DC}^{s,l} = \tilde{\rho}_{a,R}^{s,l} - \delta\hat{\rho}_{I,a}^{s,l} - \delta\hat{\rho}_{T,a}^{s} + \delta\hat{\rho}_{c}^{s,l} + \nabla\rho_{dc}^{s,l}, \qquad (10.2)$$

where the notation is as defined in Section 8.5.3. Application of only some of these corrections is also valid. However, it is essential that the same convention is adopted by both the reference station and the users. The ionosphere correction obtained from dual-frequency measurements is generally more accurate than that from DGNSS, while a troposphere model should be used for air applications as the troposphere errors vary significantly with height. After applying the differential corrections, the position may be determined using the same methods as for stand-alone GNSS (see Section 9.4). The measurement error covariance or measurement noise covariance should be adjusted to remove the variance of those errors that cancel between the user and reference, but add the reference receiver tracking noise variance.

Most LADGNSS systems adopt the Radio Technical Committee for Maritime Services (RTCM) Special Committee (SC) 104 transmission protocol. This supports a number of different messages, enabling each LADGNSS system's transmissions to be tailored to the user base and data rate, which can be as low as 50 bit s^{-1} [5, 6]. Range-rate corrections are transmitted to enable users to compensate for latency in the range corrections, while *delta corrections* are transmitted for users of old ephemeris and clock data broadcast by the satellite constellation.

Many LADGNSS stations transmit in the 283.5–325-kHz marine radio-beacon band, with coverage radii of up to 300 km. VHF and UHF band data links, cell-phone systems, radio and television broadcasts, the Internet, ELoran signals (Section 11.2.1), and Iridium (Section 11.4.1) are also used, while in cooperative positioning, differential corrections can be transmitted between peers.

Regional area DGNSS (RADGNSS) enables LADGNSS users to obtain greater accuracy by using corrections from multiple reference stations, combined using

$$\nabla \rho_{a,dc}^{s,l} = \sum_i W_i \nabla \rho_{a,dc,i}^{s,l}, \quad \sum_i W_i = 1, \qquad (10.3)$$

where the weighting factors, W_i, are determined by the user's distance from each reference station. RADGNSS may be implemented entirely within the receiver or corrections from multiple reference stations may be included in a single transmission.

Reference stations do not have to be at fixed locations. A mobile user with access to a more accurate positioning system can also estimate DGNSS corrections as described in Section 16.3.2 [7].

10.1.3 Wide Area DGNSS and Precise Point Positioning

A wide area DGNSS (WADGNSS) system aims to provide positioning to meter accuracy over a continent, such as Europe, or large country, such as the United States, using much fewer reference stations than LADGNSS or RADGNSS would require. From the user's perspective, the key difference is that corrections for the different error sources are transmitted separately. Typically, 10 or more reference stations at known locations send pseudo-range and dual-frequency ionosphere delay measurements to a master control station (MCS). The MCS then computes corrections to the GNSS system broadcast ephemeris and satellite clock parameters, together with ionosphere data, which are transmitted to the users [8–10]. The ionosphere data comprises estimates of the vertical propagation delay over a grid of pierce points; Section G.7.3 of Appendix G on the CD shows how to apply them.

WADGNSS operates on the same principle as the GNSS control segments (see Section 8.1.1 and Section G.1 of Appendix G on the CD). Actually, a stand-alone GNSS is really a global WADGNSS system as it cannot operate without the satellite ephemeris and satellite clock parameters being determined by the control segment and then broadcast by the satellites to the users.

WADGNSS is one of the functions of the SBAS systems, described in Sections 8.2.6 and 8.4.4. Other satellite-delivered WADGNSS services include NASA's Global Differential GPS System [11] and the commercial OmniStar [12] and StarFire [13] systems. WADGNSS data can also be transmitted to users via terrestrial radio links, cellphones, and the Internet.

Using a denser network of reference stations than the system control segments enables WADGNSS to achieve more accurate ephemeris and satellite clock calibration over the area spanned by the reference stations, while the ionosphere data is only provided for this service area. However, improvements in the accuracy of the ephemeris and satellite clock data broadcast by the GNSS satellites, together with the full advent of dual-frequency ionosphere correction for civil users, could limit the benefit of WADGNSS.

Precise point positioning (PPP) is a class of positioning techniques that combine WADGNSS with dual-frequency ionosphere delay calibration (Section 9.3.2) and carrier smoothing of the pseudo-ranges (Section 9.2.7) [12, 14]. It is most commonly used for postprocessed applications, but is also used in real-time positioning, particularly within the offshore oil and gas industry.

PPP provides decimeter-accuracy positioning after an initialization period of about 20 minutes, which is required for the carrier-smoothing of the pseudo-ranges to converge, averaging out the code-tracking errors. Real-time ephemeris and satellite clock data is available from commercial service providers. Examples include the OmniStar High Performance (HP) service [15] and StarFire [16].

Freely available *precision orbit and clock products*, used in place of the broadcast navigation message data are provided via the Internet by the International GNSS Service (IGS). This is a voluntary network of over 200 organizations in over 80 countries, operating more than 370 active reference stations. Their real-time orbit and clock data has an accuracy of around 10 cm [17], while for postprocessed applications, orbit and clock data accurate to 2.5 cm are available [18].

10.1.4 Relative GNSS

Relative GNSS (RGNSS) is used in applications, such as shipboard landing of aircraft and in-flight refueling, where the user position must be known accurately with respect to the reference station, but the position accuracy with respect to the Earth is less important. This relative position is known as a *baseline*. In RGNSS, the reference station transmits absolute pseudo-range measurements, which are then differenced with the user's pseudo-range measurements:

$$\nabla \tilde{\rho}_{ra,R}^{s,l} = \tilde{\rho}_{a,R}^{s,l} - \tilde{\rho}_{r,R}^{s,l}, \tag{10.4}$$

where the r denotes the reference station body frame Then, from (8.49) and (9.128), assuming the user and reference are close enough for the ionosphere, troposphere, and Sagnac corrections to cancel,

$$\nabla \tilde{\rho}_{ra,R}^{s,l} \approx \left| \hat{r}_{es}^{e}(\tilde{t}_{st,a}^{s,l}) - \hat{r}_{ea}^{e}(t_{sa,a}^{s,l}) \right| - \left| \hat{r}_{es}^{e}(\tilde{t}_{st,r}^{s,l}) - \hat{r}_{er}^{e}(t_{sa,r}^{s,l}) \right| + \nabla \hat{\rho}_{c}^{ra}(t_{sa,a}^{s,l}) + \delta\rho_{ra,\varepsilon}^{s,l+}, \tag{10.5}$$

where $\nabla \hat{\rho}_{c}^{ra}(t_{sa,a}^{s,l})$ is the relative receiver clock error. By analogy with (9.141), the weighted least-squares ECEF-frame relative position solution is then

$$\begin{pmatrix} \hat{r}_{ra}^{e+} \\ \nabla \hat{\rho}_{c}^{ra+} \end{pmatrix} = \begin{pmatrix} \hat{r}_{ea}^{e+} - \hat{r}_{er}^{e+} \\ \nabla \hat{\rho}_{c}^{ra+} \end{pmatrix} = \begin{pmatrix} \hat{r}_{ra}^{e-} \\ \nabla \hat{\rho}_{c}^{ra-} \end{pmatrix} + \left(\mathbf{H}_{G}^{e\mathrm{T}} \mathbf{C}_{\rho}^{\nabla -1} \mathbf{H}_{G}^{e} \right)^{-1} \mathbf{H}_{G}^{e\mathrm{T}} \mathbf{C}_{\rho}^{\nabla -1} \begin{pmatrix} \nabla \tilde{\rho}_{ra,R}^{1} - \nabla \hat{\rho}_{ra,R}^{1-} \\ \nabla \tilde{\rho}_{ra,R}^{2} - \nabla \hat{\rho}_{ra,R}^{2-} \\ \vdots \\ \nabla \tilde{\rho}_{ra,R}^{m} - \nabla \hat{\rho}_{ra,R}^{m-} \end{pmatrix}, \tag{10.6}$$

where $\hat{r}_{ra}^{e-}$ and $\nabla \hat{\rho}_{c}^{ra-}$ are, respectively, the predicted relative position and clock offset; the measurement matrix, $\mathbf{H}_{G}^{e}$, is as given by (9.144); $\mathbf{C}_{\rho}^{\nabla}$ is the differential measurement error covariance matrix; and the predicted pseudo-range difference is

$$\nabla \hat{\rho}_{ra,R}^{j-} \approx \left| \hat{r}_{es}^{e-}(\tilde{t}_{st,a}^{j}) - \hat{r}_{ea}^{e-}(\tilde{t}_{sa,a}^{j}) \right| - \left| \hat{r}_{es}^{e-}(\tilde{t}_{st,r}^{j}) - \hat{r}_{er}^{e-}(\tilde{t}_{sa,r}^{j}) \right| + \nabla \hat{\rho}_{c}^{ra-}(t_{sa,a}^{j}). \tag{10.7}$$

where j denotes the combination of a satellite, s, and signal, l, from that satellite. Similarly, for an EKF-based solution, the relative position and velocity, $\hat{\mathbf{r}}_{ra}^e$ and $\hat{\mathbf{v}}_{ra}^e$, are estimated instead of their absolute counterparts.

10.2 Real-Time Kinematic Carrier-Phase Positioning and Attitude Determination

When the user and reference station are relatively close, the accuracy of code-based differential GNSS is determined by the tracking and multipath errors. However, in range terms, carrier phase tracking is much less noisy and exhibits smaller multipath errors than code tracking. Therefore, by performing relative positioning with carrier measurements as well as code measurements, centimeter accuracy is potentially attainable. However, the inherent ambiguity in carrier-based ranging measurements, due to successive waveforms being indistinguishable from each other, must be resolved.

There are many different carrier-phase positioning techniques tailored to the needs of different applications. These may be classified into real time or postprocessed and static or dynamic. Only the real-time dynamic techniques are relevant to navigation. Carrier-phase positioning techniques that can operate in real time over moving baselines are known as real-time kinematic (RTK) positioning or kinematic carrier phase tracking (KCPT). Information on very-high-precision static positioning may be found in [19–21].

Like other forms of differential GNSS, RTK positioning may be implemented as a local-area, regional-area, or wide-area system. Local-area RTK uses a single reference station. After the ambiguities have been resolved, centimeter-accuracy positioning can be achieved using a single frequency with baselines of up to about 20 km. However, accuracy is degraded with longer baselines due to decorrelation of the ionosphere and troposphere propagation errors (see Section 10.1.1). For longer baselines, dual-frequency operation and troposphere modeling is required.

Regional-area RTK is known as network RTK and enables a given accuracy to obtained using more widely spaced reference stations. In the virtual reference station technique, measurements from multiple reference stations are interpolated to create a virtual reference station close to the user, compensating for most of the spatial decorrelation effects [22]. Network RTK reference data services are typically provided commercially by survey equipment manufacturers. The reference stations are publicly operated in many countries with their data freely available over the Internet for postprocessed applications (albeit at a lower rate in some cases).

Finally, wide-area RTK is known as PPP-RTK. By adding additional information to a basic PPP service (see Section 10.1.3), the integer wavelength ambiguities may be resolved, improving the precision and reducing the time required for initialization [23, 24]. Network-based ionosphere and troposphere error estimates are also provided.

This section begins by describing the principles of positioning using the accumulated delta range measurements, often loosely known as carrier phase. A single-epoch navigation solution is then presented, followed by discussions of more efficient ambiguity resolution techniques exploiting signal geometry and using multiple frequencies. Finally, the use of GNSS ADR measurements for attitude determination

10.2.1 Principles of Accumulated Delta Range Positioning

ADR measurements (see Section 9.2.7) have a common phase reference for all signals of the same type and an integer wavelength ambiguity for each signal that remains constant provided that carrier-phase tracking is maintained continuously without cycle slips. When a Costas carrier-phase discriminator is used, due to the presence of navigation data bits, the carrier phase measurement can be half a cycle out. Normally, the user equipment corrects for this using the sign of known bits in the navigation data message. When this is not done, the wavelength ambiguity can take half-integer values as well as integer values.

By analogy with (8.47) and (8.48), the raw ADR measurement, $\tilde{\Phi}_{a,R}^{s,l}$, may be expressed in terms of the true range, r_{as}, the wavelength ambiguity, $N_a^{s,l}$, and various error sources:

$$\tilde{\Phi}_{a,R}^{s,l} = r_{as} + N_a^{s,l}\lambda_{ca}^l + \delta\Phi_{I,a}^{s,l} + \delta\rho_{T,a}^s - \delta\rho_c^{s,l} + \delta\rho_c^a - \delta\Phi_b^{s,l} + \delta\Phi_b^{a,l} + \delta\Phi_{p,a}^{s,l} + \delta\Phi_{M,a}^{s,l} + w_{\Phi,a}^{s,l},$$

(10.8)

where $\delta\Phi_{I,a}^{s,l}$, $\delta\rho_{T,a}^s$, $\delta\rho_c^{s,l}$, and $\delta\rho_c^a$ are, the range errors due to, respectively, ionosphere propagation of the carrier, troposphere propagation, the satellite clock, and the receiver clock, as already defined; $\delta\Phi_b^{s,l}$ and $\delta\Phi_b^{a,l}$ are the range errors due to, respectively, the satellite and receiver phase biases; $\delta\Phi_{p,a}^{s,l}$ is the range error due to the line-of-sight-dependent phase wind-up error; $\delta\Phi_{M,a}^{s,l}$ is the range error due to carrier-phase multipath and NLOS reception; and $w_{\Phi,a}^{s,l}$ is the carrier-phase tracking range error.

Different authors use different conventions for the satellite and receiver phase biases. Here they are defined as the delays in the carrier phase with respect to the code that are independent of the line of sight and occur within the transmitter or receiver hardware and software and the transmit or receive antenna. Delays that affect code and carrier equally are absorbed into the clock bias terms. The receiver phase bias varies between different GLONASS FDMA frequencies. This must be calibrated in order to use GLONASS FDMA signals for precise positioning [25].

The phase wind-up error is a lag in the received carrier phase with respect to the code that depends on the relative orientation of the transmit and receive antennas [19]. This arises because GNSS signals are circularly polarized. A rotation of either antenna within its plane changes the measured carrier phase by one cycle per complete antenna rotation regardless of the line of sight. Consequently, the phase wind-up due to the orientation of the receive antenna within its own plane is common to all signals received so may be absorbed into the receiver phase bias, $\delta\Phi_b^{a,l}$. Similarly, the phase wind-up due to the orientation of the satellite antenna within its own plane may be absorbed into the satellite phase bias, $\delta\Phi_b^{s,l}$. The remaining phase wind-up error, $\delta\Phi_{p,a}^{s,l}$, is dependent on the orientation of the LOS vector with respect to the planes of the two antennas. This cancels between the user and reference receivers when the planes of their antennas are parallel (e.g., they are both horizontal) and the baseline between them is short enough for the LOS vectors to common satellites to be effectively parallel.

To determine the range from an ADR measurement, all of the other terms in (10.8) must be accounted for, except for the carrier-phase tracking error, which remains. The satellite clock error and phase bias, together with most of the ionosphere and troposphere errors, are eliminated using differential GNSS. In a wide-area or PPP-RTK implementation, satellite phase bias information is transmitted to users [23, 24]. The multipath error may be minimized using some of the techniques described in Section 10.4. The receiver clock offset and phase bias are common to all signals of the same type, so may be estimated as part of the navigation solution. However, in local-area and regional-area RTK, the ADR measurements are usually double differenced (Section 7.1.4.4), which eliminates the receiver phase bias and clock offset.

When the reference station transmits absolute measurements, the double-differenced pseudo-range and ADR measurements are

$$\begin{aligned} \nabla\Delta\tilde{\rho}_{ra,R}^{ts,l} &= \tilde{\rho}_{a,R}^{s,l} - \tilde{\rho}_{a,R}^{t,l} - \tilde{\rho}_{r,R}^{s,l} + \tilde{\rho}_{r,R}^{t,l} \\ \nabla\Delta\tilde{\Phi}_{ra,R}^{ts,l} &= \tilde{\Phi}_{a,R}^{s,l} - \tilde{\Phi}_{a,R}^{t,l} - \tilde{\Phi}_{r,R}^{s,l} + \tilde{\Phi}_{r,R}^{t,l} \end{aligned} \quad (10.9)$$

where s and t denote the body frames of two different satellites and r is the reference receiver body frame. If corrections are transmitted by the reference station, the double-differenced measurements are

$$\begin{aligned} \nabla\Delta\tilde{\rho}_{ra,R}^{ts,l} &= \tilde{\rho}_{a,R}^{s,l} - \tilde{\rho}_{a,R}^{t,l} + \nabla\tilde{\rho}_{dc}^{s,l} - \nabla\tilde{\rho}_{dc}^{t,l} \\ \nabla\Delta\tilde{\Phi}_{ra,R}^{ts,l} &= \tilde{\Phi}_{a,R}^{s,l} - \tilde{\Phi}_{a,R}^{t,l} + \nabla\tilde{\Phi}_{dc}^{s,l} - \nabla\tilde{\Phi}_{dc}^{t,l} \end{aligned} \quad (10.10)$$

where $\nabla\tilde{\Phi}_{dc}^{s,l}$ and $\nabla\tilde{\Phi}_{dc}^{t,l}$ are the differential ADR corrections for the two signals.

This leaves the double-differenced integer wavelength ambiguity to be determined, a process known as *ambiguity resolution*. The simplest method is to start with the user and reference antennas at known locations. The ambiguity is then estimated using

$$\nabla\Delta\hat{N}_{ra}^{ts,l} = \frac{1}{\lambda_{ca}^l}\left(\nabla\Delta\tilde{\Phi}_{ra,R}^{ts,l} - |\hat{\mathbf{r}}_{is}^i - \hat{\mathbf{r}}_{ia}^i| + |\hat{\mathbf{r}}_{it}^i - \hat{\mathbf{r}}_{ia}^i| + |\hat{\mathbf{r}}_{is}^i - \hat{\mathbf{r}}_{ir}^i| - |\hat{\mathbf{r}}_{it}^i - \hat{\mathbf{r}}_{ir}^i|\right) \quad (10.11)$$

and fixed by rounding $\nabla\Delta\hat{N}_{ra}^{ts,l}$ to the nearest integer (or half integer) to give $\nabla\Delta\check{N}_{ra}^{ts,l}$. This is sometimes known as receiver initialization [12].

Note that it is the fixing of the wavelength ambiguity to an integer or half integer that enables higher precision positioning to be achieved using the ADR measurements. Otherwise, the accuracy is no better than techniques based on carrier-smoothed code.

When the user equipment is not initialized at a known location, the pseudo-range measurements may be used to aid ambiguity resolution. Figure 10.2 shows the probability distributions of double-differenced pseudo-range measurements obtained from code and carrier. The carrier-based measurements are ambiguous but more precise. Combining the two reduces the number of possible values of the integer wavelength ambiguity to the order of 10.

10.2 Real-Time Kinematic Carrier-Phase Positioning and Attitude Determination

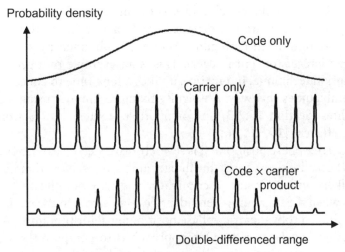

Figure 10.2 Double-differenced range probability distributions from code and carrier measurements.

From (8.47) and (8.48), a pseudo-range measurement may be expressed as

$$\tilde{\rho}_{a,R}^{s,l} = r_{as} + \delta\rho_{I,a}^{s,l} + \delta\rho_{T,a}^{s} - \delta\rho_{c}^{s,l} + \delta\rho_{c}^{a} + \delta\rho_{M,a}^{s,l} + w_{\rho,a}^{s,l}. \qquad (10.12)$$

By subtracting the double-differenced pseudo-range measurements from the corresponding ADR and applying (9.81), (10.8), (10.9), and (10.12), the double-differenced ambiguity may be expressed as

$$\nabla\Delta N_{ra}^{ts,l} \approx \frac{1}{\lambda_{ca}^{l}}\begin{pmatrix} \nabla\Delta\tilde{\Phi}_{ra,R}^{ts,l} - \nabla\Delta\tilde{\rho}_{ra,R}^{ts,l} + 2\nabla\Delta\delta\rho_{I,ra}^{ts,l} + \nabla\Delta\delta\rho_{M,ra}^{ts,l} \\ +\nabla\Delta w_{\rho,ra}^{ts,l} - \nabla\Delta\delta\Phi_{\phi,ra}^{ts,l} - \nabla\Delta\delta\Phi_{M,ra}^{ts,l} - \nabla\Delta w_{\Phi,ra}^{ts,l} \end{pmatrix}, \qquad (10.13)$$

where $\nabla\Delta\delta\rho_{I,ra}^{ts,l}, \nabla\Delta\delta\rho_{M,ra}^{ts,l}, \nabla\Delta w_{\rho,ra}^{ts,l}, \nabla\Delta\delta\Phi_{\phi,ra}^{ts,l}, \nabla\Delta\delta\Phi_{M,ra}^{ts,l}$, and $\nabla\Delta w_{\phi,ra}^{ts,l}$ are the double-differenced range errors due to, respectively, the ionosphere modulation delay, code multipath error, code tracking error, LOS-dependent phase wind-up, carrier multipath error, and carrier tracking error. The standard deviation of this will typically be several wavelengths. The ionosphere and phase-wind-up errors are minimized by minimizing the baseline between the user and reference receivers and keeping their antenna planes parallel. The effects of the tracking and multipath errors may be reduced by time averaging.

The double-differenced ambiguity may therefore be estimated using

$$\nabla\Delta\hat{N}_{ra}^{ts,l} = \frac{1}{n\lambda_{ca}^{l}}\sum_{k=1}^{n}\left(\nabla\Delta\tilde{\Phi}_{ra,R,k}^{ts,l} - \nabla\Delta\tilde{\rho}_{ra,R,k}^{ts,l}\right), \qquad (10.14)$$

where k denotes the epoch and n is the number of epochs used. Once the uncertainty in the estimate has dropped below a certain fraction of a wavelength, the corresponding fixed ambiguity, $\nabla\Delta \hat{N}_{ra}^{ts,l}$, is obtained by rounding $\nabla\Delta \hat{N}_{ra}^{ts,l}$ to the nearest integer (or half integer). This is an example of a geometry-free ambiguity resolution technique. In practice, it takes a long time to collect sufficient data to fix the ambiguities this way. Ambiguity-fixing techniques using signals from multiple satellites (Section 10.2.3) and/or multiple frequencies (Section 10.2.4) are much more efficient [19, 20, 26].

Carrier-phase positioning is severely disrupted by carrier cycle slips (see Section 9.2.4) because they change the integer ambiguities. Note that half-integer cycle slips are likely with a Costas discriminator and may be followed by half-cycle corrections when the user equipment identifies sign errors in the navigation data message. Therefore, a robust carrier-phase positioning algorithm must incorporate cycle slip detection and correction. One approach is to compare the step change between successive relative carrier phase or double-differenced measurements with values predicted from the range-rate measurements [27] or velocity solution.

10.2.2 Single-Epoch Navigation Solution Using Double-Differenced ADR

Using the weighted least-squares method described in Section 9.4.1, the ECEF-frame relative position of the user antenna with respect to the reference antenna, $\hat{\mathbf{r}}_{ra}^{e+}$, may be estimated from the double-differenced ADR measurements and fixed ambiguities using

$$\hat{\mathbf{r}}_{ra}^{e+} = \hat{\mathbf{r}}_{ra}^{e-} + \left(\mathbf{H}_G^{\Delta e \mathrm{T}} \mathbf{C}_\Phi^{\nabla\Delta -1} \mathbf{H}_G^{\Delta e}\right)^{-1} \mathbf{H}_G^{\Delta e \mathrm{T}} \mathbf{C}_\Phi^{\nabla\Delta -1} \begin{pmatrix} \nabla\Delta \tilde{\Phi}_{ra,R}^{t1,l} - \nabla\Delta \check{N}_{ra}^{t1,l} \lambda_{ca}^l - \nabla\Delta \hat{r}_{ra}^{t1-} \\ \nabla\Delta \tilde{\Phi}_{ra,R}^{t2,l} - \nabla\Delta \check{N}_{ra}^{t2,l} \lambda_{ca}^l - \nabla\Delta \hat{r}_{ra}^{t2-} \\ \vdots \\ \nabla\Delta \tilde{\Phi}_{ra,R}^{tm,l} - \nabla\Delta \check{N}_{ra}^{tm,l} \lambda_{ca}^l - \nabla\Delta \hat{r}_{ra}^{tm-} \end{pmatrix}, \quad (10.15)$$

where $\hat{\mathbf{r}}_{ra}^{e-}$ is the predicted relative position; $\nabla\Delta \hat{r}_{ra}^{ts-}$ is the predicted double-differenced range, given by

$$\nabla\Delta \hat{r}_{ra}^{ts-} = \left| \hat{\mathbf{r}}_{es}^e\left(\tilde{t}_{st,a}^{s,l}\right) - \hat{\mathbf{r}}_{ea}^{e-}\left(\tilde{t}_{sa,a}^{s,l}\right) \right| - \left| \hat{\mathbf{r}}_{et}^e\left(\tilde{t}_{st,a}^{t,l}\right) - \hat{\mathbf{r}}_{ea}^{e-}\left(\tilde{t}_{sa,a}^{s,l}\right) \right| - \left| \hat{\mathbf{r}}_{es}^e\left(\tilde{t}_{st,r}^{s,l}\right) - \hat{\mathbf{r}}_{er}^{e-}\left(\tilde{t}_{sa,r}^{s,l}\right) \right|$$
$$+ \left| \hat{\mathbf{r}}_{et}^e\left(\tilde{t}_{st,r}^{t,l}\right) - \hat{\mathbf{r}}_{er}^{e-}\left(\tilde{t}_{sa,r}^{s,l}\right) \right|; \quad (10.16)$$

$\mathbf{H}_G^{\Delta e}$ is the measurement matrix for measurements differenced across satellites; and $\mathbf{C}_\Phi^{\nabla\Delta}$ is the double-differenced ADR measurement error covariance matrix, which is not diagonal. These are given by

$$\mathbf{H}_G^{\Delta e} = \mathbf{D}_G \mathbf{H}_{rG}^e, \qquad \mathbf{C}_\Phi^{\nabla\Delta} = \mathbf{D}_G \mathbf{C}_\Phi^{\nabla} \mathbf{D}_G^\mathrm{T}, \quad (10.17)$$

where the corresponding measurement matrix for single-satellite measurements, $\mathbf{H}_{rG}^e$, and the differencing matrix, $\mathbf{D}_G$, are given by

10.2 Real-Time Kinematic Carrier-Phase Positioning and Attitude Determination

$$\mathbf{H}_{rG}^e = \begin{pmatrix} -u_{a1,x}^e & -u_{a1,y}^e & -u_{a1,z}^e \\ -u_{a2,x}^e & -u_{a2,y}^e & -u_{a2,z}^e \\ \vdots & \vdots & \vdots \\ -u_{am,x}^e & -u_{am,y}^e & -u_{am,z}^e \\ -u_{at,x}^e & -u_{at,y}^e & -u_{at,z}^e \end{pmatrix}_{\mathbf{r}_{ea}^e = \hat{\mathbf{r}}_{ea}^{e-}}, \quad \mathbf{D}_G = \begin{pmatrix} 1 & 0 & \cdots & 0 & -1 \\ 0 & 1 & \cdots & 0 & -1 \\ \vdots & \vdots & \ddots & \vdots & \vdots \\ 0 & 0 & \cdots & 1 & -1 \end{pmatrix} \quad (10.18)$$

and $\mathbf{C}_\Phi^\nabla$ is the measurement error covariance matrix for ADR measurements differenced across receivers. This accounts for tracking noise and multipath errors at both receivers, and the spatial decorrelation of the ionosphere, troposphere, and LOS-dependent phase wind-up errors.

10.2.3 Geometry-Based Integer Ambiguity Resolution

The signal geometry may be used to aid ambiguity resolution wherever the navigation solution is overdetermined (i.e., the number of double-differenced ADR measurements from different satellites exceeds the number of position and bias states estimated). This requires at least five satellites to be tracked. In this case, only certain solutions to the set of integer wavelength ambiguities will produce a consistent solution. Figure 10.3 illustrates this for the determination of a 2-D position solution from four ambiguous range measurements. Within the search area, there are lots of places where the candidate lines of position from two of the range measurements intersect, several places where three intersect, but only one place where all four intersect.

To exploit signal geometry for improving ambiguity resolution, the ambiguities for all of the double-differenced ADR measurements must be resolved together. This is what *geometry-based* ambiguity resolution techniques do.

In principle, the ambiguities can be resolved by computing navigation solutions for all possible ambiguity combinations within the search area defined by the double-differenced pseudo-ranges and then selecting the most consistent. A suitable

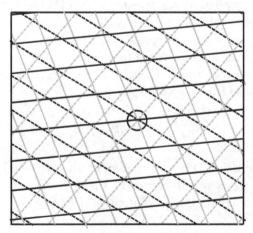

Figure 10.3 Intersection in two dimensions of the lines of position from four ambiguous range measurements.

consistency measure is the chi-square test statistic computed from the measurement residuals as described in Section 17.4.1. However, this approach is impractical as over a million possible solutions would often have to be evaluated.

The float-and-fix method is much more efficient [19, 20, 26]. First, the set of ambiguities is estimated as continuous values, known as *float ambiguities*, as part of the navigation solution. This *float solution* is calculated from the double-differenced ADR measurements and the double-differenced pseudo-ranges. Sections G.9.1 and G.9.2 of Appendix G on the CD describe single-epoch and filter-based float solutions, respectively.

Let the float solution at epoch k, $\hat{\mathbf{x}}_k^+$, and its error covariance, $\mathbf{P}_k^+$, be partitioned as

$$\hat{\mathbf{x}}_k^+ = \begin{pmatrix} \hat{\mathbf{x}}_{G,k}^+ \\ \hat{\mathbf{x}}_{N,k}^+ \end{pmatrix} \qquad \mathbf{P}_k^+ = \begin{pmatrix} \mathbf{P}_{G,k}^+ & \mathbf{P}_{GN,k}^+ \\ \mathbf{P}_{GN,k}^{+\,T} & \mathbf{P}_{N,k}^+ \end{pmatrix}, \qquad (10.19)$$

where the subscript N denotes the double-differenced wavelength ambiguity states and G denotes the position states and any additional states, such as velocity. The ambiguities are fixed by finding the set of integer values that are most consistent with the float solution. A suitable test statistic is

$$s_{N,i}^2 = \left(\mathbf{x}_{N,i} - \hat{\mathbf{x}}_{N,k}^+\right)^T \left(\mathbf{P}_{N,k}^+\right)^{-1} \left(\mathbf{x}_{N,i} - \hat{\mathbf{x}}_{N,k}^+\right), \qquad (10.20)$$

where $\mathbf{x}_{N,i}$ is the ith set of candidate integer ambiguities. This uses the float solution error covariance to normalize the distance between the fixed and float solutions. Whichever candidate set produces the smallest value of $S_{N,i}^2$ is deemed the likeliest solution.

If the search space for the integer ambiguities is defined by the individual uncertainties of the float estimates, scaled up to provide a suitable confidence region, the number of candidates to consider will be impractically large. However, the float ambiguity estimates are highly correlated with each other, enabling the search space to be reduced considerably. The most commonly used approach is the least-squares ambiguity decorrelation adjustment (LAMBDA) method [28, 29]. Sections G.9.3, G.9.4, and G.9.5 of Appendix G on the CD, respectively, describe the correlation properties of the float ambiguity estimates, the LAMBDA method, and a simple method of validating the fixed ambiguities.

When the fixed ambiguity set, $\check{\mathbf{x}}_N$, is validated, the position solution and any other estimated states are adjusted using

$$\check{\mathbf{x}}_{G,k}^+ = \hat{\mathbf{x}}_{G,k}^+ + \mathbf{P}_{GN,k}^+ \left(\mathbf{P}_{N,k}^+\right)^{-1} \left(\check{\mathbf{x}}_N - \hat{\mathbf{x}}_{N,k}^+\right). \qquad (10.21)$$

If the validation fails, the float solution is retained and further measurements are required to fix the ambiguities. Alternatively, a partial set of the wavelength ambiguities, or combinations thereof, may be fixed with the remainder left as float values [30].

When a sufficiently large number of satellites are tracked, geometry-based methods can provide an ambiguity fix from a single set of measurements, particularly for shorter baselines. Otherwise, data taken over a few minutes is required to average out the tracking errors and the effects of any multipath interference. The greater the number of satellites tracked, the quicker the ambiguities may be resolved. Changes in the signal geometry as the satellites move can also aid ambiguity resolution.

10.2.4 Multifrequency Integer Ambiguity Resolution

For short baselines between the user and reference receivers, the ionosphere propagation delay essentially cancels, so the second frequency can be used to aid ambiguity resolution. The separation between the candidate carrier-based differential range measurements is different on each frequency, so only the candidate differential ranges where the measurements on the different frequencies align need be considered. Figure 10.4 illustrates this.

A common approach to dual-frequency ambiguity resolution is to combine the ADR measurements on the two frequencies to produce *wide-lane* measurements that have much longer wavelengths, but are also noisier. The dual-frequency pseudo-ranges are combined so as to minimize the noise. The ratio of the uncertainty of the double-differenced pseudo-ranges to the wavelength is therefore reduced considerably, making the ambiguities easier to resolve. Both geometry-free and geometry-based ambiguity resolution techniques may be used [19, 20, 26]. More information on wide-lane measurements may be found in Section G.9.6 of Appendix G on the CD.

For longer baselines, the difference in ionosphere propagation delay experienced at the user and reference receivers cannot be neglected. However, when this is substantially less than a wavelength, dual-frequency ambiguity resolution can still be performed by using a geometry-free float solution, estimating double-differenced range, ionosphere delay, and ambiguities on both frequencies. Ionosphere states can also be incorporated in a geometry-based float solution. These methods may also be used for triple-frequency and multifrequency ambiguity resolution, which can operate over much longer baselines [31, 32].

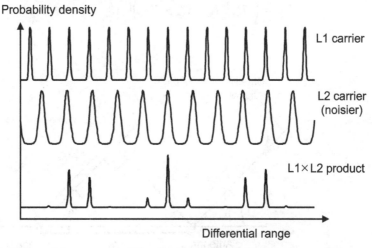

Figure 10.4 Differential range probability distributions from dual-frequency carrier measurements.

10.2.5 GNSS Attitude Determination

When a carrier-phase relative GNSS solution is obtained between a pair of antennas attached to the same vehicle, it can be used to obtain information about the host vehicle's attitude. As the baseline between the antennas is much smaller than the distance to the satellites, the line-of-sight vectors from a pair of antennas to a given satellite may be treated as parallel. Therefore, the angle, θ, between the baseline and the line of sight to satellite s is given by $\cos\theta = (r_{bs} - r_{as})/r_{ab}$, where r_{as} and r_{bs} are the ranges between satellite s and, respectively, antennas a and b and r_{ab} is the known baseline length between the two antennas, as shown in Figure 10.5. The line-of-sight vector with respect to the Earth is known, so information about the host vehicle's attitude with respect to the Earth can be obtained. This technique is known as *interferometric attitude determination* or a *GNSS compass*.

More generally, if carrier-phase GNSS is used to make a measurement of the baseline in local navigation frame axes, $\tilde{\mathbf{r}}^n_{ab}$, which may be obtained from an ECEF-frame measurement using (2.62) and (2.150), this may be related to the known body-frame baseline, $\mathbf{r}^b_{ab}$, by

$$\tilde{\mathbf{r}}^n_{ab} = \tilde{\mathbf{C}}^n_b \mathbf{r}^b_{ab}. \qquad (10.22)$$

However, this does not give a unique solution to the attitude, $\mathbf{C}^n_b$, as the component about the baseline is undetermined; only two components may be obtained from a single baseline measurement. To resolve this, a third antenna, denoted by c, must be introduced that is noncolinear with the other two antennas, providing a second baseline. Combining the two measurements [3],

$$\begin{pmatrix} \tilde{\mathbf{r}}^n_{ba} & \tilde{\mathbf{r}}^n_{bc} \end{pmatrix} = \tilde{\mathbf{C}}^n_b \begin{pmatrix} \mathbf{r}^b_{ba} & \mathbf{r}^b_{bc} \end{pmatrix}. \qquad (10.23)$$

The attitude can be obtained by adding a vector-product column and rearranging, giving:

$$\tilde{\mathbf{C}}^n_b = \begin{pmatrix} \tilde{\mathbf{r}}^n_{ba} & \tilde{\mathbf{r}}^n_{bc} & \tilde{\mathbf{r}}^n_{ba} \wedge \tilde{\mathbf{r}}^n_{bc} \end{pmatrix} \begin{pmatrix} \mathbf{r}^b_{ba} & \mathbf{r}^b_{bc} & \mathbf{r}^b_{ba} \wedge \mathbf{r}^b_{bc} \end{pmatrix}^{-1}. \qquad (10.24)$$

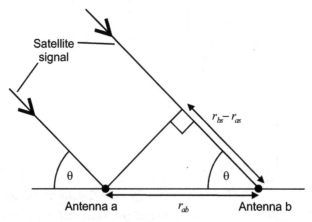

Figure 10.5 Schematic of GNSS attitude determination.

With four or more antennas, the solution is

$$\tilde{C}_b^n = \begin{pmatrix} \tilde{r}_{ba}^n & \tilde{r}_{bc}^n & \tilde{r}_{ba}^n \wedge \tilde{r}_{bc}^n & \tilde{r}_{bd}^n & \cdots \end{pmatrix} \begin{pmatrix} r_{ba}^b & r_{bc}^b & r_{ba}^b \wedge r_{bc}^b & r_{bd}^b & \cdots \end{pmatrix}^T \\ \times \left[\begin{pmatrix} r_{ba}^b & r_{bc}^b & r_{ba}^b \wedge r_{bc}^b & r_{bd}^b & \cdots \end{pmatrix} \begin{pmatrix} r_{ba}^b & r_{bc}^b & r_{ba}^b \wedge r_{bc}^b & r_{bd}^b & \cdots \end{pmatrix}^T \right]^{-1} \qquad (10.25)$$

where they are coplanar and

$$\tilde{C}_b^n = \begin{pmatrix} \tilde{r}_{ba}^n & \tilde{r}_{bc}^n & \tilde{r}_{bd}^n & \cdots \end{pmatrix} \begin{pmatrix} r_{ba}^b & r_{bc}^b & r_{bd}^b & \cdots \end{pmatrix}^T \left[\begin{pmatrix} r_{ba}^b & r_{bc}^b & r_{bd}^b & \cdots \end{pmatrix} \begin{pmatrix} r_{ba}^b & r_{bc}^b & r_{bd}^b & \cdots \end{pmatrix}^T \right]^{-1}$$

(10.26)

otherwise. Other solutions are described in [33, 34].

An attitude solution may be obtained with signals from only two GNSS satellites once the integer wavelength ambiguities have been resolved. This is because the known baseline lengths remove one degree of freedom from the baseline measurements, while the use of a common receiver design and shared receiver clock across all the antennas removes the relative phase offsets, provided the antenna cable lags are calibrated. Note also that the known baseline lengths can be used to constrain the ambiguities [35].

The attitude measurement accuracy is given by the ratio of the carrier phase baseline measurement accuracy to the baseline length. So, for a 1-m rigid baseline, measured to a 1-cm precision, the attitude measurement standard deviation is 10 mrad (about 0.6°). Longer baselines provide greater attitude measurement precision, provided that the baseline is rigid. Flexure degrades the measurement accuracy. However, short baselines convey the advantage of fewer integer ambiguity combinations to search.

As the measurement errors are noise-like, accuracy for static applications is improved by averaging over time. For dynamic applications, noise smoothing can be achieved through integration with an INS as described in Section 14.4.3.

Attitude may also be determined using a single precisely calibrated antenna by comparing the measured c/n_0 of each signal received with the antenna gain pattern. This is typically used for space applications where the antenna calibration costs can be justified, there is no signal attenuation, and the only multipath interference is due to the spacecraft body, which is predictable. The accuracy is about 200 mrad with one GNSS constellation and 140 mrad with two [36], which is considerably poorer than the interferometric method.

10.3 Interference Rejection and Weak Signal Processing

This section reviews techniques that enable GNSS user equipment to operate in a poor signal-to-noise (i.e., low C/N_0) environment. Antenna systems, receiver front-end filtering, extended range tracking, receiver sensitivity, combined acquisition and tracking, and vector tracking are discussed. Most of these techniques may be used together, while performance can also be improved through aiding and assistance as discussed in Section 10.5. The section begins by summarizing the sources of unintentional interference, deliberate jamming, and signal attenuation that cause low signal-to-noise levels.

10.3.1 Sources of Interference, Jamming, and Attenuation

Sources of unintentional interference include broadcast television, mobile satellite services, UWB communications, radar, mobile communications, car key fobs, DME/TACAN, and faulty GNSS user equipment [37–40]. Proposals in the United States to introduce mobile broadband services in spectrum immediately adjacent to the L1/E1 band raised particular concern and were eventually rejected [41]. Receivers with a higher precorrelation bandwidth (see Section 9.1.3) are typically more vulnerable to adjacent band interference. Most sources of unintentional interference can be mitigated by using GNSS signals in more than one frequency band. However, interference from solar radio bursts can affect all GNSS signals [42].

Deliberate jamming of GNSS signals has historically been a military issue. However, with the advent of GNSS-based road user charging, asset tracking, and law enforcement, it is becoming more widespread. GPS jammer designs are readily available on the Internet. Low-power jammers, sometimes called *personal privacy devices*, which will block reception of all GNSS signals within a few meters, are commercially available for less than $30 (€25). More powerful 25-W jammers, for which the suppliers claim a range of up to 300m, are also available [40].

In indoor environments, GNSS signals are typically 15–40 dB weaker than out in the open. This is due to a mixture of attenuation by the fabric of the building and rapid fading due to multipath interference between signal components of similar strength [43]. Signals can also be attenuated by foliage, with a typical attenuation of 1–4 dB per tree [44], by the human body [45, 46], by helicopter blades [47], and as a result of device masking, particularly on phone-based receivers.

When NLOS signals must be used to obtain a position solution, such as in many urban canyons, these will be attenuated on reflection and are often LHCP, reducing the antenna gain. Diffracted signals are also attenuated. For space applications, signals may be received from the low-gain regions of the GNSS satellite antenna patterns.

In space, indoor, and dense urban environments, some GNSS signals are much stronger than others, rendering intersignal interference a problem. This can manifest both as noise-like interference and cross-correlation effects (see Section 9.2.1).

GNSS is also vulnerable to spoofing, the transmission of fake signals causing user equipment to report a false position solution. Spoofing generation, detection, and mitigation methods are reviewed in [48].

10.3.2 Antenna Systems

The most effective defense against unintentional interference and deliberate jamming is a controlled-reception-pattern antenna (CRPA) system. The CRPA comprises an array of GPS antennas, mounted with their centers usually about half a wavelength apart, as illustrated by Figure 10.6. Operational CRPAs tend to comprise four or

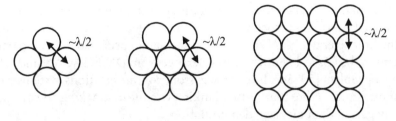

Figure 10.6 Schematic of 4-, 7-, and 16-element controlled-reception-pattern antennas.

seven elements, while larger arrays have been developed for experimental use. An antenna control unit (ACU) then varies the reception pattern of the antenna array by combining the signals from each antenna element with different gains and phases.

Early designs of CRPA system are null-steering, whereby the ACU acts to minimize the received RF power on the basis that unwanted interfering signals must always be stronger than the wanted GNSS signals, as the latter lie below thermal noise levels (see Section 8.1.2). This results in an antenna pattern with minima, or nulls, in the direction of each interfering source, improving the signal-to-noise level within the receiver by more than 20 dB [38, 49]. An n-element CRPA system can produce up to $n-1$ deliberate nulls. Incidental, or parasitic, nulls also occur. A null can sometimes coincide with a GNSS satellite line of sight, attenuating signals from that satellite. Null-steering CRPA systems offer no benefit in weak-signal environments.

More advanced CRPA systems are beam-forming, whereby an antenna pattern is created with a gain maximum in the direction of the wanted satellite signal [50]. A separate antenna pattern is formed for each satellite tracked, so the receiver must have multiple front ends. The ACU may determine the gain maxima by seeking to maximize the receiver's C/N_0 measurements. Alternatively, the CRPA attitude may be supplied by an INS and combined with satellite line-of-sight data. Beam-forming CRPA systems can potentially improve the receiver signal to noise in weak-signal environments.

CRPA systems have the drawback of being expensive, at over \$10,000 (€8,000) each, and large, with seven-element CRPAs between 14 and 30 cm in diameter. A simpler alternative for air applications is a canceller, which uses only two antennas, pointing upwards and downwards, and takes advantage of the fact that the GNSS signals generally come from above, while the interference generally comes from below [49]. Synthetic array processing may be used to achieve beam-forming with a single moving antenna element [51].

When there is significant spatial variation in signal strength due to deep multipath fading and/or attenuation, more reliable reception may be obtained by using dual antennas half a wavelength apart with separate receiver hardware and then combining the correlator outputs [52].

10.3.3 Receiver Front-End Filtering

Careful design of the receiver's AGC and ADC can improve performance in weak signal-to-noise environments. An AGC prevents interference from saturating the receiver, while a larger number of quantization levels lets more signal information through [38, 39]. Pulsed interference may be mitigated using pulse blanking, whereby the ADC output is zeroed when the interference exceeds a certain margin, improving the time-averaged C/N_0 in the baseband signal processor. This is particularly important in the L5/E5 band where interference from DME/TACAN and other sources can be a problem [53, 54]. Interference from communications systems can also occur in short bursts. Note that pulse blanking can allow tracking to continue, but disrupt navigation data message demodulation.

When the interference source has a narrower bandwidth than the wanted GNSS signal, it can be filtered by a spectral filtering technique, such as an adaptive transversal filter (ATF) [38] or frequency-domain interference suppressor (FDIS) [55]. These use an FFT to generate a frequency-domain power spectrum and then identify which frequencies are subject to interference. The components of the signal at the interfering frequencies are then attenuated. Cognitive radio techniques may be used to optimally combine pulse blanking and spectral filtering [56].

10.3.4 Extended Range Tracking

A limiting factor of conventional code and carrier tracking is the pull-in range of the discriminator functions (see Figures 9.13, 9.17, and 9.18). If the pull-in range can be extended, higher levels of tracking noise and larger dynamics response lags (see Section 9.3.3) may be tolerated before tracking lock is lost.

Carrier frequency tracking may be maintained at lower c/n_0 by adding additional low-frequency and high-frequency correlation channels, analogous to the early and late code correlation channels. These perform carrier wipeoff (see Section 9.1.4) at offset carrier frequencies (e.g., $\pm 0.443/\tau_a$) and then correlate the signal with the prompt reference code. Suitable carrier frequency discriminators are then

$$F_{DPP} = (I_{LF} - I_{HF})I_P + (Q_{LF} - Q_{HF})Q_P$$
$$F_{LHP} = (I_{LF}^2 + Q_{LF}^2) - (I_{HF}^2 + Q_{HF}^2)$$
, (10.27)

where the subscripts LF and HF denote the low-frequency and high-frequency correlator outputs, respectively.

The code tracking function pull-in range can be expanded by replacing the early, prompt, and late correlators by a correlator bank. However, feeding the outputs of an extended-range correlator bank into a discriminator function increases the noise, canceling the benefit of an extended pull-in range. A solution is to feed the outputs from a bank of correlators into a limited-window acquisition algorithm with a duty cycle spanning several correlator accumulation intervals and matched to the desired tracking-loop time constant. Implementing an FFT also extends the carrier-frequency pull-in range. In this *batch processing* approach [57, 58], a tracking loop can be formed by using the output of one acquisition cycle to center the code and Doppler search window of the following cycle via the NCOs. Figure 10.7 illustrates this. The batch processing approach enables tracking to be maintained at lower C/N_0 levels, albeit at the cost of a higher processing load. Note that larger pseudo-range and pseudo-range-rate errors than those in conventional tracking can be exhibited before the tracking lock is lost.

10.3 Interference Rejection and Weak Signal Processing

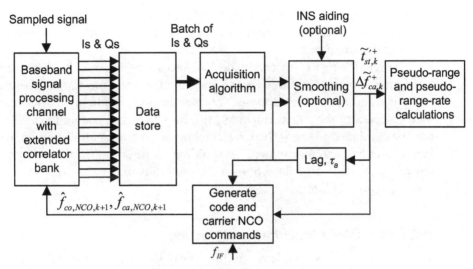

Figure 10.7 Batch-processing tracking architecture.

10.3.5 Receiver Sensitivity

Receiver sensitivity in a poor signal-to-noise environment is optimized for both acquisition and tracking by maximizing the accumulation, or coherent integration, interval of the correlator I and Q outputs. For the data-free GNSS signals, the main factor limiting the accumulation interval is the pseudo-range-rate tracking accuracy as discussed in Section 9.1.4.4. This may be enhanced using a CSAC (see Section 9.3.2) and aiding from another navigation sensor (Section 10.5.1), which also enables the noncoherent integration interval to be extended (see Section 9.2.1).

For signals modulated with a navigation data message, accumulation over more than one data-bit interval also requires the receiver to multiply the reference code or correlator outputs by a locally generated copy of the message. This is known as data wipeoff. Assisted GNSS (Section 10.5.2) and cooperative GNSS enables real-time transmission of the navigation data message from a receiver in a strong signal-to-noise environment, while the legacy GPS and GLONASS messages are regular enough for the user equipment to predict most of their content from recently stored messages, where it is able to gather this data.

A number of techniques have been developed for estimating unknown navigation data bits [59–62]. They are also applicable to unknown secondary PRN-code bits. Essentially, the candidate coherent summations with each possible data-bit combination are computed and then the set that gives the highest value of $(\Sigma I_p)^2 + (\Sigma Q_p)^2$ is used. A coherent integration time of 10 seconds has been demonstrated using a stationary antenna, assisted GPS and a high-quality reference oscillator [63].

As shown in Sections 9.2.2, 9.2.3, and 9.3.3, the code and carrier tracking-loop bandwidths are a tradeoff between noise resistance and response to dynamics, including reference oscillator noise. Thus, tracking sensitivity in a weak signal-to-noise and low-dynamics environment may be improved by reducing the tracking-loop bandwidths. This is equivalent to extending the coherent or noncoherent integration interval in acquisition mode, depending on whether the discriminator is coherent or noncoherent. The minimum c/n_0 at which tracking can be maintained varies as the inverse square of the tracking-loop bandwidth with a noncoherent code discriminator or Costas carrier discriminator, and as the inverse of the bandwidth with a coherent discriminator.

The ranging processor can be designed to adapt the tracking loop bandwidths as a function of the measured c/n_0 to maintain the optimum tradeoff between noise resistance and dynamics response. This may be done implicitly if a Kalman filter is used to perform the signal tracking [59], in which case code and carrier tracking may be combined [64, 65]. Narrow tracking-loop bandwidths may be maintained in a high-dynamics environment using aiding information from another navigation sensor (see Section 10.5.1), while a CSAC exhibits much less noise than a conventional crystal oscillator.

10.3.6 Combined Acquisition and Tracking

New GNSS satellites broadcast multiple open-access signals. The effective signal to noise is improved if these are acquired and tracked together rather than separately. Signals on the same frequency are synchronized, so the different correlator outputs may be combined to produce common acquisition test statistics and tracking discriminator functions as described in Sections 9.2.1 and 9.2.2.

Signals on different frequencies are offset due to the ionosphere propagation delay and frequency-dependent satellite and receiver biases. For acquisition, this offset is less than the code bin spacing for low-chipping-rate signals. Correlator outputs from different signals may therefore be combined to form common test statistics, provided a high-chipping-rate signal is not acquired on more than one frequency [66, 67].

Multifrequency tracking requires a Kalman filter-based estimation algorithm that estimates the code and carrier interfrequency offset, together with code phase, carrier phase, carrier frequency, and rate of frequency change on one frequency. The measurements comprise separate code and carrier discriminator functions for each signal [68].

10.3.7 Vector Tracking

In conventional GNSS user equipment, the information from the baseband signal processing channels, the Is and Qs, is filtered by the code and carrier tracking loops before being passed to the navigation processor. This smooths out noise and enables the navigation processor to be iterated at a lower rate. However, it also filters out some of the signal information. Each set of pseudo-range and pseudo-range rate measurements input to the navigation processor is derived from several successive sets of Is and Qs, but the older data has been down-weighted by the tracking loops, partially discarding it.

Once an initial navigation solution has been obtained, the signal tracking and navigation-solution determination can be combined into a single estimation algorithm, usually Kalman-filter-based. The Is and Qs are used to estimate corrections to the navigation solution, from which the NCO control commands are derived. This is known as vector tracking. It brings the benefit that all I and Q data is weighted equally in the navigation solution, reducing the impact of tracking errors. When the navigation solution is overdetermined, the tracking of each signal is aided by the

10.3 Interference Rejection and Weak Signal Processing

others and tracking lock may be maintained through a single-channel outage. Thus, a given navigation-solution precision can be obtained in a poorer signal-to-noise environment. The main drawback is an increase in processor load. When fewer than four satellites are tracked, the user equipment must revert to conventional tracking.

The simplest implementation of vector tracking is the vector delay lock loop (VDLL) [69]. This uses the navigation processor to track code, but retains independent carrier tracking loops as shown in Figure 10.8. When four GNSS signals are tracked, the states and system model are the same as for the conventional navigation filter described in Section 9.4.2. Discriminator functions are used to obtain a measurement of each code tracking error, $\tilde{x}_k^j$, from the Is and Qs as described in Section 9.2.2. Using (9.12), each pseudo-range measurement innovation may then be obtained:

$$\begin{aligned}\delta z_{\rho,k}^{j-} &= \rho_{a,C,k}^{j} - \hat{\rho}_{a,C,k}^{j-} + w_{m,\rho,k}^{j} \\ &\approx \rho_{a,R,k}^{j} - \hat{\rho}_{a,R,k}^{j-} + w_{m,\rho,k}^{j}, \\ &= -\frac{c}{f_{co}} \tilde{x}_k^j\end{aligned} \quad (10.28)$$

where w_m is the measurement noise. The pseudo-range-rate measurements are obtained from the carrier tracking loops and perform the same role as the range-rate aiding inputs to conventional code tracking loops. The measurement model described in Section 9.4.2.3 may then be applied.

When signals on more than one frequency are tracked, the ionosphere propagation delay for each satellite must be estimated as a Kalman filter state. When more than four satellites are tracked, range-bias estimation can be used to maintain code

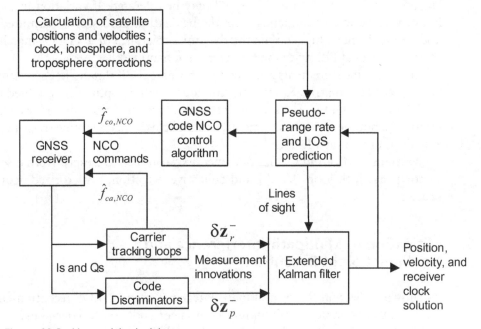

Figure 10.8 Vector delay lock loop.

tracking at the peak of each signal's correlation function. This is more important where the correlation peak is narrow, as applies to many BOC signals, and/or only one frequency is used, in which case the range biases are larger.

The code NCO commands are generated as described in Section 14.3.3.5 (for deeply coupled INS/GNSS integration). The update interval may be longer than the correlation period, τ_a, with measurement innovations averaged over this interval. However, the update rate must be at least twice the code-tracking bandwidth, B_{L_CO}, which is not constant in a Kalman filter implementation.

The vector DLL may be extended to incorporate carrier frequency tracking by deriving each pseudo-range-rate measurement innovation from a carrier-frequency discriminator (see Section 9.2.3):

$$\begin{aligned} \delta z_{r,k}^{j-} &= \dot{\rho}_{a,C,k}^{j} - \hat{\dot{\rho}}_{a,C,k}^{j-} + w_{m,r,k}^{j} \\ &\approx \dot{\rho}_{a,R,k}^{j} - \hat{\dot{\rho}}_{a,R,k}^{j-} + w_{m,r,k}^{j} \\ &= -\frac{c}{f_{ca}} \delta \tilde{f}_{ca,k}^{j} \end{aligned} \qquad (10.29)$$

This is also known as a vector delay and frequency lock loop (VDFLL) [70, 71] and is shown in Figure 10.9. Carrier NCO command generation is also described in Section 14.3.3.5. The addition of acceleration and rate-of-clock-drift Kalman filter states to enable second-order, as opposed to first-order, frequency tracking is recommended. The measurement update rate must be at least twice the carrier tracking bandwidth, so a 50-Hz update rate enables a bandwidth of up to 25 Hz.

The dynamics tolerance of a VDFLL depends on the interface between the receiver and the navigation processor [72]. If the NCO commands are accepted at 50 Hz and the acceleration estimate is used to compensate them for the control lag, an acceleration of up to 120 m s^{-1} may be tolerated. If variation in the NCO frequencies between commands from the navigation processor is permitted or the update rate is much higher, then the dynamics tolerance of a VDFLL matches that of a conventional FLL with the same tracking bandwidth.

If the c/n_0 measurements are used to determine the Kalman filter's measurement noise covariance matrix, $\mathbf{R}_G$, the measurements will be optimally weighted in both the VDLL and VDFLL. The state uncertainties (obtained from the error covariance matrix, $\mathbf{P}$), not the c/n_0 measurements, should be used to determine tracking lock as explained in Section 14.3.3.1.

Sections G.10.1 and G.10.2 of Appendix G on the CD, respectively, discuss the vector phase lock loop (VPLL) and collective detection, a vectorized acquisition process.

10.4 Mitigation of Multipath Interference and Nonline-of-Sight Reception

As shown in Section 9.3.4, multipath interference and NLOS reception can produce significant errors in GNSS user-equipment code and carrier-phase measurements. The errors caused by a given reflected signal depend on the signal type,

10.4 Mitigation of Multipath Interference and Nonline-of-Sight Reception

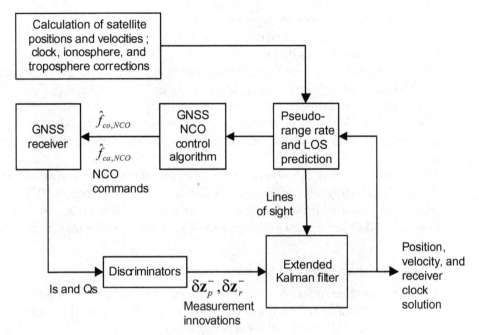

Figure 10.9 Vector delay and frequency lock loop.

antenna design, and receiver design in the case of multipath interference, but not when a single reflected signal is received in the absence of the direct signal. This section reviews a number of techniques for mitigating these errors, focusing mainly on the code tracking errors. They are divided into antenna-based, receiver-based, and navigation-processor-based techniques. Some methods only mitigate multipath interference, while others mitigate both multipath and NLOS reception. Different techniques may be deployed in parallel.

In addition, multipath mapping and some techniques for reference stations are discussed in Section G.11 of Appendix G on the CD, while some carrier-phase multipath mitigation techniques are discussed in [73].

10.4.1 Antenna-Based Techniques

The strength of reflected signals reaching a GNSS receiver can be minimized by careful design of the antenna system. GNSS signals are transmitted with RHCP, while reflection by a smooth surface generally reverses this to LHCP, assuming an angle of incidence less than Brewster's angle. Therefore, an antenna designed to be sensitive to RHCP signals, but not LHCP signals, can reduce multipath interference by about 10 dB. However, signals reflected by very rough surfaces are polarized randomly, so they are only attenuated by 3 dB by a RHCP antenna [74].

Another characteristic of multipath environments is that most reflected signals have low or negative elevation angles. A choke-ring antenna system uses a series of concentric rings, mounted on a ground plane around the antenna element, to attenuate these signals. This is too large for most navigation applications, but it can be deployed on ships.

Beam-forming CRPA systems may be used to mitigate multipath by maximizing the antenna gain for direct signals [75–77]. A seven-element CRPA system reduces the pseudo-range errors due to multipath by a factor of about 2, having a greater effect for high elevation satellites. Alternatively, beam-forming with a single moving antenna element may be achieved using synthetic array processing [78].

All of these antenna-based techniques attenuate reflected signals, reducing errors due to multipath interference. However, ranging errors due to NLOS reception are not reduced unless the signals are attenuated sufficiently to prevent acquisition and tracking.

An antenna array can also be used to measure the angle of arrival of the signals, essentially inverting interferometric attitude determination (Section 10.2.5). By comparing the measured lines of sight with those determined from the satellite and user positions, NLOS reception and severe multipath interference may be detected [79].

10.4.2 Receiver-Based Techniques

A number of techniques have been developed that mitigate multipath by increasing the resolution of the code discriminator on the basis that the higher frequency components of a GNSS signal are less impacted by moderate-delay multipath interference. However, as the power in a BPSK GNSS signal is concentrated in the low-frequency components, these techniques achieve multipath mitigation at the expense of signal-to-noise performance [80]. None of these techniques mitigate errors due to NLOS-only signal reception.

Three techniques that replace the conventional early, prompt, and late correlators have been compared in [81]. The double-delta discriminator, also known as the Leica Type A [82], strobe correlator [83], high-resolution correlator [84], and pulse-aperture correlator, adds very early (VE) and very late (VL) correlators to conventional correlators with narrow spacing and uses them to correct the discriminator function, giving

$$D_{\Delta\Delta} = (I_E^2 + Q_E^2) - (I_L^2 + Q_L^2) - \tfrac{1}{2}(I_{VE}^2 + Q_{VE}^2) + \tfrac{1}{2}(I_{VL}^2 + Q_{VL}^2). \quad (10.30)$$

The early/late slope technique, or multipath elimination technology (MET) [85], places two pairs of narrowly spaced correlation channels on each side of the correlation peak and uses these to compute the slope on each side. The prompt correlation channels are then synchronized with the point where the two slopes intersect. The e1e2 technique [86] operates on the basis that multipath interference mainly distorts the late half of the correlation function (see Figure 9.27). Therefore, it places two correlation channels on the early side of the peak and acts to maintain a constant ratio between the two.

The gated-correlator method [84] retains conventional discriminators, but blanks out both the signal and the reference code away from the chip transitions to sharpen the autocorrelation function. The superresolution method [87] simply boosts the high-frequency components by filtering the spectrum of the reference code and/or the signal prior to correlation.

The multipath-estimating delay lock loop (MEDLL) is another superresolution technique. It samples the whole combined correlation function of the direct and reflected signals using a bank of up to 48 narrowly-spaced correlators [88]. It then fits the sum of a number of idealized correlation functions to the measurements, separating the direct and reflected signal components.

The vision correlator [89] uses extra accumulators to build up a measurement of the shape of the code-chip transitions. A number of idealized chip transitions are then fitted to the measurements, enabling the relative amplitude, lag, and phase of the reflected components to be determined and the signal tracking corrected. This method has been shown to give better performance, particularly with short-delay multipath, than older techniques.

All of these multipath mitigation techniques require a large precorrelation bandwidth relative to the code chipping rate to obtain the best performance. There is much less scope to apply them to the high-chipping-rate GNSS signals, which are limited by the transmission bandwidth. However, the high-chipping-rate signals are only affected by short-delay multipath, which the receiver-based mitigation techniques do not compensate well. Hence, these multipath mitigation techniques effectively match the multipath performance of the low-chipping-rate signals to that of the high-chipping-rate signals at the expense of signal-to-noise performance. Thus, it is better to use high-chipping-rate signals where available.

When the host vehicle is moving with respect to the reflectors, reflected signals will have different range rates from the directly-received signal. Thus, by implementing extended range tracking (Section 10.3.4) in both the code-phase and Doppler shift domains, it is possible to separate out the different signal components by Doppler shift. To achieve the necessary resolution for vehicle applications, the coherent integration interval must be extended to 100 ms (see Sections 9.1.4.4 and 10.3.5) [90, 91].

10.4.3 Navigation-Processor-Based Techniques

When the user equipment is moving with respect to the reflectors, the multipath errors fluctuate rapidly due to phase changes. Therefore, their effect on pseudo-range measurements can be much reduced by smoothing the pseudo-range with carrier phase measurements. This may be done directly, using (9.74), (9.75), or (9.89), or by implementing a filtered navigation solution (Section 9.4.2), with the measurement noise covariance tuned to favor the pseudo-range-rate measurements. Vector tracking (Section 10.3.7) may also be used to minimize the impact of multipath interference and NLOS reception on the navigation solution [92].

The effect of both multipath- and NLOS-induced ranging errors on the navigation solution may be reduced by predicting the susceptibility of each measurement to these errors and weighting them accordingly within the navigation solution (see Section 9.4). A number of criteria may be used. A reduced and/or fluctuating C/N_0 is indicative of both NLOS reception and severe multipath interference [93], although there are other causes (see Section 10.3.1). Low-elevation signals are more susceptible to multipath and NLOS reception. Other factors, such as the azimuth of the signal with respect to the street direction, the distance to the nearest building, and the building heights, may also be used where the necessary data is available [94].

Better positioning performance may be obtained by detecting multipath-contaminated and NLOS measurements and either downweighting them in the navigation solution or excluding them altogether, depending on the number and quality of the other available measurements. AOA-based detection is described in Section 10.4.1.

Comparing different measurements of signals from the same satellite may be used to detect multipath interference but not NLOS reception. Multipath indicators suitable for dynamic applications include discrepancies between the change in pseudo-range between epochs and the carrier-derived delta range, discrepancies between delta-range measurements on different frequencies [95], and differences between amplitude fluctuations on the early and late correlator outputs [96]. Phase differences between the early and late correlation channels can be used to detect multipath for both static and dynamic applications; this is more robust if two frequencies are used [97].

NLOS reception and large multipath-induced errors may be detected by comparing signals from different satellites. Single-epoch consistency checks (Section 17.4) provide rapid detection of erroneous measurements, but can struggle in environments in which the majority of receivable signals are contaminated. Performance is better when C/N_0-based measurement weighting is used [98, 99]. For land applications, a terrain height database (see Section 13.2) can provide additional information to increase the robustness of consistency checking [100].

When a filtered navigation solution is implemented, measurement innovation-based outlier detection (Section 17.3) can be more sensitive than consistency checking, particularly with a multisensor navigation filter. Also, NLOS reception and multipath can be distinguished by comparing a series of innovations, with the former indicated by a bias and the latter by a larger variance than normal [101].

NLOS reception and very strong multipath interference may be detected by separately correlating the RHCP and LHCP outputs of a dual-polarization antenna and comparing the C/N_0 measurements [102].

A sky-pointing camera with a panoramic lens or an array of cameras can produce an image of the entire field of view above the receiver's masking angle. When the orientation of the camera is known, the blocked lines of sight may be determined from the image, enabling NLOS signals to be identified [103, 104]. Blocked lines of sight may also be identified using a 3-D city model. This is straightforward where the user position is known [105]; otherwise, the user position and NLOS signals must be determined jointly [106, 107].

Finally, for ships, trains, large aircraft, and reference stations, antennas may be deployed at multiple locations and their measurements compared to determine which are contaminated by multipath and/or NLOS reception.

10.5 Aiding, Assistance, and Orbit Prediction

Aiding and assistance are external sources of information that can be used to improve GNSS performance, reducing the time to first fix and/or enabling acquisition, tracking, and position computation under poor GNSS reception conditions. They are distinct from integration, whereby external information is used only for computing the navigation solution as described in Chapters 14 and 16. Here aiding is defined as information about the user position, velocity, and time, while assistance comprises information about the satellites and signals. Some authors use other definitions. Note also that the term height aiding is sometimes used to describe integration architectures that combine GNSS measurements (e.g., pseudo-ranges) with height measurements from other sources as described in Sections 13.2.7 and 16.3.2.

This section discusses aiding and assistance in turn, followed by orbit prediction techniques. Both assistance and orbit prediction enable positioning without navigation data demodulation, in which case the pseudo-ranges are ambiguous. Section G.12 of Appendix G on the CD reviews techniques for positioning with these ambiguous pseudo-ranges.

10.5.1 Acquisition and Velocity Aiding

There are two main types of aiding: acquisition and velocity. Acquisition aiding is used to reduce the search space, enabling acquisition to take place more quickly and/or at lower signal-to-noise levels (see Section 9.2.1) [108]. An approximate time and user position enables the range rate due to satellite motion with respect to the Earth to be predicted (assuming approximate satellite orbits are known). Frequency aiding from a terrestrial radio signal at a known frequency also enables the receiver clock drift to be calibrated, potentially eliminating the Doppler search altogether for a stationary user. Time and frequency aiding can also be obtained from Iridium satellite signals (see Section 11.4.1).

Position aiding is often accurate to within a few code chips. However, to enable the code-phase search region to be reduced, the receiver clock must be calibrated to within a code repetition interval (1 ms for GPS C/A code). Accurate clock calibration is inherent in the CDMA Interface Standard (IS)-95 and CDMA IS-2000 cellphone systems. Base stations of the other mobile phone systems are not synchronized, but do maintain a stable time offset. They may therefore be calibrated using GNSS user equipment that is already receiving the navigation data message, and then used to aid acquiring receivers, a technique known as fine time aiding (FTA) [109].

In cooperative positioning, acquisition aiding can be provided by another GNSS receiver nearby. As well as time, frequency, and approximate position information, pseudo-ranges, Doppler shifts, and C/N_0 information for individual satellite signals may also be provided [110].

When aiding is available from a more accurate positioning system than standalone GNSS, even if this is not continuous, it can be used to estimate GNSS range biases as described in Section 16.3.2 [7].

Velocity aiding comprises an independent user velocity solution (e.g., from an INS or other dead-reckoning sensors). This is used to aid the code and carrier tracking loops, which then track the errors in the aiding information and receiver clock instead of absolute user dynamics. This enables them to operate with lower tracking bandwidths, boosting noise resistance (see Section 9.3.3). Velocity aiding can also enable acquisition to operate with longer dwell times when the user is moving by adjusting the reference signal for user motion as well as satellite motion. Similarly, it can aid reacquisition by bridging the reference signal through outages in reception of the satellite signals. Finally, velocity aiding can enable the coherent integration or accumulation interval to be extended for both acquisition and tracking, noting that the longer this interval, the more accurate the velocity aiding must be (see Section 9.1.4.4).

Velocity aiding does not account for the changes in the receiver clock drift so oscillator noise limits the performance that can be achieved. One solution is to use a terrestrial radio signal with a known stable frequency and transmitter location for Doppler shift compensation [111]; another is to use a CSAC.

When a velocity solution used to provide tracking aiding is also calibrated using GNSS, the integration architecture must be carefully designed to avoid positive feedback destabilizing the tracking (see Section 14.1.4). One solution is to combine the tracking and integration into a single algorithm. This is known as deeply coupled integration and is analogous to vector GNSS tracking (Section 10.3.7). Deeply coupled INS/GNSS integration is described in Sections 14.1.5 and 14.3.3. Odometry and UWB measurements have also been deeply coupled with GNSS [112, 113].

When an external velocity solution is not used to aid the carrier tracking loops, it may instead be used to detect GNSS cycle slips (see Section 9.2.4). This may be done by comparing the changes in ADR with a prediction made from the velocity solution and clock drift estimate. Velocity from both inertial navigation [114] and odometry [115] has been used in cycle-slip detection.

External aiding can also be used to determine the behavioral context (see Section 16.1.10), enabling the GNSS acquisition, tracking, and navigation algorithms to be adapted to the user antenna dynamics. For pedestrian navigation, the navigation filter position uncertainty may be increased whenever a step is detected (see Section 6.4).

10.5.2 Assisted GNSS

In many poor signal-to-noise environments, GNSS signals can be acquired and tracked, but the navigation data message cannot be demodulated. The continuous carrier tracking required to download the ephemeris data can also be difficult to achieve while moving around dense urban and indoor environments due to intermittent signal blockage. Therefore, a stand-alone GNSS receiver may have to rely on out-of-date ephemeris, satellite clock, and ionosphere calibration parameters, degrading the navigation solution, while a "cold-start" navigation solution cannot be obtained at all. One solution is assisted GNSS, which uses a separate communication link to provide the information in the navigation data message [108, 116]. This also shortens the TTFF.

AGNSS is often implemented through the mobile phone system where it is also known as network assistance and incorporates acquisition aiding (see Section 10.5.1). In some phone-based AGNSS systems, the navigation solution is determined by the network instead of the user equipment. AGNSS data can also be obtained via a wireless Internet connection, while Iridium satellites (Section 11.4.1) can provide both AGNSS data and frequency aiding. Similarly, any differential GNSS system that provides absolute ephemeris and satellite clock data, instead of corrections to the broadcast parameters, is also an AGNSS system. Assistance data may also be provided cooperatively by direct communication between nearby users [110].

10.5.3 Orbit Prediction

When current ephemeris data is not available via the broadcast navigation message or a current AGNSS data link, satellite orbits more accurate than the almanac data may often be predicted from the most recently available ephemeris data. This is called extended ephemeris. Best performance requires modeling of forces such as gravitation and solar radiation pressure [117]. The orbit predictions may be calculated on a server, transmitted to the user via AGNSS and stored until required [108]. In a self-assisted model, the computations are run on the user equipment itself. However, the force modeling approach is computationally intensive. A simpler, though less accurate, option is to simply extrapolate forward the broadcast ephemeris data [118].

Satellite clock errors are not accurately predictable because they are essentially a random walk process. However, the broadcast clock parameters are useful for about three days. Within this timeframe, positions to within 100m may be computed using orbit prediction [118].

10.6 Shadow Matching

Shadow matching is a multiconstellation GNSS positioning technique based on pattern matching (see Sections 1.3.1 and 7.1.6) instead of ranging. It is intended for use in dense urban areas where conventional GNSS positioning exhibits poor accuracy, particularly in the cross-street direction. This is because even if sufficient direct LOS signals are receivable, the geometry is typically poor as shown in Figure 9.35.

Shadow matching is based on a similar principle to RSS fingerprinting (see Section 7.1.6). However, because GNSS satellites are constantly moving, it is impractical to build a database from RSS measurements. Instead, a 3-D city model is used to predict which satellites are directly visible and which are blocked by buildings [119]. Signals from many GNSS satellites will be directly receivable in some parts of a street, but not others. Consequently, by determining whether a direct signal is being received from a given satellite, and comparing this with the prediction from

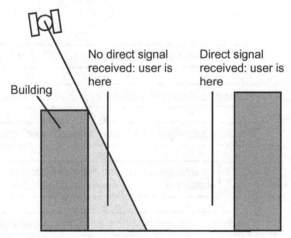

Figure 10.10 The shadow-matching concept.

the city model, the user can localize their position to within one of two areas of the street. Figure 10.10 illustrates this. By considering other satellites, the position solution may be refined further. Thus the observed signal shadowing is matched with the predicted shadowing to determine position.

Figure 10.11 shows the stages of a shadow-matching algorithm [120]. A conventional GNSS position solution, together with lines of sight and C/N_0 measurements, should be obtained first. The accuracy of the conventional solution should then be assessed to determine whether shadow-matching is necessary and, if so, to define a suitable search region. An outdoor positioning context should also be confirmed (see Section 16.1.10). After this, a set of candidate positions is determined, comprising a regularly spaced grid of points within the outdoor portion of the search region. Direct signal availability at each point is then predicted by comparing each LOS with the 3-D city model. Precomputing and storing the elevation of the building boundary as a function of azimuth at each point reduce the real-time processing load.

The next step is to evaluate the similarity between predicted and observed satellite visibility at each candidate position. A score is computed for each satellite above the elevation mask angle. When a high C/N_0 signal is received, the score is one if the signal was predicted and zero otherwise. Similarly, if no signal is received, zero is scored if signal reception was predicted and one if it was not. Intermediate scores are allocated to signals with lower C/N_0 levels as these could be NLOS, diffracted, or attenuated direct signals. The overall score for each position is simply the sum of the satellite-matching scores. The likeliest user position is that with the highest overall score.

Tests with GPS and GLONASS measurements have shown that shadow matching can correctly determine which side of a street the user is on when conventional GNSS positioning cannot [120, 121].

When only a basic road map is available, perpendicular streets in urban areas may be distinguished by comparing the azimuths of the highest C/N_0 signals with the road directions [private communication with P. Mattos, August 2011].

Problems and exercises for this chapter are on the accompanying CD.

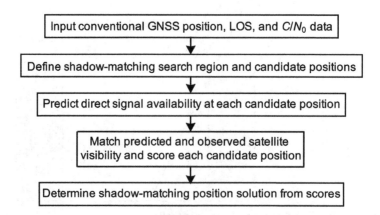

Figure 10.11 The stages of a shadow-matching algorithm.

References

[1] Lin, T., C. O'Driscoll, and G. Lachapelle, "Development of a Context-Aware Vector-Based High-Sensitivity GNSS Software Receiver," *Proc. ION ITM*, San Diego, CA, January 2011, pp. 1043–1055.

[2] Shivaramaiah, N. C., and A. G. Dempster, "Cognitive GNSS Receiver Design: Concept and Challenges," *Proc. ION GNSS 2011*, Portland, OR, September 2011, pp. 2782–2789.

[3] Costentino, R. J., et al., "Differential GPS," in *Understanding GPS: Principles and Applications*, 2nd ed., E. D. Kaplan and C. J. Hegarty, (eds.), Norwood, MA: Artech House, 2006, pp. 379–452.

[4] Parkinson, B. W., and P. K. Enge, "Differential GPS," in *Global Positioning System: Theory and Applications, Volume II*, B. W. Parkinson and J. J. Spilker, Jr., (eds.), Washington, D.C.: AIAA, 1996, pp. 3–50.

[5] Kalafus, R. M., A. J. Van Dierendonck, and N. A. Pealer, "Special Committee 104 Recommendations for Differential GPS Service," *Navigation: JION*, Vol. 33, No. 1, 1986, pp. 26–41.

[6] SC 104, *RTCM Recommended Standards for Differential GNSS Service*, Version 3.0, RTCM, Alexandria, VA, 2004.

[7] Rife, J., and X. Xiao, "Estimation of Spatially Correlated Errors in Vehicular Collaborative Navigation with Shared GNSS and Road-Boundary Measurements," *Proc. ION GNSS 2010*, Portland, OR, September 2010, pp. 1667–1677.

[8] Kee, C., B. W. Parkinson, and P. Axelrad, "Wide Area Differential GPS," *Navigation: JION*, Vol. 38, No. 2, 1991, pp. 123–145.

[9] Ashkenazi, V., et al., "Wide-Area Differential GPS: A Performance Study," *Navigation: JION*, Vol. 40, No. 3, 1993, pp. 297–319.

[10] Kee, C., "Wide Area Differential GPS," in *Global Positioning System: Theory and Applications, Volume II*, B. W. Parkinson and J. J. Spilker, Jr., (eds.), Washington, DC: AIAA, 1996, pp. 81–115.

[11] Muellerschoen, R. J., et al., "Real-Time Precise-Positioning Performance Evaluation of Single-Frequency Receivers Using NASA's Global Differential GPS System," *Proc. ION GNSS 2004*, Long Beach, CA, September 2004, pp. 1872–1880.

[12] El-Rabbany, A., *Introduction to GPS: The Global Positioning System*, 2nd ed., Norwood, MA: Artech House, 2006.

[13] Sharpe, T., R. Hatch, and F. Nelson, "John Deere's StarFire System: WADGPS for Precision Agriculture," *Proc. ION GPS 2000*, Salt Lake City, UT, September 2000, pp. 2269–2277.

[14] Bisnath, S., and Y. Gao, "Precise Point Positioning: A Powerful Technique with a Promising Future," *GPS World*, April 2009, pp. 43–50.

[15] Pocknee, S., et al., "Experiences with the OmniSTAR HP Differential Correction Service on an Autonomous Agricultural Vehicle," *Proc. ION 60th AM*, Dayton, OH, June 2004, pp. 346–353.

[16] Dixon, K., "StarFire^{TM}: A Global SBAS for Sub-Decimeter Precise Point Positioning," *Proc. ION GNSS 2006*, Fort Worth, TX, September 2006, pp. 2286–2296.

[17] Caissy, M., et al, "Coming Soon: The International GNSS Real-Time Service," *GPS World*, June 2012, pp. 52–58.

[18] "International GNSS Service," http://igscb.jpl.nasa.gov/, accessed in November 2011.

[19] Leick, A., *GPS Satellite Surveying*, 3rd ed., New York: Wiley, 2004.

[20] Hoffman-Wellenhof, B., H. Lichtenegger, and E. Wasle, *GNSS: Global Navigation Satellite Systems: GPS, GLONASS, Galileo & More*, Vienna, Austria: Springer, 2008.

[21] Rizos, C., and D. A. Grejner-Brzezinska, "Geodesy and Surveying," in *GNSS Applications and Methods*, S. Gleason and D. Gebre-Egziabher, (eds.), Norwood, MA: Artech House, 2009, pp. 347–380.

[22] Raquet, J., G. Lachapelle, and T. Melgård, "Test of a 400 km × 600 km Network of Reference Receivers for Precise Kinematic Carrier-Phase Positioning in Norway," *Proc. ION GPS-98*, Nashville, TN, September 1998, pp. 407–416.

[23] Wübbena, G., M. Schmitz, and A. Bagge, "PPP-RTK: Precise Point Positioning Using State-Space Representation in RTK Networks," *Proc. ION GNSS 2005*, Long Beach, CA, September 2005, pp. 2584–2594.

[24] Ge, M., et al., "Resolution of GPS Carrier-Phase Ambiguities in Precise Point Positioning (PPP) with Daily Observations," *Journal of Geodesy*, Vol. 82, No. 7, 2008, pp. 389–399.

[25] Yamada, H., et al., "Evaluation and Calibration of Receiver Inter-Channel Biases for RTK-GPS/GLONASS," *Proc. ION GNSS 2010*, Portland, OR, September 2010, pp. 1580–1587.

[26] Misra, P., and P. Enge, *Global Positioning System Signals, Measurements, and Performance*, 2nd ed., Lincoln, MA: Ganga-Jamuna Press, 2006.

[27] Kim, D., and R. B. Langley, "Instantaneous Real-Time Cycle-Slip Correction for Quality Control of GPS Carrier-phase Measurements," *Navigation: JION*, Vol. 49, No. 4, 2002, pp. 205–222.

[28] Hatch, R. R., "A New Three-Frequency, Geometry-Free, Technique for Ambiguity Resolution," *Proc. ION GNSS 2006*, Fort Worth, TX, September 2006, pp. 309–316.

[29] Teunissen, P. J. G, "Least Squares Estimation of Integer GPS Ambiguities," *Proc. International Association of Geodesy General Meeting*, Beijing, China, August 1993.

[30] Teunissen, P. J. G, P. J. De Jonge, and C. C. J. M. Tiberius, "Performance of the LAMBDA Method for Fast GPS Ambiguity Resolution," *Navigation: JION*, Vol. 44, No. 3, 1997, pp. 373–383.

[31] Lawrence, D. G., "A New Method for Partial Ambiguity Resolution," *Proc. ION ITM*, Anaheim, CA, January 2009, pp. 652–663.

[32] Feng, Y., and B. Li, "Three Carrier Ambiguity Resolution: Generalised Problems, Models, Methods and Performance Analysis Using Semi-Generated Triple Frequency GPS Data," *Proc. ION GNSS 2008*, Savannah, GA, September 2008, pp. 2831–2840.

[33] Cohen, C. E., "Attitude Determination," in *Global Positioning System: Theory and Applications, Volume II*, B. W. Parkinson and J. J. Spilker, Jr., (eds.), Washington, D.C.: AIAA, 1996, pp. 518–538.

[34] Farrell, J. A., and M. Barth, *The Global Positioning System and Inertial Navigation*, New York: McGraw-Hill, 1999.

[35] Teunissen, P. J. G., G. Giorgi, and P. J. Buist, "Testing of a New Single-Frequency GNSS Carrier Phase Attitude Determination Method: Land, Ship and Aircraft Experiments," *GPS Solutions*, Vol. 15, No. 1, 2011, pp. 15–28.

[36] Wang, C., R. A. Walker, and Y. Feng, "Performance Evaluation of Single Antenna GPS Attitude Algorithms with the Aid of Future GNSS Constellations," *Proc. ION GNSS 2007*, Fort Worth, TX, September 2007, pp. 883–891.

[37] Carroll, J., et al., *Vulnerability Assessment of the Transportation Infrastructure Relying on the Global Positioning System*, John A. Volpe National Transportation Systems Center report for U.S. Department of Transportation, 2001.

[38] Spilker, J. J., Jr., and F. D. Natali, "Interference Effects and Mitigation Techniques," in *Global Positioning System: Theory and Applications, Volume I*, pp. 717–771, B. W. Parkinson and J. J. Spilker, Jr. (eds.), Washington, D.C.: AIAA, 1996.

[39] Ward, P. W., J. W. Betz, and C. J. Hegarty, "Interference, Multipath and Scintillation," in *Understanding GPS: Principles and Applications*, 2nd ed., E. D. Kaplan and C. J. Hegarty, (eds.), Norwood, MA: Artech House, 2006, pp. 243–299.

[40] Thomas, M., et al., *Global Navigation Space Systems: Reliance and Vulnerabilities*, London, U.K.: Royal Academy of Engineering, 2011.

[41] Boulton, P., et al., "GPS Interference Testing: Lab, Live, and LightSquared," *Inside GNSS* July/August 2011, pp. 32–45.

[42] Cerruti, A., "Observed GPS and WAAS Signal-to-Noise Degradation Due to Solar Radio Bursts," *Proc. ION GNSS 2006*, Fort Worth, TX, September 2006, pp. 1369–1376.

[43] Haddrell, T., and A. R. Pratt, "Understanding the Indoor GPS Signal," *Proc. ION GPS 2001*, Salt Lake City, UT, September 2001, pp. 1487–1499.

[44] Spilker, J. J., Jr., "Foliage Attenuation for Land Mobile Users," in *Global Positioning System: Theory and Applications, Volume I*, B. W. Parkinson and J. J. Spilker, Jr., (eds.), Washington, D.C.: AIAA, 1996, pp. 569–583.

[45] Bancroft, J. B., et al., "Observability and Availability for Various Antenna Locations on the Human Body," *Proc. ION GNSS 2010*, Portland, OR, September 2010, pp. 2941–2951.

[46] Bancroft, J. B., et al., "GNSS Antenna-Human Body Interaction," *Proc. ION GNSS 2011*, Portland, OR, September 2011, pp. 3952–3958.

[47] Brodin, G., J. Cooper, and D. Walsh, "The Effect of Helicopter Rotors on GPS Signal Reception," *Journal of Navigation*, Vol. 58, No. 3, 2005, pp. 433–450.

[48] Jafarnia-Jahromi, A., et al., "GPS Vulnerability to Spoofing Threats and a Review of Antispoofing Techniques," *International Journal of Navigation and Observation*, 2012, Article ID 127072.

[49] Rounds, S., "Jamming Protection of GPS Receivers, Part II: Antenna Enhancements," *GPS World*, February 2004, pp. 38–45.

[50] Owen, J. I. R., and M. Wells, "An Advanced Digital Antenna Control Unit for GPS," *Proc. ION NTM*, Long Beach, CA, January 2001, pp. 402–407.

[51] Soloviev, A., and F. Van Graas, "Beam Steering in Global Positioning System Receivers Using Synthetic Phased Arrays," *IEEE Trans. of Aerospace and Electronic Systems*, Vol. 446, No. 3, 2010, pp. 1513–1521.

[52] Nielson, J., et al., "Enhanced Detection of Weak GNSS Signals Using Spatial Combining," *Navigation: JION*, Vol. 56, No. 2, 2009, pp. 83–95.

[53] Hegarty, C., et al., "Suppression of Pulsed Interference Through Blanking," *Proc. ION 56th AM*, San Diego, CA, June 2000, pp. 399–408.

[54] Anyaegbu, E., et al., "An Integrated Pulsed Interference Mitigation for GNSS Receivers," *Journal of Navigation*, Vol. 61, No. 2, 2008, pp. 239–255.

[55] Capozza, P. T., et al., "A Single-Chip Narrow-Band Frequency-Domain Excisor for a Global Positioning System (GPS) Receiver," *IEEE Journal of Solid-State Circuits*, Vol. 35, No. 3, 2000, pp. 401–411.

[56] Dafesh, P. A., R. Prabhu, and E. L. Vallés, "Cognitive Antijam Receiver System (CARS) for GNSS," *Proc. ION NTM*, San Diego, CA, January 2010, pp. 657–666.

[57] Van Graas, F., et al., "Comparison of Two Approaches for GNSS Receiver Algorithms: Batch Processing and Sequential Processing Considerations," *Proc. ION GNSS 2005*, Long Beach, CA, September 2005, pp. 200–211.

[58] Anyaegbu, E., "A Frequency Domain Quasi-Open Tracking Loop for GNSS Receivers," *Proc. ION GNSS 2006*, Fort Worth, TX, September 2006, pp. 790–798.

[59] Ziedan, N. I., *GNSS Receivers for Weak Signals*, Norwood, MA: Artech House, 2006

[60] Soloviev, A., F. Van Graas, and S. Gunawardena, "Implementation of Deeply Integrated GPS/Low-Cost IMU for Reacquisition and Tracking of Low CNR GPS Signals," *Proc. ION NTM*, San Diego, CA, January 2004, pp. 923–935.

[61] Ziedan, N. I., and J. L. Garrison, "Unaided Acquisition of Weak GPS Signals Using Circular Correlation or Double-Block Zero Padding," *Proc. IEEE PLANS*, Monterey, CA, April 2004, pp. 461–470.

[62] Psiaki, M. L., and H., Jung, "Extended Kalman Filter Methods for Tracking Weak GPS Signals," *Proc. ION GPS 2002*, Portland OR, September 2002, pp. 2539–2553.

[63] Watson, W., et al., "Investigating GPS Signals Indoors with Extreme High-Sensitivity Detection Techniques," *Navigation: JION*, Vol. 52, No. 4, 2005, pp. 199–213.

[64] Mongrédien, C., M. E. Cannon, and G. Lachapelle, "Performance Evaluation of Kalman Filter Based Tracking for the New GPS L5 Signal," *Proc. ION GNSS 2007*, Fort Worth, TX, September 2007, pp. 749–758.

[65] Kim, K. -H., et al., "The Adaptive Combined Receiver Tracking Filter Design for High Dynamic Situations," *Proc. IEEE/ION PLANS*, Monterey, CA, May 2008, pp. 203–209.

[66] Ioannides, R. T., L. E. Aguado, and G. Brodin, "Coherent Integration of Future GNSS Signals," *Proc. ION GNSS 2006*, Fort Worth, TX, September 2006, pp. 1253–1268.

[67] Gernot, C., K. O'Keefe, and G. Lachapelle, "Assessing Three New GPS Combined L1/L2C Acquisition Methods," *IEEE Trans. on Aerospace and Electronic Systems*, Vol. 47, No. 3, 2011, pp. 2239–2247.

[68] Gernot, C., K. O'Keefe, and G. Lachapelle, "Combined L1/L2 Kalman Filter-Based Tracking Scheme for Weak Signal Environments," *GPS Solutions*, Vol. 15, No. 4, 2011, pp. 403–414.

[69] Spilker, J. J., Jr., "Fundamentals of Signal Tracking Theory," in *Global Positioning System: Theory and Applications, Volume I*, B. W. Parkinson and J. J. Spilker, Jr., (eds.), Washington, D.C.: AIAA, 1996, pp. 245–327.

[70] Lashley, M., and D. M. Bevly, "Vector Delay/Frequency Lock Loop Implementation and Analysis," *Proc. ION ITM*, Anaheim, CA, January 2009, pp. 1073–1086.

[71] Bhattacharyya, S., and D. Gebre-Egziabher, "Development and Validation of Parametric Models for Vector Tracking Loops," *NAVIGATION: JION*, Vol. 57, No. 4, 2010, pp. 275–295.

[72] Groves, P. D., and C. J. Mather, "Receiver Interface Requirements for Deep INS/GNSS Integration and Vector Tracking," *Journal of Navigation*, Vol. 63, No. 3, 2010, pp. 471–489.

[73] Lau, L., and P. Cross, "Investigations into Phase Multipath Mitigation Techniques for High Precision Positioning in Difficult Environments," *Journal of Navigation*, Vol. 60, No. 1, 2007, pp. 95–105.

[74] Braasch, M. S., "Multipath Effects," in *Global Positioning System: Theory and Applications, Volume I*, B. W. Parkinson and J. J. Spilker, Jr., (eds.), Washington, D.C.: AIAA, 1996, pp. 547–568.

[75] Brown, A., and N. Gerein, "Test Results from a Digital P(Y) Code Beamsteering Receiver for Multipath Minimization," *Proc. ION 57th AM*, Albuquerque, NM, June 2001, pp. 872–878.

[76] Weiss, J. P., et al., "Analysis of P(Y) Code Multipath for JPALS LDGPS Ground Station and Airborne Receivers," *Proc. ION GNSS 2004*, Long Beach, CA, September 2004, pp. 2728–2741.

[77] McGraw, G. A., et al., "GPS Multipath Mitigation Assessment of Digital Beam Forming Antenna Technology in a JPALS Dual Frequency Smoothing Architecture," *Proc. ION NTM*, San Diego, CA, January 2004, pp. 561–572.

[78] Draganov, S., M. Harlacher, and L. Haas, "Multipath Mitigation via Synthetic Aperture Beamforming," *Proc. ION GNSS 2009*, Savannah, GA, September 2009, pp. 1707–1715.

[79] Keshvadi, M. H., A. Broumandan, and G. Lachapelle, "Analysis of GNSS Beamforming and Angle of Arrival Estimation in Multipath Environments," *Proc. ION ITM*, San Diego, CA, January 2011, pp. 427–435.

[80] Pratt, A. R., "Performance of Multi-Path Mitigation Techniques at Low Signal to Noise Ratios," *Proc. ION GNSS 2004*. Long Beach, CA, September 2004, pp. 43–53.

[81] Irsigler, M., and B. Eissfeller, "Comparison of Multipath Mitigation Techniques with Consideration of Future Signal Structures," *Proc. ION GPS/GNSS 2003*, Portland, OR, September 2003, pp. 2584–2592.

[82] Hatch, R. R., R. G. Keegan, and T. A. Stansell, "Leica's Code and Phase Multipath Mitigation Techniques," *Proc. ION NTM*, January 1997, pp. 217–225.

[83] Garin, L., and J.-M., Rousseau, "Enhanced Strobe Correlator Multipath Rejection for Code and Carrier," *Proc. ION GPS-97*, Kansas, MO, September 1997, pp. 559–568.

[84] McGraw. G. A., and M. S. Braasch, "GNSS Multipath Mitigation Using Gated and High Resolution Correlator Concepts," *Proc. ION GPS-99*, Nashville, TN, September 1999, pp. 333–342.

[85] Townsend, B., and P. Fenton, "A Practical Approach to the Reduction of Pseudorange Multipath Errors in a L1 GPS Receiver," *Proc. ION GPS-94*, Salt Lake City, UT, September 1994, pp. 143–148.

[86] Mattos, P. G., "Multipath Elimination for the Low-Cost Consumer GPS," *Proc. ION GPS-96*, Kansas, MO, September 1996, pp. 665–672.

[87] Weill, L. R., "Application of Superresolution Concepts to the GPS Multipath Mitigation Problem," *Proc. ION NTM*, Long Beach, CA, January 1998, pp. 673–682.

[88] Townsend, B. R., "Performance Evaluation of the Multipath Estimating Delay Lock Loop," *Proc. ION NTM*, Anaheim, CA, January 1995.

[89] Fenton, P. C., and J. Jones, "The Theory and Performance of NovAtel Inc's Vision Correlator," *Proc. ION GNSS 2005*, Long Beach, CA, September 2005, pp. 2178–2186.

[90] Soloviev, A., and F. van Graas, "Utilizing Multipath Reflections in Deeply Integrated GPS/INS Architecture for Navigation in Urban Environments," *Proc. IEEE/ION PLANS*, Monterey, CA, May 2008, pp. 383–393.

[91] Xie, P., M. G. Petovello, and C. Basnayake, "Multipath Signal Assessment in the High Sensitivity Receivers for Vehicular Applications," *Proc. ION GNSS 2011*, Portland, OR, pp. 1764–1776.

[92] Lashley, M., and D. M. Bevly, "Comparison in the Performance of the Vector Delay/Frequency Lock Loop and Equivalent Scalar Tracking Loops in Dense Foliage and Urban Canyon," *Proc. ION GNSS 2011*, Portland, OR, September 2011, pp. 1786–1803.

[93] Viandier, N., et al., "GNSS Performance Enhancement in Urban Environment Based on Pseudo-Range Error Model," *Proc. IEEE/ION PLANS*, Monterey, CA, May 2008, pp. 377–382.

[94] Mendoume, I., et al., "GNSS Positioning Enhancement Based on Statistical Modeling in Urban Environment," *Proc. ION GNSS 2010*, Portland, OR, September 2010, pp. 2221–2227.

[95] Kashiwayanagi, T., et al., "Novel Algorithm to Exclude Multipath Satellites by Dual Frequency Measurements in RTK-GPS," *Proc. ION GNSS 2007*, Fort Worth, TX, September 2007, pp. 1741–1747.

[96] Mattos, P. G., *Multipath Indicator to Enhance RAIM and FDE in GPS/GNSS Systems*, Patent Application No. 11112819.5, filed July 2011.

[97] Mubarak, O. M., and A. G. Dempster, "Analysis of Early Late Phase in Single- and Dual-Frequency GPS Receivers for Multipath Detection," *GPS Solutions*, Vol. 14, No. 4, 2010, pp. 381–388.

[98] Jiang, Z., et al., "Multi-Constellation GNSS Multipath Mitigation Using Consistency Checking," *Proc. ION GNSS 2011*, Portland, OR, pp. 3889–3902.

[99] Jiang, Z., and P. Groves, "GNSS NLOS and Multipath Error Mitigation Using Advanced Multi-Constellation Consistency Checking with Height Aiding," *Proc. ION GNSS 2012*, Nashville, TN, September 2012, pp. 79–88.

[100] Iwase, T., N. Suzuki, and Y. Watanabe, "Estimation and Exclusion of Multipath Range Error for Robust Positioning," *GPS Solutions*, 2012, DOI 10.1007s/10291-012-0260-1.

[101] Spangenberg, M., et al., "Detection of Variance Changes and Mean Value Jumps in Measurement Noise for Multipath Mitigation in Urban Navigation," *Navigation: JION*, Vol. 57, No. 1, pp. 35–52.

[102] Jiang, Z., and P. D. Groves, "NLOS GPS Signal Detection Using a Dual-Polarization Antenna," *GPS Solutions*, Accepted for publication December 2012, DOI 10.1007/s10291-012-0305.

[103] Marais, J., M. Berbineau, and M. Heddebaut, "Land Mobile GNSS Availability and Multipath Evaluation Tool," *IEEE Trans. on Vehicular Technology*, Vol. 54, No. 5, 2005, pp. 1697–1704.

[104] Meguro, J., et al., "GPS Multipath Mitigation for Urban Area Using Omnidirectional Infrared Camera," *IEEE Trans. on Intelligent Transportation Systems*, Vol. 10, No. 1, 2009, pp. 22–30.

[105] Obst, M., S. Bauer, and G. Wanielik, "Urban Multipath Detection and Mitigation with Dynamic 3D Maps for Reliable Land Vehicle Localization," *Proc. IEEE/ION PLANS*, Myrtle Beach, SC, April 2012, pp. 685–691.

[106] Groves, P. D., et al., "Intelligent Urban Positioning Using Multi-Constellation GNSS with 3D Mapping and NLOS Signal Detection," *Proc. ION GNSS 2012*, Nashville, TN, September 2012, pp. 458–472.

[107] Bourdeau, A., M. Sahmoudi, and J. -Y. Tourneret, "Tight Integration of GNSS and a 3D City Model for Robust Positioning in Urban Canyons, *Proc. ION GNSS 2012*, Nashville, TN, September 2012, pp. 1263–1269.

[108] Van Diggelen, F., *A-GPS: Assisted GPS, GNSS, and SBAS*, Norwood, MA: Artech House, 2009.

[109] Pratt, T., R. Faragher, and P. Duffett-Smith, "Fine Time Aiding and Pseudo-Synchronisation of GSM Networks," *Proc. ION NTM*, Monterey, CA, January 2006, pp. 167–173.

[110] Garello, R., et al., "Peer-to-Peer Cooperative Positioning Part 1: GNSS-Aided Acquisition," *Inside GNSS*, March/April 2012, pp. 55–63.

[111] Wesson, K. D., et al., "Opportunistic Frequency Stability Transfer for Extending the Coherence Time of GNSS Receiver Clocks," *Proc. ION GNSS 2010*, Portland, OR, September 2010, pp. 2959–2968.

[112] Li, T., et al., "Real-Time Ultra-Tight Integration of GPS L1/L2C and Vehicle Sensors," *Proc. ION ITM*, San Diego, CA, January 2011, pp. 725–736.

[113] Chan, B., and M. G. Petovello, "Collaborative Vector Tracking of GNSS Signals with Ultra-Wideband Augmentation in Degraded Signal Environments," *Proc. ION ITM*, San Diego, CA, January 2011, pp. 404–413.

[114] Colombo, O. L., U. V. Bhapkar, and A. G. Evans, "Inertial-Aided Cycle-Slip Detection/Correction for Precise, Long-Baseline Kinematic GPS," *Proc. ION GPS-99*, Nashville, TN, September 1999, pp. 1915–1921.

[115] Song, J., et al., "Odometer-Aided Real Time Cycle Slip Detection Algorithm for Land Vehicle Users," *Proc. ION ITM*, San Diego, CA, January 2011, pp. 326–335.

[116] Bullock, J. B., et al., "Integration of GPS with Other Sensors and Network Assistance," in *Understanding GPS: Principles and Applications*, 2nd ed., E. D. Kaplan and C. J. Hegarty, (eds.), Norwood, MA: Artech House, 2006, pp. 459–558.

[117] Stacey, P., and M. Ziebart, "Long-Term Extended Ephemeris Prediction for Mobile Devices," *Proc. ION GNSS 2011*, Portland, OR, September 2011, pp. 3235–3244.

[118] Mattos, P. G., "Hotstart Every Time: Compute the Ephemeris on the Mobile," *Proc. ION GNSS 2008*, Savannah, GA, September 2008, pp. 204–211.

[119] Groves, P. D., "Shadow Matching: A New GNSS Positioning Technique for Urban Canyons," *Journal of Navigation*, Vol. 64, No. 3, 2011, pp. 417–430.

[120] Wang, L., P. D. Groves, and M. K. Ziebart, "GNSS Shadow Matching: Improving Urban Positioning Accuracy Using a 3D City Model with Optimized Visibility Prediction Scoring," *Proc. ION GNSS 2012*, Nashville, TN, September 2012, pp. 423–437.

[121] Groves, P. D., L. Wang, and M. K. Ziebart, "Shadow Matching: Improved GNSS Accuracy in Urban Canyons," *GPS World*, February 2012, pp. 14–29.

CHAPTER 11
Long- and Medium-Range Radio Navigation

This chapter describes the main features of long- and medium-range radio positioning systems other than GNSS, building on the principles described in Chapter 7 and focuses mainly on self-positioning. Most, but not all, of these systems use terrestrial base stations. Note that measurements from different types of signal can be combined to form a position solution in locations where there are insufficient signals of one type.

Section 11.1 describes aircraft navigation systems, including DME. Sections 11.2 and 11.3 describe Enhanced Loran and phone positioning, respectively. Finally, Section 11.4 summarizes positioning using Iridium, marine radio beacons, and television and radio broadcasts, and discusses generic radio positioning techniques.

Further positioning systems are described in Appendix F on the CD. Section F.2 describes radio determination satellite services (RDSS), including the Beidou Position Reporting Service; Section F.3 describes landing guidance systems for aircraft; Section F.4 describes a number of radio tracking systems, including the Enhanced Position Location Reporting System (EPLRS), Datatrak, and the Deep Space Network (DSN); Section F.5 summarizes phone positioning terminology; and Section F.6 provides further information on positioning using television and radio broadcasts. Furthermore, descriptions of some historical radio navigation systems, including Omega, Decca Navigator, and Loran A–D, may be found in Appendix K on the CD.

11.1 Aircraft Navigation Systems

A number of radio navigation systems have been developed specifically for aviation use and are thus optimized for that context. These systems use VHF and UHF signals, which rely on line-of-sight propagation. Due to a combination of Earth curvature and terrestrial obstructions, these signals propagate much further where the transmitter or receiver is in the air. Figure 11.1 shows how the radio horizon (see Section 7.4.1) varies with receiver height, assuming a transmit antenna height of 100m above the terrain.

Most of the technologies described here predate GNSS and are retained to back up and supplement GNSS so that the demanding integrity, continuity, and availability requirements (see Section 17.6) of many aviation applications may be met. DME is described first, followed by range-bearing systems, nondirectional beacons, and JTIDS/MIDS Relative Navigation. The section concludes with a discussion of possible future systems. In addition, Section F.3 of Appendix F on the CD describes

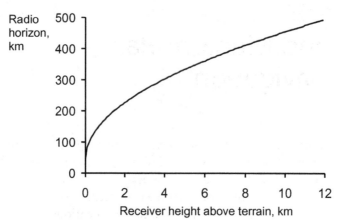

Figure 11.1 Radio horizon as a function of receiver height for a transmit antenna height of 100m above terrain.

the Instrument Landing System (ILS) and Microwave Landing System (MLS), which are used for guiding aircraft approaches to runways.

11.1.1 Distance Measuring Equipment

Distance Measuring Equipment is a medium-range two-way ranging system (see Section 7.1.4.5), providing horizontal positioning only [1–3]. Mutually-incompatible DME systems were developed by a number of countries during and after World War II. An international standard was adopted by the International Civil Aviation Organization (ICAO) in 1959, which forms the basis of the current DME system. DME was originally designed to operate as part of a range-bearing system (see Section 11.1.2). However, stand-alone DME operation using two or more base stations is now the recommended mode of operation. This is sometimes referred to as DME/DME and a number of new DME-only stations have been introduced to improve coverage. Note from Section 7.1.4 that prior or additional information is needed to resolve an ambiguity where only two base stations are used.

The standard service radius of a DME base station is 240 km at aircraft altitudes between 5.5 and 13.7 km above ground level, reducing to 185 km at altitudes of 4.4–5.5 and 13.7–18.3 km, and to 74 km at altitudes of 0.3–4.4 km [4]. Modern equipment will actually operate up to a range of 500 km [5] (radio horizon permitting), albeit with reduced accuracy.

11.1.1.1 Signal Structure and Ranging Protocol

DME signals are vertically polarized and comprise double pulses, which are easier to distinguish from pulse interference than single pulses. Ranging is initiated by the user equipment, known as an interrogator, transmitting a double pulse. The DME base station or beacon, known as a transponder, then broadcasts a double pulse on a separate frequency 50 μs or 56 μs after receiving the interrogator's signal. Range is calculated by the interrogator as described in Section 7.1.4.5.

11.1 Aircraft Navigation Systems

Table 11.1 DME Mode Characteristics

Mode	X	Y	W	Z
Pulse interval (interrogation) (μs)	12	36	24	21
Pulse interval (reply) (μs)	12	30	24	15
Transponder reply delay (μs)	50	56	50	56
Number of channels	126	126	20	80
Interrogation frequencies (MHz)	1,025–1,150	1,025–1,150	1,042–1,080 (even only)	1,041–1,080 and 1,104–1,143
Reply frequencies (MHz)	962–1,024 and 1,151–1,213	1,025–1,150	979–1,017 (odd only)	1,041–1,080 and 1,104–1,143

DME operates using 252 carrier frequencies at a 1-MHz separation within the 960–1,215-MHz aeronautical radionavigation band. In the normal DME operating mode, the bandwidth is 300–400 kHz and the pulse shape is Gaussian with an FWHM amplitude of 3.5±0.5 μs [3]. DME was originally designed as an FDMA only system. However, a limited amount of CDMA in the form of four modes, known as X, Y, W, and Z, was subsequently introduced to increase capacity. Most stations operate in X mode. A DME channel thus comprises a pair of frequencies and a mode. Table 11.1 lists the mode characteristics [6]. In all cases, the interrogation and reply frequencies are 63 MHz apart.

Each DME transponder is designed to serve 100 users at a time, although newer transponders can handle more users. If too many users attempt to use a transponder, it will respond to the strongest interrogation signals. When signals from two interrogators are received in close succession, typically within about 100 μs, the transponder can only respond to the earlier signal. Random intervals between successive interrogations prevent repeated clashes between any pair of users.

As each DME transponder transmits pulses in response to many users, the interrogator must identify which are in response to its own signals. Initially, the range to the transponder is unknown, so the interrogator operates in search mode, essentially an acquisition process, where it may emit up to 150 pulse pairs per second. It attempts to detect a response at a number of fixed intervals from transmission (depending on the available processing power), changing these intervals every few pulses. When the interval corresponds to the response time, pulses will be received in response to most interrogations. Otherwise, pulses will only be received occasionally as the responses to the other users are uncorrelated with the interrogator's transmissions. Figure 11.2 illustrates this. Once the response time has been found, the interrogator switches to track mode, dropping its interrogation rate to within 30 pulse pairs per second and only processing responses close to the predicted response time. Modern user equipment can typically acquire a DME transponder within 1 second. When multiple DME transponders are acquired and tracked, the pulse rate limits are applied to the total interrogator output as opposed to the output per transponder.

Every 30–40 seconds, each DME station transmits an identification message for about 3 seconds. During this time, it stops responding to interrogations. DME tracking loop time constants are sufficiently long for this gap to be bridged using previous measurements. An interrogator will typically switch to acquisition mode

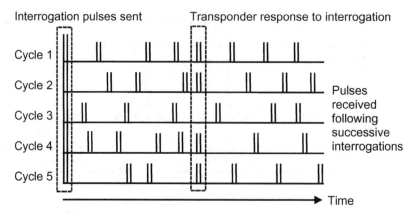

Figure 11.2 DME pulses received following successive interrogations. (*After:* [1].)

after 10 seconds without responses from a base station. This must be accounted for when integrating DME with other navigation sensors.

Each base station is equipped with a monitoring interrogator that switches the station to a standby transponder or shuts it down in the event of a fault.

11.1.1.2 Position Determination

DME only provides a horizontal position solution. However, the basic 2-D positioning algorithm presented in Section 7.3.3 will not provide an accurate solution for two reasons. First, the curvature of the Earth is significant for medium-range systems. Second, the height difference of the aircraft and the base station must be accounted for. Because of this, DME measurements are known as slant ranges. Figure 11.3 illustrates this. Therefore, 3-D positioning equations (see Section 9.4) should be used with the aircraft height, obtained from an altimeter (Section 6.2), treated as a known parameter.

Assuming an overdetermined solution with m measurements, the measured ECEF-frame Cartesian position of the user antenna, $\tilde{\mathbf{r}}_{ea}^e$, , is obtained by solving equations of the form

$$\tilde{r}_{aj,C} = \sqrt{\left(\mathbf{r}_{ej}^e - \tilde{\mathbf{r}}_{ea}^e\right)^{\mathrm{T}}\left(\mathbf{r}_{ej}^e - \tilde{\mathbf{r}}_{ea}^e\right)} + \delta r_{aj,\varepsilon}^+ \quad j \in 1,2,\ldots,m, \qquad (11.1)$$

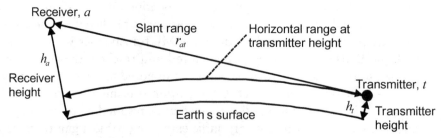

Figure 11.3 Relationship between slant range and horizontal range.

11.1 Aircraft Navigation Systems

where $\tilde{r}_{aj,C}$ is the jth slant range measurement, corrected for the estimated troposphere propagation delay (see Section 7.4.1), $\mathbf{r}^e_{ej}$ is the jth transmit antenna position and $\delta r^+_{aj,\varepsilon}$ is the jth measurement residual, defined by (7.34). The Sagnac effect (see Section 8.5.3) is neglected here as it introduces errors of less than a meter. Cartesian ECEF position is obtained from latitude, longitude, and height using (2.112).

Predicted slant ranges are obtained from the predicted user position, $\hat{\mathbf{r}}^{e-}_{ea}$, using

$$\hat{r}^-_{aj} = \sqrt{\left(\mathbf{r}^e_{ej} - \hat{\mathbf{r}}^{e-}_{ea}\right)^T \left(\mathbf{r}^e_{ej} - \hat{\mathbf{r}}^{e-}_{ea}\right)} \quad j \in 1,2,\ldots,m, \quad (11.2)$$

Subtracting the predicted ranges from the measured ranges and applying a first-order Taylor expansion about the predicted user position gives

$$\begin{pmatrix} \tilde{r}_{a1,C} - \hat{r}^-_{a1,C} \\ \tilde{r}_{a2,C} - \hat{r}^-_{a2,C} \\ \vdots \\ \tilde{r}_{am,C} - \hat{r}^-_{am,C} \end{pmatrix} = \mathbf{H}^{L\lambda}_R \begin{pmatrix} \tilde{L}_a - \hat{L}^-_a \\ \tilde{\lambda}_a - \hat{\lambda}^-_a \end{pmatrix} + \begin{pmatrix} \delta r^+_{a1,\varepsilon} \\ \delta r^+_{a2,\varepsilon} \\ \vdots \\ \delta r^+_{am,\varepsilon} \end{pmatrix}, \quad (11.3)$$

where the linearization errors are included in the residuals, and the measurement matrix, $\mathbf{H}^{L\lambda}_R$, is given by

$$\begin{aligned}
\mathbf{H}^{L\lambda}_R &= \begin{pmatrix} \frac{\partial r_{a1}}{\partial x^e_{ea}} & \frac{\partial r_{a1}}{\partial y^e_{ea}} & \frac{\partial r_{a1}}{\partial z^e_{ea}} \\ \frac{\partial r_{a2}}{\partial x^e_{ea}} & \frac{\partial r_{a2}}{\partial y^e_{ea}} & \frac{\partial r_{a2}}{\partial z^e_{ea}} \\ \vdots & \vdots & \vdots \\ \frac{\partial r_{am}}{\partial x^e_{ea}} & \frac{\partial r_{am}}{\partial y^e_{ea}} & \frac{\partial r_{am}}{\partial z^e_{ea}} \end{pmatrix}_{\mathbf{r}^e_{ea}=\hat{\mathbf{r}}^{e-}_{ea}} \times \begin{pmatrix} \frac{\partial x^e_{ea}}{\partial L_a} & \frac{\partial x^e_{ea}}{\partial \lambda_a} \\ \frac{\partial y^e_{ea}}{\partial L_a} & \frac{\partial y^e_{ea}}{\partial \lambda_a} \\ \frac{\partial z^e_{ea}}{\partial L_a} & \frac{\partial z^e_{ea}}{\partial \lambda_a} \end{pmatrix}_{(L_a,\lambda_a)=(\hat{L}^-_a,\hat{\lambda}^-_a)} \\
&= \begin{pmatrix} -\hat{u}^{e-}_{a1,x} & -\hat{u}^{e-}_{a1,y} & -\hat{u}^{e-}_{a1,z} \\ -\hat{u}^{e-}_{a2,x} & -\hat{u}^{e-}_{a2,y} & -\hat{u}^{e-}_{a2,z} \\ \vdots & \vdots & \vdots \\ -\hat{u}^{e-}_{am,x} & -\hat{u}^{e-}_{am,y} & -\hat{u}^{e-}_{am,z} \end{pmatrix} \begin{pmatrix} -\left(R_E(\hat{L}^-_a) + \hat{h}_a\right)\sin\hat{L}^-_a \cos\hat{\lambda}^-_a & -\left(R_E(\hat{L}^-_a) + \hat{h}_a\right)\cos\hat{L}^-_a \sin\hat{\lambda}^-_a \\ -\left(R_E(\hat{L}^-_a) + \hat{h}_a\right)\sin\hat{L}^-_a \sin\hat{\lambda}^-_a & \left(R_E(\hat{L}^-_a) + \hat{h}_a\right)\cos\hat{L}^-_a \cos\hat{\lambda}^-_a \\ \left[(1-e^2)R_E(\hat{L}^-_a) + \hat{h}_a\right]\cos\hat{L}^-_a & 0 \end{pmatrix}
\end{aligned}$$
$$(11.4)$$

where e is the eccentricity of the Earth ellipsoid, $\partial R_E/\partial L_a$ is neglected, and the jth line-of-sight unit vector is given by

$$\hat{\mathbf{u}}^{e-}_{aj} = \frac{\mathbf{r}^e_{ej} - \hat{\mathbf{r}}^{e-}_{ea}}{\sqrt{\left(\mathbf{r}^e_{ej} - \hat{\mathbf{r}}^{e-}_{ea}\right)^T \left(\mathbf{r}^e_{ej} - \hat{\mathbf{r}}^{e-}_{ea}\right)}}. \quad (11.5)$$

Using an iterated least-squares algorithm (see Section 7.3.3), the position solution is updated using

$$\begin{pmatrix} \hat{L}_a^+ \\ \hat{\lambda}_a^+ \end{pmatrix} = \begin{pmatrix} \hat{L}_a^- \\ \hat{\lambda}_a^- \end{pmatrix} + \left(\mathbf{H}_R^{L\lambda\mathrm{T}} \mathbf{H}_R^{L\lambda}\right)^{-1} \mathbf{H}_R^{L\lambda\mathrm{T}} \begin{pmatrix} \tilde{r}_{a1,C} - \hat{r}_{a1,C}^- \\ \tilde{r}_{a2,C} - \hat{r}_{a2,C}^- \\ \vdots \\ \tilde{r}_{am,C} - \hat{r}_{am,C}^- \end{pmatrix}. \qquad (11.6)$$

Equations (11.2) and (11.4)–(11.6) are then iterated until the required degree of convergence has been obtained. A weighted ILS or filtered position determination algorithm may also be used (see Section 9.4).

11.1.1.3 Error Sources and Position Accuracy

DME ranging errors vary considerably with the distance between user and base station, with most error sources increasing with distance. Except at close range, the dominant error source is measurement noise; this has greater impact at longer range due to the reduced signal strength. Measurement noise affects the timing of the received signal in both transponder and interrogator as discussed in Section 7.4.3. With modern equipment, signal timing should be accurate to 0.1 μs in each direction, corresponding to 15m of range error per direction [2]. However, an old interrogator or transponder may only be accurate to 1 μs [3]. The current specification for transponder-induced range errors is 92.5m (1σ) for users within the 240-km standard service radius [4].

Scale factor errors (biases proportional to the range) of up to 10 ppm can occur due to interrogator oscillator biases. Variation in the tropospheric refractive index with the weather leads to scale factor errors of around 25 ppm [7]. Multipath can be significant for users at short range and low altitude, with ranging errors of up to 20m [7], depending on the design and siting of the base station antenna. Finally, base station survey errors can be of the order of 5m if old data is used [7]. With modern equipment, total DME ranging errors are around 100m (1σ) at mid-range.

As discussed in Section 7.4.5, the overall position error also depends on the signal geometry. Further dilution of precision occurs due to the interrogator and transponders not being coplanar. When all elevation angles are equal, the additional DOP factor is $1/\cos\theta_{nu}^{at}$, where θ_{nu}^{at} is the elevation angle [8]. Horizontal position errors also arise from errors in the assumed interrogator height.

11.1.1.4 Future Developments

To increase capacity and introduce a datalink, the addition of a passive ranging signal, transmitted by the DME base stations, has been proposed [9]. This could comprise 500 pulse pairs per second, combining a known synchronization sequence with data transmitted using pulse position modulation. Using passive ranging alone would decrease coverage and availability as at least three signals would be required. The positioning accuracy is also much more sensitive to signal geometry in passive

11.1 Aircraft Navigation Systems

ranging than in two-way ranging as explained in Section 7.4.5. Therefore, a hybrid system is proposed in which each user would make fewer two-way ranging measurements than at present, enabling more users to be served simultaneously [10].

DME ranges have traditionally been measured using the half amplitude point on the first pulse, which is less affected by multipath interference than later parts of the waveform. Correlation-based ranging and curve-fitting to the incoming signal both offer reduced tracking noise and multipath interference. However, they would require a standardized pulse shape to be implemented [11].

Although DME pulses are not coherent, the underlying carrier is continuous. Therefore, by measuring the carrier phase each time a pulse is received, the carrier of the signal from the base station to the aircraft can be tracked using a PLL or FLL in a manner similar to that of GNSS. For passive ranging, this enables the tracking and multipath errors to be reduced using carrier smoothing of the pseudo-range measurements in the same way as for GNSS (see Section 9.2.7) [12].

11.1.2 Range-Bearing Systems

Range-bearing positioning systems combine two-way ranging (Section 7.1.4.5) with angular positioning (Section 7.1.5), enabling a full horizontal position solution to be obtained using a single base station. Figure 11.4 illustrates this.

DME was originally designed as the ranging component of the Tactical Air Navigation (TACAN) range-bearing system, deployed by U.S. and NATO military forces. For civil range-bearing navigation, it is paired with VHF omnidirectional radio range, while VORTAC beacons serve both VOR/DME and TACAN users. In all cases, the range and bearing beacons are collocated.

VOR predates DME with international standardization achieved in 1949. Beacons transmit in the 108–118-MHz band. Each of the 200 channels is paired with a DME channel and the coverage radius is the same as for DME. Each VOR is modulated with a 30-Hz AM signal, a 30-Hz FM signal on a subcarrier, an identification code, and an optional voice signal. The relative phase of the AM and FM signals varies with azimuth. By measuring this, VOR receivers can obtain their bearing from the transmitter, generally with respect to magnetic north [1–3]. The accuracy is 1–2° (1σ), which corresponds to a 4–8-km position accuracy at a range of 240 km.

TACAN transmits bearing information on the same frequency as the DME transponder responses [1]. The amplitude of all DME pulses transmitted by the

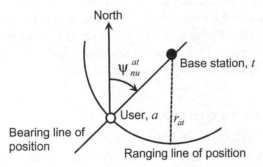

Figure 11.4 Ranging and angular positioning in the horizontal plane with a single base station.

transponder beacon is modulated with an azimuth-dependent signal that rotates at 15 Hz. North-reference pulse codes are emitted once a cycle, whenever the maximum amplitude is directed east. These are interspersed with eight auxiliary-reference pulse codes per cycle. Thus, 135 reference pulse codes per second are transmitted, which take priority over responses to interrogators. Users determine their bearing by comparing the amplitude and timing of the regular pulses with those of the reference pulses. TACAN bearings are slightly more accurate than those from VOR.

RSBN, a Russian acronym for radio engineers' system of short-range navigation, is a system similar to VOR, DME and TACAN. Coverage and accuracy are similar, although bearing information is with respect to true north. RSBN uses the 116–118-MHz band for angular positioning and the 873–1,001-MHz band for both ranging and angular positioning. It is used in Eastern Europe, Russia, and surrounding countries by both military and civil users, though some airlines have switched to VOR and DME.

Approximate user latitude and longitude may be obtained from any of the range-bearing systems using

$$\tilde{L}_a \approx L_t + \frac{\sqrt{\tilde{r}_{at,C}^2 - (\hat{h}_a - h_t)^2}\cos(\tilde{\psi}_{mu}^{ta} + \hat{\alpha}_{nm})}{R_N(L_t) + h_t},$$
$$\tilde{\lambda}_a \approx \lambda_t + \frac{\sqrt{\tilde{r}_{at,C}^2 - (\hat{h}_a - h_t)^2}\sin(\tilde{\psi}_{mu}^{ta} + \hat{\alpha}_{nm})}{[R_E(L_t) + h_t]\cos L_t} \quad (11.7)$$

where $\tilde{r}_{at,C}$ is the slant range measurement, $\tilde{\psi}_{mu}^{ta}$ is the bearing from the transmitter to the user with respect to magnetic north, $\hat{\alpha}_{nm}$ is the magnetic declination (see Section 6.1.1), $L_t, \lambda_t,$ and h_t are the beacon latitude, longitude, and height, and $\hat{h}_a$ is the user height. The Earth's curvature is neglected here as the resulting error is much less than that from the bearing measurement.

Range-bearing navigation systems are typically used by aircraft flying along airways linking the beacons. They navigate in terms of bearing and slant range from the beacon rather than using latitude and longitude (or other Earth-referenced coordinates). With all aircraft following the same route and maintaining safe along-track separation, accurate cross-track positioning is not required. However, airways are being replaced by area navigation (RNAV) routing, for which current range-bearing systems are not sufficiently accurate. Many VOR and land-based TACAN beacons are therefore being withdrawn [4, 13]. A minimal network of VOR transmitters will be retained until there is sufficient coverage from multiple DME or other terrestrial systems (see Section 11.1.5), while sea-based TACAN will remain until GNSS-based shipboard landing technology is mature.

11.1.3 Nondirectional Beacons

Nondirectional beacons (NDBs) are the oldest and simplest form of radio positioning system. They broadcast omnidirectional signals with a simple Morse identification.

Most aeronautical NDBs broadcast between 190 and 530 kHz, with a few in the 530–1,750-kHz band. A direction-finding receiver (see Section 7.1.5) may measure a bearing to within 5°. A very rough position fix may be obtained from two beacons. However, aircraft tend to use the bearing measurements to fly towards or away from the beacon. There is a vertical null in the beacon's transmission pattern. By detecting this when flying directly overhead, an aircraft may obtain a proximity position fix [1]. Dedicated NDBs are being phased out, with many already decommissioned at the time of this writing. A few will be retained in remote areas where DME signals are unavailable [4, 13]. Marine radio beacons (Section 11.4.2) and AM broadcast stations (Section 11.4.3) are also used as NDBs.

11.1.4 JTIDS/MIDS Relative Navigation

The relative navigation (RelNav) function of the Joint Tactical Information Distribution System and Multi-functional Information Distribution System is an example of a relative navigation chain (see Section 7.1.2). It is used by NATO aircraft, which communicate using Link 16 signals in the 960–1,215-MHz band [14, 15]. Each participant broadcasts a ranging signal every 3–12 seconds with a range of about 500 km. Participants are time synchronized so passive ranging is used. The position accuracy, integrated with inertial navigation, is 30–100m, depending on how far down the chain the user is.

11.1.5 Future Air Navigation Systems

At the time of this writing, the United States Federal Aviation Administration (FAA) was developing alternative position, navigation, and timing (APNT) technologies to provide a fully modernized backup to GNSS [9]. Three options, and combinations thereof, were under consideration: enhancements to DME (Section 11.1.1.4), automatic dependent surveillance–broadcast (ADS-B) multilateration, and a pseudolite system.

ADS-B multilateration is a remote-positioning system whereby aircraft positions are determined on the ground from signals transmitted by the aircraft. The position solutions are then uploaded to the aircraft.

A long-range pseudolite system (see Section 12.1) would use ground-based transmitters of FDMA or CDMA GNSS-like signals in the 960–1,215-MHz band. In Europe, the future L-Band Digital Aeronautical Communication System (LDACS) has been proposed for this [16].

11.2 Enhanced Loran

Enhanced long-range navigation (ELoran, often written as eLoran) is the latest in a series of long-range, one-way ranging systems. It is intended to provide a backup to GNSS for marine positioning and critical timing applications with very high availability and integrity requirements. It is also suitable for aircraft nonprecision approach. ELoran is being implemented as an internationally coordinated series of improvements to the previous Loran-C system, described in Section K.4 of Appendix K on the CD. These commenced in the mid-1990s and were ongoing at the time of this writing [17]. The ELoran name was adopted around 2004.

ELoran uses the 90–110-kHz band in the low-frequency region of the spectrum. As this is very different to the spectrum used by GNSS, common failure modes are minimized. Transmitters are synchronized to UTC (independently of GNSS), so positioning is by passive ranging (Section 7.1.4.1). For historical reasons, the transmitters are grouped into chains, each comprising one master station and two to five secondary stations. Some transmitters, known as dual rates, belong to two chains [1–3, 18].

Ground-wave propagation is employed that, at low frequencies, enables long range to be achieved independently of altitude. Signals are receivable at ranges of 1,000–2,000 km over land (depending on the terrain) and up to 3,000 km over an all-sea path [2, 3, 19]. Sky-wave signals travel further, but are unreliable and inaccurate. They are much stronger at night when sky-wave interference can reduce the useful range of the ground wave by about 30%, depending on the geomagnetic latitude and solar weather [3]. A major advantage of Loran signals, compared to GNSS, is that they penetrate well into valleys, urban canyons, and buildings, even basements. LF signals are also very difficult to jam over a wide area.

The position accuracy requirement for ELoran is about 150m (1σ) generally and 5m (1σ) for areas, such as harbors and waterways, for which differential corrections are provided [20].

By 2012, prototype ELoran signals were transmitted in North West Europe, South Korea, and Saudi Arabia. There were also proposals to upgrade Russia's Chayka system, which is almost identical to Loran, to Enhanced Chayka (EChayka). In addition, there was legacy Loran-C infrastructure available in the United States and several other countries that could be used for ELoran or a new positioning system using the same LF spectrum [21].

This section describes the ELoran signals, followed by the user equipment processing, the position computation, and the error sources. It then concludes with a summary of differential Loran (DLoran).

11.2.1 Signals

Loran signals are all transmitted on a 100-kHz carrier with a 20-kHz double-sided bandwidth and vertical polarization. Stations within a chain transmit in turn, a form of TDMA. Figure 11.5 shows the signals received from one chain [2]. Each transmission comprises a group of eight 500-μs pulses, starting 1 ms apart, with master stations adding an additional pulse 2 ms after the eighth pulse (1 ms for Chayka).

Each transmitter repeats its pulse group transmission at a constant interval between 40 and 99.99 ms, known as the group repetition interval (GRI). The GRI, together with the nominal emission delays (between transmitters within the chain), must be sufficient to avoid pulses from different transmitters in a chain overlapping anywhere within the coverage area. The GRI is different for each chain so it is used as identification. Multiples of 10 μs are used in Europe and India with multiples of 100 μs used elsewhere. Signals from different Loran chains can potentially interfere. Careful selection of the GRIs keeps the repetition intervals of the cross-chain interference

11.2 Enhanced Loran

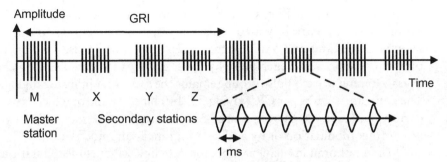

Figure 11.5 Loran signals received from one chain. (*After:* [2].)

patterns in excess of 10 seconds. Furthermore, modern Loran user equipment can predict which pulses will be subject to interference from other Loran stations, so can ignore them or even subtract a replica of the unwanted signal. Signals more than 40 dB weaker than the strongest available signal can thus be tracked [19].

The received signal pulses can be distorted by sky waves from the same transmitter, a form of multipath interference. A single-reflection sky wave is lagged by 35–500 μs with respect to the ground wave, while a multihop sky wave may be lagged by up to 4 ms. At night and at long range, a sky wave can be up to 20 dB stronger than the corresponding ground wave. Receivers take multiple samples of each pulse and then process them to separate out the ground-wave component [22].

The polarity of each pulse within a group is varied to produce a phase code, which repeats every two groups, known as the phase code interval (PCI). The phase codes are selected such that when signals differing by more than 1 ms are correlated, their product averages to zero over the PCI [3]. This minimizes interference from multihop sky-wave propagation of the same signal. It also reduces the effects of interference from Loran transmitters in other chains. Secondary stations use a different phase code to the master. The two phase codes are orthogonal; their product averages to zero over the PCI regardless of the offset between them. The combination of different GRIs and pulse codes is a form of CDMA.

The signal for each Loran pulse, illustrated by Figure 11.6 may be described by [1, 3]

$$s(t_{sa}) = \begin{matrix} A\left(\dfrac{t_{sa}-t_0}{\tau_p}\right)^2 \exp\left[-2\dfrac{(t_{sa}-t_0)}{\tau_p}\right]\sin\left[2\pi(f_{ca}+\Delta f_{ca})(t_{sa}-t_0)+\phi_{PC}+\phi_{ECD}\right] & t_{sa} \geq t_0 \\ 0 & t_{sa} < t_0 \end{matrix}$$

(11.8)

where t_{sa} is the arrival time, A is a constant proportional to the signal amplitude, t_0 is the arrival time of the beginning of the pulse, τ_p is the pulse time constant (usually 65 μs), f_{ca} is the carrier frequency, Δf_{ca} is the Doppler shift, $\phi_{PC} \in {0,\pi}$ is the phase code, and ϕ_{ECD} is the envelope to cycle difference (ECD). The ECD occurs due to the dispersive nature of ground-wave propagation and varies with distance from the transmitter and ground conductivity. Dispersion can also affect the pulse shape. Chayka pulses are shorter than Loran pulses [1], except in joint Loran/Chayka chains.

An innovation of ELoran is the incorporation of a data link. In Europe, the Eurofix system is currently used. This was originally developed to carry GNSS differential corrections on Loran-C signals [22]. Eurofix uses pulse position modulation, offsetting the timing of each pulse by 0, +1, or −1 μs to provide a data channel. The average timing of each pulse group remains the same to minimize ranging distortion. The data rate is 7 bits per GRI, giving 70–140 bit s^{-1}, some of which is used for error correction. However, there is a need to increase the date rate, either by modifying the Eurofix modulation or by adding additional data modulation.

In a full ELoran implementation, the data link will include the station identification, an almanac of transmitter locations, timing information, authentication data, differential-Loran data (see Section 11.2.4), and integrity alerts [20]. ELoran already incorporates infrastructure-based integrity monitoring (Section 17.5). However, the transmitter is currently switched off when a fault is detected.

11.2.2 User Equipment and Positioning

Modern Loran user equipment will track all receivable signals, which can be more than 30. This is known as all-in-view operation. Antennas may be magnetic (H) field or electric (E) field. E-field antennas perform better on ships, provided that a radome is used to reduce precipitation static interference [private communication with P. Williams, August 2012].

H-field antennas eliminate precipitation static interference altogether [19, 23], and are less sensitive to man-made interference, such as lighting. H-field antennas for receipt of vertically polarized signals are directional within the horizontal plane. Consequently, a pair of orthogonally-mounted antennas must be used to ensure good reception of all signals. They may be used as part of a goniometer system for direction finding (see Section 7.1.5), which may be used to determine host vehicle heading. This is sometimes referred to as a "Loran compass." The goniometer system may also be used to minimize interference.

Signal timing is performed by correlating each incoming signal with an internally-generated replica as described in Section 7.3.2. Both the carrier phase and the

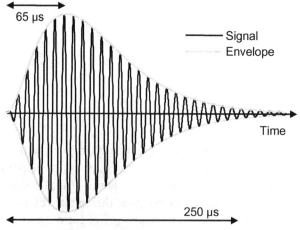

Figure 11.6 Loran pulse shape.

pulse envelope are tracked, with the envelope effectively used to resolve the range ambiguity of the carrier. Corrected pseudo-range measurements, $\tilde{\rho}^t_{a,C}$, are obtained from the measured TOA, $\tilde{t}^t_{sa,a}$, using

$$\tilde{\rho}^t_{a,C} = \left(\tilde{t}^t_{sa,a} - t^t_{st,a}\right)c + \Delta\rho^t_{ASF,a}, \qquad (11.9)$$

where $t^t_{st,a}$ is the time of transmission, c is the effective propagation speed over saltwater, and $\Delta\rho^t_{ASF,a}$ is a (negative) propagation delay correction, known as an additional secondary factor (ASF).

The ASF is analogous to the various corrections applied to GNSS pseudo-ranges (see Sections 8.5.3 and 9.3). It models variation in the signal propagation speed, primarily over land, where it is about 0.5% lower than over water. Thus, for a 1,000-km land path, the ASF is around 5 km. The propagation speed varies with terrain conductivity. This can change with the seasons. For example, summer-to-winter variations in the ASFs of over 100m can occur when land freezes and thaws [24]. The propagation speed also varies with terrain roughness; signals take longer to propagate across valleys and mountains than over flat terrain.

ELoran user equipment derives its ASF corrections from databases. ASF measurements, interpolated using modeling, are used to provide corrections that reduce the range biases to within 100m [25]. Higher-precision ASF corrections may be obtained for areas of interest, such as airports and harbors, by using a higher-resolution grid than applied generally [26]. The seasonal variations in ASFs may also be incorporated in the databases. However, the accuracy is limited by variations in the timing and severity of the seasons each year.

For land and marine users, the signal measured by the user equipment is the ground wave, which follows a great circle path. Consequently, the pseudo-range may be expressed in terms of the transmitter latitude, L_t, and longitude, λ_t, and the user latitude, $\hat{L}_a$, and longitude, $\hat{\lambda}_a$, solution by

$$\tilde{\rho}^t_{a,C} = \hat{s}\left(L_t, \lambda_t; \hat{L}_a, \hat{\lambda}_a\right) + \delta\hat{\rho}^a_c + \delta\rho^{t+}_{a,\varepsilon}, \qquad (11.10)$$

where $\delta\hat{\rho}^a_c$ is the estimated receiver clock offset, $\delta\rho^{t+}_{a,\varepsilon}$ is the measurement residual for the signal from transmitter t, and s is the geodesic, the shortest distance between the transmitter and receiver across the surface of the Earth. Note that the Sagnac effect is accounted for within the ASF correction.

The geodesic is approximately [3]

$$\hat{s}\left(L_t, \lambda_t; \hat{L}_a, \hat{\lambda}_a\right) \approx \hat{u}\left(L_t, \lambda_t; \hat{L}_a, \hat{\lambda}_a\right)$$

$$\times \sqrt{\frac{1}{2}\left(R_N^2(L_t) + R_N^2(\hat{L}_a)\right) + \frac{\sin^2\psi^{at}_{n(a)s}}{2}\left[\frac{\cos^2\hat{L}_a}{\cos^2 L_t}\left(R_E^2(L_t) - R_N^2(L_t)\right) + R_E^2(\hat{L}_a) - R_N^2(\hat{L}_a)\right]^2},$$

$$(11.11)$$

where R_N and R_E are, respectively, the north–south and east–west great circle radii of curvature, given by (2.105) and (2.106), μ is the angle subtended by the geodesic, given by

$$\hat{u}\left(L_t, \lambda_t; \hat{L}_a, \hat{\lambda}_a\right) \approx \arccos\left[\sin L_t \sin \hat{L}_a + \cos L_t \cos \hat{L}_a \cos\left(\lambda_t - \hat{\lambda}_a\right)\right], \quad (11.12)$$

and $\hat{\psi}_{n(a)s}^{at}$ is the azimuth, with respect to true north, of the geodesic from the user antenna to transmitter t, determined at the user antenna and given by

$$\hat{\psi}_{n(a)s}^{at} \approx \begin{matrix} \arcsin\left(\dfrac{\sin\left(\lambda_t - \hat{\lambda}_a\right)\cos L_t}{\sin\left[u\left(L_t, \lambda_t; \hat{L}_a, \hat{\lambda}_a\right)\right]}\right) & L_t \geq \hat{L}_a \\[2ex] \pi - \arcsin\left(\dfrac{\sin\left(\lambda_t - \hat{\lambda}_a\right)\cos L_t}{\sin\left[u\left(L_t, \lambda_t; \hat{L}_a, \hat{\lambda}_a\right)\right]}\right) & L_t < \hat{L}_a \end{matrix}. \quad (11.13)$$

More accurate formulae may be found in [27].

At least three pseudo-range measurements are needed to solve for the latitude, longitude, and clock offset. With more measurements, the solution is overdetermined. A solution may be obtained by least-squares or using a Kalman filter in analogy with the GNSS navigation solution, described in Section 9.4.

A least-squares solution from m measurements is

$$\begin{pmatrix} \hat{L}_a^+ \\ \hat{\lambda}_a^+ \\ \delta\hat{\rho}_c^{a+} \end{pmatrix} = \begin{pmatrix} \hat{L}_a^- \\ \hat{\lambda}_a^- \\ \delta\hat{\rho}_c^{a-} \end{pmatrix} + \begin{pmatrix} 1/R_N(\hat{L}_a^-) & 0 & 0 \\ 0 & 1/\left[R_E(\hat{L}_a^-)\cos \hat{L}_a^-\right] & 0 \\ 0 & 0 & 1 \end{pmatrix} \left(\mathbf{H}_R^{nC\,T} \mathbf{H}_R^{nC}\right)^{-1} \mathbf{H}_R^{nC\,T} \begin{pmatrix} \tilde{\rho}_{a,C}^1 - \hat{\rho}_{a,C}^{1-} \\ \tilde{\rho}_{a,C}^2 - \hat{\rho}_{a,C}^{2-} \\ \vdots \\ \tilde{\rho}_{a,C}^m - \hat{\rho}_{a,C}^{m-} \end{pmatrix},$$
(11.14)

where $\hat{\rho}_{a,C}^{j-}$ is the predicted value of the jth pseudo-range measurement and the measurement matrix is

$$\mathbf{H}_R^{nC} = \begin{pmatrix} -\cos\hat{\psi}_{n(a)s}^{a1-} & -\sin\hat{\psi}_{n(a)s}^{a1-} & 1 \\ -\cos\hat{\psi}_{n(a)s}^{a2-} & -\sin\hat{\psi}_{n(a)s}^{a2-} & 1 \\ \vdots & \vdots & \vdots \\ -\cos\hat{\psi}_{n(a)s}^{am-} & -\sin\hat{\psi}_{n(a)s}^{am-} & 1 \end{pmatrix}. \quad (11.15)$$

For airborne users, the navigation solution is more complex as the signal propagation is a mixture of ground wave and line of sight. At a given latitude and longitude, the signal has to propagate further to reach an airborne user. However, line-of-sight propagation is faster than ground-wave propagation. In practice, a height-dependent ASF database is used in conjunction with the land and marine position determination method.

11.2.3 Error Sources

The main cause of biases in Loran pseudo-range measurements is variations in the ASFs that are not accounted for by the database. These are both spatial and temporal, with variations in seasonal effects from year to year being a significant factor. ASFs vary over time due to the weather with variations over the course of a day of order 3–10m (1σ) [19, 28]. ASFs are typically less accurate for airborne users due to the greater complexity of mixed ground-wave and line-of-sight propagation and the variation of the troposphere refractivity with the weather. To fully calibrate the time-varying biases, either differential Loran (Section 11.2.4) or Loran integrated with other positioning systems, such as GNSS (see Section 16.3), must be used.

Like any radio system, Loran signals are subject to multipath. Due to the long wavelength, this arises only from large objects, such as mountains, bridges, and transmission lines. When high-resolution databases are used, the effect of multipath can be incorporated into the ASF corrections as it is correlated over several hundred meters and exhibits little time variation as the transmitters are fixed. However, absorption and reradiation of signals by conducting wires can be a problem, especially in urban areas, introducing phase distortions as well as multipath effects; in principle, this may also be calibrated. Thus, the ranging accuracy obtainable from ELoran depends on the context of the application.

Random errors in Loran measurements are due to transmitter and receiver timing jitter and RF noise arising from atmospheric electrical noise, man-made interference, sky-wave propagation, interference from signals on frequencies adjacent to the Loran band, and interference from the other Loran signals, known as cross-rate interference. With a strong signal, a modern transmitter, and modern user equipment, the range error standard deviation due to noise can be as low as 1.5m [23], while the noise with a weak signal can be around 100m. Thus, with all-in-view user equipment, it is important to weight each measurement appropriately in the navigation solution. As with GNSS (see Section 9.3.3), there is a design tradeoff between noise performance and dynamics response. The signal tracking algorithms in Loran user-equipment typically have time constants of several seconds, so significant position errors can arise from delays in responding to changes in host vehicle velocity. Therefore, the user equipment design should be carefully matched to the application.

As discussed in Section 7.4.5, the dilution of precision can be used to predict the position accuracy from the pseudo-range accuracy. With pseudo-range measurements, the DOPs are given by

$$\begin{pmatrix} D_N^2 & \cdot & \cdot \\ \cdot & D_E^2 & \cdot \\ \cdot & \cdot & D_T^2 \end{pmatrix} = \left(\mathbf{H}_R^{nC\,T} \mathbf{H}_R^{nC} \right)^{-1}, \qquad (11.16)$$

where $\mathbf{H}_R^{nC}$ is given by (11.15). In practice, the position accuracy is substantially degraded if all of the signals come from similar directions, as can happen towards the edge of an ELoran system's coverage area. The effect of signal geometry is diminished where the user equipment is equipped with a precision clock, such as a CSAC (see Section 9.1.2), that has been calibrated independently (e.g., from GNSS).

11.2.4 Differential Loran

Like differential GNSS (Section 10.1), differential Loran is designed to eliminate time-varying signal propagation delays and time-of-transmission errors by measuring these at a reference station at a known location and then transmitting corrections to the users. The corrections comprise the difference between the smoothed pseudo-range measurements and their nominal values. A database is used to account for the spatial variation in propagation delays as normal. This is specifically designed for use with a particular reference station to ensure that the sum of the differential and spatial corrections gives the correct total ASF correction [29].

Differential corrections and monitor station locations may be transmitted using Eurofix or a private link. Because of variation in the ground-wave propagation characteristics over different terrain, the spatial decorrelation is higher than for DGNSS, so best results are obtained within about 30 km of the reference station [17]. However, the temporal decorrelation is lower so lower update rates may be used for the corrections. A sub-5m (1σ) positioning accuracy may be obtained where good geometry is available. An operational DLoran service has been implemented at several ports in the United Kingdom [30].

Note that DLoran is not suited to most aviation applications due to the impracticality of measuring the variation of the ASFs with height in real time.

11.3 Phone Positioning

Mobile phone, or cellphone, positioning was originally developed primarily to meet mandatory emergency caller location requirements, commonly known as E911 in the United States and E112 in Europe. It is also used for location-based services and personal navigation. A number of methods may be used. However, this section focuses on positioning using the phone signals themselves.

Mobile phones use frequencies mainly in the 800–960-MHz, 1,710–2,170-MHz, and 2,490–2,690-MHz ranges, with the exact bands varying between countries. There are two main second generation (2G) standards, offering digital audio communication and limited data. The Global System for Mobile communication (GSM) is used in Europe and many other countries, while CDMA IS-95 is the main 2G standard in the United States. The Universal Mobile Telecommunication System (UMTS) is the international third generation (3G) standard, offering higher data communication rates. CDMA IS-2000, or simply CDMA2000, is a competing 3G standard that evolved from CDMA IS-95. All of these systems combine FDMA with frequency sharing. GSM uses TDMA, dividing each 200-kHz-wide channel into eight timeslots, each allocated to a different user. The other systems use CDMA to share channels between users, with channel widths of 1.25 MHz for IS-95 and IS-2000 and 5 MHz for UMTS [5].

There are two fourth generation (4G) standards, offering the highest data communication rates. These are Long-Term Evolution (LTE) Advanced and IEEE 802.16m, also known as Mobile Worldwide Interoperability for Microwave Access (WiMAX) Release 2 and Wireless Metropolitan Access Network (MAN)-Advanced. Both standards use orthogonal frequency division multiple access (OFDMA), which combines OFDM with FDMA. In LTE, different numbers of subcarriers are allocated, depending on the data rate required, giving a bandwidth between 1.4 and 20 MHz [31]. IEEE 802.16m signals have a 10-MHz bandwidth [32].

Both self-positioning, using the downlink (or forward link) signals from the base stations, and remote positioning, using the uplink signals from the phones, may be implemented [5]. The proximity, ranging, and pattern-matching positioning methods may all be used. Remote positioning may also use angular positioning by performing direction finding using an antenna array at the base station. However, direction finding at the phone will often give the direction of a reflecting surface rather than the base station, while direction-finding antenna arrays are too large to fit on a phone.

Phones communicate mainly with the base station serving the cell they are within. However, they also exchange control signals with the base stations of neighboring cells to facilitate handover. The downlink control signals are available to all receivers, not just those that are subscribed to the relevant network, so may be treated as signals of opportunity and used for positioning without the cooperation of the network provider.

A multinetwork SOOP approach greatly increases the number of base stations available for positioning. However, the operators do not make the locations of their base stations publicly available. Regulators provide site locations in many countries, but these are only accurate to within 100m and do not include the cell IDs of the base stations, which can change.

The remainder of this section discusses the proximity, pattern-matching, and ranging positioning methods, while Section F.5 of Appendix F on the CD summarizes some phone positioning terminology, which varies between standards.

11.3.1 Proximity and Pattern Matching

Cell ID is the simplest form of mobile phone positioning and simply reports that the user is somewhere within the coverage area of the serving base station. However, coverage radii can vary from 1 km in urban areas to 35 km in rural areas, so accuracy is poor. Cells designed to serve large numbers of users may be divided by bearing from the base station into up to six sectors, each with their own channels, improving cell ID accuracy [5].

Cell ID can be enhanced using containment intersection (see Section 7.1.3 and Figure 7.4) [31] or pattern matching (Section 7.1.6) [33]. For multinetwork SOOP positioning, these methods have the advantage of not requiring knowledge of the base station locations, only the areas where the signals are receivable. Positioning algorithms must be robust against signal reception being blocked by local obstructions, including the user's body, while the phone is within the nominal reception area of a base station.

Pattern-matching methods are further enhanced by using the received signal strength instead of simple signal availability [34, 35]. Again, signal attenuation by the user's body and other obstacles must be accounted for. A positioning accuracy of 50m (1σ) can typically be achieved using all available phone signals in an urban environment with a 30-m grid spacing [36].

11.3.2 Ranging

All CDMA IS-95, IS-2000, and IEEE 802.16m base stations and some LTE base stations are synchronized to UTC. Therefore, phones can perform passive ranging (Section 7.1.4.1) using the downlink. Control signals from at least four base stations are required for a unique latitude and longitude solution. Three stations can be used with prior position information, while a line of position can be obtained with two base stations. Ranging using the IS-95 or IS-2000 uplink is impractical because the uplink signal power is reduced when a phone is close to the serving base station to prevent interference to signals from other phones [5].

GSM and UMTS base stations are not normally time synchronized and LTE may also operate in an asynchronous mode. Therefore, differential positioning (Section 7.1.4.3) is normally used. Methods using reference stations have largely been replaced by the matrix method, in which pseudo-range measurements from multiple phones are pooled [37]. As discussed in Section 7.1.4.3, a 2-D position solution requires at least four base stations and three phones (or five stations and two phones). For both approaches, either uplink or downlink signals may be used and position solutions may be computed in either the phone or the network.

Base stations use high-quality oscillators to maintain frequency stability. Consequently, the clocks are relatively stable, drifting by between 2 μs and 300 μs over 3 days, depending on the design [38]. Thus, once the clock offsets have been calibrated, single-phone position solutions may be obtained by passive ranging. As well as the matrix method, clock offset calibration may be performed using signals from already-calibrated base stations or from other positioning technologies, as described in Section 16.3.2. Furthermore, matrix positioning may be performed by visiting multiple locations in close succession using a single phone.

In SOOP positioning, the base station positions may also be determined by performing ranging measurements at several different locations, the positions of which may be determined using signals from transmitters at known locations [39]. SLAM-based techniques may be used when there are insufficient known signals, but dead reckoning is available. In cooperative positioning, transmitter position and clock offset information may be shared between peers.

In computing a position solution, the curvature of the Earth may be neglected as the ensuing errors at mobile phone ranges are small compared to the measurement errors. For the best horizontal positioning accuracy, the height difference between base station and phone should be accounted for. A terrain height database may be used to estimate the height of phones outdoors. However, 2-D positioning algorithms are typically used in practice as the error from neglecting height is usually much smaller than the signal propagation errors.

The position accuracy is typically 50–100m for GSM, IS-95 and IS-2000, and 25–50m for UMTS, depending on the signal propagation environment and signal geometry [5, 37, 40]. Ranging errors due to multipath interference and NLOS signal reception can be several hundred meters in urban and indoor environments [41]. The wider bandwidth of UMTS and LTE signals enables better multipath resolution, but does not affect the errors due to NLOS reception.

A major limitation of phone signal ranging is that there are not always sufficient base stations within range to determine a position solution. For GSM-Rail (GSM-R), which is used for voice and data communication across rail networks, the base station geometry allows positioning along the track, but not generally across it [42].

GSM and UMTS phones also perform a two-way ranging measurement to synchronize their TDMA slot with the serving base station, providing a circular line of position accurate to about 500m for GSM [43] and 35m for UMTS. This may be used to enhance proximity positioning where a full position solution from ranging is unavailable.

For maximum availability, phone-signal positioning should be used as part of a multisensor navigation system rather than stand alone.

11.4 Other Systems

This section describes Iridium positioning, marine radio beacons, AM and FM radio broadcasts, digital television and radio, and, finally, generic radio positioning. Radio and television broadcasts make convenient signals of opportunity because the modulation formats and most transmitter locations (to a few tens of meters) are publicly known.

11.4.1 Iridium Positioning

Iridium is a satellite communications system. Its constellation comprises 66 low Earth orbit (LEO) satellites distributed among six orbital planes inclined at 86.4° to the equator. The orbital radius is 7,158 km, corresponding to an altitude of about 780 km. At most locations on Earth, either one or two satellites are visible at any given time; more are visible in polar regions. The orbital period is 100 minutes and a satellite is typically visible to a given user for about 9 minutes. The 1,616–1,626.5-MHz band is used for two-way user-satellite communications. A mixture of TDMA and FDMA is used with each channel 41.67 kHz wide [44].

The Boeing Timing and Location (BTL) service adds a ranging service to the 1,626.104-MHz Iridium paging signal, comprising 23.32-ms QPSK bursts every 1–1.5 seconds whenever the satellite is receivable within the BTL service coverage area. The BTL signals are much stronger than GNSS signals, enabling an attenuation that is 15–20 dB greater to be tolerated before reception is lost. In practice, a BTL position solution can be obtained in any building above ground level [45].

Using only the BTL signals, a position accurate to 30–100m may be obtained in about 30 seconds using Doppler positioning (Section 7.1.7). A single Iridium satellite is sufficient for this because the line-of-sight changes rapidly. The receiver clock offset and drift are also calibrated using passive ranging. Therefore, BTL may be used to aid GNSS acquisition and tracking as described in Section 10.5.1. The BTL service was launched in the United States in 2012 and could easily be extended worldwide.

It has also been proposed to use Iridium differential carrier-phase ranging to aid GNSS carrier-phase ambiguity resolution and integrity monitoring [46].

11.4.2 Marine Radio Beacons

Marine radio beacons broadcast omnidirectional signals in the 283.5–325-kHz band. They are located along the coast in order to provide coverage over sea. Ranges of up to 300 km are typical. The beacons were originally used for direction finding (see Section 7.1.5) with a simple Morse identification, like NDBs. Under good conditions, accuracies of about 2° can be obtained. Refraction of the ground wave at the land-sea boundary can bend the signal path. The effect of this on positioning is minimized by siting transmitters as close to the coast as possible. However, significant positioning errors can occur when a signal path crosses a peninsula [47].

Since the late 1990s, marine radio beacons in many countries have been used to transmit LADGNSS information (see Section 10.1.2), with additional transmitters installed to provide coverage to inland areas. It has been proposed that the transmitters be synchronized to UTC to enable them to be used for passive ranging as part of a backup to GNSS [48, 49], a concept known as R mode. Precise ranging measurements would be obtained from the carrier phase with the modulation used to resolve the ambiguity. ASF corrections to account for variations in propagation speed with terrain could be provided in a similar manner to ELoran (Section 11.2), with a database of the spatial variations and differential corrections to account for temporal variations. The effective coverage area would be reduced at night due to sky-wave interference.

R mode could also incorporate passive ranging from fixed Automatic Identification System (AIS) beacons. AIS is a VHF communication system used by vessels to transmit their position, velocity, destination, and other information. Fixed beacons act as hazard markers on rocks and sandbanks and can also be deployed on buoys. AIS is a TDMA system with each station typically transmitting 10 times a minute. The range of each transmission is around 60 km.

As with Loran, equipping the user with a precision clock, such as a CSAC, that has been calibrated independently, can compensate for poor signal geometry in cases where the R-mode signals come from similar directions.

11.4.3 AM Radio Broadcasts

AM radio broadcasts are the original signals of opportunity, having been used for direction finding from soon after the start of regular broadcasts in the 1920s. From the 1990s, several positioning systems using differential ranging have been developed, noting that AM radio transmitters are not usually synchronized unless they are part of a single frequency network (SFN). The carrier may be separated from the modulation simply by using a narrowband filter. Ranging measurements obtained from the carrier phase are more precise than those from the modulation. However, they are subject to an integer wavelength ambiguity.

In the MF and LF broadcasting bands, the wavelength ranges from 175m to 2 km. The integer ambiguity may be resolved by three methods: starting at a known position [50], modulation-based ranging [51], and consistency checking between signals [52]. Consistency checking requires more signals than the other methods. A position accuracy within 10m can be achieved using differential carrier-phase positioning in the MF band [50, 52]. However, this requires either the baseline between the user and reference to be limited to a few kilometers or extensive calibration of terrain- and frequency-dependent ground wave propagation speeds and azimuth-dependent phase biases at the transmit antenna [52].

Absorption and reradiation of signals by conducting wires can completely change the carrier phase in and around buildings and transmission lines, rendering conventional carrier-phase positioning useless in these locations [52]. These perturbations result in near-field position-dependent phase differences between the electric-field and magnetic-field components of the signal. These phase differences can be measured by user equipment with separate electric-field and magnetic-field antennas and used for positioning by pattern matching [53].

Note that several countries are phasing out AM broadcasting.

11.4.4 FM Radio Broadcasts

Positioning using the RSS pattern-matching method (see Section 7.1.6) has been performed using FM radio broadcasts. An accuracy of tens of meters has been achieved outdoors [54], while the indoor accuracy, assuming that the building is known, is a few meters [55, 56]. Signals in the 88–108-MHz FM broadcast band are much less affected by the human body than higher-frequency signals so the RSS at a particular location is more stable. There is also less spatial variation, which limits precision but also reduces ambiguity.

11.4.5 Digital Television and Radio

Each digital television transmission, known as a multiplex, incorporates multiple television programs and other data. Multiplex channel widths are 6, 7, or 8 MHz (including a guard interval), depending on the country. Frequencies range from 45 to 900 MHz, although most countries use 174–223 MHz and 470–854 MHz. The different standards are described in Section F.6.1 of Appendix F on the CD.

To correctly demodulate digital signals, the receiver must be time synchronized to the transmitter. This is done by transmitting sequences of data that are known to the receiver and can be timed using a correlation, acquisition, and tracking process (see Section 7.3.2). These synchronization signals may be used for range measurement [57, 58].

Ranging accuracy using television signals can vary considerably. When direct line-of-sight reception is available, an accuracy of around 10m, depending on multipath conditions, is achievable using either differential ranging with a reference receiver at a known location or passive ranging with accurately synchronized transmitters [57, 58]. As television signals are usually received at low elevation angles, line-of-sight reception is often blocked in urban areas and indoors. Using reflected signals introduces ranging errors of around 100m [58–60]. Performing passive ranging using signals from poorly synchronized transmitters can introduce ranging errors of several kilometers [59].

In Europe, it is common for all television signals serving a given area to be transmitted from the same mast. In countries without a mast-sharing culture, such as the United States, transmission masts are often clustered together on a hill or a group of tall buildings. When all signals come from the same direction, it is not possible to separate position and timing information using passive or differential ranging (see Section 7.4.5). However, because ranging uses known symbol sequences, it can operate with much lower signal strengths than those required for television content reception. This enables additional ranging measurements to be obtained from signals providing television coverage to neighboring areas.

Digital Audio Broadcasting (DAB) and its derivatives form the terrestrial digital radio standard for Europe and several other countries. DAB transmitters form SFNs, so are tightly time synchronized, with offsets from UTC varying between transmitters to prevent destructive interference. Passive ranging may be performed where these offsets are known. Otherwise, differential positioning must be used. A positioning accuracy of about 150m has been achieved using a prototype system [61]. Further information on DAB and other digital radio systems is provided in Section F.6.2 of Appendix F on the CD.

11.4.6 Generic Radio Positioning

Positioning by proximity, direction finding, and RSS pattern matching can be performed using any identifiable signals. However, most ranging techniques require some knowledge of the signal structure in order to measure the time of flight. By performing differential positioning in the signal domain instead of the range domain, any modulated FDMA signal can potentially be used for ranging without any knowledge of how that signal is modulated.

The downconverted and filtered signal is sampled by both the user and reference receivers at an agreed rate over an agreed period. No demodulation is necessary. The samples are then transmitted from one of the receivers to the other, where the two sets of samples are correlated over a range of time offsets. The time offset corresponding to the correlation peak is the TDOA of the signal between the two receivers, subject to the relative clock offset between them. There must be sufficient time synchronization between the receivers for the two sampling windows to have a significant overlap; this can be accomplished using the data link between them. This modulation correlation technique has been demonstrated using AM radio broadcasts [51].

It is not necessary to correlate the whole signal, so the data-link bandwidth may be traded off against ranging accuracy. One way of reducing the data rate is to use spectral compression processing (SCP) [62].

Problems and exercises for this chapter are on the accompanying CD.

References

[1] Uttam, B. J., et al., "Terrestrial Radio Navigation Systems," in *Avionics Navigation Systems*, 2nd ed., M. Kayton and W. R. Fried, (eds.), New York: Wiley, 1997, pp. 99–177.

[2] Enge, P., et al., "Terrestrial Radionavigation Technologies," *Navigation: JION*, Vol. 42, No. 1, 1995, pp. 61–108.

[3] Forssell, B., *Radionavigation Systems*, Norwood, MA: Artech House, 2008 (originally published 1991).

[4] Anon., *2010 Federal Radionavigation Plan*, U.S. Departments of Defense, Homeland Security, and Transportation, 2010.

[5] Bensky, A., *Wireless Positioning Technologies and Applications*, Norwood, MA: Artech House, 2008.

[6] RTCA Sub-committee 149, *Minimum Operational Performance Standards for Airborne Distance Measuring Equipment (DME) Operating Within the Radio Frequency Range of 960–1215*, RTCA DO-189, September 1985.

[7] Latham, R. W., and R. S. Townes, "DME Errors," *Navigation: JION*, Vol. 22, No. 4, 1975, pp. 332–342.

[8] Tran, M., "DME/DME Accuracy," *Proc. ION NTM*, San Diego, CA, January 2008, pp. 443–451.

[9] Lo, S. C., et al., "Alternative Position Navigation & Timing (APNT) Based on Existing DME and UAT Ground Signals," *Proc. ION GNSS 2011*, Portland, OR, September 2011, pp. 3309–3317.

[10] Lo, S. C., and P. K. Enge, "Signal Structure Study for a Passive Ranging System using Existing Distance Measuring Equipment (DME)," *Proc. ION ITM*, Newport Beach, CA, January 2012, pp. 97–107.

[11] Li, K., and W. Pelgrum, "Optimal Time-of-Arrival Estimation for Enhanced DME," *Proc. ION GNSS 2011*, Portland, OR, September 2011, pp. 3493–3502.

[12] Li, K., and W. Pelgrum, "Flight Test Performance of Enhanced DME (eDME)," *Proc. ION ITM*, Newport Beach, CA, January 2012, pp. 131–141.

[13] *Navigation Application & NAVAID Infrastructure Strategy for the ECAC Area Up to 2020*, Edition 2.9, Eurocontrol, May 2008.

[14] Fried, W. R., J. A., Kivett, and E. Westbrook, "Terrestrial Integrated Radio Communication-Navigation Systems," in *Avionics Navigation Systems*, 2nd ed., M. Kayton and W. R. Fried, (eds.), New York: Wiley, 1997, pp. 283–312.

[15] Ranger, J. F. O., "Principles of JTIDS Relative Navigation," *Journal of Navigation*, Vol. 49, No. 1, 1996, pp. 22–35.

[16] Belabbas, B., et al., "LDACS1 for an Alternate Positioning Navigation and Time Service," *Proc. 5th European Workshop on GNSS Signals and Signal Processing*, Toulouse, France, December 2011.

[17] Shmihluk, K., et al., "Enhanced LORAN Implementation and Evaluation for Timing and Frequency," *Proc. ION 61st AM*, Cambridge, MA, June 2005, pp. 379–385.

[18] Lo, S., and B. Peterson, "Integrated GNSS and Loran Systems," in *GNSS Applications and Methods*, S. Gleason and D. Gebre-Egziabher, (eds.), Norwood, MA: Artech House, 2009, pp. 269–289.

[19] Roth, G. L., and P. W. Schick, "New Loran Capabilities Enhance Performance of Hybridized GPS/Loran Receivers," *Navigation: JION*, Vol. 46, No. 4, 1999, pp. 249–260.

[20] *Enhanced Loran (eLoran) Definition Document*, International Loran Association, October 2007.

[21] Helwig, A., et al., "Low Frequency (LF) Solutions for Alternative Positioning, Navigation, Timing, and Data (APNT&D) and Associated Receiver Technology," *Proc. NAV 10*, London, U.K., November-December 2010.

[22] Offermans, G. W. A., and A. W. S. Helwig, *Integrated Navigation System Eurofix: Vision, Concept, Design, Implementation & Test*, Ph.D. Thesis, Delft University, 2003.

[23] Roth, G. L., et al., "Performance of DSP: Loran/H-Field Antenna System and Implications for Complementing GPS," *Navigation: JION*, Vol. 49, No. 2, 2002, pp. 81–90.

[24] Lo, S. C., et al., "Developing and Validating the Loran Temporal ASF Bound Model for Aviation," *Navigation: JION*, Vol. 56, No. 1, 2009, pp. 9–21.

[25] Williams, P., and D. Last, "Mapping the ASFs of the North West European Loran-C System," *Journal of Navigation*, Vol. 53, No. 2, 2000, pp. 225–235.

[26] Hartnett, R., G. Johnson, and P. Swaszek, "Navigating Using an ASF Grid for Harbor Entrance and Approach," *Proc. ION 60th AM*, Dayton, OH, June 2004, pp. 200–210.

[27] Vincenty, T., "Direct and Inverse Solutions of Geodesics on the Ellipsoid with Application of Nested Equations," *Survey Review*, Vol. 23, No. 176, 1975, pp. 88–93.

[28] Samaddar, S. N., "Weather Effect on LORAN-C Propagation," *Navigation: JION*, Vol. 27, No. 1, 1980, pp. 39–53.

[29] Hargreaves, C., P. Williams, and M. Bransby, "ASF Quality Assurance for eLoran," *Proc. IEEE/ION PLANS*, Myrtle Beach, SC, April 2012, pp. 1169–1174.

[30] Williams, P., G. Shaw, and C. Hargreaves, "GLA Maritime eLoran Activities in 2011 (and beyond!)," *Proc. ENC 2011*, London, U.K., November-December 2011.

[31] Kangas, A., I. Siomina, and T. Wigren, "Positioning in LTE," in *Handbook of Position Location: Theory, Practice, and Advances*, S. A. Zekavat and R. M. Buehrer, (eds.), New York: Wiley, 2012, pp. 1081–1127.

[32] Tseng, P. -H., and K. -T. Feng, "Cellular-Based Positioning for Next-Generation Telecommunication Systems," in *Handbook of Position Location: Theory, Practice, and Advances*, S. A. Zekavat and R. M. Buehrer, (eds.), New York: Wiley, 2012, pp. 1055–1079.

[33] Bshara, M., et al., "Robust Tracking in Cellular Networks Using HMM Filters and Cell-ID Measurements," *IEEE Trans. on Vehicular Technology*, Vol. 60, No. 3, 2011, pp. 1016–1024.

[34] Laitinen, H., J. Lahteenmaki, and T. Nordstrom, "Database Correlation Method for GSM location," *Proc. IEEE 53rd Vehicular Technology Conference*, Rhodes, Greece, May 2001, pp. 2504–2508.

[35] Chen, M. Y., et al., "Practical Metropolitan-Scale Positioning for GSM Phones," *Proc. Ubicomp 2006*, Irvine, CA, September 2006, pp. 225–242.

[36] Bhattacharrya, T., et al., "Location by Database: Radio-Frequency Pattern Matching," *GPS World*, June 2012, pp. 8–12.

[37] Duffett-Smith, P. J., and P. Hansen, "Precise Time Transfer in a Mobile Radio Terminal," *Proc. ION NTM*, San Diego, CA, January 2005, pp. 1101–1106.

[38] Couronneau, N., and P. J. Duffett-Smith, "Experimental Evaluation of Fine-Time Aiding in Unsynchronized Networks," *Proc. ION GNSS 2012*, Nashville, TN, September 2012, pp. 711–716.

[39] Pesnya, K. M., et al., "Tightly-Coupled Opportunistic Navigation for Deep Urban and Indoor Positioning," *Proc. ION GNSS 2011*, Portland OR, September 2011, pp. 3605–3616.

[40] Kim, H. S., et al., "Performance Analysis of Position Location Methods Based on IS-801 standard," *Proc. ION GPS 2000*, Salt Lake City, UT, September 2000, pp. 545–553.

[41] Faragher, R. M., and P. J. Duffett-Smith, "Measurements of the Effects of Multipath Interference on Timing Accuracy in a Cellular Radio Positioning System," *IET Radar, Sonar, and Navigation*, Vol. 4, No. 6, 2010, pp. 818–824.

[42] Faragher, R. M., *Lost in Space: The Science of Navigation (Without GPS)*, presentation by BAE Systems at University College London, January 2012.

[43] Kitching, T. D., "GPS and Cellular Radio Measurement Integration," *Journal of Navigation*, Vol. 53, No. 3, 2000, pp. 451–463.

[44] Pratt, S. R., et al., "An Operational and Performance Overview of the IRIDIUM Low Earth Orbit Satellite System," *IEEE Communications Surveys*, Second Quarter 1999, pp. 2–10.

[45] Whelan, D., G. Gutt, and P. Enge, "Boeing Timing & Location: An Indoor Capable Time Transfer and Geolocation System," *5th Stanford University Symposium on Position, Navigation, and Timing*, Menlo Park, CA, November 2011.

[46] Joerger, M., et al., "Analysis of Iridium-Augmented GPS for Floating Carrier Phase Positioning," *Navigation: JION*, Vol. 57, No. 2, 2010, pp. 137–160.

[47] Tetley, L., and D. Calcutt, *Electronic Aids to Navigation*, London, U.K.: Edward Arnold, 1986.

[48] Johnson, G. W., et al., "Beacon-Loran Integrated Navigation Concept (BLINC): An Integrated Medium Frequency Ranging System," *Proc. ION GNSS 2007*, Fort Worth, TX, September 2007, pp. 1101–1110.

[49] Oltmann, J. -H., and M. Hoppe, "Maritime Terrestrial Augmentation and Backup Radio Navigation Systems: State of the Art and Future Developments," Presentation by Wasser- und Schiffahrtsverwaltung des Bundes, 2008.

[50] Duffett-Smith, P. J., and G. Woan, "The CURSOR Radio Navigation and Tracking System," *Journal of Navigation*, Vol. 45, No. 2, 1992, pp. 157–165.

[51] Webb, T. A., et al., "A New Differential Positioning Technique Applicable to Generic FDMA Signals of Opportunity," *Proc. ION GNSS 2011*, Portland OR, September 2011, pp. 3527–3538.

[52] Hall, T. D., *Radiolocation Using AM Broadcast Signals*, Ph.D. thesis, Cambridge, MA: Massachusetts Institute of Technology, September 2002.

[53] Schantz, H. G., "Theory and Practice of Near-Field Electromagnetic Ranging," *Proc. ION ITM*, Newport Beach, CA, January 2012, pp. 978–985.

[54] Fang, S. -H., et al., "Is FM a RF-Based Positioning Solution in a Metropolitan-Scale Environment? A Probabilistic Approach with Radio Measurements Analysis," *IEEE Trans. on Broadcasting*, Vol. 55, No. 3, 2009, pp. 577–588.

[55] Moghtadaiee, V., A. G. Dempster, and S. Lim, "Indoor Localization Using FM Radio Signals: A Fingerprinting Approach," *Proc. Indoor Positioning and Indoor Navigation*, Guimarães, Portugal, September 2011.

[56] Popleteev, A., V. Osmani, and O. Mayora, "Investigation of Indoor Localization with Ambient FM Radio Stations," *Proc. IEEE International Conference on Pervasive Computing and Communications*, Lugano, Switzerland, March 2012, pp. 171–179.

[57] Rabinowitz, M., and J. J. Spilker, Jr., "A New Positioning System Using Television Synchronization Signals," *IEEE Trans. on Broadcasting*, Vol. 51, No. 1, 2005, pp. 51–61.

[58] Kovář, P., and F. Vejražka, "Multi System Navigation Receiver," *Proc. IEEE/ION PLANS*, Monterey, CA, May 2008, pp. 860–864.

[59] Do, J.-Y., M. Rabinowitz, and P. Enge, "Multi-Fault Tolerant RAIM Algorithm for Hybrid GPS/TV Positioning," *Proc. ION NTM*, San Diego, CA, January 2007, pp. 788–797.

[60] Thevenon, P., et al., "Positioning Using Mobile TV Based on the DVB-SH Standard," *Navigation: JION*, Vol. 58, No. 2, 2011, pp. 71–90.

[61] Palmer, D., et al., "Radio Positioning Using the Digital Audio Broadcasting (DAB) Signal," *Journal of Navigation*, Vol. 64, No. 1, 2011, pp. 45–59.

[62] Mathews, M. B., P. F. Macdoran, and K. L. Gold, "SCP Enabled Navigation Using Signals of Opportunity in GPS Obstructed Environments," *Navigation: JION*, Vol. 58, No. 2, 2011, pp. 91–110.

CHAPTER 12
Short-Range Positioning

This chapter describes the main features of short-range radio positioning systems, building on the principles described in Chapter 7. Acoustic, ultrasound, infrared, optical, and magnetic positioning systems that operate on the same principles as radio positioning are also described here. These systems generally have ranges of less than 3 km. The emphasis is on positioning systems used for navigation, not tracking, although many of the technologies described here may be used for both.

Section 12.1 discusses pseudolites, including GNSS-based pseudolites and repeaters, the Indoor Messaging System (IMES), Locata, and Terralite XPS. Section 12.2 describes ultrawideband positioning, and Section 12.3 covers positioning using short-range communication systems, such as WLAN, WPAN, RFID, Bluetooth low energy (BLE), and dedicated short-range communication (DSRC). Section 12.4 describes acoustic positioning for use underwater. Finally, Section 12.5 summarizes a number of other positioning technologies.

Many short-range positioning systems operate in the international 2.4–2.5 GHz industrial, scientific, and medical (ISM) band. Low-power transmissions are permitted within ISM bands without a license. Other ISM bands include 433.05–434.79 MHz in Europe, Africa, and parts of Asia, 902–928 MHz in the Americas, and 5.725–5.875 GHz internationally.

12.1 Pseudolites

A pseudolite (a contraction of pseudo-satellite) is a ground-based, ship-based, or airborne transmitter of GNSS-like signals (see Chapter 8). The operational principles are the same as for GNSS. A key advantage of pseudolite positioning is that user equipment hardware may be shared with GNSS positioning, reducing costs. The principal drawback is that where CDMA is used for terrestrial positioning, the signals from nearby transmitters can block reception of signals from distant transmitters. This is known as the *near-far problem* and limits pseudolite technology to short-range applications.

Pseudolites were originally deployed for user equipment testing during the GPS development phase when the satellite constellation was limited. They were subsequently proposed for improving integrity and ambiguity resolution in single-constellation GBAS (Section 8.2.6) [1]. More recently, they have been used for mitigating GNSS signal shadowing in dense urban areas, open-cast mines, and harbors within mountainous areas [2].

This section discusses in-band pseudolites, Locata and Terralite XPS, and IMES, while GNSS repeaters are discussed in Section G.13 of Appendix G on the CD.

12.1.1 In-Band Pseudolites

Transmitting pseudolite signals on the same frequencies and with the same modulation as GNSS signals minimizes user equipment costs as common front ends and baseband signal processors may be used. Only software enhancements to GNSS user equipment are required to support the different ranging codes and navigation message formats of pseudolite signals, noting that satellite ephemeris parameters are not suitable for conveying the position of a ground-based or airborne transmitter [3]. In-band pseudolites used to supplement GNSS should be synchronized to GNSS system time. This can be done via receipt of GNSS signals. Alternatively, differential positioning with a reference station may be used.

In-band pseudolite systems must be designed to avoid disruption to GNSS signal reception. The near-far problem is particularly severe for GPS C/A codes due to the relatively short repetition length of the code; interference due to cross-correlation can occur where received signal strengths differ by more than 21 dB (see Section 9.1.4). L1/E1 is also the band of choice for single-frequency GNSS user equipment. Consequently, many countries have banned pseudolites (and GNSS repeaters) in this band. GNSS signals in the L5/E5 band are less susceptible to the near-far problem, making it more suitable for pseudolite operation. Another way of mitigating the near-far problem is to pulse the pseudolite signals on and off [1, 4]. As pseudolite signals do not pass through the ionosphere, it is not necessary to use multiple frequencies.

Where multiple pseudolites are required within a localized area, infrastructure costs may be reduced by using a common signal generator and applying different time delays to each transmitter [5]. Mutual interference will be no greater than that with separate PRN codes provided the received signals always differ by more than two code chips throughout the reception area.

12.1.2 Locata and Terralite XPS

Locata and Terralite XPS are proprietary pseudolite-based positioning systems. Locata, designed primarily for surveying, operates in the 2.4–2.48-GHz ISM band [6]. Each transmitter, known as a Locatalite, broadcasts a 10 Mchip s^{-1} DSSS ranging code and also receives the signals from the other Locatalites in the network, known as a Locatanet. Each Locatalite uses GNSS to determine its own position and the received Locata signals for time synchronization with respect to the master. The near-far problem is solved by using TDMA as well as CDMA (i.e., each Locatalite transmits in turn). However, interference from other users of the ISM band can occur.

Terralite XPS is designed for positioning within deep open-cast mines. It operates on a similar principle to Locata, but uses the 9.5–10-GHz X-band [7].

Both systems have ranges of 2–10 km, depending on the degree of signal obscuration. When direct line-of-sight reception of sufficient signals is available, a horizontal positioning accuracy for static users of a few centimeters may be achieved using both code and carrier measurements (and assuming a constant troposphere refractive index). Vertical accuracy is typically much poorer due to signal geometry.

12.1.3 Indoor Messaging System

IMES is an indoor positioning system proposed for implementation in Japan as part of QZSS (Section 8.2.5) [8]. It uses L1-band C/A code transmitters with a range of 3m. A very low power limits the interference to GPS. Interference is further limited by a frequency offset of ±8.2 kHz, which is equivalent to the Doppler shift when the pseudo-range rate is ±1,560 m s^{-1}. IMES transmitters are not time synchronized and no ranging is performed. Instead, users simply decode the navigation message, repeated every few seconds, to obtain a proximity position fix (Section 7.1.3) and other location information.

12.2 Ultrawideband

Ultrawideband signals are formally defined as signals with an absolute bandwidth of at least 500 MHz or a fractional bandwidth (bandwidth divided by carrier frequency) of 20%, where the bandwidth is double-sided and bounds a continuous region within which the PSD is within 10 dB of the maximum [9, 10].

The main attraction of UWB signals for positioning is multipath resolution (see Section 7.4.2). For example, if the signal bandwidth is 1 GHz, multipath components with a differential path delay of 0.3m or more may be separately resolved. This enables a much greater ranging accuracy to be obtained within indoor and urban environments where signals typically follow multiple paths from transmitter to receiver and the direct path is often severely attenuated by walls, leaving it much weaker than some of the reflected signals. A gigahertz-region signal is typically attenuated by about 10 dB by an internal wall, 20 dB by an external wall, and 30 dB by a concrete floor [11].

Dedicated spectrum for UWB transmissions is not available. Consequently, UWB signals must share spectrum with other users. To avoid causing interference to these users, UWB transmissions must have a very low PSD. The maximum average PSD allowed for unlicensed UWB transmissions is –41.3 dBmW/MHz (7.4×10^{-14} W Hz^{-1}) [9, 10]. This corresponds to a maximum power of 74 μW for a 1 GHz-bandwidth signal. Different countries permit UWB communications and ranging in different parts of the spectrum as shown in Figure 12.1. Some countries additionally require use of detect and avoid (DAA) technology in some bands; this continuously detects narrowband signals and minimizes the transmitted UWB power within the conflicting spectrum.

To obtain useful coverage with a very low PSD, spread spectrum techniques (see Section 7.2.1) are used. Assuming a PSD close to the maximum, the range depends on the bandwidth-to-data-rate ratio, which is typically high for UWB signals. For example, a free-space range of about 1 km is achievable with a bandwidth-to-data-rate ratio of about 10^5 [11]. UWB ranging is typically performed using known data sequences, maximizing the bandwidth-to-data-rate ratio. This improves the sensitivity of the receiver (compared to communications use), allowing higher precision timing measurements to be made. It also enables the detection of the direct-path signal needed for ranging in cases where it is highly attenuated. Positioning performance is better when a higher proportion of the overall UWB system's duty cycle is used for ranging.

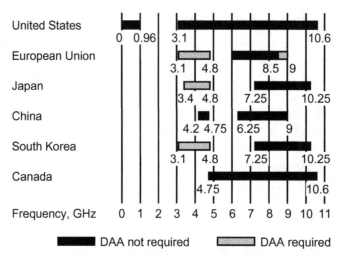

Figure 12.1 UWB spectrum allocations in selected countries.

A key application of UWB positioning is the navigation and tracking of emergency and military personnel inside buildings without having to rely on base stations within those buildings [12, 13]. It has also been demonstrated for ranging between road vehicles in a relative positioning system [14] and is used for indoor asset tracking [9].

The remainder of this section describes UWB modulation schemes, signal timing, and positioning.

12.2.1 Modulation Schemes

Three different types of modulation have been used for UWB ranging: impulse radio (IR), orthogonal frequency division multiplex, and frequency-hopping direct-sequence spread spectrum.

IR systems transmit a series of subnanosecond pulses, which are inherently ultrawideband. In practical UWB systems, these are modulated onto one or more carriers [15]. Pulse intervals are typically much larger than pulse durations. Variation of the relative polarity and timing of the pulses is used both to convey data and as a spread spectrum technique. A benefit of IR is that simple receivers may be used that simply square the incoming signal and detect the pulses, although correlation-based receivers (see Section 7.3.2) are more sensitive [10].

OFDM comprises multiple carriers, typically about 100 for UWB, each modulated with a combination of data and a DSSS code (see Section 7.2.1). OFDM modulation has two main benefits over IR [10]. First, its spectrum is close to flat, enabling it to make very efficient use of a given channel as shown in Figure 12.2. Second, an OFDM signal does not have to occupy a continuous block of spectrum, enabling frequencies occupied by in-band narrowband signals to be avoided, minimizing mutual interference. This is crucial for spectrum subject to DAA restrictions.

Unmodulated OFDM is known as multicarrier (MC). Ranging may be performed by measuring the received phase difference between successive carriers, while CDMA may be achieved using known initial phase offsets on each carrier [16]. Note that

12.2 Ultrawideband

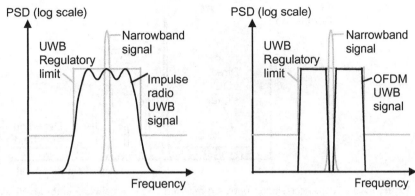

Figure 12.2 Comparison of IR and OFDM signal spectra.

there is a range ambiguity which is inversely proportional to the carrier spacing. For example, a 300-kHz spacing produces a 1-km ambiguity.

An FH-DSSS signal comprises a single frequency-hopping carrier, modulated with a combination of data and a DSSS code. Different transmitters use different frequency-hopping sequences to minimize interference. For example, the Thales Research and Technology (TRT) UWB positioning system uses a signal with a 20 Mchip s^{-1} chipping rate that hops over 1.25 GHz of spectrum within a 1-ms cycle, giving a signal bandwidth over 1 ms of 1.25 GHz [11]. FH-DSSS modulation is almost as spectrum efficient as OFDM and also has the capability to avoid in-band narrowband signals.

There are three standard protocols for UWB communications [9, 10]. The Ecma-368 standard is for high-rate communications over a range of a few meters. It uses OFDM modulation and supports data rates of 50–480 Mbit s^{-1}. The IEEE 802.15.4a and 802.15.4f UWB standards are for lower-rate communications over a range of a few tens of meters. They uses IR modulation and support data rates of 0.11–27 Mbit s^{-1}. All three standards incorporate protocols for two-way ranging (Section 7.1.4.5). However, only 802.15.4f was designed specifically to support positioning. A transmission protocol optimized for communications does not necessarily offer the best ranging performance (and vice versa). Many UWB positioning systems therefore use proprietary protocols.

12.2.2 Signal Timing

In a typical UWB positioning environment, the signal will follow multiple paths between transmitter and receiver. Consequently, a correlation of the received signal with an internally-generated replica of the transmitted signal will produce multiple peaks. The direct-path signal is often not the strongest signal, so it is assumed to correspond to the earliest arriving correlation peak above a certain threshold and within a certain time window of the largest peak. Figure 12.3 illustrates this.

A typical UWB positioning receiver acquires and tracks the strongest component of the received signals. In addition, a bank of parallel correlators, maintained at fixed time offsets from the peak of the strongest signal, is used to measure the correlation

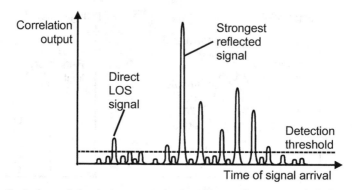

Figure 12.3 Typical correlation between received and internally-generated UWB signals in a multipath environment.

profile. From this, the time of arrival or round-trip time of the direct LOS signal is measured; this is used for positioning [9].

There are a number of reasons why a UWB receiver may select the wrong peak. First, the direct-path signal may be too weak, falling below the detection threshold. Second, there may be interference from another UWB signal of the same type, known as multiple access interference (MAI); this is a particular problem with the IEEE 802.15.4a protocol [9]. Finally, side lobes of a larger correlation peak can be mistaken for the direct-path signal if the signal waveform and spreading sequence are not designed with care [9].

Statistical tests using the detected correlation peaks of the multipath components can be used to estimate whether the direct LOS signal is receivable and distinguishable [17]. When sufficient ranging measurements are available, innovation filtering (Section 17.3.1) and consistency checks (Section 17.4) can be used to identify measurements of nonline-of-sight signals. A NLOS range measurement can be used to define a containment area (see Section 7.1.3) as it will always exceed the direct range.

Assuming correct selection of the direct-path component, the signal timing error depends on the receiver and transmitter timing resolution and stability, the signal-to-noise level, narrowband interference, building material dielectric properties and dispersion in the receive and transmit antennas [9, 13]. Antenna dispersion may be mitigated by modifying the waveform of the internally-generated signals used for correlation [10].

12.2.3 Positioning

UWB positioning systems may use base-station-to-mobile ranging, mobile-to-base-station ranging, or two-way ranging, depending on the application [18]. However, two-way ranging conveys three significant benefits: one less base station is needed, base stations need not be time synchronized, and there are fewer constraints on base station placement for optimizing signal geometry (see Section 7.4.5). As Figures 7.24 and 7.27 illustrated, it is not necessary to surround a user with base stations on all sides to obtain optimum position accuracy with two-way ranging, whereas it is with one-way ranging [13]. A two-way ranging protocol also allows ranging

12.2 Ultrawideband

between mobiles and base stations to be easily supplemented by peer-to-peer ranging (Section 7.1.2) in order to extend coverage [12].

As UWB provides short-range high-precision positioning, height differences between transmitters and receivers are significant and there will often be sufficient signal geometry to determine these from the UWB signals. Consequently, a 3-D positioning algorithm should be used. A local-tangent-plane coordinate frame, denoted by l, is a suitable reference and resolving frame. A position solution, $\tilde{\mathbf{r}}_{la}^{l}$, for the user antenna, a, may be obtained using equations of the form

$$\tilde{r}_{aj,C} = \sqrt{\left(\mathbf{r}_{lj}^{l} - \tilde{\mathbf{r}}_{la}^{l}\right)^{\mathrm{T}}\left(\mathbf{r}_{lj}^{l} - \tilde{\mathbf{r}}_{la}^{l}\right)} + \delta r_{aj,\varepsilon}^{+}, \tag{12.1}$$

where j denotes the jth base station antenna frame, $\tilde{r}_{aj,C}$ is the measured two-way range between the user and the jth base station with any necessary corrections applied, and $\delta r_{aj,\varepsilon}^{+}$ is the jth residual.

Using an ILS algorithm (see Section 7.3.3), a user position estimate, $\hat{\mathbf{r}}_{la}^{l}$, may be obtained from m measurements by iterating

$$\hat{\mathbf{r}}_{la}^{l} = \hat{\mathbf{r}}_{la}^{l-} + \left(\mathbf{H}_{R}^{l\mathrm{T}}\mathbf{H}_{R}^{l}\right)^{-1}\mathbf{H}_{R}^{l\mathrm{T}}\begin{pmatrix}\tilde{r}_{a1,C} - \hat{r}_{a1}^{-} \\ \tilde{r}_{a2,C} - \hat{r}_{a2}^{-} \\ \vdots \\ \tilde{r}_{am,C} - \hat{r}_{am}^{-}\end{pmatrix}, \tag{12.2}$$

where $\hat{\mathbf{r}}_{la}^{l-}$ is the predicted user position, $\mathbf{H}_{R}^{l}$ is the measurement matrix, given by

$$\mathbf{H}_{R}^{l} = \begin{pmatrix} -(x_{l1}^{l} - \hat{x}_{la}^{l-})/\hat{r}_{a1}^{-} & -(y_{l1}^{l} - \hat{y}_{la}^{l-})/\hat{r}_{a1}^{-} & -(z_{l1}^{l} - \hat{z}_{la}^{l-})/\hat{r}_{a1}^{-} \\ -(x_{l2}^{l} - \hat{x}_{la}^{l-})/\hat{r}_{a2}^{-} & -(y_{l2}^{l} - \hat{y}_{la}^{l-})/\hat{r}_{a2}^{-} & -(z_{l2}^{l} - \hat{z}_{la}^{l-})/\hat{r}_{a2}^{-} \\ \vdots & \vdots & \vdots \\ -(x_{lm}^{l} - \hat{x}_{la}^{l-})/\hat{r}_{am}^{-} & -(y_{lm}^{l} - \hat{y}_{la}^{l-})/\hat{r}_{am}^{-} & -(z_{lm}^{l} - \hat{z}_{la}^{l-})/\hat{r}_{am}^{-} \end{pmatrix}, \tag{12.3}$$

and $\hat{r}_{aj}^{-}$ is the jth predicted range, given by

$$\hat{r}_{aj}^{-} = \sqrt{\left(\mathbf{r}_{lj}^{l} - \hat{\mathbf{r}}_{la}^{l-}\right)^{\mathrm{T}}\left(\mathbf{r}_{lj}^{l} - \hat{\mathbf{r}}_{la}^{l-}\right)}, \tag{12.4}$$

A Kalman filter-based position solution may also be implemented (see Section 9.4.2). Note that better results may be obtained using a UKF measurement model (Section 3.4.2) because of the short range between transmitters and receivers.

UWB positioning performance varies from system to system. TRT have reported an accuracy of about 5 cm without intervening walls between transceivers and about 1m with intervening walls. The range is about 100m with one external wall or two internal walls between transceivers and about 30m with an external wall and an internal wall or with a concrete floor [11].

12.3 Short-Range Communications Systems

Positioning using preexisting short-range communications infrastructure and standard user equipment is attractive because it minimizes the costs of both implementation and operation. Solutions may be self-positioning, implemented in user equipment, or remote positioning, implemented in a network server (see Section 7.1.1). However, the use of equipment and communications protocols that were not designed with positioning in mind can impose significant performance constraints. WLAN, WPAN, RFID, Bluetooth low energy, and DSRC are discussed in turn.

12.3.1 Wireless Local Area Networks (Wi-Fi)

Wireless local area network technology provides computer networking in the 2.4–2.5-GHz ISM band and at various frequencies, depending on country, within the 4.9–5.9-GHz range. It is standard in smartphones. Most WLAN implementations correspond to one of the IEEE 802.11 family of standards and are commonly known as Wi-Fi (a contraction of wireless fidelity). Base stations, known as access points (APs), are situated in homes, offices, and public areas, such as cafés and airports. Each AP is identified by a unique code. WLAN signals have a range of up to 100m, though attenuation by walls and buildings usually reduces this to a few tens of meters. The signal bandwidth is 22 MHz and either OFDM or DSSS modulation is used, depending on the standard.

Practical WLAN positioning systems use either proximity or pattern matching. A number of commercial positioning services are available using both methods. Ranging and angular positioning are also possible in principle. Each method is discussed in turn.

Proximity positioning (Section 7.1.3) identifies which APs are within range of the user. It then uses a database of the AP locations to determine a user position, typically within 20–30m. The leading commercial provider is Skyhook Wireless, which supplies Apple and Dell among others and produces the Loki tool. An independent system is operated by Google. The databases cover major towns and cities worldwide. Only preexisting APs are used, so they are effectively treated as signals of opportunity. Service availability thus depends on an AP being within range, so coverage tends to be better in urban areas than in suburban areas.

Databases may be generated using a moving vehicle that scans every AP within range, a practice known as "wardriving." A proprietary algorithm is then used to estimate the AP positions from the vehicle position solution and WLAN signal strength data [19]. An alternative approach is crowdsourcing, whereby smartphone users with a good GNSS position solution upload the WLAN APs they can receive to a central server that builds the database. APs can move location, including from one city to another, between database updates. Therefore, users should exercise caution in using a position solution obtained using only one AP.

A position solution can be obtained from WLAN RSS measurements using pattern matching as described in Section 7.1.6. An accuracy of 1–5m can typically be obtained, depending on the number of APs receivable, database grid spacing, and the mixture of measurement data and signal propagation modeling used in generating the database [20, 21]. Commercial systems include the Ekahau Real-Time Location System (RTLS) [22] and Aeroscout MobileView. Note, however, that WLAN RSS measurements are not standardized, so different chipsets produce different results from the same signals. Also, a device can take about 5 seconds to complete a new AP scan, while more APs than required for providing an Internet service are typically needed for the best accuracy.

Ranging using WLAN signals presents a number of challenges. WLAN transmissions are not time synchronized, so differential or two-way ranging must be used (see Sections 7.1.4.3 and 7.1.4.5). Standard Wi-Fi equipment can only time signals to a resolution of 1 μs, corresponding to a range of 300m. Multiple measurements are thus needed for accurate ranging. To obtain a meter-level resolution, measurements must be made over about 10 seconds [10]. Furthermore, in indoor and urban environments, the direct-path signal is often attenuated by walls, while reflected signals can be stronger. The signal bandwidth is not sufficient to distinguish multipath components easily. Therefore, the accuracy of timing-based WLAN positioning is typically limited to around 10m with standard user equipment [23]. A submeter positioning accuracy may be obtained by performing differential ranging with special high-sampling-rate user and reference receivers and using super-resolution techniques to separate out the components of multipath-contaminated signals [24].

Angular positioning (Section 7.1.5) may also be implemented using the IEEE 802.11n WLAN standard if the APs are modified to incorporate antenna arrays. A positioning accuracy of around 2m has been reported using this method [25].

12.3.2 Wireless Personal Area Networks

Wireless personal area networks are designed for peer-to-peer communication between mobile users, although they can also be used for communication between mobile users and a base station. Consequently, WPANs are suitable for implementing cooperative, or, collaborative positioning concepts. These include relative positioning (Section 7.1.2), sharing of GNSS assistance data (see Section 10.5.2) and exchange of spatial data for use by map matching algorithms (Section 13.1).

There are six main WPAN standards [10]. Bluetooth (IEEE 802.15.1), ZigBee (IEEE 802.15.4), IEEE 802.15.3, and the chirp spread spectrum version of IEEE 802.15.4a all use the 2.4–2.5-GHz ISM band. The other two standards use UWB signals, so are covered by Section 12.2; they are Ecma-368 and the UWB version of IEEE 802.15.4a. This section focuses on Bluetooth and ZigBee.

Bluetooth is the dominant WPAN standard worldwide and is used in many consumer applications. The majority of Bluetooth devices are in power class 2, providing a range of up to 10m (class 1 devices have a 100-m range and class 3 devices have a 1-m range). All practical Bluetooth positioning uses the proximity method (Section 7.1.3). Time-based ranging is not supported by standard Bluetooth hardware, while RSS-based ranging is rendered impractical by a combination of variable transmission power and a standard RSS indication protocol that outputs the same value for any RSS within an optimal 20-dB range [10].

A further impediment to Bluetooth positioning is that it can take a few seconds to establish a Bluetooth connection. For vehicle navigation, this is typically longer than that for which Bluetooth devices will be within range. Even for pedestrian navigation, there may be insufficient time to connect to all Bluetooth devices within range, limiting the use of proximity by intersection of containment areas (see Section 7.1.3).

ZigBee is a low-power low-data-rate WPAN standard commonly used in wireless sensor networks. The range of ZigBee signals is typically 20–30m and connections can be established in 15 ms. As well as proximity positioning, some ZigBee hardware incorporates a capability to determine position to an accuracy of a few meters from RSS-derived range measurements (see Section 7.1.4.6) [10], while a position accuracy of about 2m (1σ) may be obtained using RSS pattern matching (Section 7.1.6) with ceiling-mounted base stations [26].

12.3.3 Radio Frequency Identification

RFID tags are used to identify objects in a similar manner to barcodes, but without the need for a direct line of sight between tag and reader. Typical applications include tracking of consignments, library books, and medical equipment; building access control; and public transport ticketing (e.g., the Oyster system in London).

Passive RFID tags are powered from the reader's RF signal using electromagnetic induction. They only cost a few cents, but their memory is limited and they respond only with an identification code. Active tags have their own batteries, enabling them to store and transmit more information, but they costs tens of dollars or euros each. Frequencies used by RFID include 125–134.2 kHz, 140–148.5 kHz, 13.56 MHz, 863–870 MHz in Europe, 902–928 MHz in the Americas, and the ISM bands. Passive tags have a range of about 0.5m in the LF band and a few meters at 13.56 MHz [10]. Active tags typically operate over a few tens of meters [27].

RFID positioning can operate in a remote-positioning or self-positioning configuration (see Section 7.1.1). For tracking, remote positioning, where the tags are mobile and the readers fixed, is normally used whereas, for navigation, self-positioning, where the tags are fixed and the readers mobile, is more common. The proximity positioning method (Section 7.1.3) is normally used with passive RFID tags due to their short range. Passive tags cannot transmit their locations so a database must be stored by the mobile user or accessed via a separate datalink.

Longer-range active-tag positioning uses RSS pattern matching techniques (Section 7.1.6) to obtain an accuracy of a few meters. As RFID readers (and tags) use directional antennas, performance can be improved by measuring RSS in multiple directions [27].

12.3.4 Bluetooth Low Energy

Bluetooth low energy is a feature of Bluetooth 4.0 technology that is likely to become standard equipment in smartphones. However, it is closer to an active RFID system than a WPAN. It is designed for rapid communication of small amounts of data. Power consumption is minimized through a low duty cycle, with tags operating in short bursts, enabling batteries to last several years. A connection may be established in 3 ms, while the range is about 50m.

Proximity positioning to room-level precision may be obtained by installing a basic BLE tag in each room. Installation costs are minimized by running on battery power.

Nokia's high-accuracy indoor positioning system uses angular positioning by nonisotropic transmission (see Section 7.1.5). A ceiling-mounted HAIP beacon broadcasts multiple highly directional BLE signals. The user's position is then determined by which of these signals is received. The accuracy varies from about 0.3m in office spaces to 1m in large indoor areas such as shopping centers, airports, and train stations [28].

12.3.5 Dedicated Short-Range Communication

DSRC is used for vehicle-to-vehicle (V2V) and vehicle-to-infrastructure (V2I) communications for intelligent transportation system applications in the 5.725–5.925-GHz band. Standards and frequencies vary between countries. DSRC provides a suitable data link for cooperative positioning between vehicles. In principle, it could also be used for relative positioning. However, with a range of several hundred meters, it is unsuited to proximity positioning. RSS pattern matching requires a predictable environment, RSS-based ranging is not accurate enough, and TOF-based ranging would require modifications to the user equipment and protocols [29].

12.4 Underwater Acoustic Positioning

Radio navigation signals do not propagate underwater. Instead, submarines, ROVs, AUVs, and sometimes individual divers use sound for underwater positioning. Four different self-positioning methods are used for acoustic positioning, as shown in Figure 12.4.

In long baseline (LBL) positioning, three or more transponders are placed at known locations on the bed below the water [30]. The baselines between these range from several hundred meters to a few kilometers. Two-way ranging (Section 7.1.4.5) must be used to determine a 3-D user position as, with the transponders roughly in a plane, there is insufficient signal geometry (see Sections 7.4.5 and 9.4.3) to separate the user time offset from one of the spatial dimensions (the direction of which varies with the user position). The vertical positioning geometry is also poor when the vessel requiring positioning is close to the sea bed. Note that there is no need to use a fourth transponder to resolve position ambiguity as only one of the two possible solutions will be within the water; the other will be underground.

Two-way ranging is instigated by the user equipment. A transducer aboard the host vehicle sends out a burst of digitally modulated sound, known as a ping. Frequencies of up to 40 kHz are used. An LBL transponder then replies with a similar ping, a fixed interval after receiving the user's ping.

LBL systems must be calibrated to determine the positions of the transponders. Ranging between the transponders can determine their relative positions. However, range measurements to a ship at several known positions are required to determine the absolute positions. Conversely, LBL can be used for positioning surface vessels as well as underwater vehicles.

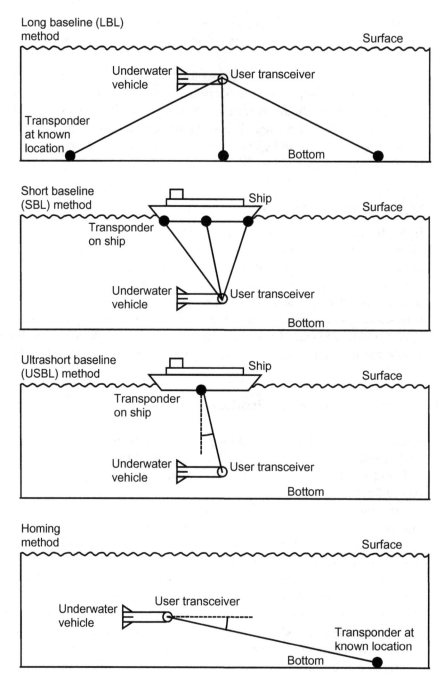

Figure 12.4 Underwater acoustic positioning methods.

Short baseline (SBL) positioning is similar to LBL positioning except that the transponders are located on the underside of a nearby ship. This architecture is suited to positioning underwater vehicles (and divers) relatively close to the ship because positioning accuracy is degraded by poor signal geometry if the distance from the mobile user to the ship greatly exceeds the baselines between the shipboard transponders.

12.4 Underwater Acoustic Positioning

Ultra-short baseline (USBL) positioning, also known as super-short baseline (SSBL) positioning, uses a single directional transponder located on the underside of the ship [30, 31]. This comprises multiple transducers with an ultra-short baseline between them of around 0.1m. AOA measurements are made (at the ship) by comparing the phase of the acoustic signals arriving at each antenna in analogy with GNSS attitude determination (Section 10.2.5). Two-way ranging is initiated by the ship transponder with the mobile unit responding. The mobile vehicle position is thus calculated on board the ship, an example of remote positioning. Note that accurate position determination from the AOA measurements requires an accurate ship attitude solution. Long and ultra-short baseline (LUSBL) positioning combines the LBL and USBL techniques, using both shipboard and seabed transponders.

The final method is homing, whereby the moving vehicle is equipped with a directional acoustic transponder and both two-way ranging and AOA measurements are made using a sequence of single transponders at known locations [32].

Determining the two-way range from an acoustic RTT measurement requires knowledge of the speed of sound in water. This is around 1,500 m s^{-1}. However, it varies, depending on temperature, depth, and salinity. Therefore, assuming a constant value can result in scale factor errors of a few percent. The speed of sound must therefore be carefully measured. One way of doing this is to measure the range between a transponder at a known location and a known position on the surface. Some LBL systems also use ranging between the fixed transponders.

Acoustic positioning systems are vulnerable to interference from underwater noise, while excessive aeration of the water can severely attenuate the signals. Therefore, they cannot be relied upon to provide continuous positioning and are typically used as part of an integrated navigation system.

The maximum range of a transponder operating in the 10–15-kHz band is 10 km [30]. With a speed of sound of around 1,500 m s^{-1}, the round-trip time between sending the initial ping and receiving the returned ping can therefore exceed 10 seconds. Vehicle motion during this time can be significant and must be accounted for in the position determination. Both underwater vehicles and ships are typically equipped with dead reckoning so they can determine short-term position changes relatively accurately.

Positioning equations of the following form should be solved for LBL and SBL ranging measurements

$$\left(\Delta \tilde{t}_{rt,a}^t - \tau_r^t\right)c_w = \sqrt{\left[\mathbf{r}_{lt}^l - \mathbf{r}_{la}^l\left(t_{st,t}^a\right)\right]^T \left[\mathbf{r}_{lt}^l - \mathbf{r}_{la}^l\left(t_{st,t}^a\right)\right]} + \sqrt{\left[\mathbf{r}_{lt}^l - \mathbf{r}_{la}^l\left(t_{sa,a}^t\right)\right]^T \left[\mathbf{r}_{lt}^l - \mathbf{r}_{la}^l\left(t_{sa,a}^t\right)\right]}, \quad (12.5)$$

where $\Delta \tilde{t}_{rt,a}^t$ is the measured RTT between the mobile user and the transponder t, τ_r^t is the transponder response time, c_w is the speed of sound in water, $\mathbf{r}_{lt}^l$ is the transponder position with respect to and resolved in a local-tangent-plane coordinate frame, $\mathbf{r}_{la}^l$ is the user antenna position, $t_{st,t}^a$ is the time of transmission of the outgoing signal, and $t_{sa,a}^t$ the time of arrival of the incoming signal. For USBL positioning, the surface ship motion must be accounted for.

A well-calibrated LBL or homing acoustic positioning system can achieve a positioning accuracy of 0.2–2m [30, 32]. USBL positioning accuracy varies from around 0.2m at close range to about 10m at the maximum range of about 3 km [30].

GNSS-equipped intelligent buoys (GIBs) are used for remote acoustic positioning. An underwater vehicle or diver transmits to a network of beacons on the surface, with position determined at a network server.

12.5 Other Positioning Technologies

This section briefly reviews a number of other short-range positioning technologies, divided into radio, ultrasound, infrared, optical, and magnetic categories.

12.5.1 Radio

A number of other positioning technologies using proximity (Section 7.1.3) are in use. Radio signposts, used for bus networks [33], and balises, used for rail navigation [34], are intended to be used alongside dead reckoning. A balise is a track-mounted transponder, powered by induction from a passing train. It can provide positioning to submeter accuracy.

Proprietary signals can be used for timing-based ranging (Section 7.1.4) in the ISM bands [35], overcoming some of the limitations of the standard WLAN and WPAN protocols.

12.5.2 Ultrasound

Ultrasound is sound at frequencies above the upper threshold for human hearing. Because of the slow speed of sound in air (around 340 m s^{-1}), ultrasound can provide a very high-ranging resolution for a given timing resolution compared to radio methods. For example, ultrasound at a frequency of 40 kHz has a wavelength of less than a centimeter. However, high-accuracy ranging requires line-of-sight signal propagation. Sensors also react to ambient ultrasonic noise, including jangling keys, slamming doors, malfunctioning fluorescent lights, and other ultrasound signals, so outlier detection is required [22].

Because of the line-of-sight requirement, most ultrasonic positioning systems operate indoors with a separate set of beacons deployed in each room. Examples include MIT Cricket [36], Active Bat [37], Dolphin [38], and 3D-LOCUS [39]. Cricket and Active Bat use radio for time synchronization of the ultrasound signals and to convey beacon location information to the mobile units. Cricket, Active Bat, and Dolphin have positioning accuracies of a few centimeters, while subcentimeter positioning has been demonstrated with 3D-LOCUS. In general, denser beacon networks lead to more precise positioning.

An ultrasonic system for the relative positioning (Section 7.1.2) of land vehicles has also been demonstrated, giving submeter accuracy [40].

12.5.3 Infrared

Infrared technology, similar to that used by remote controls, can be used for proximity positioning (Section 7.1.3) at low cost. The range using typical consumer technology is up to 30m, though this can be reduced significantly in the presence of strong sunlight or ambient heat [22]. An example is Active Badge, used for locating people within buildings [41].

12.5.4 Optical

The AOA of a light signal may be measured using three orthogonal low-cost photodiodes. These measure the intensity of the incident light, which is a function of the angle of incidence at each detector. Light signals may be transmitted from multiple light-emitting diodes (LEDs), which may be amplitude modulated to enable the different signals to be distinguished. A positioning accuracy of about 5% of the interbeacon distance has been achieved using two LED beacons [42].

12.5.5 Magnetic

Both the magnitude and the direction of an artificially generated magnetic field vary with position. Therefore, by measuring these, a 3-D user position may be deduced from a single source. In order to distinguish this magnetic signal from the Earth's magnetic field and other man-made sources, it must be modulated. Frequencies below 100 Hz fall within the bandwidth of a typical magnetometer and the mains power frequencies of 50 Hz and 60 Hz should be avoided. A range of about 20m is achievable and magnetic signals can penetrate deep inside buildings and even be received underground [43].

Problems and exercises for this chapter are on the accompanying CD.

References

[1] Elrod, B. D., and A. J. Van Dierendonck, "Pseudolites," in *Global Positioning System: Theory and Applications, Volume II*, B. W. Parkinson and J. J. Spilker, Jr., (eds.), Washington, D.C.: AIAA, 1996, pp. 51–79.

[2] Grant, A., et al., "MARUSE: Demonstrating the Use of Maritime Galileo Pseudolites," *Proc. ION GNSS 2007*, Fort Worth, TX, September 2007, pp. 1923–1930.

[3] Kim, D, et al., "Design of Efficient Navigation Message Format for UAV Pseudolite Navigation System," *IEEE Trans. AES*, Vol. 44, No. 4, 2008, pp. 1342–1355,

[4] O'Driscoll, C., D. Borio, and J. Fortuny-Guasch, "Investigation of Pulsing Schemes for Pseudolite Applications," *Proc. ION GNSS 2011*, Portland, OR, September 2011, pp. 3480–3492.

[5] Im, S.-H., and G.-I. Jee, "Feasibility Study of Pseudolite Techniques Using Signal Transmission Delay and Code Offset," *Proc. ION ITM*, Anaheim, CA, January 2009, pp. 798–803.

[6] Barnes, J., et al., "High Accuracy Positioning Using Locata's Next Generation Technology," *Proc. ION GNSS 2005*, Long Beach, CA, September 2005, pp. 2049–2056.

[7] Zimmerman, K. R., et al., "A New GPS Augmentation Solution: Terralite™ XPS System for Mining Applications and Initial Experience," *Proc. ION GNSS 2005*, Long Beach, CA, September 2005, pp. 2775–2788.

[8] Manandhur, D., et al., "Development of Ultimate Seamless Positioning System Based on QZSS IMES," *Proc. ION GNSS 2008*, Savannah, GA, September 2008, pp. 1698–1705.

[9] Sahinoglu, Z., S. Gezici, and I. Guvenc, *Ultra-Wideband Positioning Systems: Theoretical Limits, Ranging Algorithms, and Protocols*, New York: Cambridge University Press, 2008.

[10] Bensky, A., *Wireless Positioning Technologies and Applications*, Norwood, MA: Artech House, 2008.

[11] Harmer, D., "Indoor Positioning," *Proc. NAV '09—Positioning & Location*, Nottingham, U.K., November 2009.

[12] Harmer, D., et al., "EUROPCOM: Emergency Ultrawideband Radio for Positioning and Communications," *Proc. IEEE International Conference on Ultra-Wideband*, Hannover, Germany, September 2008, pp. 85–88.

[13] Michalson, W. R., A. Navalekar, and H. K. Parikh, "Error Mechanisms in Indoor Positioning Systems Without Support from GNSS," *Journal of Navigation*, Vol. 62, No. 2, 2009, pp. 239–249.

[14] Petovello, M. G., et al., "Demonstration of Inter-Vehicle UWB Ranging to Augment DGPS for Improved Relative Positioning," *Proc. ION GNSS 2010*, Portland, OR, September 2010, pp. 1198–1209.

[15] Yu., H., "Long-Range High-Accuracy UWB Ranging for Precise Positioning," *Proc. ION GNSS 2006*, Fort Worth, TX, September 2006, pp. 83–94.

[16] Cyganski, D., J. Orr, and W. R. Michalson, "Performance of a Precision Indoor Positioning System Using a Multi-Carrier Approach," *Proc. ION NTM*, San Diego, CA, January 2004, pp. 175–180.

[17] Guvenc, I., C. C. Chong, and F. Watanabe, "NLOS Identification and Mitigation for UWB Localization Systems," *Proc. IEEE Wireless Communications Networking Conference*, Hong Kong, China, March 2007, pp. 3488–3492.

[18] Kang, D., et al., "A Simple Asynchronous UWB Position Location Algorithm Based on Single Round-Trip Transmission," *Proc. International Conference on Advanced Communication Technology*, February 2006, pp. 1458–1461.

[19] Jones, K., L. Liu, and F. Alizadeh-Shabdiz, "Improving Wireless Positioning with Look-Ahead Map-Matching," *Proc. MobiQuitous 2007*, Philadelphia, PA, February 2008, pp. 1–8.

[20] Eissfeller, B., et al., "Indoor Positioning Using Wireless LAN Radio Signals," *Proc. ION GNSS 2004*, Long Beach, CA, September 2004, pp. 1936–1947.

[21] Hatami, A., and K. Pahlavan, "A Comparative Performance Evaluation of RSS-Based Positioning Algorithms Used in WLAN Networks," *Proc. IEEE Wireless Communications and Networking Conference*, March 2005, pp. 2331–2337.

[22] Kolodziej, K. W., and J. Hjelm, *Local Positioning Systems: LBS Applications and Services*, Boca Raton, FL: CRC/Taylor and Francis, 2006.

[23] Galler, S., et al., "Analysis and Practical Comparison of Wireless LAN and Ultra-Wideband Technologies for Advanced Localization," *Proc. IEEE/ION PLANS*, San Diego, CA, April 2006, pp. 198–203.

[24] Nur, K., et al., "A New Time Estimation Technique for High Accuracy Indoor WLAN," *Proc. European Navigation Conference*, London, U.K., November 2011.

[25] Wong, C. M., G. G. Messier, and R. Klukas, "Evaluating Measurement-Based AOA Indoor Location Using WLAN Infrastructure," *Proc. ION GNSS 2007*, Fort Worth, TX, September 2007, pp. 1139–1145.

[26] Hsu, L.-T., W.-M. Tsai, and S.-S. Jan, "Development of a Real Time Indoor Location Based Service Test Bed," *Proc. ION GNSS 2010*, Portland, OR, September 2010, pp. 1175–1183.

[27] Fu, Q., and G. Retscher, "Active RFID Trilateration and Location Fingerprinting Based on RSSI for Pedestrian Navigation," *Journal of Navigation*, Vol. 62, No. 2, 2009, pp. 323–340.

[28] Kalliola, K., "High Accuracy Indoor Positioning Based on BLE," Nokia Research Center Presentation, April 27, 2011.

[29] Allen, J. W., and D. M. Bevly, "Performance Evaluation of Range Information Provided by Dedicated Short Range Communication (DSRC) Radios," *Proc. ION GNSS 2010*, Portland, OR, September 2010, pp. 1631–1635.

[30] *High Precision Acoustic Positioning—HiPAP*, Kongsberg product description, accessed March 2010.

[31] Jalving, B., and K. Gade, "Positioning Accuracy for the Hugin Detailed Seabed Mapping UUV," *Proc. IEEE Oceans '98*, 1998, pp. 108–112.

[32] Butler, B., and R. Verrall, "Precision Hybrid Inertial/Acoustic Navigation System for a Long-Range Autonomous Underwater Vehicle," *Navigation: JION*, Vol. 48, No. 1, 2001, pp. 1–12.

[33] El-Gelil, M. A., and A. El-Rabbany, "Where's My Bus? Radio Signposts, Dead Reckoning and GPS," *GPS World*, June 2004, pp. 68–72.

[34] Mirabadi, A., F. Schmid, and N. Mort, "Multisensor Integration Methods in the Development of a Fault-Tolerant Train Navigation System," *Journal of Navigation*, Vol. 56, No. 3, 2003, pp. 385–398.

[35] Hedley, M., D. Humphrey, and P. Ho, "System and Algorithms for Accurate Indoor Tracking Using Low-Cost Hardware," *Proc. IEEE/ION PLANS*, Monterey, CA, May 2008, pp. 633–639.

[36] Priyantha, N. B., *The Cricket Indoor Location System*, Ph.D. Thesis, Massachusetts Institute of Technology, 2005.

[37] Harter, A., et al., "The Anatomy of a Context-Aware Application," *Proc. Mobicom '99*, Seattle, WA, August 1999.

[38] Hazas, M., and A. Hopper, "Broadband Ultrasonic Location Systems for Improved Indoor Positioning," *IEEE Trans. on Mobile Computing*, Vol. 5, No. 5, 2006, pp. 536–547.

[39] Prieto, J. C., et al., "Performance Evaluation of 3D-LOCUS Advanced Acoustic LPS," *IEEE Trans. on Instrumentation and Measurement*, Vol. 58, No. 8, 2009, pp. 2385–2395.

[40] Henderson, H. P., Jr., and D. M. Bevly, "Relative Position of UGVs in Constrained Environments Using Low Cost IMU and GPS Augmented with Ultrasonic Sensors," *Proc. IEEE/ION PLANS*, Monterey, CA, May 2008, pp. 1269–1277.

[41] Want, R., et al., "The Active Badge Location System," *ACM Transactions on Information Systems*, Vol. 10, No. 1, 1992, pp. 91–102.

[42] Arafa, A., X. Jin, and R. Klukas, "A Differential Photosensor for Indoor Optical Wireless Positioning," *Proc. ION GNSS 2011*, Portland, OR, September 2011, pp. 1758–1763.

[43] Blankenbach, J., and A, Norrdine, "Position Estimation Using Artificial Generated Magnetic Fields," *Proc. Indoor Positioning and Indoor Navigation*, Zurich, Switzerland, September 2010.

CHAPTER 13
Environmental Feature Matching

Environmental feature-matching techniques can determine the user's position by measuring features of the environment and comparing them with a database in the same way that a person would compare features with a map. Features may be manmade, such as roads, buildings, and street furniture, or geophysical, such as terrain height and the Earth's gravitational and magnetic fields. Figure 13.1 shows the main components of a feature-matching position-fixing system. The proximity, ranging, angular, and pattern-matching positioning methods described in Section 1.3.1 may all be used with environmental features, with different position-fixing methods suited to different features. Note that the environmental features used as landmarks for proximity, ranging, and angular positioning are identified using pattern-matching techniques. However, this process is referred to a feature matching to avoid confusion with the pattern-matching positioning method.

Most practical position-fixing systems using environmental features require position aiding from another navigation system for initialization purposes. This is analogous to acquisition aiding in radio positioning systems and is used to determine which region of the feature database to search. Limiting the database search area minimizes the computational load and the number of instances in which there are multiple possible matches between the measured features and those in the database.

Databases are typically preloaded into the feature-matching system with updates often available via the Internet. Alternatively, local feature data may be downloaded via a mobile data link as the user enters the relevant area, a form of network assistance. Feature data may also be exchanged between participants in a cooperative (or collaborative) positioning system. Feature matching is inherently context dependent as different types of feature will be encountered in different environments. It is therefore important to match the database and sensor(s) to the application.

Environmental features can also be used for dead reckoning. Successive measurements of the same feature(s) are compared to determine the motion of the host vehicle as shown in Figure 13.2. Some feature-matching systems can operate in both position-fixing and dead-reckoning modes, with dead reckoning used where there are insufficient matches between the observed features and the database for a position fix. However, not all of the sensors used to measure environmental features are suitable for dead reckoning as this requires the same features to be measured more than once. This can be achieved using an imaging sensor, a movable sensor, or multiple sensors on the same vehicle.

Environmental feature matching and tracking can fail to provide navigation information if there are insufficient features in the environment or the database. Ambiguous features, such as parallel roads or a group of similar houses, can also cause problems. There are three main ways of dealing with ambiguous matches:

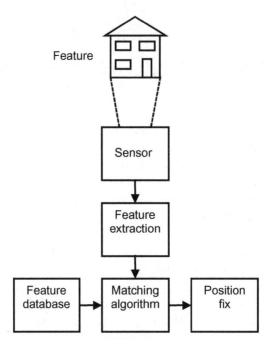

Figure 13.1 A generic feature-matching position-fixing system.

selecting one of the candidates, rejecting all of the candidates, and considering multiple possibilities until there is sufficient information to resolve the ambiguity. The latter approach is most robust, but imposes the highest processing load. The incorporation of ambiguous feature matches into a multisensor integration algorithm is discussed in Section 16.3.5. False position fixes can also result from temporary features of the environment, such as a marquee or a vehicle; an out-of-date database; or obstruction of the sensor. Therefore, any navigation system using feature matching or tracking should have the capability to detect faults and recover from them, as described in Chapter 17.

Ambiguity can be reduced by combining multiple environmental features into a location signature and matching them together. A location signature may include different categories of feature. It may also combine observations from different locations, provided their relative positions may be determined from a velocity solution. Similarly, if suitable environmental features are not available continuously, a velocity solution may be required to bridge the gaps between position fixes. Velocity can sometimes be obtained from the feature-matching system's own dead-reckoning mode or predicted from recent position fixes. However, aiding from external sensors, such as inertial navigation or odometry, is often required. Thus, environmental-feature-based position fixing is normally implemented as part of a multisensor integrated navigation system.

Environmental feature matching is a core component of most simultaneous localization and mapping systems alongside mobile mapping and dead-reckoning (by feature tracking or other means). SLAM builds its own environmental features database, using its dead-reckoning navigation solution to determine the approximate position of the features. On revisiting a feature, feature matching is used both to

13.1 Map Matching

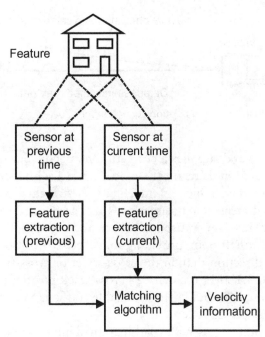

Figure 13.2 A generic feature-matching dead-reckoning system.

update the host-vehicle position solution and to refine the feature position estimate. Over several visits to the features, the database is improved. SLAM is discussed further in Section 16.3.6. It is typically implemented using an imaging sensor on an autonomous vehicle. However, the concept is broader. SLAM may also be implemented cooperatively, sharing data between users.

Section 13.1 describes road, rail, and pedestrian map-matching techniques. Section 13.2 describes terrain-referenced navigation (TRN) for air, marine, and land applications, and also terrain database height aiding. Section 13.3 describes image-based navigation, including stellar navigation, using a range of different sensors. Finally, Section 13.4 discusses other feature-matching techniques, focusing on gravity gradiometry, magnetic field variations, and celestial X-ray sources. Further information is provided in Appendix H on the CD.

13.1 Map Matching

Road vehicles usually travel on roads, trains always travel on rails, and pedestrians do not walk through walls. Map matching, also known as map aiding or snap to map, is used in land applications to correct the integrated navigation solution by applying these constraints. It is inherently context dependent as the normal behavior of the host vehicle or user is embedded in the rules of the map-matching algorithm. It combines aspects of the proximity and pattern-matching positioning methods.

Map matching is most commonly used in road vehicle navigation, generally integrated with GNSS and dead reckoning [1]. The map-matching algorithm compares the input position solution from the rest of the navigation system with the roads

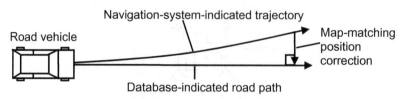

Figure 13.3 Map-matching position correction for a road vehicle.

in its database and supplies a correction perpendicular to the road direction if the navigation solution drifts off the road. Figure 13.3 illustrates this.

While the host vehicle is traveling in a straight line, map matching can only provide one-dimensional positioning. Turns are needed to obtain a 2-D fix as Figure 13.4 shows. However, in urban canyons and cuttings, GNSS satellite visibility and geometry are often poor, providing much better positioning in the along-street or along-track direction than in the cross-street or cross-track direction (see Sections 9.4.3 and 10.3.1). Map matching's cross-track positioning is thus complementary. Map matching does not currently enable traffic lane identification when there are multiple lanes per direction.

This section begins with road map matching. Digital road maps are described first, followed by road link identification and positioning. Rail and pedestrian map matching are then discussed.

13.1.1 Digital Road Maps

A digital road map is a type of GIS and is stored in a vector format. The centerline of each road is represented as a series of straight-line segments, known as links, which are joined by nodes. Curves are typically approximated by a series of straight lines. Divided highways (dual carriageways) are typically represented as parallel pairs of links. Nodes are stored as two dimensional coordinates, typically in projected form (see Section 2.4.5), while links are stored as the IDs of the nodes at each end [2, 3]. Some road maps also include direction restrictions (i.e., one-way streets), turn restrictions, and numbers of lanes. Figure 13.5 shows some examples.

A road map describes the centerline of each road. However, vehicles travel in individual lanes 2.4–3.7m wide. Consequently, matching the vehicle position to the centerline of the road introduces a bias-like error. This lane bias varies between 1m for a narrow residential street to 5.4m for the outer lanes of a four-lane road. Noise-like errors are introduced by approximating curves to straight lines, while surveying errors will also be present.

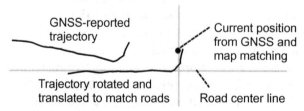

Figure 13.4 Two-dimensional map matching over a vehicle turn.

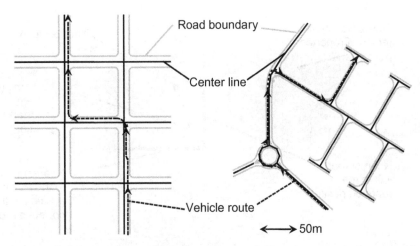

Figure 13.5 Example road maps of a U.S. urban area (left) and a U.K. suburban area (right).

13.1.2 Road Link Identification

The key to successful map matching is the correct identification of which road link within the map database the vehicle is on. The simplest technique, known as *point-to-curve* matching, just selects the nearest road link to the position input from the rest of the navigation system (e.g., GNSS or GNSS integrated with dead reckoning). This can work well in rural areas where the road separation is much greater than the uncertainty bounds of the input navigation solution. However, in urban areas, where the road network is denser and GNSS performance can be poor, this can often produce errors, as Figure 13.6 illustrates.

Road link identification can be improved by also matching the direction of travel. However, this can still lead to a false match, particularly where the roads are arranged in a grid pattern, as shown in Figure 13.6. Traffic-rule information, such as one-way streets and illegal turns, can also help. However, reliable road-segment identification requires *curve-to-curve* matching techniques, which match a series of consecutive reported positions to the map database [3]. This enables connectivity information to be used; thus successive vehicle positions should be either on the same road link or on directly connected links. If there is insufficient time for the vehicle to travel from one link to another between position fixes, then at least one of those links must be incorrect.

Most link ID algorithms adopt a multiple-hypothesis approach whereby all links within a search region defined by the uncertainty bounds of the input position fix are considered as candidates. The search region should represent a confidence level of at least 99% (see Appendix B on the CD) and account for directional differences in the input position uncertainty. Each candidate link is given a score based on proximity, direction, and connectivity. Scores may be determined purely heuristically [2] or based on fuzzy logic [4], Dempster-Schafer theory (also known as belief theory or evidence theory) [5], or Bayesian inference [6]. Scores from successive position fixes are combined according to the link connectivity. Link hypotheses scoring below a certain threshold are eliminated, while surviving hypotheses are carried forward

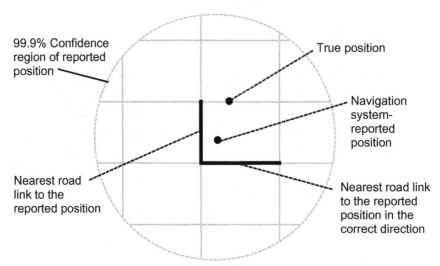

Figure 13.6 False road identification in an urban area.

to the next position fix. The process continues until only one hypothesis survives, indicating the correct link. Figure 13.7 illustrates the stages of a suitable algorithm.

A likelihood score, Λ_k^i, for the ith candidate link at the kth epoch may be determined using

$$\Lambda_k^i = \Lambda_{p,k}^i \Lambda_{d,k}^i \sum_j c_j^i \Lambda_{n,k-1}^j, \tag{13.1}$$

where $\Lambda_{p,k}^i$ is the proximity likelihood, $\Lambda_{d,k}^i$ is the direction likelihood, $\Lambda_{n,k-1}^j$ is the normalized likelihood for the jth link at the previous epoch, and c_j^i is the connectivity from link j to link i. Note that the set of links considered at the previous epoch may differ from that considered at the current epoch.

The proximity likelihood is

$$\Lambda_{p,k}^i = \exp\left[-\frac{1}{2}\begin{pmatrix} \hat{x}_{bm}^{p,i} & \hat{y}_{bm}^{p,i} \end{pmatrix}\left(\mathbf{C}_{xy\_bm}^p\right)^{-1}\begin{pmatrix} \hat{x}_{bm}^{p,i} \\ \hat{y}_{bm}^{p,i} \end{pmatrix}\right]_k, \tag{13.2}$$

where $\left(\hat{x}_{bm}^{p,i}, \hat{y}_{bm}^{p,i}\right)$ is the line from the input position, b, to the nearest point on the ith link, m, resolved along the axes of a planar coordinate frame, p, and $\mathbf{C}_{xy\_bm}^p$ is the corresponding error covariance matrix, given by

$$\mathbf{C}_{xy\_bm}^p = \mathbf{C}_\gamma^p \mathbf{P}_b^\gamma \mathbf{C}_p^\gamma + \mathbf{R}_m, \tag{13.3}$$

where $\mathbf{P}_b^\gamma$ is the input position error covariance matrix, resolved about the axes of a generic frame, γ; $\mathbf{C}_\gamma^p$ is the γ-frame-to-p-frame coordinate transformation matrix (which will have only two rows) and $\mathbf{R}_m$ is the map-link position error covariance, which will typically be a 2×2 diagonal matrix.

13.1 Map Matching

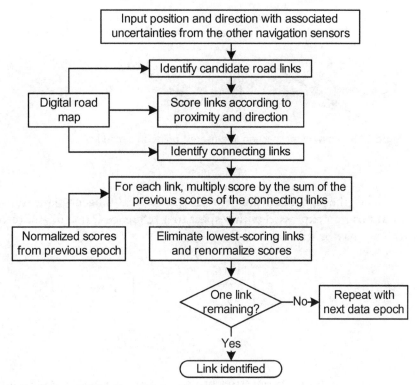

Figure 13.7 Stages of a link identification algorithm.

To determine m, the position of the point, q, where the normal from b intersects the road link must first be calculated. This is shown in Figure 13.8. If the link starts at point s and finishes at point f, then the intersection point may be determined using

$$\mathbf{r}_{bq}^T \mathbf{r}_{sf} = 0 \qquad (13.4)$$

and

$$\mathbf{r}_{bq} \wedge \mathbf{r}_{sf} = \mathbf{r}_{bs} \wedge \mathbf{r}_{sf}, \qquad (13.5)$$

where the link ID is omitted for convenience.

Solving these, noting that the z component of all planar coordinates is zero, gives

$$\hat{x}_{pq}^p = \hat{x}_{pb}^p + \frac{\left(\hat{x}_{bs}^p y_{sf}^p - \hat{y}_{bs}^p x_{sf}^p\right) y_{sf}^p}{x_{sf}^{p\,2} + y_{sf}^{p\,2}}, \quad \hat{y}_{pq}^p = \hat{y}_{pb}^p + \frac{\left(\hat{y}_{bs}^p x_{sf}^p - \hat{x}_{bs}^p y_{sf}^p\right) x_{sf}^p}{x_{sf}^{p\,2} + y_{sf}^{p\,2}}, \qquad (13.6)$$

where

$$\begin{aligned}
\hat{x}_{bs}^p &= x_{ps}^p - \hat{x}_{pb}^p, & x_{sf}^p &= x_{pf}^p - x_{ps}^p \\
\hat{y}_{bs}^p &= y_{ps}^p - \hat{y}_{pb}^p, & y_{sf}^p &= y_{pf}^p - y_{ps}^p
\end{aligned} \qquad (13.7)$$

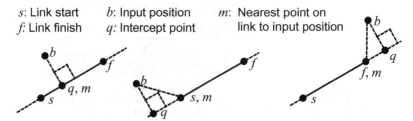

Figure 13.8 Point of intersection and nearest point on a road link.

When the planar coordinate frame, p, is a local-tangent-plane frame, a Cartesian input position expressed with respect to a reference frame β, and resolved about a frame γ may be transformed using

$$\begin{pmatrix} \hat{x}^p_{pb} \\ \hat{y}^p_{pb} \end{pmatrix} = \begin{pmatrix} x^p_{p\beta} \\ y^p_{p\beta} \end{pmatrix} + \mathbf{C}^p_\gamma \hat{\mathbf{r}}^\gamma_{\beta b}$$

$$= \mathbf{C}^p_\gamma \left(\hat{\mathbf{r}}^\gamma_{\beta b} - \mathbf{r}^\gamma_{\beta p} \right)$$

(13.8)

When the frame p comprises projected coordinates, the conversion of the input position is more complex; Section C.4 of Appendix C on the CD describes this for the transverse Mercator projection. Datum conversion may also be required; this is discussed in Section C.1 of Appendix C on the CD.

When the intersection point lies on the link (i.e., between the start and finish points), the nearest point on the road link, m, is equal to the intersection point, q. Otherwise, m is equal to the start or finish point, whichever is nearer. This is also illustrated in Figure 13.8. Thus,

$$\left(\hat{x}^p_{pm}, \hat{y}^p_{pm} \right) = \begin{array}{ll} \left(\hat{x}^p_{pq}, \hat{y}^p_{pq} \right) & 0 < \mu < 1 \\ \left(x^p_{ps}, y^p_{ps} \right) & \mu \leq 0 \\ \left(x^p_{pf}, y^p_{pf} \right) & 1 \leq \mu \end{array},$$

(13.9)

where

$$\mu = \frac{\left(\hat{x}^p_{pq} - x^p_{ps} \right) \operatorname{sign}\left(x^p_{sf} \right) + \left(\hat{y}^p_{pq} - y^p_{ps} \right) \operatorname{sign}\left(y^p_{sf} \right)}{\sqrt{x^{p\,2}_{sf} + y^{p\,2}_{sf}}}.$$

(13.10)

Finally,

$$\begin{pmatrix} \hat{x}^{p,i}_{bm} \\ \hat{y}^{p,i}_{bm} \end{pmatrix} = \begin{pmatrix} \hat{x}^{p,i}_{pm} \\ \hat{y}^{p,i}_{pm} \end{pmatrix} - \begin{pmatrix} \hat{x}^p_{pb} \\ \hat{y}^p_{pb} \end{pmatrix}.$$

(13.11)

The direction likelihood is

$$\Lambda^i_{d,k} = \exp\left[-\frac{(\hat{\psi}_{nb} - \psi^i_{nL})^2}{2(\sigma^2_{\psi\_nb} + \sigma^2_{\psi\_nLi})}\right]_k, \quad (13.12)$$

where $\hat{\psi}_{nb}$ is the input heading from the rest of the navigation system, ψ^i_{nL} is the road link direction, expressed as a bearing, σ_{ψ_nb} is the input heading uncertainty, and σ_{ψ_nLi} is the expected standard deviation of the difference between the vehicle heading and the direction of the correct road link. σ_{ψ_nLi} may be modeled as a function of yaw rate, where a suitable measurement is available. For a one-way street, $\hat{\psi}_{nb} - \psi^i_{nL}$ must be limited to the range $-\pi$ to π by adding or subtracting multiples of 2π. For a two-way street, a vehicle heading in the opposite direction to the road link is a good match, so $\hat{\psi}_{nb} - \psi^i_{nL}$ must be limited to the range $-\pi/2$ to $\pi/2$ by adding or subtracting multiples of π.

The road-link direction is given by

$$\psi^i_{nL} = \arctan_2\left(r^{n,i}_{sf,E}, r^{n,i}_{sf,N}\right), \quad (13.13)$$

where

$$\mathbf{r}^{n,i}_{sf} = \mathbf{C}^n_p \begin{pmatrix} x^{p,i}_{sf} \\ y^{p,i}_{sf} \end{pmatrix} \quad (13.14)$$

and $\mathbf{C}^n_p$ is the two-column coordinate transformation matrix from the planar frame to a North, East, Down frame. Note that this will vary with position where the planar coordinates are projected (see Section 2.4.5).

The accuracy of the input heading will vary, and the uncertainty used in computing the likelihood must reflect this. In particular, a heading derived from the vehicle trajectory (Section 6.1.4) is unreliable at low speed and unavailable when the vehicle is stationary. The direction likelihood should be omitted from (13.1) where the heading uncertainty exceeds a certain threshold [3].

During a turn, the heading may not match any link well, while several links will have good proximity matches. This information can be used to match the vehicle's position to the intersection, particularly if the turning motion can be detected by a yaw-axis gyro or differential odometry.

A simple link ID algorithm will set the connectivity, c^i_j, to one if links i and j connect at a common node (or are the same) and to zero otherwise. However, direction of travel should also be considered. The vehicle should be moving away from the common node on the current link and towards it on the previous link. Values between zero and one may be used where connectivity is uncertain. The connectivity should also be reduced by a large factor where turn restrictions or one-way street regulations would be violated. A more sophisticated algorithm could also use the vehicle's speed to estimate how far it could have traveled between epochs.

In (13.1), the implicit assumption is made that map-matching errors are independent between epochs. In practice, there will be significant correlation due to bias-like errors in both the input navigation solution and the map database. Accounting for these in the link identification algorithm will improve reliability at the expense of increased complexity.

After likelihoods have been calculated for all candidate links, those below a certain threshold (e.g., 0.01) should be deleted from the candidate list. To prevent the likelihood score for each link getting progressively smaller over successive epochs, a renormalization process is required. If the vehicle is assumed to be on one of the remaining links, then the probability of each link is simply its likelihood divided by the sum of the likelihoods of the remaining candidates.

However, a vehicle may not be on any of the roads in the database at all. It could be parking, or driving on a new road or a private driveway. A map-matching algorithm must therefore consider the off-road hypothesis [7]. When a vehicle is off the known road network, it is better to produce no map-matching measurements than a wrong fix.

The off-road hypothesis may be accounted for by imposing a maximum value on the renormalization coefficient, k_{N_max}. The renormalized likelihoods are thus given by

$$\Lambda_{n,k}^i = k_N \Lambda_k^i, \qquad k_N = \min\left(\frac{1}{\sum_j \Lambda_k^j}, k_{N\_\max}\right), \qquad (13.15)$$

where deleted links are omitted from the summation.

An alternative method of identifying the correct road link is to match a series of input positions to the map using similar techniques to batch-processed TRN (Section 13.2.2). This is known as topological map matching and works by matching the shape of the trajectory to the road network.

Once a correct road link has been identified with high confidence, the vehicle may be assumed to remain on the current road until either an intersection (junction) is encountered or the likelihood score for the current link drops below a certain threshold, in which case link identification is reactivated [3].

13.1.3 Road Positioning

A road map-matching algorithm can output position information to a route guidance algorithm and/or a navigation integration algorithm. These have different requirements.

The route guidance algorithm requires the vehicle's location within the vector-based road network map; it does not require a Cartesian or curvilinear position. Thus, it should be provided with the road link that the vehicle is on, its position within that road link and its direction of travel. This information is derived from both the map matching and the other navigation sensors, such as GNSS. A position fix should only be output when there is high confidence in the link identification; otherwise, incorrect directions could be supplied to the driver.

The position within the road link is that of the nearest point on the link to the input position, denoted m in Section 13.1.2. Its distance from the start of the link is given by

$$r_{sm} = \sqrt{\left(\hat{x}_{pm}^p - x_{ps}^p\right)^2 + \left(\hat{y}_{pm}^p - y_{ps}^p\right)^2}. \qquad (13.16)$$

The integration algorithm maintains a 3-D navigation solution using information from all navigation sensors. The map-matching measurements should be independent of those from the other navigation sensors. Thus, they should comprise only information derived from the map matching itself, not information input to the map-matching algorithm. The independent information provided by map matching is the position of the road links. Thus, a suitable output to the integration algorithm is a line fix. This is a one-dimensional position measurement, noting that map matching only produces one-dimensional information from a single input position. The processing of line-fix measurements by the integration algorithm is described in Section 16.3.4.

A line fix may be expressed as the bearing describing the road link direction, ψ_{nL}, given by (13.13), and the position of the road link's start point, (x_{ps}^p, y_{ps}^p), which must be converted to the form used by the integration algorithm. When p is a local-tangent-plane frame, a Cartesian position may be obtained using

$$\begin{aligned}\hat{\mathbf{r}}_{\beta s}^{\gamma} &= \mathbf{r}_{\beta p}^{\gamma} + \mathbf{C}_p^{\gamma} \begin{pmatrix} \hat{x}_{ps}^p \\ \hat{y}_{ps}^p \end{pmatrix} \\ &= \mathbf{C}_p^{\gamma} \begin{pmatrix} \hat{x}_{ps}^p - x_{p\beta}^p \\ \hat{y}_{ps}^p - y_{p\beta}^p \end{pmatrix}.\end{aligned} \qquad (13.17)$$

Conversion of projected coordinates and datum conversion are discussed in Appendix C on the CD.

Caution should be exercised in using an integrated navigation solution that incorporates map-matching measurements to provide the position and vehicle-direction input to the map-matching algorithm. This introduces positive feedback and could result in the link ID algorithm falsely assuming that the vehicle is continuing along the current road when it has actually turned off.

A navigation system incorporating map matching can achieve a horizontal position accuracy of about 10m (1σ) [3].

13.1.4 Rail Map Matching

Rail map matching is similar to road map matching. Rail networks are typically simpler than road networks. The train is always constrained to move on the track and there is no cross-track uncertainty, so a track map can effectively constrain rail navigation to a one-dimensional problem. Line fixes may be supplied to the navigation integration algorithm [8].

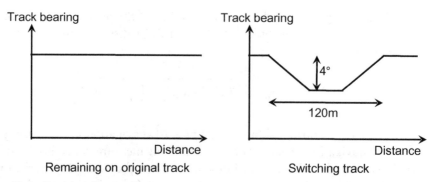

Figure 13.9 Track heading profiles at a switch (points).

Parallel tracks are always represented separately on rail maps and their centerlines can be as close as 3.3m in some countries. For many applications, such as signaling, it is essential to know which track the train is on. Initial track determination requires a precision input position from differential GNSS or a balise. Movements between parallel tracks at switches (points) can be detected by matching the heading changes measured by a yaw-axis gyro or IMU with the candidate track profiles in the database [9]. Figure 13.9 illustrates some examples. The along-track position should be adjusted to obtain the best fit to each profile. Profile matching may also be used to obtain along-track position information on curved track, while straight stretches of track may be used to calibrate the gyro or IMU.

13.1.5 Pedestrian Map Matching

For pedestrian navigation, the user position is nowhere near as constrained as for road vehicles and trains. A pedestrian cannot always be assumed to be moving along the direction of a street or a corridor. Therefore, more flexible map matching is required. One approach is obstacle-based map matching. Figure 13.10 illustrates some examples. If the navigation solution and database indicate the user is walking perpendicularly through a wall, the position solution needs to be shifted to the nearest doorway, while if the user appears to be drifting through a wall to their side, the direction of travel needs correcting. A useful map database need not contain every wall and obstacle.

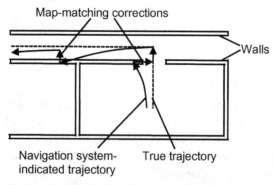

Figure 13.10 Application of obstacle-based map matching to pedestrian navigation.

The preceding discussion assumes the navigation solution is sufficiently accurate for the map-matching algorithm to determine the correct path. More sophisticated approaches use a particle filter (Section 3.5) to maintain a distribution of dead-reckoning navigation hypotheses covering different positions, headings, and solution drifts. Whenever a hypothesis violates the map (e.g., by requiring passage through a wall), it is deleted. This continues until only the correct hypothesis remains [10–12]. This approach enables indoor positioning using only inertial sensors and a map. However, it is processor intensive and it can be difficult to obtain a unique fix in a building with repeating features.

An alternative to obstacle-based map matching is to use a database of permitted routes, shifting the position solution to the nearest route. The database can be built cooperatively, by combining routes traveled by multiple users [13]. However, this may omit routes that are permitted but unpopular.

Context (Section 1.4.2) may also be used for pedestrian map matching. Indoor/outdoor context detection, described in Section 16.1.10, may be used to constrain the position solution to the indoor or outdoor region of the map. Furthermore, different classes of human motion may be identified using inertial sensors as described in Section 6.4. This enables steps, stairs, elevators, and escalators to be identified and matched with a map to determine position [14].

For pedestrians walking along corridors, the position solution is constrained to a network of links and nodes, enabling a similar approach to road map matching to be adopted. This also applies to sidewalk networks in countries where road crossing is only permitted at certain locations [15].

Most walking within a typical building is along one of four cardinal directions matching the orientations of the external walls. These directions can be determined from a city model; an internal building map is not required. The same applies to many urban sidewalk networks. Cardinal heading aiding corrects the heading drift inherent in foot-mounted inertial navigation and PDR using step detection by aligning the measured trajectory to the nearest cardinal direction [16]. Figure 13.11 illustrates this. A heading measurement innovation is computed as follows:

$$\delta z_{\psi,k} = \psi_c - \arctan_2\left(\Delta \hat{r}^n_{eb,E}, \Delta \hat{r}^n_{eb,N}\right), \quad (13.18)$$

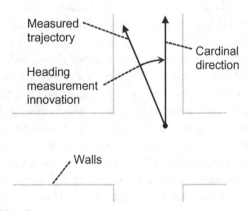

Figure 13.11 Cardinal heading measurement.

where ψ_c is the bearing of the nearest cardinal direction and $\Delta \hat{\mathbf{r}}_{eb}^n$ is the change in position measured over the past few steps. The heading measurement may then be processed as described in Section 15.1.2. Note that the sensor assembly is not assumed to be aligned with the direction of travel.

Cardinal heading aiding is not applicable all the time. People may depart from a cardinal direction to avoid an obstacle or cross an open area. One approach is to only perform a cardinal heading update where the measured trajectory is straight and within a certain threshold (e.g., 10°) of one of the cardinal directions [16]. Another approach is to limit the size of the correction applied at each epoch [17].

13.2 Terrain-Referenced Navigation

Terrain-referenced navigation determines position using pattern matching by comparing a series of terrain height measurements with a database. The terrain may be land or sea bed. Terrain-referenced navigation is also known as terrain-aided navigation (TAN), terrain-contour navigation (TCN), terrain-contour matching (TCM), terrain-based navigation (TBN), and terrain-matching navigation (TMN). Furthermore, the term terrain-referenced navigation is sometimes used to describe a broader family of terrain-based, feature-matching techniques.

The first TRN system for air use was developed in the 1950s using analog technology [18]. Systems currently available commercially include Terrain Profile Matching (TERPROM) [19, 20] and Terrain Contour Matching (TERCOM) [21]. They are used for terrain collision avoidance as well as navigation.

Conventional airborne TRN systems determine terrain height by using a radar altimeter (Section 6.2.3) to measure height above terrain and then subtracting this from the host vehicle's height solution. Figure 13.12 illustrates this. The height above terrain may also be measured using a video camera by comparing the host aircraft velocity (assumed to be known) with the rate at which features flow across the image (see Section 13.3.5) [22].

A number of terrain height databases, known as digital terrain models (DTMs) or digital elevation models (DEMs), of varying resolutions and coverage areas are available [23]. For the optimum tradeoff between accuracy and data storage, the database resolution should match that of the radalt measurements. Most military TRN systems use Digital Terrain Elevation Data (DTED), collated by the NGA. Level 1 is the lowest DTED resolution used for TRN; this has a grid spacing of about 100m and requires about 2 GB of memory for a whole Earth database [24]. Note that databases tend to use orthometric rather than geodetic height (see Section 2.4.4).

In theory, three terrain height measurements are needed to obtain a 3-D position fix from TRN. However, as the measurement and database errors are often large compared to the short-distance terrain height variation, many more are usually needed for a unique match between the measurements and database. Over flat terrain and when flying over water, TRN cannot provide horizontal position fixes at all, only height information. The methods for obtaining position from the measurements may be grouped into two categories: sequential and batch processing. These are described next, followed by a discussion of performance and error sources, and a summary of TRN techniques using laser ranging to obtain higher precision.

13.2 Terrain-Referenced Navigation

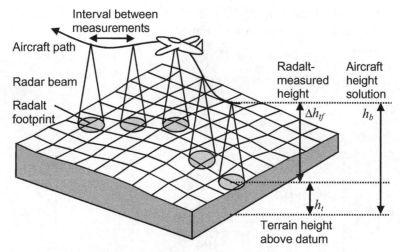

Figure 13.12 Terrain height measurement using a radar altimeter.

In more recent years, TRN techniques using sonar to determine the height above the sea bed have become established for marine applications, particularly underwater vehicles. TRN for land applications using a barometric altimeter (Section 6.2.1) has also been investigated. A DTM may also be used for height aiding. These topics conclude the section.

13.2.1 Sequential Processing

In sequentially processed TRN, used in TERPROM and Sandia Inertial TAN (SITAN) [25], an extended Kalman filter (Section 3.4.1) is used to estimate the position error of the host vehicle's navigation system. Each radar altimeter measurement, $\Delta \tilde{h}_{tr}$, is then processed separately to refine the position error estimate. The principal advantage of this method is a much lower processor load compared to batch-processed pattern-matching techniques.

The measurement innovation comprises the difference between the measured and database-indicated terrain height. Thus

$$\delta z_T = \hat{h}_b - \begin{pmatrix} 0 & 0 & 1 \end{pmatrix} \hat{C}_b^n \mathbf{l}_{br}^b - \Delta \tilde{h}_{tr} - h_{t,D}\left(\hat{L}_b, \hat{\lambda}_b\right), \tag{13.19}$$

where $\hat{L}_b$, $\hat{\lambda}_b$, and $\hat{h}_b$ comprise the host navigation system's corrected latitude, longitude, and height solution; $\mathbf{l}_{br}^b$ is the lever arm from the host-vehicle body frame to the radalt antenna; and $h_{t,D}$ is the geodetic height from the database.

Defining the state vector as

$$\mathbf{x}^n = \begin{pmatrix} \delta L_b \\ \delta \lambda_b \\ \delta h_b \\ \vdots \end{pmatrix}, \quad \mathbf{x}^\gamma = \begin{pmatrix} \delta \mathbf{r}_{\gamma b}^\gamma \\ \vdots \end{pmatrix} \quad \gamma \in i,e,l, \tag{13.20}$$

where the latitude, longitude, and height errors, δL_b, $\delta \lambda_b$, and δh_b are defined by (5.108) and the Cartesian position error is defined by (5.107), the measurement matrix is

$$\mathbf{H}_T^n = \begin{pmatrix} \dfrac{\partial h_{t,D}(\hat{L}_b, \hat{\lambda}_b)}{\partial L} & \dfrac{\partial h_{t,D}(\hat{L}_b, \hat{\lambda}_b)}{\partial \lambda} & 1 & 0 \end{pmatrix}$$

$$\mathbf{H}_T^\gamma = \begin{bmatrix} \begin{pmatrix} \dfrac{\partial h_{t,D}(\hat{L}_b, \hat{\lambda}_b)}{\partial L} & 0 & 0 \\ 0 & \dfrac{\partial h_{t,D}(\hat{L}_b, \hat{\lambda}_b)}{\partial \lambda} & 0 \\ 0 & 0 & 1 \end{pmatrix} \mathbf{T}_{r(n)}^p \mathbf{C}_\gamma^n & 0 \end{bmatrix}, \quad \gamma \in i, e, l \qquad (13.21)$$

where $\mathbf{T}_{r(n)}^p$ is given by (2.119).

The key to successful operation of this method is correct determination of the terrain gradients, $\partial h_{t,D}/\partial L$ and $\partial h_{t,D}/\partial \lambda$ [26]. This is constrained by the database accuracy. However, the main limitation is that the gradient is calculated below the estimated host vehicle location, not the actual location. If the direction of the slopes at these locations differs, the estimated position can diverge from the truth instead of converging with it. In practice, sequential-processing TRN using an EKF does not work well when the combined position and database errors exceed a few hundred meters.

Two solutions to this problem have been developed. One is to replace the EKF with a nonlinear non-Gaussian estimation algorithm, such as a Viterbi algorithm, [27], point-mass filter [28], or particle filter (Section 3.5) [29, 30]. This removes the dependency on the terrain gradient at one point, but substantially increases the processor load. The other approach is to operate separate tracking and acquisition modes as in typical GNSS user equipment [31]. The EKF is the tracking mode and operates when the position uncertainty is below its operational limit. The acquisition mode operates otherwise and may be a batch-processing algorithm (Section 13.2.2) or parallel filters, such as a MMAE filter bank (see Section 3.4.4) [32]. With a filter bank, each filter is offset in latitude and longitude and uses its own terrain gradient estimates. The filter which is closest to the truth exhibits the smallest measurement innovations over time.

13.2.2 Batch Processing

In batch-processed TRN, also known as template matching or terrain contour matching, a series of typically 5–16 terrain height measurements, forming a location signature, are processed together. Each radalt measurement is tagged with the host-vehicle navigation system's position estimate, $\hat{L}_b$, $\hat{\lambda}_b$, and $\hat{h}_b$. Subtracting the radalt measurement from the vehicle height at each point and then linking the points together produces a 3-D *transect* of positions. This is equivalent to linking together the radalt footprints shown in Figure 13.12. Positioning is then performed in two stages. First, a probability distribution as a function of a position error, common to all points of the transect, is obtained by fitting the transect to the terrain height database, essentially a correlation process. Second, a position correction is obtained from the distribution [33].

Obtaining a probability distribution as a continuous function of position is not practical, so a grid is used. The extent of the grid is typically matched to the 3σ error bounds of the host-vehicle navigation solution, while the spacing depends on the database resolution, terrain correlation length, and processing capacity. For an m-point transect, offset from the estimated host-vehicle position by ΔL, $\Delta\lambda$, and Δh, the measurement innovation vector is

$$\delta \mathbf{z}_T(\Delta L, \Delta\lambda, \Delta h) = \begin{pmatrix} \hat{h}_{b,1} + \Delta h - \begin{pmatrix} 0 & 0 & 1 \end{pmatrix} \hat{\mathbf{C}}_{b,1}^n \mathbf{l}_{br}^b - \Delta \tilde{h}_{tr,1} - h_{t,D}\left(\hat{L}_{b,1} + \Delta L, \hat{\lambda}_{b,1} + \Delta\lambda\right) \\ \hat{h}_{b,2} + \Delta h - \begin{pmatrix} 0 & 0 & 1 \end{pmatrix} \hat{\mathbf{C}}_{b,2}^n \mathbf{l}_{br}^b - \Delta \tilde{h}_{tr,2} - h_{t,D}\left(\hat{L}_{b,2} + \Delta L, \hat{\lambda}_{b,2} + \Delta\lambda\right) \\ \vdots \\ \hat{h}_{b,m} + \Delta h - \begin{pmatrix} 0 & 0 & 1 \end{pmatrix} \hat{\mathbf{C}}_{b,m}^n \mathbf{l}_{br}^b - \Delta \tilde{h}_{tr,m} - h_{t,D}\left(\hat{L}_{b,m} + \Delta L, \hat{\lambda}_{b,m} + \Delta\lambda\right) \end{pmatrix}, \quad (13.22)$$

where the notation is as defined in the previous section. Using a normalized mean square difference statistic, the likelihood of a particular offset, given the measurements, is [33]

$$\Lambda(\Delta L, \Delta\lambda, \Delta h) = \exp\left(-\tfrac{1}{2} \delta \mathbf{z}_T^\mathrm{T} \mathbf{R}_T^{-1} \delta \mathbf{z}_T\right), \quad (13.23)$$

where the measurement noise covariance, $\mathbf{R}_T$, accounts for radalt measurement noise and database errors. Correlated measurement noise between different elements of the transect can occur as a result of radalt footprint errors (see Section 6.2.3) and host-vehicle velocity errors, so the matrix is not necessarily diagonal.

A 3-D likelihood grid can be impractical due to the processor load it imposes. The vertical component may be eliminated by differencing successive points in the transect [33]. Alternatively, the height offset for each latitude and longitude offset may be optimized by performing a least-squares fit, giving [34]

$$\Delta h(\Delta L, \Delta\lambda,) = \frac{1}{m} \sum_{i=1}^m \left[\Delta \tilde{h}_{tr,i} + h_{t,D}\left(\hat{L}_{b,i} + \Delta L, \hat{\lambda}_{b,i} + \Delta\lambda\right) - \hat{h}_{b,i} + \begin{pmatrix} 0 & 0 & 1 \end{pmatrix} \hat{\mathbf{C}}_{b,i}^n \mathbf{l}_{br}^b \right]. \quad (13.24)$$

Once a likelihood grid has been obtained, it is used to determine a position fix. Sometimes, the likelihood function has a single dominant peak, clearly denoting the host vehicle position. However, this is not always the case. Figure 13.13 shows a typical likelihood surface over a 2-D grid, where the measurement–database fit is ambiguous. There is more than one peak and significant levels of noise, making it difficult to determine the correct position offset from which to generate a fix. The maximum likelihood point is not necessarily the correct fix due to noise. Alternatively,

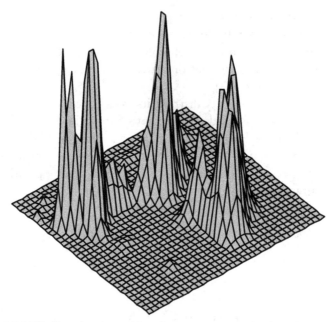

Figure 13.13 A TRN likelihood surface where the measurement–database fit is ambiguous. (*After:* [35].)

fitting a bivariate Gaussian distribution to the surface will overweight some parts of the solution space and downweight others. For example, the means of the distribution can fall between the peaks where the likelihood is actually very low. A number of solutions to this problem of determining fixes from irregular likelihood surfaces have been developed; they are discussed in order of increasing processor load.

The simplest solution is to use transects sufficiently long that ambiguous fits rarely occur; at least 3 km is typically required [33]. However, long transects reduce the update rate, increasing the effect of host navigation system velocity error on the integrated position solution. Velocity errors can also smear the likelihood surface as they cause the best position correction for the end of the transect to diverge from that for the beginning.

The next simplest approach, known as the stockpot algorithm for robust TAN (SPARTAN), fits a Gaussian distribution to the likelihood surface, but then adds the residuals of that fit to the likelihood surface from the next transect [33].

Another method is to fit multiple Gaussian distributions to the likelihood function. A simple way of doing this is to divide the likelihood grid into signal and noise cells and then fit a Gaussian function to each contiguous clump of signal cells [34]. A better performing, but processor-intensive, approach is to perform an iterative fit of a weighted sum of Gaussian functions to the likelihood surface. In either case, multiple position-fix hypotheses, each with an associated covariance and probability score are passed to the integration algorithm (see Sections 3.4.5 and 16.3.5).

The most sophisticated batch-processing TRN algorithms modify the integration filter to accept the whole likelihood surface as a measurement. IGMAP [35] multiplies the integration algorithm's prior position error distribution by the TRN likelihood surface and then fits one or more Gaussian functions to the product. The particle filter (Section 3.5) [36] and other Monte Carlo estimation algorithms, such as the Metropolis-coupled Monte Carlo Markov chain (MCMCMC) method [37], do away with Gaussian approximations altogether.

13.2.3 Performance

Under optimum conditions, conventional TRN operates with horizontal position errors of around 50m (1σ). Performance depends on the roughness of the terrain below the host vehicle. Established TRN systems typically require terrain with an RMS gradient of at least 3% and height varying in all directions to operate. Greater roughness is needed to obtain the best performance. More sophisticated TRN algorithms offer better performance over low roughness terrain [30, 34] and will operate with a lower minimum roughness. Performance can also be limited where the terrain is too rough due to variation in terrain height within the radalt footprint causing large errors in measuring the distance to the ground directly below the aircraft (see Section 6.2.3). Thus, to ensure continuous availability of TRN throughout a flight, the trajectory must be carefully selected [38].

A number of other factors affect TRN performance. The velocity accuracy of the inertial or dead-reckoning solution determines the accuracy of the assumed distance between successive radalt measurements and the degree to which TRN noise may be smoothed in the integrated navigation solution. The host-vehicle speed is important as the information available from TRN depends on the terrain correlation length; with a faster host vehicle, information is obtained at a faster rate. Performance degrades with increasing height above terrain, as a larger radalt footprint leads to noisier measurements. However, the fundamental limits to TRN performance are the database accuracy and the radalt beamwidth. Newer terrain height databases provide greater accuracy and resolution than DTED Level 1 [23], while interferometric radalts provide narrower beams than conventional units [24]. Together, these enable an accuracy of around 3m to be achieved [18].

13.2.4 Laser TRN

The problem of a large sensor footprint can be eliminated by replacing the radalt with a laser rangefinder. With a conventional TRN algorithm, this reduces average position errors by up to 50% [39]. However, by scanning the laser from side to side as shown in Figure 13.14, many more data points can be obtained for a given distance traveled [18, 40, 41]. This is known as light detection and ranging (LIDAR) or laser detection and ranging (LADAR). Using the host vehicle's integrated velocity and attitude solution, each data point is transformed from a time, range, and scanning angle to a relative position on the terrain from which the laser was reflected. The measurements can then be compared with the database using a batch-processing algorithm (Section 13.2.2). With a DTED level 1 database, the ambiguity in the matches is largely removed, but the accuracy, integrated with an aircraft-grade INS, is only about 30m (1σ) [40].

To get the full benefit from using a laser scanner, a much higher database resolution is needed, with a 5-m grid spacing and submeter accuracy. This enables a

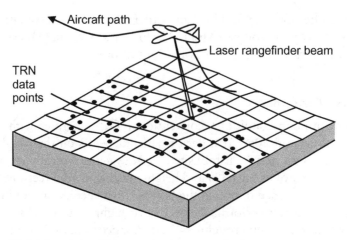

Figure 13.14 TRN data points with a scanning laser rangefinder.

horizontal position accuracy within 10m (1σ) to be obtained [18, 41]. With a 1-m database grid spacing and integration with an aviation-grade INS, an accuracy of about 2m (1σ) per axis is obtainable [18, 42]. With a large number of data points, the processor load for conventional batch processing is very high. Solutions include subsampling the laser data to match the terrain correlation length and gradient-based matching algorithms.

Correlating the profiles measured using two laser scanners pointing in fore and aft directions enables the aircraft velocity to be determined by dead reckoning over terrain where a database is not available [43]. The greater the separation of the two sensor footprints, the greater the time interval between successive profiles of the same terrain and the higher the velocity resolution for a given sensor resolution.

A limitation of laser rangefinders is that they have shorter ranges than radalts and are sensitive to the weather. However, their higher precision is most useful for low-altitude operations, such as landing.

13.2.5 Sonar TRN

Ships, submarines, AUVs, and ROVs can use a multibeam sonar echo-sounder to measure the range from the vehicle to a number of points on the sea bed or river bed. The relative positions of these points, obtained from the sonar measurements, is known as a *bathymetric profile*. TRN may then be performed by matching this to a suitable database using batch-processing methods. This is sometimes known as bottom-contour navigation (BCN). As with other forms of TRN, performance depends on the terrain height variation and sensor resolution. The highest resolution sensors can measure more than 10,000 points simultaneously. The positioning accuracy can be as good as 1m but is typically around 10m [44–46].

If no database is available, dead reckoning may be used by comparing successive overlapping bathymetric profiles to obtain the velocity, $\mathbf{v}_{eb}^b$, resolved along body-frame axes.

13.2.6 Barometric TRN

For land vehicles and pedestrians outdoors, the navigation system generally maintains a constant height above terrain, so a barometric altimeter (Section 6.2.1), with the bias calibrated (Section 16.2.2), may be used to measure terrain height. Thus, in principle, terrain-referenced navigation may be performed [47, 48]. However, as the host vehicle or user speed is much lower, a high-resolution database is needed to capture sufficient terrain height variation. Further research is needed to determine the range of terrain over which such a system would operate. Note also that this technique is context dependent and is not suited to use indoors.

13.2.7 Terrain Database Height Aiding

Terrain database height aiding uses a DTM to determine the height of a land vehicle or outdoor pedestrian from latitude and longitude or projected horizontal coordinates, interpolating as necessary. For marine applications, a tidal model may be used to estimate the height of sea level above or below the geoid. A geoid model may be required to convert orthometric height to geodetic height. Height aiding may simply be used to determine the height from a well-known horizontal position solution. It can also be used to test the validity of a 3-D position solution by determining whether the horizontal and vertical components match.

Another way of using height aiding is to enable 4-D positioning with GNSS in cases where only three pseudo-range measurements are available or the signal geometry is poor. Height aiding is in the form of a virtual range measurement that defines a spherical surface of position. When on the surface of water or the terrain is flat, the center of the Earth may be taken as the origin of the virtual measurement, while the range is simply the magnitude of the ECEF-frame Cartesian position, determined from an approximate latitude and longitude and the database-derived height using (2.112). For sloping terrain, better performance is obtained by fitting the SOP from the virtual range measurement to the DTM at the approximate latitude and longitude. Section 16.3.2 shows how to incorporate a virtual range measurement into the integrated navigation solution.

When the uncertainty of the input latitude and longitude is significant, the height aiding should be iterated. Thus, the latitude and longitude obtained from the GNSS and height aiding measurements should be used to determine a more accurate height aiding measurement which is then used to calculate a more accurate latitude and longitude. Several iterations may be required. As with any pattern-matching technique, there is scope for false and ambiguous matches as there may be more than one point on the DTM's terrain surface that is consistent with the GNSS measurements.

When the latitude and longitude are initially unknown, the height aiding measurement should be set to the equatorial radius with the origin at the center of the Earth. The ellipsoid radius, given by (2.137), should be used for the second iteration with the terrain height database used on subsequent iterations. Figure 13.15 illustrates the process. This terrain height aiding technique is also used with RDSS (Section F.1 of Appendix F on the CD), Doppler positioning, and short-range UWB and pseudolite positioning. A comparison of different height-aiding methods may be found in [49].

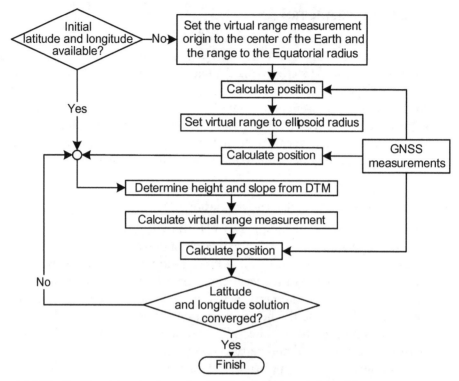

Figure 13.15 Position determination using GNSS and database height aiding.

13.3 Image-Based Navigation

Image-based navigation systems use a sensor to obtain a 2-D or 3-D image of the surrounding environment. Sensors for imaging terrestrial features are typically downward looking for air applications and forward looking for land and sea applications. By comparing the images with a database of known environmental features, information on the position and orientation of the sensor, sometimes known as its *pose*, may be obtained. By comparing successive images, information about the translational and rotational motion of the sensor may be obtained, a form of dead reckoning. The same sensor may be used for both position fixing and dead reckoning.

Like TRN, image-based navigation does not work everywhere as suitable distinct features, generally man-made, are needed for matching. However, more features tend to be found on flat terrain, over which TRN performance is poor, so the two techniques are complementary. Image matching does not work over water.

In this section, imaging sensors are reviewed first, followed by a discussion of feature comparison methods. Position fixing using individual image features and by matching the whole-image is described next. The two main dead reckoning approaches are then discussed. These are visual odometry, which compares successive images, and tracking of individual features over several images. Finally, stellar navigation is described.

Only *inside-out* positioning, whereby the mobile user observes the environment, is considered here. However, for tracking applications, imaging sensors are often used in an *outside-in* configuration, whereby the environment observes mobile targets.

13.3.1 Imaging Sensors

Imaging sensors may be classed as bearing-elevation, range-bearing, or 3-D, depending on the type of image they produce. In all cases, the alignment of the sensor with respect to the host vehicle and the lever arm between the sensor and host-vehicle body frames must be carefully calibrated to enable the vehicle navigation solution to be derived from the sensor pose.

The oldest imaging sensor is the human eye. Combined with a gimbaled telescope, this can be used to obtain precise bearings and elevations of environmental features, a technique still used on ships.

A camera also provides bearing-elevation information. It is a passive sensor as it only receives information. A standard video camera, offering a resolution between 640×480 and 1,920×1,080 pixels, is suitable for navigation. Alternatively, an infrared camera can work well at night or under poor illumination and is less affected by rain and clouds.

The camera must be calibrated to match each pixel to a bearing and elevation from the camera boresight. Assuming an ideal camera, the line-of-sight vector of a feature, f, with respect to the navigation system body frame, b, is given by [50]

$$\mathbf{u}^b_{bf} = \frac{1}{\sqrt{x^c_{cf}{}^2 + y^c_{cf}{}^2 + F^2}} \mathbf{C}^b_c \begin{pmatrix} -x^c_{cf} \\ -y^c_{cf} \\ F \end{pmatrix}, \qquad (13.25)$$

where (x^c_{cf}, y^c_{cf}) is the position of the feature within the camera's detector, or image plane, as shown in Figure 13.16, F is the focal length of the camera lens and $\mathbf{C}^b_c$ is the camera-to-body-frame coordinate transformation matrix, assumed to be known. In practice, a more sophisticated model will be required to account for lens distortion

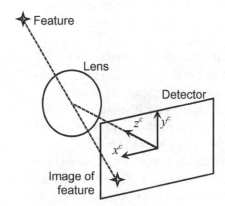

Figure 13.16 Image of a feature within the camera frame.

and other systematic errors. The line-of-sight vector is related to the bearing, ψ_{bu}^{bf}, and elevation, θ_{bu}^{bf}, by

$$\mathbf{u}_{bf}^{b} = \begin{pmatrix} \cos\theta_{bu}^{bf}\cos\psi_{bu}^{bf} \\ \cos\theta_{bu}^{bf}\sin\psi_{bu}^{bf} \\ -\sin\theta_{bu}^{bf} \end{pmatrix}, \quad \begin{aligned} \theta_{bu}^{bf} &= -\arcsin\left(u_{bf,z}^{b}\right) \\ \psi_{bu}^{bf} &= \arctan_{2}\left(u_{bf,y}^{b}, u_{bf,x}^{b}\right) \end{aligned}. \quad (13.26)$$

For navigation, a camera with a wide field of view (FOV) is typically fixed to the host vehicle body, a strapdown configuration. This enables multiple features to be observed simultaneously. For air applications, a zoom lens may be used to maintain a constant area of terrain within view as the camera height varies. Alternatively, a narrow FOV camera may be mounted on gimbals. This is known as an electro-optic (EO) sensor and provides precise observations of a single feature.

A number of methods may be used to add range information to a camera image. For land vehicles, the camera may be assumed to be a fixed height above the ground, while the height of an aircraft above ground maybe determined using a radalt (Section 6.2.3) or by differencing the host-vehicle height with a terrain height database. More generally, the size of a known feature may be used to scale an image or the image velocity may be compared with an independent measure of the host vehicle velocity, where available, which is equivalent to observing features from multiple positions. Range may be measured directly using a stereo camera [51] or by defocusing the image using a coded aperture [52].

For accurate range determination, an active sensor must be used, based on radar, sonar, or LADAR/LIDAR. Active sensors transmit signals and measure the time taken for the reflections of those signals to return to the sensor. They are more expensive and require more power than passive sensors, while for military applications, they are nonstealthy. Radar and LIDAR are both scanning sensors so the motion of the host vehicle over the scanning period may require compensation.

Imaging radar is a range-bearing sensor and is standard equipment on ships and many aircraft. Shipboard radar uses a large rotating antenna and performs a scan over 360° in azimuth every 1.5–3 seconds. The beamwidth is 1°–2° horizontally and 20° vertically, while a typical range resolution is 2–16m. The range is between 2 and 40 km, depending on the size of the target [53]. Many buoys and shoreline features are equipped with racons; these transmit identification codes whenever they receive a radar signal.

Airborne radar can detect objects up to 300 km away. The raw azimuth resolution is lower than for ships because there is insufficient space for a large antenna. However, processing techniques such as Doppler beam sharpening and synthetic aperture radar (SAR), take advantage of the rapid motion of the aircraft to greatly improve the resolution of the images. Using SAR, a ground resolution as good as 1.5×1.5m may be obtained. However, the size of a high-resolution radar image is limited by the available processing power [54, 55].

Laser scanners provide a much higher resolution than radar with a subcentimeter range resolution. However, the range is much shorter. Single-axis scanners provide range-bearing images when mounted horizontally and can scan over 180°

in a few tens of milliseconds. Three-dimensional images may be obtained from a moving vehicle by scanning perpendicularly to the direction of travel and combining the measurements from successive scans. However, as no point is scanned more than once, this cannot be used for feature tracking. Dual-axis laser scanners are also available. They can scan over a complete hemisphere in a few seconds, but have yet to be used for navigation.

Flash LADAR/LIDAR is a 3-D imaging system that emits wide-angle laser pulses and detects the returns using a sensor array, measuring range to within a centimeter. This eliminates the motion blur associated with scanning systems. At the time of writing, the angular resolution is 128×128 over a FOV of around 40×40°, so each image comprises 16,384 data points in 3-D space, known as a point cloud. The update rate is typically 30 Hz or 60 Hz [56].

With an active sensor, TRN (Section 13.2) and image-based navigation may be performed using the same sensor. This reduces hardware costs, removes relative alignment errors, and enables range data to be used to aid detection of image boundaries [40].

Image-based navigation is of particular interest for autonomous vehicles and robots as these already possess suitable sensors for route guidance and obstacle avoidance.

13.3.2 Image Feature Comparison

There are a number of methods for comparing features between camera images. The simplest is area correlation. This compares each pixel of either a pair of whole images or a selected region of the images to determine the degree of correspondence. This is repeated at different offsets between images in both directions to find the alignment with the most matching pixels.

Area correlation also requires matching of the scaling and rotation of the images. If this cannot be done using external information, scaling and rotation dimensions must be added to the search process. This substantially increases both the processor load and the probability of false matches due to noise and ambiguity. If the camera is not looking directly downwards, a rotation search about three axes may be required.

For image-database matching, comparing full color or even grayscale images is not effective due to seasonal variations, changes in lighting conditions, and day/night contrast inversion of infrared images. Furthermore, the captured and stored images may have been obtained using different types of sensor. Instead, images are processed to extract boundaries as illustrated by Figure 13.17 [57, 58]. This also has the advantage of reducing the correlation processing load and database storage requirements.

Straight-line feature matching imposes a lower processing than area correlation. Data storage is also efficient as only the start and end points are needed. Suitable line features in aerial images include roads, rail tracks, buildings, and field boundaries [58, 59]. Figure 13.17 shows an example of line features extracted from an image. Absolute or relative position information is obtained by determining the offsets between lines of similar orientation that appear in an image and a feature database or in two images. Two-dimensional positioning information may be obtained from two nonparallel lines or from a single line fully contained with an image. However,

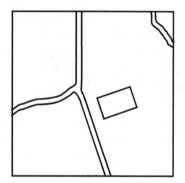 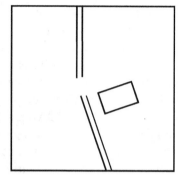

Figure 13.17 Example of boundary features (left) and line features (right) extracted from an aerial image.

for a line that extends beyond the image boundary, positioning information may only be obtained along the image direction perpendicular to the line. Thus, an image-matching system might sometimes produce a line fix (see Section 16.3.4) instead of a position fix.

As line features are very simple and often occur in parallel, ambiguity is a problem; an image line may match several possible lines in a feature database or another image. Furthermore, a single line in a database may correspond to multiple lines in an image. For example, a rail track could be represented as two lines or one. Therefore, all possible matches must be considered and consistency checks (Section 17.4) used to determine the combination of features that gives the most consistent image-database or image-image offset. When there is insufficient information to do this, multiple fix hypotheses may be output. Section 16.3.5 discusses the handling of ambiguous measurements in multisensor integrated navigation. When the orientation and/or scaling of an image are unknown, the ambiguities are much more difficult to resolve and many more line features are required. In practice, the problem can become intractable.

Another way of comparing images is to use small-feature descriptors. First, several tens of small features within an image, comprising blocks of pixels that contain a lot of variation, are selected. These are often based around corners. For each feature, a descriptor is then constructed that is insensitive to changes in camera viewpoint, including scale, rotation, and illumination. Commonly used descriptors include the scale-invariant feature transform (SIFT) [60], speeded-up robust feature (SURF) [61], and binary robust independent elementary features (BRIEF) [62] methods. These descriptors can then be used to match objects between images taken from different viewpoints. Ambiguity is resolved by requiring the descriptors of several small features of a given object to match.

Repetition of similar features at different locations, causing ambiguous feature matches, can be a problem, particularly in indoor and urban environments. One solution is to introduce unique features by placing optical tags, such as QR codes or other matrix barcodes, into the environment [63]. Figure 13.18 shows some examples.

Finally, text from signs can be extracted from camera images using standard optical character recognition (OCR) software. This has the advantage of being easy to store and process. Standard pictographic signs, such as road signs, could also be

Figure 13.18 Some examples of optical tags.

used [64]. Some signs are highly distinctive, whereas others are found in many different places.

Airborne radar can produce an image similar to that from a camera so the same feature comparison techniques apply. However, for radar and single-axis laser scanner measurements in the horizontal plane, only the nearest object at each bearing is measured, so less information is available. Distant objects are typically compared as point targets. A nearby feature, such as a wall may be compared as a line feature and its corner as a point feature [65].

Point clouds produced by 3-D laser-based sensors are also limited to the nearest object in each direction. However, there is much more information than provided by a 2-D scan. Individual points in different scans will fall on different parts of the surrounding objects so cannot be compared directly. Instead, planes, lines, and corner features are extracted. For example, the translation and rotation between two point clouds can be uniquely-determined by matching three nonparallel planes [56]. Again, where there is an ambiguous correspondence between planes in different point clouds, additional planes must be used and consistency checks applied to determine the best match.

Finally, feature identification can also be performed by a human operator, where available, noting that human vision is still more reliable than machine vision.

13.3.3 Position Fixing Using Individual Features

As discussed in the chapter introduction, practical position fixing using environmental features requires an approximate position solution and associated uncertainty to be input in order to constrain the region of the feature database to be searched. A suitable search area is defined by the 3σ uncertainty bounds of this position solution, which may be input as aiding information (see Section 16.3.7) or predicted from previous fixes.

Versions of the proximity positioning method described in Section 1.3.1 may be used. An approximate position may be obtained from a single environmental feature matched with the database. More precise positioning requires multiple features, averaging their positions or using containment intersection. However, ranging and angular positioning generally provide greater accuracy.

Obtaining an accurate and complete position fix from a single matched environmental feature requires the range, bearing, and elevation of the feature from the sensor as shown in Figure 13.19. The sensor measures the bearing and elevation of the feature with respect to its boresight. However, the azimuth, elevation, and roll

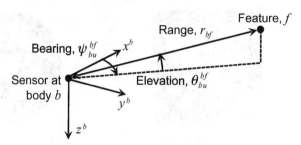

Figure 13.19 Position determination from the range, bearing, and elevation of a single environmental feature.

angle of the boresight with respect to the coordinate frame of the feature database are also required. The orientation of the sensor boresight with respect to the navigation system body frame is determined by calibrating the sensor at installation, while the orientation of the body frame with respect to the database frame is determined from the host vehicle attitude solution.

The user position with respect to a local tangent-plane frame, l, is

$$\mathbf{r}_{lb}^{l} = \mathbf{r}_{lf}^{l} - \mathbf{C}_{b}^{l}\left[\mathbf{u}_{bf}^{b} r_{cf} + \mathbf{l}_{bc}^{b}\right], \tag{13.27}$$

where $\mathbf{r}_{lf}^{l}$ is the position of the feature, r_{cf} is the range, $\mathbf{C}_{b}^{l}$ is the body-to-local-tangent-plane coordinate transformation matrix, and $\mathbf{l}_{bc}^{b}$ is the body-to-sensor lever arm.

When range measurements are not available, the bearing and elevation measurements can still contribute to an integrated navigation solution (see Section 16.3.3). A unique position fix may be obtained using two environmental features as shown in Figure 13.20, provided that the feature and sensor positions are not collinear. Position determination is the same as for angular positioning using radio signals (Section 7.1.5). Best precision is obtained when the lines of sight are perpendicular. However, this is not always possible to achieve with a single imaging sensor.

When the sensor orientation is not accurately known, it may be determined using a third feature as shown in Figure 13.21 for two-dimensional positioning. Alternatively, the relative bearing and relative elevation of two matched features can still contribute to an integrated navigation solution. Conversely, where the sensor

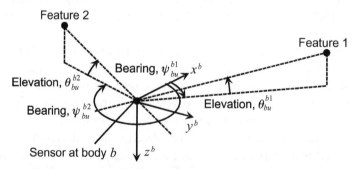

Figure 13.20 Position determination from the bearings and elevations of two environmental features.

13.3 Image-Based Navigation

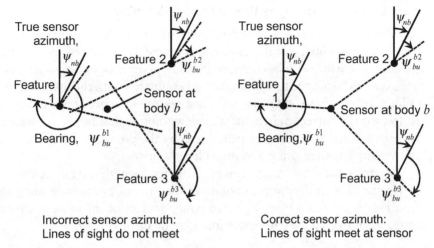

Figure 13.21 Determination of sensor boresight azimuth from bearing measurements of three environmental features.

position is known, bearing and elevation measurements to two environmental features may be used to determine the sensor and host vehicle attitude.

When an image feature is incompletely matched with the database (e.g., a line with unknown endpoints), only a linear combination of the range, bearing, and elevation (or just the bearing and elevation) may be determined. Therefore, more features are required to obtain a full position fix. Additional features may also be used to identify erroneous matches using consistency checks (Section 17.4).

With shipboard radar or a single-axis laser scanner operating within the horizontal plane, a horizontal position fix may be obtained from the range and absolute bearing to a single identified environmental feature. Alternatively, range and uncalibrated bearing measurements to two identified features enable determination of the horizontal position and sensor boresight azimuth. A different approach is to measure the range to the nearest object in every direction and then use a map comprising a Boolean grid of points that are either occupied or unoccupied to determine position by pattern matching; no identification of individual features is required [66].

Systematic error sources in image feature matching include database errors, sensor alignment and calibration errors, and uncompensated scaling variations. Noise-like errors arise from focusing errors and the resolution limits of the sensor, database, and feature-comparison algorithms. The statistical parameters assumed in the image-matching algorithms must be tuned to account for these errors and appropriate uncertainties associated with the output position information. Also, where a position solution is computed from features matched at different times, any motion of the vehicle between image captures must be accounted for.

Section H.1 of Appendix H on the CD describes an example image feature-matching system, Continuous Visual Navigation (CVN), an airborne system accurate to 10–20m when integrated with inertial navigation.

A special case of image feature matching is road marking detection [67]. This improves the cross-track positioning accuracy obtainable with road map matching (Section 13.1), enabling traffic lane identification. It can also be used as a lane departure warning system.

13.3.4 Position Fixing by Whole-Image Matching

An alternative to matching individual features within an image is whole-image matching. Each image in the database is stored as a series of small-feature descriptors, some of which will be common to overlapping images. Small-feature descriptors are extracted from the sensed image and compared with the images in the database. Whichever database image has the most feature matches above a certain threshold is deemed the correct match and the viewpoint of that image is assumed to be the sensor position [68]. Thus, positioning is by the proximity method. This technique can be implemented using a mobile phone camera.

Separate database images are required for each viewing direction from a given viewpoint. Therefore the technique can also provide sensor heading information. Accuracy is limited by the degree to which images of the same environmental features taken from different viewpoints may be distinguished.

13.3.5 Visual Odometry

Dead reckoning using successive 2-D camera images is sometimes known as visual odometry. The first step is the matching of environmental features between images. A common approach is optical flow [69], whereby one image is divided into 100–1,000 equal-sized tiles, each of which is matched with the other image using area correlation to determine the motion of that tile's centroid. Alternatives include area correlation of smaller tiles centered on corners [70], matching small-feature descriptors [71, 72], and matching irregular boundary features [73]. Figure 13.22 illustrates the resulting motion vectors.

Matching image features that lie within a common plane in the environment may be related using the homography matrix, **H**:

$$\begin{pmatrix} x_{cf}^c(+) \\ y_{cf}^c(+) \\ 1 \end{pmatrix} = \mathbf{H} \begin{pmatrix} x_{cf}^c(-) \\ y_{cf}^c(-) \\ 1 \end{pmatrix}, \qquad (13.28)$$

where the subscripts (–) and (+) denote the old and new images, respectively. **H** may be obtained from four or more matched features using general least-squares

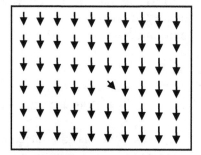 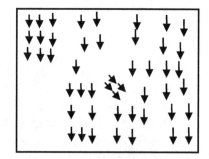

Figure 13.22 Vectors showing motion between successive downward-looking images using optical flow (left) and small-feature matching (right). The host aircraft is turning left and the anomaly in the center is due to another aircraft flying below.

estimation. When the orientation of the plane, denoted by the unit normal, $\mathbf{n}_p^l$, and its normal distance from the camera, d_{pc}, are both known, the motion of the camera, sometimes known as egomotion, may be deduced using [70]

$$\mathbf{H} = \mathbf{C}_{c(-)}^{c(+)} + \mathbf{C}_l^c \frac{\Delta \mathbf{r}_{lc}^l \mathbf{n}_p^{l\,\mathrm{T}}}{d_{pc}} \mathbf{C}_c^l, \qquad (13.29)$$

where $\Delta \mathbf{r}_{lc}^l$ is the change in camera position, $\mathbf{C}_{c(-)}^{c(+)}$ is the coordinate transformation matrix from the old to new the camera frames and $\mathbf{C}_l^c$ is the coordinate transformation matrix from the local-tangent-plane frame to the camera frame. Separation of the rotation and translation of the camera can be difficult and is impossible where the image and object planes are parallel. One solution is to use inertial sensors to measure the rotation. For a land vehicle, motion constraints may be used (see Section 15.4.1) [69]. A further option is to use multiple cameras pointing in perpendicular directions [74].

In most cases, the ground is used as the reference plane for dead reckoning. With a downward-looking camera, most of the features will be on the ground and outlier detection should be used to remove mismatched and moving features, including clouds. With a forward-looking camera, the ground will typically occupy the central part of the bottom of the image and its separation from other objects in view is more challenging [69]. Visual odometry with a camera is accurate to within 2% of distance traveled [70].

With a forward-looking camera, the vanishing point of line features in the boresight direction may be used to determine sensor rotation. This point is given by the intersection of lines that are neither vertical, nor perpendicular to the vertical as shown in Figure 13.23. The translation of the vanishing point within the image plane determines the pitch and yaw motion while the rotation of features about the vanishing point determines the roll [75].

Using a 2-D laser scanner in the horizontal plane, the horizontal velocity, resolved in body-frame axes, and the yaw rotation rate may be determined by matching two

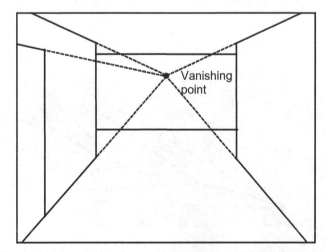

Figure 13.23 The vanishing point in a forward-looking image with line features extracted.

or more line features between successive scans [76, 77]. Line features typically correspond to walls and an accuracy within 1% of distance traveled may be achieved. Using Flash LADAR or a 3-D scanner, full 3-D velocity and angular rate, resolved in body-frame axes, are obtained by matching three planes between successive scans [56].

13.3.6 Feature Tracking

To track individual environmental features across several successive images or scans, as shown in Figure 13.24, their positions must be estimated as extra states in a filtered navigation solution, typically alongside user position, velocity, and attitude [65, 78, 79]. The first few measurements are used to determine the feature position with subsequent measurements maintaining the user position solution. With a bearing-elevation imaging sensor, the feature must be observed from different directions to fully determine its position.

This technique is often used with gimbaled EO sensors that only observe one feature at a time. However, multiple features must be tracked to separately determine the velocity and angular rate of the host vehicle. The number of simultaneously tracked features is limited by the processing capacity. In basic feature tracking, feature position states are deleted once the relevant feature is out of view. However, in SLAM, they are retained in case the feature is revisited in the future.

Environmental feature tracking is usually integrated with other navigation technologies, such as inertial navigation, and can be used to calibrate their errors. Section 16.3.6 describes the incorporation of feature-tracking states and measurements into a multisensor integration algorithm.

13.3.7 Stellar Navigation

Stellar, or celestial, navigation has been used for centuries. Today, it is used as a backup to GNSS for ships in mid-ocean and in remote areas where other signals are unavailable. It is also a cost-effective positioning technique for the exploration of other planets. Here, only modern methodology, based on stellar imagery, is described.

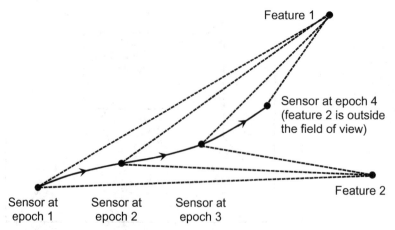

Figure 13.24 Successive observations of tracked features.

13.3 Image-Based Navigation

Historical celestial navigation techniques are described in Sections K.7.1 to K.7.3 of Appendix K on the CD.

Stellar imagery measures the line-of-sight vectors to a number of stars and then compares them with a star catalog to determine the attitude of the sensor body with respect to the CIRS inertial frame, a form of angular positioning. This is commonly used for attitude determination, particularly on satellites and other spacecraft. Further information may then be used to determine the position.

A star imager, or strapdown star tracker, comprises a charge-coupled-device (CCD) camera fixed to the host vehicle body. The resolution is about 1,000×1,000 pixels, while the field of view varies from about 8° to 40° [80]. It will work in daylight as well as at night and a baffle may be used to reduce ambient light from outside the field of view. Infrared sensors are less sensitive to cloud cover than those operating at visible wavelengths. The camera is deliberately defocused to spread the image of each star over several pixels. A *centroiding* algorithm can then determine the center of the star's image to subpixel precision [50, 81]. Accuracy depends on the camera's sensitivity and the level of noise. A high specification sensor can measure the attitude of a star to within about 5 μrad (3×10^{-4}°) in space. However, accuracy is limited to about 50 μrad (3×10^{-3}°) within the atmosphere.

A typical star catalog will store line-of-sight data for at least 6,000 stars. Given the correct time, it can apply corrections for precession and nutation of the Earth's spin axis, and for stellar parallax and aberration [82]. Provided at least four stars have been measured, they may be identified without any prior position or orientation information. The pyramid method compares the angles between the lines of sight of the measured stars with those in the catalog to find the best match [83].

The coordinates of a star feature, f, are normally expressed as a right ascension, α_{if}, and a declination, δ_{if}. The line-of-sight vector, resolved about the CIRS inertial frame, is then

$$\mathbf{u}_{if}^{i} = \begin{pmatrix} \cos\delta_{if}\cos\alpha_{if} \\ \cos\delta_{if}\sin\alpha_{if} \\ \sin\delta_{if} \end{pmatrix}. \tag{13.30}$$

This is then related to the line-of-sight vector of the star image centroid observed by the camera (see Section 13.3.1) by

$$\mathbf{u}_{if}^{i} = \mathbf{C}_{b}^{i}\mathbf{u}_{bf}^{b} \qquad f \in 1,2,\ldots, \tag{13.31}$$

where $\mathbf{C}_{b}^{i}$ is the body-to-inertial-frame coordinate transformation matrix, which is to be determined. As attitude has three independent components, while line of sight has two components, measurements from at least two stars are required. In practice, around 10 stars are typically used with the optimal attitude solution determined using least-squares estimation [50, 83].

Applying (2.144), the attitude with respect to an ECEF-frame, $\mathbf{C}_{b}^{e}$, may be obtained where time is known, and the attitude with respect to a local navigation frame, $\mathbf{C}_{b}^{n}$, obtained where latitude and longitude are also known. Stellar imagery can therefore be used to calibrate out gyro drift in inertial navigation. With an aviation-grade INS, horizontal position errors may essentially be constrained within about 500m indefinitely (see Section 5.7.2) [82, 84]. Such systems have been deployed in high-level reconnaissance aircraft and long-range bombers.

Stellar navigation determines latitude and longitude using inertially referenced attitude from stellar imagery; a time source; and roll and pitch obtained from inertial navigation, an AHRS, accelerometer leveling, tilt sensors, or horizon sensing [85]. The ECEF-to-wander-azimuth-frame coordinate transformation matrix is given by

$$\mathbf{C}_e^w = \mathbf{C}_b^w \mathbf{C}_i^b \mathbf{C}_e^i, \tag{13.32}$$

where

$$\mathbf{C}_b^w = \begin{bmatrix} \cos\theta_{nb} & \sin\phi_{nb}\sin\theta_{nb} & \cos\phi_{nb}\sin\theta_{nb} \\ 0 & \cos\phi_{nb} & -\sin\phi_{nb} \\ -\sin\theta_{nb} & \sin\phi_{nb}\cos\theta_{nb} & \cos\phi_{nb}\cos\theta_{nb} \end{bmatrix} \tag{13.33}$$

and the latitude and longitude are obtained using (5.61). A heading solution is not required. With accelerometers accurate to 100 μg, the latitude and longitude may be determined to within about 600m. To obtain higher accuracy, it is necessary to use a precision gravity model as well as higher-quality leveling sensors [85].

13.4 Other Feature-Matching Techniques

This section discusses navigation using other types of environmental features, measured using different types of sensor. In principle, any measurable feature of the environment may be used for navigation. There are three criteria that determine the suitability of a given class of feature. First, there should be spatial variation, so the feature must either be a property of the environment that varies spatially or an identifiable object that is only found in certain places. The distance over which observations of that feature are spatially correlated determines the precision with which position may be determined, while the degree of similarity between observations made at separate locations determines the likelihood of ambiguous position fixes.

The second criterion is temporal variation. A feature that varies unpredictably with time is much less useful than one that is constant or varying in a predictable way. If the temporal variation of a feature is much greater than the spatial variation, it is of little use for navigation. Slow temporal variation may be mitigated by regularly updating the database.

The final criterion is practicality. The size, weight, cost, and power consumption of the sensor required to measure the feature must be appropriate for the application concerned, as must the data processing and storage requirements for feature identification and position determination.

In this section, gravity gradiometry, magnetic field variations, and celestial X-ray sources are discussed in turn. In addition, road texture, smells and particulates, ambient sound, microclimate, background radiation, and sferics are discussed in Section H.2 of Appendix H on the CD.

13.4.1 Gravity Gradiometry

The acceleration due to gravity varies slightly over a few kilometers due to variations in the terrain height because higher terrain contains more matter, from which gravitational attraction arises. It also varies over tens of kilometers as a result of geological formations as different materials have different densities.

In principle, position may be determined using pattern matching by measuring the acceleration due to gravity and matching a succession of measurements with a high-resolution gravity database. However, in practice, better pattern-matching performance is obtained using the spatial gravity gradient, $\partial \mathbf{g}/\partial \mathbf{r}$. This is because the signal to noise of the local variations is greater in gradient measurements, which are also less susceptible to disturbance from vehicle motion. The spatial gravity gradient, $\partial \mathbf{g}/\partial \mathbf{r}$, has nine components. However, only five of these are independent. The three diagonal components sum to a constant, while because g is itself the spatial derivative of the gravity potential, V_g, the two spatial differentiation operations commute, so $\partial \mathbf{g}/\partial \mathbf{r}$ is a symmetric matrix.

A gravity gradiometer measures each component of the gravity gradient by differencing the outputs of a pair of accelerometers with parallel sensitive axes. The host vehicle acceleration and the constant component of gravity cancel out. The instrument biases may be cancelled by mounting the accelerometers on a rotating disc and processing the outputs accordingly [86]. However, this does limit the bandwidth of the gravity gradient measurements. A typical rotation rate is 0.25 Hz, giving a sensor bandwidth of 0.06 Hz [87].

Gravity gradiometry is commonly deployed aboard military submarines alongside a high-performance INS and depth sensor, limiting the position error to about 400m [86]. It has the advantage over sonar of maintaining covertness. The instrument is around a cubic meter in size and has a mass of several hundred kilograms.

In principle, gravity gradiometry can also be used for aircraft positioning. Over land, a terrain height database can be used to calculate the higher spatial-frequencies of the gravity distribution [88]. Gravity gradiometry is also one of the few environmental feature-matching techniques that can operate over water. However, the resolution achievable with a submarine gradiometer on an aircraft is limited to about 2 km. This is partly due to bandwidth limitations and partly because much greater sensitivity is needed in the air as the size of the gravity gradient diminishes as the cube of the distance from the terrain features producing that gradient [87].

New gradiometer designs, based on superconducting accelerometer technology, offer the potential of meter-accuracy gravity-inertial navigation [89]. However, these devices require cooling to 4K and are as large as conventional gradiometers. Cold-atom interferometry (see Section E.1.2 of Appendix E on the CD) can provide high-accuracy gravity gradiometry in a smaller package without external cooling [90].

Gravity gradiometry is also used for airborne surveying and mineral prospecting, and has been proposed for navigating spacecraft in the vicinity of other planets [89].

13.4.2 Magnetic Field Variation

The Earth's magnetic field (see Section 6.1.1) varies with location, so if the host vehicle attitude, including heading, can be determined by other means, the position can be obtained by comparing the measured inclination and dip angles with a global magnetic model. For orbital spacecraft, the Earth's magnetic field is regular, enabling position to be determined within a few hundred meters [91].

For air and marine applications, local anomalies, which may be time varying, limit the position accuracy achievable from the global geomagnetic field to a few kilometers. However, by measuring the variations in the magnetic field over a few kilometers and comparing this with a database using pattern matching, positions accurate to about a kilometer can be obtained for low-altitude aircraft [92]. Similar accuracies should be achievable for ships and submarines.

For pedestrian and road vehicle navigation, the local magnetic anomalies may be used as features for pattern matching with a database to determine position. A series of measurements of the either the 3-D magnetic flux density or its overall magnitude are made over successive steps (detected using inertial sensors). The measurements over this transect are then compared with a database, which either may be grid-based [93] or may comprise known routes [94, 95], to determine the position and direction of travel. Matching of 3-D magnetic flux density must account for the possibility of significant errors in the user heading solution. When there is sufficient local magnetic field variation, submeter accuracies may be achieved.

13.4.3 Celestial X-Ray Sources

Pulsars are rotating neutron stars that produce regular stable pulses over a wide spectrum that may be measured using directional X-ray detectors. Although the distance to each pulsar and the times of transmission are unknown, the lines of sight have been determined to a high precision. Therefore, differential ranging may be used to determine the relative position of a spacecraft with respect to a reference station. As pulsars are highly stable, updates from the reference station may be infrequent. However, the signals are very weak, so observation times of several hours are needed for kilometer-level accuracy. Pulsar signals repeat every few milliseconds so the ambiguity is of the order of 1,000 km.

Stronger celestial X-ray sources are also available, but their signals are irregular. Consequently, differential ranging requires much more information to be transmitted over the data link. X-ray signals may also be used for stellar imagery (Section 13.3.7). Despite many decades of research, X-ray navigation is still experimental. Its most promising potential application is missions to planets in the outer solar system [96, 97].

Problems and exercises for this chapter are on the accompanying CD.

References

[1]   Zhao, Y., *Vehicle Location and Navigation Systems*, Norwood, MA: Artech House, 1997.
[2]   Taylor, G., and G. Blewitt, *Intelligent Positioning: GIS-GPS Unification*, New York: Wiley, 2006.

[3]   Quddus, M. A., *High Integrity Map Matching Algorithms for Advanced Transport Telematics Applications*, Ph.D. Thesis, Imperial College London, 2006.

[4]   Quddus, M. A., W. Y. Ochieng, and R. B. Noland, "Current Map-Matching Algorithms for Transport Applications: State-of-the-Art and Future Research Directions," *Transportation Research Part C*, Vol. 15, No. 5, 2007, pp. 312–328.

[5]   Yang, D., B. Cai, and Y. Yuan, "An Improved Map-Matching Algorithm Used in Vehicle Navigation System," *Proc. IEEE Intelligent Transportation Systems Conference*, Shanghai, China, October 2003, pp. 1246–1250.

[6]   Pyo, J. -S., D. -H. Shin, and T. -Y. Sung, "Development of a Map Matching Method Using the Multiple Hypothesis Technique," *Proc. IEEE Intelligent Transportation Systems Conference*, Oakland, CA, August 2001, pp. 23–27.

[7]   Jagadeesh, G. R., T. Srikathan, and X. D. Zhang, "A Map Matching Method for GPS Based Real-Time Vehicle Location," *Journal of Navigation*, Vol. 57, No.3, 2004, pp. 429–440.

[8]   Zheng, Y., and P. Cross, "Integrated GNSS with Different Accuracy of Track Database for Safety-Critical Railway Control Systems," *GPS Solutions*, Vol. 16, No. 2, 2012, pp. 169–179.

[9]   Mueller, T., et al., "Design and Testing of a Robust High Speed Rail Prototype GPS Locomotive Location System," *Proc. ION GNSS 2004*, Long Beach, CA, September 2004, pp. 729–740.

[10] Woodman, O., and R. Harle, "Pedestrian Localisation for Indoor Environments," *Proc. UbiComp '08*, Seoul, Korea, September 2008.

[11] Beauregard, S., Widyawan, and M. Klepal, "Indoor PDR Performance Enhancement Using Minimal Map Information and Particle Filters," *Proc. IEEE/ION PLANS*, Monterey, CA, May 2008, pp. 141–147.

[12] Blanchart, P., L. He, and F. Le Gland, "Information Fusion for Indoor Localization," *Proc. 12th International Conference on Information Fusion*, Seattle, WA, July 2009, pp. 2083–2090.

[13] Robertson, P., M. Garcia Puyol, and M. Angermann, "Collaborative Pedestrian Mapping of Buildings Using Inertial Sensors and FootSLAM," *Proc. ION GNSS 2011*, Portland, OR, September 2011, pp. 1366–1377.

[14] Gusenbauer, D., C. Isert, and J. Krösche, "Self-Contained Indoor Positioning on Off-the-Shelf Mobile Devices," *Proc. Indoor Positioning and Indoor Navigation (IPIN)*, Zürich, Switzerland, September 2010.

[15] Ren, M., and H. A. Karimi, "Movement Pattern Recognition Assisted Map Matching for Pedestrian/Wheelchair Navigation," *Journal of Navigation*, Vol. 65, No. 4, 2012, pp. 617–633.

[16] Abdulrahim, K., et al., "Aiding Low Cost Inertial Navigation with Building Heading for Pedestrian Navigation," *Journal of Navigation*, Vol. 64, No. 2, 2011, pp. 219–233.

[17] Aggarwal, P., et al., "Map Matching and Heuristic Elimination of Gyro Drift for Personal Navigation Systems in GPS-Denied Conditions," *Measurement Science and Technology*, Vol. 22, No. 2, 2011, paper 025205.

[18] Campbell, J. L., *Application of Airborne Laser Scanner: Aerial Navigation*, Ph.D. Dissertation, Ohio University, 2006.

[19] Robins, A. J., "Recent Developments in the TERPROM Integrated Navigation System," *Proc. ION 44th AM*, Annapolis, MD, June 1988.

[20] Cowie, M., N. Wilkinson, and R. Powlesland, "Latest Development of the TERPROM® Digital Terrain System (DTS)," *Proc. IEEE/ION PLANS*, Monterey, CA, May 2008, pp. 1219–1229.

[21] Golden, J. P., "Terrain Contour Matching (TERCOM): A Cruise Missile Guidance Aid," *Image Processing for Missile Guidance*, SPIE Vol. 238, 1980, pp. 10–18.

[22] Xhang, J., W. Hu, and Y. Wu, "Novel Technique for Vision-Based UAV Navigation," *IEEE Trans. on Aerospace and Electronic Systems*, Vol. 47, No. 4, 2011, pp. 2731–2741.

[23] El-Sheimy, N., C. Valeo, and A. Habib, *Digital Terrain Modeling: Acquisition, Manipulation, and Applications*, Norwood, MA: Artech House, 2005.

[24] Perrett, M., and J. Krempasky, "Terrain Aiding for Precision Navigation in Heavy GPS Jamming," *Proc. ION GPS 2001*, Salt Lake City, UT, September 2001, pp. 924–931.

[25] Boozer, D. D., and J. R. Fellerhoff, "Terrain-Aided Navigation Test Results in the AFTI/F-16 Aircraft," *Navigation: JION*, Vol. 35, No. 2, 1988, pp. 161–175.

[26] Yu, P., Z. Chen, and J. C. Hung, "Performance Evaluation of Six Terrain Stochastic Linearization Techniques for TAN," *Proc. IEEE National Aerospace and Electronics Conference*, Dayton, OH, May 1991, pp. 382–388.

[27] Enns, R., and D. Morrell, "Terrain-Aided Navigation Using the Viterbi Algorithm," *Journal of Guidance, Control and Dynamics*, Vol. 18, No. 6, 1995, pp. 1444–1449.

[28] Bergman, N., L. Ljung, and F. Gustafsson, "Terrain Navigation Using Bayesian Statistics," *IEEE Control Systems Magazine*, June 1999, pp. 33–40.

[29] Nordlund, P. -J., and F. Gustafsson, "Recursive Estimation of Three-Dimensional Aircraft Position Using Terrain Aided Positioning," *Proc. IEEE International Conference on Acoustics, Speech and Signal Processing*, Orlando, FL, 2002, pp. II-1121–1124.

[30] Metzger, J., and G. F. Trommer, "Improvement of Modular Terrain Navigation Systems by Measurement Decorrelation," *Proc. ION 59th AM*, Albuquerque, NM, June 2003, pp. 353–362.

[31] Vaman, D., and P. Oonincx, "Exploring a GPS Inspired Acquisition and Tracking Concept for Terrain Referenced Navigation," *Proc. ION ITM*, San Diego, CA, January 2010, pp. 459–466.

[32] Hollowell, J., "HELI/SITAN: A Terrain Referenced Navigation Algorithm for Helicopters," *Proc. IEEE PLANS*, Las Vegas, NV, March 1990, pp. 616–625.

[33] Runnalls, A. R., "A Bayesian Approach to Terrain Contour Navigation," *Proc. NATO AGARD 40th Guidance and Control Panel Symposium*, May 1985, paper 43.

[34] Groves, P. D., R. J. Handley, and A. R. Runnalls, "Optimising the Integration of Terrain Referenced Navigation with INS and GPS," *Journal of Navigation*, Vol. 59, No. 1, 2006, pp. 71–89.

[35] Runnalls, A. R., P. D. Groves, and R. J. Handley, "Terrain-Referenced Navigation Using the IGMAP Data Fusion Algorithm," *Proc. ION 61st Annual Meeting*, Boston, MA, June 2005, pp. 976–987.

[36] Nordlund, P. -J., and F. Gustafsson, "Marginalized Particle Filter for Accurate and Reliable Terrain-Aided Navigation," *IEEE Trans. on Aerospace and Electronic Systems*, Vol. 45, No. 4, 2009, pp. 1385–1399.

[37] Runnalls, A. R., and R. J. Handley, "The 'Gold Standard' Navigator," *Proc. Eurofusion '98*, Great Malvern, U.K., November 1998, pp. 77–82.

[38] Bar-Gill, A., P. Ben-Ezra, and I. Y. Bar-Itzack, "Improvement of Terrain-Aided Navigation in a Trajectory Optimization," *IEEE Trans. on Control Systems Technology*, Vol. 2, No. 4, 1994, pp. 336–342.

[39] Neregård, F., et al., "Saab TERNAV, an Algorithm for Real Time Terrain Navigation and Results from Flight Trials Using a Laser Altimeter," *Proc. ION GNSS 2006*, Fort Worth, TX, September 2006, pp. 1136–1145.

[40] Handley, R. J., et al., "Future Terrain Referenced Navigation Techniques Exploiting Sensor Synergy," *Proc. GNSS 2003, ENC*, Graz, Austria, April 2003.

[41] Campbell, J. L., M. Uijt de Haag, and F. van Graas, "Terrain Referenced Positioning Using Airborne Laser Scanner," *Navigation: JION*, Vol. 52, No. 4, 2005, pp. 189–197.

[42] Campbell, J. L., M. Uijt de Haag, and F. van Graas, "Terrain Referenced Precision Approach Guidance Proof-of-Concept Flight Test Results," *Navigation: JION*, Vol. 54, No. 1, 2007, pp. 21–29.

[43] Vadlamanai, A. K., and M. Uijt de Haag, "Use of Laser Range Scanners for Precise Navigation in Unknown Environments," *Proc. ION GNSS 2006*, Fort Worth, TX, September 2006. pp. 1104–1114.

[44] Lucido, L., et al., "A Terrrain Referenced Underwater Positioning Using Sonar Bathymetric Profiles and Multiscale Analysis," *Proc. MTS/IEEE Oceans '96*, Fort Lauderdale, FL, September 1996, pp. 417–421.

[45] Strauss, O., F. Comby, and M. J. Aldon, "Multibeam Sonar Image Matching for Terrain-Based Underwater Navigation," *Proc. MTS/IEEE Oceans '99*, September 1999, pp. 882–887.

[46] Nygren, I., "Robust and Efficient Terrain Navigation of Underwater Vehicles," *Proc. IEEE/ION PLANS*, Monterey, CA, May 2008, pp. 923–932.

[47] Soehren, W., and W., Hawkinson, "A Prototype Personal Navigation System," *Proc. IEEE/ION PLANS*, San Diego, CA, April 2006, pp. 539–546.

[48] Sönmez, T., and H. E. Bingöl, "Modeling and Simulation of a Terrain Aided Inertial Navigation Algorithm for Land Vehicles," *Proc. IEEE/ION PLANS*, Monterey, CA, May 2008, pp. 1046–1052.

[49] Amt, J. R., and J. F. Raquet, "Positioning for Range-Based Land Navigation Systems Using Surface Topography," *Proc. ION GNSS 2006*, Fort Worth, TX, September 2006, pp. 1494–1505.

[50] Christian, J. A., and E. G. Lightsey, "High-Fidelity Measurement Models for Optical Spacecraft Navigation," *Proc. ION GNSS 2009*, Savannah, GA, September 2009, pp. 1486–1503.

[51] Veth, M. J., *Fusion of Imaging and Inertial Sensors for Navigation*, Ph.D. Dissertation, Air Force Institute of Technology, 2006.

[52] Morrison, J. R, J. F. Raquet, and M. J. Veth, "Performance Evaluation of a Vision Aided Inertial Navigation System Augmented with a Coded Aperture," *Proc. ION ITM*, Anaheim, CA, January 2009, pp. 61–73.

[53] Norris, A., *Integrated Bridge Systems Vol. 1: Radar and AIS*, London, U.K.: The Nautical Institute, 2008.

[54] Pearson, J. O., T. S. Abbott, Jr., and R. H. Jeffers, "Mapping and Multimode Radars in KF," in *Avionics Navigation Systems*, 2nd ed., M. Kayton and W. R. Fried, (eds.), New York: Wiley, 1997, pp. 503–550.

[55] Bevington, J. E., and C. A. Marttila, "Precision Aided Inertial Navigation Using SAR and Digital Map Data," *Proc. IEEE PLANS*, Las Vegas, NV, March 1990, pp. 490–496.

[56] Campbell, J. L., et al., "Flash-LADAR Inertial Navigator Aiding," *Proc. IEEE/ION PLANS*, San Diego, CA, April 2006, pp. 677–683.

[57] Sim, D.-G., et al., "Hybrid Estimation of Navigation Parameters from Aerial Image Sequence," *IEEE Trans. on Image Processing*, Vol. 8, No. 3, 1999, pp. 429–435.

[58] Handley, R. J., J. P. Abbott, and C. R. Surawy, "Continuous Visual Navigation: An Evolution of Scene Matching," *Proc. ION NTM*, Long Beach, CA, January 1998, pp. 217–224.

[59] Handley, R. J., L. Dack, and P. McNeil, "Flight Trials of the Continuous Visual Navigation System," *Proc. ION NTM*, Long Beach, CA, January 2001, pp. 185–192.

[60] Lowe, D. G., "Distinctive Image Features from Scale-Invariant Keypoints," *International Journal of Computer Vision*, Vol. 60, No. 2, 2004, pp. 91–110.

[61] Bay, H., et al., "SURF: Speeded Up Robust Features," *Computer Vision and Image Understanding*, Vol. 10, 2008, pp. 346–359.

[62] Calonder, M., et al., "BRIEF: Binary Robust Independent Features," *Lecture Notes in Computer Science*, Vol. 6314, 2010, pp. 778–792.

[63] Harle, R., and A. Hopper, "Cluster Tagging: Robust Fiducial Tracking for Smart Environments," *Proc. Second International Workshop on Location- and Context-Awareness*, Dublin, Ireland, May 2006, pp. 14–29.

[64] Walter, D., *Feasibility Study of Novel Environmental Feature Mapping to Bridge GNSS Outages*, University College London, 2012.

[65] Hirokawa, R., et al., "Autonomous Vehicle Navigation with Carrier Phase DGPS and Laser-Scanner Augmentation," *Proc. ION GNSS 2004*, Long Beach, CA, September 2004, pp. 1115–1123.

[66] Ma, Y., and J. B. McKitterick, "Range Sensor Aided Inertial Navigation Using Cross Correlation on the Evidence Grid," *Proc. IEEE/ION PLANS*, Myrtle Beach, SC, April 2012, pp. 422–427.

[67] Allen, J. W., and D. M. Bevly, "Relating Local Vision Measurements to Global Navigation Satellite Systems Using Waypoint Based Maps," *Proc. IEEE/ION PLANS*, Palm Springs, CA, May 2010, pp. 1204–1211.

[68] Hide, C., T. Botterill, and M. Andreotti, "An Integrated IMU, GNSS and Image Recognition Sensor for Pedestrian Navigation," *Proc. ION GNSS 2009*, Savannah, GA, September 2009, pp. 527–537.

[69] Giachetti, A., M. Campani, and V. Torre, "The Use of Optical Flow for Road Navigation," *IEEE Trans. on Robotics and Automation*, Vol. 14, No. 1, 1998, pp. 34–48.

[70] Hide, C., T. Botterill, and M. Andreotti, "Vision-Aided IMU for Handheld Pedestrian Navigation," *Proc. ION GNSS 2010*, Portland, OR, September 2010, pp. 534–541.

[71] Wang, J. J., S. Kodagoda, and G. Dissanayake, "Vision Aided GPS/INS System for Robust Land Vehicle Navigation," *Proc. ION GNSS 2009*, Savannah, GA, September 2009, pp. 600–609.

[72] De Agostino, M., et al., "GIMPhI: A Novel Vision-Based Navigation Approach for Low Cost MMS," *Proc. IEEE/ION PLANS*, Palm Springs, CA, May 2010, pp. 1238–1244.

[73] Sanchiz, J. M., and F. Pla, "Feature Correspondence and Motion Recovery in Vehicle Planar Navigation," *Pattern Recognition*, Vol. 32, 1999, pp. 1961–1977.

[74] Soloviev, A., et al., "Integrated Multi-Aperture Sensor and Navigation Fusion," *Proc. ION GNSS 2009*, Savannah, GA, September 2009, pp. 759–766.

[75] Kessler, C., et al., "Vision-Based Attitude Estimation for Indoor Navigation Using Vanishing Points and Lines," *Proc. IEEE/ION PLANS*, Palm Springs, CA, May 2010, pp. 310–318.

[76] Joerger, M., and B. Pervan, "Range-Domain Integration of GPS and Laser-Scanner Measurements for Outdoor Navigation," *Proc. ION GNSS 2006*, Fort Worth, TX, September 2006, pp. 1115–1123.

[77] Soloviev, A., D. Bates, and F. van Graas, "Tight Coupling of Laser Scanner and Inertial Measurements for a Fully Autonomous Relative Navigation Solution," *Navigation: JION*, Vol. 54, No. 3, 2007, pp. 189–205.

[78] Hoshizaki, T., et al., "Performance of Integrated Electro-Optical Navigation Systems," *Navigation: JION*, Vol. 51, No. 2, 2004, pp. 101–121.

[79] Pachter, M., A. Porter, and M. Polat, "INS-Aiding Using Bearings-Only Measurements of an Unknown Ground Object," *Navigation: JION*, Vol. 53, No. 1, 2006, pp. 1–19.

[80] Liebe, C. C., "Star Trackers for Attitude Determination," *IEEE AES Magazine*, June 1994, pp. 10–16.

[81] Woodbury, D. P., et al., "Stellar Positioning System (Part II): Improving Accuracy During Implementation," *Navigation: JION*, Vol. 57, No. 1, 2010, pp. 13–24.

[82] Knobbe, E. J., and G. N. Haas, "Celestial Navigation," in *Avionics Navigation Systems*, 2nd ed., M. Kayton and W. R. Fried, (eds.), New York: Wiley, 1997, pp. 551–596.

[83] Mortari, D., et al., "The Pyramid Star Identification Technique," *Navigation: JION*, Vol. 51, No. 3, 2004, pp. 171–183.

[84] Levine, S., R. Dennis, and K. L. Bachman, "Strapdown Astro-Inertial Navigation Utilizing the Optical Wide-Angle Lens Startracker," *Navigation: JION*, Vol. 37, No. 4, 1990, pp. 347–362.

[85] Parish, J. J., et al., "Stellar Positioning System (Part I): An Autonomous Position Determination Solution," *Navigation: JION*, Vol. 57, No. 1, 2010, pp. 1–12.

[86] Jircitano, A., and D. E. Dosch, "Gravity Aided Inertial Navigation System (GAINS)," *Proc. ION 47th AM*, June 1991, pp. 221–229.

[87] Gleason, D. M., "Passive Airborne Navigation and Terrain Avoidance Using Gravity Gradiometry," *Journal of Guidance, Control and Dynamics*, Vol. 18, No. 6, 1995, pp. 1450–1458.

[88] Affleck, C. A., and A. Jircitano, "Passive Gravity Gradiometer Navigation System," *Proc. IEEE PLANS*, Las Vegas, NV, March 1990, pp. 60–66.

[89] Richeson, J. A., *Gravity Gradiometer Aided Inertial Navigation Within Non-GNSS Environments*, Ph.D. Dissertation, University of Maryland, 2008.

[90] Brown, D., et al., "Atom Interferometric Gravity Sensor System," *Proc. IEEE/ION PLANS*, Myrtle Beach, SC, April 2012, pp. 30–37.

[91] Zuo, W., and F. Song, "An Autonomous Navigation Scheme Using Global Positioning System/Geomagnetism Integration for Small Satellites," *Proc. Institute of Mechanical Engineers*, Vol. 214G, No. 4, 2000, pp. 207–215.

[92] Wilson, J. M., et al., "Passive Navigation Using Local Magnetic Field Variations," *Proc. ION NTM*, Monterey, CA, January 2006, pp. 770–779.

[93] Storms, W. F., and J. F. Raquet, "Magnetic Field Aided Vehicle Tracking," *Proc. ION GNSS 2009*, Savannah, GA, September 2009, pp. 1767–1774.

[94] Judd, T., and T. Vu, "Use of a New Pedometric Dead Reckoning Module in GPS Denied Environments," *Proc. IEEE/ION PLANS*, Monterey, CA, May 2008, pp. 120–128.

[95] Shockley, J. A., and J. F. Raquet, "Three-Axis Magnetometer Navigation in Suburban Areas," *Proc. ION GNSS 2012*, Nashville, TN, pp. 1607–1618.

[96] Ray, P. S., et al., "Deep Space Navigation Using Celestial X-Ray Sources," *Proc. ION NTM*, San Diego, CA, January 2008, pp. 101–109.

[97] Sheikh, S. I., et al., "Deep Space Navigation Augmentation Using Variable Celestial X-Ray Sources," *Proc. ION ITM*, Anaheim, CA, January 2009, pp. 34–48.

CHAPTER 14
INS/GNSS Integration

Inertial navigation (Chapter 5) has a number of advantages. It operates continuously, bar hardware faults, provides high-bandwidth output at at least 50 Hz, and exhibits low short-term noise. It provides effective attitude, angular rate, and acceleration measurements as well as position and velocity. It is also invulnerable to jamming and interference, and is nonradiating (which is important for military stealth). However, an inertial navigation solution requires initialization and its accuracy then degrades with time as the inertial instrument errors are integrated through the navigation equations. Furthermore, INS capable of providing effective sole-means navigation for more than a few minutes after initial alignment are expensive at around \$100,000 (€80,000).

GNSS (Chapters 8 to 10) provides a high long-term position accuracy with errors limited to a few meters (stand-alone), while user equipment is available for less than \$100 (€80). However, compared to INS, the output rate is low, typically around 10 Hz, the short-term noise of a code-based position solution is high, and standard GNSS user equipment does not measure attitude. GNSS signals are also subject to obstruction and interference, so GNSS cannot be relied upon to provide a continuous navigation solution.

The benefits and drawbacks of INS and GNSS are complementary, so by integrating them, the advantages of both technologies are combined to give a continuous, high-bandwidth, complete navigation solution with high long- and short-term accuracy. In an integrated INS/GNSS, or GNSS/INS, navigation system, GNSS measurements prevent the inertial solution drifting, while the INS smooths the GNSS solution and bridges signal outages.

INS/GNSS integration, sometimes known as hybridization, is suited to established inertial navigation applications such as ships, airliners, military aircraft, and long-range missiles. Integration with GNSS also makes inertial navigation practical with lower cost tactical-grade inertial sensors (see Chapter 4), making INS/GNSS a suitable navigation solution for light aircraft, helicopters, UAVs, short- and medium-range guided weapons, smaller boats, and potentially trains. INS/GNSS is sometimes used for road vehicles and personal navigation. However, lower cost dead-reckoning techniques, such as odometers and PDR, are often integrated with GNSS instead as discussed in Section 16.2.

Figure 14.1 shows the basic configuration of a typical INS/GNSS navigation system. The integration algorithm compares the inertial navigation solution with the outputs of GNSS user equipment and estimates corrections to the inertial position, velocity, and attitude solution, usually alongside other parameters. It is usually based on a Kalman filter, described in Chapter 3. The corrected inertial navigation solution then forms the integrated navigation solution. This architecture ensures

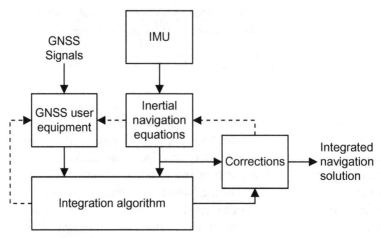

Figure 14.1 Generic INS/GNSS integration architecture.

that an integrated navigation solution is always produced, regardless of GNSS signal availability. The dotted lines in Figure 14.1 show data flows present in some systems but not others; these are discussed later.

The hardware configuration of INS/GNSS systems varies. The integration algorithm may be hosted in the INS, the GNSS user equipment, or separately. Alternatively, everything may be hosted in one unit, sometimes known as an embedded GNSS in INS (EGI) or integrated GNSS/INS (IGI). Where the inertial navigation equations and integration algorithm share the same processor, but the IMU is separate, the system is sometimes known as an integrated IMU/GNSS or GNSS/IMU. However, an IMU/GNSS is no different to an INS/GNSS.

Section 14.1 describes and compares the different INS/GNSS integration architectures. Section 14.2 discusses state selection for INS/GNSS integration Kalman filters and describes typical system models, while Section 14.3 describes measurement models. Finally, Section 14.4 discusses advanced INS/GNSS implementations, such as those using differential and carrier-phase GNSS and GNSS attitude, handling large heading errors, performing smoothing, or using advanced inertial sensor modeling. In addition, Appendix I on the CD describes several alternative formulations of INS/GNSS integration and time synchronization error estimation. MATLAB INS/GNSS integration software is also included on the CD, together with a number of demonstrations.

Figure 14.2 shows the typical stages of an INS/GNSS integration algorithm.

14.1 Integration Architectures

The architecture of an INS/GNSS integrated navigation system varies in three respects: how corrections are applied to the inertial navigation solution, what types of GNSS measurements are used, and how the GNSS user equipment is aided by the INS and integration algorithm. These are largely independent of each other. In the literature, terms such as loosely coupled, tightly coupled, ultratightly coupled, closely coupled, cascaded, deeply coupled, and deep are used to define integration architectures [1–5].

14.1 Integration Architectures

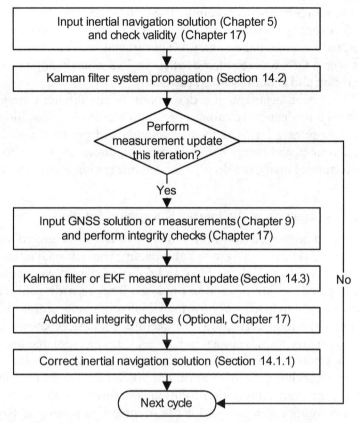

Figure 14.2 Typical stages of an INS/GNSS integration algorithm.

However, there is no commonly agreed definition of these terms. Here, the most widely used definitions are adopted.

A *loosely coupled* INS/GNSS system uses the GNSS position and velocity solution as the measurement inputs to the integration algorithm, irrespective of the type of INS correction or GNSS aiding used. It is a cascaded architecture where the GNSS user equipment incorporates a navigation filter. This is *position-domain* integration.

A *tightly coupled* INS/GNSS system uses the GNSS pseudo-range and pseudo-range-rate, delta-range, or ADR measurements as inputs to the integration algorithm, again irrespective of the type of INS correction or GNSS aiding used. This is *range-domain* integration.

A *deeply coupled* INS/GNSS system combines INS/GNSS integration and GNSS signal tracking into a single estimation algorithm. It uses the Is and Qs from the GNSS correlation channels as measurements, either directly or via discriminator functions, and generates the NCO commands used to control the reference code and carrier within the GNSS receiver (see Section 9.1.4). This is *tracking-domain* integration and is also known as deep integration.

The term *ultratightly coupled* (UTC) is used to describe both tracking-domain integration and range-domain integration with inertial aiding of the GNSS tracking loops, while the term *closely coupled* has been applied to both position-domain and range-domain integration architectures. Because of these ambiguities, both terms are avoided here.

The simplest way of combining INS and GNSS is an *uncoupled* system, whereby GNSS is simply used to reset the inertial navigation solution at intervals, often prompted by a manual command. This architecture has been applied in some aircraft where GPS was retrofitted when an INS was already installed. It is not true integration and is not discussed further.

The section begins with a description of the different methods of correcting the inertial navigation solution: open-loop correction, closed-loop correction, and total-state integration. The loosely coupled and tightly coupled architectures are then described and compared, followed by a discussion of GNSS aiding with these architectures. Finally, the deeply coupled integration architecture is described.

14.1.1 Correction of the Inertial Navigation Solution

The integrated navigation solution of an INS/GNSS integrated navigation system is the corrected inertial navigation solution. In a conventional integration architecture using an error-state Kalman filter and separate inertial navigation processing, correction may be either open-loop or closed-loop, regardless of what type of GNSS measurements are used and how the GNSS user equipment is aided.

The open-loop correction architecture, sometimes known as a feed-forward complementary filter, is shown in Figure 14.3. The estimated position, velocity, and attitude errors are used to correct the inertial navigation solution within the integration algorithm at each iteration but are not fed back to the INS. Consequently, only the integrated navigation solution contains the Kalman filter estimates and a raw INS solution is available for use in integrity monitoring (see Section 17.4.2). Either the raw INS or integrated navigation solution may be used for GNSS aiding.

The corrected inertial navigation solution, $\hat{C}_b^\gamma, \hat{v}_{\beta b}^\gamma$, and $\hat{r}_{\beta b}^\gamma$ or $\hat{p}_b$, which forms the integrated navigation solution, is obtained from the raw inertial navigation solution, $\tilde{C}_b^\gamma, \tilde{v}_{\beta b}^\gamma$, and $\tilde{r}_{\beta b}^\gamma$ or $\tilde{p}_b$, using

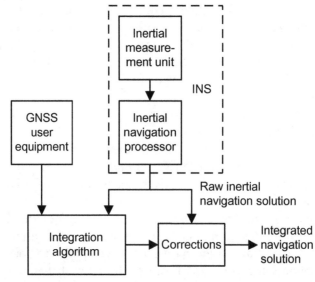

Figure 14.3 Open-loop INS correction architecture.

14.1 Integration Architectures

$$\hat{\mathbf{C}}_b^\gamma = \delta\hat{\mathbf{C}}_b^{\gamma\mathrm{T}}\tilde{\mathbf{C}}_b^\gamma, \tag{14.1}$$

$$\hat{\mathbf{v}}_{\beta b}^\gamma = \tilde{\mathbf{v}}_{\beta b}^\gamma - \delta\hat{\mathbf{v}}_{\beta b}^\gamma \tag{14.2}$$

and

$$\hat{\mathbf{r}}_{\beta b}^\gamma = \tilde{\mathbf{r}}_{\beta b}^\gamma - \delta\hat{\mathbf{r}}_{\beta b}^\gamma \tag{14.3}$$

or

$$\begin{aligned}\hat{L}_b &= \tilde{L}_b - \delta\hat{L}_b \\ \hat{\lambda}_b &= \tilde{\lambda}_b - \delta\hat{\lambda}_b \\ \hat{h}_b &= \tilde{h}_b - \delta\hat{h}_b\end{aligned} \tag{14.4}$$

where the attitude, velocity, and position errors, $\delta\hat{\mathbf{C}}_b^\gamma$, $\delta\hat{\mathbf{v}}_{\beta b}^\gamma$, $\delta\hat{\mathbf{r}}_{\beta b}^\gamma$, $\delta\hat{L}_b$, $\delta\hat{\lambda}_b$, and $\delta\hat{h}_b$, are as defined by (5.107) to (5.109), with ^ denoting a Kalman filter estimate. Note that $\delta\hat{\mathbf{C}}_b^\gamma$ is resolved about the γ-frame axes. The reference frame, β, and resolving axes, γ, are given by

$$\{\beta,\gamma\} \in \{i,i\},\{e,e\},\{e,n\}, \tag{14.5}$$

and depend on which coordinate frames are used for the inertial navigation equations (see Chapter 5).

When the small angle approximation is applicable to the attitude errors, which is often not the case with open-loop integration, (14.1) becomes

$$\hat{\mathbf{C}}_b^\gamma \approx \left(\mathbf{I}_3 - \left[\delta\hat{\boldsymbol{\psi}}_{\gamma b}^\gamma \wedge\right]\right)\tilde{\mathbf{C}}_b^\gamma, \tag{14.6}$$

where $\delta\hat{\boldsymbol{\psi}}_{\gamma b}^\gamma$ is the Kalman filter estimate of the attitude error of the INS body frame, b, with respect to frame γ, resolved about the frame γ axes. The corresponding quaternion correction is included in Section E.6.3 of Appendix E on the CD.

In the closed-loop correction architecture, sometimes known as a feedback complementary filter, GNSS is used to aid the INS via the integration algorithm. This is shown in Figure 14.4. The estimated position, velocity, and attitude errors are fed back to the inertial navigation processor, where they are used to correct the inertial navigation solution itself. The feedback may occur either on each Kalman filter iteration or at longer intervals. The Kalman filter's position, velocity, and attitude error estimates are zeroed after each set of corrections is fed back. Consequently, there is no independent uncorrected inertial navigation solution. As discussed in Section 3.2.6, a closed-loop Kalman filter minimizes the size of the states, minimizing the linearization errors in the system model.

In closed-loop integration, there is only the corrected inertial navigation solution. New corrections are applied using

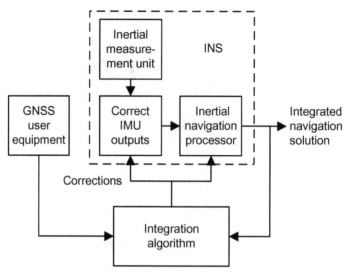

Figure 14.4 Closed-loop INS correction architecture.

$$\hat{\mathbf{C}}_b^\gamma(+) = \delta\hat{\mathbf{C}}_b^{\gamma\mathrm{T}}\hat{\mathbf{C}}_b^\gamma(-) \approx \left(\mathbf{I}_3 - \left[\delta\hat{\boldsymbol{\psi}}_{\gamma b}^\gamma \wedge\right]\right)\hat{\mathbf{C}}_b^\gamma(-), \tag{14.7}$$

$$\hat{\mathbf{v}}_{\beta b}^\gamma(+) = \hat{\mathbf{v}}_{\beta b}^\gamma(-) - \delta\hat{\mathbf{v}}_{\beta b}^\gamma \tag{14.8}$$

and

$$\hat{\mathbf{r}}_{\beta b}^\gamma(+) = \hat{\mathbf{r}}_{\beta b}^\gamma(-) - \delta\hat{\mathbf{r}}_{\beta b}^\gamma \tag{14.9}$$

or

$$\begin{aligned}\hat{L}_b(+) &= \hat{L}_b(-) - \delta\hat{L}_b \\ \hat{\lambda}_b(+) &= \hat{\lambda}_b(-) - \delta\hat{\lambda}_b\,, \\ \hat{h}_b(+) &= \hat{h}_b(-) - \delta\hat{h}_b\end{aligned} \tag{14.10}$$

where the suffixes (−) and (+) denote before and after the correction, respectively, and the small angle approximation is usually applicable to the attitude error.

In the closed-loop integration architecture, any accelerometer and gyro errors estimated by the Kalman filter are fed back to correct the IMU measurements, using (4.19) and (4.20), as they are input to the inertial navigation equations. These corrections are in addition to any that may be applied by the IMU's processor. Unlike the position, velocity, and attitude corrections, the accelerometer and gyro corrections must be applied on every iteration of the navigation equations, with feedback from the Kalman filter periodically updating the accelerometer and gyro errors. It may either feed back replacement estimates to the inertial navigation processor, or estimate residual errors and feed back perturbations to the error estimates stored by the navigation processor. In the latter case, the Kalman filter estimates are zeroed on feedback.

The choice of open- or closed-loop INS/GNSS integration is a function of both the INS quality and the integration algorithm quality. When low-grade inertial sensors are used, only the closed-loop configuration is suitable, regardless of the integration algorithm quality. This is because the raw inertial navigation solution will be of little use, while an open-loop configuration is likely to lead to large linearization errors in the Kalman filter. Conversely, when a high-quality INS is used with a low-quality integration algorithm, an open-loop configuration should be used as integrity monitoring is likely to be needed, whereas linearization errors will be small. Alternatively, a raw inertial navigation solution may be maintained in parallel to a closed-loop integrated solution. When both the INS and the integration algorithm are high quality, the open- and closed-loop configurations are both suitable.

Navigation systems where the IMU is supplied separately and the inertial navigation equations and integration algorithms share a common processor are ideally suited to closed-loop operation, as the feedback of corrections is fully under the control of the integrated navigation system designer. However, where the INS is supplied as a complete unit, closed-loop integration should be approached with caution, as it is then necessary to ensure that the corrections are sent in the form that the INS is expecting, which may not be clearly defined.

The alternative to an error-state Kalman filter in INS/GNSS integration is a total-state Kalman filter, which estimates absolute position, velocity, and attitude instead of the errors in the corresponding INS outputs. In a total-state Kalman filter, the inertial navigation equations are embedded in the system model. As these are nonlinear, the extended Kalman filter form of the system model (see Section 3.4.1) must be used. The system model is then a function of the IMU outputs [6]. Figure 14.5 shows the system architecture.

In total-state integration, the system model must be iterated at the same rate as the inertial navigation equations in an error-state implementation. However, the processor load can be limited by iterating the system propagation of the error covariance matrix, $\mathbf{P}$, at a lower rate than that of the state vector, $\mathbf{x}$. The equations processed in a total-state INS/GNSS implementation are the same as those in a closed-loop error-state implementation, so the performance will be the same. The difference lies in the software architecture.

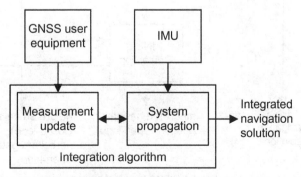

Figure 14.5 Total-state INS/GNSS integration architecture.

14.1.2 Loosely Coupled Integration

Figure 14.6 shows a loosely coupled INS/GNSS integration architecture. The GNSS position and velocity solutions (Section 9.4) are input as measurements to the integration Kalman filter, which uses them to estimate the INS errors. The integrated navigation solution is the INS navigation solution, corrected with the Kalman filter estimates of its errors. Thus, the integration is performed in the position domain, as opposed to the range domain.

Loosely coupled integration can operate with GNSS position measurements alone as the Kalman filter can infer the INS velocity error from the rate of change of position error in the same way as it infers the INS attitude and instrument errors from the changes in velocity error (see Sections 14.2.1 and 14.3.4). However, the velocity obtained by differentiating position is noisier, particularly where a code-based single-epoch GNSS position solution is used, as opposed to a carrier-smoothed-code-based or EKF-based solution. In principle, integration can also proceed using only the GNSS velocity measurements. However, this will not fully constrain the growth of the INS position error as the velocity measurement noise will be integrated up into the position state estimates. Therefore, most INS/GNSS integration algorithms use both velocity and position measurements.

The principal advantage of loosely coupled integration is simplicity. It can be used with any INS and any GNSS user equipment, making it particularly suited to retrofit applications. A further benefit is redundancy: there is usually a stand-alone

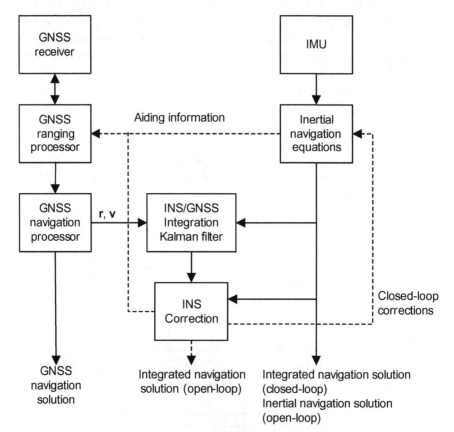

Figure 14.6 Loosely coupled INS/GNSS integration architecture.

GNSS navigation solution available in addition to the integrated solution. When open-loop INS correction is implemented, there is also an independent INS solution. This enables basic parallel solutions integrity monitoring (Section 17.4.2).

When the GNSS navigation processor is Kalman filter-based or uses carrier-smoothed pseudo-range measurements, it will output a position and velocity solution with time-correlated noise. However, an assumption of the Kalman filter used for INS/GNSS integration is that noise on the measurements it inputs is uncorrelated in time. This is a standard problem with cascaded filtered integration architectures (Section 16.1.3). As discussed in Section 3.4.3, input of measurements with time-correlated errors can disrupt Kalman filter state estimation unless the filter gain is reduced or the correlated errors are modeled as states. The correlation time of the GNSS navigation-solution errors varies and can be up to 100 seconds on the position and 20 seconds on the velocity. Selection of the integration Kalman filter gain and measurement iteration rate is therefore critical. If they are too high, the filter is liable to become unstable. However, if they are too low, the stochastic observability of the INS errors will be reduced (see Section 3.2.5). For stability, the system must be tuned so that the integration Kalman filter bandwidth is always less than that of the GNSS Kalman filter, noting that the bandwidths vary. Measurement-update intervals of 10 seconds are common in loosely coupled systems.

The cascading stability problem does not arise where the GNSS user equipment computes a single-epoch position solution without carrier smoothing of the pseudo-ranges (provided the update interval exceeds the tracking-loop time constants). However, this type of position solution is much noisier, so a low gain is still needed for the integration Kalman filter.

A further problem is that the integration filter needs to know the covariance of the GNSS navigation solution, both to ensure the correct gain is used and to estimate the accuracy of the integrated navigation solution. However, the GNSS solution covariance varies with satellite geometry and availability. With some geometries, there can be significant correlation between the position and velocity errors along different axes (see Sections 7.4.5 and 9.4.3). Furthermore, there may be significant correlation between the position error and the velocity error, depending on the GNSS navigation processor design. Few designs of GNSS user equipment output realistic error covariances; many just output DOP information, and some provide no information at all on the solution accuracy. Consequently, a conservative GNSS solution covariance must be assumed, limiting the Kalman filter gain further.

The final limitation of loosely coupled integration is that signals from at least four satellites are required to maintain a GNSS navigation solution, though an EKF-based solution can be maintained for short periods with fewer satellites. Consequently, when fewer satellites are tracked, the GNSS data cannot be used to aid the INS.

14.1.3 Tightly Coupled Integration

Figure 14.7 shows a tightly coupled INS/GNSS integration architecture. An extended Kalman filter inputs the pseudo-range and pseudo-range rate measurements from the GNSS ranging processor (Section 9.2) and uses them to estimate the INS errors, together with the GNSS receiver clock offset and bias. The integration is thus performed in the range domain. As with a loosely coupled architecture, the corrected inertial navigation solution forms the integrated navigation solution. This is an example of centralized filtered integration (Section 16.1.4) as the GNSS navigation processor is bypassed, its function subsumed into the INS/GNSS integration filter.

The pseudo-range and pseudo-range rate measurements are complementary. Pseudo-ranges come from code tracking and can be used to determine absolute position, whereas pseudo-range rates are derived from the more precise, but less robust, carrier tracking and are used to determine changes in position. It is thus standard practice to use both. GNSS user equipment often outputs Doppler shift measurements instead of pseudo-range rates; however, the latter may be calculated from the former using (9.70). With suitable modifications to the EKF design, ADR or delta-range measurements may be used instead of pseudo-range rates.

A tightly coupled integration architecture has a number of advantages. By combining the GNSS navigation processor and INS/GNSS integration algorithm into one, the filter cascading problems of the loosely coupled architecture are eliminated. Also, the variation of the GNSS solution covariance with satellite geometry and availability is accounted for implicitly. These enable higher gains to be used, increasing the stochastic observability of the INS errors. Note, however, that the Kalman filter bandwidths must still be kept within the GNSS tracking-loop bandwidths to prevent time-correlated tracking noise from contaminating the state estimates. The final advantage of tightly coupled integration is that GNSS measurement data is

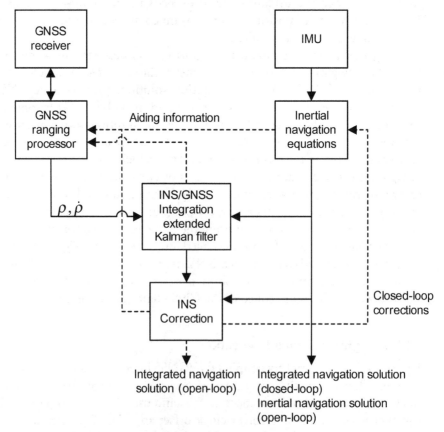

Figure 14.7 Tightly coupled INS/GNSS integration architecture.

still input and used to aid the INS when insufficient satellite signals are tracked to compute a stand-alone GNSS navigation solution.

The main limitation of tightly coupled integration is that not all designs of GNSS user equipment output the range-domain measurements required. Although there is no inherent stand-alone GNSS solution, a GNSS-only navigation solution may be maintained in parallel where required for integrity monitoring. Given the same inertial instruments and the same GNSS user equipment, a tightly coupled INS/GNSS should be both more accurate and more robust than its loosely coupled counterpart.

14.1.4 GNSS Aiding

In loosely and tightly coupled integration, the inertial navigation solution may be used to aid GNSS acquisition and tracking as discussed in Section 10.5.1 [7]. In deeply coupled integration, aiding of GNSS acquisition is the same, but tracking aiding is an inherent part of the integration architecture. When open-loop INS correction is used, either the raw or corrected inertial navigation solution may be used for GNSS acquisition and tracking aiding. The corrected solution is generally more accurate, but the raw solution is wholly independent of GNSS (after initialization), so is not subject to positive-feedback-induced errors. In tightly and deeply coupled integration, the receiver clock may also be corrected.

Acquisition aiding provides the GNSS ranging processor with the approximate position and velocity, limiting the number of cells that need to be searched to acquire the signal (see Section 9.2.1). For reacquisition, where the satellite positions and velocities are known and the receiver clock is calibrated, the number of cells to search can be very small, allowing very long dwell times in each cell. Simulations have shown that inertially aided reacquisition should be feasible at C/N_0 levels down to about 10 dB-Hz [8]. Using AGNSS (Section 10.5.2) to provide satellite information and receiver clock calibration enables a similar performance to be achieved with inertially aided initial acquisition [9].

GNSS tracking-loop bandwidths are a tradeoff between dynamics response and noise resistance (see Sections 9.2.2, 9.2.3, and 9.3.3). However, if the tracking loops are aided with the inertial navigation solution, they only have to track the receiver clock noise and the error in the INS solution, not the absolute dynamics of the user antenna. This enables narrower tracking-loop bandwidths to be used, improving noise resistance and allowing tracking to be maintained at a lower C/N_0 [10]. Integration architectures with GNSS tracking aiding are sometimes known as ultratightly coupled.

However, the downside of narrow tracking bandwidths is longer error correlation times on the GNSS measurements input to the integration algorithm. Figure 14.8 shows the code tracking time constant as a function of minimum C/N_0. This requires lower gains to be used in the Kalman filter to prevent instability (see Section 3.4.3). With an aviation-grade INS, this does not present a problem. However, with tactical-grade inertial sensors, a reduced-gain Kalman filter leads to poorer inertial calibration and larger navigation-solution errors. One solution is adaptive tightly coupled (ATC) integration [11, 12], in which the tracking-loop bandwidths are varied according to the measured c/n_0 and the measurement noise covariance in the integration algorithm adapted to the tracking bandwidths so that the Kalman

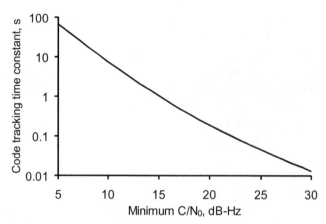

Figure 14.8 Code tracking time constant versus minimum C/N_0. [Calculated using (9.98) assuming $\sigma_x \ll 1/10$ chip, a DPP discriminator, and a time constant of $1/4B_{L\_CO}$.]

filter gains are matched to the tracking-error correlation times. ATC enables GNSS code to be tracked at a C/N_0 around 8 dB-Hz lower than a conventional tightly coupled system tuned for optimum INS calibration [11].

Inertial aiding of the code-tracking loop is normally implemented as a reversionary mode to carrier aiding. The pseudo-range rate for satellite s, estimated from the inertial navigation solution, is

$$\begin{aligned} \hat{\dot{\rho}}_R^s &= \hat{\mathbf{u}}_{as}^{i\,\mathrm{T}}\left(\hat{\mathbf{v}}_{is}^i(t_{st,a}^s) - \hat{\mathbf{v}}_{ia}^i(t_{sa,a}^s)\right) + \delta\hat{\dot{\rho}}_c^a - \delta\hat{\dot{\rho}}_c^s \\ &= \hat{\mathbf{u}}_{as}^{e\,\mathrm{T}}\left(\hat{\mathbf{v}}_{es}^e(t_{st,a}^s) - \hat{\mathbf{v}}_{ea}^e(t_{sa,a}^s)\right) + \delta\hat{\dot{\rho}}_{ie,a}^s + \delta\hat{\dot{\rho}}_c^a - \delta\hat{\dot{\rho}}_c^s, \end{aligned} \qquad (14.11)$$

where the notation is as defined in Section 8.5.3. The receiver clock drift estimate, $\delta\hat{\dot{\rho}}_c^a$, is obtained from the integration Kalman filter in tightly coupled integration and the GNSS navigation solution in loosely coupled integration. The satellite velocity, $\hat{\mathbf{v}}_{is}^i$ or $\hat{\mathbf{v}}_{es}^e$, is obtained from the navigation data message (see Section 8.5.2). The line-of-sight vector, $\hat{\mathbf{u}}_{as}^i$ or $\hat{\mathbf{u}}_{as}^e$, is given by (8.40) or (8.41), the Sagnac correction, $\delta\hat{\dot{\rho}}_{ie,a}^s$, is given by (8.46), and the satellite clock drift correction, $\delta\hat{\dot{\rho}}_c^s$, obtained by differentiating (9.77). The GNSS antenna velocity is, from (2.165),

$$\hat{\mathbf{v}}_{\beta a}^\gamma = \hat{\mathbf{v}}_{\beta b}^\gamma + \hat{\mathbf{C}}_b^\gamma\left(\hat{\boldsymbol{\omega}}_{ib}^b \wedge \mathbf{l}_{ba}^b\right), \qquad (14.12)$$

where $\mathbf{l}_{ba}^b$ is the lever arm from the IMU to the antenna, resolved about the IMU body frame and β and γ are defined by (14.5). When carrier tracking is lost, the inertial aiding information must also be used to control the carrier NCO to maintain signal coherence over the correlator accumulation interval in the receiver and GNSS ranging processor (see Section 9.1.4.4). For a 20-ms accumulation interval, the pseudo-range rate must be accurate to within about 4 m s^{-1}.

Inertial aiding can also be used to maintain synchronization of the reference code phase and carrier frequency through short signal blockages, enabling tracking to resume when the signal returns without having to undergo reacquisition first.

The key is to compensate for any loss of synchronization between loss of signal and detection of the loss of tracking lock. All-channel outages may be bridged for several tens of seconds and single-channel outages for several minutes [8].

Inertial and carrier aiding of the code tracking loop may also be implemented in parallel, in which case, a weighted average of the two pseudo-range rates should be constructed, based on the respective error standard deviations.

For inertial aiding of carrier-phase tracking, simulations have suggested that a 4 dB-Hz reduction in the minimum tolerable C/N_0 should be achievable using tactical-grade inertial sensors and a high-quality receiver oscillator [13]. However, carrier tracking aiding is challenging to implement, largely because it requires tight time synchronization for high-dynamics applications [14]. One solution is to store and retrieve the precorrelation GNSS signal samples so that inertial aiding derived from contemporaneous IMU measurements may be used; however, this requires a software receiver.

The integration algorithm's receiver-clock estimates may be fed back to correct the receiver clock itself in analogy with closed-loop INS correction. The Kalman filter states are zeroed when feedback takes place, which can occur at every iteration, at regular intervals or when the estimates exceed a predetermined threshold. Caution must be exercised in correcting for the effects of any lags in applying the clock corrections and disabling any clock feedback from the GNSS navigation processor. As no approximations are made in implementing the clock states in the Kalman filter, closed-loop receiver clock correction has no impact on performance.

The corrected inertial navigation solution can also be used to aid GNSS integrity monitoring (Chapter 17) and detection of GNSS cycle slips (see Section 10.5.1).

14.1.5 Deeply Coupled Integration

Deeply coupled, or deep, INS/GNSS integration is the INS/GNSS equivalent of combining GNSS navigation and tracking into a vector-tracking architecture, as discussed in Section 10.3.7. The terms ultratightly coupled and deeply integrated are also used to describe this architecture. Figure 14.9 shows the deeply coupled integration architecture with closed-loop INS correction. The code and carrier NCO commands are generated using the corrected inertial navigation solution, the satellite position and velocity from the navigation data message, and various GNSS error estimates. The accumulated correlator outputs from the GNSS receiver, the Is and Qs, are input via prefilters to the integration algorithm, usually Kalman filter-based, where a number of INS and GNSS errors are estimated. The corrected inertial navigation solution forms the integrated navigation solution as in the other architectures. The integration is thus performed in the tracking domain, as opposed to the range or position domain.

Compared with GNSS vector tracking, deep INS/GNSS integration has the advantage that only the errors in the INS solution need be tracked, as opposed to the absolute dynamics. This enables a lower tracking bandwidth to be used, increasing noise resistance. Deeply coupled integration can also operate with fewer than four GNSS satellites for limited periods.

Compared with tightly coupled integration, deeply coupled avoids the weighting down of the older I and Q measurement data when the pseudo-range or pseudo-range

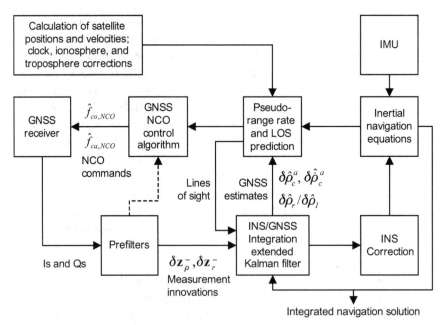

Figure 14.9 Deeply coupled INS/GNSS integration architecture (closed-loop INS correction).

rate output interval is greater than the corresponding tracking-loop time constant and avoids the need to reduce the Kalman filter gain when it is not. This enables deeply coupled integration to operate at lower C/N_0 levels. Like ATC, a deeply coupled integration algorithm can adapt to different C/N_0 levels by varying the measurement weighting. Bridging code and carrier-frequency tracking through brief signal outages is inherent in deep INS/GNSS integration. By removing the cascade between the tracking-loop filters and integration filter, deep coupling offers an optimal integration architecture [15].

For data-modulated signals, the Is and Qs must be output by the GNSS receiver at at least the navigation message symbol rate (50 Hz for the legacy GPS and GLONASS signals). A faster data rate reduces communication lags, but requires the integration algorithm to know where the navigation data-bit transitions are so that it can perform correct coherent summation of the Is and Qs. The NCO commands are usually input by the GNSS receiver at the same rate as the Is and Qs are output. The need to implement a new and much faster interface between the GNSS user equipment and integration algorithm is the main drawback of deeply coupled integration and makes it difficult to retrofit [16].

The prefilters input the Is and Qs from the receiver at its output rate, process them and then output pseudo-range and pseudo-range-rate or delta-range measurement innovations to the integration filter, usually an EKF, at a lower rate, typically 1–10 Hz. By reducing both the number of measurements input to the EKF and their rate, the processor load is reduced to a practical level. There is one prefilter per signal or satellite tracked. In some architectures, information from the prefilters is also used to correct the NCO commands; this is shown by the dotted line in Figure 14.9.

This combination of prefilters and integration filter is an example of a federated integration architecture as described in Section 16.1.5 [17]. As a key driver of deeply coupled integration is to maximize the filtering gain by removing cascades between successive filters, the zero-reset version of the federated architecture is employed. This avoids cascading errors by removing information from the prefilters when it is passed to the integration filter.

There are two classes of deeply coupled integration; coherent and noncoherent [18]. These are distinguished by the prefilter design. In noncoherent integration, the prefilters process the Is and Qs using separate code and carrier discriminator functions. In coherent integration, the prefilters estimate the code-phase, carrier-phase, and carrier-frequency tracking errors between integration EKF updates, inputting the Is and Qs directly. If the prefilters are bypassed and the Is and Qs input directly to the integration filter, the integration is coherent. However, this is not currently practical due to the much higher processing load.

Coherent integration is more accurate as it avoids discriminator nonlinearities and reduces code-tracking noise to that obtained with a coherent discriminator (see Section 9.3.3). However, it can only operate where there is sufficient signal to noise to track carrier phase, so noncoherent deeply coupled integration is more robust. The benefits of both approaches may be realized by switching the prefilters between coherent and noncoherent operation according to the signal quality and dynamics. This is analogous to switching between carrier-phase and carrier-frequency tracking in a conventional GNSS receiver as discussed in Section 9.2.3.

Tests have shown that noncoherent deep INS/GPS integration can stably track code at C/N_0 levels of 8–12 dB-Hz using a tactical-grade IMU [19–22]. This represents a margin of at least 8–12 dB-Hz over conventional tightly coupled INS/GPS with wide tracking bandwidths. Carrier-phase tracking down to 15 dB-Hz C/N_0 has been demonstrated with coherent deeply coupled integration and navigation data-bit estimation [23, 24].

The tracking sensitivity is affected by a number of factors. The quality of the IMU is important as sensor noise limits the extent to which the receiver tracking bandwidth may be narrowed. In most studies reported in the literature, a tactical-grade IMU was used. Using an aviation-grade IMU enables tracking to be maintained at a C/N_0 6 dB-Hz lower than when a tactical-grade IMU is used [21]. However, with calibrated automotive- or consumer-grade inertial sensors, the tracking sensitivity is 4 dB-Hz poorer than with tactical-grade sensors [21, 22]. This may be no more sensitive than stand-alone GNSS using vector tracking [25]. The receiver oscillator must also be of an adequate standard, particularly for coherent implementations [26].

A further limiting factor is the C/N_0 measurement algorithm (Section 9.2.6). The C/N_0 measurements (or equivalent parameters such as signal amplitude) are used to determine the correct weighting of the GNSS measurements in the integration filter (see Section 14.3.3). However, at the lowest signal-to-noise levels, C/N_0 measurements are very noisy and the design of the C/N_0 measurement algorithm has been shown to have a significant impact on tracking robustness [18].

14.2 System Model and State Selection

The system model of a Kalman filter is described in Section 3.2.3. For INS/GNSS integration, it depends on which quantities are estimated as Kalman filter states which, in turn, depends on the application, inertial sensors, integration architecture, and choice of coordinate frame. State selection is thus described first. A typical INS state-propagation model is then derived in an ECI frame, followed by descriptions of the ECEF-frame and local-navigation-frame equivalents. Modeling additional IMU errors, INS system noise modeling, and GNSS state propagation and system noise are then described. Discussion of state and error covariance initialization completes the section. In addition, wander-azimuth-frame and local-tangent-plane-frame state propagation models are described in Sections I.3 and I.4, respectively, of Appendix I on the CD.

There is no interaction between the INS and GNSS states in the system model; they only interact through the measurement model. Therefore, the system, transition, and system noise covariance matrices may be partitioned:

$$\mathbf{F} = \begin{pmatrix} \mathbf{F}_{INS} & 0 \\ 0 & \mathbf{F}_{GNSS} \end{pmatrix}, \quad \mathbf{\Phi} = \begin{pmatrix} \mathbf{\Phi}_{INS} & 0 \\ 0 & \mathbf{\Phi}_{GNSS} \end{pmatrix}, \quad \mathbf{Q} = \begin{pmatrix} \mathbf{Q}_{INS} & 0 \\ 0 & \mathbf{Q}_{GNSS} \end{pmatrix}, \quad (14.13)$$

where

$$\mathbf{x} = \begin{pmatrix} \mathbf{x}_{INS} \\ \mathbf{x}_{GNSS} \end{pmatrix}. \quad (14.14)$$

An error-state implementation is assumed here; for total-state INS/GNSS integration, the propagation of the position, velocity, and attitude errors should be replaced by the inertial navigation equations described in Chapter 5.

14.2.1 State Selection and Observability

All error-state INS/GNSS integration algorithms estimate the position and velocity errors. These may be expressed in an ECI frame as δr_{ib}^i and δv_{ib}^i, in an ECEF frame as δr_{eb}^e and δv_{eb}^e, or in a local navigation frame as δL_b, $\delta \lambda_b$, δh_b, and δv_{eb}^n. An ECI-frame implementation leads to the sparsest system matrix, which can thus be multiplied most efficiently. However, it is also good practice to match the coordinate frame used for the integration algorithm to that used for the inertial navigation equations (Chapter 5).

For all but the highest grades of INS, there is significant benefit in estimating the attitude error, which here is expressed as a small angle, resolved about the coordinate frame used for navigation, $\delta \psi_{\gamma b}^{\gamma}$. It may also be resolved about the INS body frame axes, giving $\delta \psi_{\gamma b}^{b}$ (see Section I.5 of Appendix I on the CD), or expressed as a quaternion [27] or rotation vector [2]. The small angle approximation is only suited to attitude errors in which closed-loop INS correction is applied or the INS is of a high grade. For example, applying this approximation to a 2° attitude error perturbs the state estimate by about 5%. The handling of large attitude errors is discussed in Section 14.4.4.

The choice of inertial instrument errors to estimate depends on their effect on the position, velocity, and attitude solution. If an IMU error has a significant impact on the navigation accuracy, it will be observable. Conversely, if its impact is much less than that of the random noise (Section 4.4.3), which cannot be calibrated, it will not be observable. Kalman filter state observability is discussed in general terms in Section 3.2.5. The deterministic observability of the inertial sensor errors, and also the attitude errors, depends on the user dynamics. The stochastic observability depends primarily on the IMU design; larger systematic errors are easier to observe, while high levels of random noise make systematic errors more difficult to observe. The Kalman filter measurement noise also impacts the stochastic observability of the states. Observability of INS error states is discussed more formally in [28]. Note that the observability is independent of the axes about which the states are resolved.

Except where they are very small, the accelerometer biases, $\mathbf{b}_a$, should always be estimated where the attitude errors are estimated. Conversely, the attitude errors should always be estimated where the accelerometer biases are estimated. Otherwise, the attitude error estimates are contaminated by the effects of the acceleration errors or vice versa. This is because, in INS/GNSS integration, the attitude and acceleration errors are observed through the growth in the velocity and position error they produce (see Section 14.3.4). As described in Section 5.7, both types of error lead to an initial linear growth in velocity error and quadratic growth in position error.

As (5.114) shows, the heading error only produces a velocity error when there is acceleration in the horizontal plane. Therefore, the navigation system's host must undergo significant maneuvering for the INS heading error to be observed and calibrated using GNSS measurements. When the navigation system is level and not accelerating, the vertical accelerometer bias is the only Kalman filter state that causes growth of the vertical velocity error. This makes it the most observable of the accelerometer biases and, as a consequence, an INS/GNSS navigation solution exhibits lower vertical drift than horizontal drift during GNSS outages of a few minutes.

The roll and pitch attitude errors and horizontal accelerometer biases are observed as linear combinations under conditions of constant acceleration and attitude. To fully separate the estimates of these states, the host-vehicle must turn, while a forward acceleration will separate the pitch error and forward accelerometer bias. The Kalman filter keeps a record of the correlations between its state estimates in the off-diagonal elements of its error covariance matrix, $\mathbf{P}$. This is used to separate the attitude and acceleration error estimates when their impact on the velocity error changes. Note that initialization of the roll and pitch solution using accelerometer leveling introduces correlations between the roll and pitch errors and all three accelerometer biases as described in Section 14.2.8.

The gyro biases, $\mathbf{b}_g$, are also estimated in most INS/GNSS integration algorithms. These are the only significant error sources that produce a quadratic growth in the velocity error with time. Consequently, the x- and y-axis gyro biases, which couple continuously into the velocity error (see Section 5.7.1), have better deterministic observability than the roll and pitch attitude errors and horizontal accelerometer biases as no maneuvers are required to separate them from other states. However, the stochastic observability is poorer as measurements must be made over time in order to distinguish the quadratic, linear, and random contributions to the velocity error growth. As with the attitude error, the z-axis (heading) gyro bias is more difficult to observe as a series of maneuvers that provide deterministic observability of the heading error must be performed.

Whether there is any benefit in estimating the accelerometer and gyro scale factor and cross-coupling errors (Section 4.4.2) or the gyro g-dependent biases (Section 4.4.4) depends on the accelerations and angular rates exhibited by the host vehicle as well as the magnitude of the errors. Maneuver-dependent INS error growth is discussed in Section 5.7.3. For most air, land, sea, and space applications, these errors are difficult to observe unless they are comparatively large. Exceptions are highly dynamic applications, such as motor sports, combat aircraft, and some guided weapons. Gyro scale factor and cross-coupling errors can also be significant for aircraft performing circling movements and roll-stabilized guided weapons.

A common mistake is to estimate scale factor errors, but not cross-coupling errors, as in most IMUs, they are of a similar magnitude. Estimating all of the scale factor and cross-coupling errors requires 15 states, while estimating all components of the gyro g-dependent biases requires a further nine states. Often, individual dynamics-dependent IMU errors are observable, but not the full set. Therefore, to make best use of the available processing capacity, simulations (or analytical sensitivity analysis) should be conducted in order to determine which errors should be estimated (see Section 3.3.5 and Appendix J on the CD).

Sometimes, where the observability of scale factor and cross-coupling errors is borderline, their inclusion in the Kalman filter state vector can improve the estimation of the accelerometer and gyro biases through the use of a more representative model of the IMU in the Kalman filter. Separating the biases into separate static and dynamic states (see Section 4.4.1) can have a similar effect [11]. Alternatively, these errors may be modeled as correlated system noise using a Schmidt-Kalman filter (Section D.2 of Appendix D on the CD).

The inertial navigation solution will also be subject to gravity-modeling errors. These produce acceleration biases aligned with the reference frame, as opposed to the vehicle body frame. The biases are of order 10^{-3} m s^{-2} per axis where a simple gravity model is used [1, 29] and of order 2×10^{-4} m s^{-2} per axis with a precision model [30]. However, estimation of the gravity-modeling errors as separate states is typically restricted to high-precision applications. Otherwise, their effects are absorbed by the z-axis accelerometer bias state and the components of the attitude error states resolved about the horizontal axes.

The choice of GNSS states to estimate depends on the integration architecture. In loosely coupled integration, no GNSS states are normally estimated. In tightly and deeply coupled integration, the receiver clock offset, $\delta\rho_c^a$, and drift, $\delta\dot{\rho}_c^a$, must normally be estimated as described in Section 9.4.2 for a stand-alone GNSS navigation filter. In an integrated INS/GNSS, the specific force on the receiver's reference oscillator is known, so the clock g-dependent errors may be estimated by the Kalman filter where they have an impact on system performance. Estimation of interconstellation timing biases may also be required in multiconstellation implementations.

In tightly coupled integration, GNSS measurements may be differenced across satellites. This cancels out the receiver clock errors enabling the clock states to be omitted, reducing the processor load. However, at least two satellites must then be tracked for GNSS measurements to be processable and any INS calibration to thus take place. Furthermore, the benefit of clock coasting to help limit position error growth during partial GNSS outages is lost.

In dual-frequency GNSS user equipment, the ionosphere propagation error is normally corrected by combining pseudo-range measurements made on different frequencies, as described in Section 9.3.2. Another way of performing the smoothing is to estimate the ionosphere propagation delays as Kalman filter states, inputting the pseudo-range measurements on different frequencies separately. In a tightly coupled architecture, this significantly increases the processor load without bringing performance benefits. In deeply coupled integration, it is the only way of performing dual-frequency ionosphere calibration. However, the ionosphere error states can be incorporated in the prefilters instead of the main integration filter. Either the total ionosphere propagation delays or the errors in ionosphere model predictions may be estimated.

As discussed in Section 9.4.2.1, the Kalman filter may also estimate the correlated range errors due to the ephemeris and the residual satellite clock, troposphere, and ionosphere errors. These range biases are partially observable where signals from more than four GNSS satellites are tracked. Their inclusion as states or correlated system noise leads to a more representative error covariance matrix, $\mathbf{P}$. In deeply coupled integration, range bias estimation can be used to keep the reference codes aligned with their respective signal correlation peaks where the range biases are a significant proportion of the code chip length. When range biases are not estimated, the position and clock offset uncertainties should be modified as described in Section 9.4.2.5.

If a short correlation time is modeled, range bias states could also be used to absorb NLOS reception errors, and possibly multipath errors, lessening their impact on the navigation solution.

Finally, if it is not practical to accurately measure the lever arm from the IMU to the GNSS antenna, $\mathbf{l}_{ba}^b$, this may also be estimated as Kalman filter states.

14.2.2 INS State Propagation in an Inertial Frame

A state propagation model is developed here for a Kalman filter estimating attitude, velocity, and position errors referenced to and resolved in an ECI frame, together with the accelerometer and gyro biases. The INS partition of the state vector comprises the following 15 states

$$\mathbf{x}_{INS}^i = \begin{pmatrix} \delta \boldsymbol{\psi}_{ib}^i \\ \delta \mathbf{v}_{ib}^i \\ \delta \mathbf{r}_{ib}^i \\ \mathbf{b}_a \\ \mathbf{b}_g \end{pmatrix}, \qquad (14.15)$$

where the superscript i is used to denote an ECI-frame implementation of the integration Kalman filter, consistent with the notation used in Section 9.4.2.

To obtain the INS system model, the time derivative of each Kalman filter state must be calculated. The attitude error propagation is derived first, followed by the velocity and position error propagation. The complete continuous-time system model is then presented, followed by a discussion of its discrete-time implementation in the form of a transition matrix.

In a real navigation system, the true values of the attitude, velocity, and other kinematic parameters are unknown. The best estimates available are the corrected INS-indicated parameters. Thus, for a generic parameter, y, the approximation $y \approx E(y) = \hat{y}$ must be made in deriving the system model. This also applies to the measurement model. Note that where closed-loop correction of INS errors is implemented, the raw and corrected INS outputs are the same, so for a generic parameter, y, $\tilde{y} = \hat{y}$.

14.2.2.1 Attitude Error Propagation

From (5.111), the time derivative of the small-angle attitude error, $\delta\psi_{ib}^i$, may be obtained by differentiating its coordinate transformation matrix counterpart:

$$[\delta\psi_{ib}^i \wedge] \approx \delta\dot{C}_b^i. \tag{14.16}$$

From (5.109),

$$\delta\dot{C}_b^i = \dot{\hat{C}}_b^i C_i^b + \tilde{C}_b^i \dot{C}_i^b, \tag{14.17}$$

while from (2.56)

$$\dot{C}_b^i = C_b^i \Omega_{ib}^b, \tag{14.18}$$

noting that Ω_{ib}^b is the skew-symmetric matrix of the angular rate, ω_{ib}^b. Substituting (14.18) and (14.17) into (14.16) gives

$$[\delta\psi_{ib}^i \wedge] \approx \tilde{C}_b^i \tilde{\Omega}_{ib}^b C_i^b + \tilde{C}_b^i C_i^b \Omega_{bi}^i. \tag{14.19}$$

Applying (2.51) and rearranging,

$$\tilde{C}_i^b [\delta\psi_{ib}^i \wedge] C_b^i \approx \tilde{\Omega}_{ib}^b - \Omega_{ib}^b. \tag{14.20}$$

A fundamental assumption of the Kalman filter is that the system model is a linear function of the states; otherwise a nonlinear estimation algorithm must be used. For a linear system model to be valid in practice, the product of any two Kalman filter states must be negligible. Thus, applying the approximation $\delta\psi_{ib}^i \delta\psi_{ib}^i \approx 0$,

$$\begin{aligned}\tilde{C}_i^b [\delta\psi_{ib}^i \wedge] C_b^i &\approx \hat{C}_i^b [\delta\psi_{ib}^i \wedge] \hat{C}_b^i \\ &= [(\hat{C}_i^b \delta\psi_{ib}^i) \wedge]\end{aligned}. \tag{14.21}$$

14.2 System Model and State Selection

Substituting this into (14.20), taking the components of the skew-symmetric matrices and rearranging gives

$$\begin{aligned}\delta\dot{\psi}^i_{ib} &\approx \hat{C}^i_b\left(\tilde{\omega}^b_{ib} - \omega^b_{ib}\right) \\ &= \hat{C}^i_b \delta\omega^b_{ib}\end{aligned} \quad (14.22)$$

The error in the gyro triad's angular-rate output, $\delta\omega^b_{ib}$, is given by (4.17) and (4.18), When the biases are the only gyro errors modeled as Kalman filter states, the expectation of (14.22) is

$$E\left(\delta\dot{\psi}^i_{ib}\right) \approx \hat{C}^i_b \mathbf{b}_g. \quad (14.23)$$

14.2.2.2 Velocity Error Propagation

From (5.18) and (5.19), the time derivative of the ECI-frame velocity is

$$\dot{\mathbf{v}}^i_{ib} = \mathbf{f}^i_{ib} + \gamma^i_{ib}. \quad (14.24)$$

Thus, the time derivative of the velocity error is

$$\delta\dot{\mathbf{v}}^i_{ib} = \tilde{\mathbf{f}}^i_{ib} - \mathbf{f}^i_{ib} + \tilde{\gamma}^i_{ib} - \gamma^i_{ib}. \quad (14.25)$$

The accelerometers measure specific force in body axes, so the error in $\tilde{\mathbf{f}}^i_{ib}$ is due to a mixture of accelerometer and attitude errors:

$$\tilde{\mathbf{f}}^i_{ib} - \mathbf{f}^i_{ib} = \tilde{C}^i_b \tilde{\mathbf{f}}^b_{ib} - C^i_b \mathbf{f}^b_{ib}. \quad (14.26)$$

Assuming the products of Kalman filter states may be neglected,

$$\begin{aligned}\tilde{\mathbf{f}}^i_{ib} - \mathbf{f}^i_{ib} &= \tfrac{1}{2}\left(\tilde{C}^i_b + C^i_b\right)\left(\tilde{\mathbf{f}}^b_{ib} - \mathbf{f}^b_{ib}\right) + \tfrac{1}{2}\left(\tilde{C}^i_b - C^i_b\right)\left(\tilde{\mathbf{f}}^b_{ib} + \mathbf{f}^b_{ib}\right) \\ &\approx \hat{C}^i_b\left(\tilde{\mathbf{f}}^b_{ib} - \mathbf{f}^b_{ib}\right) + \left(\tilde{C}^i_b - C^i_b\right)\hat{\mathbf{f}}^b_{ib}\end{aligned} \quad (14.27)$$

The error in the accelerometer triad's specific-force output, $\delta\mathbf{f}^b_{ib}$, is given by (4.16) and (4.18). When the biases are the only accelerometer errors modeled as Kalman filter states, the expectations of the accelerometer errors are

$$E\left(\tilde{\mathbf{f}}^b_{ib} - \mathbf{f}^b_{ib}\right) = E\left(\delta\mathbf{f}^b_{ib}\right) \approx \mathbf{b}_a. \quad (14.28)$$

Applying the small angle approximation to the attitude error, (5.109) and (5.111) give

$$\tilde{\mathbf{C}}_b^i - \mathbf{C}_b^i = (\delta\mathbf{C}_b^i - \mathbf{I}_3)\mathbf{C}_b^i \approx [\delta\boldsymbol{\psi}_{ib}^i \wedge]\mathbf{C}_b^i \qquad (14.29)$$

Turning to the gravitational term in (14.25), from Section 2.4.7, this scales with height roughly as [(2.138) repeated]:

$$\boldsymbol{\gamma}_{ib}^i \approx \frac{(r_{eS}^e(L_b))^2}{(r_{eS}^e(L_b) + h_b)^2} \boldsymbol{\gamma}_0^i(L_b),$$

where the geocentric radius at the Earth's surface, r_{eS}^e, is given by (2.137).

As the variation of gravitation with latitude is small, the effect of the latitude error on the assumed gravitational acceleration may be neglected. Making the additional assumption that $h_b \ll r_{eS}^e$ gives

$$\tilde{\boldsymbol{\gamma}}_{ib}^i - \boldsymbol{\gamma}_{ib}^i \approx -2\frac{(\tilde{h}_b - h_b)}{r_{eS}^e(\hat{L}_b)} \hat{\boldsymbol{\gamma}}_{ib}^i, \qquad (14.30)$$

where $\hat{\boldsymbol{\gamma}}_{ib}^i$ is the gravitational acceleration at the estimated position, $\hat{\mathbf{r}}_{ib}^i$. Assuming that the height error corresponds approximately to the position error component resolved along the Cartesian position vector from the center of the Earth gives

$$\tilde{\boldsymbol{\gamma}}_{ib}^i - \boldsymbol{\gamma}_{ib}^i \approx -\frac{2\hat{\boldsymbol{\gamma}}_{ib}^i}{r_{eS}^e(\hat{L}_b)} \frac{\hat{\mathbf{r}}_{ib}^{i\,\mathrm{T}}}{|\hat{\mathbf{r}}_{ib}^i|} \delta\mathbf{r}_{ib}^i. \qquad (14.31)$$

Substituting (14.28) and (14.29) into (14.27), and (14.27) and (14.31) into (14.25) gives the expectation of the time derivative of the velocity error in terms of the Kalman filter states:

$$\mathrm{E}(\delta\dot{\mathbf{v}}_{ib}^i) \approx -(\hat{\mathbf{C}}_b^i \hat{\mathbf{f}}_{ib}^b) \wedge \delta\boldsymbol{\psi}_{ib}^i - \frac{2\hat{\boldsymbol{\gamma}}_{ib}^i}{r_{eS}^e(\hat{L}_b)} \frac{\hat{\mathbf{r}}_{ib}^{i\,\mathrm{T}}}{|\hat{\mathbf{r}}_{ib}^i|} \delta\mathbf{r}_{ib}^i + \hat{\mathbf{C}}_b^i \mathbf{b}_a. \qquad (14.32)$$

14.2.2.3 Position Error Propagation

As the resolving axes and reference frame are the same, the time derivative of ECI-frame position is simply velocity, as shown by (5.22):

$$\dot{\mathbf{r}}_{ib}^i = \mathbf{v}_{ib}^i.$$

The time derivative of the position error is thus simply the velocity error:

$$\delta\dot{\mathbf{r}}_{ib}^i = \delta\mathbf{v}_{ib}^i. \qquad (14.33)$$

14.2.2.4 System and Transition Matrices

The accelerometer and gyro biases, $\mathbf{b}_a$ and $\mathbf{b}_g$, are assumed not to have a known time variation. Thus, their expected time derivatives are zero:

$$E(\dot{\mathbf{b}}_a) = 0, \qquad E(\dot{\mathbf{b}}_g) = 0. \tag{14.34}$$

Substituting (14.23) and (14.32) to (14.34) into (3.26), the system matrix, expressed in terms of 3×3 submatrices corresponding to the components of the state vector in (14.15), is

$$\mathbf{F}_{INS}^i = \begin{pmatrix} \mathbf{0}_3 & \mathbf{0}_3 & \mathbf{0}_3 & \mathbf{0}_3 & \hat{\mathbf{C}}_b^i \\ \mathbf{F}_{21}^i & \mathbf{0}_3 & \mathbf{F}_{23}^i & \hat{\mathbf{C}}_b^i & \mathbf{0}_3 \\ \mathbf{0}_3 & \mathbf{I}_3 & \mathbf{0}_3 & \mathbf{0}_3 & \mathbf{0}_3 \\ \mathbf{0}_3 & \mathbf{0}_3 & \mathbf{0}_3 & \mathbf{0}_3 & \mathbf{0}_3 \\ \mathbf{0}_3 & \mathbf{0}_3 & \mathbf{0}_3 & \mathbf{0}_3 & \mathbf{0}_3 \end{pmatrix}, \tag{14.35}$$

where

$$\mathbf{F}_{21}^i = \left[-\left(\hat{\mathbf{C}}_b^i \hat{\mathbf{f}}_{ib}^b\right) \wedge\right], \qquad \mathbf{F}_{23}^i = -\frac{2\hat{\boldsymbol{\gamma}}_{ib}^i \hat{\mathbf{r}}_{ib}^{i\,\mathrm{T}}}{r_{eS}^e(\hat{L}_b)\left|\hat{\mathbf{r}}_{ib}^i\right|}. \tag{14.36}$$

As shown in Section 3.2.3, the transition matrix, $\boldsymbol{\Phi}$, is a power-series expansion of $\mathbf{F}\tau_s$. In practice, the Kalman filter designer must determine which terms to include and which to neglect. This does not have to be a uniform truncation of the power-series (e.g., some second-order terms may be included and others neglected). Assuming that all zeroth- and first-order terms are retained, the maximum magnitude of each higher-order term should be estimated and a determination made as to whether its inclusion will have a significant effect on the integration algorithm performance. The shorter the state propagation interval, τ_s, the smaller the higher-order terms will be. To ensure that the power-series expansion converges, τ_s should be 1 second or less. When there is a long interval between measurement updates, such as in many loosely coupled implementations, it may be necessary to run the system propagation at shorter intervals or use a nonlinear filter, such as a UKF.

When the power series is limited to the first order in $\mathbf{F}\tau_s$, the transition matrix is

$$\boldsymbol{\Phi}_{INS}^i \approx \begin{bmatrix} \mathbf{I}_3 & \mathbf{0}_3 & \mathbf{0}_3 & \mathbf{0}_3 & \hat{\mathbf{C}}_b^i \tau_s \\ \mathbf{F}_{21}^i \tau_s & \mathbf{I}_3 & \mathbf{F}_{23}^i \tau_s & \hat{\mathbf{C}}_b^i \tau_s & \mathbf{0}_3 \\ \mathbf{0}_3 & \mathbf{I}_3 \tau_s & \mathbf{I}_3 & \mathbf{0}_3 & \mathbf{0}_3 \\ \mathbf{0}_3 & \mathbf{0}_3 & \mathbf{0}_3 & \mathbf{I}_3 & \mathbf{0}_3 \\ \mathbf{0}_3 & \mathbf{0}_3 & \mathbf{0}_3 & \mathbf{0}_3 & \mathbf{I}_3 \end{bmatrix}. \tag{14.37}$$

When terms of up to the third order in $\mathbf{F}\tau_s$ are included and $\left|F^i_{23,ij}\right|\tau_s^2 \ll 1$ for all i and j is assumed, it becomes

$$\boldsymbol{\Phi}^i_{INS} \approx \begin{bmatrix} \mathbf{I}_3 & \mathbf{0}_3 & \mathbf{0}_3 & \mathbf{0}_3 & \hat{\mathbf{C}}^i_b \tau_s \\ \mathbf{F}^i_{21}\tau_s & \mathbf{I}_3 & \mathbf{F}^i_{23}\tau_s & \hat{\mathbf{C}}^i_b \tau_s & \tfrac{1}{2}\mathbf{F}^i_{21}\hat{\mathbf{C}}^i_b\tau_s^2 \\ \tfrac{1}{2}\mathbf{F}^i_{21}\tau_s^2 & \mathbf{I}_3\tau_s & \mathbf{I}_3 & \tfrac{1}{2}\hat{\mathbf{C}}^i_b\tau_s^2 & \tfrac{1}{6}\mathbf{F}^i_{21}\hat{\mathbf{C}}^i_b\tau_s^3 \\ \mathbf{0}_3 & \mathbf{0}_3 & \mathbf{0}_3 & \mathbf{I}_3 & \mathbf{0}_3 \\ \mathbf{0}_3 & \mathbf{0}_3 & \mathbf{0}_3 & \mathbf{0}_3 & \mathbf{I}_3 \end{bmatrix}. \quad (14.38)$$

14.2.3 INS State Propagation in an Earth Frame

When the Kalman-filter-estimated attitude, velocity, and position errors are referenced to and resolved in an ECEF frame, the state vector becomes

$$\mathbf{x}^e_{INS} = \begin{pmatrix} \delta\boldsymbol{\psi}^e_{eb} \\ \delta\mathbf{v}^e_{eb} \\ \delta\mathbf{r}^e_{eb} \\ \mathbf{b}_a \\ \mathbf{b}_g \end{pmatrix}, \quad (14.39)$$

where the superscript e is used to denote an ECEF-frame implementation.

As shown in Section 5.3.1, the attitude propagation depends on the Earth rate as well as the gyro measurements. Determination of the attitude error derivative follows its ECI-frame counterpart from (14.16) to (14.22), giving

$$\delta\dot{\boldsymbol{\psi}}^e_{eb} \approx \hat{\mathbf{C}}^e_b \left(\tilde{\boldsymbol{\omega}}^b_{eb} - \boldsymbol{\omega}^b_{eb} \right). \quad (14.40)$$

Splitting this up into gyro measurement and Earth-rate terms, and then applying (5.109) and (5.111),

$$\begin{aligned} \delta\dot{\boldsymbol{\psi}}^e_{eb} &\approx \hat{\mathbf{C}}^e_b \delta\boldsymbol{\omega}^b_{ib} - \hat{\mathbf{C}}^e_b \left(\tilde{\mathbf{C}}^b_e - \mathbf{C}^b_e \right) \boldsymbol{\omega}^e_{ie} \\ &= \hat{\mathbf{C}}^e_b \delta\boldsymbol{\omega}^b_{ib} - \boldsymbol{\Omega}^e_{ie} \delta\boldsymbol{\psi}^e_{eb} \end{aligned}. \quad (14.41)$$

In terms of the Kalman filter states, the expectation of the rate of change of the attitude error is thus

$$\mathrm{E}\left(\delta\dot{\boldsymbol{\psi}}^e_{eb}\right) \approx -\boldsymbol{\Omega}^e_{ie}\delta\boldsymbol{\psi}^e_{eb} + \hat{\mathbf{C}}^e_b \mathbf{b}_g. \quad (14.42)$$

The rate of change of the Earth-referenced velocity is [repeating (5.35)]:

14.2 System Model and State Selection

$$\dot{\mathbf{v}}_{eb}^e = \mathbf{f}_{ib}^e + \mathbf{g}_b^e(\mathbf{r}_{eb}^e) - 2\mathbf{\Omega}_{ie}^e \mathbf{v}_{eb}^e.$$

Thus, the time derivative of the velocity error is

$$\delta\dot{\mathbf{v}}_{eb}^e = \tilde{\mathbf{f}}_{eb}^e - \mathbf{f}_{eb}^e + \mathbf{g}_b^e(\tilde{\mathbf{r}}_{eb}^e) - \mathbf{g}_b^e(\mathbf{r}_{eb}^e) - 2\mathbf{\Omega}_{ie}^e(\tilde{\mathbf{v}}_{eb}^e - \mathbf{v}_{eb}^e). \tag{14.43}$$

Compared to an ECI-frame implementation, this replaces the gravitational term with a gravity term and adds a Coriolis term. By analogy with Section 14.2.2.2, the expectation of the specific force term is

$$\mathrm{E}\left(\tilde{\mathbf{f}}_{ib}^e - \mathbf{f}_{ib}^e\right) \approx -\left(\hat{\mathbf{C}}_b^e \hat{\mathbf{f}}_{ib}^b\right) \wedge \delta\boldsymbol{\psi}_{eb}^e + \hat{\mathbf{C}}_b^e \mathbf{b}_a. \tag{14.44}$$

The gravity error may be expressed as

$$\begin{aligned}
\mathbf{g}_b^e(\tilde{\mathbf{r}}_{eb}^e) - \mathbf{g}_b^e(\mathbf{r}_{eb}^e) &= \tilde{\boldsymbol{\gamma}}_{ib}^e - \boldsymbol{\gamma}_{ib}^e - \mathbf{\Omega}_{ie}^e \mathbf{\Omega}_{ie}^e \delta\mathbf{r}_{eb}^e \\
&\approx -\frac{2\hat{\boldsymbol{\gamma}}_{ib}^e}{r_{eS}^e(\hat{L}_b)} \frac{\hat{\mathbf{r}}_{eb}^{eT}}{|\hat{\mathbf{r}}_{eb}^e|} \delta\mathbf{r}_{eb}^e,
\end{aligned} \tag{14.45}$$

where $\hat{\boldsymbol{\gamma}}_{ib}^e$ is given by (2.142), the assumption that $h_b \ll r_{eS}^e$ has been made, and the centrifugal term neglected as it is three orders of magnitude smaller near the Earth's surface.

Substituting (5.107), (14.44), and (14.45) into (14.43) gives the expectation of the time derivative of the velocity error in terms of the Kalman filter states

$$\mathrm{E}\left(\delta\dot{\mathbf{v}}_{eb}^e\right) \approx -\left(\hat{\mathbf{C}}_b^e \hat{\mathbf{f}}_{ib}^b\right) \wedge \delta\boldsymbol{\psi}_{eb}^e - 2\mathbf{\Omega}_{ie}^e \delta\mathbf{v}_{eb}^e - \frac{2\hat{\boldsymbol{\gamma}}_{ib}^e}{r_{eS}^e(\hat{L}_b)} \frac{\hat{\mathbf{r}}_{eb}^{eT}}{|\hat{\mathbf{r}}_{eb}^e|} \delta\mathbf{r}_{eb}^e + \hat{\mathbf{C}}_b^e \mathbf{b}_a. \tag{14.46}$$

The time derivative of the position error is the same as for an ECI-frame implementation. Thus,

$$\delta\dot{\mathbf{r}}_{eb}^e = \delta\mathbf{v}_{eb}^e. \tag{14.47}$$

Substituting (14.42), (14.46), (14.47), and (14.34) into (3.26), the system matrix is

$$\mathbf{F}_{INS}^e = \begin{pmatrix} -\mathbf{\Omega}_{ie}^e & \mathbf{0}_3 & \mathbf{0}_3 & \mathbf{0}_3 & \hat{\mathbf{C}}_b^e \\ \mathbf{F}_{21}^e & -2\mathbf{\Omega}_{ie}^e & \mathbf{F}_{23}^e & \hat{\mathbf{C}}_b^e & \mathbf{0}_3 \\ \mathbf{0}_3 & \mathbf{I}_3 & \mathbf{0}_3 & \mathbf{0}_3 & \mathbf{0}_3 \\ \mathbf{0}_3 & \mathbf{0}_3 & \mathbf{0}_3 & \mathbf{0}_3 & \mathbf{0}_3 \\ \mathbf{0}_3 & \mathbf{0}_3 & \mathbf{0}_3 & \mathbf{0}_3 & \mathbf{0}_3 \end{pmatrix}, \tag{14.48}$$

where

$$F^e_{21} = \left[-\left(\hat{C}^e_b \hat{f}^b_{ib}\right) \wedge\right], \quad F^e_{23} = -\frac{2\hat{\gamma}^e_{ib}}{r^e_{eS}(\hat{L}_b)} \frac{\hat{r}^{e\,T}_{eb}}{|\hat{r}^e_{eb}|}. \tag{14.49}$$

Moving from continuous to discrete time, the transition matrix, limited to the first order in $F\tau_s$, is

$$\Phi^e_{INS} \approx \begin{bmatrix} I_3 - \Omega^e_{ie}\tau_s & 0_3 & 0_3 & 0_3 & \hat{C}^e_b\tau_s \\ F^e_{21}\tau_s & I_3 - 2\Omega^e_{ie}\tau_s & F^e_{23}\tau_s & \hat{C}^e_b\tau_s & 0_3 \\ 0_3 & I_3\tau_s & I_3 & 0_3 & 0_3 \\ 0_3 & 0_3 & 0_3 & I_3 & 0_3 \\ 0_3 & 0_3 & 0_3 & 0_3 & I_3 \end{bmatrix}. \tag{14.50}$$

A version with terms of up to the third order in $F\tau_s$ is presented in Section I.1 of Appendix I on the CD. The first-order version of this system model is included in the MATLAB functions, LC_KF_Epoch and TC_KF_Epoch, on the CD.

14.2.4 INS State Propagation Resolved in a Local Navigation Frame

When the Kalman filter-estimated attitude and velocity are Earth-referenced and resolved in a local navigation frame, while the position error is expressed in terms of the latitude, longitude, and height, the state vector becomes

$$\mathbf{x}^n_{INS} = \begin{pmatrix} \delta\psi^n_{nb} \\ \delta\mathbf{v}^n_{eb} \\ \delta\mathbf{p}_b \\ \mathbf{b}_a \\ \mathbf{b}_g \end{pmatrix}, \quad \delta\mathbf{p}_b = \begin{pmatrix} \delta L_b \\ \delta \lambda_b \\ \delta h_b \end{pmatrix}, \tag{14.51}$$

where the superscript n denotes a local-navigation-frame implementation.

As shown in Section 5.4.1, the attitude propagation equations incorporate a transport-rate term in addition to the Earth-rate and gyro-measurement terms. Following its ECI-frame counterpart from (14.16) to (14.22), the attitude error derivative in a local navigation frame is

$$\delta\dot{\psi}^n_{nb} \approx \hat{C}^n_b\left(\tilde{\omega}^b_{nb} - \omega^b_{nb}\right). \tag{14.52}$$

Expanding this into gyro-measurement, Earth-rate, and transport-rate terms,

$$\delta\dot{\psi}^n_{nb} \approx \hat{C}^n_b\delta\omega^b_{ib} - \hat{C}^n_b\left(\tilde{\omega}^b_{ie} - \omega^b_{ie}\right) - \hat{C}^n_b\left(\tilde{\omega}^b_{en} - \omega^b_{en}\right). \tag{14.53}$$

14.2 System Model and State Selection

Expanding the Earth-rate and transport-rate terms, neglecting products of error states,

$$\hat{C}_b^n(\tilde{\omega}_{ie}^b - \omega_{ie}^b) + \hat{C}_b^n(\tilde{\omega}_{en}^b - \omega_{en}^b) \approx \hat{C}_b^n(\tilde{C}_n^b - C_n^b)(\hat{\omega}_{ie}^n + \hat{\omega}_{en}^n) + (\tilde{\omega}_{ie}^n - \omega_{ie}^n) + (\tilde{\omega}_{en}^n - \omega_{en}^n)$$
$$\approx \Omega_{in}^n \delta\psi_{nb}^n + (\tilde{\omega}_{ie}^n - \omega_{ie}^n) + (\tilde{\omega}_{en}^n - \omega_{en}^n).$$
(14.54)

From (2.123), the expectation of the Earth-rate error is

$$E(\tilde{\omega}_{ie}^n - \omega_{ie}^n) = -\omega_{ie}\begin{pmatrix} \sin\hat{L}_b \\ 0 \\ \cos\hat{L}_b \end{pmatrix}\delta L_b. \tag{14.55}$$

From (5.44), neglecting products of error states and the variation of the radii of curvature with latitude, the expectation of the transport-rate error is

$$E(\tilde{\omega}_{en}^n - \omega_{en}^n) \approx \begin{bmatrix} \delta v_{eb,E}^n/\left(R_E(\hat{L}_b) + \hat{h}_b\right) \\ -\delta v_{eb,N}^n/\left(R_N(\hat{L}_b) + \hat{h}_b\right) \\ -\delta v_{eb,E}^n \tan\hat{L}_b/\left(R_E(\hat{L}_b) + \hat{h}_b\right) \end{bmatrix} - \begin{pmatrix} 0 \\ 0 \\ 1 \end{pmatrix}\frac{\hat{v}_{eb,E}^n}{\left(R_E(\hat{L}_b) + \hat{h}_b\right)\cos^2\hat{L}_b}\delta L_b$$

$$+ \begin{bmatrix} -\hat{v}_{eb,E}^n/\left(R_E(\hat{L}_b) + \hat{h}_b\right)^2 \\ \hat{v}_{eb,N}^n/\left(R_N(\hat{L}_b) + \hat{h}_b\right)^2 \\ \hat{v}_{eb,E}^n \tan\hat{L}_b/\left(R_E(\hat{L}_b) + \hat{h}_b\right)^2 \end{bmatrix}\delta h_b$$
(14.56)

The rate of change of the velocity, v_{eb}^n, is [repeating (5.53)]:

$$\dot{v}_{eb}^n = f_{ib}^n + g_b^n(L_b, h_b) - \left(\Omega_{en}^n + 2\Omega_{ie}^n\right)v_{eb}^n,$$

adding a transport-rate term to the ECEF-frame equivalent. The time derivative of the velocity error is thus

$$\delta\dot{v}_{eb}^n = \tilde{f}_{ib}^n - f_{ib}^n + g_b^n(\tilde{L}_b, \tilde{h}_b) - g_b^n(L_b, h_b) - \left(\tilde{\Omega}_{en}^n + 2\tilde{\Omega}_{ie}^n\right)\tilde{v}_{eb}^n + \left(\Omega_{en}^n + 2\Omega_{ie}^n\right)v_{eb}^n. \quad (14.57)$$

By analogy with Section 14.2.2.2, the expectation of the specific force term is

$$\mathrm{E}\left(\tilde{\mathbf{f}}_{ib}^n - \mathbf{f}_{ib}^n\right) \approx -\left(\hat{\mathbf{C}}_b^n \hat{\mathbf{f}}_{ib}^b\right) \wedge \delta\boldsymbol{\psi}_{nb}^n + \hat{\mathbf{C}}_b^n \mathbf{b}_a. \tag{14.58}$$

The gravity error may be approximated to

$$\mathbf{g}_b^n\left(\tilde{L}_b, \tilde{h}_b\right) - \mathbf{g}_b^n(L_b, h_b) \approx -2\frac{g_0(\hat{L}_b)}{r_{eS}^e(\hat{L}_b)} \delta h_b \hat{\mathbf{u}}_D^n, \tag{14.59}$$

where $\hat{\mathbf{u}}_D^n$ is the local navigation frame down unit vector, $h_b \ll r_{eS}^e$ is assumed, and the centrifugal term and latitude-error dependence have been neglected.

For a linear system model to be valid, the product of any two Kalman filter states must be negligible. Therefore, the Earth-rate and transport-rate terms may be approximated to

$$\begin{aligned}\left(\tilde{\boldsymbol{\Omega}}_{en}^n + 2\tilde{\boldsymbol{\Omega}}_{ie}^n\right)\tilde{\mathbf{v}}_{eb}^n - \left(\boldsymbol{\Omega}_{en}^n + 2\boldsymbol{\Omega}_{ie}^n\right)\mathbf{v}_{eb}^n &\approx \left(\tilde{\boldsymbol{\Omega}}_{en}^n - \boldsymbol{\Omega}_{en}^n\right)\hat{\mathbf{v}}_{eb}^n + 2\left(\tilde{\boldsymbol{\Omega}}_{ie}^n - \boldsymbol{\Omega}_{ie}^n\right)\hat{\mathbf{v}}_{eb}^n + \left(\hat{\boldsymbol{\Omega}}_{en}^n + 2\hat{\boldsymbol{\Omega}}_{ie}^n\right)\left(\tilde{\mathbf{v}}_{eb}^n - \mathbf{v}_{eb}^n\right) \\ &= \left(\hat{\boldsymbol{\Omega}}_{en}^n + 2\hat{\boldsymbol{\Omega}}_{ie}^n\right)\delta\mathbf{v}_{eb}^n - \hat{\mathbf{v}}_{eb}^n \wedge \left(\tilde{\boldsymbol{\omega}}_{en}^n - \boldsymbol{\omega}_{en}^n\right) - 2\hat{\mathbf{v}}_{eb}^n \wedge \left(\tilde{\boldsymbol{\omega}}_{ie}^n - \boldsymbol{\omega}_{ie}^n\right)\end{aligned} \tag{14.60}$$

Substituting (5.107), (14.58), (14.59), and (14.60) into (14.57) gives the expectation of the time derivative of the velocity error in terms of the Kalman filter states:

$$\begin{aligned}\mathrm{E}\left(\delta\dot{\mathbf{v}}_{eb}^n\right) \approx & -\left(\hat{\mathbf{C}}_b^n \hat{\mathbf{f}}_{ib}^b\right) \wedge \delta\boldsymbol{\psi}_{nb}^n - \left(\hat{\boldsymbol{\Omega}}_{en}^n + 2\hat{\boldsymbol{\Omega}}_{ie}^n\right)\delta\mathbf{v}_{eb}^n + \hat{\mathbf{v}}_{eb}^n \wedge \mathrm{E}\left(\tilde{\boldsymbol{\omega}}_{en}^n - \boldsymbol{\omega}_{en}^n\right) \\ & + 2\hat{\mathbf{v}}_{eb}^n \wedge \mathrm{E}\left(\tilde{\boldsymbol{\omega}}_{ie}^n - \boldsymbol{\omega}_{ie}^n\right) - 2\frac{g_0(\hat{L}_b)}{r_{eS}^e(\hat{L}_b)} \hat{\mathbf{u}}_D^n \delta h_b + \hat{\mathbf{C}}_b^n \mathbf{b}_a\end{aligned} \tag{14.61}$$

From (2.111), the expectation of the time derivative of the position error, neglecting products of error states and the variation of the radii of curvature with latitude, is

$$\begin{aligned}\mathrm{E}\left(\delta\dot{L}_b\right) &= \frac{\delta v_{eb,N}^n}{R_N(\hat{L}_b) + \hat{h}_b} - \frac{\hat{v}_{eb,N}^n \delta h_b}{\left(R_N(\hat{L}_b) + \hat{h}_b\right)^2} \\ \mathrm{E}\left(\delta\dot{\lambda}_b\right) &\approx \frac{\delta v_{eb,E}^n}{\left(R_E(\hat{L}_b) + \hat{h}_b\right)\cos\hat{L}_b} + \frac{\hat{v}_{eb,E}^n \sin\hat{L}_b \delta L_b}{\left(R_E(\hat{L}_b) + \hat{h}_b\right)\cos^2\hat{L}_b} - \frac{\hat{v}_{eb,E}^n \delta h_b}{\left(R_E(\hat{L}_b) + \hat{h}_b\right)^2 \cos\hat{L}_b} \\ \delta\dot{h}_b &\approx -\delta v_{eb,D}^n\end{aligned} \tag{14.62}$$

Substituting (14.53) to (14.56), (14.61), (14.62), and (14.34) into (3.26), the system matrix is

14.2 System Model and State Selection

$$\mathbf{F}_{INS}^n = \begin{pmatrix} \mathbf{F}_{11}^n & \mathbf{F}_{12}^n & \mathbf{F}_{13}^n & \mathbf{0}_3 & \hat{\mathbf{C}}_b^n \\ \mathbf{F}_{21}^n & \mathbf{F}_{22}^n & \mathbf{F}_{23}^n & \hat{\mathbf{C}}_b^n & \mathbf{0}_3 \\ \mathbf{0}_3 & \mathbf{F}_{32}^n & \mathbf{F}_{33}^n & \mathbf{0}_3 & \mathbf{0}_3 \\ \mathbf{0}_3 & \mathbf{0}_3 & \mathbf{0}_3 & \mathbf{0}_3 & \mathbf{0}_3 \\ \mathbf{0}_3 & \mathbf{0}_3 & \mathbf{0}_3 & \mathbf{0}_3 & \mathbf{0}_3 \end{pmatrix}, \tag{14.63}$$

where

$$\mathbf{F}_{11}^n = -\left[\hat{\boldsymbol{\omega}}_{in}^n \wedge\right], \tag{14.64}$$

$$\mathbf{F}_{12}^n = \begin{bmatrix} 0 & \dfrac{-1}{R_E(\hat{L}_b) + \hat{h}_b} & 0 \\ \dfrac{1}{R_N(\hat{L}_b) + \hat{h}_b} & 0 & 0 \\ 0 & \dfrac{\tan \hat{L}_b}{R_E(\hat{L}_b) + \hat{h}_b} & 0 \end{bmatrix}, \tag{14.65}$$

$$\mathbf{F}_{13}^n = \begin{bmatrix} \omega_{ie} \sin \hat{L}_b & 0 & \dfrac{\hat{v}_{eb,E}^n}{\left(R_E(\hat{L}_b) + \hat{h}_b\right)^2} \\ 0 & 0 & \dfrac{-\hat{v}_{eb,N}^n}{\left(R_N(\hat{L}_b) + \hat{h}_b\right)^2} \\ \omega_{ie} \cos \hat{L}_b + \dfrac{\hat{v}_{eb,E}^n}{\left(R_E(\hat{L}_b) + \hat{h}_b\right)\cos^2 \hat{L}_b} & 0 & \dfrac{-\hat{v}_{eb,E}^n \tan \hat{L}_b}{\left(R_E(\hat{L}_b) + \hat{h}_b\right)^2} \end{bmatrix}, \tag{14.66}$$

$$\mathbf{F}_{21}^n = -\left[\left(\hat{\mathbf{C}}_b^n \hat{\mathbf{f}}_{ib}^b\right) \wedge\right], \tag{14.67}$$

$$\mathbf{F}_{22}^n = \begin{bmatrix} \dfrac{\hat{v}_{eb,D}^n}{R_N(\hat{L}_b) + \hat{h}_b} & -\dfrac{2\hat{v}_{eb,E}^n \tan \hat{L}_b}{R_E(\hat{L}_b) + \hat{h}_b} - 2\omega_{ie} \sin \hat{L}_b & \dfrac{\hat{v}_{eb,N}^n}{R_N(\hat{L}_b) + \hat{h}_b} \\ \dfrac{\hat{v}_{eb,E}^n \tan \hat{L}_b}{R_E(\hat{L}_b) + \hat{h}_b} + 2\omega_{ie} \sin \hat{L}_b & \dfrac{\hat{v}_{eb,N}^n \tan \hat{L}_b + \hat{v}_{eb,D}^n}{R_E(\hat{L}_b) + \hat{h}_b} & \dfrac{\hat{v}_{eb,E}^n}{R_E(\hat{L}_b) + \hat{h}_b} + 2\omega_{ie} \cos \hat{L}_b \\ -\dfrac{2\hat{v}_{eb,N}^n}{R_N(\hat{L}_b) + \hat{h}_b} & -\dfrac{2\hat{v}_{eb,E}^n}{R_E(\hat{L}_b) + \hat{h}_b} - 2\omega_{ie} \cos \hat{L}_b & 0 \end{bmatrix},$$

$$\tag{14.68}$$

$$F_{23}^n = \begin{bmatrix} -\dfrac{(\hat{v}_{eb,E}^n)^2 \sec^2 \hat{L}_b}{R_E(\hat{L}_b) + \hat{h}_b} - 2\hat{v}_{eb,E}^n \omega_{ie} \cos \hat{L}_b & 0 & \dfrac{(\hat{v}_{eb,E}^n)^2 \tan \hat{L}_b}{\left(R_E(\hat{L}_b) + \hat{h}_b\right)^2} - \dfrac{\hat{v}_{eb,N}^n \hat{v}_{eb,D}^n}{\left(R_N(\hat{L}_b) + \hat{h}_b\right)^2} \\ \begin{pmatrix} \dfrac{\hat{v}_{eb,N}^n \hat{v}_{eb,E}^n \sec^2 \hat{L}_b}{R_E(\hat{L}_b) + \hat{h}_b} + 2\hat{v}_{eb,N}^n \omega_{ie} \cos \hat{L}_b \\ -2\hat{v}_{eb,D}^n \omega_{ie} \sin \hat{L}_b \end{pmatrix} & 0 & -\dfrac{\hat{v}_{eb,N}^n \hat{v}_{eb,E}^n \tan \hat{L}_b + \hat{v}_{eb,E}^n \hat{v}_{eb,D}^n}{\left(R_E(\hat{L}_b) + \hat{h}_b\right)^2} \\ 2\hat{v}_{eb,E}^n \omega_{ie} \sin \hat{L}_b & 0 & \dfrac{(\hat{v}_{eb,E}^n)^2}{\left(R_E(\hat{L}_b) + \hat{h}_b\right)^2} + \dfrac{(\hat{v}_{eb,N}^n)^2}{\left(R_N(\hat{L}_b) + \hat{h}_b\right)^2} - \dfrac{2g_0(\hat{L}_b)}{r_{eS}^e(\hat{L}_b)} \end{bmatrix}, \tag{14.69}$$

$$F_{32}^n = \begin{bmatrix} \dfrac{1}{R_N(\hat{L}_b) + \hat{h}_b} & 0 & 0 \\ 0 & \dfrac{1}{\left(R_E(\hat{L}_b) + \hat{h}_b\right) \cos \hat{L}_b} & 0 \\ 0 & 0 & -1 \end{bmatrix}, \tag{14.70}$$

$$F_{33}^n = \begin{bmatrix} 0 & 0 & -\dfrac{\hat{v}_{eb,N}^n}{\left(R_N(\hat{L}_b) + \hat{h}_b\right)^2} \\ \dfrac{\hat{v}_{eb,E}^n \sin \hat{L}_b}{\left(R_E(\hat{L}_b) + \hat{h}_b\right) \cos^2 \hat{L}_b} & 0 & -\dfrac{\hat{v}_{eb,E}^n}{\left(R_E(\hat{L}_b) + \hat{h}_b\right)^2 \cos \hat{L}_b} \\ 0 & 0 & 0 \end{bmatrix}. \tag{14.71}$$

Moving from continuous to discrete time, the transition matrix, limited to first order in $F\tau_s$, is

$$\Phi_{INS}^n \approx \begin{bmatrix} I_3 + F_{11}^n \tau_s & F_{12}^n \tau_s & F_{13}^n \tau_s & 0_3 & \hat{C}_b^n \tau_s \\ F_{21}^n \tau_s & I_3 + F_{22}^n \tau_s & F_{23}^n \tau_s & \hat{C}_b^n \tau_s & 0_3 \\ 0_3 & F_{32}^n \tau_s & I_3 + F_{33}^n \tau_s & 0_3 & 0_3 \\ 0_3 & 0_3 & 0_3 & I_3 & 0_3 \\ 0_3 & 0_3 & 0_3 & 0_3 & I_3 \end{bmatrix}. \tag{14.72}$$

A version with terms of up to third order in $F\tau_s$ is presented in Section I.1 of Appendix I on the CD.

When consumer-grade inertial sensors are used, the gyro errors will greatly exceed the Earth rate and transport rate, enabling these terms to be neglected in

14.2 System Model and State Selection

the system model [31]. The approximated system matrix is similar to its ECI-frame counterpart:

$$\mathbf{F}_{INS}^{\prime n} \approx \begin{pmatrix} \mathbf{0}_3 & \mathbf{0}_3 & \mathbf{0}_3 & \mathbf{0}_3 & \hat{\mathbf{C}}_b^n \\ \mathbf{F}_{21}^n & \mathbf{0}_3 & \mathbf{F}_{23}^{\prime n} & \hat{\mathbf{C}}_b^n & \mathbf{0}_3 \\ \mathbf{0}_3 & \mathbf{F}_{32}^n & \mathbf{0}_3 & \mathbf{0}_3 & \mathbf{0}_3 \\ \mathbf{0}_3 & \mathbf{0}_3 & \mathbf{0}_3 & \mathbf{0}_3 & \mathbf{0}_3 \\ \mathbf{0}_3 & \mathbf{0}_3 & \mathbf{0}_3 & \mathbf{0}_3 & \mathbf{0}_3 \end{pmatrix} \quad \mathbf{F}_{23}^{\prime n} = -\frac{2g_0(\hat{L}_b)}{r_{eS}^e(\hat{L}_b)} \begin{pmatrix} 0 & 0 & 0 \\ 0 & 0 & 0 \\ 0 & 0 & 1 \end{pmatrix}, \quad (14.73)$$

Section I.2 of Appendix I on the CD presents an alternative formulation of local-navigation-frame INS/GNSS integration with Cartesian position error states, while Section I.3 presents a wander-azimuth-frame implementation.

14.2.5 Additional IMU Error States

Where estimated, the scale-factor, cross-coupling, and g-dependent errors are typically modeled with no deterministic time variation. Thus,

$$E(\dot{\mathbf{M}}_a) = \mathbf{0}_3, \qquad E(\dot{\mathbf{M}}_g) = \mathbf{0}_3, \qquad E(\dot{\mathbf{G}}_g) = \mathbf{0}_3, \quad (14.74)$$

where $\mathbf{M}_a$ is the accelerometer scale-factor and cross-coupling error matrix, $\mathbf{M}_g$ is the gyro scale-factor and cross-coupling error matrix, and $\mathbf{G}_g$ is the gyro g-dependent error matrix, as defined in Section 4.4.

The additional IMU error states lead to additional terms in the attitude and velocity error models as follows:

$$E(\delta\dot{\boldsymbol{\psi}}_{\gamma b}^\gamma) \to E(\delta\dot{\boldsymbol{\psi}}_{\gamma b}^\gamma) + \hat{\mathbf{C}}_b^\gamma \mathbf{M}_g \hat{\boldsymbol{\omega}}_{ib}^b + \hat{\mathbf{C}}_b^\gamma \mathbf{G}_g \hat{\mathbf{f}}_{ib}^b, \quad (14.75)$$

$$E(\delta\dot{\mathbf{v}}_{\beta b}^\gamma) \to E(\delta\dot{\mathbf{v}}_{\beta b}^\gamma) + \hat{\mathbf{C}}_b^\gamma \mathbf{M}_a \hat{\mathbf{f}}_{ib}^b, \quad (14.76)$$

where γ and β are defined by (14.5).

When the accelerometer scale-factor and cross-coupling errors and gyro g-dependent errors are not estimated as Kalman filter states, the product of these errors with the reaction to gravity (i.e., the average specific force) will perturb the accelerometer and gyro bias estimates. This need not be a problem, particularly for low-dynamics applications, provided the initial uncertainties of the bias states are sufficient.

The accelerometer and gyro biases may be split into separate static and dynamic states using [repeating (4.13)]

$$\begin{aligned} \mathbf{b}_a &= \mathbf{b}_{as} + \mathbf{b}_{ad} \\ \mathbf{b}_g &= \mathbf{b}_{gs} + \mathbf{b}_{gd} \end{aligned}.$$

The static bias states, $\mathbf{b}_{as}$ and $\mathbf{b}_{gs}$, are modeled with zero known time variation:

$$E(\dot{\mathbf{b}}_{as}) = 0, \qquad E(\dot{\mathbf{b}}_{gs}) = 0, \tag{14.77}$$

whereas the dynamic bias states, $\mathbf{b}_{ad}$ and $\mathbf{b}_{gd}$, should be modeled as exponentially correlated fixed-variance first-order Markov processes (see Section B.4.3 of Appendix B on the CD). Their deterministic time variation is then

$$E(\dot{\mathbf{b}}_{ad}) = -\frac{\mathbf{b}_{ad}}{\tau_{bad}}, \qquad E(\dot{\mathbf{b}}_{gd}) = -\frac{\mathbf{b}_{gd}}{\tau_{bgd}}, \tag{14.78}$$

where τ_{bad} and τ_{bgd} are the correlation times of the dynamic accelerometer and gyro biases, respectively. Note that these states will absorb vibration rectification errors (see Section 4.4.5) as well as variation in the sensor biases themselves.

Modeling of selected higher order IMU errors is described in [3].

When gravity-modeling errors are estimated, they should also be modeled as exponentially correlated fixed-variance first-order Markov processes with velocity-dependent correlation times to account for the fact that these errors are spatially, as opposed to temporally, correlated. When a precision gravity model is used, the local variation in gravity not accounted for by the model has a standard deviation of about 2×10^{-4} m s^{-2} per axis and a correlation length of about 40 km [30].

14.2.6 INS System Noise

The main sources of system noise on the inertial navigation solution are random walk of the velocity error due to noise on the accelerometer specific-force measurements and random walk of the attitude error due to noise on the gyro angular-rate measurements. IMU random noise is discussed in Section 4.4.3. In addition, where separate accelerometer and gyro dynamic bias states are not estimated, the in-run variation of the accelerometer and gyro biases (Section 4.4.1) can be approximated as white noise.

From (3.43), the system noise covariance is obtained by integrating the power spectral densities of these noise sources over the state propagation interval, accounting for the deterministic system model. Thus, assuming 15 states are estimated as defined by (14.15), (14.39), or (14.51), the INS system noise covariance matrix, $\mathbf{Q}_{INS}$, is

$$\mathbf{Q}^{\gamma}_{INS} = \int_0^{\tau_s} \exp(\mathbf{F}^{\gamma}_{INS} t') \begin{pmatrix} S_{rg}\mathbf{I}_3 & \mathbf{0}_3 & \mathbf{0}_3 & \mathbf{0}_3 & \mathbf{0}_3 \\ \mathbf{0}_3 & S_{ra}\mathbf{I}_3 & \mathbf{0}_3 & \mathbf{0}_3 & \mathbf{0}_3 \\ \mathbf{0}_3 & \mathbf{0}_3 & \mathbf{0}_3 & \mathbf{0}_3 & \mathbf{0}_3 \\ \mathbf{0}_3 & \mathbf{0}_3 & \mathbf{0}_3 & S_{bad}\mathbf{I}_3 & \mathbf{0}_3 \\ \mathbf{0}_3 & \mathbf{0}_3 & \mathbf{0}_3 & \mathbf{0}_3 & S_{bgd}\mathbf{I}_3 \end{pmatrix} \exp(\mathbf{F}^{\gamma}_{INS}{}^T t') \, dt' \qquad \gamma \in i,e,n, \tag{14.79}$$

14.2 System Model and State Selection

where S_{rg}, S_{ra}, S_{bad}, and S_{bgd} are the power spectral densities of, respectively, the gyro random noise, accelerometer random noise, accelerometer bias variation, and gyro bias variation, and it is assumed that all gyros and all accelerometers have equal noise characteristics.

Assuming the PSDs are constant and substituting in (14.35), (14.48), and (14.63) give an exact system noise covariance of

$$\mathbf{Q}_{INS}^{\gamma} = \begin{pmatrix} \mathbf{Q}_{11} & \mathbf{Q}_{21}^{\gamma\,\mathrm{T}} & \mathbf{Q}_{31}^{\gamma\,\mathrm{T}} & \mathbf{0}_3 & \tfrac{1}{2}S_{bgd}\tau_s^2\hat{\mathbf{C}}_b^{\gamma} \\ \mathbf{Q}_{21}^{\gamma} & \mathbf{Q}_{22}^{\gamma} & \mathbf{Q}_{32}^{\gamma\,\mathrm{T}} & \tfrac{1}{2}S_{bad}\tau_s^2\hat{\mathbf{C}}_b^{\gamma} & \tfrac{1}{3}S_{bgd}\tau_s^3\mathbf{F}_{21}^{\gamma}\hat{\mathbf{C}}_b^{\gamma} \\ \mathbf{Q}_{31}^{\gamma} & \mathbf{Q}_{32}^{\gamma} & \mathbf{Q}_{33}^{\gamma} & \mathbf{Q}_{34}^{\gamma} & \mathbf{Q}_{35}^{\gamma} \\ \mathbf{0}_3 & \tfrac{1}{2}S_{bad}\tau_s^2\hat{\mathbf{C}}_{\gamma}^{b} & \mathbf{Q}_{34}^{\gamma\,\mathrm{T}} & S_{bad}\tau_s\mathbf{I}_3 & \mathbf{0}_3 \\ \tfrac{1}{2}S_{bgd}\tau_s^2\hat{\mathbf{C}}_{\gamma}^{b} & \tfrac{1}{3}S_{bgd}\tau_s^3\mathbf{F}_{21}^{\gamma\,\mathrm{T}}\hat{\mathbf{C}}_{\gamma}^{b} & \mathbf{Q}_{35}^{\gamma\,\mathrm{T}} & \mathbf{0}_3 & S_{bgd}\tau_s\mathbf{I}_3 \end{pmatrix}, \quad \gamma \in i,e,n, \tag{14.80}$$

where

$$\mathbf{Q}_{11} = \left(S_{rg}\tau_s + \tfrac{1}{3}S_{bgd}\tau_s^3\right)\mathbf{I}_3$$

$$\mathbf{Q}_{21}^{\gamma} = \left(\tfrac{1}{2}S_{rg}\tau_s^2 + \tfrac{1}{4}S_{bgd}\tau_s^4\right)\mathbf{F}_{21}^{\gamma}$$

$$\mathbf{Q}_{22}^{\gamma} = \left(S_{ra}\tau_s + \tfrac{1}{3}S_{bad}\tau_s^3\right)\mathbf{I}_3 + \left(\tfrac{1}{3}S_{rg}\tau_s^3 + \tfrac{1}{5}S_{bgd}\tau_s^5\right)\mathbf{F}_{21}^{\gamma}\mathbf{F}_{21}^{\gamma\,\mathrm{T}}$$

$$\mathbf{Q}_{31}^{i/e} = \left(\tfrac{1}{3}S_{rg}\tau_s^3 + \tfrac{1}{5}S_{bgd}\tau_s^5\right)\mathbf{F}_{21}^{i/e}$$

$$\mathbf{Q}_{31}^{n} = \left(\tfrac{1}{3}S_{rg}\tau_s^3 + \tfrac{1}{5}S_{bgd}\tau_s^5\right)\mathbf{T}_{r(n)}^{p}\mathbf{F}_{21}^{n}$$

$$\mathbf{Q}_{32}^{i/e} = \left(\tfrac{1}{2}S_{ra}\tau_s^2 + \tfrac{1}{4}S_{bad}\tau_s^4\right)\mathbf{I}_3 + \left(\tfrac{1}{4}S_{rg}\tau_s^4 + \tfrac{1}{6}S_{bgd}\tau_s^6\right)\mathbf{F}_{21}^{i/e}\mathbf{F}_{21}^{i/e\,\mathrm{T}}$$

$$\mathbf{Q}_{32}^{n} = \left(\tfrac{1}{2}S_{ra}\tau_s^2 + \tfrac{1}{4}S_{bad}\tau_s^4\right)\mathbf{T}_{r(n)}^{p} + \left(\tfrac{1}{4}S_{rg}\tau_s^4 + \tfrac{1}{6}S_{bgd}\tau_s^6\right)\mathbf{T}_{r(n)}^{p}\mathbf{F}_{21}^{n}\mathbf{F}_{21}^{n\,\mathrm{T}} \qquad \gamma \in i,e,n$$

$$\mathbf{Q}_{33}^{i/e} = \left(\tfrac{1}{3}S_{ra}\tau_s^3 + \tfrac{1}{5}S_{bad}\tau_s^5\right)\mathbf{I}_3 + \left(\tfrac{1}{5}S_{rg}\tau_s^5 + \tfrac{1}{7}S_{bgd}\tau_s^7\right)\mathbf{F}_{21}^{i/e}\mathbf{F}_{21}^{i/e\,\mathrm{T}}$$

$$\mathbf{Q}_{33}^{n} = \left(\tfrac{1}{3}S_{ra}\tau_s^3 + \tfrac{1}{5}S_{bad}\tau_s^5\right)\mathbf{T}_{r(n)}^{p}\mathbf{T}_{r(n)}^{p} + \left(\tfrac{1}{5}S_{rg}\tau_s^5 + \tfrac{1}{7}S_{bgd}\tau_s^7\right)\mathbf{T}_{r(n)}^{p}\mathbf{F}_{21}^{n}\mathbf{F}_{21}^{n\,\mathrm{T}}\mathbf{T}_{r(n)}^{p}$$

$$\mathbf{Q}_{34}^{i/e} = \tfrac{1}{3}S_{bad}\tau_s^3\hat{\mathbf{C}}_b^{\gamma}$$

$$\mathbf{Q}_{34}^{n} = \tfrac{1}{3}S_{bad}\tau_s^3\mathbf{T}_{r(n)}^{p}\hat{\mathbf{C}}_b^{n}$$

$$\mathbf{Q}_{35}^{i/e} = \tfrac{1}{4}S_{bgd}\tau_s^4\mathbf{F}_{21}^{i/e}\hat{\mathbf{C}}_b^{\gamma}$$

$$\mathbf{Q}_{35}^{n} = \tfrac{1}{4}S_{bgd}\tau_s^4\mathbf{T}_{r(n)}^{p}\mathbf{F}_{21}^{n}\hat{\mathbf{C}}_b^{n}$$

(14.81)

and $\mathbf{T}_{r(n)}^{p}$ is given by (2.119).

For small propagation intervals ($\tau_s \leq 0.2$ second), the system noise covariance matrix may be approximated to

$$\mathbf{Q}_{INS}^{\gamma} \approx \mathbf{Q}_{INS}^{\prime\gamma} = \begin{pmatrix} S_{rg}\mathbf{I}_3 & \mathbf{0}_3 & \mathbf{0}_3 & \mathbf{0}_3 & \mathbf{0}_3 \\ \mathbf{0}_3 & S_{ra}\mathbf{I}_3 & \mathbf{0}_3 & \mathbf{0}_3 & \mathbf{0}_3 \\ \mathbf{0}_3 & \mathbf{0}_3 & \mathbf{0}_3 & \mathbf{0}_3 & \mathbf{0}_3 \\ \mathbf{0}_3 & \mathbf{0}_3 & \mathbf{0}_3 & S_{bad}\mathbf{I}_3 & \mathbf{0}_3 \\ \mathbf{0}_3 & \mathbf{0}_3 & \mathbf{0}_3 & \mathbf{0}_3 & S_{bgd}\mathbf{I}_3 \end{pmatrix} \tau_s, \quad \gamma \in i,e,n. \quad (14.82)$$

This approximation may also be used in conjunction with (3.46) for longer propagation intervals. This version of the system noise covariance model is included in the MATLAB functions, LC_KF_Epoch and TC_KF_Epoch, on the CD.

If σ_{ra} is the standard deviation of the noise on the accelerometer specific-force measurements and σ_{rg} is the standard deviation of the noise on the gyro angular-rate measurements, then the PSDs of the accelerometer and gyro noise are

$$S_{ra} = \sigma_{ra}^2 \tau_i, \quad S_{rg} = \sigma_{rg}^2 \tau_i, \quad (14.83)$$

where τ_i is the interval between the input of successive accelerometer and gyro outputs to the inertial navigation equations. Note that vibration can significantly increase the effective sensor noise, particularly for MEMS sensors. The PSDs should either be sufficiently large to account for the maximum likely level of vibration or modeled as a function of the vibration level measured by the IMU.

Similarly, the bias variation PSDs are

$$S_{bad} = \frac{\sigma_{bad}^2}{\tau_{bad}}, \quad S_{bgd} = \frac{\sigma_{bgd}^2}{\tau_{bgd}}, \quad (14.84)$$

where σ_{bad} and σ_{bgd} are the standard deviations of the accelerometer and gyro dynamic biases, respectively. These should account for the VREs (see Section 4.4.5) as well as variation in the sensor biases themselves. Vibration-dependent models are an option.

Sufficient system noise must be modeled on the horizontal attitude error and vertical accelerometer bias states to account for the spatial variation of the gravity-modeling errors where these are not estimated as states.

When errors such as the accelerometer and gyro scale factor, cross-coupling, and g-dependent errors are not modeled either as states or correlated system noise, their effects may be rather crudely approximated by increasing the modeled accelerometer and gyro random noise. A similar approach may be adopted for higher-order sensor errors. To account for these errors more effectively, sensor noise PSDs may be modeled as functions of acceleration and angular rate. Note that the product of these errors sources with the reaction to gravity is usually absorbed by the bias states. To maintain Kalman filter stability, these white noise approximations must overbound the true impact on the Kalman filter states. When the higher-order IMU errors are estimated as states, a certain level of system noise on the states should be modeled to account for any in-run variation.

The system noise variance for a state, x_{mi}, modeled as an exponentially correlated fixed-variance first-order Markov process is

$$Q_{mi} = \left[1 - \exp(-2\tau_s/\tau_{mi})\right]\sigma_{mi}^2, \tag{14.85}$$

where σ_{mi} is the standard deviation of the state and τ_{mi} is the correlation time. This is applicable to estimation of the accelerometer and gyro dynamic biases.

The preceding discussion assumes that the PSDs are independent of frequency as a Kalman filter assumes that all noise sources are white. However, the true PSDs will vary with frequency. The assumed PSDs are therefore approximations.

An estimate of the system noise covariance can be obtained from the sensor manufacturer's specifications and knowledge of the dynamics and vibration likely to be encountered for the application in question. However, for best results, a semi-empirical approach should be adopted. This is particularly important where system noise is used to account for unmodeled systematic errors. Testing with data recorded under a range of representative conditions should be conducted to enable the overall integrated INS/GNSS performance to be compared with different system noise coefficients. Laboratory tests of the IMU under representative dynamics and vibration may also be conducted where suitable equipment is available.

Advanced system noise modeling is discussed in Section 14.4.5.

14.2.7 GNSS State Propagation and System Noise

When the receiver clock offset and drift are estimated, the state dynamics from (9.146) are

$$\frac{\partial}{\partial t}\delta\rho_c^a = \delta\dot{\rho}_c^a, \qquad \mathrm{E}\left(\frac{\partial}{\partial t}\delta\dot{\rho}_c^a\right) = 0, \tag{14.86}$$

giving

$$\Phi_{GNSS} = \begin{pmatrix} 1 & 0 \\ \tau_s & 1 \end{pmatrix}. \tag{14.87}$$

As described in Section 9.4.2.2, the exact system noise covariance for these states is

$$Q_{GNSS} = \begin{pmatrix} S_{c\phi}^a \tau_s + \frac{1}{3}S_{cf}^a \tau_s^3 & \frac{1}{2}S_{cf}^a \tau_s^2 \\ \frac{1}{2}S_{cf}^a \tau_s^2 & S_{cf}^a \tau_s \end{pmatrix}, \tag{14.88}$$

where S_{cf}^a is the receiver clock frequency drift PSD (in units of m^2 s^{-3}) and $S_{c\phi}^a$ is the phase drift PSD (in units of m^2 s^{-1}). For small propagation intervals and for use in conjunction with (3.46), this may be approximated to

$$\mathbf{Q}_{GNSS} \approx \mathbf{Q}'_{GNSS} = \begin{pmatrix} S^a_{c\phi}\tau_s & 0 \\ 0 & S^a_{cf}\tau_s \end{pmatrix}. \tag{14.89}$$

Typical values for a TCXO are $S^a_{cf} \approx 0.04$ m² s⁻³ and $S^a_{c\phi} \approx 0.01$ m² s⁻¹ [32]. This GNSS system model is included in the MATLAB functions, LC_KF_Epoch and TC_KF_Epoch, on the CD.

When the clock g-dependent error coefficients, $\mathbf{s}^a_{cg}$, are also estimated, the expectation of the clock offset derivative is

$$\mathrm{E}\left(\frac{\partial}{\partial t}\delta\rho^a_c\right) = \delta\dot{\rho}^a_c + \hat{\mathbf{f}}^{b\mathrm{T}}_{ib}\mathbf{s}^a_{cg}, \tag{14.90}$$

where it is assumed that the axes of the receiver's reference oscillator are fixed with respect to the IMU. The g-dependent error coefficients may be assumed constant, so there is no physical requirement to model system noise on these states. However, a small amount of system noise is useful for maintaining numerical stability.

When they are included, ionosphere and range-bias states should be modeled as exponentially correlated fixed-variance first-order Markov processes as described by (3.3), with the system noise given by (14.85). The standard deviation of ionosphere states should be modeled as a function of satellite elevation angle and local time as discussed in Section 9.3.2. A correlation time of around 30 minutes is suitable. An elevation-dependent standard deviation is also suitable for the range-bias states, while the choice of correlation time depends on whether they are intended primarily to capture the ephemeris, satellite clock, ionosphere, and troposphere errors, demanding a long correlation time, or the multipath and NLOS errors, demanding a much smaller correlation time. When both types of errors are important, separate states can be used.

14.2.8 State Initialization

In an error-state implementation of INS/GNSS integration, the state estimates should generally be initialized at zero. An exception is the GNSS receiver clock offset and drift estimates in tightly- and deeply-coupled integration, which should be initialized using a GNSS navigation solution (either single-epoch or filtered, see Section 9.4). This is to avoid numerical problems arising from large disparities between the different state uncertainties.

The initial error covariance, $\mathbf{P}_0$, should reflect the state uncertainties and any covariance between state residuals as accurately as possible. When the position, velocity, and receiver clock solution are initialized from GNSS, the GNSS solution error covariance matrix, where available, should be used to initialize the relevant components of $\mathbf{P}_0$. Otherwise, when the satellite lines of sight are known, the error covariance may be estimated from the geometry matrix, $\mathbf{H}_G$, (see Section 9.4.3) using

$$\mathrm{E}\left[\begin{pmatrix} \delta\delta\mathbf{r}^\gamma_{\beta b} \\ \delta\delta\rho^a_c \end{pmatrix}\begin{pmatrix} \delta\delta\mathbf{r}^\gamma_{\beta b} \\ \delta\delta\rho^a_c \end{pmatrix}^\mathrm{T}\right] = \left(\mathbf{H}^{\gamma\mathrm{T}}_G\mathbf{H}^\gamma_G\right)\sigma^2_\rho, \qquad \mathrm{E}\left[\begin{pmatrix} \delta\delta\mathbf{v}^\gamma_{\beta b} \\ \delta\delta\dot{\rho}^a_c \end{pmatrix}\begin{pmatrix} \delta\delta\mathbf{v}^\gamma_{\beta b} \\ \delta\delta\dot{\rho}^a_c \end{pmatrix}^\mathrm{T}\right] = \left(\mathbf{H}^{\gamma\mathrm{T}}_G\mathbf{H}^\gamma_G\right)\sigma^2_r, \tag{14.91}$$

14.2 System Model and State Selection

where σ_p and σ_r are, respectively, the estimated pseudo-range and pseudo-range rate error standard deviations, the leading δ denotes the state residual (see Section 3.2.1), and β and γ are given by (14.5). The choice of σ_p and σ_r depends on the GNSS user equipment design and operating environment. Suitably conservative values for most applications are $\sigma_p = 5\text{m}$ and $\sigma_r = 0.05 \text{ m s}^{-1}$.

Another common approach is to simply assume equal values for each component of the position and velocity uncertainty and zero off-diagonal covariance. Typical values are 10m for the clock offset uncertainty and each component of the position uncertainty and 0.1 m s^{-1} for the clock drift uncertainty and each component of the velocity uncertainty. For local-navigation-frame implementations, uncertainties a factor of two larger may be assumed for the vertical components.

From (2.118), the curvilinear position error covariance may be determined using

$$E\left[\begin{pmatrix} \delta\delta\mathbf{p}_b \\ \delta\delta\rho_c^a \end{pmatrix}\begin{pmatrix} \delta\delta\mathbf{p}_b \\ \delta\delta\rho_c^a \end{pmatrix}^T\right] = \begin{pmatrix} \mathbf{T}_{r(n)}^p & 0_{3\times1} \\ 0_{1\times3} & 1 \end{pmatrix} E\left[\begin{pmatrix} \delta\delta\mathbf{r}_{eb}^n \\ \delta\delta\rho_c^a \end{pmatrix}\begin{pmatrix} \delta\delta\mathbf{r}_{eb}^n \\ \delta\delta\rho_c^a \end{pmatrix}^T\right]\begin{pmatrix} \mathbf{T}_{r(n)}^p & 0_{3\times1} \\ 0_{1\times3} & 1 \end{pmatrix} \tag{14.92}$$

where $\mathbf{T}_{r(n)}^p$ is given by (2.119).

The position, velocity, and receiver clock states are easy to observe and converge quickly. Consequently, integration algorithm performance is not especially sensitive to the initial uncertainties of these states, provided they are not excessively large or small. The initial attitude and IMU error uncertainties have more impact.

Manufacturers' specifications may be used to guide the selection of the initial IMU error uncertainties. However, experimental tests should also be carried out as the sensor specifications are not always complete and the effective sensor performance can vary significantly with the vibration environment. Typical errors for different types of inertial sensor are discussed in Section 4.4.

Heading initialization techniques are described in Sections 5.6.2, 6.1, 15.1, and 15.2.1. Different techniques suit different applications and the accuracy varies with the environment and sensor quality. When the initial heading uncertainty exceeds about 2° or 35 mrad, large heading error Kalman filter implementations (Section 14.4.4) should be considered.

Roll and pitch attitude is typically initialized by accelerometer leveling (Section 5.6.2). Therefore, the initial roll and pitch uncertainties in radians are approximately the same as the accelerometer bias uncertainties in g units. However, the leveling process introduces correlation between the attitude errors and accelerometer biases. From (5.102), the initial roll and pitch errors, respectively, $\delta\phi_{nb}$ and $\delta\theta_{nb}$, may be expressed in terms of the accelerometer biases using

$$\begin{aligned} \delta\phi_{nb} &= \frac{f_{ib,z}^b b_{a,y} - f_{ib,y}^b b_{a,z}}{f_{ib,y}^{b\,2} + f_{ib,z}^{b\,2}} \\ \delta\theta_{nb} &= \frac{\left(f_{ib,y}^{b\,2} + f_{ib,z}^{b\,2}\right)b_{a,x} - f_{ib,x}^b f_{ib,y}^b b_{a,y} - f_{ib,x}^b f_{ib,z}^b b_{a,z}}{\left(f_{ib,x}^{b\,2} + f_{ib,y}^{b\,2} + f_{ib,z}^{b\,2}\right)\sqrt{f_{ib,y}^{b\,2} + f_{ib,z}^{b\,2}}} \end{aligned} \tag{14.93}$$

From (2.59), the small-angle attitude errors may be expressed in terms of the roll, pitch, and heading errors using

$$\delta\boldsymbol{\psi}_{\gamma b}^{\gamma} = \mathbf{C}_b^{\gamma} \mathbf{T}_{\psi}^{\omega} \begin{pmatrix} \delta\phi_{nb} \\ \delta\theta_{nb} \\ \delta\psi_{nb} \end{pmatrix}, \quad \mathbf{T}_{\psi}^{\omega} = \begin{pmatrix} 1 & 0 & -\sin\theta_{nb} \\ 0 & \cos\phi_{nb} & \sin\phi_{nb}\cos\theta_{nb} \\ 0 & -\sin\phi_{nb} & \cos\phi_{nb}\cos\theta_{nb} \end{pmatrix}. \quad (14.94)$$

Therefore, the initial attitude error and accelerometer bias covariance, accounting for the error correlations, is

$$E\left[\begin{pmatrix} \delta\delta\boldsymbol{\psi}_{\gamma b}^{\gamma} \\ \delta\mathbf{b}_a \end{pmatrix} \begin{pmatrix} \delta\delta\boldsymbol{\psi}_{\gamma b}^{\gamma} \\ \delta\mathbf{b}_a \end{pmatrix}^T \right] = \begin{pmatrix} \mathbf{C}_b^{\gamma}\mathbf{T}_{\psi}^{\omega} & \mathbf{0}_3 \\ \mathbf{0}_3 & \mathbf{I}_3 \end{pmatrix} \mathbf{A} \begin{pmatrix} \sigma_{ba}^2 \mathbf{I}_3 & \mathbf{0}_{3\times 1} \\ \mathbf{0}_{1\times 3} & \sigma_{\delta\psi}^2 \end{pmatrix} \mathbf{A}^T \begin{pmatrix} \mathbf{T}_{\psi}^{\omega T}\mathbf{C}_{\gamma}^b & \mathbf{0}_3 \\ \mathbf{0}_3 & \mathbf{I}_3 \end{pmatrix}$$

$$\mathbf{A} = \begin{pmatrix} 0 & \dfrac{f_{ib,z}^b}{f_{ib,y}^{b\,2} + f_{ib,z}^{b\,2}} & \dfrac{-f_{ib,y}^b}{f_{ib,y}^{b\,2} + f_{ib,z}^{b\,2}} & 0 \\ \dfrac{f_{yz}}{f_{xyz}^2} & \dfrac{-f_{ib,x}^b f_{ib,y}^b}{f_{xyz}^2 f_{yz}} & \dfrac{-f_{ib,x}^b f_{ib,z}^b}{f_{xyz}^2 f_{yz}} & 0 \\ 0 & 0 & 0 & 1 \\ 1 & 0 & 0 & 0 \\ 0 & 1 & 0 & 0 \\ 0 & 0 & 1 & 0 \end{pmatrix}, \quad \begin{aligned} f_{yz} &= \sqrt{f_{ib,y}^{b\,2} + f_{ib,z}^{b\,2}} \\ f_{xyz}^2 &= f_{ib,x}^{b\,2} + f_{ib,y}^{b\,2} + f_{ib,z}^{b\,2} \end{aligned}$$

(14.95)

where $\sigma_{\delta\psi}$ is the initial heading uncertainty and σ_{ba} is the standard deviation of each accelerometer bias.

When the INS is initialized using transfer alignment (Section 15.1) or quasi-stationary alignment (Section 15.2), the final error covariance matrix from the alignment algorithm should be used to initialize the relevant components of the INS/GNSS integration error covariance matrix. This should be increased to account for the error covariance of the alignment reference. When open-loop INS correction is implemented, the state estimates must also be transferred across.

14.3 Measurement Models

The measurement model of a Kalman filter is described in Section 3.2.4. In INS/GNSS integration, the differences between measurements output by the GNSS user equipment and predictions of those measurements from the inertial navigation solution are used to update the state vector. Which measurements are used depends on the integration architecture, so the loosely coupled measurement model is described first, followed by the tightly coupled model and models for deeply coupled integration.

14.3 Measurement Models

The section concludes with a discussion of how the attitude and instrument errors are estimated. A number of alternative formulations of the loosely coupled and tightly coupled measurement models may be found in Appendix I on the CD. The MATLAB functions on the CD, Loosely_coupled_INS_GNSS and Tightly_coupled_INS_GNSS, simulate loosely coupled and tightly coupled integration, respectively.

Consider a measurement, $\tilde{m}_G$, output by the GNSS user equipment and a prediction of that measurement, $\tilde{m}_I$, obtained from the raw inertial navigation solution (and the GNSS navigation data message, where appropriate). Estimates of the errors in these measurements, $\delta \hat{m}_G$ and $\delta \hat{m}_I$, can then be obtained from the Kalman filter state vector. There are then two ways in which these can legitimately be assembled into a Kalman filter measurement, z_G, and estimate thereof, $\hat{z}_G^-$. These are

$$z_G = \tilde{m}_G - \tilde{m}_I, \qquad \hat{z}_G^- = \delta \hat{m}_G - \delta \hat{m}_I \qquad (14.96)$$

and

$$z_G = \tilde{m}_G, \qquad \hat{z}_G^- = \tilde{m}_I - \delta \hat{m}_I + \delta \hat{m}_G. \qquad (14.97)$$

When closed-loop correction of the INS is implemented, the predicted measurement is obtained from the corrected inertial navigation solution and becomes $\hat{m}_I$. It may also be convenient to do this in an open-loop architecture, noting that $\hat{m}_I = \tilde{m}_I - \delta \hat{m}_I$. The options for the measurement and its estimate are then

$$z_G = \tilde{m}_G - \hat{m}_I, \qquad \hat{z}_G^- = \delta \hat{m}_G. \qquad (14.98)$$

and

$$z_G = \tilde{m}_G, \qquad \hat{z}_G^- = \hat{m}_I + \delta \hat{m}_G. \qquad (14.99)$$

For tightly coupled integration, an extended Kalman filter (Section 3.4.1) is needed for (14.97) and (14.99), but not generally for (14.96) and (14.98). However, the measurement innovation, δz_G^-, is the same in all cases:

$$\begin{aligned} \delta z_G^- &= \tilde{m}_G - \delta \hat{m}_G - \tilde{m}_I + \delta \hat{m}_I \\ &= \tilde{m}_G - \delta \hat{m}_G - \hat{m}_I \end{aligned} \qquad (14.100)$$

and may be computed directly. Therefore the distinction between a standard Kalman filter and an EKF is one of semantics rather than implementation. Here, the convention of expressing the measurement innovation directly is adopted. Note that the subscript G is used to distinguish GNSS measurement innovations from other types in multisensor integration (Chapter 16).

The INS-derived and GNSS-derived measurement data must have the same time of validity. Otherwise, contributions to the measurement innovations caused by time synchronization errors will corrupt the state estimates. This may be achieved by storing the inertial navigation solution and retrieving the version at the GNSS time of validity as described in Section 3.3.4. However, this requires the IMU and GNSS measurements to be synchronized to the same time base. The INS–GNSS timing offset may be determined by comparing the INS and GNSS velocity and/or position solutions during host vehicle maneuvers. When the offset is fixed, a manual comparison of graphs can be very effective. Fixed offsets will typically occur where the INS or IMU timing is synchronized to the GNSS user equipment. For variable offsets, iterative least squares may be used to estimate position, velocity and time offsets from a short sequence of data prior to initialization of the integration algorithm. The time offset may also be estimated as a Kalman filter state as described in Section I.6 of Appendix I on the CD.

Before inputting GNSS measurements to the integration algorithm, they should be tested for faults as described in Chapter 17. For high-integrity applications, measurements may be delayed to enable testing over multiple epochs before acceptance. However, this is at the expense of accuracy as inertial error growth between the time of validity of the measurements and the time when they are processed is only partially corrected. The processing of delayed measurements in a Kalman filter is described in Section 3.3.4.

14.3.1 Loosely Coupled Integration

Loosely coupled INS/GNSS integration uses the GNSS user equipment's position and velocity solution. Therefore, the measurement innovation vector comprises the difference between the GNSS and corrected inertial position and velocity solutions, accounting for the lever arm from the INS to the GNSS antenna, l_{ba}^b, which is assumed here to be well known. The coordinate frames for the measurement innovation should match those for the state vector. Thus,

$$\delta \mathbf{z}_{G,k}^{i-} = \begin{pmatrix} \hat{\mathbf{r}}_{iaG}^i - \hat{\mathbf{r}}_{ib}^i - \hat{\mathbf{C}}_b^i \mathbf{l}_{ba}^b \\ \hat{\mathbf{v}}_{iaG}^i - \hat{\mathbf{v}}_{ib}^i - \hat{\mathbf{C}}_b^i \left(\hat{\boldsymbol{\omega}}_{ib}^b \wedge \mathbf{l}_{ba}^b \right) \end{pmatrix}_k^-, \quad (14.101)$$

$$\delta \mathbf{z}_{G,k}^{e-} = \begin{pmatrix} \hat{\mathbf{r}}_{eaG}^e - \hat{\mathbf{r}}_{eb}^e - \hat{\mathbf{C}}_b^e \mathbf{l}_{ba}^b \\ \hat{\mathbf{v}}_{eaG}^e - \hat{\mathbf{v}}_{eb}^e - \hat{\mathbf{C}}_b^e \left(\hat{\boldsymbol{\omega}}_{ib}^b \wedge \mathbf{l}_{ba}^b \right) + \boldsymbol{\Omega}_{ie}^e \hat{\mathbf{C}}_b^e \mathbf{l}_{ba}^b \end{pmatrix}_k^-, \quad (14.102)$$

and

$$\delta \mathbf{z}_{G,k}^{n-} = \begin{pmatrix} \hat{\mathbf{p}}_{aG} - \hat{\mathbf{p}}_b - \hat{\mathbf{T}}_{r(n)}^p \hat{\mathbf{C}}_b^n \mathbf{l}_{ba}^b \\ \hat{\mathbf{v}}_{eaG}^n - \hat{\mathbf{v}}_{eb}^n - \hat{\mathbf{C}}_b^n \left(\hat{\boldsymbol{\omega}}_{ib}^b \wedge \mathbf{l}_{ba}^b \right) + \hat{\boldsymbol{\Omega}}_{ie}^n \hat{\mathbf{C}}_b^n \mathbf{l}_{ba}^b \end{pmatrix}_k^-, \quad (14.103)$$

where the subscript k denotes the measurement update iteration, superscripts i, e, and n respectively denote ECI-frame, ECEF-frame, and local-navigation-frame implementations, the subscript G denotes GNSS indicated; $\boldsymbol{\Omega}_{ie}^e$ and $\boldsymbol{\Omega}_{ie}^n$ are given by

14.3 Measurement Models

(5.25) and (5.41), respectively; and from (2.119), the Cartesian-to-curvilinear position change transformation matrix, $\hat{T}^p_{r(n)}$, is

$$\hat{T}^p_{r(n)} = \begin{pmatrix} \dfrac{1}{R_N(\hat{L}_b) + \hat{h}_b} & 0 & 0 \\ 0 & \dfrac{1}{\left(R_E(\hat{L}_b) + \hat{h}_b\right)\cos\hat{L}_b} & 0 \\ 0 & 0 & -1 \end{pmatrix}. \quad (14.104)$$

From (5.109) and (5.111),

$$\tilde{C}^\gamma_b \approx \left(I_3 + \left[\delta\psi^\gamma_{\gamma b} \wedge\right]\right)C^\gamma_b. \quad (14.105)$$

Substituting the attitude error states with their residual, $\delta\delta\psi^\gamma_{\gamma b}$ [see (3.4)],

$$\hat{C}^\gamma_b \approx \left(I_3 + \left[\delta\delta\psi^\gamma_{\gamma b} \wedge\right]\right)C^\gamma_b. \quad (14.106)$$

Therefore,

$$\hat{C}^\gamma_b l^b_{ba} \approx C^\gamma_b l^b_{ba} - \left[\left(C^\gamma_b l^b_{ba}\right) \wedge\right]\delta\delta\psi^\gamma_{\gamma b} \quad (14.107)$$

and

$$\begin{aligned}\hat{C}^\gamma_b\left(\hat{\omega}^b_{ib} \wedge l^b_{ba}\right) &\approx C^\gamma_b\left(\hat{\omega}^b_{ib} \wedge l^b_{ba}\right) - \left[\left\{C^\gamma_b\left(\hat{\omega}^b_{ib} \wedge l^b_{ba}\right)\right\} \wedge\right]\delta\delta\psi^\gamma_{\gamma b}\\ &\approx C^\gamma_b\left(\omega^b_{ib} \wedge l^b_{ba}\right) - \left[\left\{C^\gamma_b\left(\hat{\omega}^b_{ib} \wedge l^b_{ba}\right)\right\} \wedge\right]\delta\delta\psi^\gamma_{\gamma b} - C^\gamma_b\left(l^b_{ba} \wedge \delta b_g\right)\end{aligned}$$
$$(14.108)$$

The measurement matrix is given by (3.90):

$$H_k = \left.\dfrac{\partial h(x, t_k)}{\partial x}\right|_{x=\hat{x}^-_k} = \left.\dfrac{\partial z(x, t_k)}{\partial x}\right|_{x=\hat{x}^-_k}.$$

Taking the measurement vector (14.101) and the state vector (14.15), noting that no GNSS states are estimated in loosely coupled integration, and making use of (14.107) and (14.108), the measurement matrix for ECI-frame loosely coupled INS/GNSS integration is

$$H^i_{G,k} = \begin{pmatrix} H^i_{r1} & 0_3 & -I_3 & 0_3 & 0_3 \\ H^i_{v1} & -I_3 & 0_3 & 0_3 & H^i_{v5} \end{pmatrix}_k, \quad (14.109)$$

where

$$\begin{aligned}
\mathbf{H}_{r1}^i &\approx \left[\left(\hat{\mathbf{C}}_b^i \mathbf{l}_{ba}^b\right) \wedge\right] \\
\mathbf{H}_{v1}^i &\approx \left[\left\{\hat{\mathbf{C}}_b^i \left(\hat{\boldsymbol{\omega}}_{ib}^b \wedge \mathbf{l}_{ba}^b\right)\right\} \wedge\right], \\
\mathbf{H}_{v5}^i &\approx \hat{\mathbf{C}}_b^i \left[\mathbf{l}_{ba}^b \wedge\right]
\end{aligned} \qquad (14.110)$$

noting that the corrected INS-indicated attitude, $\hat{\mathbf{C}}_b^i$, is the best available estimate of the true attitude, $\mathbf{C}_b^i$.

Similarly, from (14.102), (14.39), (14.107), and (14.108), the ECEF-frame loosely coupled measurement matrix is

$$\mathbf{H}_{G,k}^e = \begin{pmatrix} \mathbf{H}_{r1}^e & \mathbf{0}_3 & -\mathbf{I}_3 & \mathbf{0}_3 & \mathbf{0}_3 \\ \mathbf{H}_{v1}^e & -\mathbf{I}_3 & \mathbf{0}_3 & \mathbf{0}_3 & \mathbf{H}_{v5}^e \end{pmatrix}_k, \qquad (14.111)$$

where

$$\begin{aligned}
\mathbf{H}_{r1}^e &\approx \left[\left(\hat{\mathbf{C}}_b^e \mathbf{l}_{ba}^b\right) \wedge\right] \\
\mathbf{H}_{v1}^e &\approx \left[\left\{\hat{\mathbf{C}}_b^e \left(\hat{\boldsymbol{\omega}}_{ib}^b \wedge \mathbf{l}_{ba}^b\right) - \boldsymbol{\Omega}_{ie}^e \hat{\mathbf{C}}_b^e \mathbf{l}_{ba}^b\right\} \wedge\right]. \\
\mathbf{H}_{v5}^e &\approx \hat{\mathbf{C}}_b^e \left[\mathbf{l}_{ba}^b \wedge\right]
\end{aligned} \qquad (14.112)$$

From (14.103), (14.51), (14.107), and (14.108), the loosely coupled measurement matrix for an Earth-referenced local-navigation-frame implementation is

$$\mathbf{H}_{G,k}^n = \begin{pmatrix} \mathbf{H}_{r1}^n & \mathbf{0}_3 & -\mathbf{I}_3 & \mathbf{0}_3 & \mathbf{0}_3 \\ \mathbf{H}_{v1}^n & -\mathbf{I}_3 & \mathbf{0}_3 & \mathbf{0}_3 & \mathbf{H}_{v5}^n \end{pmatrix}_k, \qquad (14.113)$$

where

$$\begin{aligned}
\mathbf{H}_{r1}^n &\approx \hat{\mathbf{T}}_{r(n)}^p \left[\left(\hat{\mathbf{C}}_b^n \mathbf{l}_{ba}^b\right) \wedge\right] \\
\mathbf{H}_{v1}^n &\approx \left[\left\{\hat{\mathbf{C}}_b^n \left(\hat{\boldsymbol{\omega}}_{ib}^b \wedge \mathbf{l}_{ba}^b\right) - \hat{\boldsymbol{\Omega}}_{ie}^n \hat{\mathbf{C}}_b^n \mathbf{l}_{ba}^b\right\} \wedge\right]. \\
\mathbf{H}_{v5}^n &\approx \hat{\mathbf{C}}_b^n \left[\mathbf{l}_{ba}^b \wedge\right]
\end{aligned} \qquad (14.114)$$

Except where the lever arm is very large, the Earth-rotation terms in (14.102), (14.103), (14.112), and (14.114) may be neglected.

In practice, the coupling of the attitude errors and gyro biases into the measurements through the lever arm terms is also weak. These states are mainly estimated through the change in the velocity error as described in Section 14.3.4. Therefore, the measurement matrices can often be approximated to

$$\mathbf{H}_{G,k}^{i/e/n} \approx \begin{pmatrix} \mathbf{0}_3 & \mathbf{0}_3 & -\mathbf{I}_3 & \mathbf{0}_3 & \mathbf{0}_3 \\ \mathbf{0}_3 & -\mathbf{I}_3 & \mathbf{0}_3 & \mathbf{0}_3 & \mathbf{0}_3 \end{pmatrix}_k, \qquad (14.115)$$

noting that the remaining lever arm terms in the measurement innovations, (14.101) to (14.103), must not be neglected. This approximate version of the ECEF-frame measurement model is included in the MATLAB function, LC_KF_Epoch, on the CD.

Measurement noise arises due to a combination of factors such as RF and thermal noise, dynamics response lag, multipath interference, and NLOS signal reception. This may then be smoothed within the GNSS navigation processor by an EKF or carrier-smoothing algorithm, reducing the magnitude of the noise, but increasing its correlation time. Ideally, the measurement noise covariance matrix, $\mathbf{R}_G$, should be based on the error covariance of the GNSS navigation solution (e.g., $\mathbf{P}$ for an EKF), enabling the GNSS data to be weighted according to the GNSS user equipment's own level of confidence. However, this information is rarely output in practice. The measurement noise covariance is often assumed to be constant, but is better modeled as a function of the measured carrier power to noise density, $\tilde{c}/\tilde{n}_0$, satellite signal geometry, and acceleration. Acceleration is relevant because the effects of INS-GNSS time synchronization errors are larger under high dynamics. The measurement noise covariance must also account for any flexure in the lever arm between the INS and the GNSS antenna, which can also be excited by dynamics.

When the measurement-update interval is shorter than the correlation time of the noise on the GNSS position and velocity solution, the measurement noise covariance assumed in the integration Kalman filter should be increased to downweight the measurements. Note that reducing the measurement-update interval can better capture the host-vehicle dynamics, improving the observability of some of the attitude and instrument error states.

A suitable value for a component of the measurement noise covariance is thus the variance of the position or velocity error multiplied by the ratio of the error correlation time to the measurement update interval. Typical values for a 1-Hz update interval are $(2\text{--}20\text{m})^2$ for position and $(0.1\text{--}1 \text{ m s}^{-1})^2$ for velocity, with the larger values used where GNSS reception is relatively poor, such as in urban areas.

In local-navigation-frame implementations of the loosely coupled measurement model, mixing latitude and longitude in radians with height in meters can cause numerical problems in the $\left(\mathbf{H}_k \mathbf{P}_k^- \mathbf{H}_k^T + \mathbf{R}_k\right)$ matrix inversion step of the measurement update. One solution is to rescale the latitude and longitude components, in which case, the measurement innovation, $\delta \mathbf{z}_{G,k}^{n-}$, and measurement matrix, $\mathbf{H}_{G,k}^n$, become

$$\delta \mathbf{z}_{G,k}^{n-} = \begin{pmatrix} \mathbf{S}_p \left(\hat{\mathbf{p}}_{aG} - \hat{\mathbf{p}}_b - \hat{\mathbf{T}}_{r(n)}^p \hat{\mathbf{C}}_b^n \mathbf{l}_{ba}^b \right) \\ \hat{\mathbf{v}}_{eaG}^n - \hat{\mathbf{v}}_{eb}^n - \hat{\mathbf{C}}_b^n \left(\hat{\boldsymbol{\omega}}_{ib}^b \wedge \mathbf{l}_{ba}^b \right) + \hat{\boldsymbol{\Omega}}_{ie}^n \hat{\mathbf{C}}_b^n \mathbf{l}_{ba}^b \end{pmatrix}_k, \qquad (14.116)$$

$$\mathbf{H}_{G,k}^{n} = \begin{pmatrix} \mathbf{S}_p \mathbf{H}_{r1}^{n} & \mathbf{0}_3 & -\mathbf{S}_p & \mathbf{0}_3 & \mathbf{0}_3 \\ \mathbf{H}_{v1}^{n} & -\mathbf{I}_3 & \mathbf{0}_3 & \mathbf{0}_3 & \mathbf{H}_{v5}^{n} \end{pmatrix}_k , \qquad (14.117)$$

$$\approx \begin{pmatrix} \mathbf{0}_3 & \mathbf{0}_3 & -\mathbf{S}_p & \mathbf{0}_3 & \mathbf{0}_3 \\ \mathbf{0}_3 & -\mathbf{I}_3 & \mathbf{0}_3 & \mathbf{0}_3 & \mathbf{0}_3 \end{pmatrix}_k$$

where $\mathbf{S}_p$ is the curvilinear position scaling matrix, given by

$$\mathbf{S}_p = \begin{pmatrix} s_{L\lambda} & 0 & 0 \\ 0 & s_{L\lambda} & 0 \\ 0 & 0 & 1 \end{pmatrix}, \qquad (14.118)$$

where a suitable value for $s_{L\lambda}$ is 10^3, which converts the latitude and longitude components of the measurement innovation from radians to milliradians.

14.3.2 Tightly Coupled Integration

Tightly coupled INS/GNSS integration uses the GNSS ranging processor's pseudo-range and pseudo-range-rate measurements, obtained from code and carrier tracking, respectively. The measurement model is thus based on that of the GNSS navigation filter, described in Section 9.4.2.3. The measurement innovation vector comprises the differences between the GNSS measured pseudo-range and pseudo-range rates and values predicted from the corrected inertial navigation solution at the same time of validity, estimated receiver clock offset and drift, and navigation-data-indicated satellite positions and velocities. Thus,

$$\delta \mathbf{z}_{G,k}^{-} = \begin{pmatrix} \delta \mathbf{z}_{\rho,k}^{-} \\ \delta \mathbf{z}_{r,k}^{-} \end{pmatrix}, \quad \begin{aligned} \delta \mathbf{z}_{\rho,k}^{-} &= \left(\tilde{\rho}_{a,C}^{1} - \hat{\rho}_{a,C}^{1-}, \tilde{\rho}_{a,C}^{2} - \hat{\rho}_{a,C}^{2-}, \cdots \tilde{\rho}_{a,C}^{m} - \hat{\rho}_{a,C}^{m-} \right)_k \\ \delta \mathbf{z}_{r,k}^{-} &= \left(\tilde{\dot{\rho}}_{a,C}^{1} - \hat{\dot{\rho}}_{a,C}^{1-}, \tilde{\dot{\rho}}_{a,C}^{2} - \hat{\dot{\rho}}_{a,C}^{2-}, \cdots \tilde{\dot{\rho}}_{a,C}^{m} - \hat{\dot{\rho}}_{a,C}^{m-} \right)_k \end{aligned} . \qquad (14.119)$$

For the jth measurement, comprising signal l from satellite s, the corrected pseudo-range and pseudo-range-rate measurements, $\tilde{\rho}_{a,C}^{j} \equiv \tilde{\rho}_{a,C}^{s,l}$ and $\tilde{\dot{\rho}}_{a,C}^{j} \equiv \tilde{\dot{\rho}}_{a,C}^{s,l}$, are given by [repeating (8.49)]

$$\tilde{\rho}_{a,C}^{s,l} = \tilde{\rho}_{a,R}^{s,l} - \delta \hat{\rho}_{I,a}^{s,l} - \delta \hat{\rho}_{T,a}^{s} + \delta \hat{\rho}_{c}^{s,l}$$
$$\tilde{\dot{\rho}}_{a,C}^{s,l} = \tilde{\dot{\rho}}_{a,R}^{s,l} + \delta \hat{\dot{\rho}}_{c}^{s}$$.

Their estimated counterparts are given by [repeating (9.161)]

$$\hat{\rho}_{a,C,k}^{j-} = \sqrt{\left[\hat{\mathbf{r}}_{ij}^{i}\left(\tilde{t}_{st,a,k}^{j}\right) - \hat{\mathbf{r}}_{ia,k}^{i-} \right]^{\mathrm{T}} \left[\hat{\mathbf{r}}_{ij}^{i}\left(\tilde{t}_{st,a,k}^{j}\right) - \hat{\mathbf{r}}_{ia,k}^{i-} \right]} + \delta \hat{\rho}_{c,k}^{a-}$$
$$\hat{\dot{\rho}}_{a,C,k}^{j-} = \hat{\mathbf{u}}_{aj,k}^{i-\mathrm{T}} \left[\hat{\mathbf{v}}_{ij}^{i}\left(\tilde{t}_{st,a,k}^{j}\right) - \hat{\mathbf{v}}_{ia,k}^{i-} \right] + \delta \hat{\dot{\rho}}_{c,k}^{a-}$$

14.3 Measurement Models

for an ECI-frame calculation and [repeating (9.164)]

$$\hat{\rho}_{a,C,k}^{j-} = \sqrt{\left[\hat{\mathbf{r}}_{ej}^{e}\left(\tilde{t}_{st,a,k}^{j}\right) - \hat{\mathbf{r}}_{ea,k}^{e-}\right]^{\mathrm{T}}\left[\hat{\mathbf{r}}_{ej}^{e}\left(\tilde{t}_{st,a,k}^{j}\right) - \hat{\mathbf{r}}_{ea,k}^{e-}\right]} + \delta\hat{\rho}_{c,k}^{a-} + \delta\rho_{ie}^{j}$$

$$\dot{\hat{\rho}}_{a,C,k}^{j-} = \hat{\mathbf{u}}_{aj,k}^{e-\mathrm{T}}\left[\hat{\mathbf{v}}_{ej}^{e}\left(\tilde{t}_{st,a,k}^{j}\right) - \hat{\mathbf{v}}_{ea,k}^{e-}\right] + \delta\dot{\hat{\rho}}_{c,k}^{a-} + \delta\dot{\rho}_{ie}^{j}$$

or [repeating (9.165)]

$$\hat{\rho}_{a,C,k}^{j-} = \sqrt{\left[\mathbf{C}_{e}^{I}\left(\tilde{t}_{st,a,k}^{j}\right)\hat{\mathbf{r}}_{ej}^{e}\left(\tilde{t}_{st,a,k}^{j}\right) - \hat{\mathbf{r}}_{ea,k}^{e-}\right]^{\mathrm{T}}\left[\mathbf{C}_{e}^{I}\left(\tilde{t}_{st,a,k}^{j}\right)\hat{\mathbf{r}}_{ej}^{e}\left(\tilde{t}_{st,a,k}^{j}\right) - \hat{\mathbf{r}}_{ea,k}^{e-}\right]} + \delta\hat{\rho}_{c,k}^{a-}$$

$$\dot{\hat{\rho}}_{a,C,k}^{j-} = \hat{\mathbf{u}}_{as,j}^{e-\mathrm{T}}\left[\mathbf{C}_{e}^{I}\left(\tilde{t}_{st,a,k}^{j}\right)\left(\hat{\mathbf{v}}_{ej}^{e}\left(\tilde{t}_{st,a,k}^{j}\right) + \mathbf{\Omega}_{ie}^{e}\hat{\mathbf{r}}_{ej}^{e}\left(\tilde{t}_{st,a,k}^{j}\right)\right) - \left(\hat{\mathbf{v}}_{ea,k}^{e-} + \mathbf{\Omega}_{ie}^{e}\hat{\mathbf{r}}_{ea,k}^{e-}\right)\right] + \delta\dot{\hat{\rho}}_{c,k}^{a-}$$

for an ECEF-frame calculation, where the transmission time used to calculate $\hat{\mathbf{r}}_{ej}^{e}$ is given by [repeating (9.125)]

$$\tilde{t}_{st,a}^{s,l} = \tilde{t}_{sa,a}^{s,l} - \left(\tilde{\rho}_{a,R}^{s,l} + \delta\hat{\rho}_{c}^{s,l}\right)/c.$$

However, $\mathbf{C}_{e}^{I}$ is calculated using (8.36) as a function of the range between the user antenna and satellite, not the pseudo-range, because the I frame is synchronized with the ECEF frame at the true time of signal arrival, not the measured time.

The position and velocity of the user antenna are obtained from the inertial navigation solution using

$$\begin{aligned}\hat{\mathbf{r}}_{ia}^{i} &= \hat{\mathbf{r}}_{ib}^{i} + \hat{\mathbf{C}}_{b}^{i}\mathbf{l}_{ba}^{b} \\ \hat{\mathbf{v}}_{ia}^{i} &= \hat{\mathbf{v}}_{ib}^{i} + \hat{\mathbf{C}}_{b}^{i}\left(\hat{\boldsymbol{\omega}}_{ib}^{b} \wedge \mathbf{l}_{ba}^{b}\right)\end{aligned} \tag{14.120}$$

$$\begin{aligned}\hat{\mathbf{r}}_{ea}^{e} &= \hat{\mathbf{r}}_{eb}^{e} + \hat{\mathbf{C}}_{b}^{e}\mathbf{l}_{ba}^{b} \\ \hat{\mathbf{v}}_{ea}^{e} &= \hat{\mathbf{v}}_{eb}^{e} + \hat{\mathbf{C}}_{b}^{e}\left(\hat{\boldsymbol{\omega}}_{ib}^{b} \wedge \mathbf{l}_{ba}^{b}\right) + \mathbf{\Omega}_{ie}^{e}\hat{\mathbf{C}}_{b}^{e}\mathbf{l}_{ba}^{b}\end{aligned} \tag{14.121}$$

or

$$\hat{\mathbf{r}}_{ea}^{e} = \begin{pmatrix} \left(R_{E}(\hat{L}_{b}) + \hat{h}_{b}\right)\cos\hat{L}_{b}\cos\hat{\lambda}_{b} \\ \left(R_{E}(\hat{L}_{b}) + \hat{h}_{b}\right)\cos\hat{L}_{b}\sin\hat{\lambda}_{b} \\ \left[(1-e^{2})R_{E}(\hat{L}_{b}) + \hat{h}_{b}\right]\sin\hat{L}_{b} \end{pmatrix} + \hat{\mathbf{C}}_{n}^{e}\hat{\mathbf{C}}_{b}^{n}\mathbf{l}_{ba}^{b},$$

$$\hat{\mathbf{v}}_{ea}^{e} = \hat{\mathbf{C}}_{n}^{e}\hat{\mathbf{v}}_{eb}^{n} + \hat{\mathbf{C}}_{n}^{e}\hat{\mathbf{C}}_{b}^{n}\left(\hat{\boldsymbol{\omega}}_{ib}^{b} \wedge \mathbf{l}_{ba}^{b}\right) + \mathbf{\Omega}_{ie}^{e}\hat{\mathbf{C}}_{n}^{e}\hat{\mathbf{C}}_{b}^{n}\mathbf{l}_{ba}^{b}$$

(14.122)

where $\mathbf{\Omega}_{ie}^{e}$ and $\hat{\mathbf{C}}_{n}^{e}$ are given by (5.25) and (2.150), respectively, and the Earth-rotation terms may be neglected except where the lever arm is very large.

For tightly coupled integration, the state vector typically comprises the inertial states, receiver clock offset, and clock drift. Thus,

$$\mathbf{x}^\gamma = \begin{pmatrix} \mathbf{x}^\gamma_{INS} \\ \delta\rho^a_c \\ \delta\dot\rho^a_c \end{pmatrix}, \qquad (14.123)$$

where the inertial state vectors given by (14.15), (14.39), and (14.51) are assumed.

The measurement matrix is given by (3.90) and can be expressed in terms of submatrices as

$$\mathbf{H}^\gamma_{G,k} = \begin{pmatrix} \dfrac{\partial \mathbf{z}_\rho}{\partial \delta \boldsymbol{\psi}^\gamma_{\gamma b}} & \mathbf{0}_{m,3} & \dfrac{\partial \mathbf{z}_\rho}{\partial \delta \mathbf{r}^\gamma_{\gamma b}} & \mathbf{0}_{m,3} & \mathbf{0}_{m,3} & \dfrac{\partial \mathbf{z}_\rho}{\partial \delta \rho^a_c} & \mathbf{0}_{m,1} \\[6pt] \dfrac{\partial \mathbf{z}_r}{\partial \delta \boldsymbol{\psi}^\gamma_{\gamma b}} & \dfrac{\partial \mathbf{z}_r}{\partial \delta \mathbf{v}^\gamma_{\gamma b}} & \dfrac{\partial \mathbf{z}_r}{\partial \delta \mathbf{r}^\gamma_{\gamma b}} & \mathbf{0}_{m,3} & \dfrac{\partial \mathbf{z}_\rho}{\partial \mathbf{b}_g} & \mathbf{0}_{m,1} & \dfrac{\partial \mathbf{z}_r}{\partial \delta \dot\rho^a_c} \end{pmatrix}_{\mathbf{x}=\hat{\mathbf{x}}^-_k} \quad \gamma \in i,e \qquad (14.124)$$

or

$$\mathbf{H}^n_{G,k} = \begin{pmatrix} \dfrac{\partial \mathbf{z}_\rho}{\partial \delta \boldsymbol{\psi}^n_{nb}} & \mathbf{0}_{m,3} & \dfrac{\partial \mathbf{z}_\rho}{\partial \delta \mathbf{p}_b} & \mathbf{0}_{m,3} & \mathbf{0}_{m,3} & \dfrac{\partial \mathbf{z}_\rho}{\partial \delta \rho^a_c} & \mathbf{0}_{m,1} \\[6pt] \dfrac{\partial \mathbf{z}_r}{\partial \delta \boldsymbol{\psi}^n_{nb}} & \dfrac{\partial \mathbf{z}_r}{\partial \delta \mathbf{v}^n_{eb}} & \dfrac{\partial \mathbf{z}_r}{\partial \delta \mathbf{p}_b} & \mathbf{0}_{m,3} & \dfrac{\partial \mathbf{z}_\rho}{\partial \mathbf{b}_g} & \mathbf{0}_{m,1} & \dfrac{\partial \mathbf{z}_r}{\partial \delta \dot\rho^a_c} \end{pmatrix}_{\mathbf{x}=\hat{\mathbf{x}}^-_k}. \qquad (14.125)$$

The differentials may be calculated analytically or numerically by perturbing the state estimates and calculating the change in estimate pseudo-range and pseudo-range rate. The dependence of the measurement innovations on the attitude error and of the pseudo-range-rate measurements on the position and gyro errors is weak, so a common approximation to the analytical solution is

$$\mathbf{H}^\gamma_{G,k} \approx \left(\begin{array}{ccccccc} \mathbf{0}_{1,3} & \mathbf{0}_{1,3} & \mathbf{u}^{\gamma\,T}_{a1} & \mathbf{0}_{1,3} & \mathbf{0}_{1,3} & 1 & 0 \\ \mathbf{0}_{1,3} & \mathbf{0}_{1,3} & \mathbf{u}^{\gamma\,T}_{a2} & \mathbf{0}_{1,3} & \mathbf{0}_{1,3} & 1 & 0 \\ \vdots & \vdots & \vdots & \vdots & \vdots & \vdots & \vdots \\ \mathbf{0}_{1,3} & \mathbf{0}_{1,3} & \mathbf{u}^{\gamma\,T}_{am} & \mathbf{0}_{1,3} & \mathbf{0}_{1,3} & 1 & 0 \\ \hline \mathbf{0}_{1,3} & \mathbf{u}^{\gamma\,T}_{a1} & \mathbf{0}_{1,3} & \mathbf{0}_{1,3} & \mathbf{0}_{1,3} & 0 & 1 \\ \mathbf{0}_{1,3} & \mathbf{u}^{\gamma\,T}_{a2} & \mathbf{0}_{1,3} & \mathbf{0}_{1,3} & \mathbf{0}_{1,3} & 0 & 1 \\ \vdots & \vdots & \vdots & \vdots & \vdots & \vdots & \vdots \\ \mathbf{0}_{1,3} & \mathbf{u}^{\gamma\,T}_{am} & \mathbf{0}_{1,3} & \mathbf{0}_{1,3} & \mathbf{0}_{1,3} & 0 & 1 \end{array} \right)_{\mathbf{x}=\hat{\mathbf{x}}^-_k} \quad \gamma \in i,e \qquad (14.126)$$

or

$$\mathbf{H}_{G,k}^n \approx \left(\begin{array}{ccccccc} \mathbf{0}_{1,3} & \mathbf{0}_{1,3} & \mathbf{h}_{\rho p}^{1\,\mathrm{T}} & \mathbf{0}_{1,3} & \mathbf{0}_{1,3} & 1 & 0 \\ \mathbf{0}_{1,3} & \mathbf{0}_{1,3} & \mathbf{h}_{\rho p}^{2\,\mathrm{T}} & \mathbf{0}_{1,3} & \mathbf{0}_{1,3} & 1 & 0 \\ \vdots & \vdots & \vdots & \vdots & \vdots & \vdots & \vdots \\ \mathbf{0}_{1,3} & \mathbf{0}_{1,3} & \mathbf{h}_{\rho p}^{m\,\mathrm{T}} & \mathbf{0}_{1,3} & \mathbf{0}_{1,3} & 1 & 0 \\ \hline \mathbf{0}_{1,3} & \mathbf{u}_{a1}^{n\,\mathrm{T}} & \mathbf{0}_{1,3} & \mathbf{0}_{1,3} & \mathbf{0}_{1,3} & 0 & 1 \\ \mathbf{0}_{1,3} & \mathbf{u}_{a2}^{n\,\mathrm{T}} & \mathbf{0}_{1,3} & \mathbf{0}_{1,3} & \mathbf{0}_{1,3} & 0 & 1 \\ \vdots & \vdots & \vdots & \vdots & \vdots & \vdots & \vdots \\ \mathbf{0}_{1,3} & \mathbf{u}_{am}^{n\,\mathrm{T}} & \mathbf{0}_{1,3} & \mathbf{0}_{1,3} & \mathbf{0}_{1,3} & 0 & 1 \end{array} \right)_{\mathbf{x}=\hat{\mathbf{x}}_k^-}, \quad (14.127)$$

where

$$\mathbf{h}_{\rho p}^j = \begin{pmatrix} \left(R_N(\hat{L}_b) + \hat{h}_b\right) u_{aj,N}^n \\ \left(R_E(\hat{L}_b) + \hat{h}_b\right) \cos \hat{L}_b u_{aj,E}^n \\ -u_{aj,D}^n \end{pmatrix}, \quad (14.128)$$

where the line-of-sight unit vector, $\mathbf{u}_{aj}^\gamma$, is as defined in Section 8.5.3. Note that where the position error is large (e.g., when fewer than four GNSS satellites are tracked), the error in resolving the line-of-sight vectors along north, east, and down can be significant. The approximate version of the ECEF-frame measurement model is included in the MATLAB function, TC_KF_Epoch, on the CD.

The measurement noise covariance, $\mathbf{R}_G$, accounts for GNSS tracking errors, multipath variations, satellite clock noise, residual INS-GNSS synchronization errors, and flexure in the lever arm between the IMU and GNSS antenna. It should ideally be modeled as a function of $\tilde{c}/\tilde{n}_0$ and acceleration, though a constant value is often assumed. It may also be scaled to weight the measurements according to the expected range bias standard deviations and account for the effects of time-correlated noise. The matrix $\mathbf{R}_G$ is diagonal unless the pseudo-range measurements are carrier-smoothed. See Section 9.4.2.4 for more information.

An alternative to using pseudo-range rate measurements is to use carrier-smoothed pseudo-ranges (see Section 9.2.7) and estimate the time-correlated tracking errors as additional Kalman filter states. The pseudo-range rates are then inferred from the changes in the carrier-smoothed pseudo-ranges. In terms of processor load, this is typically more efficient when a vector measurement update is used and less efficient with a sequential measurement update. Otherwise, using carrier-smoothed pseudo-ranges does not significantly impact performance.

Another alternative is to use the delta ranges, which are the changes in ADR. The delta-range measurements are less noisy than the pseudo-range rates where the measurement-update interval is much longer than the carrier tracking-loop correlation time. This applies with a 1-second update interval and carrier phase tracking. However, using delta range adds complexity to the Kalman filter, requiring either the use of delayed position error states or the incorporation of back-propagation through the system model into the measurement model as described in Section 3.3.4 and in Section G.8.3 of Appendix G on the CD [31, 33].

As discussed in Section 14.2.1, pseudo-range and pseudo-range-rate or delta-range measurements may also be differenced between satellites [31], in which case the receiver clock states are omitted. Using the superscript Δ to denote a differenced measurement, the differenced measurement innovation, measurement matrix, and measurement noise covariance matrix are related to their undifferenced counterparts by

$$\delta \mathbf{z}_{G,k}^{\Delta-} = \begin{pmatrix} \mathbf{D}_G & 0 \\ 0 & \mathbf{D}_G \end{pmatrix} \delta \mathbf{z}_{G,k}^{-}$$

$$\mathbf{H}_{G,k}^{\Delta \gamma} = \begin{pmatrix} \mathbf{D}_G & 0 \\ 0 & \mathbf{D}_G \end{pmatrix} \mathbf{H}_{G,k}^{\gamma} \qquad \gamma \in i,e,n \ , \qquad (14.129)$$

$$\mathbf{R}_{G,k}^{\Delta} = \begin{pmatrix} \mathbf{D}_G & 0 \\ 0 & \mathbf{D}_G \end{pmatrix} \mathbf{R}_{G,k} \begin{pmatrix} \mathbf{D}_G & 0 \\ 0 & \mathbf{D}_G \end{pmatrix}^{\mathrm{T}}$$

where $\mathbf{D}_G$ is the differencing matrix, which is given by (10.18) in cases where a common reference satellite is used. Note that the measurement noise covariance matrix, $\mathbf{R}_{G,k}^{\Delta}$, is not diagonal.

Some GNSS receivers periodically implement 1-ms clock corrections, causing jumps of 3×10^5 m in their pseudo-range measurements. Tightly coupled integration algorithms using these receivers must detect these jumps and correct the clock offset estimate, $\delta \hat{\rho}_c^a$, using the algorithm shown in Figure 9.31.

14.3.3 Deeply Coupled Integration

The measurement model of a deep INS/GNSS integration and tracking algorithm comprises both the measurement model of the integration filter and the prefilters, implemented for each signal or satellite. The integration filter measurement model is described first, followed by descriptions of the noncoherent and coherent prefilters, a discussion on multisignal prefiltering, and finally, a description of receiver NCO control.

14.3.3.1 Integration Filter Measurement Model

The integration filter estimates the INS and receiver clock errors, and optionally, ionosphere or range-bias states. The measurement update rate is typically between 1 and 10 Hz. This must be set high enough to track the receiver clock noise and capture the host-vehicle dynamics.

14.3 Measurement Models

Pseudo-range and pseudo-range rate measurement innovations are generated by the prefilters and input directly to the integration Kalman filter. Thus, the integration filter measurement model is essentially the same as for tightly coupled integration as described in Section 14.3.2.

Extra columns of the measurement matrix, $\mathbf{H}_{G,k}$, are required where range bias and/or ionosphere propagation delay states are estimated. For the pseudo-range rate measurements, these columns are all zero. For the pseudo-range measurements, the range bias elements are equal to 1 where measurement corresponds to the same satellite as the range bias state and to 0 otherwise. Similarly, the ionosphere elements are equal to $\left(f_{ca}^L/f_{ca}^l\right)^2$ where the pseudo-range measurement corresponds to the same satellite as the ionosphere state and to 0 otherwise; f_{ca}^l is the carrier frequency of the measurement and f_{ca}^L is a reference carrier frequency, typically the L1 frequency. Determination of the integration filter measurement noise covariance, $\mathbf{R}_G$, is discussed with the prefilters.

Measurements from a given signal should only be accepted where the tracking of that signal is in lock; otherwise, that signal must be reacquired. As deeply coupled integration can bridge short-term GNSS signal outages, lock detection cannot be c/n_0-based. Code tracking lock is determined by comparing the pseudo-range uncertainties, obtained by resolving the position-error and clock-offset state uncertainties along each user-satellite line of sight, with the code chip length. For determining carrier-frequency tracking lock, the pseudo-range rate uncertainties may be obtained from the velocity-error and clock-drift state uncertainties and compared against the tolerance given by (9.25); this is inversely proportional to the coherent integration, or accumulation, interval, τ_a. When there is insufficient c/n_0 for tracking to be maintained, $\mathbf{R}_G$ will be large, causing the state uncertainties to grow.

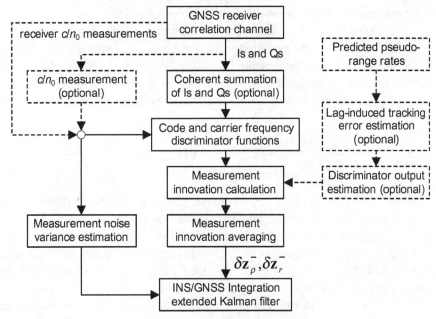

Figure 14.10 Noncoherent prefilter for deeply coupled integration.

14.3.3.2 Noncoherent Prefilters

Figure 14.10 shows the stages of the noncoherent prefilter. Coherent summation of the Is and Qs over the navigation data bit interval may be implemented where the receiver generates Is and Qs at a faster rate and the timing of the data bit transitions is known. Similarly, when navigation-data wipe-off (see Section 10.3.5) or a data-free signal is used, coherent I and Q summation may be performed over longer intervals. As discussed in Section 9.1.4.4, increasing the coherent summation interval improves signal-to-noise performance, but reduces the carrier-frequency tracking tolerance. One option is to vary the coherent summation interval as a function of the uncertainty in the predicted pseudo-range rate.

Following any coherent summation, the Is and Qs are used to form discriminator measurements as in a conventional GNSS ranging processor, described in Sections 9.2.2 and 9.2.3. Pseudo-range and pseudo-range-rate measurement innovations can then be obtained from the discriminator outputs using

$$\delta z_{\rho,k}^{j-} = -\frac{c}{f_{co}}\left(\tilde{x}_k^j - \hat{x}_k^j\right)$$
$$\delta z_{rk}^{j-} = -\frac{c}{f_{ca}}\left(\delta \tilde{f}_{ca,k}^j - \delta \hat{f}_{ca,k}^j\right), \quad (14.130)$$

where $\tilde{x}_k^j$ is the code discriminator output for channel j at iteration k, $\delta \tilde{f}_{ca,k}^j$ is the corresponding carrier-frequency discriminator output, and $\hat{x}_k^j$ and $\delta \hat{f}_{ca,k}^j$ are their estimated counterparts. It is assumed that the code discriminator outputs are normalized using $\tilde{c}/\tilde{n}_0$. Note that the estimated discriminator outputs are zero where there are no significant lags in applying the NCO commands (e.g., where a software receiver is used or NCO commands are applied at the IMU output rate). What constitutes a significant lag depends on the host-vehicle dynamics.

Unless a long coherent integration interval is used, the measurement innovations given by (14.130) are averaged (noncoherently) over successive iterations to reduce the integration filter processing load. Note that averaging should always be used in preference to undersampling, as it gives a better signal-to-noise level.

The noise variances of the pseudo-range and pseudo-range rate measurements are

$$\mathrm{E}\left(w_{m,\rho,k}^{j}{}^2\right) = \frac{c^2 \sigma^2(N_D D)}{f_{co}^2 n_D} + R_{\rho D,k}^j, \quad \mathrm{E}\left(w_{m,r,k}^{j}{}^2\right) = \frac{c^2 \sigma^2(N_F F)}{f_{ca}^2 n_D} + R_{rD,k}^j. \quad (14.131)$$

where $\sigma^2(N_D D)$ and $\sigma^2(N_F F)$ are, respectively, the variances of the code and carrier-frequency discriminators, given in Section 9.3.3; n_D is the number of successive discriminator measurements averaged to form the measurement innovations; and $R_{\rho D,k}^j$ and $R_{rD,k}^j$ are empirically-determined terms to account for the effects of uncompensated NCO control lags (see Section 14.3.3.5), which may be proportional to the squares of the line-of-sight acceleration and/or jerk.

In theory, a noncoherent prefilter can also process carrier phase measurements, obtained from conventional carrier-phase discriminators. However, in practice, a coherent prefilter is used.

14.3.3.3 Coherent Prefilters

Figure 14.11 shows the main elements of a coherent prefilter for deep INS/GNSS integration. At its core, is a tracking filter that estimates the code-phase, carrier-phase, and carrier-frequency tracking errors for the relevant signal. This may be an EKF or a nonlinear estimation algorithm [34–36]. To maintain accurate carrier phase estimates, the tracking filter measurement update must be performed at at least 50 Hz. Therefore, coherent summation of the Is and Qs prior to the tracking filter is only performed where the receiver outputs them at a much higher rate. In coherent implementations of deeply coupled integration, pseudo-range and pseudo-range rate or delta-range measurements are typically output to the integration filter at 1 or 2 Hz.

All coherent prefilters estimate the code tracking error, x^j, the carrier phase tracking error, $\delta\phi_{ca}^j$ [given by (9.11)], and the carrier frequency tracking error, δf_{ca}^j. The derivative of the carrier frequency tracking error, $\delta \dot{f}_{ca}^j$, and the signal amplitude or carrier power to noise density may also be estimated [34, 35].

When the derivative of the carrier frequency tracking error is estimated, the deterministic system model for this state and the carrier frequency tracking error is

$$\frac{\partial}{\partial t}\delta f_{ca}^j = \delta \dot{f}_{ca}^j, \qquad E\left(\frac{\partial}{\partial t}\delta \dot{f}_{ca}^j\right) = 0. \qquad (14.132)$$

Otherwise, the deterministic system model is

$$E\left(\frac{\partial}{\partial t}\delta f_{ca}^j\right) = 0. \qquad (14.133)$$

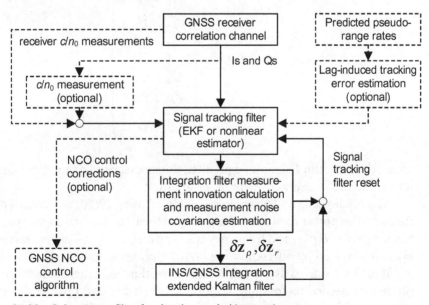

Figure 14.11 Coherent prefilter for deeply coupled integration.

When the prefilter frequency tracking error estimate is not used to modify the NCO commands, the deterministic system model for the code and carrier-phase tracking errors is

$$\delta \dot{x}^j = \frac{f_{co}}{f_{ca}} \delta f_{ca}^j, \qquad \delta \dot{\phi}_{ca}^j = 2\pi \delta f_{ca}^j. \qquad (14.134)$$

Otherwise, the deterministic system model for the state estimates becomes

$$\mathrm{E}\left(\delta \dot{x}^j\right) = 0, \qquad \mathrm{E}\left(\delta \dot{\phi}_{ca}^j\right) = 0 \qquad (14.135)$$

as the known frequency tracking error should be compensated through the NCO command corrections. However, (14.134) should be used for the propagation of the prefilter error covariance, $\mathbf{P}_P^j$.

When necessary, control inputs can be added to the system model to account for known tracking errors due to lags in applying the IMU-measured dynamics to the NCO commands in the GNSS receiver.

System noise of variance $f_{ca}^2 S_c^a \tau_{Ps}/c^2$ must be modeled on the frequency tracking error to account for receiver clock noise, where τ_{Ps} is the prefilter system propagation interval and S_c^a is as defined in Section 14.2.7. Additional system noise, modeled on $\delta \dot{f}_{ca}^j$ where estimated and δf_{ca}^j, otherwise may be needed to account for the effects of uncompensated NCO control lags; its variance may be proportional to the squares of the line-of-sight acceleration and/or jerk.

In a coherent prefilter, the differences between the measured Is and Qs from the GNSS receiver and their predicted values form the measurement innovations. Thus,

$$\delta \mathbf{z}_{P,k'}^{j-} = \begin{pmatrix} \tilde{I}_E^j - \hat{I}_E^{j-} \\ \tilde{I}_P^j - \hat{I}_P^{j-} \\ \tilde{I}_L^j - \hat{I}_L^{j-} \\ \tilde{Q}_E^j - \hat{Q}_E^{j-} \\ \tilde{Q}_P^j - \hat{Q}_P^{j-} \\ \tilde{Q}_L^j - \hat{Q}_L^{j-} \end{pmatrix}_{k'}, \qquad (14.136)$$

where the subscript P denotes a prefilter, the superscript j denotes the signal to which it applies, and k' denotes the prefilter iteration.

The predicted Is and Qs may be calculated using (9.10) as a function of the measured carrier power to noise density, $\tilde{c}/\tilde{n}_0$, estimated code tracking error, $\hat{x}_{k'}^j$, carrier tracking error, $\delta \hat{\phi}_{ca,k'}^j$, and frequency tracking error, $\delta \hat{f}_{ca,k'}^j$. The measurement matrix, $\mathbf{H}_{P,k'}^j$, is obtained by differentiating (9.10) with respect to the Kalman filter states.

If an EKF is used as the estimation algorithm, accurate estimates of the carrier phases are needed to compute both the predicted Is and Qs and the measurement matrices; without them, the tracking filter will not work. One solution to this problem

14.3 Measurement Models

is to use a particle filter as each particle has its own carrier-phase estimate [37]. A particle filter can thus tolerate much higher carrier-phase tracking errors before losing lock. However, the processor load is much higher.

When navigation-data wipe-off or data-free signals are used, the incorporation of the Is and Qs into the estimation algorithm as measurements maintains coherent summation over a varying interval, determined by the filter gain. This optimizes the signal-to-noise performance (see Section 9.1.4.4). Consequently, coherent summation of the Is and Qs prior to their input to the tracking filter has little impact on performance [26]. When there are unknown data bits that are not estimated (see Section 10.3.5), coherent integration is limited to the data-bit intervals and the integration algorithm must use the sign of one of the Is and Qs to correct for the data bits. Thus,

$$\delta \mathbf{z}_{P,k'}^{j-} = \begin{pmatrix} \tilde{I}_E^j \tilde{D}_{k'}^j - \hat{I}_E^{j-} \\ \tilde{I}_P^j \tilde{D}_{k'}^j - \hat{I}_P^{j-} \\ \tilde{I}_L^j \tilde{D}_{k'}^j - \hat{I}_L^{j-} \\ \tilde{Q}_E^j \tilde{D}_{k'}^j - \hat{Q}_E^{j-} \\ \tilde{Q}_P^j \tilde{D}_{k'}^j - \hat{Q}_P^{j-} \\ \tilde{Q}_L^j \tilde{D}_{k'}^j - \hat{Q}_L^{j-} \end{pmatrix}_k \quad \tilde{D}_{k'}^j = \begin{array}{ll} \dfrac{\text{sign}(\tilde{I}_P^j)}{\text{sign}(\hat{I}_P^{j-})} & |\tilde{I}_P^j| \geq |\tilde{Q}_P^j| \\[1em] \dfrac{\text{sign}(\tilde{Q}_P^j)}{\text{sign}(\hat{Q}_P^{j-})} & |\tilde{I}_P^j| < |\tilde{Q}_P^j| \end{array}. \quad (14.137)$$

When an AGC is used in the receiver front end, the measurement noise should be constant (with the signal amplitude varying as a function of c/n_0). A suitable value for the measurement noise covariance matrix [assuming the measurement order defined in (14.136)] is

$$\mathbf{R}_P^j = \sigma_{IQ}^2 \begin{pmatrix} 1 & (1-d/2)^2 & (1-d)^2 & 0 & 0 & 0 \\ (1-d/2)^2 & 1 & (1-d/2)^2 & 0 & 0 & 0 \\ (1-d)^2 & (1-d/2)^2 & 1 & 0 & 0 & 0 \\ 0 & 0 & 0 & 1 & (1-d/2)^2 & (1-d)^2 \\ 0 & 0 & 0 & (1-d/2)^2 & 1 & (1-d/2)^2 \\ 0 & 0 & 0 & (1-d)^2 & (1-d/2)^2 & 1 \end{pmatrix}, \quad (14.138)$$

where σ_{IQ} is the noise standard deviation of the accumulated correlator outputs and d is the early–late correlator spacing in code chips. The off-diagonal terms are nonzero because the noise is partially correlated between the three I measurements and between the three Q measurements from each channel. The noise standard deviation varies with the receiver design so must be determined empirically. It may be necessary to introduce a $\tilde{c}/\tilde{n}_0$ dependence if the AGC reaches minimum gain at a c/n_0 level where signals may still be tracked; this must also be determined empirically.

Pseudo-range and pseudo-range-rate measurement innovations for output to the integration filter can be obtained from the tracking filter state estimates using

$$\delta z_{\rho,k}^{j-} = -\frac{c}{f_{co}}\hat{x}_{k'}^{j}, \qquad \delta z_{rk}^{j-} = -\frac{c}{f_{ca}}\delta \hat{f}_{ca,k'}^{j}. \qquad (14.139)$$

As a federated zero-reset architecture (Section 16.1.5.3) [17] is implemented for the code-phase and carrier-frequency states in the tracking filters, these states are zeroed when data is output to the integration filter to ensure that no information is held simultaneously by two sequential filters. The corresponding state uncertainties must be reset to the code-phase and carrier-frequency tracking uncertainties derived from the integration filter. Off-diagonal elements of the error covariance matrix, $\mathbf{P}_P^j$, pertaining to the reset states are zeroed. When estimated, $\delta \dot{f}_{ca}^j$ should be similarly reset. However, the carrier-phase error state is not reset as this information is required for processing future I and Q measurements and is not passed onto the integration filter.

The elements of the integration filter measurement noise covariance, $\mathbf{R}_G$, for the relevant signal should be determined from the tracking filter error covariance, $\mathbf{P}_P^j$, noting that, with a coherent prefilter, the pseudo-range and pseudo-range rate measurement noise will be correlated. Thus,

$$E\left[\begin{pmatrix} w_{m,\rho,k}^j \\ w_{m,r,k}^j \end{pmatrix}\begin{pmatrix} w_{m,\rho,k}^j \\ w_{m,r,k}^j \end{pmatrix}^T\right] = \begin{pmatrix} \frac{c}{f_{co}} & 0 \\ 0 & \frac{c}{f_{ca}} \end{pmatrix} E\left[\begin{pmatrix} \delta x_{k'}^j \\ \delta \delta f_{ca,k'}^j \end{pmatrix}\begin{pmatrix} \delta x_{k'}^j \\ \delta \delta f_{ca,k'}^j \end{pmatrix}^T\right]\begin{pmatrix} \frac{c}{f_{co}} & 0 \\ 0 & \frac{c}{f_{ca}} \end{pmatrix}.$$

(14.140)

An alternative to the tracking filter described here is to use the batch-processing acquisition and tracking algorithm described in Section 10.3.4 for the prefilters in coherent deep INS/GNSS integration [38].

14.3.3.4 Multisignal Prefiltering

When the receiver processes GNSS signals on more than one frequency, it is more efficient to process all measurements from a given satellite using the same prefilter. In this case, the prefilter must maintain an estimate of the ionosphere propagation delay which is used by the NCO control algorithm.

In a noncoherent prefilter, separate discriminators should be computed for each signal with ionosphere-corrected combinations (see Section 9.3.2) used to form the measurement innovations output to the integration filter. The differences between the discriminators for different frequencies should be used as the measurement inputs to an ionosphere delay estimation algorithm; this could be a simple fixed-gain loop filter.

One way of configuring a multifrequency coherent prefilter is to implement a separate tracking filter for each signal. At the output and reset stage, ionosphere-corrected combinations of the tracking filter estimates can then be used to form

14.3.3.5 GNSS Receiver NCO Control

In deeply coupled INS/GNSS integration, the GNSS receiver NCO commands are generated by the navigation processor. By analogy with (9.39), the code NCO command for the jth signal is

$$\hat{f}^j_{co,NCO,k+1} = f_{co}\left[1 - \frac{\left(\hat{\rho}^{j-}_{a,R,k+1} - \hat{\rho}^{j-}_{a,R,k}\right)}{c\tau_N}\right], \quad (14.141)$$

where τ_N is the NCO-command update interval and the pseudo-range estimate is given by

$$\begin{aligned}\hat{\rho}^j_{a,R} &= \left|\hat{\mathbf{r}}^i_{ij}\left(\tilde{t}^j_{st,a}\right) - \hat{\mathbf{r}}^i_{ia}\right| + \delta\hat{\rho}^a_c + \delta\hat{\rho}^j_{I,a} + \delta\hat{\rho}^j_{T,a} - \delta\hat{\rho}^j_c + \delta\hat{\rho}^j_r \\ &= \left|\hat{\mathbf{C}}^I_e\left(\tilde{t}^j_{st,a}\right)\hat{\mathbf{r}}^e_{ej}\left(\tilde{t}^j_{st,a}\right) - \hat{\mathbf{r}}^e_{ea}\right| + \delta\hat{\rho}^a_c + \delta\hat{\rho}^j_{I,a} + \delta\hat{\rho}^j_{T,a} - \delta\hat{\rho}^j_c + \delta\hat{\rho}^j_r\end{aligned}, \quad (14.142)$$

where the notation is as defined in Section 8.5.3, except for $\delta\hat{\rho}^j_r$, which is the Kalman-filter-estimated residual range bias (see Section 14.2.1). The receiver clock offset estimate, $\delta\hat{\rho}^a_c$, is obtained from the integration Kalman filter. The satellite position, $\hat{\mathbf{r}}^\gamma_{\gamma j}$, is obtained from the navigation data message as described in Section 8.5.2. The satellite clock correction, $\delta\hat{\rho}^j_c$, is given by (9.77), the troposphere error estimate, $\delta\hat{\rho}^j_{T,a}$, is obtained from a model, and the ionosphere error estimate, $\delta\hat{\rho}^j_{I,a}$, is obtained from a model and/or the relevant prefilter or the integration Kalman filter. Troposphere and ionosphere models are presented in Section 9.3.2 and in Section G.7 of Appendix G on the CD. The GNSS antenna position is, from (2.162),

$$\hat{\mathbf{r}}^\gamma_{\gamma a} = \hat{\mathbf{r}}^\gamma_{\gamma b} + \hat{\mathbf{C}}^\gamma_b \mathbf{l}^b_{ba} \quad \gamma \in i,e, \quad (14.143)$$

where $\hat{\mathbf{r}}^e_{eb}$ may be calculated from $\hat{L}_b$, $\hat{\lambda}_b$, and $\hat{h}_b$ using (2.112) where required.

Assuming the signal and reference carrier phases are not synchronized, the carrier NCO command is given by

$$\hat{f}^j_{ca,NCO,k+1} = f_{IF} - \frac{f_{ca}}{c}\hat{\dot{\rho}}^{j-}_{a,R,k+1}, \quad (14.144)$$

where the pseudo-range rate estimated from the inertial navigation solution is given by (14.11), noting that the receiver clock drift estimate is obtained from the integration Kalman filter.

There can be a significant lag between the time of validity of the inertial navigation solution used to generate the NCO commands and the application of those commands in the GNSS receiver. Contributions to this lag include the length of the IMU sampling window; the time taken to output the IMU samples to the navigation processor, update the inertial navigation solution, and generate the NCO commands; the time taken to communicate the NCO commands to the GNSS receiver and then apply them; and the length of the GNSS signal correlation window [16]. This NCO control lag can cause significant tracking errors. However, it can be mitigated by using the estimated pseudo-range acceleration to predict the NCO commands forward in time from the inertial navigation solution time of validity to the signal correlation time. The range acceleration is calculated as shown in Section G.4.1 of Appendix G on the CD. The pseudo-range acceleration is assumed to be the same as the rate of change of the clock drift is not easily predictable. If the receiver does not permit variation of the NCO frequencies over the correlation interval, the NCO control algorithm should supply average values, determined using the pseudo-range acceleration.

When the navigation system is subject to significant jerk, an acceleration valid over the IMU sampling interval may be out of date by the time of the GNSS signal correlation interval when it is applied. However, by the time the I and Q measurements from that correlation interval are processed by the prefilters, further IMU measurements will have been processed. Consequently, corrections may be applied within the prefilters as discussed in Sections 14.3.3.2 and 14.3.3.3.

NCO control lags may be eliminated by using a software receiver, which enables the GNSS signal samples to be stored until the contemporaneous IMU measurements have been processed.

For most applications, an NCO command update rate of 50 Hz is sufficient when data latency compensation is applied, while a 100-Hz update rate is recommended without latency compensation. Exceptions are applications with extreme dynamics or vibration where the acceleration is liable to exceed 100 m s^{-2} or the jerk 100 m s^{-3}, noting that oscillatory jerk is not expected to cause loss of tracking where the position amplitude is much less than a GNSS carrier wavelength [16].

14.3.4 Estimation of Attitude and Instrument Errors

In INS/GNSS integration, the measurement innovations input to the Kalman filter are based on position and velocity or pseudo-range and pseudo-range rate. As (14.115), (14.126), and (14.127) show, the direct coupling of these measurements to the attitude-error and instrument-error states is usually negligible. Yet the Kalman filter still estimates these errors and how it does this is not intuitive. Here an explanation is presented.

As explained in Section 14.2.1, the attitude and instrument errors are observed through the growth in the velocity error they produce, with the attitude errors and accelerometer biases inducing a linear growth in the velocity error and gyro biases inducing a quadratic growth. The corresponding growths in position error are quadratic and cubic, respectively. This coupling of the states is represented by the system model (Sections 14.2.2–14.2.5). Each time the error covariance matrix, $\mathbf{P}$, is propagated through the system model [using (3.15) or (3.46)], information on the correlations between the residual attitude and instrument errors and the residual position and velocity errors is built up in the off-diagonal elements of **P**. This enables the measurement model to estimate corrections to the attitude and instrument error estimates from measurements of the position and velocity errors (or linear combinations thereof).

When a measurement update is performed, the error covariance matrix, **P**, is used in the calculation of the Kalman gain matrix, **K**, using (3.21). The products of the off-diagonal elements of **P** with the elements of the measurement matrix, $\mathbf{H}_G$, coupling the position and velocity errors to the measurements give elements of **K** coupling the measurements to the attitude and instrument errors. Thus, when a state-vector update, (3.24), is performed, the attitude and instrument error estimates are updated alongside the position and velocity error estimates and any GNSS states.

14.4 Advanced INS/GNSS Integration

This section collects together a number of advanced INS/GNSS integration topics. Integration of INS with differential, carrier-phase, and multiantenna GNSS; modeling large heading errors; advanced IMU error modeling; and smoothing are discussed.

14.4.1 Differential GNSS

Differential GNSS, described in Section 10.1, improves position accuracy by calibrating out much of the temporally and spatially-correlated biases in the pseudo-range measurements due to ephemeris prediction errors, residual satellite clock errors, ionospheric refraction, and sometimes tropospheric refraction.

The architectures for integrating DGNSS with INS are essentially the same as for stand-alone GNSS. Differential corrections are applied to the pseudo-range measurements. In loosely coupled integration, this occurs within the GNSS user equipment. For tightly coupled integration, the pseudo-range measurements may be corrected by the GNSS ranging processor or within the integration algorithm's measurement model. For deeply coupled integration, the differential corrections are applied within the NCO control algorithm.

The measurement noise covariance, $\mathbf{R}_G$, due to tracking noise, multipath, time-synchronization errors, and lever arm flexure is the same as for stand-alone GNSS as these error sources are unchanged by the application of differential corrections. However, the standard deviation of the range biases, whether modeled as states (see Sections 14.2.1 and 14.2.7) or as additional position and clock-offset uncertainty (see Section 9.4.2.5), will be smaller and should be reduced to a value commensurate with the accuracy of the overall differential corrections. This will also affect any weighting of $\mathbf{R}_G$ to account for variation in the range bias standard deviations between measurements. The noise on the differential corrections will typically be much smaller than that on the user receiver's pseudo-range measurements due to the use of smoothing at the reference station.

14.4.2 Carrier-Phase Positioning

As described in Section 10.2, real-time centimeter-accuracy positioning can be obtained by comparing GNSS ADR measurements with those made by user equipment at a precisely surveyed base station. Integration with INS, as well as bridging the position solution through GNSS outages, can also bridge the ambiguity resolution process through outages of up to about a minute [39, 40] and aid the detection and correction of cycle slips [41].

Loosely coupled integration of carrier-phase GNSS with INS is the same as integration of stand-alone or differential GNSS, except that the measurement noise covariance, $\mathbf{R}_G$, must be adjusted to account for the smaller GNSS position errors and their different time correlation properties. The position error variance will typically be of order $(0.01–0.05\text{m})^2$. However, to determine $\mathbf{R}_G$, this should be multiplied by the ratio of the error correlation time to the measurement update time interval, as for code-based GNSS position. The error correlation time will vary with the GNSS user equipment design.

Tightly coupled INS/GNSS integration may be performed independently of ambiguity resolution by inputting double-differenced ADR measurements, with the integer wavelength ambiguities corrected, to the integration algorithm [42, 43]. These may be converted to ambiguity-corrected ADR measurements differenced across satellites only using

$$\Delta \breve{\tilde{\Phi}}_{a,R}^{ts,l} = \nabla \Delta \tilde{\Phi}_{ra,R}^{ts,l} - \nabla \Delta \Phi \breve{N}_{ra}^{ts,l} \lambda_{ca}^{l} + \left| \mathbf{C}_e^l(\tilde{t}_{st,r}^{s,l}) \hat{\mathbf{r}}_{es}^e(\tilde{t}_{st,r}^{s,l}) - \mathbf{r}_{er}^e \right| - \left| \mathbf{C}_e^l(\tilde{t}_{st,r}^{s,l}) \hat{\mathbf{r}}_{et}^e(\tilde{t}_{st,l}^{t,l}) - \mathbf{r}_{er}^e \right|, \quad (14.145)$$

where the notation is as defined in Section 10.2 and the reference station position, $\mathbf{r}_{er}^e$, is assumed to be known. The measurement innovation vector is then

$$\delta \mathbf{z}_{\Phi,k}^- = \begin{pmatrix} \Delta \breve{\tilde{\Phi}}_{a,R}^{t1,l} - \hat{r}_{a1}^- + \hat{r}_{at}^- \\ \Delta \breve{\tilde{\Phi}}_{a,R}^{t2,l} - \hat{r}_{a2}^- + \hat{r}_{at}^- \\ \vdots \\ \Delta \breve{\tilde{\Phi}}_{a,R}^{tm,l} - \hat{r}_{am}^- + \hat{r}_{at}^- \end{pmatrix}_k, \quad (14.146)$$

where

$$\begin{aligned}
\hat{r}_{as,k}^- &= \sqrt{\left[\hat{\mathbf{r}}_{is}^i(\tilde{t}_{st,a,k}^s) - \hat{\mathbf{r}}_{ia,k}^{i-}\right]^T \left[\hat{\mathbf{r}}_{is}^i(\tilde{t}_{st,a,k}^s) - \hat{\mathbf{r}}_{ia,k}^{i-}\right]} \\
&= \sqrt{\left[\hat{\mathbf{r}}_{es}^e(\tilde{t}_{st,a,k}^s) - \hat{\mathbf{r}}_{ea,k}^{e-}\right]^T \left[\hat{\mathbf{r}}_{es}^e(\tilde{t}_{st,a,k}^s) - \hat{\mathbf{r}}_{ea,k}^{e-}\right]} + \delta \rho_{ie}^s \\
&= \sqrt{\left[\mathbf{C}_e^l(\tilde{t}_{st,a,k}^s) \hat{\mathbf{r}}_{es}^e(\tilde{t}_{st,a,k}^s) - \hat{\mathbf{r}}_{ea,k}^{e-}\right]^T \left[\mathbf{C}_e^l(\tilde{t}_{st,a,k}^s) \hat{\mathbf{r}}_{es}^e(\tilde{t}_{st,a,k}^s) - \hat{\mathbf{r}}_{ea,k}^{e-}\right]}
\end{aligned} \quad (14.147)$$

14.4 Advanced INS/GNSS Integration

Note there is no benefit in using pseudo-range or pseudo-range rate measurements alongside ADR measurements, though the pseudo-ranges are used for the wavelength ambiguity determination.

The state vector is the same as for conventional tightly coupled integration except that the receiver clock states are omitted. Assuming the state vector defined by (14.15), (14.39), or (14.51), the measurement matrix is

$$\mathbf{H}_{\Phi,k}^{\Delta\gamma} \approx \begin{pmatrix} \mathbf{0}_{1,3} & \mathbf{0}_{1,3} & \left(\mathbf{u}_{a1}^{\gamma} - \mathbf{u}_{at}^{\gamma}\right)^{\mathrm{T}} & \mathbf{0}_{1,3} & \mathbf{0}_{1,3} \\ \mathbf{0}_{1,3} & \mathbf{0}_{1,3} & \left(\mathbf{u}_{a2}^{\gamma} - \mathbf{u}_{at}^{\gamma}\right)^{\mathrm{T}} & \mathbf{0}_{1,3} & \mathbf{0}_{1,3} \\ \vdots & \vdots & \vdots & \vdots & \vdots \\ \mathbf{0}_{1,3} & \mathbf{0}_{1,3} & \left(\mathbf{u}_{am}^{\gamma} - \mathbf{u}_{at}^{\gamma}\right)^{\mathrm{T}} & \mathbf{0}_{1,3} & \mathbf{0}_{1,3} \end{pmatrix}_{\mathbf{x}=\hat{\mathbf{x}}_k^-} \quad \gamma \in i,e \quad (14.148)$$

or

$$\mathbf{H}_{\Phi,k}^{\Delta n} \approx \begin{pmatrix} \mathbf{0}_{1,3} & \mathbf{0}_{1,3} & \left(\mathbf{h}_{\rho p}^{1} - \mathbf{h}_{\rho p}^{t}\right)^{\mathrm{T}} & \mathbf{0}_{1,3} & \mathbf{0}_{1,3} \\ \mathbf{0}_{1,3} & \mathbf{0}_{1,3} & \left(\mathbf{h}_{\rho p}^{2} - \mathbf{h}_{\rho p}^{t}\right)^{\mathrm{T}} & \mathbf{0}_{1,3} & \mathbf{0}_{1,3} \\ \vdots & \vdots & \vdots & \vdots & \vdots \\ \mathbf{0}_{1,3} & \mathbf{0}_{1,3} & \left(\mathbf{h}_{\rho p}^{m} - \mathbf{h}_{\rho p}^{t}\right)^{\mathrm{T}} & \mathbf{0}_{1,3} & \mathbf{0}_{1,3} \end{pmatrix}_{\mathbf{x}=\hat{\mathbf{x}}_k^-}, \quad (14.149)$$

where $\mathbf{h}_{\rho p}^{s}$ is given by (14.128).

The measurement noise covariance matrix is nondiagonal and given by

$$\mathbf{R}_{\Phi,k}^{\nabla\Delta} = \mathbf{D}_G \mathbf{R}_{\Phi,k}^{\nabla} \mathbf{D}_G^{\mathrm{T}}, \quad (14.150)$$

where $\mathbf{R}_{\Phi,k}^{\nabla}$ is a diagonal matrix representing the noise on the single-satellite ADR measurements differenced between user and reference. It will typically be at centimeter level and should account for reference station as well as user measurement noise. The range biases may often be neglected.

In both the loosely coupled and tightly coupled integration architectures, the corrected INS position solution may be used to aid GNSS ambiguity resolution. However, both the integration and ambiguity resolution algorithms must be carefully tuned to avoid positive-feedback problems. Alternatively, where the ambiguity resolution algorithm is Kalman filter-based, it may be combined with the integration algorithm into a single algorithm, estimating both the INS errors and the float carrier wavelength ambiguities. This is described in Section I.7.1 of Appendix I on CD and also applies to deeply coupled integration.

14.4.3 GNSS Attitude

GNSS attitude determination uses relative carrier-phase positioning between antennas mounted on the same vehicle as described in Section 10.2.5. GNSS attitude is very noisy, but does not drift, making it highly complementary to INS attitude and a solution to the heading calibration problem that occurs for some INS/GNSS applications. By combining INS with multiantenna GNSS, a precise and stable attitude solution may be obtained. A full GNSS attitude solution requires three or more antennas. However, two antennas is sufficient for INS/GNSS in cases where conventional INS/GNSS meets the roll and pitch accuracy requirements [44].

For loosely coupled integration of GNSS attitude, an additional measurement innovation is added:

$$\mathbf{I}_3 + \left[\delta \mathbf{z}_{\psi,k}^{\gamma-} \wedge\right] \approx \tilde{\mathbf{C}}_{aG,k}^{\gamma} \mathbf{C}_b^a \hat{\mathbf{C}}_{\gamma,k}^{b-} \qquad \gamma \in i,e,n, \qquad (14.151)$$

where $\tilde{\mathbf{C}}_{aG}^{\gamma}$ is the GNSS attitude measurement, and the relative orientation of the INS and GNSS body frames, $\mathbf{C}_b^a$, is assumed to be known. The measurement matrix is

$$\mathbf{H}_{\psi k}^{i/e/n} = \begin{pmatrix} -\mathbf{I}_3 & \mathbf{0}_3 & \mathbf{0}_3 & \mathbf{0}_3 & \mathbf{0}_3 \end{pmatrix}, \qquad (14.152)$$

assuming the state vector defined by (14.15), (14.39), or (14.51).

Tightly coupled integration of GNSS attitude may be performed by inputting measurements of ADR differenced between antennas [45]. Note that this differencing eliminates the position and velocity information, leaving only attitude. Measurements may also be differenced across satellites to eliminate any residual timing and phase biases between antennas, noting that a common receiver clock is typically used for multi-antenna GNSS.

When measurements comprise ADRs double-differenced across satellites and between antennas a and A, the tightly coupled measurement innovations are

$$\delta \mathbf{z}_{\psi,k}^{\gamma-} = \begin{pmatrix} \nabla\Delta\tilde{\Phi}_{aA,R}^{t1,l} - \nabla\Delta\breve{N}_{aA}^{t1,l}\lambda_{ca}^l + \left(\mathbf{u}_{a1}^{\gamma} - \mathbf{u}_{at}^{\gamma}\right)^{\mathrm{T}} \hat{\mathbf{C}}_b^{\gamma} \mathbf{r}_{aA}^b \\ \nabla\Delta\tilde{\Phi}_{aA,R}^{t2,l} - \nabla\Delta\breve{N}_{aA}^{t2,l}\lambda_{ca}^l + \left(\mathbf{u}_{a2}^{\gamma} - \mathbf{u}_{at}^{\gamma}\right)^{\mathrm{T}} \hat{\mathbf{C}}_b^{\gamma} \mathbf{r}_{aA}^b \\ \vdots \\ \nabla\Delta\tilde{\Phi}_{aA,R}^{tm,l} - \nabla\Delta\breve{N}_{aA}^{tm,l}\lambda_{ca}^l + \left(\mathbf{u}_{am}^{\gamma} - \mathbf{u}_{at}^{\gamma}\right)^{\mathrm{T}} \hat{\mathbf{C}}_b^{\gamma} \mathbf{r}_{aA}^b \end{pmatrix}_k \qquad \gamma \in i,e,n, \qquad (14.153)$$

where $\nabla\Delta\tilde{\Phi}_{aA,R}^{ts,l}$ is the double-differenced ADR, given by

$$\nabla\Delta\tilde{\Phi}_{aA,R}^{ts,l} = \tilde{\Phi}_{A,R}^{s,l} - \tilde{\Phi}_{A,R}^{t,l} - \tilde{\Phi}_{a,R}^{s,l} + \tilde{\Phi}_{a,R}^{t,l}, \qquad (14.154)$$

$\nabla\Delta\breve{N}_{aA}^{ts,l}$ is the fixed double-differenced integer wavelength ambiguity estimate, and $\mathbf{r}_{aA}^a$ is the position of antenna A with respect to antenna a, resolved about INS body-frame axes, b. When the error in the corrected inertial attitude solution, $\hat{\mathbf{C}}_b^\gamma$, is relatively small, the correct integer ambiguities will be those that minimize the measurement innovations. The measurement matrix is

$$\mathbf{H}_{\psi k}^\gamma = \begin{pmatrix} (\mathbf{u}_{a1}^\gamma - \mathbf{u}_{at}^\gamma)^{\mathrm{T}}\left[(\hat{\mathbf{C}}_b^\gamma \mathbf{r}_{aA}^b)\wedge\right] & \mathbf{0}_{1\times 3} & \mathbf{0}_{1\times 3} & \mathbf{0}_{1\times 3} & \mathbf{0}_{1\times 3} \\ (\mathbf{u}_{a2}^\gamma - \mathbf{u}_{at}^\gamma)^{\mathrm{T}}\left[(\hat{\mathbf{C}}_b^\gamma \mathbf{r}_{aA}^b)\wedge\right] & \mathbf{0}_{1\times 3} & \mathbf{0}_{1\times 3} & \mathbf{0}_{1\times 3} & \mathbf{0}_{1\times 3} \\ \vdots & \vdots & \vdots & \vdots & \vdots \\ (\mathbf{u}_{am}^\gamma - \mathbf{u}_{at}^\gamma)^{\mathrm{T}}\left[(\hat{\mathbf{C}}_b^\gamma \mathbf{r}_{aA}^b)\wedge\right] & \mathbf{0}_{1\times 3} & \mathbf{0}_{1\times 3} & \mathbf{0}_{1\times 3} & \mathbf{0}_{1\times 3} \end{pmatrix}_{\hat{\mathbf{x}} = \hat{\mathbf{x}}_k^-} \quad \gamma \in i, e, n, \quad (14.155)$$

again assuming the previous state vector defined by (14.15), (14.39), or (14.51). The measurement noise covariance is as given by (14.150). Note that ADR difference measurements between a second pair of antennas must be added to provide three-axis GNSS attitude measurements.

Note that both the loosely coupled and tightly coupled forms of GNSS attitude integration may be combined with either loosely coupled or tightly coupled integration of the GNSS range measurements.

Another form of tightly coupled attitude integration, which is not compatible with loosely coupled position and velocity integration, processes separate code- and carrier-phase-derived range-domain measurements from each antenna in a single EKF. The EKF's measurement matrix performs the differencing of carrier-phase-derived measurements between antennas that is required to obtain the attitude information. Note that the appropriate lever arms must be modeled and the attitude-error components of the measurement matrix, $\mathbf{H}_G$, must not be neglected [42].

The accuracy of INS/GNSS attitude determination depends on the quality of the inertial sensors and the antenna separation. Longer lever arms produce more precise GNSS attitude measurements, but can be subject to flexure. One solution to the flexure problem is to measure it using the main IMU, additional inertial sensors at the antennas, and strain gauges [46].

The inertial attitude solution can be used to aid the GNSS wavelength ambiguity resolution process, either by constraining the search space or by estimating the float ambiguities in the integration algorithm as described in Section I.7.2 of Appendix I on the CD. With a short baseline, the inertial attitude alone may be sufficient resolve the ambiguities.

14.4.4 Large Heading Errors

In the examples of INS/GNSS Kalman filters presented in Sections 14.2 and 14.3, the small angle approximation is applied to the attitude errors. This is usually valid for the roll and pitch attitude, which may be observed through leveling (Section 5.6.2). However, the heading (or azimuth) is more difficult to initialize. Only the higher grades of INS are capable of gyrocompassing, while magnetic heading is subject to environmental anomalies and heading derived from the GNSS trajectory is subject to sideslip-induced errors.

When consumer-grade inertial sensors are used, all three components of the attitude error can be relatively large, particularly during GNSS outages or when signal reception is poor.

When the small angle approximation is applied to the attitude components resolved about the horizontal axes, but not the heading, the attitude-error coordinate transformation matrix may be expressed as

$$
\begin{aligned}
\delta\mathbf{C}_b^n &\approx \begin{pmatrix} \cos\delta\psi_{nb,D}^n & -\sin\delta\psi_{nb,D}^n & \delta\psi_{nb,N}^n \sin\delta\psi_{nb,D}^n + \delta\psi_{nb,E}^n \cos\delta\psi_{nb,D}^n \\ \sin\delta\psi_{nb,D}^n & \cos\delta\psi_{nb,D}^n & -\delta\psi_{nb,N}^n \cos\delta\psi_{nb,D}^n + \delta\psi_{nb,E}^n \sin\delta\psi_{nb,D}^n \\ -\delta\psi_{nb,E}^n & \delta\psi_{nb,N}^n & 1 \end{pmatrix} \\
&= \begin{pmatrix} \cos\delta\psi_{nb,D}^n & -\sin\delta\psi_{nb,D}^n & 0 \\ \sin\delta\psi_{nb,D}^n & \cos\delta\psi_{nb,D}^n & 0 \\ 0 & 0 & 1 \end{pmatrix} \left(\mathbf{I}_3 + \left[\begin{pmatrix} \delta\psi_{nb,N}^n \\ \delta\psi_{nb,E}^n \\ 0 \end{pmatrix} \wedge \right] \right)
\end{aligned}, \quad (14.156)
$$

noting that the down component of the attitude error, $\delta\psi_{nb,D}^n$, is the heading error, $\delta\psi_{nb}$.

Using (14.156), the system model is no longer a linear function of the error states, a key requirement of Kalman filtering. One solution is to replace the heading error state with sine and cosine terms, $\sin\delta\psi_{nb}$ and $\cos\delta\psi_{nb}$ [47] or $\delta\sin\psi_{nb}$ and $\delta\cos\psi_{nb}$ [48, 49]. These enable INS/GNSS integration and other forms of fine alignment (Section 5.6.3) to take place with no prior knowledge of heading.

When the heading error is very large, the products of the heading error state(s) with other states are no longer negligible, so a system model of at least second order is required. With a consumer-grade IMU, this can also apply to the roll and pitch. An extended Kalman filter (Section 3.4.1) is not suitable as it linearizes the propagation of the error covariance matrix, **P**, through which the attitude errors are observed (Section 14.3.4). Alignment with large heading errors has been demonstrated using an unscented Kalman filter (Section 3.4.2) [50] and a number of other nonlinear filters [51], noting that nonlinear filtering is only needed for the system propagation phase [52]. However, other research has shown that a conventional approach with closed-loop INS correction works equally well for initial attitude uncertainties up to 30° [53]. Where the initial heading is completely unknown, a particle filter (Section 3.5) may be used.

When either a particle filter or the system propagation phase of a UKF is implemented, total-state integration should be used as discussed in Section 14.1.1. In this case, separate inertial navigation equations are implemented for each particle or sigma point. Separate corrections are also applied to the IMU outputs. Each system propagation phase will typically comprise tens of inertial navigation processing cycles using successive IMU outputs. Figure 14.12 illustrates this for a UKF. GNSS state estimates may be propagated in one step.

14.4 Advanced INS/GNSS Integration

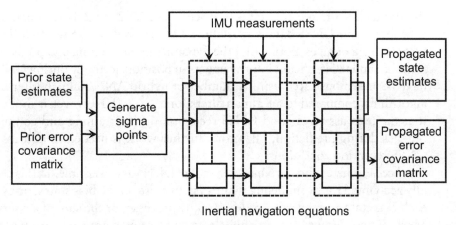

Figure 14.12 Total-state system propagation of INS states using a UKF.

14.4.5 Advanced IMU Error Modeling

Low-grades of IMU, particularly those using MEMS sensors, can exhibit high levels of noise. To optimize the Kalman filter gain, as discussed in Section 3.3.1, it is important to match the assumed sensor noise to its true value. Unfortunately, the manufacturer's specification may not be an accurate guide due both to variation in noise performance between individual sensors and to variation in effective noise levels with the vibration environment (see Sections 4.4.3 and 4.4.5).

One solution is to use an adaptive Kalman filter to vary the assumed system noise covariance according to the measurement innovations as described in Section 3.4.4. Both the innovation-based and multiple-model adaptive estimation techniques have been shown to speed up the rate of convergence of the state estimates with their true counterparts [54–57]. Interestingly, these algorithms tend to select high levels of system noise initially, leading to high Kalman filter gains and faster convergence, and then switch to lower system noise after convergence, producing lower gains, more stable state estimates, and smaller state uncertainties. Note that IAE-derived covariances may require rescaling to account for time-correlated measurement noise. Using an ANN to estimate the system noise covariance as a function of velocity increments and attitude increments [58] and using reinforcement learning to automate the manual tuning process [59] have also been shown to improve performance.

When the dominant vibration modes are known, or determined from the IMU data, they can be incorporated into the Kalman filter. Modeling the sensor noise as correlated by second-order Markov processes using a Schmidt-Kalman filter (Section D.2 of Appendix D on the CD) has been shown to improve alignment performance for missiles in an air-carriage environment [60].

MEMS IMUs can also exhibit complex higher-order systematic and slowly time-varying errors that are difficult to model using a Kalman filter. However, these errors can be modeled using an ANN alongside a conventional Kalman filter estimating the standard 15 INS error states (see Sections 14.2.2–14.2.4). The neural network is trained while GNSS data is available and then predicts the residual INS position errors during outages, noting that the dynamics in the training and prediction phases must be similar. Significant improvements in position accuracy during GNSS outages have been demonstrated using a number of hybrid ANN/Kalman filter integration algorithms, compared to a Kalman filter alone [61–65]. However, it should be noted that the tests assumed good GNSS reception during training and complete GNSS outages during prediction. Therefore, further work is needed to address degraded GNSS reception conditions.

Accelerometer and gyro biases (Section 4.4.1) vary over time. This is conventionally accounted for by modeling white system noise on the bias states, or by modeling each bias state as either first-order Markov processes or the sum of a constant and a Markov state. However, these models are only a rough approximation of the sensor behavior. Performance improvements have been demonstrated by representing the biases as second- and third-order autoregressive models, tuned to each sensor type [66, 67], while a frequency-domain approach [68] avoids the need to make a priori assumptions about the time variations and can improve the speed of convergence.

All of these methods of modeling IMU errors require more processing capacity, and were relatively immature at the time of writing. However, they demonstrate the scope to improve upon the conventional system models described in Section 14.2.

14.4.6 Smoothing

For many applications, such as surveying, geo-referencing, vehicle testing, and military ranges, the navigation solution is required for analysis after the event. In these cases, the INS errors can be calibrated using GPS measurements taken after the time of interest as well as before. A standard Kalman filter will not do this; the solution is to use a Kalman smoother, as described in Section 3.4.6. Whether smoothing significantly improves performance depends on the application. It is useful when it is not practical to undergo a period of INS calibration before the data set of interest or where a heading solution is required and the heading error is difficult to observe. However, it has the biggest impact where GNSS signal availability is relatively poor, such as in urban areas, particularly where carrier-phase accuracy is required [69] or a low-grade IMU is used [70]. Smoothing effectively halves the period of INS drift during GNSS outages, reducing the maximum position error by a factor of up to 4.

Problems and exercises for this chapter are on the accompanying CD.

References

[1] Titterton, D. H., and J. L. Weston, *Strapdown Inertial Navigation Technology*, 2nd ed., Stevenage, U.K.: IEE, 2004.

[2] Grewal, M. S., L. R. Weill, and A. P. Andrews, *Global Positioning Systems, Inertial Navigation, and Integration*, 2nd ed., New York: Wiley, 2007.

[3] Farrell, J. A., *Aided Navigation*, New York: McGraw-Hill, 2008.

[4] Phillips, R. E., and G. T. Schmidt, "GPS/INS Integration," *Proc. NATO AGARD MSP Lecture Series on "System Implications and Innovative Applications of Satellite Navigation,"* LS-207, Paris, France, July 1996.

[5] Greenspan, R. L., "GPS and Inertial Integration," in *Global Positioning System: Theory and Applications, Volume II*, B. W. Parkinson and J. J. Spilker, Jr., (eds.), Washington, D.C.: AIAA, 1996, pp. 187–220.

[6] Wagner, J. F., G. Kasties, and M. Klotz, "An Alternative Filter Approach to Integrate Satellite Navigation and Inertial Sensors," *Proc. ION NTM*, Santa Monica, CA, January 1997, pp. 141–150.

[7] Cox, D. B., Jr., "Integration of GPS with Inertial Navigation Systems," *Navigation: JION*, Vol. 25, No. 2, 1978, pp. 236–245.

[8] Groves, P. D., and D. C. Long, "Inertially-Aided GPS Signal Re-Acquisition in Poor Signal to Noise Environments and Tracking Maintenance Through Short Signal Outages," *Proc. ION GNSS 2005*, Long Beach, CA, September 2005, pp. 2408–2417.

[9] Kubrak, D., et al., "Improvement of GNSS Signal Acquisition Using Low-Cost Inertial Sensors," *Proc. ION GNSS 2008*, Savannah GA, September 2008, pp. 2145–2155.

[10] Alban, S., et al., "Performance Analysis and Architectures for INS-Aided GPS Tracking Loops," *Proc. ION NTM*, Anaheim, CA, January 2003, pp. 611–622.

[11] Groves, P. D., and D. C. Long, "Combating GNSS Interference with Advanced Inertial Integration," *Journal of Navigation*, Vol. 58, No. 3, 2005, pp. 419–432.

[12] Groves, P. D. and D. C. Long, "Adaptive Tightly Coupled, a Low Cost Alternative Anti-Jam INS/GPS Integration Technique," *Proc. ION NTM*, Anaheim, CA, January 2003, pp. 429–440.

[13] Gebre-Egziabher, D., et al., "Sensitivity and Performance Analysis of Doppler-Aided GPS Carrier-Tracking Loops," *Navigation: JION*, Vol. 52, No. 2, 2005, pp. 49–60.

[14] Bye, C. T., G. L. Hartmann, and A. Killen, "Inertial and GPS Technology Advances on the GGP Program," *Proc. ION 53rd AM*, June 1997, pp. 639–648.

[15] Copps, E. M., et al., "Optimal Processing of GPS Signals," *Navigation: JION*, Vol. 27, No. 3, 1980, pp. 171–182.

[16] Groves, P. D., and C. J. Mather, "Receiver Interface Requirements for Deep INS/GNSS Integration and Vector Tracking," *Journal of Navigation*, Vol. 63, No. 3, 2010, pp. 471–489.

[17] Carlson, N. A., "Federated Filter for Distributed Navigation and Tracking Applications," *Proc. ION 58th AM*, Albuquerque, NM, June 2002, pp. 340–353.

[18] Groves, P. D., C. J. Mather and A. A. Macaulay, "Demonstration of Non-Coherent Deep INS/GPS Integration for Optimized Signal to Noise Performance," *Proc. ION GNSS 2007*, Fort Worth, TX, September 2007, pp. 2627–2638.

[19] Gustafson, D., J. Dowdle, and K. Flueckiger, "A Deeply Integrated Adaptive GPS-Based Navigator with Extended Range Code Tracking," *Proc. IEEE PLANS*, San Diego, CA, March 2000, pp. 118–124.

[20] Buck, T. M., J. Wilmore, and M. J. Cook, "A High G, MEMS Based, Deeply Integrated, INS/GPS, Guidance, Navigation and Control Flight Management Unit," *Proc. IEEE/ION PLANS*, San Diego, CA, April 2006, pp. 772–794.

[21] Mather, C. J., A. A. Macaulay, and A. R. Pratt, "Performance Evaluation of Ultra Tightly Coupled INS/GPS Integration Using a C/A Code GPS Receiver," *Proc. European Navigation Conference*, London, U.K., November 2011.

[22] Kiesel, S., et al., "Discriminator Weighting and Performance of a Deeply Coupled GPS/INS System at Low CNo," *Proc. ION ITM*, San Diego, CA, January 2011, pp. 858–867.

[23] Soloviev, A., S. Gunawardena, and F. Van Graas, "Deeply Integrated GPS/Low-Cost IMU for Low CNR Signal Processing: Concept Description and In-Flight Demonstration," *Proc. ION GNSS 2004*, Long Beach, CA, September 2004, pp. 1598–1608.

[24] Soloviev, A., S. Gunawardena, and F. Van Graas, "Deeply Integrated GPS/Low-Cost IMU for Low CNR Signal Processing: Flight Test Results and Real Time Implementation," *Navigation: JION*, Vol. 55, No. 1, 2008, pp. 1–13.

[25] Lashley, M., D. M. Bevly, and J. Y. Hung, "Impact of Carrier to Noise Power Density, Platform Dynamics, and IMU Quality on Deeply Integrated Navigation," *Proc. IEEE/ION PLANS*, Monterey, CA, May 2008, pp. 9–16.

[26] O'Driscoll, C., M. G. Petovello, and G. Lachapelle, "Impact of Extended Coherent Integration Times on Weak Signal RTK in an Ultra-Tight Receiver," *Proc. NAV 08 (Royal Institute of Navigation)*, London, U.K., October 2008.

[27] Jun, W., et al., "Quaternion-Based Attitude Estimation Using Multiple GPS Antennas, MEMS IMU," *Proc. ION GPS/GNSS 2003*, Portland, OR, September 2003, pp. 480–488.

[28] Rhee, I., M. F. Abdel-Hafez, and J. L. Speyer, "Observability of an Integrated GPS/INS During Manoeuvres," *IEEE Trans. on Aerospace and Electronic Systems*, Vol. 40, No. 2, 2004, pp. 526–535 and 1421.

[29] Britting, K. R., *Inertial Navigation Systems Analysis*, New York: Wiley, 1971 (Republished by Norwood, MA: Artech House, 2010).

[30] Farrell, J. L., *GNSS Aided Navigation and Tracking*, Baltimore, MD: American Literary Press, 2007.

[31] Farrell, J. L., "GPS/INS—Streamlined," *Navigation: JION*, Vol. 49, No. 4, 2002, pp. 171–182.

[32] Brown, R. G., and P. Y. C. Hwang, *Introduction to Random Signals and Applied Kalman Filtering*, 3rd ed., New York: Wiley, 1997.

[33] Wendel, J., T. Obert, and G. F. Trommer, "Enhancement of a Tightly Coupled GPS/INS System for High Precision Attitude Determination of Land Vehicles," *Proc. ION 59th AM*, Albuquerque, NM, June 2003, pp. 200–208.

[34] Sennott, J. W., and D. Senffner, *Navigation Receiver with Coupled Signal-Tracking Channels*, U.S. Patent 5,343,209, granted 1994.

[35] Abbott, A. S., and W. E. Lillo, *Global Positioning System and Inertial Measuring Unit Ultra-tight Coupling Method*, U.S. Patent 6,516,021, granted 2003.

[36] Beser, J., et al., "Trunav: A Low-Cost Guidance/Navigation Unit Integrating a SAASM-Based GPS and MEMS IMU in a Deeply Coupled Mechanization," *Proc. ION GPS 2002*, Portland OR, September 2002, pp. 545–555.

[37] Sivananthan, S., and J. Weitzen, "Improving Optimality of Deeply Coupled Integration of GPS and INS," *Proc. ION ITM*, Anaheim, CA, January 2009, pp. 426–433.

[38] Van Graas, F., et al., "Comparison of Two Approaches for GNSS Receiver Algorithms: Batch Processing and Sequential Processing Considerations," *Proc. ION GNSS 2005*, Long Beach, CA, September 2005, pp. 200–211.

[39] Petovello, M. G., M. E. Cannon, and G. Lachapelle, "Benefits of Using a Tactical Grade IMU for High-Accuracy Processing," *Navigation: JION*, Vol. 51, No. 1, 2004, pp. 1–12.

[40] Zhang, H. T., M. G. Petovello, and M. E. Cannon, "Performance Comparison of Kinematic GPS Integrated with Different Tactical Level IMUs," *Proc. ION NTM*, San Diego, CA, January 2005, pp. 243–254.

[41] Colombo, O. L., U. V. Bhapkar, and A. G. Evans, "Inertial-Aided Cycle-Slip Detection/Correction for Precise, Long-Baseline Kinematic GPS," *Proc. ION GPS-99*, Nashville, TN, September 1999, pp. 1915–1921.

[42] Farrell, J. A. and M. Barth, *The Global Positioning System and Inertial Navigation*, New York: McGraw-Hill, 1999.

[43] Lorga, J. F. M., Q. P. Chu, and J. A. Mulder, "Tightly Coupled IMU/GPS Carrier-Phase Navigation System," *Proc. ION NTM*, Anaheim, CA, January 2003, pp. 385–396.

[44] Tazartes, D., et al., "Synergistic Interferometric GPS-INS," *Proc. ION NTM*, Anaheim, CA, January 1995, pp. 657–671.

[45] Hirokawa, R., and T. Ebinuma, "A Low-Cost Tightly Coupled GPS/INS for Small UAVs Augmented with Multiple GPS Antennas," *Navigation: JION*, Vol. 56, No. 1, 2009, pp. 35–44.

[46] Wagner, J. F., and G. Kasties, "Modelling the Vehicle Kinematics as Key Element for the Design of Integrated Navigation System," *Proc. IAIN World Congress*, Berlin, Germany, October 2003.

[47] Scherzinger, B. M., "Inertial Navigator Error Models for Large Heading Uncertainty," *Proc. IEEE PLANS*, Atlanta, GA, April 1996, pp. 477–484.

[48] Rogers, R. M., "IMU In-Motion Alignment Without Benefit of Attitude Initialization," *Navigation: JION*, Vol. 44, No. 3, 1997, pp. 301–311.

[49] Han, S., and J. Wang, "A Novel Initial Alignment Scheme for Low-Cost INS Aided by GPS for Land Vehicle Applications," *Journal of Navigation*, Vol. 63, No. 4, 2010, pp. 663–680.

[50] Shin, E. -H., and N. El-Sheimy, "An Unscented Kalman Filter for In-Motion Alignment of Low-Cost IMUs," *Proc. IEEE PLANS*, Monterey, CA, April 2004, pp. 273–279.

[51] Fujioka, S., et al., "Comparison of Nonlinear Filtering Methods for INS/GPS In-Motion Alignment," *Proc. ION GNSS 2005*, Long Beach, CA, September 2005, pp. 467–477.

[52] Xing, Z., and D. Gebre-Egziabher, "Comparing Non-Linear Filters for Aided Inertial Navigators," *Proc. ION ITM*, Anaheim, CA, January 2009, pp. 1048–1053.

[53] Wendel, J., et al., "A Performance Comparison of Tightly Coupled GPS/INS Navigation Systems Based on Extended and Sigma Point Kalman Filters," *Navigation: JION*, Vol. 53, No. 1, 2006, pp. 21–31.

[54] Mohammed, A. H., and K. P. Schwarz, "Adaptive Kalman Filtering for INS/GPS," *Journal of Geodesy*, Vol. 73, 1999, pp. 193–203.

[55] Wang, J., M. Stewart, and M. Tsakiri, "Online Stochastic Modelling for INS/GPS Integration," *Proc. ION GPS '99*, Nashville, TN, September 1999, pp. 1887–1895.

[56] Hide, C., T. Moore, and M. Smith, "Adaptive Kalman Filtering for Low Cost INS/GPS," *Journal of Navigation*, Vol. 56, No. 1, 2003, pp. 143–152.

[57] Hide, C., T. Moore, and M. Smith, "Multiple Model Kalman Filtering for GPS and Low-Cost INS Integration," *Proc. ION GNSS 2004*, Long Beach, CA, September 2004, pp. 1096–1103.

[58] Lee, J. K., and C. Jekeli, "Neural Network Aided Adaptive Filtering and Smoothing for an Integrated INS/GPS Unexploded Ordnance Geolocation System," *Journal of Navigation*, Vol. 63, No. 2, 2010, pp. 251–268.

[59] Goodall, C., X. Niu, and N. El-Sheimy, "Intelligent Tuning of a Kalman Filter for INS/GPS Navigation Applications," *Proc. ION GNSS 2007*, Fort Worth, TX, September 2007, pp. 2121–2128.

[60] Wendel, J., and G. F. Trommer, "An Efficient Method for Considering Time Correlated Noise in GPS/INS Integration," *Proc. ION NTM*, San Diego, CA, January 2004, pp. 903–911.

[61] Kaygisiz, B. H., I. Erkmen, and A. M. Erkmen, "GPS/INS Enhancement for Land Navigation Using Neural Network," *Journal of Navigation*, Vol. 57, No. 2, 2004, pp. 297–310.

[62] El-Sheimy, N., W. Abdel-Hamid, and G. Lachapelle, "An Adaptive Neuro-Fuzzy Model for Bridging GPS Outages in MEMS-IMU/GPS Land Vehicle Navigation," *Proc. ION GNSS 2004*, Long Beach, CA, September 2004, pp. 1088–1095.

[63] Goodall, C., N. El-Sheimy, and K.-W. Chiang, "The Development of a GPS/MEMS INS Integrated System Utilizing a Hybrid Processing Architecture," *Proc. ION GNSS 2005*, Long Beach, CA, September 2005, pp. 1444–1455.

[64] Abdelazim, T., W. Abdel-Hamid, and N. El-Sheimy, "A Genetic Fuzzy and Kalman Filtering Model for MEMS-IMU/GPS Integration," *Proc. ION ITM*, Anaheim, CA, January 2009, pp. 609–616.

[65] Aggarwal, P., et al., *MEMS-Based Integrated Navigation*, Norwood, MA: Artech House, 2010.

[66] Nassar, S., et al., "Modeling Inertial Sensor Errors Using Autoregressive (AR) Models," *Proc. ION NTM*, Anaheim, CA, January 2003, pp. 116–125

[67] Nassar, S., and N. El-Sheimy, "A Combined Algorithm of Improving INS Error Modeling and Sensor Measurements for Accurate INS/GPS Navigation," *GPS Solutions*, Vol. 10, No. 1, 2006, pp. 29–39.

[68] Soloviev, A., and F. van Graas, "Investigation into Performance Characteristics of Frequency Domain INS Calibration Procedure Under Noisy GPS Environments," *Proc. ION GPS 2002*, Portland, OR, September 2002, pp. 1454–1463.

[69] Shin, E. -H., and N. El-Sheimy, "Optimizing Smoothing Computation for Near-Real-Time GPS Measurement Gap Filling in INS/GPS Systems," *Proc. ION GPS 2002*, Portland, OR, September 2002, pp. 1434–1441.

[70] Hide, C., and T. Moore, "GPS and Low Cost INS Integration for Positioning in the Urban Environment," *Proc. ION GNSS 2005*, Long Beach, CA, September 2005, pp. 1007–1015.

CHAPTER 15
INS Alignment, Zero Updates, and Motion Constraints

This chapter describes a number of Kalman filter-based methods of calibrating and constraining INS errors. Transfer alignment and quasi-stationary alignment are fine alignment methods (Section 5.6.3) that improve the calibration of an INS attitude solution and inertial sensor errors between initialization (Section 5.6) and use of the inertial navigation solution. In transfer alignment, described in Section 15.1, measurements from a nearby reference navigation system are used to initialize, align, and calibrate an INS over a few seconds or minutes while in motion. In quasi-stationary alignment, described in Section 15.2, the fact that the INS is roughly stationary with respect to the Earth is used as an alignment reference. Coarse quasi-stationary alignment may be used for heading initialization of an aviation- or marine-grade INS, while fine quasi-stationary alignment follows when the heading is approximately known.

A zero velocity update (ZVU or ZUPT), described in Section 15.3, operates on the same principle as fine quasi-stationary alignment. It is used in land applications to correct the navigation solution and calibrate the INS errors whenever the host vehicle (or pedestrian) stops. A zero angular rate update (ZARU), may be similarly applied when the host is known not to be rotating.

Finally, motion constraints, also known as nonholonomic constraints, exploit constraints in the freedom of movement of the host vehicle or pedestrian to correct and calibrate INS errors. These are described in Section 15.4 and are often used with a partial IMU (Section 5.9).

The techniques described in this chapter are based on similar principles to INS/GNSS integration, described in Chapter 14. With the exception of coarse quasi-stationary alignment, they may be implemented as additional measurement models within a common Kalman filter or EKF as the system models are very similar.

15.1 Transfer Alignment

Transfer alignment is used to initialize, align, and calibrate an INS in motion. Applications include:

- Guided weapons and UAVs launched from aircraft and ships;
- Aircraft, AUVs, and ROVs launched from ships;
- Torpedoes launched from ships and submarines;
- Sensor-pointing IMUs on all types of vehicle, for which alignment runs continuously.

The discussion here focuses on airborne transfer alignment, sometimes known as in-flight alignment, a term also applied to alignment from GNSS.

Figure 15.1 illustrates the most challenging airborne transfer alignment environment, whereby the weapon, UAV, or sensor pod containing the aligning INS is mounted on a wing pylon and the reference navigation system, an INS or integrated INS/GNSS, is mounted in the aircraft body. This maximizes the linear and angular motion of the lever arm between the two navigation systems. Flexure of both the wing and pylon occurs when the aircraft maneuvers and as the loading changes over time due to fuel consumption and the launching of other weapons or UAVs. Vibration occurs due to turbulence and the transmission of engine vibration from the aircraft. More benign environments include a shoulder pylon, closer to the aircraft body, and a trapeze inside the aircraft, though flexure and vibration still occur in both cases.

Transfer alignment comprises up to three phases: a "one-shot" initialization, a measurement-matching phase, and a reinitialization immediately prior to the launch of the aligning INS's host vehicle [1]. The one-shot phase initializes the aligning INS with the reference navigation system's position, velocity, and attitude solution, corrected with the estimated lever arm between and relative orientation of the two navigation systems, and predicted forward to compensate the data transmission lag. This is good enough for position initialization. However, flexure and vibration can limit the attitude initialization accuracy to about 2°, which, in turn, can lead to position error growth in excess of 500m over the first minute (see Section 5.7.1).

The measurement-matching phase compares the aligning-INS and reference navigation solutions using a Kalman filter (Chapter 3) to estimate corrections to the aligning-INS navigation solution and calibrate the IMU errors. The duration of this phase can vary from 2 seconds to many minutes, depending on the application. Best performance requires at least 2 minutes. Figure 15.2 illustrates this with continuous closed-loop correction of the INS (see Section 14.1.1). Conventional measurement matching uses only linear measurements, while rapid alignment also uses angular measurements. Both are described next, followed by a discussion of different types of reference navigation system. Sections I.3.5 and I.5.4 of Appendix I on the CD, describe implementations resolved in the wander-azimuth-frame and with body-frame-resolved attitude errors, respectively.

At the reinitialization phase, the aligning-INS position solution is reset from the aircraft's integrated position solution. Furthermore, if the error covariance matrix, **P**, is used to initialize a subsequent integration Kalman filter, it must be reset to account for the errors of the reference navigation system. This is because these

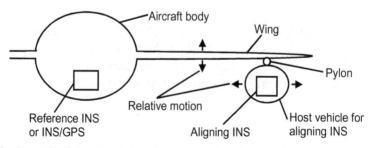

Figure 15.1 Example of an airborne transfer alignment environment.

15.1 Transfer Alignment 629

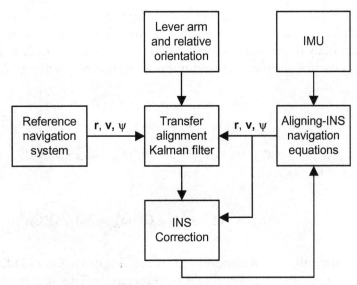

Figure 15.2 Architecture of transfer-alignment measurement-matching phase.

errors are inherited by the aligning INS, but are not accounted for by $\mathbf{P}$ during the measurement-matching phase.

15.1.1 Conventional Measurement Matching

Conventional transfer alignment uses only linear measurement matching. In contrast to many forms of INS/GNSS integration, either position or velocity measurements are used, not both. Position and velocity measurements comprise the same information, as one is the integral of the other. Using velocity measurements generally leads to faster estimation of the attitude and IMU errors as position is a further integration step from them. However, at low measurement-update rates, around 1 Hz, position measurements can perform better due to the averaging of measurement noise due to lever-arm vibration, Using time-averaged velocity measurements combines the benefits of both measurement types [2]. Faster measurement update or averaging rates also reduce state estimation biases caused by synchronization of the vibration and measurement-update cycles obstructing the averaging out of vibration-induced noise [3].

The system model and state selection for transfer alignment is essentially the same as for the INS states in INS/GNSS integration, described in Section 14.2. However, where velocity matching is used with closed-loop correction of the aligning-INS navigation solution, the position error states may be omitted as the position error growth over a few minutes of transfer alignment will be small. Position is also reset at the reinitialization phase. The error in the assumed lever arm between the aligning INS and reference navigation system is sometimes estimated. However, it is difficult to observe, while lever-arm errors of up to 0.5m are easily tolerated.

The measurement model is analogous to that for loosely coupled INS/GNSS integration (Section 14.3.1). The measurement innovation comprises the difference between the reference navigation system and corrected aligning-INS velocity solutions, accounting for the lever arm from the reference to the aligning INS, l_{rb}^b, which is assumed here to be well known. Note that r denotes the reference-navigation-system body frame. For the implementations resolved in ECI, ECEF, and local navigation frames, the measurement innovations are, respectively,

$$\delta z_{V,k}^{i-} = \left[\hat{v}_{irR}^i - \hat{v}_{ib}^i + \hat{C}_b^i (\hat{\omega}_{ib}^b \wedge l_{rb}^b) \right]_k, \quad (15.1)$$

$$\delta z_{V,k}^{e-} = \left[\hat{v}_{erR}^e - \hat{v}_{eb}^e + \hat{C}_b^e (\hat{\omega}_{ib}^b \wedge l_{rb}^b) - \Omega_{ie}^e \hat{C}_b^e l_{rb}^b \right]_k, \quad (15.2)$$

and

$$\delta z_{V,k}^{n-} = \left[\hat{v}_{erR}^n - \hat{v}_{eb}^n + \hat{C}_b^n (\hat{\omega}_{ib}^b \wedge l_{rb}^b) - \hat{\Omega}_{ie}^n \hat{C}_b^n l_{rb}^b \right]_k, \quad (15.3)$$

where the subscript R denotes reference-navigation-system-indicated and Ω_{ie}^e and Ω_{ie}^n are given by (5.25) and (5.41), respectively. The aligning-INS and reference velocity solutions must be time synchronized as described in Section 3.3.4. The time offset may also be estimated as a Kalman filter state as described in Section I.6.3 of Appendix I on the CD.

Defining the state vector as

$$\mathbf{x}^\gamma = \begin{pmatrix} \delta \psi_{\gamma b}^\gamma \\ \delta \mathbf{v}_{\beta b}^\gamma \\ \mathbf{b}_a \\ \mathbf{b}_g \\ \vdots \end{pmatrix} \quad \{\beta, \gamma\} \in \{i, i\}, \{e, e\}, \{e, n\}, \quad (15.4)$$

where $\delta \psi_{\gamma b}^\gamma$ is the attitude error, $\delta \mathbf{v}_{\beta b}^\gamma$ is the velocity error, $\mathbf{b}_a$ is the accelerometer biases, and $\mathbf{b}_g$ is the gyro biases of the aligning INS, the measurement matrix is

$$\mathbf{H}_{V,k}^\gamma = \begin{pmatrix} \mathbf{H}_{v1}^\gamma & -\mathbf{I}_3 & \mathbf{0}_3 & \mathbf{H}_{v4}^\gamma & 0 \end{pmatrix}_k, \quad \gamma \in i, e, n, \quad (15.5)$$

where

$$\begin{aligned} \mathbf{H}_{v1}^i &\approx -\left[\{ \hat{C}_b^i (\hat{\omega}_{ib}^b \wedge l_{rb}^b) \} \wedge \right] \\ \mathbf{H}_{v4}^i &\approx -\hat{C}_b^i [l_{rb}^b \wedge] \end{aligned}, \quad (15.6)$$

$$\begin{aligned} \mathbf{H}_{v1}^e &\approx -\left[\{ \hat{C}_b^e (\hat{\omega}_{ib}^b \wedge l_{rb}^b) - \Omega_{ie}^e \hat{C}_b^e l_{rb}^b \} \wedge \right] \\ \mathbf{H}_{v4}^e &\approx -\hat{C}_b^e [l_{rb}^b \wedge] \end{aligned}, \quad (15.7)$$

and

15.1 Transfer Alignment

$$\begin{aligned} \mathbf{H}_{v1}^n &\approx -\left[\left\{\hat{\mathbf{C}}_b^n\left(\hat{\boldsymbol{\omega}}_{ib}^b \wedge \mathbf{l}_{rb}^b\right) - \hat{\boldsymbol{\Omega}}_{ie}^n \hat{\mathbf{C}}_b^n \mathbf{l}_{rb}^b\right\} \wedge\right] \\ \mathbf{H}_{v4}^n &\approx -\hat{\mathbf{C}}_b^n\left[\mathbf{l}_{rb}^b \wedge\right] \end{aligned} \quad (15.8)$$

The coupling of the attitude errors and gyro biases into the measurements through the lever arm is weak, so a suitable approximation is

$$\mathbf{H}_{V,k}^{\gamma} \approx \begin{pmatrix} \mathbf{0}_3 & -\mathbf{I}_3 & \mathbf{0}_3 & \mathbf{0}_3 & \mathbf{0} \end{pmatrix}_k \quad \gamma \in i, e, n. \quad (15.9)$$

The measurement noise arises mainly from lever-arm vibration and time-synchronization errors. It thus depends on the host aircraft and where the aligning INS is mounted. For optimum performance, a suitable value for the measurement noise covariance matrix, $\mathbf{R}_V$, should be determined for each scenario. However, in practice, a worst-case value is often assumed.

The attitude and IMU errors are determined mainly from the time evolution of the velocity error as explained in Section 14.3.4. The observability of these errors from velocity-matching measurements is the same as for INS/GNSS integration, discussed in Section 14.2.1. To observe the heading error and separate the roll and pitch errors from the horizontal accelerometer biases, the host vehicle must undergo significant maneuvering, including turns. Without this, alignment performance is significantly degraded. An s-weave alignment maneuver is typically performed [4]. Note that the heading calibration degrades slightly between the end of the maneuver and completion of transfer alignment.

15.1.2 Rapid Transfer Alignment

Rapid transfer alignment algorithms add attitude-measurement matching to the linear-measurement matching of conventional transfer alignment. This is designed to speed up the estimation of attitude errors, enabling a transfer alignment to take place within 10 seconds [5], noting that longer periods are still needed to calibrate the IMU errors. Rapid transfer alignment reduces the maneuver requirement, removing the need for turns if a wing rock (pair of equal and opposite rolls) is performed. It also prevents subsequent degradation of the heading alignment.

Rapid transfer alignment was first demonstrated on a helicopter [6]. Although there are significant levels of vibration, the maneuver-dependent lever-arm flexure is limited. However, where the aligning INS is mounted on a wing pylon of a fixed-wing aircraft, the relative orientation of the aligning INS and the reference can change significantly as the aircraft maneuvers due to wing flexure. The roll relative orientation during roll maneuvers is particularly affected and can change by a few degrees, severely biasing some of the attitude and IMU-error estimates if the Kalman filter does not model the flexure [7, 8]. This happens because the forces in the aircraft body frame are significantly different in roll maneuvers to those in level flight and coordinated turns, as Figure 15.3 illustrates. The simplest solution to this problem is to limit attitude-measurement matching to the heading component as this is least affected by flexure, while the heading attitude error is

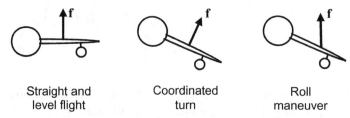

Figure 15.3 Specific forces on an aircraft wing during different maneuvers.

most difficult to observe using conventional transfer alignment. The heading measurement innovation is simply

$$\delta z^-_{\psi,k} = \left(\hat{\psi}_{nr} - \hat{\psi}_{nb} + \psi_{rb}\right)_k, \tag{15.10}$$

where $\hat{\psi}_{nb}$ is the aligning-INS-indicated heading, $\hat{\psi}_{nr}$ is the reference-indicated heading, and the relative heading, ψ_{rb}, is assumed to be known. $\delta z^-_{\psi,k}$ must be limited to the range $-\pi$ to π by adding or subtracting 2π as appropriate. Assuming the state vector defined by (15.4), the measurement matrix is

$$\mathbf{H}^n_{\psi,k} = \left(\begin{pmatrix} 0 & 0 & -1 \end{pmatrix} \quad \mathbf{0}_{1\times 3} \quad \mathbf{0}_{1\times 3} \quad \mathbf{0}_{1\times 3} \quad 0 \right)_k, \tag{15.11}$$

or

$$\mathbf{H}^\gamma_{\psi,k} = \left(-\mathbf{C}^n_{\gamma 3,1:3} \quad \mathbf{0}_{1\times 3} \quad \mathbf{0}_{1\times 3} \quad \mathbf{0}_{1\times 3} \quad 0 \right)_k \quad \gamma \in i,e. \tag{15.12}$$

Alternatively, three-component attitude matching can be made to work by estimating the coefficients of the relative orientation's variation with specific force as additional Kalman filter states [8]. The full attitude-matching measurement innovation, $\delta \mathbf{z}^{\gamma-}_{A,k}$, is given by

$$\mathbf{I}_3 + \left[\delta \mathbf{z}^{\gamma-}_{A,k} \wedge\right] = \hat{\mathbf{C}}^\gamma_r \hat{\mathbf{C}}^r_b \hat{\mathbf{C}}^b_\gamma \quad \gamma \in i,e,n, \tag{15.13}$$

where the estimated relative orientation, assumed to be a small angle, is given by [8]

$$\hat{\mathbf{C}}^r_b = \mathbf{I}_3 + [\hat{\boldsymbol{\psi}}_{rb} \wedge]$$

$$\hat{\boldsymbol{\psi}}_{rb} = \hat{\boldsymbol{\psi}}_{rb,s} + \begin{pmatrix} 0 & \hat{\eta}_{xy} & \hat{\eta}_{xz} \\ \hat{\eta}_{yx} & 0 & \hat{\eta}_{yz} \\ \hat{\eta}_{zx} & \hat{\eta}_{zy} & 0 \end{pmatrix} \begin{pmatrix} f^b_{ib,x} \\ f^b_{ib,y} \\ f^b_{ib,z} + g \end{pmatrix}, \tag{15.14}$$

where $\psi_{rb,s}$ is the static relative orientation, $\boldsymbol{\eta} = \{\eta_{xy}, \eta_{xz}, \eta_{yx}, \eta_{yz}, \eta_{zx}, \eta_{zy}\}$ are the flexure coefficients, and g is the acceleration due to gravity. Defining the state vector as

$$\mathbf{x}^\gamma = \begin{pmatrix} \delta\boldsymbol{\psi}^\gamma_{\gamma b} \\ \delta\mathbf{v}^\gamma_{\beta b} \\ \mathbf{b}_a \\ \mathbf{b}_g \\ \boldsymbol{\psi}_{rb,s} \\ \boldsymbol{\eta} \\ \vdots \end{pmatrix} \quad \{\beta,\gamma\} \in \{i,i\},\{e,e\},\{e,n\}, \tag{15.15}$$

the attitude measurement matrix is

$$\mathbf{H}^\gamma_{A,k} = \begin{pmatrix} -\mathbf{I}_3 & \mathbf{0}_3 & \mathbf{0}_3 & \mathbf{0}_3 & \hat{\mathbf{C}}^\gamma_b & \mathbf{H}^\gamma_{a6} & 0 \end{pmatrix}_k \quad \gamma \in i,e,n, \tag{15.16}$$

where

$$\mathbf{H}^\gamma_{a6} = \hat{\mathbf{C}}^\gamma_b \begin{pmatrix} f^b_{ib,y} & f^b_{ib,z}+g & 0 & \mathbf{0}_3 & 0 & 0 \\ 0 & 0 & f^b_{ib,x} & f^b_{ib,z}+g & 0 & 0 \\ 0 & 0 & 0 & 0 & f^b_{ib,x} & f^b_{ib,y} \end{pmatrix} \quad \gamma \in i,e,n. \tag{15.17}$$

With a tactical-grade IMU, the posttransfer-alignment position drift is typically within 10m per axis over the first 60 seconds, excluding errors inherited from the reference navigation system [1, 9]. A large boost maneuver immediately after the launch of the aligning INS's host vehicle can degrade this as it increases the coupling of attitude and accelerometer scale factor errors into the position and velocity solution (see Section 5.7.3).

15.1.3 Reference Navigation System

An integrated INS/GNSS navigation system provides a more accurate reference for transfer alignment than a stand-alone INS, as the velocity solution does not drift. It can also provide a better reference than stand-alone GNSS, as the velocity is less noisy and attitude measurements are available [10]. However, an INS/GNSS velocity can exhibit a transient when reacquisition of GNSS following a long outage leads to a large correction to the integrated navigation solution. Such a transient can disrupt transfer alignment as the Kalman filter wrongly attributes the velocity change to attitude and IMU errors in the aligning INS [8]. There are three main options for resolving this.

The simplest solution is to use a stand-alone INS as the reference. A velocity correction from the host aircraft's INS/GNSS solution is then applied alongside the position reset at the reinitialization phase at the end of transfer alignment. The degradation of the attitude and IMU-error estimates is minimal. However, if the vertical channel of the reference INS is baro-aided (see Section 6.2.1), the vertical velocity is subject to transients which are difficult to model in the transfer-alignment Kalman filter. Therefore, velocity-measurement matching is limited to the horizontal components. This degrades the vertical navigation performance of the aligned INS by about a factor of 2.

The optimal solution to the reference transient problem is to transmit to the aligning INS the velocity (and attitude) corrections applied to the reference INS. At the aligning INS, they are applied directly to the navigation solution, bypassing the Kalman filter. Consequently, no transients are seen in the transfer alignment measurement innovations [8]. The third approach is to implement innovation sequence monitoring, as described in Sections 17.3.2 and 17.3.3 [8].

15.2 Quasi-Stationary Alignment

Quasi-stationary alignment operates only when the INS is approximately stationary with respect to the Earth. This information is used as a reference, against which the velocity, attitude, and IMU errors are calibrated. Vibratory motion and small displacements due to human activity, such as loading, boarding, and fuelling, are treated as measurement noise. Quasi-stationary alignment may also be used on a ship provided that the assumed measurement noise is large enough to account for the oscillatory motion induced by the sea state and that the update rate is not a harmonic of the sea-state frequency.

Three types of measurement may be used: specific force, velocity, and position displacement or integrated velocity. In a low-vibration environment, specific-force measurements [11] provide faster estimation of the attitude and IMU errors, as the measurements are fewer integration steps away from these errors. However, in a high-vibration environment, position-displacement [12] or integrated velocity [13] measurements give better performance, as the standard deviation of the position displacement is correctly modeled as a constant, rather than growing with time. Use of integrated specific force measurements has also been proposed [14].

Coarse alignment is described first, followed by fine alignment.

15.2.1 Coarse Alignment

Where the heading is unknown at the start of the quasi-stationary alignment, it can be determined during the alignment using indirect gyrocompassing, provided the gyros are sufficiently accurate, as discussed in Section 5.6.2. With a known pitch and roll, but initially unknown heading, a wander-azimuth coordinate frame (see Section 2.1.6) must be used. The algorithm described below uses an error-state Kalman filter with position displacement measurements.

The inertial navigation equations may be simplified as the average Earth-referenced velocity is zero. Furthermore, the north component of the acceleration due to gravity is neglected as it is very small near the Earth's surface and the direction of the wander-azimuth frame axes with respect to north is initially unknown. The attitude may be updated using

15.2 Quasi-Stationary Alignment

$$\hat{\mathbf{C}}_b^w(+) \approx \hat{\mathbf{C}}_b^w(-)\left(\mathbf{I}_3 + \mathbf{\Omega}_{ib}^b\right)$$

$$-\omega_{ie}\tau_i \begin{pmatrix} 0 & \sin L_b & -\sin\psi_{nw}\cos L_b \\ -\sin L_b & 0 & -\cos\psi_{nw}\cos L_b \\ \sin\psi_{nw}\cos L_b & \cos\psi_{nw}\cos L_b & 0 \end{pmatrix} \hat{\mathbf{C}}_b^w(-), \quad (15.18)$$

with orthogonalization and normalization as described in Section 5.5.1. The velocity and position displacement are then updated using

$$\mathbf{v}_{eb}^w(+) \approx \mathbf{v}_{eb}^w(-) + \mathbf{f}_{ib}^w \tau_i + \begin{pmatrix} 0 \\ 0 \\ g_{b,0}^n(L_b, h_b) \end{pmatrix} \tau_i \quad (15.19)$$

and

$$\Delta \mathbf{r}_{eb}^w(+) = \Delta \mathbf{r}_{eb}^w(-) + \frac{\tau_i}{2}\left(\mathbf{v}_{eb}^w(-) + \mathbf{v}_{eb}^w(+)\right). \quad (15.20)$$

The latitude, L_b, and height, h_b, obtained from a position initialization procedure (Section 5.6.1) are assumed constant. It is also assumed that the specific-force and angular-rate measurements are corrected using the IMU-error estimates and that closed-loop correction of the position displacement, velocity, attitude, and wander angle, ψ_{nw}, from the Kalman filter takes place (see Sections 3.2.6 and 14.1.1).

The sine and cosine of the wander angle are treated as separate parameters, both in the navigation equations and the Kalman filter, in order to maintain linearity in the latter. As both have zero-mean distributions, they can each be initialized at zero. Once confident estimates of both have been obtained, they may be made consistent by applying a common scaling factor to achieve the relationship $\cos^2\psi_{nw} + \sin^2\psi_{nw} = 1$.

A suitable Kalman filter state vector, assuming operation away from the poles, is thus

$$\mathbf{x}^w = \begin{pmatrix} \delta\mathbf{\psi}_{wb}^w \\ \delta\mathbf{v}_{eb}^w \\ \delta\mathbf{r}_{eb}^w \\ \delta\sin\psi_{nw} \\ \delta\cos\psi_{nw} \\ \mathbf{b}_a \\ \mathbf{b}_g \\ \vdots \end{pmatrix}, \quad (15.21)$$

where $\delta\mathbf{\psi}_{wb}^w$, $\delta\mathbf{v}_{eb}^w$, and $\delta\mathbf{r}_{eb}^w$ are the attitude, velocity, and position errors (see Section 5.7); $\delta\sin\psi_{nw}$ and $\delta\cos\psi_{nw}$ are the errors in the sine and cosine of the wander angle;

and $\mathbf{b}_a$ and $\mathbf{b}_g$ are the accelerometer and gyro biases. The state-propagation equations that form the system model may be simplified from those described in Section 14.2 by assuming $\mathbf{v}_{eb}^w \approx 0$, giving

$$\mathrm{E}\left(\delta\dot{\boldsymbol{\psi}}_{wb}^w\right) = -\left[\hat{\boldsymbol{\omega}}_{ie}^w \wedge\right]\delta\boldsymbol{\psi}_{wb}^w + \omega_{ie}\cos\hat{L}_b\left(\begin{pmatrix}0\\1\\0\end{pmatrix}\delta\sin\psi_{nw} - \begin{pmatrix}1\\0\\0\end{pmatrix}\delta\cos\psi_{nw}\right) + \hat{\mathbf{C}}_b^w\mathbf{b}_g$$

$$\mathrm{E}\left(\delta\dot{\mathbf{v}}_{eb}^w\right) \approx -\left[\left(\hat{\mathbf{C}}_b^w\hat{\mathbf{f}}_{ib}^b\right)\wedge\right]\delta\boldsymbol{\psi}_{wb}^w + \hat{\mathbf{C}}_b^w\mathbf{b}_a$$

$$\mathrm{E}\left(\delta\dot{\mathbf{r}}_{eb}^w\right) = \delta\mathbf{v}_{eb}^w$$

$$\mathrm{E}\left(\frac{\partial\delta\sin\psi_{nw}}{\partial t}\right) = 0 \quad \mathrm{E}\left(\frac{\partial\delta\cos\psi_{nw}}{\partial t}\right) = 0$$

$$\mathrm{E}\left(\dot{\mathbf{b}}_a\right) = 0 \quad \mathrm{E}\left(\dot{\mathbf{b}}_g\right) = 0$$

(15.22)

where

$$\hat{\boldsymbol{\omega}}_{ie}^w = \omega_{ie}\begin{pmatrix}\cos\hat{\psi}_{nw}\cos\hat{L}_b\\-\sin\hat{\psi}_{nw}\cos\hat{L}_b\\\sin\hat{L}_b\end{pmatrix}. \tag{15.23}$$

The wander-angle errors prevent compensation of the attitude, $\mathbf{C}_b^w$, for the rotation of the Earth. It is the propagation of these errors through the system model that enables the Kalman filter to calibrate the wander angle and hence the heading, $\psi_{nb} = \psi_{nw} + \psi_{wb}$, from the position-displacement measurements (see Section 14.3.4).

The position-displacement measurements are always zero, so the measurement innovation for quasi-stationary alignment is simply

$$\delta\mathbf{z}_{Q,k}^{w-} = -\Delta\hat{\mathbf{r}}_{eb,k}^w, \tag{15.24}$$

while the measurement matrix is

$$\mathbf{H}_{Q,k}^w = \begin{pmatrix}\mathbf{0}_3 & \mathbf{0}_3 & -\mathbf{I}_3 & \mathbf{0}_{1,3} & \mathbf{0}_{1,3} & \mathbf{0}_3 & \mathbf{0}_3 & 0\end{pmatrix}. \tag{15.25}$$

The measurement noise covariance represents the variance of the position displacement due to vibration and disturbance. Depending on the relationship between the vibration frequency and measurement-update rate, it may be necessary to treat the measurement noise as time correlated (see Section 3.4.3). Thus, $\mathbf{R}_Q$, is best determined empirically for each application.

It typically takes between 30 seconds and 60 seconds to determine the heading to within about 2°, enabling the small angle approximation to be made, and a few minutes to obtain the heading accuracy permitted by the gyro bias.

15.2.2 Fine Alignment

When the heading is known within a few degrees, quasi-stationary fine alignment can proceed using standard inertial navigation equations (Sections 5.2–5.5) and a Kalman filter state vector and system model of the same form as INS/GNSS and multisensor integration (Chapters 14 and 16). Thus, the same Kalman filter used in the navigation phase can also accept the stationarity or zero-velocity measurements during the initial alignment phase.

For alignment operating over a few minutes, position-displacement measurements enable the best modeling of the vibration. The measurement innovation is thus

$$\delta z_{Q,k}^{n-} = \hat{\mathbf{p}}_b(t_0) - \hat{\mathbf{p}}_b(t) \tag{15.26}$$

in a local navigation frame,

$$\delta z_{Q,k}^{e-} = \hat{\mathbf{r}}_{eb}^e(t_0) - \hat{\mathbf{r}}_{eb}^e(t) \tag{15.27}$$

in an ECEF frame and

$$\delta z_{Q,k}^{i-} = \begin{pmatrix} \cos\omega_{ie}(t-t_0) & -\sin\omega_{ie}(t-t_0) & 0 \\ \sin\omega_{ie}(t-t_0) & \cos\omega_{ie}(t-t_0) & 0 \\ 0 & 0 & 1 \end{pmatrix} \hat{\mathbf{r}}_{ib}^i(t_0) - \hat{\mathbf{r}}_{ib}^i(t) \tag{15.28}$$

in an ECI frame, where t_0 is the alignment start time. If the state vector is

$$\mathbf{x}^n = \begin{pmatrix} \delta\psi_{nb}^n \\ \delta\mathbf{v}_{eb}^n \\ \delta\mathbf{p}_b \\ \mathbf{b}_a \\ \mathbf{b}_g \\ \vdots \end{pmatrix}, \quad \mathbf{x}^\gamma = \begin{pmatrix} \delta\psi_{\gamma b}^\gamma \\ \delta\mathbf{v}_{\gamma b}^\gamma \\ \delta\mathbf{r}_{\gamma b}^\gamma \\ \mathbf{b}_a \\ \mathbf{b}_g \\ \vdots \end{pmatrix} \quad \gamma \in i,e, \tag{15.29}$$

where the curvilinear position error, $\delta\mathbf{p}_b$, is defined by (14.51), the measurement matrix is

$$\mathbf{H}_{Q,k} = \begin{pmatrix} \mathbf{0}_3 & \mathbf{0}_3 & -\mathbf{I}_3 & \mathbf{0}_3 & \mathbf{0}_3 & 0 \end{pmatrix}. \tag{15.30}$$

A local-navigation-frame implementation with Cartesian position error and a wander-azimuth-frame implementation are described in Sections I.2.6 and I.3.6 of Appendix I on the CD, respectively.

15.3 Zero Updates

If the host vehicle or user is often stationary during navigation, zero velocity updates may be used to maintain INS alignment and calibration. In these instances, the INS may only be stationary briefly, so velocity measurements are generally better than position displacement measurements. ZVUs are particularly useful in poor GNSS signal environments as often found in urban areas. They are used for land vehicle navigation [15] in the absence of odometry (which provides velocity all of the time). For pedestrian navigation, ZVUs may be combined with PDR using step detection (Section 6.4) as well as with inertial navigation [16]. For pedestrian navigation with a shoe-mounted IMU, a ZVU may be performed on every step. When combined with a method for reducing heading drift, this enables inertial navigation with a low position-drift rate to be performed with very-low-cost consumer-grade inertial sensors [17–20]. Other applications of ZVUs include robotics, helicopter navigation (during touchdowns), and inertial surveying.

Stationary-condition detection is discussed first, followed by a description of the zero velocity update. The section concludes with a description of the zero angular rate update. ZVUs and ZARUs are often performed together, but may also be implemented independently.

15.3.1 Stationary-Condition Detection

ZVUs must only be performed when the navigation system is stationary. For surveying applications, stationarity is typically indicated by the operator. However, for navigation, automated detection is needed. This is an example of context detection (see Section 1.4.2) and must be tailored to the individual application.

Stationary-condition detection for pedestrians is based on the fact that most parts of the body are constantly accelerating and decelerating during walking and running. For example, the foot acceleration can reach 30 m s^{-2} during walking and 60 m s^{-2} during running. A test quantity, $\left\| \mathbf{f}_{ib}^b \right\| - g(L_b, h_b)$, determined from the accelerometer measurements, is compared with a threshold. When the test quantity is below the threshold throughout a moving time window, the navigation system is assumed to be stationary. When body-mounted sensors are used, a detection window of about 0.5 second is suitable for determining that the pedestrian as a whole is stationary [16].

For foot-mounted inertial sensors, the same test quantity may be used, but a shorter detection window is required for determining the stance phase of each step (i.e., when the foot is on the ground). For walking, a 0.2-second window and 1.5 m s^{-2} threshold is suitable, while running and jogging require a shorter window and higher threshold [20]. Stance-phase detection has also been demonstrated using the incremental change in the inertial sensor outputs [18] and the variance of the resultant specific force, $|\mathbf{f}_{ib}^b|$ [19, 21]. Detection techniques based on the magnitude of the angular rate and using both gyro and accelerometer measurements have also been proposed [22]. During the stance phase of crawling motion, the foot can wobble significantly, which makes both stationarity detection and the application of ZVUs more difficult.

In land vehicles, the horizontal velocity solution itself is often used for stationary-condition detection. This is compared with a threshold, again over a suitable time window. The detection threshold must reflect the quality of the sensors. For mobile mapping using an aviation-grade IMU, a threshold as low as 0.0075 m s^{-1} per axis may be used [15], while for navigation using consumer-grade sensors, a 0.5 m s^{-1} threshold is more realistic [23]. The detection threshold may also be scaled with the velocity solution uncertainty.

Acceleration-based stationarity detection is more difficult for land vehicles as the acceleration can be low when the vehicle is moving. However, a vehicle typically undergoes less vibration when stationary. Therefore, the standard deviation of the accelerometer measurements can be used to confirm velocity-based stationarity detection [23]. Using frequency-domain filtering to eliminate most of the engine-induced vibration increases the sensitivity of accelerometer-based stationary-condition detection, enabling it to be used on its own [24].

As land vehicles do not turn when they are stationary, additional tests may be applied using the yaw angular rate [15]. A magnetometer may also be used to confirm that a vehicle is stationary as the measured magnetic flux density along each body-frame axis will then be constant, regardless of whether there are any local magnetic anomalies.

For robotics applications, the guidance and control system may inform the navigation system when the robot is stationary. Furthermore, the navigation system may request that the guidance and control system stops the robot for a ZVU when the position or velocity uncertainty exceeds a certain threshold [25].

During longer stationary periods, it is best to trigger multiple ZVUs. This keeps the inertial error calibration up to date and enables the averaging out of noise due to residual motion.

15.3.2 Zero Velocity Update

The measurement innovation for a ZVU is

$$\delta z^{\gamma-}_{ZV,k} = -\hat{v}^{\gamma}_{eb,k} \quad \gamma \in e, n \tag{15.31}$$

or

$$\delta z^{i-}_{ZV,k} = \mathbf{\Omega}^i_{ie}\hat{r}^i_{ib,k} - \hat{v}^i_{ib,k} \tag{15.32}$$

and the measurement matrix, assuming a state vector defined by (15.29), is

$$\mathbf{H}^{\gamma}_{ZV,k} = \begin{pmatrix} \mathbf{0}_3 & -\mathbf{I}_3 & \mathbf{0}_3 & \mathbf{0}_3 & \mathbf{0}_3 & 0 \end{pmatrix} \quad \gamma \in e, n \tag{15.33}$$

or

$$\mathbf{H}^{i}_{ZV,k} = \begin{pmatrix} \mathbf{0}_3 & -\mathbf{I}_3 & \mathbf{\Omega}^i_{ie} & \mathbf{0}_3 & \mathbf{0}_3 & 0 \end{pmatrix}. \tag{15.34}$$

The measurement noise covariance describes the variance and covariance of the nominally-zero velocity due to vibration and disturbances. Vehicles may be disturbed by wind gusts, other vehicles passing, people moving around inside, loading, and unloading. Similarly, a nominally-stationary pedestrian may be moving on the spot. The amount of residual motion during a ZVU will depend on what the stationary-condition detection algorithm allows, so there should be a relationship between the assumed measurement noise covariance and the detection threshold(s).

Although ZVUs do not provide absolute position information, the Kalman filter system model builds up information on the correlation between the velocity and position errors in the off-diagonal elements of the error covariance matrix, **P**. This enables a ZVU to correct most of the position drift since the last measurement update, ZVU or otherwise [16].

In principle, the heading error and yaw gyro bias may also be calibrated using ZVUs (see Section 14.3.4). However, in practice, they are weakly observable, particularly for pedestrian applications. Therefore, better performance is often obtained using an algorithm, such as a UKF, that permits large heading errors as discussed in Section 14.4.4 [26].

For extended stationary periods, zero position displacement measurements, described in Section 15.2.5, may be used instead of zero velocity measurements.

15.3.3 Zero Angular Rate Update

A zero angular rate update is useful for applications with low-performance gyros. It is applicable to a partial IMU (Section 5.9) and a stand-alone gyro (Section 6.1.3) as well as to a full IMU. The context for ZARUs is different from that for ZVUs. Therefore, ZARUs and ZVUs are sometimes performed separately and sometimes performed together.

For land vehicle applications, a zero angular rate may be assumed whenever the vehicle is stationary so a ZARU may be performed whenever a ZVU is performed. When odometry (Section 6.3) is available, it may be used to trigger a ZARU when the vehicle is stationary.

A ZARU may also be performed when the vehicle is moving at a constant heading. This may be detected by comparing three parameters with thresholds: the standard deviation of the recent yaw-rate gyro measurements, the standard deviation of the steering-angle commands, and the yaw rate obtained from differential odometry (Section 6.3.2) [27].

The use of ZARUs in pedestrian applications should be approached with caution. The residual angular motion of a stationary person's body can be much larger than the gyro errors, while a foot can rotate during the stance phase. Therefore, ZARUs should not be automatically performed each time stationarity is detected. Instead, there should be additional tests to determine whether a ZARU should be performed alongside a ZVU. The IMU should be stationary for longer than the stance phase of walking and a rotation test should be applied. This could be based on the standard deviation of the gyro measurements as pedestrians rarely rotate at a constant rate. A magnetometer could also be used as the measured magnetic flux density along each body-frame axis is normally constant when the instrument is both stationary and nonrotating.

The measurement innovation for a full-IMU ZARU is

$$\delta \mathbf{z}^-_{ZA,k} = -\hat{\boldsymbol{\omega}}^b_{ib,k}. \qquad (15.35)$$

Assuming a state vector defined by (15.29), the measurement matrix is simply

$$\mathbf{H}_{ZA,k} = \begin{pmatrix} \mathbf{0}_3 & \mathbf{0}_3 & \mathbf{0}_3 & \mathbf{0}_3 & -\mathbf{I}_3 & 0 \end{pmatrix}, \qquad (15.36)$$

noting that this is independent of the coordinate frames used for the position, velocity, and attitude states.

The measurement noise covariance represents the variance of the nominally-zero angular rate due to vibration and disturbances. For a stationary land vehicle, there will be less disturbance about the yaw axis than about the roll and pitch axes, which should be reflected in the assumed measurement noise covariance. A larger measurement noise should be assumed for a moving vehicle.

ZARUs are only useful where the gyro errors are significant compared to the angular disturbances. Thus, ZARUs are sometimes applied in the yaw axis only and are not typically used with high-performance gyros.

15.4 Motion Constraints

Motion constraints exploit limitations in the freedom of movement of the host vehicle or pedestrian in order to correct and calibrate INS errors. They thus contribute additional information to the navigation solution based on the operating context as discussed in Section 1.4.2. Motion constraints are also known as nonholonomic constraints, which means that they introduce dependency of the state estimates on their previous values (in addition to that already accounted for by the system model). This section describes land vehicle motion constraints and then discusses constraints for pedestrians and for ships and boats.

15.4.1 Land Vehicle Constraints

Normal land vehicle motion is subject to two constraints. The velocity of the vehicle is zero along the rotation axis of any of its wheels and is also zero in the direction perpendicular to the road or rail surface [28, 29]. Note that, because of frame rotation, zero velocity does not necessarily imply zero acceleration. Selecting the rear wheels, these constraints may be expressed as

$$\begin{pmatrix} 0 & 1 & 0 \\ 0 & 0 & 1 \end{pmatrix} \mathbf{v}^r_{er} = \mathbf{0}, \qquad (15.37)$$

where r denotes the rear-wheel body frame. The top row of each equation applies to the transverse constraint and the bottom row to the perpendicular constraint. Applying (2.165) and assuming that the coordinate transformation matrix from the IMU body frame, b, to the r frame, $\mathbf{C}^r_b$, is constant, the constraints become

$$\begin{pmatrix} 0 & 1 & 0 \\ 0 & 0 & 1 \end{pmatrix} \mathbf{C}_b^r \left(\mathbf{C}_\gamma^b \mathbf{v}_{\beta b}^\gamma - \boldsymbol{\omega}_{ib}^b \wedge \mathbf{l}_{rb}^b \right) \approx 0, \qquad \{\beta,\gamma\} \in \{i,i\},\{e,e\},\{e,n\}, \qquad (15.38)$$

where $\mathbf{l}_{rb}^b$ is the lever arm from the r frame to the b frame and the Earth-rotation and transport-rate terms have been neglected.

This vehicle velocity constraint can be applied as a Kalman filter measurement update with measurement innovation

$$\delta \mathbf{z}_{VC,k}^{\gamma-} = -\begin{pmatrix} 0 & 1 & 0 \\ 0 & 0 & 1 \end{pmatrix} \hat{\mathbf{C}}_b^r \left(\hat{\mathbf{C}}_\gamma^b \hat{\mathbf{v}}_{\beta b}^\gamma - \hat{\boldsymbol{\omega}}_{ib}^b \wedge \mathbf{l}_{rb}^b \right), \qquad \{\beta,\gamma\} \in \{i,i\},\{e,e\},\{e,n\}. \qquad (15.39)$$

This is an example of a pseudo-measurement. Assuming a state vector defined by (15.29) and neglecting the coupling of the attitude errors and gyro biases into the measurements through the lever arm, the measurement matrix may be approximated to

$$\mathbf{H}_{VC,k}^\gamma \approx \begin{pmatrix} \mathbf{0}_{2\times 3} & -\begin{pmatrix} 0 & 1 & 0 \\ 0 & 0 & 1 \end{pmatrix} \hat{\mathbf{C}}_b^r \hat{\mathbf{C}}_\gamma^b & \mathbf{0}_{2\times 3} & \mathbf{0}_{2\times 3} & \mathbf{0}_{2\times 3} & \mathbf{0} \end{pmatrix}_k \qquad \gamma \in i,e,n. \qquad (15.40)$$

This is equivalent to a ZVU applied along only two axes. However, the motion-constraint measurements may be applied continuously; no stationary-condition detection is required. In principle, a motion-constraint measurement update may be performed whenever the navigation solution is updated. However, a lower rate is sufficient, commensurate with the Kalman filter processing cycles for other types of measurement.

Correct application of the motion constraint requires knowledge of the attitude, $\mathbf{C}_b^r$, and position, $\mathbf{l}_{rb}^b$, of the IMU with respect to the rear-wheel rotation axis. If these cannot be measured with sufficient precision at installation, they can be estimated as Kalman filter states [30]. The relative heading, ψ_{rb}, can also be determined by detecting starting and stopping motion using the accelerometers [31].

The measurement noise covariance must account for the differences between the true and assumed vehicle motion. Causes include sideslip along the wheel rotation axes during turns, zero-mean vertical motion due to the vehicle's suspension system, and engine vibration in both directions [28]. The time correlation of these effects may be significant where the measurement update rate is high.

Sideslip invalidates the transverse velocity constraint and can occur when a vehicle turns at an angular rate of more than about 0.05 rad s^{-1}. Applying this constraint during significant sideslip therefore biases the velocity solution. Consequently, better performance is obtained if the transverse motion-constraint measurements are omitted whenever the yaw rate exceeds a predefined threshold [32]. This does not affect the perpendicular velocity constraint. Large discrepancies between the true and assumed vehicle motion can also occur during skids or when a wheel loses

15.4 Motion Constraints

contact with the ground; these may be addressed using measurement innovation filtering (Section 17.3.1).

Motion-constraint measurements provide a means of correcting the errors in the zero-value IMU pseudo-measurements that may be used as substitutes for the missing sensor outputs in partial IMUs (Section 5.9) [33]. In this case, the INS system noise covariance (Section 14.2.6) must be modified to model errors in the sensor pseudo-measurements instead of real sensor noise. This will vary from axis to axis. The system noise covariance matrix under the small-propagation-interval approximation, given by (14.82), becomes

$$Q_{INS}^{\gamma} \approx Q_{INS}^{\prime\gamma} = \begin{pmatrix} Q_{11} & 0_3 & 0_3 & 0_3 & 0_3 \\ 0_3 & Q_{22} & 0_3 & 0_3 & 0_3 \\ 0_3 & 0_3 & 0_3 & 0_3 & 0_3 \\ 0_3 & 0_3 & 0_3 & S_{bad}I_3 & 0_3 \\ 0_3 & 0_3 & 0_3 & 0_3 & S_{bgd}I_3 \end{pmatrix} \tau_s, \quad \gamma \in i,e,n \qquad (15.41)$$

where

$$Q_{11} = C_b^{\gamma} \begin{pmatrix} S_{\omega x} & 0 & 0 \\ 0 & S_{\omega y} & 0 \\ 0 & 0 & S_{\omega z} \end{pmatrix} C_{\gamma}^b, \quad Q_{22} = C_b^{\gamma} \begin{pmatrix} S_{fx} & 0 & 0 \\ 0 & S_{fy} & 0 \\ 0 & 0 & S_{fz} \end{pmatrix} C_{\gamma}^b, \quad \gamma \in i,e,n.$$

(15.42)

where $S_{\omega x}$, $S_{\omega y}$, and $S_{\omega z}$ are the angular rate PSDs and S_{fx}, S_{fy}, and S_{fz} are the specific force PSDs. When a real sensor is used, the PSD is the same as the corresponding sensor noise PSD, given by (14.83), whereas where a pseudo-measurement is used, the PSD is that of the actual vehicle motion. In both cases a frequency-independent approximation of the true frequency-dependent PSD must be used.

For indoor robot navigation, the further assumption may be made that the floor is flat. This enables additional constraints to be applied to the roll and pitch angles [34].

15.4.2 Pedestrian Constraints

For pedestrian navigation, the heading drift may be constrained by mounting IMUs on both feet and, in addition to performing ZVUs, measure the range between them to calibrate both yaw-axis gyro biases [17]. This has been demonstrated using ultrasonic ranging [35]. A simpler constraint-based approach has been demonstrated in which a fixed body-frame displacement between the two IMUs is assumed each time the stance phase of one foot is detected (see Section 15.3.1) and processed as a relative position measurement [36].

Pedestrians tend to walk in approximately straight lines, particularly in indoor and urban environments where walls and roads are in the way. Therefore, another constraint that can be applied is to assume that a person's heading is constant whenever the change in the INS heading solution over a certain interval is below a predetermined threshold [37]. This is most applicable to indoor environments and to urban areas with a grid-based street layout.

Indoors, the floor is usually flat, so the height may be assumed constant unless the height change between steps or epochs is sufficient for the pedestrian to be on a staircase [38]. Escalators, elevators, and ramps must also be distinguished.

15.4.3 Ship and Boat Constraint

Ships and boats remain on the surface of the water. Consequently, an inertial navigation solution can be constrained by applying zero-position-displacement measurements, as described in Section 15.2.2, in the vertical direction only. The measurement noise covariance should be sufficient to account for the oscillatory heave motion induced by the sea state.

Problems and exercises for this chapter are on the accompanying CD.

References

[1] Groves, P. D., "Optimising the Transfer Alignment of Weapon INS," *Journal of Navigation*, Vol. 56, No. 3, 2003, pp. 323–335.

[2] Spalding, K., "An Efficient Rapid Transfer Alignment Filter," *Proc. AIAA Guidance, Navigation and Control Conference*, Hilton Head Island, SC, August 1992, pp. 1276–1286.

[3] Wendel, J., and G. Trommer, "Impact of Mechanical Vibrations on the Performance of Integrated Navigation Systems and an Optimal IMU Specification," *Proc. ION 57th AM*, Albuquerque, NM, June 2001, pp. 614–621.

[4] Titterton, D. H., and J. L. Weston, *Strapdown Inertial Navigation Technology*, 2nd ed., Stevenage, U.K.: IEE, 2004.

[5] Kain, J. E., and J. R. Cloutier, "Rapid Transfer Alignment for Tactical Weapon Applications," *Proc. AIAA Guidance, Navigation and Control Conference*, Boston, MA, August 1989, pp. 1290–1300.

[6] Graham, W., and K. Shortelle, "Advanced Transfer Alignment for Inertial Navigators (A-Train)," *Proc. ION NTM*, Anaheim, CA, January 1995, pp. 113–124.

[7] Rogers, R. M., *Applied Mathematics in Integrated Navigation Systems*, Reston: VA, AIAA, 2000.

[8] Groves, P. D., G. G. Wilson, and C. J. Mather, "Robust Rapid Transfer Alignment with an INS/GPS Reference," *Proc. ION NTM*, San Diego, CA, January 2002, pp. 301–311.

[9] Graham, W. R., K. J. Shortelle, and C. Rabourn, "Rapid Alignment Prototype (RAP) Flight Test Demonstration," *Proc. ION NTM*, Long Beach, CA, January 1998, pp. 557–568.

[10] Groves, P. D., C. A. Littlefield, and D. C. Long, "The Need for Transfer Alignment in a GPS Jamming Environment and Optimization for MEMS IMU," *Proc. ION GNSS 2004*, Long Beach, CA, September 2004, pp. 775–783.

[11] Farrell, J. A., and M. Barth, *The Global Positioning System and Inertial Navigation*, New York: McGraw-Hill, 1999.

[12] Savage, P. G., *Strapdown Analytics, Parts 1 and 2*, Maple Plain, MN: Strapdown Associates, 2000.

[13] Hua, C., "Gyrocompass Alignment with Base Motions: Result for a 1nmi/h INS.GPS System," *Navigation: JION*, Vol. 47, No. 2, 2000, pp. 65–74.

[14] Gu, D., et al., "Coarse Alignment for Marine SINS Using Gravity in the Inertial Frame as Reference," *Proc. IEEE/ION PLANS*, Monterey, CA, May 2008, pp. 961–965.

[15] Grejner-Brzezinska, D. A., Y. Yi, and C. K. Toth, "Bridging Gaps in Urban Canyons: The Benefits of ZUPTS," *Navigation: JION*, Vol. 48, No. 4, 2001, pp. 217–225.

[16] Mather, C. J., P. D. Groves, and M. R. Carter, "A Man Motion Navigation System Using High Sensitivity GPS, MEMS IMU and Auxiliary Sensors," *Proc. ION GNSS 2006*, Fort Worth, TX, September 2006, pp. 2704–2714.

[17] Brand, T. J., and R. E., Phillips, "Foot-to-Foot Range Measurements as an Aid to Personal Navigation," *Proc. ION 59th AM*, Albuquerque, NM, June 2003, pp. 113–125.

[18] Foxlin, E., "Pedestrian Tracking with Shoe-Mounted Inertial Sensors," *IEEE Computer Graphics and Applications Magazine*, November/December 2005. pp. 38–46.

[19] Godha, S., G. Lachapelle, and M. E. Cannon, "Integrated GPS/INS System for Pedestrian Navigation in a Signal Degraded Environment," *Proc. ION GNSS 2006*, Fort Worth, TX, September 2006, pp. 2151–2164.

[20] Groves, P. D., et al., "Inertial Navigation Versus Pedestrian Dead Reckoning: Optimizing the Integration," *Proc. ION GNSS 2007*, Fort Worth, TX, September 2007, pp. 2043–2055.

[21] Kwakkel, S. P., G. Lachapelle, and M. E. Cannon, "GNSS Aided In Situ Human Lower Limb Kinematics During Running," *Proc. ION GNSS 2008*, Savannah, GA, September 2008, pp. 1388–1397.

[22] Skog, I., J.-O. Nilsson, and P. Händel, "Evaluation of Zero-Velocity Detectors for Foot-Mounted Inertial Navigation Systems," *Proc. Indoor Positioning and Indoor Navigation (IPIN)*, Zurich, Switzerland, September 2010.

[23] Aggarwal, P., et al., *MEMS-Based Integrated Navigation*, Norwood, MA: Artech House, 2010.

[24] Ramanandan, A., et al., "Detection of Stationarity in an Inertial Navigation System," *Proc. ION GNSS 2010*, Portland, OR, September 2010, pp. 238–244.

[25] Venable, D. T., et al., "Performance Evaluation of Coupling Between Vehicle Guidance and Vision Aided Navigation," *Proc. ION GNSS 2009*, Savannah, GA, September 2009, pp. 826–834.

[26] Zampella, F., et al., "Unscented Kalman Filter and Magnetic Angular Rate Update (MARU) for an Improved Pedestrian Dead-Reckoning," *Proc. IEEE/ION PLANS*, Myrtle Beach, SC, April 2012, pp. 129–139.

[27] Basnayake, C., "A Novel Yaw Rate Sensor Bias Error Containment Method Using Existing Vehicle Sensors," *Proc. ION GNSS 2009*, Savannah, GA, September 2009, pp. 555–563.

[28] Dissanayake, G., et al., "The Aiding of a Low-Cost Strapdown Inertial Measurement Unit Using Vehicle Model Constraints for Land Vehicle Applications," *IEEE Trans. on Robotics and Automation*, Vol. 17, No. 5, 2001, pp. 731–747.

[29] Niu, X., S. Nassar, and. N., El-Sheimy, "An Accurate Land-Vehicle MEMS IMU/GPS Navigation System Using 3D Auxiliary Velocity Updates," *Navigation: JION*, Vol. 54, No. 3, 2007, pp. 177–188.

[30] Wu. Y., C. Goodall, and N. El-Sheimy, "Self-Calibration for IMU/Odometer Land Navigation: Simulation and Test Results," *Proc. ION ITM*, San Diego, CA, January 2010, pp. 839–849.

[31] Vinande, E., P. Axelrad, and D. Akos, "Mounting-Angle Estimation for Personal Navigation Devices," *IEEE Trans. on Vehicular Technology*, Vol. 59, No. 3, 2010, pp. 1129–1138.

[32] Ryan, J., and D. Bevly, "Robust Ground Vehicle Constraints for Aiding Stand Alone INS and Determining Inertial Sensor Errors," *Proc. ION ITM*, Newport Beach, CA, January 2012, pp. 374–384.

[33] El-Sheimy, N., "The Potential of Partial IMUs for Land Vehicle Navigation," *Inside GNSS*, Spring 2008, pp. 16–25.

[34] Chen, C., et al., "Low Cost IMU Based Indoor Mobile Robot Navigation with the Assist of Odometry and Wi-Fi Using Dynamic Constraints," *Proc. IEEE/ION PLANS*, Myrtle Beach, SC, April 2012, pp. 1274–1279.

[35] Laverne, M., et al., "Experimental Validation of Foot to Foot Range Measurements in Pedestrian Tracking," *Proc. ION GNSS 2011*, Portland, OR, September 2011, pp. 1386–1393.

[36] Bancroft, J. B., et al., "Twin IMU-HSGPS Integration for Pedestrian Navigation," *Proc. ION GNSS 2008*, Savannah, GA, September 2008, pp. 1377–1387.

[37] Borenstein, J., L. Ojeda, and S. Kwanmuang, "Heuristic Reduction of Gyro Drift for Personnel Tracking Systems," *Journal of Navigation*, Vol. 62, No. 1, 2009, pp. 41–58.

[38] Abdulrahim, K., et al., "Using Constraints for Shoe Mounted Indoor Pedestrian Navigation," *Journal of Navigation*, Vol. 65, No. 1, 2012, pp. 15–28.

CHAPTER 16
Multisensor Integrated Navigation

This chapter describes how dead reckoning, terrestrial radio navigation, and environmental feature-matching navigation systems may be integrated with INS, GNSS, and each other. It follows on from the description of INS/GNSS integration in Chapter 14.

Different combinations of navigation sensors are suited to different applications, depending on the environment, dynamics, budget, accuracy requirements, and the degree of robustness or integrity required. For commercial airliners and most military aircraft, INS and GNSS form the core of the navigation system, with further sensors, such as magnetic compass, barometric and radar altimeters, and DME/VOR/TACAN providing enhanced robustness. Road vehicles typically combine GNSS with map matching and may also use odometers or wheel speed sensors, inertial sensors, and a magnetic compass. Autonomous vehicles and robots typically incorporate cameras and/or laser scanners for route guidance and collision avoidance, so these are often used for navigation, usually in addition to dead-reckoning sensors and radio navigation. Radio signals do not penetrate far through water, so underwater vehicles use inertial navigation, various sonar and acoustic devices, a depth sensor, and sometimes gravity gradiometry. A detailed discussion of the requirements for different navigation applications and typical multisensor solutions may be found in Chapter 18.

Section 16.1 describes and compares the different architectures that may be used to integrate measurements from multiple navigation systems. Section 16.2 then discusses the integration issues and describes system and measurement models for dead reckoning, attitude, and height measurements. Similarly, Section 16.3 describes the integration of position fixes, range measurements, angular measurements, and line fixes from terrestrial radio navigation and environmental feature matching. Integration of INS and GNSS into multisensor integration architectures is essentially the same as for INS/GNSS integration, described in Chapter 14. Integration is also known as data fusion, sensor fusion, and hybridization.

16.1 Integration Architectures

There are many ways of combining information from multiple navigation systems. The design of the integration architecture is a tradeoff between maximizing the accuracy and robustness of the navigation solution, minimizing the complexity, and optimizing the processing efficiency. It must also account for the different characteristics of the various navigation technologies, combine the different types of measurement that they provide and supply them with suitable aiding information.

Dead-reckoning systems, including inertial navigation and feature tracking, measure motion resolved along body-frame axes. This must be converted to external resolving axes (e.g., north, east, and down) in order to update the position. Dead-reckoning position solutions drift over time, but may be calibrated using position-fixing systems.

Proximity-based position fixes may have large associated uncertainties or may comprise irregularly shaped coverage areas. Both ranging systems (e.g., GNSS) and angular positioning systems require a minimum number of signals or landmarks to determine a position solution, but may also contribute individual measurements to the integrated navigation solution. Pattern-matching systems generate scores for an array of candidate positions based on the matching of measurements with a database. These scores may be used to calculate a position fix or input directly to the integration algorithm.

Environmental feature-matching systems require an approximate position solution to determine which region of their database to search; this can also speed up signal acquisition in radio positioning. Pattern-matching systems, such as TRN, that must make measurements at multiple locations to determine position unambiguously also require a velocity input.

Many navigation sensors exhibit biases and other systematic errors, which can be calibrated in an integrated navigation system. However, such calibration can result in faults in one navigation sensor contaminating the calibration of another. Fault detection and integrity monitoring are described in Chapter 17 and can be implemented for all multisensor integration architectures. However, integrity monitoring is more processor intensive for some integration architectures than for others.

The design of an integrated navigation system can be severely constrained by the need to combine equipment from different manufacturers. When raw sensor or ranging measurements are not available, the systems integrator may have to work with a "black box" navigation solution with no information about its error characteristics, such as covariances, correlation times, or even uncertainty in many cases. Such systems are also limited in terms of what feedback information they may accept.

This section describes the different integration architectures and discusses their benefits and drawbacks. Integration architectures can be cascaded or centralized and single-epoch or filtered. The cascaded single-epoch, centralized single-epoch, cascaded filtered, centralized filtered, and federated filtered architectures are described in turn, including the difference between the total-state and error-state implementations of the filtered architectures. This is followed by a discussion of hybrid integrated architectures. More information on total-state and error-state Kalman filtering is then presented, including prediction and timing. The section concludes with discussions of primary and reversionary moding and context-adaptive moding, including context detection.

16.1.1 Cascaded Single-Epoch Integration

Cascaded single-epoch, or epoch-by-epoch, integration, shown in Figure 16.1, is the simplest way of combining information from different navigation systems. Each subsystem, denoted by index i, provides a position or position and velocity solution, $\hat{\mathbf{x}}_i$, and an associated error covariance matrix, $\mathbf{P}_{ii}$, which accounts for both bias-like and noise-like errors. In principle, each subsystem may use either position fixing or

16.1 Integration Architectures

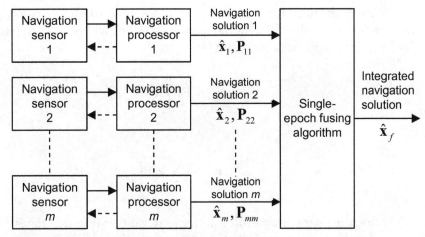

Figure 16.1 Cascaded single-epoch integration architecture.

dead reckoning. A position-fixing algorithm may be either single-epoch or filtered; the latter is then known as a local filter.

The subsystem navigation solutions are combined with a single-epoch, or snapshot, fusing algorithm, analogous to the single-epoch GNSS positioning algorithm described in Section 9.4.1. This typically uses least-squares estimation. Therefore, the integrated navigation solution, $\hat{\mathbf{x}}_f$, is simply the weighted least-squares combination of the m individual solutions, given by

$$\hat{\mathbf{x}}_f = \left(\mathbf{H}^\mathrm{T}\mathbf{P}^{-1}\mathbf{H}\right)^{-1}\mathbf{H}^\mathrm{T}\mathbf{P}^{-1}\begin{pmatrix}\hat{\mathbf{x}}_1\\\hat{\mathbf{x}}_2\\\vdots\\\hat{\mathbf{x}}_m\end{pmatrix}, \qquad \mathbf{H} = \begin{pmatrix}\mathbf{I}_n\\\mathbf{I}_n\\\vdots\\\mathbf{I}_n\end{pmatrix}, \qquad (16.1)$$

where the measurement matrix, $\mathbf{H}$, comprises m $n{\times}n$ identity matrices, where n is the number of components of $\mathbf{x}$, and $\mathbf{P}$ is the error covariance of the individual navigation solutions, an $mn{\times}mn$ matrix given by

$$\mathbf{P} = \begin{pmatrix}\mathbf{P}_{11} & \mathbf{P}_{12} & \cdots & \mathbf{P}_{1m}\\\mathbf{P}_{21} & \mathbf{P}_{22} & \cdots & \mathbf{P}_{2m}\\\vdots & \vdots & \ddots & \vdots\\\mathbf{P}_{m1} & \mathbf{P}_{m2} & \cdots & \mathbf{P}_{mm}\end{pmatrix}. \qquad (16.2)$$

The covariance of the fused navigation solution is

$$\mathbf{P}_{ff} = \left(\mathbf{H}^\mathrm{T}\mathbf{P}^{-1}\mathbf{H}\right)^{-1}. \qquad (16.3)$$

When each subsystem uses different information to obtain its navigation solution, the errors of the different navigation solutions will be uncorrelated, so $\mathbf{P}_{ij} = 0$ for $i \neq j$. This simplifies (16.1) and (16.3) to

$$\hat{\mathbf{x}}_f = \mathbf{P}_{ff} \sum_{i=1}^{m} \mathbf{P}_{ii}^{-1} \hat{\mathbf{x}}_i, \qquad (16.4)$$

$$\mathbf{P}_{ff} = \left(\sum_{i=1}^{m} \mathbf{P}_{ii}^{-1} \right)^{-1}. \qquad (16.5)$$

If the subsystem navigation solutions are valid at different times, they must be interpolated to a common time of validity prior to fusion. The faster the host vehicle, the more precise the time synchronization must be.

The cascaded single-epoch integration architecture is suited to black box navigation systems as the fusion algorithm requires no knowledge of how the navigation system errors vary with time and there is no feedback. However, to optimally combine the different navigation solutions, accurate error covariance information is needed. Figure 16.2 shows that neglecting the off-diagonal elements of $\mathbf{P}$ for a particular subsystem causes the accuracy to be overestimated in one direction and underestimated in another. Using the off-diagonal elements also allows incomplete navigation solutions to be fused. However, this requires the navigation systems to output the information matrix, $\mathbf{P}^{-1}$, instead of the error covariance matrix, as an incomplete navigation solution has infinite uncertainty in one or two directions.

When a navigation system outputs no uncertainty information, its error covariance must be estimated by the fusing algorithm. This can be a problem for radio navigation systems, such as GNSS, where the accuracy varies with the number of transmitters used, their geometry, and the signal-to-noise ratios.

Cascaded single-epoch integration has the advantages of simplicity and a low processor load. As the subsystems are completely independent, it also facilitates integrity monitoring, allowing consistency checks (Section 17.4.1) to be used. However, it has fundamental limitations. It is unsuited to integration of dead-reckoning systems, such as inertial navigation, as it offers no inherent means of calibrating the position drift. Instead, as the dead-reckoning position degrades, it is simply weighted out of

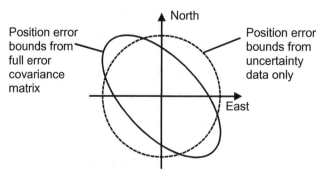

Figure 16.2 Effect of off-diagonal error covariance on position error bounds.

the integrated navigation solution. If ad hoc feedback is added to correct the position drift, the dead-reckoning position solution may no longer be treated as independent and its error correlation with the other subsystems must then be accounted for.

16.1.2 Centralized Single-Epoch Integration

In a centralized single-epoch integration architecture, shown in Figure 16.3, ranging, bearing, elevation, and line-fix measurements from different subsystems at the same epoch are combined using least-squares estimation. This essentially combines the single-system single-epoch position algorithms described in Sections 7.3.3, 9.4.1, 11.1.1.2, 11.2.2, and 12.2.3 into a multisensor positioning algorithm. For optimal weighting of each subsystem, the error standard deviation of each measurement is needed.

A centralized architecture has the advantage over cascaded integration that subsystem measurements may still contribute to the integrated navigation solution when there is insufficient information available to compute position using that subsystem alone.

In centralized single-epoch integration, no information is carried forward from one epoch to the next, not even within the individual subsystems as can happen in cascaded single-epoch integration. This has three main drawbacks. First, there is no navigation solution in cases where there is insufficient information to determine it from the current set of measurements; it cannot be estimated using previous measurements. Second, there is no capacity to calibrate subsystem systematic errors using data from multiple epochs. Finally, dead-reckoning sensors cannot easily contribute to the position solution (they can contribute to the velocity solution). However, there are integrity benefits. The measurements from the subsystems are fully independent, facilitating measurement consistency checking (Section 17.4.1). Furthermore, faulty measurements are not carried forward from one epoch to the next. Therefore, fault isolation (see Chapter 17) is easy to achieve.

Centralized single-epoch integration is best suited to applications (outside navigation) that neither require a continuous navigation solution nor incorporate dead-reckoning sensors.

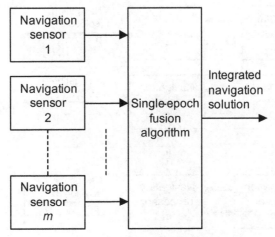

Figure 16.3 Centralized single-epoch integration architecture.

16.1.3 Cascaded Filtered Integration

Figure 16.4 shows a total-state cascaded filtered integration architecture. This is similar to the cascaded single-epoch architecture, but with the single-epoch fusing algorithm replaced by a Kalman filter (Chapter 3). This estimates the navigation solution and can also estimate errors in the navigation subsystems. As a Kalman filter-based integration algorithm retains past information from the constituent navigation systems, it can maintain calibration of the position solution drift of an INS or other dead-reckoning system and can calibrate other errors. The term total-state thus applies to the position, velocity, and attitude, not the complete state vector. Filtered integration can also handle different sensors providing measurements at different times of validity more easily (e.g., by using its velocity estimate to predict forward position information from one subsystem to the time of validity of another subsystem's position measurement).

Figure 16.5 shows an error-state cascaded filtered integration architecture. The integrated navigation solution is that of a dead-reckoning reference system, such as an INS, corrected using estimates of its position, velocity, and attitude error made by the Kalman filter integration algorithm. Open- and closed-loop correction of the reference system is discussed in Sections 14.1.1 and 16.1.8. The other subsystems are aiding systems as they aid the reference system via the integration algorithm. This brings the advantage that the integrated navigation solution may be updated at a faster rate than that at which the Kalman filter is iterated, reducing the processor load. An example of cascaded filtered integration is loosely coupled INS/GNSS integration, described in Section 14.1.2, where a GNSS navigation filter is used. When an error-state Kalman filter integrates two dead-reckoning systems, one is integrated as the reference and the other is integrated as an aiding system. Inertial navigation is normally integrated as the reference.

Although not shown in the figures, corrections from the integration algorithm may be fed back to any of the navigation processors if their software accepts them. As with other forms of closed-loop correction, the estimates of the fed-back states should be zeroed within the Kalman filter following the feedback. The reference

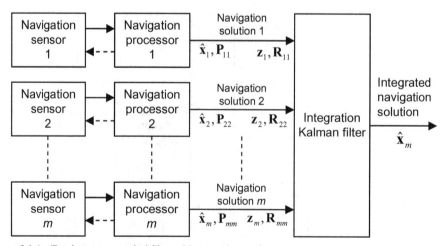

Figure 16.4 Total-state cascaded filtered integration architecture.

16.1 Integration Architectures

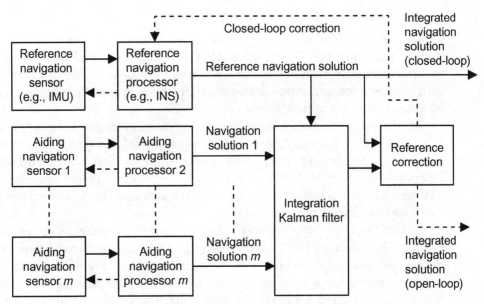

Figure 16.5 Error-state cascaded filtered integration architecture.

or integrated navigation solution may also be fed back (e.g., to aid GNSS signal tracking or determine the search area for environmental feature identification or pattern matching). However, feedback reduces the independence between navigation systems, making it more difficult to detect faults by comparing navigation solutions.

In the total-state implementation, the navigation solution of each of the navigation systems forms the Kalman filter measurement vector, $\tilde{z}$. In the error-state implementation, the measurement vector comprises the difference between the aiding and reference system navigation solutions. Section 14.3.1 shows how this is implemented for GNSS measurements integrated with INS. The implementation for other navigation systems is similar. Note that any data lags must be compensated as discussed in Section 3.3.4. Measurements from different subsystems do not have to be processed simultaneously.

A fundamental assumption of Kalman filtering is that the navigation-system errors comprise a mixture of systematic errors, estimated as states, and white noise, modeled as system noise for a reference system and measurement noise otherwise. However, if Kalman filters or smoothing filters are used in the individual subsystem navigation processors, time-correlated noise is introduced. The integration Kalman filter must account for this, as discussed in Section 3.4.3, to prevent instability. The simplest method is to increase the assumed measurement noise covariance, $\mathbf{R}$, reducing the filter gain. Thus, when black box navigation systems are integrated in a cascaded architecture, their error characteristics must be determined across all operational conditions to ensure that the integration Kalman filter is correctly tuned. If this is not practical, a cascaded integration architecture should not be used. It is also difficult to handle incomplete subsystem navigation solutions in a cascaded architecture.

16.1.4 Centralized Filtered Integration

Figures 16.6 and 16.7, respectively, show the total-state and error-state implementations of the centralized filtered integration architecture. A total-state filter is suited to integrating position-fixing systems, whereas an error-state filter is suitable where inertial navigation or another dead-reckoning sensor is used. Subsystem errors may be estimated in either implementation.

In contrast to the cascaded architecture, sensor measurements rather than navigation solutions are generally input to the integration Kalman filter. Radio navigation systems typically provide ranging measurements; thus, tightly coupled INS/GNSS integration (Section 14.1.3) is an example of centralized filtered integration. This enables a navigation system to contribute to the integrated navigation solution when there are insufficient signals to form its own solution.

A basic Kalman filter is rarely suited to centralized integration. An extended Kalman filter (Section 3.4.1) is required to process ranging measurements because the measurement model is nonlinear. For short-range positioning systems where the position uncertainty is significant compared to the range measured, an unscented Kalman filter (Section 3.4.2) may be needed [1]. Finally, where ambiguous environmental features are used for positioning, a multiple-hypothesis Kalman filter (Section 3.4.5) or a particle filter (Section 3.5) may be required. For centralized integration, the term "Kalman filter" should therefore be taken to mean "Kalman filter-based estimation algorithm."

Either IMU or INS measurements are acceptable in a centralized architecture, as the inertial navigation equations do not incorporate any smoothing or estimation algorithm. Other dead-reckoning systems and environmental feature-matching systems may also output either the sensor or navigation measurements, provided they do not pass through smoothing or estimation algorithms.

Processing of GNSS measurements is described in Section 9.4.2 for a total-state Kalman filter and in Section 14.3.2 for an error-state Kalman filter with an INS reference. Processing of dead-reckoning measurements is described in Section 16.2, while processing of terrestrial radio navigation and environmental feature-matching measurements is described in Section 16.3. Correct handling of data lags (see Section 3.3.4) is particularly important in centralized integration.

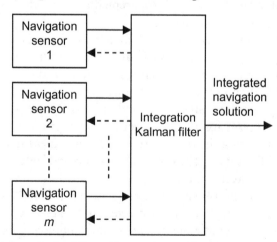

Figure 16.6 Total-state centralized filtered integration architecture.

16.1 Integration Architectures

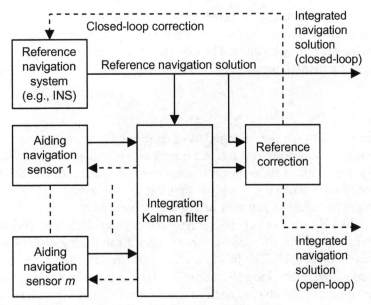

Figure 16.7 Error-state centralized filtered integration architecture.

In a centralized filtered integration architecture, the systematic errors and noise sources of all of the navigation sensors are modeled in the same Kalman filter-based estimation algorithm. This ensures that all error correlations are accounted for, all measurements optimally weighted, and the maximum information used to calibrate each error. Furthermore, the elimination of Kalman filter cascades enables higher gains to be used before there is a stability risk. Thus, the centralized filtered integration architecture provides the optimal navigation solution in terms of accuracy and robustness. However, this is contingent on having the necessary information to model all sensors correctly, requiring careful design.

With all of the error sources modeled in one place, the principal disadvantage of centralized filtered integration is a high processor load. With no independent subsystem navigation solutions available, processor-intensive parallel filters (Section 17.4.2) are needed for applications with demanding integrity requirements. As centralized integration requires raw sensor measurements and information about their error characteristics, it is not compatible with black box navigation systems.

16.1.5 Federated Filtered Integration

In a federated integration architecture, filtered error-state integration is broken down into stages. In the first stage, a reference inertial or other dead-reckoning navigation system is separately integrated with each of the aiding navigation systems in a bank of local error-state Kalman filters, EKFs, and/or UKFs. The reference navigation solution may be computed prior to input to the local filters or separately within each local filter from common sensor outputs. The latter approach enables the local filters to separately feedback closed-loop corrections to the reference navigation solution [2]. Each local filter's integration with its navigation sensors may be either centralized or cascaded and this need not be the same for every filter.

In the second stage of federated integration, the local filter outputs are combined to produce the integrated navigation solution. There are a number of different ways in which this may be done. The no-reset, fusion-reset, zero-reset, and cascaded versions of federated integration are described here.

16.1.5.1 No Reset

Figure 16.8 shows the federated no-reset (FNR) integration architecture, in which the navigation solutions and reference-system error estimates from the local filters are combined with a single-epoch fusing algorithm using (16.4) and (16.5) [3]. Changes in the raw reference navigation solution can be used to propagate the integrated navigation solution between fusing algorithm updates.

The FNR architecture is useful for integrating black box navigation systems that accept a common INS-aiding input without the tuning difficulties inherent in the cascaded approach. The local filter outputs are also suited to integrity monitoring by consistency checking (Section 17.4.1). Closed-loop corrections from the fusing algorithm can only be fed back to the reference system if they are accepted by all of the local filters; otherwise, local filter estimation would be disrupted by unmodeled transients. However, useful open-loop integration of an INS requires high-quality sensors.

A problem with the FNR architecture is that, as the reference navigation system is common to all of the local filters, their navigation solutions are no longer independent, so $\mathbf{P}_{ij} \neq 0$. Consequently, the weighting of the local filter solutions is suboptimal and the error covariance of the integrated navigation solution, $\mathbf{P}_{ff}$, is overoptimistic. A work-around solution to this is to overestimate the initial error covariance and system noise covariance of the reference-system states in the local filters by a factor of m, the number of local filters [4], though this does cause the Kalman gains for these states to be overestimated.

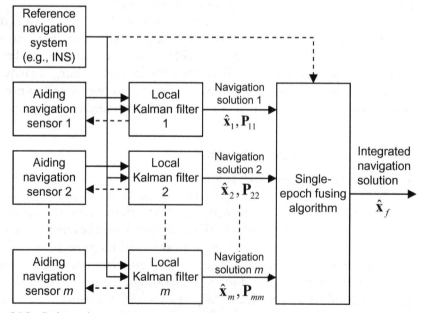

Figure 16.8 Federated no-reset integration architecture (error-state).

16.1 Integration Architectures

16.1.5.2 Fusion Reset

The federated fusion-reset (FFR) architecture, shown in Figure 16.9, feeds back the state estimates and error covariance from a single-epoch fusing algorithm to the local filters, where they replace the corresponding states and error covariance matrix elements, the latter scaled up by a factor of m [5]. This allows calibration information to be shared between all subsystems at a lower processing load than centralized integration without cascading Kalman filters. The local filters may be implemented using parallel processors. However, the problem then arises of how to model the correlations between the common and local states in each local filter; one solution makes use of a conventional Kalman filter in parallel with each local filter [6].

The FFR architecture shares many of the drawbacks of the centralized architecture: it is incompatible with black box navigation systems and does not provide independent subsystem solutions for integrity monitoring. It is also more complex than the centralized architecture and gives poorer performance [7].

16.1.5.3 Zero Reset

The federated zero-reset (FZR) architecture, shown in Figure 16.10, uses a Kalman filter to integrate the outputs of the local filters. However, Kalman filter cascading is avoided by zeroing all local filter states after they are input to the master Kalman filter, with the corresponding elements of the error covariance matrix set to their initialization values [3]. State estimates that are not passed to the master filter may be retained in the local filters without being reset. This zero reset prevents the same data being input to the master filter more than once. Unlike in the FNR and FFR architectures, measurements from different local filters may be processed by the master filter at different times.

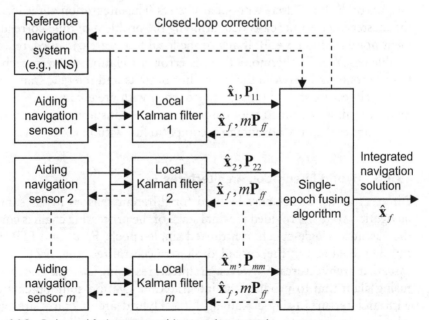

Figure 16.9 Federated fusion-reset architecture (error-state).

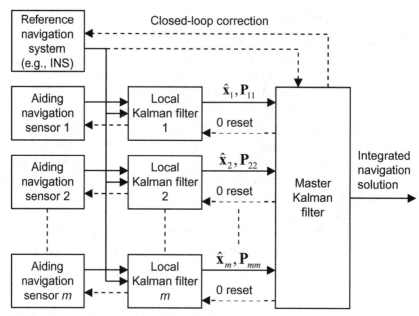

Figure 16.10 Federated zero-reset architecture (error-state).

The FZR architecture can be useful where there is a need to process measurements at a faster rate than it is practical to run the integration algorithm. It has found use in coherent deeply coupled INS/GPS integration (Section 14.3.3), with the local filters performing the signal tracking and the master filter the integration.

16.1.5.4 Cascaded

Figure 16.11 shows a federated architecture with a Kalman filter integrating the outputs of the local filters without any resets. The integration of the local filters with the master filter is thus cascaded, bringing the problem of time-correlated measurement noise (Section 3.4.3). If any of the local filters are cascaded, this introduces a double cascade. Furthermore, there is error correlation between each of the local filter outputs and between the local filter outputs and reference navigation system. This architecture should thus be approached with great caution as the master Kalman filter must be tuned very carefully to avoid instability and produce realistic uncertainties. Its only advantage is compatibility with black box local filters.

16.1.6 Hybrid Integration Architectures

Different architectures may be used for different sensors in the same integrated navigation system, provided the final stage of the processing chain is common. Thus, the cascaded single-epoch, centralized single-epoch, FNR, and FFR architectures can be mixed as can the, cascaded filtered, centralized filter, FZR, and federated-cascaded architectures. However, architectures using a single-epoch least-squares fusing algorithm to produce the integrated navigation solution cannot be mixed with architectures using a Kalman filter. Hybrid architectures are typically used where constraints in the design of the constituent navigation systems prevent use of

16.1 Integration Architectures

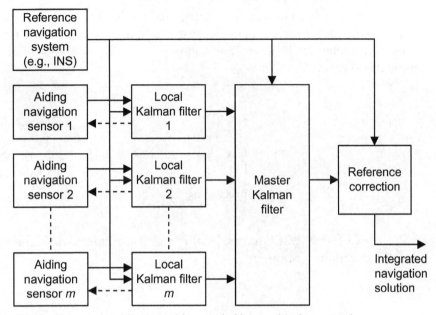

Figure 16.11 Federated architecture with cascaded integration (error-state).

the desired architecture in all cases. Figure 16.12 depicts an example integration of INS with centralized GNSS, cascaded DME/TACAN, and federated-cascaded TRN.

16.1.7 Total-State Kalman Filter Employing Prediction

Total-state integration employing prediction is applicable to centralized, master, and local Kalman filter-based estimation algorithms integrating only positioning systems. It is also suitable for use with dead-reckoning sensors where the velocity measurement noise exceeds the variation in the host vehicle's velocity between measurements.

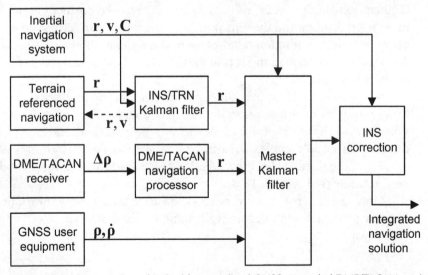

Figure 16.12 Hybrid integration of INS with centralized GNSS, cascaded DME/TACAN, and federated-cascaded TRN.

The state vector comprises an integrated navigation solution, $\mathbf{x}^{\gamma}_{Nav}$, together with error states for each of the subsystems integrated, $\mathbf{x}_{Sensor-i}$. Thus, the term total-state only applies to the navigation solution states. When n sensors are combined, the state vector is

$$\mathbf{x}^{\gamma} = \begin{pmatrix} \mathbf{x}^{\gamma}_{Nav} \\ \mathbf{x}_{Sensor-1} \\ \mathbf{x}_{Sensor-2} \\ \vdots \\ \mathbf{x}_{Sensor-n} \end{pmatrix} \qquad \gamma \in i, e, n. \qquad (16.6)$$

The ECI-frame, ECEF-frame, local-navigation-frame, and local-tangent-plane-frame navigation solutions are

$$\mathbf{x}^{i}_{Nav} = \begin{pmatrix} \mathbf{r}^{i}_{ib} \\ \mathbf{v}^{i}_{ib} \end{pmatrix}, \quad \mathbf{x}^{e}_{Nav} = \begin{pmatrix} \mathbf{r}^{e}_{eb} \\ \mathbf{v}^{e}_{eb} \end{pmatrix}, \quad \mathbf{x}^{n}_{Nav} = \begin{pmatrix} L_b \\ \lambda_b \\ h_b \\ \mathbf{v}^{n}_{eb} \end{pmatrix}, \quad \mathbf{x}^{l}_{Nav} = \begin{pmatrix} \mathbf{r}^{l}_{lb} \\ \mathbf{v}^{l}_{lb} \end{pmatrix}. \qquad (16.7)$$

When dead-reckoning sensors are integrated, attitude states are added. Alternatively, total-state position and velocity integration may be combined with error-state attitude integration.

In total-state integration, the position solution is predicted forward in time using the velocity solution. This is useful for smoothing noise in navigation system measurements, aiding time synchronization, and bridging gaps in navigation system measurements, (e.g., in tunnels). The accuracy of the predicted solution depends on the host-vehicle dynamics, so its uncertainty must be correctly modeled to ensure that the predictions and the sensor measurements are correctly weighted in the navigation solution. Prediction takes place in the Kalman filter's system model (Section 3.2.3). This is described in Section 9.4.2.2 for a GNSS navigation filter and is also applicable generally.

For high-dynamics applications, navigation solution prediction may be enhanced by adding acceleration states to the Kalman filter. Further improvements can sometimes be made by adding force modeling. For most air, land, and marine applications, the forces are too complex for this to be practical. An exception is some ballistic missiles and guided shells. However, for space applications, force modeling can be very accurate (see Section 18.8).

Most sensor errors should be modeled as either a random walk or a first-order Markov process. For a random walk state, x_{ri}, the state dynamics and system noise variance are

$$E\left(\frac{\partial x_{ri}}{\partial t}\right) = 0, \quad Q_{ri} = S_{ri}\tau_s, \qquad (16.8)$$

where τ_s is the state-propagation interval, and S_{ri} a frequency-independent approximation of the PSD of the noise producing the random walk. The state dynamics and system noise variance of an exponentially-correlated fixed-variance first-order Markov process are given by (3.3) and (14.85). Error states are not necessarily estimated for all constituent navigation systems.

The Kalman filter, EKF, and UKF measurement models are described in Sections 3.2.4, 3.4.1, and 3.4.2, respectively. When navigation systems output measurements with the same time of validity, they may be processed by the Kalman filter together. Otherwise, they must be processed separately. By analogy with (14.96) to (14.100), the measurement innovation vector for each sensor is of the form

$$\delta \mathbf{z}_i^- = \tilde{\mathbf{m}}_i - \delta \hat{\mathbf{m}}_i^- - \hat{\mathbf{m}}_{Nav}^-, \tag{16.9}$$

where $\tilde{\mathbf{m}}_i$ is the vector of measurements output by sensor i, (e.g., a position solution or a set of range measurements), $\delta \hat{\mathbf{m}}_i^-$ is the Kalman-filter-estimated error in those measurements, calculated from $\hat{\mathbf{x}}_{Sensor-i}^-$, and $\hat{\mathbf{m}}_{Nav}^-$ is the value of $\mathbf{m}$ estimated from the predicted navigation solution, $\hat{\mathbf{x}}_{Nav}^{\gamma-}$. Note that the system model must be used to predict forward the estimated navigation solution to the measurement time of validity prior to the measurement update. The measurement matrix is then

$$\mathbf{H}_{i,k}^{\gamma} = \left(-\frac{\partial \mathbf{m}}{\partial \mathbf{x}_{Nav}^{\gamma}} \quad 0 \quad \frac{\partial \mathbf{m}}{\partial \mathbf{x}_{Sensor-i}} \quad 0 \right)_{\mathbf{x}^{\gamma} = \hat{\mathbf{x}}_k^{\gamma-}}, \tag{16.10}$$

noting that $\partial \mathbf{m}/\partial \mathbf{x}_{Sensor-j}$ is a null matrix for $j \neq i$. Measurement models for GNSS are described in Sections 9.4.2.3 and 14.3, while those for dead reckoning are presented in Section 16.2, and those for terrestrial radio navigation and environmental feature matching are described in Section 16.3.

A problem that can arise in multisensor navigation systems is that, due to variation in the duty cycles and processing delays between different sensors, measurements may not become available in the order of the time of validity. Measurements may also be delayed to enable testing for faults over multiple epochs (see Chapter 17) before acceptance. The simplest solution is to delay processing of those measurements with the shorter lags so that measurements may be processed in the time-of-validity order. However, this will delay the integrated navigation solution. Another option is to use the Kalman filter system model to propagate between the times of validity of the measurements and the state estimates as described in Section 3.3.4.

16.1.8 Error-State Kalman Filter

In error-state multisensor integration, the integrated navigation solution comprises the navigation solution of a reference navigation system, corrected using measurements from the other constituent navigation systems. The reference system must be an INS or another dead-reckoning system. When noninertial dead reckoning is used, a sensor measuring velocity or distance traveled must be combined with attitude measurement sensors so that a navigation solution resolved about an ECI, ECEF, local-navigation, or local-tangent-plane frame may be maintained. The reference solution takes the place of the predicted solution of a total-state Kalman filter-based estimation algorithm.

The state vector comprises error states for the reference navigation system, $\mathbf{x}^\gamma_{Ref}$, and for each of the aiding subsystems, $\mathbf{x}_{Sensor-i}$. With n aiding sensors, it is

$$\mathbf{x}^\gamma = \begin{pmatrix} \mathbf{x}^\gamma_{Ref} \\ \mathbf{x}_{Sensor-1} \\ \mathbf{x}_{Sensor-2} \\ \vdots \\ \mathbf{x}_{Sensor-n} \end{pmatrix} \qquad \gamma \in i, e, n, l. \tag{16.11}$$

The reference navigation system error states comprise attitude, velocity, and position errors, defined in Section 5.7, together with reference sensor errors, $\mathbf{x}_{Sensor-Ref}$. Thus, for ECI-frame, ECEF-frame, local-navigation-frame and local-tangent-plane-frame error-state implementations,

$$\mathbf{x}^i_{Ref} = \begin{pmatrix} \delta\boldsymbol{\psi}^i_{ib} \\ \delta\mathbf{v}^i_{ib} \\ \delta\mathbf{r}^i_{ib} \\ \mathbf{x}_{Sensor-Ref} \end{pmatrix}, \quad \mathbf{x}^e_{Ref} = \begin{pmatrix} \delta\boldsymbol{\psi}^e_{eb} \\ \delta\mathbf{v}^e_{eb} \\ \delta\mathbf{r}^e_{eb} \\ \mathbf{x}_{Sensor-Ref} \end{pmatrix}, \quad \mathbf{x}^n_{Ref} = \begin{pmatrix} \delta\boldsymbol{\psi}^n_{nb} \\ \delta\mathbf{v}^n_{eb} \\ \delta\mathbf{p}_b \\ \mathbf{x}_{Sensor-Ref} \end{pmatrix}, \quad \mathbf{x}^l_{Ref} = \begin{pmatrix} \delta\boldsymbol{\psi}^l_{lb} \\ \delta\mathbf{v}^l_{lb} \\ \delta\mathbf{r}^l_{lb} \\ \mathbf{x}_{Sensor-Ref} \end{pmatrix},$$

$$\tag{16.12}$$

where $\delta\mathbf{p}_b$ is defined by (14.51). The INS system model is described in Section 14.2, while dead-reckoning system models are discussed in Section 16.2. The system models for the aiding subsystems are the same as for total-state integration.

By analogy with total-state integration, the measurement innovation vector and measurement matrix for each aiding subsystem are of the form

$$\delta\mathbf{z}^-_i = \tilde{\mathbf{m}}_i - \delta\hat{\mathbf{m}}^-_i - \hat{\mathbf{m}}^-_{Ref}, \tag{16.13}$$

and

$$\mathbf{H}^\gamma_{i,k} = \begin{pmatrix} -\dfrac{\partial \mathbf{m}}{\partial \mathbf{x}^\gamma_{Ref}} & 0 & \dfrac{\partial \mathbf{m}}{\partial \mathbf{x}_{Sensor-i}} & 0 \end{pmatrix}_{\mathbf{x}^\gamma = \hat{\mathbf{x}}^{\gamma-}_k}, \tag{16.14}$$

where $\hat{\mathbf{m}}^-_{Ref}$ is the value of m estimated from the corrected reference navigation solution. As discussed in Section 16.1.7, measurements from different subsystems may be delayed by different amounts. Time synchronization of the measurement innovations may be performed by using a stored reference navigation solution at the measurement time of validity, as described in Section 3.3.4. Measurements with different times of validity may be processed together, providing the appropriate reference navigation solution is used in each case. For long delays, the Kalman filter system model may also be used to propagate between the measurement and state-estimate times of validity (see Section 3.3.4).

The measurement models for the aiding sensors are the same as for total-state integration. For position and velocity errors, there is a sign change. Thus,

$$\frac{\partial \mathbf{m}}{\partial \delta \mathbf{v}_{\beta\alpha}^{\gamma}} = -\frac{\partial \mathbf{m}}{\partial \mathbf{v}_{\beta\alpha}^{\gamma}}, \qquad \frac{\partial \mathbf{m}}{\partial \delta \mathbf{r}_{\beta\alpha}^{\gamma}} = -\frac{\partial \mathbf{m}}{\partial \mathbf{r}_{\beta\alpha}^{\gamma}}, \qquad \frac{\partial \mathbf{m}}{\partial \delta \mathbf{p}_b} = -\frac{\partial \mathbf{m}}{\partial \mathbf{p}_b}. \qquad (16.15)$$

Following a measurement update, a dead-reckoning reference navigation solution is corrected in the same way as an inertial navigation solution, as described in Section 14.1.1, with both open- and closed-loop correction options available.

Processing measurement data out of a time-of-validity order is not a problem in an error-state Kalman filter, provided the time of validity of the reference navigation solution does not lag behind that of any aiding subsystem. However, the problem, described in Section 3.3.4, of repeated application of the same closed-loop corrections to the reference navigation solution can be magnified in a multisensor system in which different sensors exhibit different duty cycles and processing lags. Thus, careful Kalman filter design is needed to maintain stability.

As discussed in Section 14.1.1, total-state Kalman filters that incorporate the navigation equations of a reference navigation system in their system models are mathematically equivalent to the combination of the corresponding error-state Kalman filter and reference system navigation equations. Therefore, they are not discussed separately here.

16.1.9 Primary and Reversionary Moding

In some integration architectures, a subsystem, known as a *reversionary system*, only contributes to the integrated navigation solution when the output of another subsystem, known as a *primary system*, is either unavailable or considered unreliable. Typically, a more accurate technology, such as GNSS, would be integrated as a primary system and a less accurate technology, such as ELoran, as a reversionary system. The integrated system thus has two modes: a *primary mode* using the primary system and a *reversionary mode* using the reversionary system. In the primary mode, the primary subsystem solution may be used to calibrate errors in the reversionary subsystem. The calibration algorithm may be Kalman filter-based or an ANN. Figure 16.13 illustrates this.

More complex integration architectures may also be assembled in which some subsystems are used in the primary mode, some are used in the reversionary mode, and some are used in both modes. Figure 16.14 illustrates an example in which GNSS is a primary system, ELoran is a reversionary system, and inertial navigation is a common sensor.

A reversionary subsystem may also be used to initialize the primary subsystem, forming a tiered position fix. A rough position fix can aid GNSS signal acquisition (see Section 10.5.1) by reducing the search space. Other ranging-based positioning systems can benefit in the same way. Pattern-matching positioning techniques require

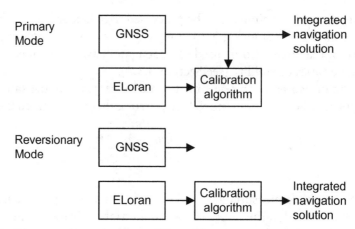

Figure 16.13 Primary and reversionary moding architecture.

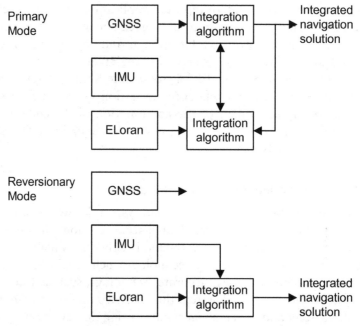

Figure 16.14 Integration of GNSS, ELoran, and inertial navigation in a primary and reversionary moding error-state architecture.

an approximate position to define their search areas. Similarly, position-fixing techniques that incorporate environmental feature identification typically need a coarse position fix to determine which region of the feature database to search and limit ambiguous matches.

In summary, primary and reversionary moding may be used to integrate a coarse, but robust and unambiguous, positioning technology with another technology that is more accurate, but less robust and/or ambiguous. This is equivalent to combining modulation-based and carrier-based measurements of a radio signal.

16.1.10 Context-Adaptive Moding

As discussed in Section 1.4.2 and elsewhere, many navigation techniques are context dependent, relying on prior knowledge of the environment and the host vehicle or user behavior. Different types of radio signal are available in different environments, particularly indoors. The appropriate tuning of a total-state Kalman filter (e.g., a GNSS navigation filter) depends on the user dynamics. Different map-matching algorithms must be used for road vehicles, trains, pedestrians outdoors, and pedestrians indoors (see Section 13.1). Certain methods of processing IMU data, such as PDR using step detection and the application of vehicle motion constraints, are also context-dependent.

This is important for multicontext applications. An example would be a handheld navigation device that is required to operate indoors, outdoors, and inside a car. For best performance, the navigation system must adapt to its context. Figure 16.15 outlines the architecture of a context-adaptive (or cognitive) navigation system. The integrated navigation solution and the sensor measurements are input to a context detection algorithm. This then selects a mode, consisting of a set of sensor-processing and integration algorithms, appropriate to the context. Some algorithms would only be used in one mode while others would be used in multiple modes, but with different tunings. On mode changes, some state estimates and their associated error covariance are passed from the old algorithm to the new while others are reset.

Behavioral context-detection algorithms must determine whether the navigation system is on a vehicle or a pedestrian and whether it is moving or stationary. It is also useful to distinguish different types of vehicle and different classes of vehicle and pedestrian motion.

Vehicle types can be identified by their motion. Aircraft travel much faster than boats and most land vehicles, and can significantly vary their height. Conversely, pedestrians are slower than nearly all types of vehicle and can suddenly stop or change

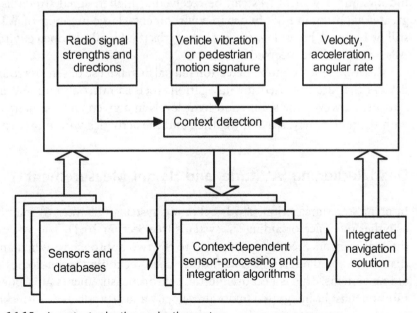

Figure 16.15 A context-adaptive navigation system.

direction. Vehicles can also be identified by their vibration signatures, comprising both engine vibration and vibration induced by air turbulence, sea-state motion, or road-surface irregularity.

If a vehicle is equipped with a short-range communications system, a mobile device can use that to determine whether it is in the vicinity of a vehicle and the identity of that vehicle. However, further information is needed to determine whether or not the device is inside the vehicle.

Pedestrian motion exhibits distinct signatures that can be identified using inertial sensors as described in Section 6.4. These motion signatures depend on the type of activity, (e.g., walking or running) and the location of the sensors on the body; they can also vary between individuals.

Stationarity detection for both pedestrians and land vehicles (including robots) is described in Section 15.3. When a navigation system is stationary for an extended period, most of the subsystems should be switched off to save power as the navigation solution will not change. A vehicle-mounted navigation system will either be switched between active and standby modes by the host vehicle or will only receive power when the host vehicle is in use. However, an autonomous navigation system, must switch itself to standby mode and then determine when to reactivate itself using minimal sensor information. A single low-cost accelerometer may be used for simple motion detection. A mobile device may also reactivate the navigation system when the user starts to interact with it or uses an application that requires position.

Environmental context may be determined using radio signals, such as GNSS or Wi-Fi. In an open environment, strong GNSS signals will be received from all directions. In an urban environment, fewer strong signals will be received and only from certain directions. Inside a building, nearly all GNSS signals will be much weaker than outside. It should also be possible to determine that the receive antenna is inside a vehicle.

Wi-Fi signals essentially vary with the environment in the opposite way to GNSS. Indoors, more WLAN APs can be received at higher signal strengths and there is greater variation in RSS. In urban environments, large numbers of WLAN APs can still be received, but at lower signal strengths [8]. Finally, in open environments, few APs, if any, will be received.

Environmental context detection may also make use of sensors found on mobile devices that are not used for navigation, such as microphones and ambient light sensors. However, the sound and light levels in a given environment will also vary with the time of day, season, weather, and the activities within that environment.

16.2 Dead Reckoning, Attitude, and Height Measurement

In error state-integration, dead-reckoning systems (Chapter 6) may be integrated as either reference or aiding subsystems (see Section 16.1). The system models for the sensor errors are largely common to the two approaches. When an INS is used, that should be the reference, with other dead-reckoning technologies integrated as aiding systems. This is because inertial sensor measurements are updated at a high rate and must be integrated to obtain velocity and attitude. A dead-reckoning system can also be a constituent subsystem in total-state integration, integrated in the same

16.2 Dead Reckoning, Attitude, and Height Measurement

way as an aiding system in error-state integration. A filtered integration architecture is assumed throughout this section.

A dead-reckoning reference may be two- or three-dimensional. However, it must incorporate both an attitude or heading measurement system, and a distance traveled or velocity measurement system. When the velocity/distance measurement is horizontal only, the reference may also include a separate height measurement. However, where multiple dead-reckoning systems provide measurements of the same type (i.e., angular or linear), only one can be the reference. The techniques for handling large INS heading errors discussed in Section 14.4.4 are also relevant to other dead-reckoning systems.

Attitude and velocity/distance measurements from aiding sensors may be combined to provide a velocity/distance measurement resolved about an ECI, ECEF, local-navigation, or local-tangent-plane frame, as described in Sections 6.3 to 6.5. However, treating these as separate measurements allows integrity monitoring, such as measurement innovation filtering (Section 17.3.1), to be performed separately so that measurements from one sensor may be accepted when those from the other are rejected.

Any dead-reckoning system may be aided by feeding back estimates of its systematic errors from the integration algorithm. These may then be used to correct the dead-reckoning sensor measurements. As discussed in Section 16.1, Kalman filter estimates that are fed back to a subsystem as closed-loop corrections should be zeroed. Therefore, once a correction is received by the dead-reckoning subsystem, it should be applied to all subsequent sensor measurements, while further corrections should be added to the existing corrections (as opposed to displacing them).

A dead-reckoning system that computes its own navigation solution may also be aided through initialization and correction of that navigation solution. The position may be initialized directly from a position-fixing system, while corrections are normally supplied by the integration algorithm. Corrections are applied in the same way as for inertial navigation (see Section 14.1.1).

Conversely, dead reckoning, or a combination of navigation sensors, may be used to aid GNSS acquisition and tracking in the same manner as inertial navigation, as described in Section 14.1.4. Aiding of other position-fixing systems using dead reckoning is discussed in Section 16.3.7.

Integration of attitude measurements is described first, followed by height and depth measurements, odometry, PDR using step detection, and Doppler radar/sonar. Integration of other dead-reckoning sensors, including visual odometry and terrain-referenced dead reckoning, follows similar principles. In addition, Appendix I on the CD presents alternative implementations with a Cartesian local-navigation-frame position error or body-frame-resolved attitude error, and shows how the time synchronization error may be estimated as a Kalman filter state.

16.2.1 Attitude

Measurements from the attitude sensors described in Section 6.1 may be integrated as either aiding sensors or as part of a reference navigation system. Integration of magnetic heading is described first, followed by other types of heading measurement, accelerometer leveling measurements, and the Kalman filter implementation of AHRS sensor integration.

16.2.1.1 Magnetic Heading

When a magnetic compass (Section 6.1.1) is integrated as an aiding sensor using a Kalman filter, the measurement innovation is

$$\delta z^-_{M,k} = \left(\tilde{\psi}_{mb} - \delta\hat{\psi}_{mb} + \hat{\alpha}_{nE} - \hat{\psi}_{nb}\right)^-_k, \tag{16.16}$$

where $\tilde{\psi}_{mb}$ is the magnetic heading; $\delta\hat{\psi}_{mb}$ is the magnetic heading error, estimated from the magnetic-compass Kalman filter states, $\mathbf{x}_M$; $\hat{\alpha}_{nE}$ is the declination angle of the Earth's magnetic field, obtained from a magnetic model; and $\hat{\psi}_{nb}$ is the integrated heading solution, comprising the corrected INS, AHRS, gyro, or differential-odometer-indicated heading.

If the state vector is defined as

$$\mathbf{x}^\gamma = \begin{pmatrix} \delta\boldsymbol{\psi}^\gamma_{\gamma b} \\ \vdots \\ \mathbf{x}_M \end{pmatrix} \qquad \gamma \in i,e,n,l, \tag{16.17}$$

where $\delta\boldsymbol{\psi}^\gamma_{\gamma b}$ is the overall attitude error, the measurement matrix is

$$\mathbf{H}^\gamma_{M,k} = \left(-\frac{\partial \tilde{\psi}_{nb}}{\partial \delta\boldsymbol{\psi}^\gamma_{\gamma b}} \quad 0 \quad \frac{\partial \tilde{\psi}_{mb}}{\partial \mathbf{x}_M} \right)_{\mathbf{x}^\gamma = \hat{\mathbf{x}}^{\gamma-}_k}. \tag{16.18}$$

When the small angle approximation applies,

$$\frac{\partial \tilde{\psi}_{nb}}{\partial \delta\boldsymbol{\psi}^\gamma_{\gamma b}} \approx \begin{pmatrix} 0 & 0 & 1 \end{pmatrix} \mathbf{C}^n_\gamma. \tag{16.19}$$

The choice of magnetometer errors to estimate as Kalman filter states varies with the application. If the hard- and soft-iron equipment magnetism is adequately calibrated by the magnetic compass software and does not vary after that calibration, no further calibration is needed in the integration algorithm. Otherwise, the Kalman filter may estimate body-frame hard- and soft-iron magnetism, $\mathbf{b}_m$ and $\mathbf{M}_m$, as defined by (6.10), or heading-domain hard- and soft-iron calibration coefficients, c_{h1}, c_{h2}, c_{s1}, and c_{s2}, as defined by (6.11). Often, only the hard-iron magnetism need be estimated. To observe the heading-domain equipment-magnetism coefficients, the host vehicle must perform turns, while to observe the full body-frame magnetism, maneuvers about the roll and pitch axes are also required. Thus, for marine and land vehicle applications, the heading-domain model is better.

When the heading-domain calibration coefficients are estimated, the magnetic compass error state vector is

$$\mathbf{x}_M = \begin{pmatrix} c_{h1} \\ c_{h2} \\ c_{s1} \\ c_{s2} \end{pmatrix}, \tag{16.20}$$

16.2 Dead Reckoning, Attitude, and Height Measurement

the measurement matrix coefficients are given by

$$\left.\frac{\partial \tilde{\psi}_{mb}}{\partial \mathbf{x}_M}\right|_{\mathbf{x}^{\gamma}=\hat{\mathbf{x}}_k^{\gamma-}} = \begin{pmatrix} -\sin\tilde{\psi}_{mb} & -\cos\tilde{\psi}_{mb} & -\sin 2\tilde{\psi}_{mb} & -\cos 2\tilde{\psi}_{mb} \end{pmatrix}_k, \quad (16.21)$$

and the estimated magnetic heading error is

$$\delta\hat{\psi}_{mb} = -(\hat{c}_{b1}\sin\tilde{\psi}_{mb} + \hat{c}_{b2}\cos\tilde{\psi}_{mb} + \hat{c}_{s1}\sin 2\tilde{\psi}_{mb} + \hat{c}_{s2}\cos 2\tilde{\psi}_{mb}). \quad (16.22)$$

The Kalman filter may also estimate regional and temporal deviations in the declination angle from that given by the magnetic model. To observe this, the observability of the heading error from positioning measurements through the growth in the velocity error must be strong (see Sections 14.2.1 and 14.3.4).

For land applications, magnetic heading measurements are often distorted by local magnetic anomalies as discussed in Section 6.1.1. Because of this, a magnetic compass should not be integrated as a reference sensor. Furthermore, to avoid contamination of the integrated navigation solution, anomalous magnetometer measurements should be identified and rejected. This can be done using measurement innovation filtering (Section 17.3.1) and/or by comparing the magnetic flux density measurements with thresholds as discussed in Section 6.1.1.

Whenever the external magnetic field is constant, even in the presence of magnetic anomalies, the variation in three-axis magnetometer measurements will be due to sensor rotation. Consequently, the magnetometer may be used to measure angular rate, which may be used to calibrate gyro errors in a magnetic angular rate update (MARU) [9]. MARUs can be performed whenever the magnetometer is detected to be stationary (see Section 15.3.1). For foot-mounted inertial navigation of pedestrians, a MARU can thus be performed during each stance phase of walking. More generally, a MARU can be triggered whenever a quasi-static field (QSF) is detected from the time variation of the magnitude of the measured magnetic flux density [10].

16.2.1.2 Heading Measurement Integration

Heading may be obtained by integrating the output of a strapdown yaw-axis gyro (Section 6.1.3), assuming it is approximately level. In an error-state integration architecture, this should be integrated as the reference heading, with other sensor(s) integrated as aiding measurements. The gyro bias may be estimated as a Kalman filter state. It is observed through the change in the heading error estimate (see Section 14.3.4) and the system model is the same as for INS integration (see Section 14.2). The gyro bias may also be observed directly in a zero angular rate update (Section 15.3.3). In a total-state integration architecture, the gyro-indicated yaw rate is input by the integration algorithm as a measurement.

Heading measurements may be obtained directly from a marine gyrocompass (Section 6.1.2) or a land vehicle trajectory (Section 6.1.4). These may be used as a reference in the absence of a gyro, but may also be integrated as aiding sensors. These heading measurements exhibit time-correlated errors. For integration as the reference, the error is modeled by the integrated heading error, $\delta\psi_{nb}$, the variation of which is modeled as system noise.

For integration as an aiding measurement, a sensor heading bias, b_ψ, should be estimated. A suitable measurement innovation is

$$\delta z^-_{\psi,k} = \left(\tilde{\psi}_{nb,\psi} - \hat{\psi}_{nb} - \hat{b}_\psi\right)^-_k, \qquad (16.23)$$

where $\hat{\psi}_{nb}$ is the corrected integrated heading solution and $\tilde{\psi}_{nb,\psi}$ is the heading sensor measurement. $\delta z^-_{\psi,k}$ must be limited to the range $-\pi$ to π by adding or subtracting 2π as appropriate. Defining the state vector as

$$\mathbf{x}^\gamma = \begin{pmatrix} \delta\psi_{\gamma b} \\ b_\psi \\ \vdots \end{pmatrix} \qquad \gamma \in n,l, \qquad (16.24)$$

the measurement matrix is

$$\mathbf{H}^\gamma_{\psi,k} = \begin{pmatrix} -1 & 1 & 0 \end{pmatrix}. \qquad (16.25)$$

A heading reference may also be maintained by integrating differential odometry measurements (Section 6.3.2); odometry integration is discussed in Section 16.2.3.

The output of an integrated heading measurement system (Section 6.1.5) may be integrated with other navigation sensors as a single measurement, forming a cascaded architecture (Section 16.1.3). However, integrating the constituent sensors within a centralized architecture (Section 16.1.4) enables more representative error modeling.

16.2.1.3 Accelerometer Leveling

Accelerometer leveling is integrated as an aiding system wherever another source of attitude is available (otherwise, the roll and pitch are computed directly as described in Section 6.1.6). The measurement innovations comprise the difference between the corrected accelerometer-indicated roll and pitch (subscript A) and the integrated roll and pitch solution (no subscript):

$$\delta \mathbf{z}^-_{L,k} = \begin{pmatrix} \hat{\phi}_{nbA} - \hat{\phi}_{nb} \\ \hat{\theta}_{nbA} - \hat{\theta}_{nb} \end{pmatrix}^-_k, \qquad (16.26)$$

where, from (5.101), the leveling measurements may be corrected with estimates of the accelerometer bias, $\mathbf{b}_a$, using

$$\begin{aligned} \hat{\phi}_{nbA} &= \arctan_2\left[-\hat{f}^b_{ib,y}, -\hat{f}^b_{ib,z}\right] \\ \hat{\theta}_{nbA} &= \arctan\left[\frac{\hat{f}^b_{ib,x}}{\sqrt{\hat{f}^{b\,2}_{ib,y} + \hat{f}^{b\,2}_{ib,z}}}\right] \end{aligned}. \qquad (16.27)$$

16.2 Dead Reckoning, Attitude, and Height Measurement

where

$$\hat{\mathbf{f}}_{ib}^b = \tilde{\mathbf{f}}_{ib}^b - \hat{\mathbf{b}}_a. \tag{16.28}$$

The system model for the accelerometer biases is the same as for INS accelerometers (see Section 14.2). The biases are observable where another source of attitude information, such as gyros or GNSS attitude, is available. When gyros are used, the navigation system must also be rotated about the roll and pitch axes and the gyro biases must not be too large.

Defining the state vector as

$$\mathbf{x}^\gamma = \begin{pmatrix} \delta \boldsymbol{\psi}_{\gamma b}^\gamma \\ \mathbf{b}_a \\ \vdots \end{pmatrix} \qquad \gamma \in i,e,n,l, \tag{16.29}$$

the measurement matrix is

$$\mathbf{H}_{L,k}^\gamma = \begin{pmatrix} -\dfrac{\partial \phi_{nb}}{\partial \delta \boldsymbol{\psi}_{\gamma b}^\gamma} & \dfrac{\partial \phi_{nbA}}{\partial \mathbf{f}_{ib}^b} & 0 \\ -\dfrac{\partial \theta_{nb}}{\partial \delta \boldsymbol{\psi}_{\gamma b}^\gamma} & \dfrac{\partial \theta_{nbA}}{\partial \mathbf{f}_{ib}^b} & 0 \end{pmatrix}_{\mathbf{x}^\gamma = \hat{\mathbf{x}}_k^{\gamma-}}, \tag{16.30}$$

where

$$\begin{pmatrix} \partial \phi_{nbA}/\partial \mathbf{f}_{ib}^b \\ \partial \theta_{nbA}/\partial \mathbf{f}_{ib}^b \end{pmatrix} \approx \begin{pmatrix} 0 & \dfrac{f_{ib,z}^b}{{f_{ib,y}^b}^2 + {f_{ib,z}^b}^2} & \dfrac{-f_{ib,y}^b}{{f_{ib,y}^b}^2 + {f_{ib,z}^b}^2} \\ \dfrac{f_{yz}}{f_{xyz}^2} & \dfrac{-f_{ib,x}^b f_{ib,y}^b}{f_{xyz}^2 f_{yz}} & \dfrac{-f_{ib,x}^b f_{ib,z}^b}{f_{xyz}^2 f_{yz}} \end{pmatrix}, \quad \begin{aligned} f_{yz} &= \sqrt{{f_{ib,y}^b}^2 + {f_{ib,z}^b}^2} \\ f_{xyz}^2 &= {f_{ib,x}^b}^2 + {f_{ib,y}^b}^2 + {f_{ib,z}^b}^2 \end{aligned}$$

$$\tag{16.31}$$

and, where the small angle approximation applies,

$$\begin{pmatrix} \partial \phi_{nb}/\partial \delta \boldsymbol{\psi}_{\gamma b}^\gamma \\ \partial \theta_{nb}/\partial \delta \boldsymbol{\psi}_{\gamma b}^\gamma \end{pmatrix} \approx \begin{pmatrix} 1 & 0 & 0 \\ 0 & 1 & 0 \end{pmatrix} \mathbf{C}_\gamma^b. \tag{16.32}$$

The roll and pitch measurement noise covariance is

$$\mathbf{R}_L = E\left[\begin{pmatrix} \delta\phi_{nbA} \\ \delta\theta_{nbA} \end{pmatrix}\begin{pmatrix} \delta\phi_{nbA} \\ \delta\theta_{nbA} \end{pmatrix}^T\right]$$

$$= \begin{pmatrix} \partial\phi_{nbA}/\partial\mathbf{f}_{ib}^b \\ \partial\theta_{nbA}/\partial\mathbf{f}_{ib}^b \end{pmatrix}_{\mathbf{x}^\gamma=\hat{\mathbf{x}}_k^{\gamma-}} \begin{pmatrix} \sigma_{a,x}^2 & 0 & 0 \\ 0 & \sigma_{a,y}^2 & 0 \\ 0 & 0 & \sigma_{a,z}^2 \end{pmatrix}\begin{pmatrix} \partial\phi_{nbA}/\partial\mathbf{f}_{ib}^b \\ \partial\theta_{nbA}/\partial\mathbf{f}_{ib}^b \end{pmatrix}^T_{\mathbf{x}^\gamma=\hat{\mathbf{x}}_k^{\gamma-}}, \quad (16.33)$$

where $\sigma_{a,x}^2$, $\sigma_{a,y}^2$, and $\sigma_{a,z}^2$ are the variances of the accelerometer measurement errors (see Section 4.4), excluding the biases, which are estimated as states. The noise contribution to the variances should be inversely proportional to the number of successive accelerometer measurements averaged to produce the leveling measurements, noting that the optimum averaging interval depends on both the accelerometer noise and the host vehicle dynamics. The systematic error contribution accounts for scale-factor, cross-coupling, and higher-order errors. It should be modeled as a function of acceleration, where known, as the coupling of these error sources with gravity is typically absorbed into the bias terms.

16.2.1.4 Attitude and Heading Reference System

When the sensors in an AHRS (Section 6.1.8) are integrated by Kalman filter, the gyros should be treated as the reference system, with the other sensors integrated as aiding measurements as described in Sections 16.2.1.1 and 16.2.1.3. A suitable state vector is

$$\mathbf{x}^\gamma = \begin{pmatrix} \delta\boldsymbol{\psi}_{\gamma b}^\gamma \\ \mathbf{b}_a \\ \mathbf{b}_g \\ \mathbf{x}_M \end{pmatrix} \quad \gamma \in i,e,n,l, \quad (16.34)$$

where the attitude error, $\delta\boldsymbol{\psi}_{\gamma b}^\gamma$, and gyro biases, $\mathbf{b}_g$, should be modeled in the same way as for an INS (see Section 14.2).

An AHRS attitude solution may be integrated with other navigation sensors either as a single unit, forming a cascaded architecture (Section 16.1.3), or as its constituent sensors within a centralized architecture (Section 16.1.4). The centralized architecture enables more representative attitude error modeling, but can require more processing power.

When an AHRS is integrated with a positioning system, such as GNSS, the position measurements may be used to estimate an acceleration correction to the leveling measurement [11]. However, as pitch and roll correction is an inherent part of the integration of inertial navigation with position-fixing systems (see Sections 14.2.1 and 14.3.4), a more efficient solution is simply to process low-grade IMU measurements in INS mode when positioning measurements are available. AHRS leveling measurements are then used only during positioning system outages as a reversionary mode [12].

16.2.2 Height and Depth

Barometric, radar, or sonar height and depth measurements (Section 6.2) may be integrated as reference or aiding sensors. When they form the reference, their systematic errors are accounted for by the navigation solution height and vertical velocity error states of the Kalman filter. When they are integrated as aiding sensors, dedicated error states may be required.

For a baro or depth pressure sensor operating close to the surface, only the bias need be estimated. This is observable whenever another source of height information, such as GNSS, is available. For air and underwater applications, a scale factor error should also be estimated; this requires vertical motion to observe. For a radar altimeter, error states are not needed unless the terrain height database is biased, while for sonar altimeter measurements, estimation of a scale factor error can account for errors in the assumed speed of sound.

For a baro or depth pressure sensor integrated as an aiding sensor with a bias, b_b, and scale factor error, s_b, estimated, the measurement innovation is

$$\delta z_{B,k}^- = \left[\tilde{h}_{bB} - \hat{b}_b - \hat{h}_b(1 + \hat{s}_b) \right]_k^-, \tag{16.35}$$

where $\tilde{h}_{bB}$ is the barometric height measurement and $\hat{h}_b$ is the integrated geodetic height solution. If the state vector is

$$\mathbf{x} = \begin{pmatrix} \delta h_b \\ \vdots \\ b_b \\ s_b \end{pmatrix}, \tag{16.36}$$

the measurement matrix is

$$\mathbf{H}_{B,k} \approx \begin{pmatrix} -1 & 0 & 1 & \hat{h}_{b,k}^- \end{pmatrix}, \tag{16.37}$$

neglecting the products of the error states. The same model applies for radar and sonar altimeters, but with fewer error states.

The baro measurement noise variance, R_B, comprises a constant term to account for the measurement noise, scaled to account for time correlation, and a term proportional to the square of the vertical velocity to account for the response lag. For radar and sonar, the assumed measurement noise covariance, R_B, should be varied with the radar footprint size and terrain roughness.

As the bias and scale factor error of a barometric altimeter depends primarily on the weather, multiple users in a given area will experience similar errors. Consequently, estimates of these errors may be shared between participants in a cooperative (or collaborative) positioning system.

16.2.3 Odometry

Integration of odometer and wheel speed sensor measurements (Section 6.3) requires the estimation of the scale factor errors as Kalman filter states. When the speed of individual wheels is measured, a scale factor error state is needed for each wheel used. These may be attributed to each wheel as s_{orL}, s_{orR}, s_{ofL}, and s_{ofR}, where r denotes rear, f denotes forward, L denotes left, and R denotes right. Alternatively, an average and difference scale factor error may be estimated for each pair:

$$\begin{aligned} s_{or} &= \tfrac{1}{2}(s_{orL} + s_{orR}), & s_{\Delta or} &= s_{orL} - s_{orR} \\ s_{of} &= \tfrac{1}{2}(s_{ofL} + s_{ofR}), & s_{\Delta of} &= s_{ofL} - s_{ofR} \end{aligned} \quad (16.38)$$

The average scale factor error is observable whenever independent position fixes are available and the vehicle has moved since the last fix. For the difference scale factor error to be observable, the vehicle must also make turns. Note that errors in the assumed track widths, T_r and T_f, are absorbed by the corresponding difference scale factor error states, $s_{\Delta or}$ and $s_{\Delta of}$, rather than being separately observable.

As tire wear is the main long-term cause of scale factor variation and heating is the main short-term cause, the scale factor errors of each pair of wheels will be highly correlated. For single-wheel scale factor states, this must be modeled with nonzero off-diagonal elements of the initial error covariance and system noise covariance matrices, P_0^+ and Q. The average and difference scale factor error states are uncorrelated with each other. The initial uncertainties and system noise should be larger for the average states. Where transmission-shaft measurements are used, only s_{or} or s_{of} is estimated.

Considering first the integration of odometry as part of the reference navigation system, a pitch correction should be applied to the raw odometry measurements, as described in Section 6.3, so that the position increments are in the horizontal plane. Additional system noise should be modeled where the pitch is unknown. The effect of the scale factor errors on the navigation solution is represented within the system model [13]. Considering rear wheel sensors only with s_{or} and $s_{\Delta or}$ estimated, together with position and attitude or heading errors, but not velocity error, the state dynamics are

$$\delta \dot{\mathbf{r}}_{\gamma b}^{\gamma} = \mathbf{C}_n^{\gamma} \begin{pmatrix} 1 & 0 & 0 \\ 0 & 1 & 0 \\ 0 & 0 & 0 \end{pmatrix} \mathbf{C}_{\gamma}^{n} \mathbf{v}_{\gamma b}^{\gamma} s_{or} \quad \gamma \in i, e \quad (16.39)$$

or

$$\begin{aligned} \delta \dot{L}_b &= \frac{v_{eb,N}^n}{R_N(L_b) + h_b} s_{or} \\ \delta \dot{\lambda}_b &= \frac{v_{eb,E}^n}{[R_E(L_b) + h_b]\cos L_b} s_{or} \end{aligned}, \quad (16.40)$$

16.2 Dead Reckoning, Attitude, and Height Measurement

and, from (6.33),

$$\delta\dot{\psi}^\gamma_{\gamma b} \approx \mathbf{C}^\gamma_{n1:3,3}\left(\dot{\psi}_{nb}s_{or} + \frac{v_{er}}{T_r}s_{\Delta or}\right), \qquad \gamma \in i,e,n, \tag{16.41}$$

where $\mathbf{C}^\gamma_{n1:3,3}$ is the third column of $\mathbf{C}^\gamma_n$, T_r is the rear track width, R_N and R_E are given by (2.105) and (2.106), Earth rate and transport rate terms are neglected, and it is assumed that odometry measurements are only used in the horizontal plane.

When odometry is incorporated as an aiding sensor or total-state integration is used, there are a number of different ways of processing individual wheel-speed measurements. They may be combined into a velocity or distance traveled in a local-navigation, ECEF or ECI frame. They may be expressed as a speed or distance traveled in the body-frame forward direction, together with a heading rate or change. Alternatively, separate measurements for each wheel may be processed. Each provides the same information. When a transmission-shaft sensor is used, only the forward speed or distance traveled is available [14].

In all cases, the assumption of zero velocity along each wheel axle is commonly made, an example of a nonholonomic motion constraint (Section 15.4.1). Instantaneous velocity measurements are noisy, so if velocity, rather than distance, is used, it should be averaged over the time since the last odometry measurement update.

In this example, the odometry measurements comprise the rear-wheel forward speed, $\tilde{v}_{erO}$, and yaw rate, $\tilde{\dot{\psi}}_{nbO}$, averaged over the interval $t - \tau_o$ to t, together with the assumption of a zero cross-track speed (i.e., the motion constraint). For the aiding sensor case, it is assumed that no pitch corrections (see Section 6.3) are applied so the measurement model is in the plane of the road surface, not the horizontal plane. It is also assumed that the axes of the rear-wheel frame are aligned with the body frame (i.e., $\mathbf{C}^r_b = \mathbf{I}_3$). The measurement innovation is thus

$$\delta\mathbf{z}^-_{O,k} = \begin{pmatrix} \tilde{v}_{erO}(1-\hat{s}_{or}) - \hat{v}_{er} \\ \tilde{\dot{\psi}}_{nbO}(1-\hat{s}_{or}) - \dfrac{\hat{v}_{er}}{T_r}\hat{s}_{\Delta or} - \overline{\hat{\dot{\psi}}_{nb}\cos\hat{\theta}_{nb}} \\ -\hat{v}_c \end{pmatrix}_k, \tag{16.42}$$

where $\hat{v}_{er}, \hat{\dot{\psi}}_{nb}$, and $\hat{v}_c$ are, respectively, the predicted rear-wheel forwards speed, yaw rate, and cross-track speed from the corrected navigation solution. They are obtained from a store of the reference velocity and attitude using

$$\hat{v}_{er} = \frac{1}{\tau_o}\int_{t-\tau_o}^{t}\begin{pmatrix} 1 & 0 & 0 \end{pmatrix}\left[\hat{\mathbf{C}}^b_\gamma(t')\hat{\mathbf{v}}^\gamma_{\beta b}(t') + \left(\boldsymbol{\omega}^b_{\beta b}(t') \wedge \mathbf{l}^b_{br}\right)\right]dt', \tag{16.43}$$

$$\overline{\hat{\dot{\psi}}_{nb}\cos\hat{\theta}_{nb}} = \frac{1}{\tau_o^2}\left[\hat{\psi}_{nb}(t) - \hat{\psi}_{nb}(t-\tau_o)\right]\int_{t-\tau_o}^{t}\cos\hat{\theta}_{nb}\,dt', \tag{16.44}$$

and

$$\hat{v}_c = \frac{1}{\tau_o} \int_{t-\tau_o}^{t} \begin{pmatrix} 0 & 1 & 0 \end{pmatrix} \left[\hat{C}_\gamma^b(t') \hat{v}_{\beta b}^\gamma(t') + \left(\omega_{\beta b}^b(t') \wedge l_{br}^b \right) \right] dt', \quad (16.45)$$

where

$$\{\beta,\gamma\} \in \{i,i\},\{e,e\},\{e,n\}. \quad (16.46)$$

Defining the state vector as

$$\mathbf{x}^\gamma = \begin{pmatrix} \delta \psi_{\gamma b}^\gamma \\ \delta \mathbf{v}_{\beta b}^\gamma \\ \vdots \\ \mathbf{b}_g \\ \vdots \\ s_{or} \\ s_{\Delta or} \end{pmatrix}, \quad (16.47)$$

noting that the gyro bias is only estimated where the reference navigation system includes gyros, the measurement matrix is

$$\mathbf{H}_{O,k}^\gamma \approx \begin{pmatrix} \mathbf{H}_{O11}^\gamma & \mathbf{H}_{O12}^\gamma & 0 & 0 & 0 & \hat{v}_{er} & 0 \\ \mathbf{H}_{O21}^\gamma & 0 & 0 & -\frac{\cos\hat{\theta}_{nb}}{\tau_o}\begin{pmatrix} 0 & 0 & 1 \end{pmatrix}\hat{C}_b^n & 0 & \hat{\psi}_{nb} & \frac{\hat{v}_{er}}{T_r} \\ \mathbf{H}_{O31}^\gamma & \mathbf{H}_{O32}^\gamma & 0 & 0 & 0 & 0 & 0 \end{pmatrix}_{\hat{\mathbf{x}}^\gamma = \hat{\mathbf{x}}_k^{\gamma-}}, \quad (16.48)$$

where

$$\mathbf{H}_{O11}^\gamma = -\frac{1}{\tau_o} \int_{t-\tau_o}^{t} \begin{pmatrix} 1 & 0 & 0 \end{pmatrix} \hat{C}_\gamma^b(t') \left[\hat{v}_{\beta b}^\gamma(t') \wedge \right] dt'$$

$$\mathbf{H}_{O12}^\gamma = -\frac{1}{\tau_o} \int_{t-\tau_o}^{t} \begin{pmatrix} 1 & 0 & 0 \end{pmatrix} \hat{C}_\gamma^b(t') \, dt'$$

$$\mathbf{H}_{O21}^\gamma \approx \frac{1}{\tau_o^2} [\hat{\psi}_{nb}(t) - \hat{\psi}_{nb}(t-\tau_o)] \int_{t-\tau_o}^{t} \sin\hat{\theta}_{nb} \begin{pmatrix} 0 & \cos\hat{\phi}_{nb} & \sin\hat{\phi}_{nb} \end{pmatrix} \hat{C}_\gamma^b(t') \, dt'. \quad (16.49)$$

$$\mathbf{H}_{O31}^\gamma = -\frac{1}{\tau_o} \int_{t-\tau_o}^{t} \begin{pmatrix} 0 & 1 & 0 \end{pmatrix} \hat{C}_\gamma^b(t') \left[\hat{v}_{\beta b}^\gamma(t') \wedge \right] dt'$$

$$\mathbf{H}_{O32}^\gamma = -\frac{1}{\tau_o} \int_{t-\tau_o}^{t} \begin{pmatrix} 0 & 1 & 0 \end{pmatrix} \hat{C}_\gamma^b(t') \, dt'$$

16.2 Dead Reckoning, Attitude, and Height Measurement

See Section I.8.1 of Appendix I on the CD for the derivation of the attitude error columns of $\mathbf{H}_{O,k}^{\gamma}$. When there is an unknown misalignment between the rear-wheel frame and the body frame, the yaw component may be estimated as an additional Kalman filter state as described in Section I.8.2 of Appendix I on the CD.

The odometry measurement noise covariance may be expressed as

$$\mathbf{R}_{O,k} = \begin{pmatrix} 0.5 & 0.5 & 0 \\ 1/T_r & -1/T_r & 0 \\ 0 & 0 & 1 \end{pmatrix} \begin{pmatrix} \sigma_{orL}^2 & 0 & 0 \\ 0 & \sigma_{orR}^2 & 0 \\ 0 & 0 & \sigma_{cmc}^2 \end{pmatrix} \begin{pmatrix} 0.5 & 1/T_r & 0 \\ 0.5 & -1/T_r & 0 \\ 0 & 0 & 1 \end{pmatrix}, \quad (16.50)$$

where σ_{orL}^2 and σ_{orR}^2 are the rear-left and rear-right WSS noise variances and σ_{cmc}^2 is the cross-track motion-constraint noise variance. The main source of WSS noise is sensor quantization. A higher measurement noise covariance should be assumed wherever the pitch is unknown (this is included in $\hat{\mathbf{C}}_{\gamma}^{b}$). Motion constraint noise is discussed in Section 15.4.

Integration of front-wheel speed measurements is more complex as the front-wheel frame is displaced from the body frame by the steering angle, ψ_{bf}, as shown in Section 6.3. The integration algorithm must thus input the steering angle and rate thereof as well as the odometer measurements. The steering-angle bias and scale factor may be estimated as Kalman filter states where necessary.

Land vehicles are subject to wheel slip, resulting in occasional erroneous odometry measurements. Therefore, measurement innovation filtering (Section 17.3.1) should be used to reject suspect measurements.

16.2.4 Pedestrian Dead Reckoning Using Step Detection

Integration of measurements from PDR using step detection (Section 6.4) requires the estimation of the coefficients of the step length estimation model or corrections to their default values, together with the boresight angle, ψ_{bb}, where this is unknown or variable. The PDR-estimated step length, Δr_P, may be modeled as [repeating (6.43)]

$$\Delta r_P = c_{P0} + \frac{c_{P1}}{\tau_P} + c_{P2}\sigma_f^2 + c_{P3}\hat{\theta}_{nb},$$

where τ_P is the duration of the current step, σ_f^2 is the variance of the specific force measurements, $\hat{\theta}_{nb}$ is the estimated angle of the slope, and the coefficients of the step length estimation model are

$$\mathbf{c}_P = \begin{pmatrix} c_{P0} \\ c_{P1} \\ c_{P2} \\ c_{P3} \end{pmatrix}. \quad (16.51)$$

Observability of the first model coefficient, c_{P0}, simply requires the availability of a more accurate source of position information. Coefficients c_{P1} and c_{P2} require variation in the pedestrian's walking pace to be observed, while coefficient c_{P3} requires variation in the terrain slope. The observability of the boresight angle is contingent on turns being made.

When PDR using step detection is integrated as part of the reference navigation system, the effect of the PDR states on the navigation solution is represented within the system model. If a Kalman filter system propagation step occurs whenever a step is detected, the expectation of the position error propagates as

$$E\left(\delta \mathbf{r}_{\gamma b,k}^{\gamma-}\right) = \begin{aligned} &\delta \mathbf{r}_{\gamma b,k-1}^{\gamma+} + \mathbf{C}_n^\gamma \begin{pmatrix} \cos(\hat{\psi}_{nb} + \hat{\psi}_{bb}) \\ \sin(\hat{\psi}_{nb} + \hat{\psi}_{bb}) \\ 0 \end{pmatrix} \begin{pmatrix} 1 & \frac{1}{\tau_P} & \sigma_f^2 & \hat{\theta}_{nb} \end{pmatrix} \delta \mathbf{c}_{P,k-1}^- \\ &+ \mathbf{C}_n^\gamma \begin{pmatrix} -\sin(\hat{\psi}_{nb} + \hat{\psi}_{bb}) \\ \cos(\hat{\psi}_{nb} + \hat{\psi}_{bb}) \\ 0 \end{pmatrix} \Delta \hat{r}_P \left(\delta \psi_{nb,k-1}^- + \delta \psi_{bb,k-1}^-\right) \end{aligned}, \quad (16.52)$$

or

$$E\left(\delta \mathbf{p}_{b,k}^-\right) = \begin{aligned} &\delta \mathbf{p}_{b,k-1}^+ + \mathbf{T}_{r(n)}^p \begin{pmatrix} \cos(\hat{\psi}_{nb} + \hat{\psi}_{bb}) \\ \sin(\hat{\psi}_{nb} + \hat{\psi}_{bb}) \\ 0 \end{pmatrix} \begin{pmatrix} 1 & \frac{1}{\tau_P} & \sigma_f^2 & \hat{\theta}_{nb} \end{pmatrix} \delta \mathbf{c}_{P,k-1}^- \\ &+ \mathbf{T}_{r(n)}^p \begin{pmatrix} -\sin(\hat{\psi}_{nb} + \hat{\psi}_{bb}) \\ \cos(\hat{\psi}_{nb} + \hat{\psi}_{bb}) \\ 0 \end{pmatrix} \Delta \hat{r}_P \left(\delta \psi_{nb,k-1}^- + \delta \psi_{bb,k-1}^-\right) \end{aligned}, \quad (16.53)$$

where $\delta \mathbf{c}_p = \tilde{\mathbf{c}}_p - \mathbf{c}_p$ is the error in the PDR-assumed model coefficients, $\Delta \hat{r}_p$ is the step length estimated using the corrected model coefficients, $\delta \psi_{bb} = \tilde{\psi}_{bb} - \psi_{bb}$ is the error in the PDR-assumed boresight angle, and $\delta \psi_{nb}$ is the heading error, which may be obtained from the attitude error using

$$\delta \hat{\psi}_{nb} \approx \begin{pmatrix} 0 & \frac{\sin \hat{\phi}_{nb}}{\cos \hat{\theta}_{nb}} & \frac{\cos \hat{\phi}_{nb}}{\cos \hat{\theta}_{nb}} \end{pmatrix} \hat{\mathbf{C}}_\gamma^b \delta \hat{\psi}_{\gamma b}^\gamma, \quad \gamma \in i,e,n,l. \quad (16.54)$$

System noise of a few centimeters standard deviation should be added to the position error states to account for variations in the step length not accounted for by the models.

When PDR using step detection is integrated as an aiding measurement or in a total-state architecture, the measurement may be resolved in either the body frame

16.2 Dead Reckoning, Attitude, and Height Measurement

or the frame used for resolving the position solution. A motion constraint of zero sideways motion may also be applied. The measurement innovation, resolved about north and east axes and applying the sideways motion constraint, is

$$\delta \mathbf{z}_{P,k}^- = \begin{pmatrix} \Delta \hat{r}_P \cos(\hat{\psi}_{nb} + \hat{\psi}_{bb}) - \Delta r_N' \\ \Delta \hat{r}_P \sin(\hat{\psi}_{nb} + \hat{\psi}_{bb}) - \Delta r_E' \end{pmatrix}_k, \quad (16.55)$$

where $\Delta \hat{r}_p$ is calculated using the estimated step length coefficients, $\hat{c}_p$, and

$$\begin{pmatrix} \Delta r_N' \\ \Delta r_E' \end{pmatrix} = \int_{t-\tau_P}^{t} \begin{pmatrix} 1 & 0 & 0 \\ 0 & 1 & 0 \end{pmatrix} \hat{C}_\gamma^n(t') \hat{v}_{\beta b}^\gamma(t') \, dt', \quad (16.56)$$

$$\{\beta, \gamma\} \in \{i,i\}, \{e,e\}, \{e,n\}, \{l,l\} \quad (16.57)$$

Defining the state vector as

$$\mathbf{x}^\gamma = \begin{pmatrix} \delta \boldsymbol{\psi}_{\gamma b}^\gamma \\ \delta \mathbf{v}_{\beta b}^\gamma \\ \vdots \\ \delta \mathbf{c}_P \\ \psi_{bb} \end{pmatrix}, \quad (16.58)$$

the measurement matrix is

$$\mathbf{H}_{P,k}^\gamma = \begin{pmatrix} \mathbf{H}_{P1}^\gamma & \mathbf{H}_{P2}^\gamma & 0 & \mathbf{H}_{P4}^\gamma & \mathbf{H}_{P5}^\gamma \end{pmatrix}_{\hat{x}^\gamma = \hat{x}_k^{\gamma-}}, \quad (16.59)$$

where

$$\mathbf{H}_{P1}^\gamma \approx \begin{pmatrix} -\sin(\hat{\psi}_{nb} + \hat{\psi}_{bb}) \\ \cos(\hat{\psi}_{nb} + \hat{\psi}_{bb}) \end{pmatrix} \Delta \hat{r}_P \begin{pmatrix} 0 & \dfrac{\sin \hat{\phi}_{nb}}{\cos \hat{\theta}_{nb}} & \dfrac{\cos \hat{\phi}_{nb}}{\cos \hat{\theta}_{nb}} \end{pmatrix} \hat{C}_\gamma^b - \int_{t-\tau_P}^{t} \begin{pmatrix} 1 & 0 & 0 \\ 0 & 1 & 0 \end{pmatrix} \hat{C}_\gamma^n(t') \left[\hat{v}_{\beta b}^\gamma(t') \wedge \right] dt',$$

$$(16.60)$$

$$\mathbf{H}_{P2}^\gamma \approx - \begin{pmatrix} 1 & 0 & 0 \\ 0 & 1 & 0 \end{pmatrix} \hat{C}_\gamma^n(t) \tau_P, \quad (16.61)$$

$$\mathbf{H}_{P4}^\gamma = \begin{pmatrix} \cos(\hat{\psi}_{nb} + \hat{\psi}_{bb}) \\ \sin(\hat{\psi}_{nb} + \hat{\psi}_{bb}) \end{pmatrix} \begin{pmatrix} 1 & \dfrac{1}{\tau_P} & \sigma_f^2 & \hat{\theta}_{nb} \end{pmatrix}. \quad (16.62)$$

$$\mathbf{H}_{P5}^{\gamma} = \begin{pmatrix} -\sin(\hat{\psi}_{nb} + \hat{\psi}_{bb}) \\ \cos(\hat{\psi}_{nb} + \hat{\psi}_{bb}) \end{pmatrix} \Delta \hat{r}_P. \qquad (16.63)$$

See Section I.8.1 of Appendix I on the CD for the derivation of $\mathbf{H}_{P1}^{\gamma}$.

The measurement noise covariance, $\mathbf{R}_P$, should account for variations in the step length and boresight unaccounted for by the models, but not for errors in the model coefficients. The unmodeled step length variation is typically a few centimeters.

The step length model parameters are tailored for individual users. Therefore, their estimates should be carried over between periods of operation by the same user, but reset when a different user takes over. It is useful to retain model coefficient estimates for multiple users.

Measurement innovation filtering (Section 17.3.1) should be used to reject suspected erroneous measurements. These may occur due to false step detection or incorrect motion classification.

In total-state integration, an alternative to inputting PDR measurements is to use step detection to adjust the amount of system noise on the position and/or velocity states. One way of implementing this is to increase the position uncertainty whenever a step is detected. This is useful in cases where the direction of walking cannot easily be determined.

16.2.5 Doppler Radar and Sonar

Integration of Doppler radar and sonar measurements (Section 6.5) requires estimation of a single scale factor error state, s_D, by the Kalman filter. For radar, this accounts for the effect of scattering over water, while for sonar, it accounts for errors in the assumed speed of sound. The scale factor error is observable wherever sufficiently accurate velocity or acceleration measurements are available from another sensor.

For radar over water, the surface velocity is also estimated when measurements from a positioning system, such as GNSS, are available. To observe both the surface velocity and the scale factor error, the host aircraft must undergo a velocity-changing maneuver.

The scale factor error will change suddenly when the host aircraft switches between overflying land and overflying water. This should be detected and the state estimate reset. Alternatively, the integration algorithm can maintain separate land and water scale factor error estimates, switching between them as appropriate, or only activate the scale factor error state when overflying water.

When there is a significant unknown misalignment between the sensor and host-vehicle body frames, this may be estimated as a set of three additional Kalman filter states representing the rotation from one body frame to the other. This is analogous to the INS attitude error states. Individual beam misalignments may also be estimated as states, where necessary.

When Doppler is integrated as part of the reference navigation system, the effect of the scale factor error on the navigation solution is accounted for in the system model. In this example, surface velocity, $\mathbf{v}_{es}^{\gamma}$, is also estimated, but not misalignments. The state dynamics are

16.2 Dead Reckoning, Attitude, and Height Measurement

$$\delta \dot{\mathbf{r}}_{ib}^i = s_D (\mathbf{v}_{ib}^i - \mathbf{v}_{ie}^i) + \mathbf{v}_{es}^i, \qquad (16.64)$$

$$\delta \dot{\mathbf{r}}_{eb}^e = s_D \mathbf{v}_{eb}^e + \mathbf{v}_{es}^e, \qquad (16.65)$$

or

$$\begin{pmatrix} \delta \dot{L}_b \\ \delta \dot{\lambda}_b \\ \delta \dot{h}_b \end{pmatrix} = \mathbf{T}_{r(n)}^p (s_D \mathbf{v}_{eb}^n + \mathbf{v}_{es}^n), \qquad (16.66)$$

where $\mathbf{T}_{r(n)}^p$ is given by (14.104).

When Doppler is incorporated as an aiding sensor or total-state integration is used, the Doppler velocity may be resolved about the body frame or each beam processed as a separate measurement. For a body-frame Doppler velocity measurement, $\tilde{\mathbf{v}}_{ebD}^b$, when the beam alignment is known, the measurement innovation is

$$\delta \mathbf{z}_{D,k}^- = \left[\tilde{\mathbf{v}}_{ebD}^b (1 - \hat{s}_D) - \hat{\mathbf{C}}_\gamma^b (\hat{\mathbf{v}}_{eb}^\gamma - \hat{\mathbf{v}}_{es}^\gamma) \right]_k^- \qquad \gamma \in e, n. \qquad (16.67)$$

Defining the state vector as

$$\mathbf{x}^\gamma = \begin{pmatrix} \delta \boldsymbol{\psi}_{\gamma b}^\gamma \\ \delta \mathbf{v}_{eb}^\gamma \\ \vdots \\ s_D \\ \mathbf{v}_{es}^\gamma \end{pmatrix} \qquad \gamma \in e, n, \qquad (16.68)$$

the measurement matrix is

$$\mathbf{H}_{D,k}^\gamma = \left(-\hat{\mathbf{C}}_\gamma^b \left[(\hat{\mathbf{v}}_{eb}^\gamma - \hat{\mathbf{v}}_{es}^\gamma) \wedge \right] \quad -\hat{\mathbf{C}}_\gamma^b \quad 0 \quad \hat{\mathbf{C}}_\gamma^b \hat{\mathbf{v}}_{eb}^\gamma \quad \hat{\mathbf{C}}_\gamma^b \right)_{\hat{\mathbf{x}}^\gamma = \hat{\mathbf{x}}_k^{\gamma-}}. \qquad (16.69)$$

Note that this is an EKF measurement model as $\mathbf{H}_{D,k}^\gamma$ is a function of $\hat{\mathbf{v}}_{es}^\gamma$. Sections I.8.1 and I.8.2 of Appendix I on the CD describe, respectively, the derivation of the attitude error columns of $\mathbf{H}_{D,k}^\gamma$ and the estimation of the misalignment between the Doppler sensor and reference body frames.

The Doppler measurement noise covariance, $\mathbf{R}_D$, should be modeled as a function of the velocity and the height above terrain as discussed in Section 6.5. When Doppler is integrated as the reference, this noise should be modeled as system noise on the position error states.

Occasional erroneous measurements can occur due to animals or vehicles interrupting the beam. Measurement innovation filtering (Section 17.3.1) should be used to reject these.

16.2.6 Visual Odometry and Terrain-Referenced Dead Reckoning

Visual odometry (Section 13.3.5) and terrain-referenced dead reckoning (Sections 13.2.4 and 13.2.5) produce measurements of Earth-referenced velocity, resolved about the vehicle body frame, $\mathbf{v}_{eb}^b$. Measurements may be 2-D or 3-D depending on the sensor configuration. Two-dimensional measurements may be integrated in the same way as wheel-based odometry measurements (see Section 16.2.3), while 3-D measurements may be treated like Doppler radar and sonar measurements (Section 16.2.5).

Velocity scale factor errors can arise due to errors in determining the distance from the sensor to the reference features. For land-vehicle and pedestrian applications, the sensor's height above ground may be estimated as a state [15]. Errors can also arise due to sensor misalignment, which may be estimated as described in Section I.8.2 of Appendix I on the CD.

16.3 Position-Fixing Measurements

This section describes the integration of position-fixing measurements, including range, bearing, elevation, and line-fix measurements. These may be obtained from terrestrial radio navigation systems (Chapters 11 and 12); from acoustic, ultrasound, and infrared position-fixing systems (Sections 12.4 and 12.5); and from environmental feature-matching systems, such as map matching, TRN, terrain database height aiding, and image-based navigation (Chapter 13). Position may be determined from a wide range of different types of environmental features and signals, including SOOP.

Position-fixing technologies may be incorporated as constituent navigation systems in total-state filtered integration and single-epoch integration, or as aiding systems in error-state filtered integration, but not as the reference in error-state filtered integration.

The integration architectures for terrestrial position-fixing systems are essentially the same as for GNSS (see Section 14.1). Loosely coupled integration uses position measurements, which may be obtained by any positioning method, including proximity, ranging, angular positioning, and pattern matching (see Section 7.1). In tightly coupled integration, ranging measurements (e.g., pseudo-range, two-way-range, or differential range), angular measurements (e.g., bearing and elevation), line-fix measurements, or any combination are used.

The integration of position, ranging, angular, and line-fix measurements are described in turn. This is followed by a discussion of methods for handling ambiguous measurements arising from positioning systems that use pattern-matching or feature identification. Feature tracking and mapping, including SLAM, is then described. Finally, the section concludes by discussing the aiding of position-fixing systems using the integrated navigation solution. In addition, Appendix I on the CD presents alternative implementations with a Cartesian local-navigation-frame position error, and shows how the time synchronization error may be estimated as a Kalman filter state.

To avoid repeating equations, a is used throughout this section to denote the body frame of both a user radio antenna and a feature-matching sensor.

16.3.1 Position Measurement Integration

Position measurement integration is an example of loosely coupled integration, forming part of a cascaded integration architecture. It has the advantage of being able to operate with any radio-navigation user equipment or environmental feature-matching system. However, measurements are only available when there is sufficient information to generate a position solution using a given positioning technology. A position-fixing subsystem may use only one type of signal or environmental feature, or it may use multiple types of signal or feature.

Single-epoch integration of position measurements is as described in Section 16.1.1. The remainder of this subsection focuses on the filtered integration of position measurements.

Most radio-navigation position solutions exhibit a slowly varying bias. Similarly, environmental feature-matching fixes may be subject to a slowly-varying database bias. When the integrated navigation system inputs measurements from a more accurate positioning system, such as GNSS, this bias will be observable and may be calibrated by estimating it as Kalman filter states. For example, this can significantly improve the accuracy of ELoran [16]. Otherwise, it may be necessary to model the bias as time-correlated measurement noise (see Section 3.4.3) and/or increase the error covariance of the integrated navigation solution to account for it.

For radio navigation, the bias will exhibit a step change each time the combination of signals used to calculate the navigation solution changes. For environmental feature matching, bias step changes can occur at database boundaries. The Kalman filter should respond by increasing the uncertainty of the bias states, where estimated, and the position states otherwise.

A curvilinear position measurement innovation is

$$\begin{aligned}\delta z_{R,k}^{n-} &= S_p \left(\tilde{p}_{bR} - \hat{p}_b - \hat{b}_R^p \right)_k^- \\ &= S_p \left(\tilde{p}_{aR} - \hat{p}_b - \hat{T}_{r(n)}^p \hat{C}_b^n l_{ba}^b - \hat{b}_R^p \right)_k^- ,\end{aligned} \qquad (16.70)$$

where k denotes the iteration, $\tilde{p}_{bR} = \left(\tilde{L}_{bR}, \tilde{\lambda}_{bR}, \tilde{h}_{bR} \right)$ is the terrestrial-radio-navigation or environmental feature-matching subsystem's curvilinear position solution for the body frame, b, $\tilde{p}_{aR} = \left(\tilde{L}_{aR}, \tilde{\lambda}_{aR}, \tilde{h}_{aR} \right)$ is the subsystem curvilinear position solution of the user antenna or feature-matching sensor, $\hat{p}_b = \left(\hat{L}_b, \hat{\lambda}_b, \hat{h}_b \right)$ is the integrated position solution at the same time of validity, l_{ba}^b is the body-to-antenna or body-to-sensor lever arm, $\hat{T}_{r(n)}^p$ is given by (14.104), $\hat{b}_R^p$ is the estimated curvilinear position solution bias, and S_p is the curvilinear position scaling matrix given by (14.118).

Similarly, the ECI-frame, ECEF-frame, and local-tangent-plane-frame measurement innovations are given by

$$\begin{aligned}\delta z_{R,k}^{\gamma-} &= \left(\tilde{r}_{\gamma bR}^{\gamma} - \hat{r}_{\gamma b}^{\gamma} - \hat{b}_R^{\gamma} \right)_k^- \\ &= \left(\tilde{r}_{\gamma aR}^{\gamma} - \hat{r}_{\gamma b}^{\gamma} - \hat{C}_b^{\gamma} l_{ba}^b - \hat{b}_R^{\gamma} \right)_k^- ,\end{aligned} \qquad \gamma \in i,e,l, \qquad (16.71)$$

where $\tilde{\mathbf{r}}_{\gamma bR}^{\gamma}$ is the subsystem Cartesian position solution for b, $\tilde{\mathbf{r}}_{\gamma aR}^{\gamma}$ is the Cartesian position solution of the user antenna or feature-matching sensor, $\hat{\mathbf{r}}_{\gamma b}^{\gamma}$ is the integrated position solution, and $\hat{\mathbf{b}}_{R}^{\gamma}$ is the estimated Cartesian position solution bias, resolved about the γ-frame axes.

Note that some subsystems that require the integrated navigation solution to be fed back to them (see Section 16.3.7) output position corrections, $\tilde{\mathbf{p}}_{bR} - \hat{\mathbf{p}}_b$ or $\tilde{\mathbf{r}}_{\gamma bR}^{\gamma} - \hat{\mathbf{r}}_{\gamma b}^{\gamma}$, instead of fixes.

Many terrestrial radio navigation and environmental feature-matching systems omit the height component, so the curvilinear or local-tangent-plane-frame position may be 2-D. It may also be necessary to convert the position fix to a common datum as described in Section C.1 of Appendix C on the CD, particularly for short-range systems.

Defining the state vector as

$$\mathbf{x}^n = \begin{pmatrix} \mathbf{p}_b \\ \vdots \\ \mathbf{b}_R^p \end{pmatrix}, \quad \mathbf{x}^{\gamma} = \begin{pmatrix} \mathbf{r}_{\gamma b}^{\gamma} \\ \vdots \\ \mathbf{b}_R^{\gamma} \end{pmatrix}, \quad \gamma \in i,e,l \qquad (16.72)$$

for total-state integration or

$$\mathbf{x}^n = \begin{pmatrix} \delta\mathbf{p}_b \\ \vdots \\ \mathbf{b}_R^p \end{pmatrix}, \quad \mathbf{x}^{\gamma} = \begin{pmatrix} \delta\mathbf{r}_{\gamma b}^{\gamma} \\ \vdots \\ \mathbf{b}_R^{\gamma} \end{pmatrix}, \quad \gamma \in i,e,l \qquad (16.73)$$

for error-state integration, the measurement matrix is

$$\begin{aligned} \mathbf{H}_{R,k}^{n} &= \mathbf{S}_p \begin{pmatrix} k_R \mathbf{I} & 0 & \mathbf{I} \end{pmatrix}, \\ \mathbf{H}_{R,k}^{\gamma} &= \begin{pmatrix} k_R \mathbf{I} & 0 & \mathbf{I} \end{pmatrix}, \quad \gamma \in i,e,l \end{aligned} \qquad (16.74)$$

where k_R is 1 for total-state integration and -1 for error-state integration.

For angular positioning systems, including many image-based techniques, and systems performing pattern-matching at a distance (e.g., TRN), the navigation system's attitude error can significantly impact the position measurement so should be accounted for in the integration algorithm. Extending the error-state state vector to

$$\mathbf{x}^n = \begin{pmatrix} \delta\mathbf{p}_b \\ \vdots \\ \mathbf{b}_R^p \\ \delta\boldsymbol{\psi}_{nb}^n \end{pmatrix}, \quad \mathbf{x}^{\gamma} = \begin{pmatrix} \delta\mathbf{r}_{\gamma b}^{\gamma} \\ \vdots \\ \mathbf{b}_R^{\gamma} \\ \delta\boldsymbol{\psi}_{\gamma b}^{\gamma} \end{pmatrix}, \quad \gamma \in i,e,l, \qquad (16.75)$$

16.3 Position-Fixing Measurements

the measurement matrix becomes

$$\mathbf{H}_{R,k}^{n} = \mathbf{S}_{p}\left(\begin{array}{ccc} -\mathbf{I} & 0 & \mathbf{I} & -\left[\left(\mathbf{C}_{b}^{n}\left\{\mathbf{u}_{bf}^{b}r_{af} + \mathbf{l}_{ba}^{b}\right\}\right)\wedge\right] \end{array}\right)_{\mathbf{x}^{n}=\hat{\mathbf{x}}_{k}^{n-}},$$

$$\mathbf{H}_{R,k}^{\gamma} = \left(\begin{array}{ccc} -\mathbf{I} & 0 & \mathbf{I} & -\left[\left(\mathbf{C}_{b}^{\gamma}\left\{\mathbf{u}_{bf}^{b}r_{af} + \mathbf{l}_{ba}^{b}\right\}\right)\wedge\right] \end{array}\right)_{\mathbf{x}^{\gamma}=\hat{\mathbf{x}}_{k}^{\gamma-}}, \quad \gamma \in i,e,l \qquad (16.76)$$

where $\mathbf{u}_{bf}^{b}$ is the line-of-sight vector from the body frame to feature f (see Section 13.3.1) and r_{af} is the range from the sensor to the same feature. For TRN and other downward-pointing systems, $\mathbf{u}_{bf}^{b} \approx (0\ 0\ 1)^{T}$.

The measurement noise covariance, $\mathbf{R}_{R}$, should be based on the error covariance of the subsystem position solution, when available. Otherwise, for radio ranging and angular positioning, it may be modeled as a function of signal-to-noise level, signal geometry, and host vehicle dynamics. In either case, it should be scaled to account for the time correlation of the noise. This is more important when the subsystem computes a filtered navigation solution than where a single-epoch position solution is used. The discussion of GNSS measurement noise covariance in Section 14.3.1 is largely applicable to other radio navigation systems.

For proximity positioning, the position-fix error distribution will be non-Gaussian and will vary with local topography, base-station antenna pattern, and signal-to-noise level. For position measurements input to a Kalman filter-based estimation algorithm, the measurement noise covariance, $\mathbf{R}_{R}$, should be a Gaussian distribution that overbounds the true error distribution. Alternatively, a non-Gaussian estimation algorithm, such as a particle filter (Section 3.5) can be used. For a given base station, the variation in the proximity position error with time will match the host vehicle velocity. Thus, in principle, trajectory information from another navigation sensor could be used to constrain a proximity fix to a smaller region.

For environmental feature matching, the measurement noise covariance, $\mathbf{R}_{R}$, will depend on sensor resolution, database resolution, feature distinctiveness, and feature-sensor range.

Section I.9.1 of Appendix I on the CD describes the addition of radio-navigation velocity measurements to loosely coupled integration.

16.3.2 Ranging Measurement Integration

Ranging measurement integration is an example of tightly coupled integration. The integration is in the range domain and the architecture is always centralized. Tightly coupled integration has the advantage of being able to incorporate radio-navigation, acousic ranging, and environmental feature-matching measurements, including terrain database height aiding, when there is insufficient information to calculate a single-technology position solution. This is particularly useful where a position solution can be obtained by combining range measurements from different position-fixing technologies. A further benefit is that the biases are easier to model in the range domain than the position domain [16, 17]. Most of this section applies to both filtered and single-epoch centralized integration.

For pseudo-range measurements from passive ranging (Section 7.1.4.1), the measurement innovation is

$$\delta \mathbf{z}_{\bar{R},k} = \begin{pmatrix} \tilde{\rho}_{a,C}^1 - \hat{\rho}_{a,C}^{1-} \\ \tilde{\rho}_{a,C}^2 - \hat{\rho}_{a,C}^{2-} \\ \vdots \\ \tilde{\rho}_{a,C}^m - \hat{\rho}_{a,C}^{m-} \end{pmatrix}_k, \qquad (16.77)$$

where m is the number of transmitters used, $\tilde{\rho}_{a,C}^j$ is the measured pseudo-range for transmitter (or tracking channel) j, corrected for any predictable errors (e.g., ELoran ASFs), and $\hat{\rho}_{a,C}^{j-}$ is the prior estimate thereof, calculated using

$$\hat{\rho}_{a,C,k}^{j-} = \left[\sqrt{\left(\hat{\mathbf{r}}_{\gamma j}^\gamma - \hat{\mathbf{r}}_{\gamma b}^\gamma - \hat{\mathbf{C}}_b^\gamma \mathbf{l}_{ba}^b\right)^T \left(\hat{\mathbf{r}}_{\gamma j}^\gamma - \hat{\mathbf{r}}_{\gamma b}^\gamma - \hat{\mathbf{C}}_b^\gamma \mathbf{l}_{ba}^b\right)} + \delta \hat{\rho}_c^a + \hat{b}_r^j \right]_k, \qquad \gamma \in i,e,l \qquad (16.78)$$

where $\delta \rho_c^a$ is the clock offset, $\mathbf{r}_{\gamma j}^\gamma$ is the position of transmitter j, b_r^j is its range bias, where estimated, and the other terms are as defined in Section 16.3.1. It is assumed that the range is short enough for the Sagnac effect to be neglected. In tightly coupled systems, the receiver clock can be shared between different radio navigation systems.

For range measurements from two-way ranging (Section 7.1.4.5), RSS-derived ranging (Section 7.1.4.6), or image feature matching (Section 13.3.3), and virtual range measurements from terrain database height aiding (Section 13.2.7), the measurement innovation is

$$\delta \mathbf{z}_{\bar{R},k} = \begin{pmatrix} \tilde{r}_{a1,C} - \hat{r}_{a1,C}^- \\ \tilde{r}_{a2,C} - \hat{r}_{a2,C}^- \\ \vdots \\ \tilde{r}_{am,C} - \hat{r}_{am,C}^- \end{pmatrix}_k, \qquad (16.79)$$

where $\tilde{r}_{aj,C}$ is the corrected measured range for transmitter, environmental feature, or virtual ranging origin j and $\hat{r}_{aj,C}^-$ is the prior estimate thereof, given by

$$\hat{r}_{aj,C,k}^- = \left[\sqrt{\left(\hat{\mathbf{r}}_{\gamma j}^\gamma - \hat{\mathbf{r}}_{\gamma b}^\gamma - \hat{\mathbf{C}}_b^\gamma \mathbf{l}_{ba}^b\right)^T \left(\hat{\mathbf{r}}_{\gamma j}^\gamma - \hat{\mathbf{r}}_{\gamma b}^\gamma - \hat{\mathbf{C}}_b^\gamma \mathbf{l}_{ba}^b\right)} + \hat{b}_r^j \right]_k, \qquad \gamma \in i,e,l \qquad (16.80)$$

where $\mathbf{r}_{\gamma j}^\gamma$ is the position of transmitter, environmental feature, or virtual ranging origin j.

The user, transmitter, and feature positions may be converted from curvilinear to Cartesian using (2.112) where necessary. It may also be necessary to convert the transmitter and/or environmental feature positions to a common datum as described in Section C.1 of Appendix C on the CD. For height aiding, the origin of the virtual

16.3 Position-Fixing Measurements

range measurements may be the center of the Earth, in which case, $\hat{\mathbf{r}}_{ij}^i = \hat{\mathbf{r}}_{ej}^e = 0$ and $\hat{\mathbf{r}}_{lj}^l = -\hat{\mathbf{r}}_{el}^l$.

For pseudo-range measurements, the state vector may be defined as

$$\mathbf{x}^\gamma = \begin{pmatrix} \mathbf{r}_{\gamma b}^\gamma \\ \vdots \\ \mathbf{b}_r \\ \delta\rho_c^a \\ \delta\dot{\rho}_c^a \end{pmatrix} \quad \gamma \in i,e,l, \qquad \mathbf{x}^n = \begin{pmatrix} \mathbf{p}_b \\ \vdots \\ \mathbf{b}_r \\ \delta\rho_c^a \\ \delta\dot{\rho}_c^a \end{pmatrix} \qquad (16.81)$$

for total-state integration or

$$\mathbf{x}^\gamma = \begin{pmatrix} \delta\mathbf{r}_{\gamma b}^\gamma \\ \vdots \\ \mathbf{b}_r \\ \delta\rho_c^a \\ \delta\dot{\rho}_c^a \end{pmatrix} \quad \gamma \in i,e,l \qquad \mathbf{x}^n = \begin{pmatrix} \delta\mathbf{p}_b \\ \vdots \\ \mathbf{b}_r \\ \delta\rho_c^a \\ \delta\dot{\rho}_c^a \end{pmatrix} \qquad (16.82)$$

for error-state integration, where $\delta\dot{\rho}_c^a$ is the receiver clock drift. These state vectors may be used for two-way radio ranging, RSS-derived range, environmental feature range, and virtual range measurements with the clock states omitted.

To use signals with unknown timing offsets that may not be synchronized to UTC, particularly signals of opportunity, it is necessary to estimate the timing offset before the transmitter may be used for positioning. These timing offsets may be estimated using the range bias states. It is not usually necessary to estimate the clock drifts as these are usually small because frequency stability is required. Instead, the clock drift should be modeled with an appropriate level of system noise; adaptive estimation (Section 3.4.4) may be used where this is unknown. When groups of transmitters are synchronized with each other, a single bias state may be used to estimate their common timing offset. For terrain database height aiding, effective range biases may occur where the host vehicle height above terrain is uncertain or the database is biased.

Range bias states should only be estimated where they are observable. As with loosely coupled integration, this will be the case where measurements from a more accurate positioning system are available. The range biases will also be partially observable when measurements from more than four landmarks (three for 2-D positioning) are used. In cooperative (or collaborative) positioning, range bias estimates and their associated uncertainties may be shared between peers. Thus, if a range bias is not observable, it may be possible to obtain an estimate from another user who can observe it.

In the case of some environmental features and SOOP, the feature or transmitter position may be unknown. Section 16.3.6 describes how this may be estimated.

The measurement matrix for pseudo-range measurements is approximately

$$\mathbf{H}_{R,k}^{\gamma} \approx \begin{pmatrix} -k_R \mathbf{u}_{a1}^{\gamma\,T} & 0 & \mathbf{h}_b^{1\,T} & 1 & 0 \\ -k_R \mathbf{u}_{a2}^{\gamma\,T} & 0 & \mathbf{h}_b^{2\,T} & 1 & 0 \\ \vdots & \vdots & \vdots & \vdots & \vdots \\ -k_R \mathbf{u}_{am}^{\gamma\,T} & 0 & \mathbf{h}_b^{m\,T} & 1 & 0 \end{pmatrix}_{\mathbf{x}^{\gamma} = \hat{\mathbf{x}}_k^{\gamma-}} \quad \gamma \in i,e,l \qquad (16.83)$$

or

$$\mathbf{H}_{R,k}^{n} \approx \begin{pmatrix} -k_R \mathbf{h}_{pp}^{1\,T} & 0 & \mathbf{h}_b^{1\,T} & 1 & 0 \\ -k_R \mathbf{h}_{pp}^{2\,T} & 0 & \mathbf{h}_b^{2\,T} & 1 & 0 \\ \vdots & \vdots & \vdots & \vdots & \vdots \\ -k_R \mathbf{h}_{pp}^{m\,T} & 0 & \mathbf{h}_b^{m\,T} & 1 & 0 \end{pmatrix}_{\mathbf{x}^{\gamma} = \hat{\mathbf{x}}_k^{\gamma-}}, \qquad (16.84)$$

where $\mathbf{u}_{aj}^{\gamma}$ is the line-of-sight vector from the user antenna to transmitter j,

$$\mathbf{h}_{pp}^{j} = \begin{pmatrix} (R_N(L_b) + h_b) u_{aj,N}^{n} \\ (R_E(L_b) + h_b) \cos \hat{L}_b u_{aj,E}^{n} \\ -u_{aj,D}^{n} \end{pmatrix}, \qquad (16.85)$$

$$\mathbf{h}_b^{j\,T} = \begin{pmatrix} \delta_{1j} & \delta_{2j} & \cdots & \delta_{nj} \end{pmatrix}, \qquad (16.86)$$

where δ is the Kronecker delta function, and k_R is 1 for total-state integration and −1 for error-state integration (as previously). These measurement matrices may be used for two-way ranging, RSS-derived ranges, ranges to matched features, and virtual ranges from height aiding if the clock-state columns are omitted or zeroed.

For filtered integration of radio and acoustic positioning, the measurement noise covariance, $\mathbf{R}_R$, accounts for receiver tracking errors, multipath variations, transmitter clock noise, residual time-synchronization errors, and antenna lever arm flexure. For environmental feature matching, $\mathbf{R}_R$, will depend on sensor resolution and database resolution. It may also be affected by the feature's distinctiveness and its distance from the sensor. For terrain database height aiding, $\mathbf{R}_R$ should account for uncertainties in the database itself, the input latitude and longitude, and the interpolation process. The matrix $\mathbf{R}_R$ is normally diagonal and should be scaled by the ratio of the error correlation time to the measurement update interval.

For single-epoch integration, the biases are not normally estimated so the measurement error covariance, $\mathbf{C}_R$, accounts for the range biases as well as the measurement noise.

Section I.9.2 of Appendix I on the CD describes the addition of pseudo-range rate or range rate measurements to tightly coupled integration.

16.3 Position-Fixing Measurements

For measurements of the TDOA across transmitters (Section 7.1.4.2), the measurement innovation with a common reference is

$$\delta \mathbf{z}_{R,k}^{\Delta-} = \begin{pmatrix} \Delta \tilde{\rho}_{a,C}^{r1} - \hat{r}_{a1,C}^- + \hat{r}_{ar,C}^- \\ \Delta \tilde{\rho}_{a,C}^{r2} - \hat{r}_{a2,C}^- + \hat{r}_{ar,C}^- \\ \vdots \\ \Delta \tilde{\rho}_{a,C}^{rm} - \hat{r}_{am,C}^- + \hat{r}_{ar,C}^- \end{pmatrix}_k, \qquad (16.87)$$

where $\Delta \tilde{\rho}_{a,C}^{rj}$ is the corrected measured pseudo-range for transmitter j minus that for the reference transmitter, r. The state vectors defined in (16.81) and (16.82) may be used with the clock states omitted. Note that the range biases are relative to that of the reference signal. The measurement matrix is approximately

$$\mathbf{H}_{R,k}^\gamma \approx \begin{pmatrix} k_R \left(\mathbf{u}_{ar}^\gamma - \mathbf{u}_{a1}^\gamma \right)^T & 0 & \mathbf{h}_b^{1\,T} \\ k_R \left(\mathbf{u}_{ar}^\gamma - \mathbf{u}_{a2}^\gamma \right)^T & 0 & \mathbf{h}_b^{2\,T} \\ \vdots & \vdots & \vdots \\ k_R \left(\mathbf{u}_{ar}^\gamma - \mathbf{u}_{am}^\gamma \right)^T & 0 & \mathbf{h}_b^{m\,T} \end{pmatrix}_{\mathbf{x}^\gamma = \hat{\mathbf{x}}_k^{\gamma-}} \quad \gamma \in i,e,l \qquad (16.88)$$

or

$$\mathbf{H}_{R,k}^n \approx \begin{pmatrix} k_R \left(\mathbf{h}_{pp}^r - \mathbf{h}_{pp}^1 \right)^T & 0 & \mathbf{h}_b^{1\,T} \\ k_R \left(\mathbf{h}_{pp}^r - \mathbf{h}_{pp}^2 \right)^T & 0 & \mathbf{h}_b^{2\,T} \\ \vdots & \vdots & \vdots \\ k_R \left(\mathbf{h}_{pp}^r - \mathbf{h}_{pp}^m \right)^T & 0 & \mathbf{h}_b^{m\,T} \end{pmatrix}_{\mathbf{x}^\gamma = \hat{\mathbf{x}}_k^{\gamma-}}. \qquad (16.89)$$

The measurement noise covariance matrix is not diagonal; it is given by

$$\mathbf{R}_{R,k}^\Delta = \mathbf{D}_{R,k} \mathbf{R}_{R,k} \mathbf{D}_{R,k}^T, \qquad (16.90)$$

where $\mathbf{R}_{R,k}$ is the measurement noise covariance of the undifferenced pseudo-ranges, an $(m + 1) \times (m + 1)$ matrix with the last row and column corresponding to the reference transmitter, and $\mathbf{D}_{R,k}$ is the $m \times (m + 1)$ differencing matrix, defined as

$$\mathbf{D}_{R,k} = \begin{pmatrix} 1 & 0 & \cdots & 0 & -1 \\ 0 & 1 & \cdots & 0 & -1 \\ \vdots & \vdots & \ddots & \vdots & \vdots \\ 0 & 0 & \cdots & 1 & -1 \end{pmatrix}. \qquad (16.91)$$

Note that a set of TDOA measurements does not have to have a common reference. TDOA measurements may also be differenced across receivers (Section 7.1.4.3) or doubly differenced across transmitters and receivers (Section 7.1.4.4).

The ranging measurement integration must be adapted to the characteristics of the individual positioning technology. For short-range systems, such as pseudolites, UWB, acoustic ranging, and image matching in urban and indoor environments, a UKF measurement update (Section 3.4.2) should be considered [1], in which case measurement innovations must be calculated for each sigma point. For two-way acoustic ranging (Section 12.4), the motion of the host vehicle during signal propagation should be accounted for by averaging the user position in (16.80).

For systems that provide only two-dimensional positioning, such as DME, ELoran, mobile phones, most signals of opportunity, radar and some laser scanners, only the two horizontal components of the user position states are incorporated within the measurement model. The height may be estimated using other sensors or omitted from the state vector altogether. In determining the prior estimated pseudo-range, (16.78), or range, (16.80), the relative height of the transmitter and receiver is either neglected or estimated using whatever information is available, depending on the signal geometry and user requirements. An estimated relative height should always be used in air applications. For a system, such as ELoran, that uses ground-wave propagation, the square root term in (16.78) and (16.80) must be replaced by the geodesic, given by (11.11). DME ranging measurements should not be used during the transmission of the 3-second identification message as they are predicted from earlier measurements at this time.

The measurements of height above terrain in a TRN system (Section 13.2) are a type of ranging measurement. For centralized integration of a sequential TRN system, the difference between the measured and database-indicated terrain height should be input directly to the integration Kalman filter using the measurement model described in Section 13.2.1.

16.3.3 Angular Measurement Integration

Angular measurement integration is an example of tightly coupled integration, which is always centralized. The integration is in the angle domain. For angular positioning using azimuth and elevation measurements (Section 7.1.5), the measurement innovations in either a filtered or single-epoch architecture are

$$\delta \mathbf{z}_{\bar{\psi},k} = \begin{pmatrix} \tilde{\psi}_{lu}^{a1} - \hat{\psi}_{lu}^{a1-} \\ \tilde{\psi}_{lu}^{a2} - \hat{\psi}_{lu}^{a2-} \\ \vdots \\ \tilde{\psi}_{lu}^{am} - \hat{\psi}_{lu}^{am-} \end{pmatrix}_k, \quad \delta \mathbf{z}_{\bar{\theta},k} = \begin{pmatrix} \tilde{\theta}_{lu}^{a1} - \hat{\theta}_{lu}^{a1-} \\ \tilde{\theta}_{lu}^{a2} - \hat{\theta}_{lu}^{a2-} \\ \vdots \\ \tilde{\theta}_{lu}^{aM} - \hat{\theta}_{lu}^{aM-} \end{pmatrix}_k, \quad (16.92)$$

where $\tilde{\psi}_{lu}^{aj}$ and $\tilde{\theta}_{lu}^{aj}$ are, respectively, the measured azimuth and elevation for transmitter or feature j and $\hat{\psi}_{lu}^{aj-}$ and $\hat{\theta}_{lu}^{aj-}$ are the prior estimates thereof, given by

$$\hat{\psi}_{lu}^{aj-} = \arctan_2\left[\left(\hat{y}_{lj}^l - \hat{y}_{la}^l\right), \left(\hat{x}_{lj}^l - \hat{x}_{la}^l\right)\right]_k + \hat{b}_{\psi,k}^{j-}, \quad (16.93)$$

$$\hat{\theta}_{lu}^{aj-} = -\arctan\left(\frac{\hat{z}_{lj}^l - \hat{z}_{la}^l}{\sqrt{\left(\hat{x}_{lj}^l - \hat{x}_{la}^l\right)^2 + \left(\hat{y}_{lj}^l - \hat{y}_{la}^l\right)^2}}\right)_k^- + \hat{b}_{\theta,k}^{j-}, \qquad (16.94)$$

where b_ψ^j and b_θ^j are, respectively the azimuth and elevation biases for transmitter or feature j, where estimated, and a local-tangent-plane coordinate frame is assumed. Spatially-varying biases can occur due to refraction and diffraction of radio signals and radar; they are not significant for optical and infrared feature matching. Azimuth measurement innovations must be limited to the range $-\pi$ to π by adding or subtracting 2π as appropriate.

The user, transmitter, and feature positions may be converted from curvilinear to Cartesian using (2.112) where necessary. It may also be necessary to convert the transmitter and/or environmental feature positions to a common datum as described in Section C.1 of Appendix C on the CD. Section 16.3.6 describes how transmitter and feature positions may be estimated in cases where they are unknown.

Defining the state vector as

$$\mathbf{x}^l = \begin{pmatrix} \mathbf{r}_{lb}^l \\ \mathbf{b}_\psi \\ \mathbf{b}_\theta \\ \vdots \end{pmatrix} \qquad (16.95)$$

for total-state integration or

$$\mathbf{x}^l = \begin{pmatrix} \delta\mathbf{r}_{lb}^l \\ \mathbf{b}_\psi \\ \mathbf{b}_\theta \\ \vdots \end{pmatrix} \qquad (16.96)$$

for error-state integration, the measurement matrices are approximately

$$\mathbf{H}_{\psi,k}^l \approx \begin{pmatrix} k_\psi \mathbf{h}_{\psi r}^{1\,\mathrm{T}} & \mathbf{h}_b^{1\,\mathrm{T}} & 0 & 0 \\ k_\psi \mathbf{h}_{\psi r}^{2\,\mathrm{T}} & \mathbf{h}_b^{2\,\mathrm{T}} & 0 & 0 \\ \vdots & \vdots & \vdots & \vdots \\ k_\psi \mathbf{h}_{\psi r}^{m\,\mathrm{T}} & \mathbf{h}_b^{m\,\mathrm{T}} & 0 & 0 \end{pmatrix}_{\hat{\mathbf{x}}^l = \hat{\mathbf{x}}_k^{l-}}, \qquad (16.97)$$

$$\mathbf{H}_{\theta,k}^l \approx \begin{pmatrix} k_\theta \mathbf{h}_{\theta r}^{1\,\mathrm{T}} & 0 & \mathbf{h}_b^{1\,\mathrm{T}} & 0 \\ k_\theta \mathbf{h}_{\theta r}^{2\,\mathrm{T}} & 0 & \mathbf{h}_b^{2\,\mathrm{T}} & 0 \\ \vdots & \vdots & \vdots & \vdots \\ k_\theta \mathbf{h}_{\theta r}^{M\,\mathrm{T}} & 0 & \mathbf{h}_b^{M\,\mathrm{T}} & 0 \end{pmatrix}_{\hat{\mathbf{x}}^l = \hat{\mathbf{x}}_k^{l-}}, \qquad (16.98)$$

where k_ψ and k_θ are 1 for total-state integration and -1 for error-state integration,

$$\mathbf{h}_{\psi r}^{j\ \mathrm{T}} = \left(\frac{\hat{y}_{lj}^l - \hat{y}_{la}^l}{\left(\hat{x}_{lj}^l - \hat{x}_{la}^l\right)^2 + \left(\hat{y}_{lj}^l - \hat{y}_{la}^l\right)^2} \quad -\frac{\hat{x}_{lj}^l - \hat{x}_{la}^l}{\left(\hat{x}_{lj}^l - \hat{x}_{la}^l\right)^2 + \left(\hat{y}_{lj}^l - \hat{y}_{la}^l\right)^2} \quad 0 \right), \quad (16.99)$$

$$\mathbf{h}_{\theta r}^{j\ \mathrm{T}} = \left(-\frac{\left(\hat{x}_{lj}^l - \hat{x}_{la}^l\right)\left(\hat{z}_{lj}^l - \hat{z}_{la}^l\right)}{\left(\Delta_{xyz}^j\right)^2 \Delta_{xy}^j} \quad -\frac{\left(\hat{y}_{lj}^l - \hat{y}_{la}^l\right)\left(\hat{z}_{lj}^l - \hat{z}_{la}^l\right)}{\left(\Delta_{xyz}^j\right)^2 \Delta_{xy}^j} \quad \frac{\Delta_{xy}^j}{\left(\Delta_{xyz}^j\right)^2} \right), \quad (16.100)$$

$$\begin{aligned} \left(\Delta_{xyz}^j\right)^2 &= \left(\hat{x}_{lj}^l - \hat{x}_{la}^l\right)^2 + \left(\hat{y}_{lj}^l - \hat{y}_{la}^l\right)^2 + \left(\hat{z}_{lj}^l - \hat{z}_{la}^l\right)^2 \\ \Delta_{xy}^j &= \sqrt{\left(\hat{x}_{lj}^l - \hat{x}_{la}^l\right)^2 + \left(\hat{y}_{lj}^l - \hat{y}_{la}^l\right)^2} \end{aligned}, \quad (16.101)$$

and $\mathbf{h}_b$ is given by (16.86). Note that the lever arm from the body-frame origin to the user antenna, $\mathbf{l}_{ba}^b$, has been neglected.

When the azimuth is obtained from a system, such as VOR (Section 11.1.2), that transmits bearing-varying signals, the measurements must be adjusted for the difference between the north reference at the transmitter (which may be magnetic north) and the x-axis of the l frame. Note that if the l frame axes are aligned with north, east, and down at the user location, they will not be exactly aligned with those directions at the transmitter location.

When azimuth and elevation are obtained from radio direction finding by the user equipment or from image-based navigation, the measurements will be made with respect to the body frame not the local-tangent-plane frame and must be transformed using the attitude solution:

$$\begin{aligned} \tilde{\theta}_{lu}^{aj} &= -\arcsin\left(\tilde{u}_{lj,z}^l\right) \\ \tilde{\psi}_{lu}^{aj} &= \arctan_2\left(\tilde{u}_{lj,y}^l, \tilde{u}_{lj,x}^l\right) \end{aligned}, \quad \tilde{\mathbf{u}}_{lj}^l = \hat{\mathbf{C}}_b^l \begin{pmatrix} \cos\tilde{\theta}_{bu}^{aj}\cos\tilde{\psi}_{bu}^{aj} \\ \cos\tilde{\theta}_{bu}^{aj}\sin\tilde{\psi}_{bu}^{aj} \\ -\sin\tilde{\theta}_{bu}^{aj} \end{pmatrix}, \quad (16.102)$$

where $\tilde{\psi}_{bu}^{aj}$ and $\tilde{\theta}_{bu}^{aj}$ are, respectively, the measured bearing and elevation with respect to the body frame for transmitter or feature j. When the xy planes of the b and l frames are parallel, this simplifies to

$$\tilde{\theta}_{lu}^{aj} = \tilde{\theta}_{bu}^{aj}, \qquad \tilde{\psi}_{lu}^{aj} = \tilde{\psi}_{bu}^{aj} + \hat{\psi}_{lb}^{aj}. \quad (16.103)$$

In this case, the attitude errors must be included in the state vector. Assuming error-state integration,

16.3 Position-Fixing Measurements

$$\mathbf{x}^l = \begin{pmatrix} \delta \mathbf{r}_{lb}^l \\ \mathbf{b}_\psi \\ \mathbf{b}_\theta \\ \delta \boldsymbol{\psi}_{lb}^l \\ \vdots \end{pmatrix}, \qquad (16.104)$$

and the measurement matrices become

$$\mathbf{H}_{\psi,k}^l \approx \begin{pmatrix} -\mathbf{h}_{\psi r}^{1\,\mathrm{T}} & \mathbf{h}_b^{1\,\mathrm{T}} & 0 & \mathbf{h}_{\psi\psi}^{1\,\mathrm{T}} & 0 \\ -\mathbf{h}_{\psi r}^{2\,\mathrm{T}} & \mathbf{h}_b^{2\,\mathrm{T}} & 0 & \mathbf{h}_{\psi\psi}^{2\,\mathrm{T}} & 0 \\ \vdots & \vdots & \vdots & \vdots & \vdots \\ -\mathbf{h}_{\psi r}^{m\,\mathrm{T}} & \mathbf{h}_b^{m\,\mathrm{T}} & 0 & \mathbf{h}_{\psi\psi}^{m\,\mathrm{T}} & 0 \end{pmatrix}_{\hat{\mathbf{x}}^l = \hat{\mathbf{x}}_k^{l-}}, \qquad (16.105)$$

$$\mathbf{H}_{\theta,k}^l \approx \begin{pmatrix} -\mathbf{h}_{\theta r}^{1\,\mathrm{T}} & 0 & \mathbf{h}_b^{1\,\mathrm{T}} & \mathbf{h}_{\theta\psi}^{1\,\mathrm{T}} & 0 \\ -\mathbf{h}_{\theta r}^{2\,\mathrm{T}} & 0 & \mathbf{h}_b^{2\,\mathrm{T}} & \mathbf{h}_{\theta\psi}^{2\,\mathrm{T}} & 0 \\ \vdots & \vdots & \vdots & \vdots & \vdots \\ -\mathbf{h}_{\theta r}^{M\,\mathrm{T}} & 0 & \mathbf{h}_b^{M\,\mathrm{T}} & \mathbf{h}_{\theta\psi}^{M\,\mathrm{T}} & 0 \end{pmatrix}_{\hat{\mathbf{x}}^l = \hat{\mathbf{x}}_k^{l-}}, \qquad (16.106)$$

where

$$\mathbf{h}_{\psi\psi}^{j\,\mathrm{T}} = \begin{pmatrix} \tan\hat{\theta}_{lu}^{aj-}\cos\hat{\psi}_{lu}^{aj-} & \tan\hat{\theta}_{lu}^{aj-}\sin\hat{\psi}_{lu}^{aj-} & 1 \end{pmatrix}, \qquad (16.107)$$

$$\mathbf{h}_{\theta\psi}^{j\,\mathrm{T}} = \begin{pmatrix} -\sin\hat{\psi}_{lu}^{aj-} & \cos\hat{\psi}_{lu}^{aj-} & 0 \end{pmatrix}. \qquad (16.108)$$

When only the heading or azimuth error is significant, including in 2-D systems, the first two components of $\delta\boldsymbol{\psi}_{lb}^l$ are omitted and the heading-error column is ones in the azimuth measurement matrix and zeros in the elevation measurement matrix. When the heading is completely unknown, it may be eliminated from the azimuth measurement model by differencing measurements of different features or transmitters as described in Section I.9.3 of Appendix I on the CD, provided that the roll and pitch are known.

Measurements from imaging sensors may also be input as image-plane coordinates or pixels, in which case the conversion to bearing and elevation (see Section 13.3.1) is included within the integration algorithm's measurement model [18].

For filtered integration, the measurement noise covariance, $\mathbf{R}_{\psi\theta}$, is determined by the resolution of the user equipment, regardless of whether the angular information is obtained from signal demodulation, radio direction finding or image matching. For image matching, feature distinctiveness is also an issue.

For single-epoch integration, the biases are not normally estimated so the measurement error covariance, $C_{\psi\theta}$, accounts for the azimuth and elevation biases as well as the measurement noise.

16.3.4 Line Fix Integration

Environmental feature-matching systems, including map matching and some image-based techniques, can sometimes determine that a user is somewhere along a line, but not where along that line they are. This provides a one-dimensional measurement in the horizontal plane that may be specified in terms of the bearing of the line, $\tilde{\psi}_{nL}$, and the position, $\mathbf{r}_{lf}^{l}$ or $\tilde{\mathbf{p}}_{f}$, of an arbitrary point, f, on that line. The perpendicular distance from the estimated user position, $\mathbf{r}_{lb}^{l}$ or $\hat{\mathbf{p}}_{b}$, to the line forms the measurement innovation, δz_L, while the bearing of the line is used in the measurement matrix. Figure 16.16 illustrates this. The lever arm, database bias, and sensor alignment errors are neglected in this discussion.

The measurement innovation may be calculated using

$$\delta z_{L,k} = \left[\begin{pmatrix} -\sin\tilde{\psi}_{nL} & \cos\tilde{\psi}_{nL} & 0 \end{pmatrix} \left(\tilde{\mathbf{r}}_{lf}^{l} - \hat{\mathbf{r}}_{lb}^{l} \right) \right]_k^-, \qquad (16.109)$$

where the local tangent-plane frame axes are aligned with north, east, and down or

$$\delta z_{L,k} = \left[\begin{pmatrix} -\sin\tilde{\psi}_{nL} & \cos\tilde{\psi}_{nL} & 0 \end{pmatrix} \hat{\mathbf{T}}_{p}^{r(n)} \left(\tilde{\mathbf{p}}_{f} - \hat{\mathbf{p}}_{b} \right) \right]_k^-, \qquad (16.110)$$

where $\mathbf{T}_{p}^{r(n)}$ is given by (2.120).

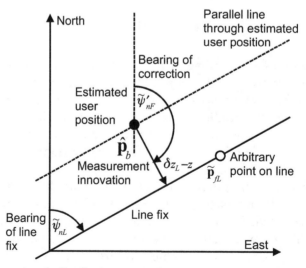

Figure 16.16 Geometry of a line-fix measurement.

16.3 Position-Fixing Measurements

The measurement matrix is then

$$\mathbf{H}^l_{L,k} = k_L \begin{bmatrix} -\sin\tilde{\psi}_{nL} & \cos\tilde{\psi}_{nL} & 0 \end{bmatrix}_k, \qquad (16.111)$$

where the state vector is simply $\mathbf{r}^l_{lb}$ or $\delta\mathbf{r}^l_{lb}$, and

$$\mathbf{H}^n_{L,k} = k_L \begin{bmatrix} \begin{pmatrix} -\sin\tilde{\psi}_{nL} & \cos\tilde{\psi}_{nL} & 0 \end{pmatrix} \hat{\mathbf{T}}^{r(n)}_p \end{bmatrix}_k, \qquad (16.112)$$

where the state vector is $\mathbf{p}_b$ or $\delta\mathbf{p}_b$, noting that k_L is 1 for total-state integration and -1 for error-state integration.

As many environmental feature-matching systems require the integrated position solution (see Section 16.3.7), the line fix may instead be output as a position correction, $\delta\tilde{z}_L$, and the bearing of that correction, $\tilde{\psi}'_{nL}$, also shown in Figure 16.16. In this case, the measurement innovation is simply the correction and the measurement matrix is

$$\mathbf{H}^l_{L,k} = -k_L \begin{bmatrix} \cos\tilde{\psi}'_{nL} & \sin\tilde{\psi}'_{nL} & 0 \end{bmatrix}_k \qquad (16.113)$$

or

$$\mathbf{H}^n_{L,k} = -k_L \begin{bmatrix} \begin{pmatrix} \cos\tilde{\psi}'_{nL} & \sin\tilde{\psi}'_{nL} & 0 \end{pmatrix} \hat{\mathbf{T}}^{r(n)}_p \end{bmatrix}_k. \qquad (16.114)$$

As with other environmental feature-matching measurements, the measurement noise covariance in filtered integration, R_L, will depend on sensor and database resolution, feature distinctiveness, and feature-sensor range.

16.3.5 Handling Ambiguous Measurements

In systems that use the pattern-matching positioning method or identify environmental features from images, there is not always a clear match between the measurements and the database. Measurement and database resolution limitations can make it difficult to distinguish nearby features, while environments often contain repeating features. For example, an image-matching system may see multiple rail tracks but only have a single line on its database. Due to these ambiguities, positioning systems based on environmental feature matching or radio-signal pattern matching all produce occasional wrong measurements. To prevent these from corrupting the integrated navigation solution in a filtered architecture, measurement-innovation-based fault detection and integrity-monitoring techniques, described in Section 17.3, should always be used.

A more robust approach is for the subsystem to output a multiple-hypothesis measurement, comprising a position, covariance, and probability for each of the most likely matches between the measurement and database. There should also be a null hypothesis that none of the matches is correct. A Kalman filter-based integration algorithm can use innovation filtering (Section 17.3.1) to reject any hypotheses totally inconsistent with the prior navigation solution and then incorporate the remaining hypotheses using the best-fix, weighted-fix, or multiple-hypothesis filtering method, as described in Section 3.4.5. Alternatively, a particle filter (Section 3.5) can input all measurement hypotheses as a combined probability distribution. It can also input the complete probability distribution of a pattern-matching measurement.

With TRN operating over low-roughness terrain with a conventional radalt sensor, using a weighted-fix integration filter makes the navigation solution more robust against false TRN fixes than a simple best-fix approach, but additional improvements have not been obtained with multiple-hypothesis filtering [19]. However, when false matches can be correlated over successive measurements, such as in CVN (see Section H.1 of Appendix H on the CD), multiple-hypothesis filtering is critical for optimum performance [20].

When multiple pattern-matching subsystems are integrated, each producing a position fix from an array of matching scores for different candidate positions, ambiguity may be reduced by combining the matching scores from the different subsystems and then using them to produce a joint position fix. Matching scores from the same system at different epochs may also be combined if a velocity solution is available (e.g., from dead reckoning) to propagate previous matching scores forward to account for motion between epochs. Figure 16.17 illustrates this for multiple subsystems and epochs.

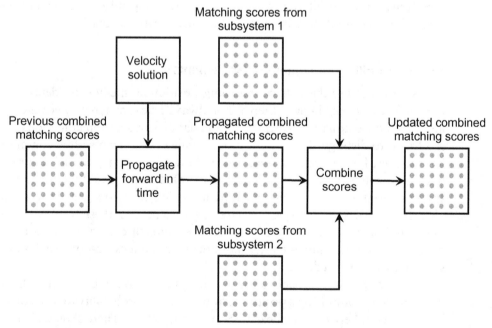

Figure 16.17 Combining matching scores from different subsystems and epochs.

16.3.6 Feature Tracking and Mapping

In a feature-tracking system, initially unknown environmental features or signals of opportunity are measured over successive epochs. The first measurements must be used to determine approximate feature positions. After that, they may be used to help correct the integrated navigation solution. By retaining feature position estimates and identification information after moving out of the feature's range, a feature map may be built for later use. This is a form of simultaneous localization and mapping. In cooperative (or collaborative) positioning, this information may be shared between users [21, 22].

For a system estimating position, velocity, and attitude errors, together with feature positions, all with respect to and resolved along the axes of a local-tangent-plane coordinate frame, a suitable state vector is

$$\mathbf{x}^l = \begin{pmatrix} \delta \mathbf{r}_{lb}^l \\ \delta \mathbf{v}_{lb}^l \\ \delta \boldsymbol{\psi}_{lb}^l \\ \mathbf{r}_{l1}^l \\ \mathbf{r}_{l2}^l \\ \vdots \\ \mathbf{r}_{lm}^l \end{pmatrix}, \tag{16.115}$$

where $\mathbf{r}_{l1}^l, \mathbf{r}_{l2}^l, \ldots, \mathbf{r}_{lm}^l$ are the feature positions. Additional feature position states must be added each time a new feature is identified. To limit the state vector to a practical size, states for features that are no longer being observed should be deleted. In mapping mode, the position estimates and their associated error covariances are then stored. A centralized filtered integration architecture must be used to enable continuous estimation of the feature positions.

Range, azimuth, and elevation measurements may be formulated as described in Sections 16.2.2 and 16.2.3 with the database-derived feature positions replaced by the relevant state estimates. The measurement matrices for the state vector defined by (16.115) are then

$$\mathbf{H}_{R,k}^l \approx \begin{pmatrix} \mathbf{u}_{a1}^{\gamma\,\mathrm{T}} & 0 & 0 & \mathbf{u}_{a1}^{\gamma\,\mathrm{T}} & 0 & \cdots & 0 \\ \mathbf{u}_{a2}^{\gamma\,\mathrm{T}} & 0 & 0 & 0 & \mathbf{u}_{a2}^{\gamma\,\mathrm{T}} & \cdots & 0 \\ \vdots & \vdots & \vdots & \vdots & \vdots & \ddots & \vdots \\ \mathbf{u}_{am}^{\gamma\,\mathrm{T}} & 0 & 0 & 0 & 0 & \cdots & \mathbf{u}_{am}^{\gamma\,\mathrm{T}} \end{pmatrix}_{\mathbf{x}^l = \hat{\mathbf{x}}_k^{l-}} \tag{16.116}$$

$$\mathbf{H}_{\psi,k}^l \approx \begin{pmatrix} -\mathbf{h}_{\psi r}^{1\,\mathrm{T}} & 0 & \mathbf{h}_{\psi\psi}^{1\,\mathrm{T}} & -\mathbf{h}_{\psi r}^{1\,\mathrm{T}} & 0 & \cdots & 0 \\ -\mathbf{h}_{\psi r}^{2\,\mathrm{T}} & 0 & \mathbf{h}_{\psi\psi}^{2\,\mathrm{T}} & 0 & -\mathbf{h}_{\psi r}^{2\,\mathrm{T}} & \cdots & 0 \\ \vdots & \vdots & \vdots & \vdots & \vdots & \ddots & \vdots \\ -\mathbf{h}_{\psi r}^{m\,\mathrm{T}} & 0 & \mathbf{h}_{\psi\psi}^{m\,\mathrm{T}} & 0 & 0 & \cdots & -\mathbf{h}_{\psi r}^{m\,\mathrm{T}} \end{pmatrix}_{\hat{\mathbf{x}}^l = \hat{\mathbf{x}}_k^{l-}}, \tag{16.117}$$

$$\mathbf{H}^l_{\theta,k} \approx \begin{pmatrix} -\mathbf{h}^{1\,T}_{\theta r} & 0 & \mathbf{h}^{1\,T}_{\theta \psi} & -\mathbf{h}^{1\,T}_{\theta r} & 0 & \cdots & 0 \\ -\mathbf{h}^{2\,T}_{\theta r} & 0 & \mathbf{h}^{2\,T}_{\theta \psi} & 0 & -\mathbf{h}^{2\,T}_{\theta r} & \cdots & 0 \\ \vdots & \vdots & \vdots & \vdots & \vdots & \ddots & \vdots \\ -\mathbf{h}^{m\,T}_{\theta r} & 0 & \mathbf{h}^{m\,T}_{\theta \psi} & 0 & 0 & \cdots & -\mathbf{h}^{m\,T}_{\theta r} \end{pmatrix}_{\hat{\mathbf{x}}^l = \hat{\mathbf{x}}^{l-}_k}. \quad (16.118)$$

If both ranging and angular positioning are implemented, use of a UKF will minimize linearization errors for close features. However, if only ranging measurements are used, the initial distributions of the feature position estimates will be ring-shaped. This cannot easily be represented in Kalman filter-based estimation, so non-Gaussian estimation techniques, such as the particle filter (Section 3.5) are often used.

If only angular measurements are used, the initial distributions of the feature position estimate will be wedge-shaped. This can be represented in Kalman filter-based estimation by using six states, comprising the user position when the feature was first observed, the bearing and elevation of the feature from that position, and the inverse of the distance [23]. Using the inverse of the distance enables modeling of very large distance uncertainties.

In a full SLAM implementation, estimates of previous user positions and their error covariance with the feature positions are also maintained. Consequently, when the user reobserves a previously observed feature or signal, known as loop closure, data from the successive visits may be combined optimally. This can have a significant impact when some of the real-time user position estimates are poor.

SLAM can also be used to build a pattern-matching database. It has been used for mapping radio signal availability [24] and pedestrian routes within a building [25].

16.3.7 Aiding of Position-Fixing Systems

Subsystems incorporating environmental feature identification or pattern matching require knowledge of their approximate position to determine which region of the database should be searched. Position aiding is thus essential for initialization unless the starting position is known. Subsequently, using the integrated position solution (as opposed to the subsystem position solution) will enable the search region to be minimized, especially where there is insufficient information to compute an accurate subsystem position solution.

Position aiding is also useful for subsystems using radio or other types of signals as it can be used to predict which signals will be receivable, enabling them to be acquired more quickly by omitting the need for a spectrum search. Furthermore, for ranging systems, knowledge of approximate position and time speeds up the signal acquisition process by reducing the search space.

Velocity aiding from the integrated navigation solution can be used to aid acquisition and tracking of ranging signals in difficult signal-to-noise environments, effectively increasing receiver sensitivity. Deeply coupled integration (Sections 14.1.5 and 14.3.3), whereby the integration algorithm tracks the ranging signals is also possible, but has yet to be implemented for signals other than GNSS.

Velocity aiding can also be used to compensate for the effects of vehicle motion in two-way ranging systems. This is particular useful for acoustic positioning (Sections 12.4 and 12.5.2) because of the relatively low speed of sound.

For pattern-matching systems, such as TRN, that have to make measurements at multiple locations to determine position, velocity aiding is essential to enable them to combine a transect of measurements into a location signature that can be matched with a database with minimal ambiguity. This approach can be extended to any pattern-matching system to increase the distinctiveness of its location signatures.

Finally, image-based positioning systems and radio direction finding can benefit from attitude aiding to optimize sensor alignment, enabling environmental features and signals to be found more quickly.

Problems and exercises for this chapter are on the accompanying CD.

References

[1] Xing, Z., and D. Gebre-Egziabher, "Comparing Non-Linear Filters for Aided Inertial Navigators," *Proc. ION ITM*, Anaheim, CA, January 2009, pp. 1048–1053.

[2] Gu, D., Y., Qin, and N. El-Sheimy, "A New Federated Kalman Filter for Integrated Navigation System Based on SINS," *Proc. ION GNSS 2007*, Fort Worth, TX, September 2007, pp. 551–561.

[3] Carlson, N. A., and M. P. Berarducci, "Federated Kalman Filter Simulation Results," *Navigation: JION*, Vol. 41, No. 3, 1994, pp. 297–321.

[4] Carlson, N. A., "Federated Filter for Distributed Navigation and Tracking Applications," *Proc. ION 58th AM*, Albuquerque, NM, June 2002, pp. 340–353.

[5] Carlson, N. A., "Federated Square Root Filter for Decentralized Parallel Processing," *IEEE Trans. on Aerospace and Electronic Systems*, Vol. 26, No. 3, 1990, pp. 517–525.

[6] Hamilton, A. S., and B. W. Chilton, "A Flexible Federated Navigation System for Next Generation Military Aircraft," *Proc. ION 52nd AM*, Cambridge, MA, June 1996, pp. 409–415.

[7] Levy, L. J., "Suboptimality of Cascaded and Federated Kalman Filters," *Proc. ION 52nd AM*, Cambridge, MA, June 1996, pp. 399–407.

[8] Shafiee, M., K. O'Keefe, and G. Lachapelle, "Context-Aware Adaptive Extended Kalman Filtering Using Wi-Fi Signals for GPS Navigation," *Proc. ION GNSS 2011*, Portland, OR, September 2011, pp. 1305–1318.

[9] Zampella, F., "Unscented Kalman Filter and Magnetic Angular Rate Update (MARU) for an Improved Pedestrian Dead-Reckoning," *Proc. IEEE/ION PLANS*, Myrtle Beach, SC, April 2012, pp. 129–139.

[10] Afzal, M. H., V. Renaudin, and G. Lachapelle, "Use of Earth's Magnetic Field for Mitigating Gyroscope Errors Regardless of Magnetic Perturbation," *Sensors*, Vol. 11, 2011, pp. 11390–11414.

[11] Gebre-Egziabher, D., et al., "A Gyro-Free Quaternion-Based Attitude Determination System Suitable for Implementation Using Low Cost Sensors," *Proc. IEEE PLANS*, San Diego, CA, March 2000, pp. 185–192.

[12] Wendel, J., et al., "MAV Attitude Estimation Using Low-Cost MEMS Inertial Sensors and GPS," *Proc. ION 61st AM*, Cambridge, MA, June 2005, pp. 397–403.

[13] Carlson, C. R., J. C. Gerdes, and J. D. Powell, "Error Sources When Land Vehicle Dead Reckoning with Differential Wheelspeeds," *Navigation: JION*, Vol. 51, No. 1, 2004, pp. 13–27.

[14] Zhao, L., et al., "An Extended Kalman Filter Algorithm for Integrating GPS and Low Cost Dead Reckoning System Data for Vehicle Performance and Emissions Monitoring," *Journal of Navigation*, Vol. 56, No. 2, 2003, pp. 257–275.

[15] Hide, C., T. Botterill, and M. Andreotti, "Vision-Aided IMU for Handheld Pedestrian Navigation," *Proc. ION GNSS 2010*, Portland, OR, September 2010, pp. 534–541.

[16] Enge, P. K., and J. R. McCullough, "Aiding GPS with Calibrated Loran-C," *Navigation: JION Navigation*, Vol. 35, No. 4, 1988, pp. 469–482.

[17] Hide, C., et al., "Integrated GPS, LORAN-C and INS for Land Navigation Applications," *Proc. ION GNSS 2006*, Fort Worth, TX, September 2006, pp. 59–67.

[18] Chu, C. -C., et al., "Performance Comparison of Tight and Loose INS-Camera Integration," *Proc. ION GNSS 2011*, Portland, OR, September 2011, pp. 3516–3526.

[19] Groves, P. D., R. J. Handley, and A. R. Runnalls, "Optimising the Integration of Terrain Referenced Navigation with INS and GPS," *Journal of Navigation*, Vol. 59, No. 1, 2006, pp. 71–89.

[20] Handley, R. J., L. Dack, and P. McNeil, "Flight Trials of the Continuous Visual Navigation System," *Proc. ION NTM*, Long Beach, CA, January 2001, pp. 185–192.

[21] Berefelt, F., et al., "Collaborative GPS/INS Navigation in Urban Environment," *Proc. ION ITM*, San Diego, CA, pp. 1114–1125.

[22] Kassas, Z. M., and T. E. Humphreys, "Observability and Estimability of Collaborative Opportunistic Navigation with Pseudorange Measurements," *Proc. ION GNSS 2012*, Nashville, TN, September 2012, pp. 621–630.

[23] Civera, J., A. J. Davison, and J. M. Martínez Montiel, "Inverse Depth Parametrization for Monocular SLAM," *IEEE Trans. on Robotics*, Vol. 24, No. 5, October 2008, pp. 932–945.

[24] Faragher, R. M., C. Sarno, and M. Newman, "Opportunistic Radio SLAM for Indoor Navigation Using Smartphone Sensors," *Proc. IEEE/ION PLANS*, Myrtle Beach, SC, April 2012, pp. 120–128.

[25] Kaiser, S., M. Garcia-Puyol, and P. Robertson, "Maps-Based Angular PDFs Used as Prior Maps for FootSLAM," *Proc. IEEE/ION PLANS*, Myrtle Beach, SC, April 2012, pp. 113–119.

CHAPTER 17
Fault Detection, Integrity Monitoring, and Testing

Like any technology, a navigation system can occasionally produce output errors much larger than the uncertainty bounds specified for or indicated by it. This may be due to hardware or software failures or to unusual operating conditions. The different failure modes that can occur in navigation systems are discussed in Section 17.1.

To get the best performance out of a navigation system, techniques for detecting and correcting faults should be implemented. These may operate at a number of levels. *Fault detection* simply indicates that a fault is present, warning the user. *Fault detection and recovery* (FDR) identifies where the fault lies and attempts to recover from navigation solution contamination occurring prior to the detection of that fault. *Fault detection and isolation* (FDI) provides a navigation solution that is isolated from (i.e., uncontaminated by) the faulty data that has been detected. *Fault detection and exclusion* (FDE) additionally verifies that the recovered navigation solution is free from faults. Note that some authors define these terms differently.

Integrity monitoring systems perform two separate but linked functions:

- Detection and mitigation of faults;
- Determination of whether the navigation solution is safe to use, also known as solution protection.

As these functions are separate, the navigation solution may be safe to use even when a fault has been detected, provided the effects of that fault have been mitigated. Conversely, if there is insufficient information available to reliably ensure that no fault is present, the navigation solution may be unsafe to use when no fault has been detected. However, as the functions are linked, the navigation solution is more likely to be unsafe when a fault has been detected than otherwise.

Integrity monitoring is essential for safety-critical applications, which require certification that the navigation solution meets a number of performance criteria, such as the availability of a fault-free navigation solution and the probability of failing to identify a fault. However, some form of fault detection and mitigation is desirable for all applications.

Fault detection and integrity monitoring techniques may be classified either as user-based, sometimes known as sensor level, or as infrastructure-based, also known as system level. The sensor-level approaches are described first: Section 17.2 discusses fault detection through range checks; Section 17.3 describes Kalman filter innovation monitoring; and Section 17.4 describes integrity monitoring through direct consistency checks between quantities calculated from different combinations of measurements. Section 17.5 discusses the infrastructure-based or system-level approaches. The different integrity monitoring techniques are not mutually exclusive and are often combined.

Integrity monitoring is an example of hypothesis testing, introduced in Section B.5 of Appendix B on the CD, in which data is used to answer a Boolean question such as whether or not a fault is present or whether the navigation solution is safe to use. Because the data is subject to the noise and bias-like errors that occur during normal system operation, fault detection tests can produce false outcomes. A *missed detection* occurs when a fault is present but the test indicates the system is fault-free, while a *false alarm* occurs when a fault is falsely detected. This is explored further in Section 17.6, which discusses the solution protection function of integrity monitoring and the specification of a required navigation performance (RNP).

Finally, Section 17.7 discusses the different ways of testing a navigation system to verify that it meets the required performance. These include field trials, recorded data testing, laboratory testing, and software simulation.

17.1 Failure Modes

The main navigation system faults that can occur and should be detected by an integrity monitoring system are summarized for each of the main navigation technologies. The section concludes by discussing integration algorithm failure modes and context failures.

17.1.1 Inertial Navigation

Individual inertial sensor faults are due to hardware failure and can manifest as no outputs at all, null readings, repeated readings, or simply much larger errors than specified. Once a sensor fault has been detected, no further data from that sensor should be accepted. Unless there are redundant inertial sensors, this means discarding the whole inertial navigation solution. Large errors exhibited by all the inertial sensors can be an indication of a much-higher vibration environment than the system is designed for or of a mounting failure. The whole IMU or INS may also exhibit a power failure, software failure, or communications failure, in which case a reset should be attempted. Achieving FDR, FDI, or FDE with inertial navigation requires redundant hardware.

17.1.2 Dead Reckoning, Attitude, and Height Measurement

Like any other equipment, dead-reckoning systems are subject to hardware or software failure, while the individual technologies have their own short-term and long-term failure modes. Magnetic heading measurements can exhibit errors due to incorrect calibration of equipment magnetism, environmental magnetic anomalies, and problems determining the sensor orientation under acceleration. Barometric height can exhibit discontinuities in the presence of sonic booms and weather fronts. Odometers are subject to wheel slip, while PDR algorithms may be fooled by unexpected movements. Doppler radar and sonar velocity measurements exhibit errors when a moving animal or vehicle interrupts one of the beams. Air data and ships' log measurements are vulnerable to turbulence and other changes in wind velocity and water currents.

17.1.3 GNSS

GNSS failure modes may be divided into four categories: satellite faults; unusual atmospheric propagation; local channel failures, which comprise faults affecting individual channels of a single set of user equipment; and general user equipment faults. A summary may be found in [1, 2]. Faults can also be present in differential corrections and assistance data.

Satellite faults include low transmission power, clock faults, transmission of faulty navigation data, and irregular code waveforms. Irregular code waveforms, known as evil waveforms, are caused by signal-generation hardware faults and distort the signal's autocorrelation function. This results in pseudo-range measurement errors that depend on the receiver design, so do not necessarily cancel in DGNSS systems. Satellite faults are best detected through ground-based signal monitoring networks, as described in Section 17.5.

Large ionospheric or tropospheric propagation delays, or ionospheric scintillation, can occur due to solar and meteorological storms. These may or may not be detected by monitoring networks, depending on the density of the network and the locality of the storm. Storms and weather fronts can also result in larger than usual residual errors in differential and relative GNSS. Thus, unusual atmospheric propagation can manifest as local channel failures.

Other causes of local channel failures are NLOS reception, multipath interference, tracking loops in the process of losing lock, and receiver hardware and software faults affecting individual channels. Satellite faults and wide-area atmospheric problems may be detected as local channel failures where no monitoring network is used or to provide a backup to a network. FDI- and FDE-level protection against single-channel failures can usually be obtained, provided that sufficient satellites are tracked. The faulty signal is simply discarded and a navigation solution computed using the remaining signals.

General user equipment faults affect all satellite signals. Hardware failure may occur in the antenna, receiver, or reference oscillator, while the processing and communication functions may be subject to hardware or software failure. These faults can sometimes produce erroneous outputs on all channels as opposed to no output at all, so they require detection. Software faults can be recovered by resetting the user equipment. Recovery from hardware faults requires redundant hardware.

17.1.4 Terrestrial Radio Navigation

Terrestrial radio navigation systems essentially exhibit the same kinds of faults as GNSS. All systems are liable to transmitter faults, whereby incorrect signals are broadcast. The user equipment may also contain incorrect or out-of-date information about transmitter positions or predicted RSS.

Signal propagation can be affected by unusual atmospheric conditions and floods. All radio signals can be subject to multipath and/or NLOS reception, while interference or signal attenuation can affect all or some signals at a given location. These problems can cause many receivers to produce a stream of false measurements prior to detecting signal unavailability. User equipment can also exhibit hardware or software failures.

17.1.5 Environmental Feature Matching and Tracking

Feature-matching systems are inherently unreliable. Environmental features may be incorrectly identified, while for pattern-matching approaches, there will always be a possibility of false matches between the measurements and the database, especially if the database is faulty or out of date. The probability of false fixes may be traded off against the availability of fixes, but false fixes cannot be eliminated altogether. Feature-matching fixes should be validated against earlier fixes and other positioning systems, where available. Otherwise, they should be treated as provisional until they can be verified by later fixes, noting that inertial navigation or dead reckoning can be used to aid comparison of successive position fixes. Multiple-hypothesis Kalman filtering techniques are discussed in Section 3.4.5. Feature-tracking systems are similarly vulnerable to false matches between successive images. Arguably, environmental feature matching and tracking should not be used at all for applications with very high integrity requirements.

17.1.6 Integration Algorithm

In addition to processor hardware and software failure, there are three ways in which faults can manifest in a Kalman filter-based integration algorithm or navigation processor: numerical problems, poor tuning, and model failure. Numerical problems are discussed in Section 3.3.3; they should be designed out but may still occur if the Kalman filter is run for a longer period than it is designed for.

As discussed in Section 3.3.1, Kalman filter tuning is a tradeoff between convergence rate and stability. If the values assumed for the initial uncertainty, system noise, and measurement noise are overoptimistic, the errors in the state estimates are more likely to be significantly larger than the state uncertainties.

Model failures occur when the system and measurement models do not properly account for a navigation system's behavior. For example, higher-order inertial sensor errors, receiver clock g-dependent errors, or vibration-induced errors may be erroneously neglected. Alternatively, the user dynamics may be underestimated, time-synchronization errors or time-correlated noise may be unaccounted for, or the variation in GNSS measurement noise with c/n_0 may be neglected. These can all lead to state-estimation errors significantly exceeding the state uncertainties.

When they can be detected, state-estimation errors due to poor tuning or modeling may be remedied by increasing the state uncertainties (see Section 17.3.3). However, for applications with high integrity requirements, effort must be expended to model all the error sources and correctly tune the Kalman filter. Integrity monitoring should not be used as a substitute for poor design.

17.1.7 Context

Many navigation techniques are designed to operate within a certain context (see Section 1.4.2), making assumptions about the host vehicle or user's behavior and/or environment. Examples include map matching, PDR using step detection, and application of motion constraints. If the navigation system then operates outside the context for which it is designed, the assumptions made may be wrong, resulting in navigation solution errors. A context-adaptive navigation system (Section 16.1.10) can detect its operating context and select the algorithms it uses accordingly. However context-detection errors can arise, leading to errors in the navigation solution.

17.2 Range Checks

This section discusses the application of checks to validate that the sensor outputs, navigation solution(s), and Kalman filter estimates lie within reasonable ranges. Parameters deviating outside their normal operational ranges can be indicative of a fault, although the converse does not apply. Range checks will not detect all failure modes, but sensor output tests will respond immediately to gross faults, protecting the navigation solution, while navigation solution checks provide an extra layer of protection, and Kalman filter estimate checks can detect slow-building faults.

17.2.1 Sensor Outputs

A number of checks may be performed on the navigation sensor outputs, such as accelerometer and gyro measurements, GNSS pseudo-ranges and pseudo-range rates, or Is and Qs, magnetic field measurements, and radar ranges. Absolute values and step changes in measurements may be compared against the operational ranges of the sensors as specified by the manufacturer and against the expected operating range of the sensor environment. For example, all vehicles have a maximum acceleration and angular rate. However, higher values may be measured due to vibration, so, in many cases, it is more effective to apply range checks to smoothed measurements. The maximum GNSS pseudo-range rate is a function of the satellite velocity and the ranges of the user velocity and receiver clock drift. Magnetic field measurements may be compared with the Earth's magnetic field strength, though it should be noted that discrepancies are more likely to be due to environmental anomalies than sensor failure, meriting a temporary rather than permanent rejection of measurements.

Failure to produce any measurements or a stream of null measurements is an indication of sensor failure. However, faulty sensors can also produce a succession of repeated measurements. A single repeated measurement should not be treated as a fault as it can occur by chance or due to a communication glitch. The likelihood of chance repeated measurements depends on the ratio of the sensor noise to the quantization level, so the number of repetitions signifying a fault should be set accordingly.

17.2.2 Navigation Solution

Every navigation application has an operational envelope. A land vehicle or ship should always be close to the Earth's surface, and every aircraft has a maximum altitude above which it cannot fly. Similarly, every vehicle has a maximum speed with respect to the Earth. For example, civil airliners currently in service do not exceed the speed of sound, and road vehicles rarely exceed 50 m s^{-1}. Therefore, if a navigation system is indicating a position or velocity outside the appropriate operational envelope, there is probably a fault. This also applies to solutions for the GNSS receiver clock drift that significantly exceed the reference oscillator specification.

17.2.3 Kalman Filter Estimates

The Kalman filter is a powerful tool for detecting faults. In most integrated navigation systems including an IMU, the accelerometer and gyro biases are estimated as states. Therefore, if a bias estimate is several times the standard deviation specified by the manufacturer (a threshold of order 5σ is suitable), then there is likely to be a fault with the sensor. Outlying state estimates can also occur when the INS calibration is poor due to a lack of measurements or observability problems, so the current state uncertainties should be accounted for in any fault-detection test.

Range checks may also be applied to GNSS range-bias state estimates. Large estimates may indicate NLOS reception when these are modeled with short correlation times. A similar approach may be adopted for terrestrial radio navigation, while feature-matching errors can sometimes be detected through large estimates of database biases. Range checks on Kalman filter estimates of magnetic heading bias and barometric altimeter and Doppler radar/sonar scale factor errors can also be used to detect faults.

State estimate checks are essentially a form of consistency check, so they rely on redundant measurement data, though not necessarily of the same type. FDR can be achieved by rejecting further measurements from the faulty sensor or signal. However, FDI requires parallel integrated navigation solutions as described in Section 17.4.2.

17.3 Kalman Filter Measurement Innovations

The measurement innovations, δz_k^-, of a Kalman filter, defined in Section 3.2, provide an indication of whether the measurements and state estimates are consistent. Innovation filtering may be used to detect large discrepancies immediately, while innovation sequence monitoring enables smaller discrepancies to be detected over time. Both are described here, together with methods of recovering the Kalman filter estimates following fault detection.

The measurement residuals, δz_k^+, may also be used for sequence monitoring and have a smaller covariance, making them more sensitive to errors. However, in a Kalman filter, they need to be calculated specially, while the innovations and their covariance are computed as part of normal operation.

For a true Kalman filter, the measurement innovation vector is

17.3 Kalman Filter Measurement Innovations

$$\delta \mathbf{z}_k^- = \mathbf{z}_k - \mathbf{H}_k \hat{\mathbf{x}}_k^-, \tag{17.1}$$

while, for an EKF (Section 3.4.1), it is

$$\delta \mathbf{z}_k^- = \mathbf{z}_k - \mathbf{h}(\hat{\mathbf{x}}_k^-). \tag{17.2}$$

The covariance of the innovations, $\mathbf{C}_{\delta z,k}^-$, comprises the sum of the measurement noise covariance and the error covariance of the state estimates transformed into measurement space. Thus:

$$\mathbf{C}_{\delta z,k}^- = \mathbf{H}_k \mathbf{P}_k^- \mathbf{H}_k^T + \mathbf{R}_k, \tag{17.3}$$

which is the denominator of the Kalman gain.

The normalized innovations are defined as

$$y_{k,j}^- = \frac{\delta z_{k,j}^-}{\sqrt{C_{\delta z,k,j,j}^-}}. \tag{17.4}$$

In an ideal Kalman filter, these have zero-mean unit-variance Gaussian distributions (see Section B.3.2 of Appendix B on the CD), and successive values are almost independent once the estimated states have converged with their true counterparts. However, time-correlated system or measurement noise, differences between the true and modeled system noise and measurement noise covariances (e.g., due to overbounding), neglect of error sources, closed-loop feedback of state estimates to nonlinear systems, and use of an extended Kalman filter all cause departures from this [3]. Therefore, in a practical system, the statistics of the normalized innovations should always be measured before designing integrity monitoring algorithms that use them.

17.3.1 Innovation Filtering

Innovation filtering is also known as spike filtering, measurement gating, measurement editing, reasonability testing, or prefiltering, noting that the latter term is also applied to measurement averaging (see Section 3.3.2). It determines whether new measurements are consistent with previous information. The magnitude of each normalized measurement innovation, y^-, is compared with a threshold and the measurement rejected for that iteration if the threshold is exceeded. Innovation filtering is applied prior to the computation of the Kalman gain (see Section 3.2.2). When a measurement is rejected, the corresponding rows of $\mathbf{H}$ and rows and columns of $\mathbf{R}$ must be excluded from the Kalman gain calculation, (3.21), and error covariance update, (3.25), as well as the state-vector update, (3.24). With a threshold of 3, 99.73% of genuine measurements are passed by the innovations filter where the normalized innovations have a zero-mean unit-variance Gaussian distribution and the innovations covariance assumed by the Kalman filter is true. In practice, the relationship between the threshold and the false alarm rate will vary and must be assessed empirically. However, if the threshold is set too low, the state estimates will be biased towards their initialization values.

Innovation filtering is applied to the normalized innovations rather than the raw innovations because the measurement innovations vary in size under normal Kalman filter operation. They are larger when the state uncertainties are larger, following initialization or a significant gap in the measurement stream, or when the measurement noise is larger. Figure 17.1 illustrates this.

When the Kalman filter measurement update is performed sequentially (Section 3.2.7), the innovations may either be filtered together at the beginning of the measurement update process or individually prior to each step of the measurement update sequence. For the sequential approach, the test becomes more sensitive as more measurement data is incorporated. Consequently, the filtering outcome is affected by the order in which the measurements are tested. For uncorrelated measurements, the normalized innovations of the corresponding vector measurement update may be used to determine the processing order.

When a navigation sensor produces measurements in the form of position, velocity, or attitude fixes, all measurements from that sensor at a given iteration should be rejected when any component fails the innovations filter. For GNSS and other radio navigation systems supplying ranging measurements, innovation filtering should be applied independently to measurements from each satellite or transmitter. However, if a large number of measurements fail the innovations filter simultaneously, the user equipment becomes suspect, so all measurements should be rejected. When a GNSS pseudo-range measurement is rejected, the corresponding pseudo-range rate or ADR should also be rejected, but not vice versa.

Innovation filtering is a common feature of Kalman filter designs, even for applications with no formal integrity requirements. It is useful for filtering out short-term erroneous data, such as false fixes from feature-matching systems, magnetometer measurements affected by environmental magnetic anomalies, and measurements from GNSS tracking loops on the verge of loss of lock. It can also filter out spurious spikes in the measurement streams due to electrical interference, data communication errors, and timing problems. Such step changes in the measurements may be rejected without redundant measurement information where they are large enough.

A measurement repeatedly failing the innovations filter is indicative of a transient. However, this may be interpreted in a number of ways. A GNSS position solution may undergo a transient when a faulty signal is removed. Similarly, INS/GNSS velocity measurements used for transfer alignment can undergo a transient when GNSS is reacquired after an outage. A transient can also indicate that a fault has arisen. The source is sometimes ambiguous. A transient in an INS/GNSS integration filter affecting all measurements could be due to a fault in either the INS or the GNSS user equipment. Similarly, a transient in magnetometer measurements occurring just after initialization, when the confidence in previous measurements is low, or in feature-matching fixes, could indicate an error in either the new or the previous measurements. In general, the more information available, the easier it is to resolve the cause of a transient. Redundant information aids all forms of integrity monitoring.

Innovation filtering responds poorly to measurement errors that build up gradually. This is because the state estimates are contaminated by those errors before the fault becomes large enough to detect. A lack of redundant measurements keeps the

17.3 Kalman Filter Measurement Innovations

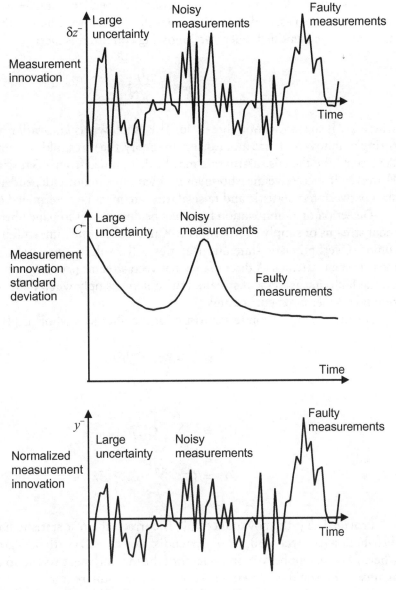

Figure 17.1 Raw and normalized measurement innovations.

measurement innovations small, so the fault is not detected at all. Otherwise, the contamination of the state estimates precludes effective fault isolation.

17.3.2 Innovation Sequence Monitoring

Smaller and slow-building discrepancies between the measurements and state estimates can be identified by forming test statistics from the last N measurements:

$$\mu_{kj} = \frac{1}{N} \sum_{i=k+1-N}^{k} y^-_{i,j}. \tag{17.5}$$

The standard deviation of the mean of N independent samples from a zero-mean unit-variance Gaussian distribution is $1/\sqrt{N}$. Therefore, a bias in the measurement innovations is identified when the following condition is met:

$$|\mu_{kj}| > \frac{T_{b\mu}}{\sqrt{N}}, \qquad (17.6)$$

where $T_{b\mu}$ is the innovation threshold. This is known as innovation sequence monitoring or innovation bias monitoring. In theory, the threshold is in units of standard deviations and the false alarm rate may be determined from a Gaussian distribution. However, if successive measurement innovations are not independent, the relationship between test statistic and false alarm rate must be determined empirically.

Detection of an innovation bias may be due to a discrepancy between measurement streams or simply overoptimistic Kalman filter state uncertainties due to poor tuning. Overoptimistic state uncertainties will result in innovation biases on many measurement streams. A discrepancy between streams produces a much larger bias on the faulty measurement stream, but sometimes only when there are at least two redundant measurement streams.

A suitable innovation test statistic for the filter as a whole is [4]

$$s^2_{\delta z,k} = \delta \mathbf{z}_\mu^{-\mathrm{T}} \mathbf{C}_\mu^{-1} \delta \mathbf{z}_\mu^-, \qquad (17.7)$$

where

$$\begin{aligned}\mathbf{C}_\mu^{-1} &= \sum_{i=k+1-N}^{k} \mathbf{C}_{\delta z,i}^{-\,-1} \\ \delta \mathbf{z}_\mu^- &= \mathbf{C}_\mu^- \sum_{i=k+1-N}^{k} \mathbf{C}_{\delta z,i}^{-\,-1} \delta \mathbf{z}_i^-\end{aligned} \qquad (17.8)$$

Provided that the covariance, $\mathbf{C}_\mu^-$, is correct, the test statistic has a chi-square distribution (see Section B.3.3 of Appendix B on the CD) with m degrees of freedom, where m is the number of components of the measurement vector, so comparing $s_{\delta z,k}$ against a threshold is sometimes known as a chi-square test.

Once a faulty measurement stream has been identified, further measurement from that stream must be rejected. Depending on the cause of the error, the measurements may recover. Therefore, computation and monitoring of the measurement innovations should continue. Remedying biased state estimates is discussed in Section 17.3.3. With a single Kalman filter, only FDR is available; FDI requires parallel filters as described in Section 17.4.2.

Selection of the sample size, N, is a tradeoff between response time and sensitivity, while selection of the threshold, $T_{b\mu}$, is a tradeoff between sensitivity and false-alarm rate. The faster the response time, the less contamination of the state estimates there will be. Multiple test statistics with different sample sizes and thresholds may be computed to enable larger biases to be detected more quickly [5]. Slowly increasing errors can sometimes be detected faster by applying an additional threshold to the rate of change of the test statistic as estimated by an additional Kalman filter [6].

17.3 Kalman Filter Measurement Innovations

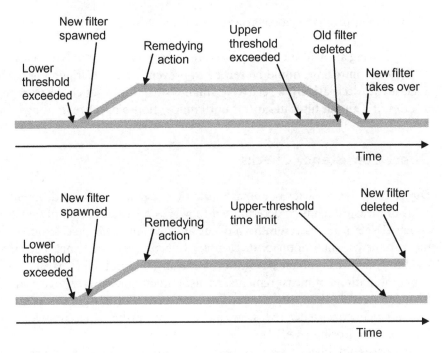

Figure 17.2 Applying parallel filters to innovation bias detection. (*From:* [8]. ©2002 QinetiQ Ltd. Reprinted with permission.)

One way of increasing the response time without increasing the false-alarm rate is to implement two thresholds with parallel filters [7]. When an innovation bias breaches the lower threshold, a parallel filter is spawned with the bias remedied and further measurements from the suspect stream rejected. If the upper threshold is then breached, the old filter is deleted, whereas if the upper threshold is not breached within a certain time, the new filter is deleted. Figure 17.2 illustrates this.

17.3.3 Remedying Biased State Estimates

When innovation biases due to overoptimistic Kalman filter state uncertainties have been detected, those uncertainties must be increased to give realistic values and optimize the Kalman gain. When the state estimates have been contaminated by erroneous measurement data or a transient that corrects a measurement stream has occurred, the Kalman filter must be made more receptive to new measurements in order to correct the estimates; this is also achieved by increasing the state uncertainties.

Increasing the state uncertainties is known as a *covariance reset* as the **P** matrix is reset or as a *Q boost*/*Q bump* because it effectively involves adding extra assumed system noise. Various reset techniques may be applied, including

- Adding appropriate noise variances to the diagonal elements of the **P** matrix;
- Multiplying the diagonal elements by a constant scaling factor;
- Multiplying the whole matrix by a scaling factor;
- Interpolating between the previous and initialization values of the matrix.

Where appropriate, a covariance reset may also be applied to a subset of the **P** matrix.

Following a covariance reset, the stored normalized residuals used for innovation sequence monitoring should be zeroed to prevent repeated triggering of the covariance reset. Alternatively, a relatively small covariance reset could be repeated over a series of Kalman filter iterations until innovation biases are no longer detected.

17.4 Direct Consistency Checks

Direct consistency checks compare quantities calculated from different combinations of measurements to determine if they are consistent. Examples include single-epoch GNSS or terrestrial radio navigation solutions, IMU accelerometer and gyro measurements, and odometer or Doppler velocity measurements. They may also be used to compare complete navigation solutions, maintained in parallel using different combinations of measurements. Measurement consistency checks are described first, followed by integrity monitoring using parallel navigation solutions. GNSS measurement consistency checks are more commonly known as receiver autonomous integrity monitoring (RAIM).

Consistency checks require redundant information (i.e., more data than is required to form a navigation solution). When at least m measurements are required to compute the quantity under comparison, $m+1$ measurements are needed for fault detection and $m+2$ measurements are needed for fault exclusion. For example, an IMU needs three accelerometers and three gyros to measure the specific force and angular rate in three dimensions. Therefore, four accelerometers and four gyros are needed for fault detection and five of each sensor for fault exclusion [9]. Similarly, four pseudo-ranges are needed to determine a GNSS position solution, so five are needed for fault detection and six for fault exclusion.

Redundancy can be achieved across different sensor types. For example, GNSS measurements can be used to determine which one of four accelerometers is faulty, enabling inertial navigation to continue using the other three. This can be easier to implement using parallel solutions. Terrain database height aiding (Section 13.2.7) can also be used for consistency checking, either by providing a virtual range measurement or by comparing the difference between the database-indicated and navigation-system-indicated heights with a threshold.

As well as the number of measurements, their geometry is also important for consistency checking. For fault detection, it must be possible to make a prediction of each measurement using the other measurements. Figure 17.3 illustrates this with a simple example of two-way ranging in two dimensions. In Figure 17.3(a), the geometry is good as each pair of measurements can be used to make a prediction of the third measurement and if all measurements have the same precision, the precision of the predictions will be the same. However, in Figure 17.3(b), the geometry is inadequate; measurements 1 and 2 can be used to form predictions of each other, but measurement 3 cannot be predicted because it is perpendicular to measurements 1 and 2, so it shares no information on the user position with them. In Figure 17.3(c), the geometry is adequate for consistency checking, but the prediction of measurement 3 will be less precise than those of measurements 1 and 2.

17.4 Direct Consistency Checks

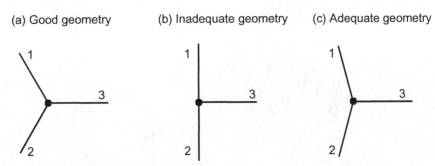

Figure 17.3 (a–c) Effect of measurement geometry on fault detection by consistency checks for two-way ranging measurements in two dimensions.

Similarly, consistency checks cannot be performed between inertial sensors with perpendicular sensitive axes, so skewed-sensor configurations tend to be used when redundancy is required [9].

For fault recovery, isolation, and exclusion, the geometry must be such that it is still possible to make a prediction of each measurement using the other measurements when one measurement is removed. This is possible in the example shown in Figure 17.4(a). However, in Figure 17.4(b), if measurement 1 is removed, measurement 3 cannot be predicted. Equivalently, if measurements 1 and 3 are found to be mutually inconsistent, it is not possible to determine which of them is faulty. The same problem occurs with measurements 2 and 4.

17.4.1 Measurement Consistency Checks and RAIM

There are four main methods for performing measurement consistency checks in general or RAIM in particular [10]. The underlying principle is the same, so they all give a similar performance.

Using the solution-separation method [11], if m measurements are available, then m different calculations of the position solution, specific force and angular rate, or velocity are made, each excluding a different measurement. Figure 17.5 illustrates this for GNSS RAIM. The differences between each calculation are then formed, and either the largest difference or the scalar average of all the differences is compared against a threshold. The test quantity exceeding the threshold denotes a fault.

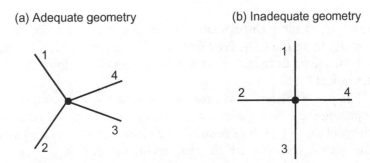

Figure 17.4 (a, b) Effect of measurement geometry on fault recovery, isolation, and exclusion by consistency checks for two-way ranging measurements in two dimensions.

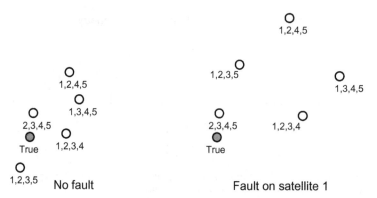

Figure 17.5 Separation of position solutions from different GNSS signal combinations.

The range comparison method [12] uses the first n measurements to calculate the position solution or other quantity under comparison. This is then used to predict the remaining $m-n$ measurements. If the measurements are consistent with the predictions, the consistency test is passed. Otherwise, a fault is declared.

The least-squares residual method first uses all the measurements, such as GNSS pseudo-ranges, to calculate a least-squares estimate of an n-component state vector, $\mathbf{x}$, such as GNSS position and clock offset, using

$$\delta \hat{\mathbf{x}}_k = \left(\mathbf{H}_k^T \mathbf{H}_k \right)^{-1} \mathbf{H}_k^T \delta \mathbf{z}_k^-, \qquad (17.9)$$

where $\delta \mathbf{z}_k^-$ is the measurement innovation vector (the actual minus predicted measurements), $\delta \hat{\mathbf{x}}_k = \hat{\mathbf{x}}_k^+ - \hat{\mathbf{x}}_k^-$ is the state innovation vector (the least-squares estimated states minus the predicted states), $\mathbf{H}_k$ is the measurement matrix, analogous to that used in a Kalman filter (see Section 3.2.4), and k denotes the epoch. Measurement residuals, $\delta \mathbf{z}_k^+$, are then computed using

$$\delta \mathbf{z}_k^+ = \delta \mathbf{z}_k^- - \mathbf{H}_k \delta \hat{\mathbf{x}}_k, \qquad (17.10)$$

These are equal to the difference between the actual measurements and those determined from the estimated states, $\hat{\mathbf{x}}_k^+$. The largest or scalar-average residual may be compared against a threshold. However, it is more common to perform a chi-square test using $\delta \mathbf{z}_k^{+T} \delta \mathbf{z}_k^+$, or the normalized test statistic

$$s_{\delta z}^2 = \delta \mathbf{z}_k^{+T} \mathbf{C}_{\delta z, k}^{+-1} \delta \mathbf{z}_k^+, \qquad (17.11)$$

where $\mathbf{C}_{\delta z,k}^+$ is the measurement residual covariance matrix (see Section D.1.3 of Appendix D on the CD). Provided that this covariance is correct, the test statistic has a chi-square distribution with $m-n$ degrees of freedom, as the residuals are not independent [13].

The parity method [14] uses the measurements to compute an exact solution comprising the n-component least-squares estimate of the comparison quantity at each epoch, $\mathbf{x}_k$ (e.g., GNSS position and clock offset), and a $m-n$-component parity vector, $\mathbf{p}_k$, the square of which, $\mathbf{p}_k^T \mathbf{p}_k$, is equal to $\delta \mathbf{z}_k^{+T} \delta \mathbf{z}_k^+$.

The thresholds for the test statistics are a tradeoff between sensitivity and false-alarm rate. For GNSS, terrestrial radio navigation, and IMU measurements, the test statistics may be averaged across successive measurement sets over time to improve both the sensitivity and false-alarm rate, making the measurement consistency checks analogous to Kalman filter innovation sequence monitoring (Section 17.3.2). When carrier-smoothed pseudo-ranges are used, this averaging is implicit. However, time averaging can increase the response time. One solution is to use multiple test statistics with different averaging times, while in some cases, the input of measurements to an integration algorithm may be delayed (see Section 3.3.4). Note that odometer and Doppler velocity measurements exhibit transitory faults due to wheel slip or moving objects in the beam, so they are not suited to averaging.

When a fault has been detected, there are a number of ways of identifying the faulty measurement stream. With the least-squares residual and parity methods, the largest measurement residual, δz^+, belongs to the faulty measurement. More generally, if there are at least two redundant measurements, the measurement consistency check may simply be repeated on m sets of $m-1$ measurements. The set excluding the faulty measurement will pass the test and exhibit the lowest test statistic.

When the measurements undergoing the consistency check are used exclusively to compute the final navigation solution, FDI is achieved simply by rejecting the faulty measurement. However, when the measurements are then input to a navigation or integration Kalman filter, or used for dead reckoning, only FDR is achievable, as undetected faulty measurement data may have been used earlier, particularly when time-averaged test statistics are used. In that case, parallel navigation solutions are needed for FDI.

GNSS RAIM using pseudo-range measurements, which may be carrier-smoothed, is sometimes known as absolute RAIM, while RAIM using delta-range measurements is known as relative RAIM. Delta-range comprises the change in carrier phase between epochs, so relative RAIM can be more sensitive to the onset of a new fault. Relative RAIM may be combined with either absolute RAIM or infrastructure-based integrity monitoring (Section 17.5), where it can compensate for the communication lag [15].

For differential carrier-phase GNSS positioning (Section 10.2), measurement consistency checking must be combined with validation of the integer wavelength ambiguity resolution to protect the integrity of the position solution [16].

17.4.2 Parallel Solutions

Parallel-solutions integrity monitoring maintains a number of parallel navigation solutions or filters, each excluding data from one sensor or radio navigation signal. Each additional navigation solution or Kalman filter is compared with the main one using a consistency test. Failure of the test indicates a fault in the sensor or signal omitted from one of the solutions. The navigation system output is then switched to the solution omitting the faulty sensor or signal. As this navigation solution has never incorporated data from the faulty source, isolation of the fault is achieved. Thus, the main benefit of parallel solutions is providing FDI. The main drawback is increased processor load.

The processor load may be reduced by implementing navigation solutions that each omit two measurement streams. However, this requires both streams to be excluded if a fault is detected on one of them.

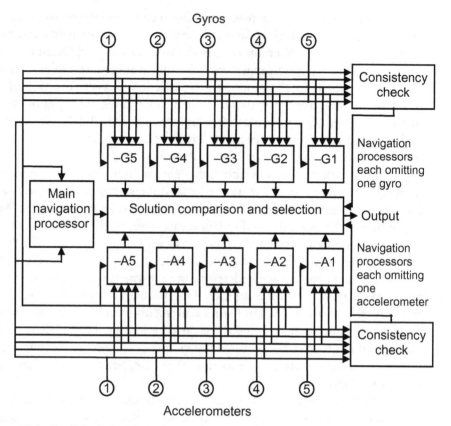

Figure 17.6 Parallel-solutions integrity monitoring applied to inertial navigation with redundant sensors.

Figure 17.6 shows how FDI may be applied to inertial navigation with redundant sensors by using multiple navigation processors. Each navigation solution, apart from the main one, omits one sensor. The processor load may be reduced by omitting accelerometer and gyro pairs. However, both sensors must then be rejected if a fault is detected.

Figure 17.7 illustrates parallel-solutions integrity monitoring for GNSS. Multiple Kalman filters are used. The main filter inputs all measurement data, while each of the other filters, known as subfilters, omits data from one of the satellites.

Figure 17.8 shows the integrity monitoring architecture for INS/GNSS without redundant inertial sensors. Closed-loop corrections to the inertial navigation equations are fed back from the main integration filter only. However, they must also be fed back to the subfilters to enable them to correct their state estimates to account for the feedback to the INS. When redundant inertial sensors are used, further subfilters based on inertial navigation solutions omitting individual sensors are added.

For multisensor integrated navigation in a centralized filtered architecture (Section 16.1.4), the INS/GNSS integrity monitoring architecture is adapted to include subfilters omitting individual sensor or signal measurements or complete position or velocity solutions from a range of different types of navigation system. In a federated architecture (Section 16.1.5), parallel-solutions integrity monitoring may

17.4 Direct Consistency Checks

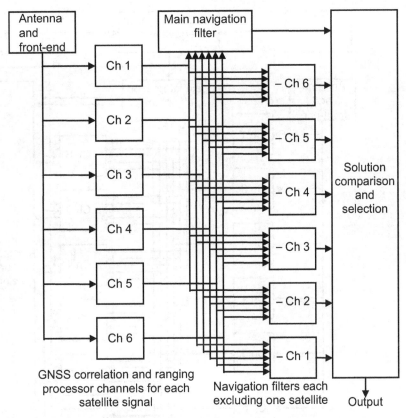

Figure 17.7 Parallel-solutions GNSS integrity monitoring.

be implemented within the local filters, while measurement consistency checks, as described in Section 17.4.1, may be applied to the set of local filter outputs prior to fusing. Note that two local filters are needed for fault detection and three for fault isolation or recovery.

When parallel navigation solutions are used for stand-alone inertial navigation or dead reckoning, consistency checks may be performed at the measurement level, as described in Section 17.4.1, with the parallel solutions simply providing fault isolation.

For parallel Kalman filters, there are several ways of applying consistency checks. The extrapolation method [4] applies innovation sequence monitoring to each Kalman filter as described in Section 17.3.2. The chi-square test statistic, (17.7), is usually computed. Measurement consistency checks (Section 17.4.1) may also be used. In both cases, when a fault is detected, the test statistics for all filters except the one omitting the faulty data will exceed the threshold.

The solution-separation method [17] compares the state estimates of each subfilter, denoted by index j, with the main filter, denoted by index 0. Assuming correct covariance information is available, a suitable chi-square test statistic for the jth subfilter is

$$s^2_{\delta x, j, k} = \left(\hat{\mathbf{x}}^+_{j,k} - \hat{\mathbf{x}}^+_{0,k} \right)^{\mathrm{T}} \mathbf{B}^{+\,-1}_{j,k} \left(\hat{\mathbf{x}}^+_{j,k} - \hat{\mathbf{x}}^+_{0,k} \right), \tag{17.12}$$

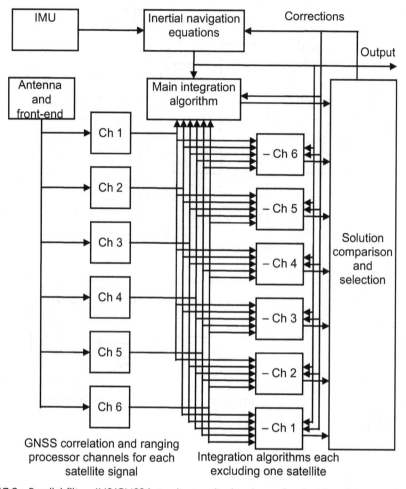

Figure 17.8 Parallel-filters INS/GNSS integrity monitoring (no redundant inertial sensors).

where $\mathbf{B}_{j,k}^+ = \mathrm{E}\left[\left(\hat{\mathbf{x}}_{j,k}^+ - \hat{\mathbf{x}}_{0,k}^+\right)\left(\hat{\mathbf{x}}_{j,k}^+ - \hat{\mathbf{x}}_{0,k}^+\right)^\mathrm{T}\right]$ is the covariance of the state vector difference. From [5],

$$\mathbf{B}_{j,k}^+ = \mathbf{P}_{j,k}^+ - \mathbf{P}_{0,k}^+, \qquad (17.13)$$

noting that the state uncertainties of the main filter should be smaller than those of the subfilters.

Once a fault has been isolated, achieving fault exclusion requires the additional step of validating that the recovered navigation solution is fault free. It is more processor efficient to do this using innovation sequence monitoring or measurement consistency checks as using solution-separation tests would require a bank of subsubfilters, each excluding two measurement streams. Subsubfilters must be used when there is a requirement to isolate two faults.

17.5 Infrastructure-Based Integrity Monitoring

Many faults are only detectable by user-equipment-based integrity monitoring. However, for radio navigation, including GNSS, faults in the transmitted signals are more easily detected by fixed monitor stations. This is because, with the user antenna at a known location and a high-performance receiver clock, the measurement data may be focused on determining the accuracy of the incoming signals. Infrastructure-based integrity monitoring thus uses fixed monitor stations to determine faults and then transmits this information to users via a data link. Many applications, such as civil aviation, require users to be notified within 10 seconds of the fault occurring.

For GNSS, errors in the transmitted signals must be distinguished from range measurement errors due to signal propagation, such as multipath or unusual ionosphere or troposphere behavior, and from faults in the monitor station itself. This is achieved by using a network of monitor stations distributed over different locations within the coverage area of the integrity monitoring service. Monitor stations may be combined with DGNSS reference stations. However, a separate network must be used where there is a requirement for independent monitoring of the DGNSS service itself. When a fault is identified, integrity alerts may be broadcast to the user via SBAS, GBAS, or another data link. This is sometimes known as a GNSS integrity channel (GIC). Also broadcast are error-bounding parameters for each signal to enable users to calculate position-solution confidence limits, including protection levels (see Section 17.6) [18]. In principle, this could also be transmitted via the GNSS signals themselves.

GNSS satellite faults detectable by monitoring stations include low transmission power, clock faults, transmission of faulty navigation data, and irregular code waveforms [19]. Satellite clock faults may manifest as sudden jumps, detected through step changes in the pseudo-range measurements, or as a high drift rate, detected by observing large pseudo-range-rate errors. Faulty navigation data is detected by comparing the ephemeris, satellite clock, and ionosphere model parameters transmitted by each satellite with those computed by the monitoring network. Signal quality monitoring algorithms detect irregular code waveforms by using receivers with multiple closely spaced correlation channels, such as the MEDLL, to observe the auto-correlation function. A vision-correlator receiver may also be used [20].

Satellite autonomous integrity monitoring (SAIM) is a concept whereby each GNSS satellite monitors its own signals for faults such as abnormal signal power, irregular code waveforms, code-carrier divergence, clock irregularities, and navigation data faults [21]. SAIM may be extended to the monitoring of the signals from neighboring satellites via intersatellite measurements [22].

Signal monitoring is also built into the infrastructure of Loran and DME. For Loran, a system area monitor at a known location checks the output of a chain of stations, while, for DME, a test interrogator is located near to the base station. When a fault is detected, the relevant ELoran transmitter or DME base station is simply shut down.

17.6 Solution Protection and Performance Requirements

For safety-critical applications, such as civil aviation and many marine applications, it is not sufficient just to implement integrity monitoring; the navigation system must be certified to meet a guaranteed level of performance. This is known as the required navigation performance and is expressed in terms of accuracy, integrity, continuity, and availability.

The accuracy requirement simply states the average accuracy, across all conditions, that the navigation system must meet and is typically expressed as 95% bounds for the radial horizontal and vertical position errors (i.e., the errors that must not be exceeded more than 5% of the time). It is usually the easiest requirement to meet.

While accuracy is concerned with the core of the error distribution, integrity is concerned with the tails. The integrity requirement is expressed in terms of a horizontal alert limit (HAL) and a vertical alert limit (VAL) that the radial horizontal and vertical position errors must not, respectively, exceed without alerting the user. Due to statistical outliers, absolute alert limits can never be enforced. Therefore, a maximum probability of missed detection, p_{md}, must also be met. This is generally specified using the *integrity risk*, which is the probability of an undetected fault, known as an *integrity failure*, occurring at any point within a specified time interval. Note that a higher alert limit combined with a lower integrity risk can be equivalent to a lower alert limit combined with a higher integrity risk.

The fault detection algorithms must meet the integrity requirement for the navigation solution to be usable, while the amount of information available for performing that fault detection will vary. Consequently, the missed detection probability of that test for a given HAL or VAL will vary. The fault detection test must be declared unavailable when it exceeds the permitted maximum. It is the role of the solution protection function of an integrity monitoring system to determine whether fault detection is available. When it is not, an integrity alert must be raised as if an uncorrectable fault had been detected.

Integrity tests are an example of hypothesis testing (Section B.5 of Appendix B on the CD). The values compared are subject to uncertainty due to noise and unknown systematic errors. In a Kalman filter, these uncertainties are represented by the error covariance, **P**, and measurement noise covariance, **R**, matrices. Thus, the true position error at the fault detection threshold of an integrity test will have a probability distribution as shown in Figure 17.9.

The integrity requirement allows position errors exceeding the alert limit to remain undetected with a probability not exceeding p_{md}. A horizontal protection level (HPL) and vertical protection level (VPL) may then be defined such that the probability of the test statistic lying below the fault detection threshold is less than p_{md} for, respectively, radial horizontal and vertical position errors exceeding that level. Therefore, if the HPL exceeds the HAL or the VPL exceeds the VAL, an integrity alert is triggered.

In any integrity monitoring system, there is a tradeoff among the protection level, the probability of missed detection, and the false alarm rate. For a given protection level, the detection threshold for the integrity-monitor test statistic determines the tradeoff between p_{md} and the false alarm probability, p_{fa}, as Figure 17.10 illustrates [23].

17.6 Solution Protection and Performance Requirements

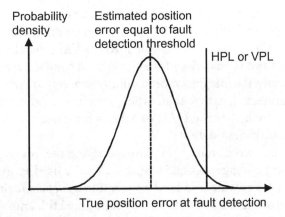

Figure 17.9 True position error distribution when the estimated position error is at the fault detection threshold.

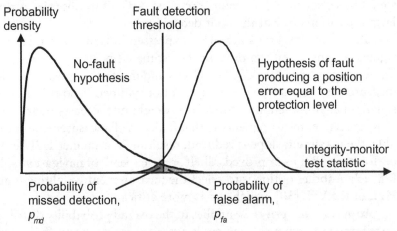

Figure 17.10 Missed detection and false alarm probability as a function of the fault detection threshold.

For a test statistic, s, and a threshold, T, p_{fa} and p_{md} are defined by

$$\int_T^\infty f_0(s)\, ds = p_{fa}, \qquad \int_{-\infty}^T f_1(s)\, ds = p_{md}, \qquad (17.14)$$

where f_0 is the fault-free probability density function (see Section B.2 of Appendix B on the CD) of the test statistic, and f_1 is the PDF given a position error equal to the protection level. To maintain a constant missed detection probability and false alarm rate, the detection threshold and protection level must vary. When approximations are made in calculating the HPL and VPL, it is essential to ensure that the assumed error distribution overbounds its true counterpart to prevent computation of over-optimistic values of the HPL and VPL, which would constitute a safety risk [24].

When FDE is implemented, an integrity alert is not automatically triggered when a fault is detected. It is triggered if the exclusion fails (i.e., the recovered navigation solution fails the fault detection test). Once a fault is detected and isolated, the HPL and VPL are recalculated for the navigation solution excluding the faulty satellite. If the new HPL or VPL exceeds the HAL or VAL, the alert is triggered. Otherwise, the integrity test is passed and the navigation solution remains safe to use.

Essentially, the integrity monitoring system reports whether or not the HAL and VAL are protected, rather than whether or not a fault has occurred. This applies regardless of whether the HAL or VAL is breached due to a genuine fault, a false alarm, or insufficient data.

There are two different ways of specifying the integrity risk requirement: specific risk and average risk [25]. Specific risk is used in aviation, whereas average risk is used for marine and land applications. A *specific integrity risk* must be met for every possible fault. Thus, for any fault, the HPL and VPL determined from the PDF of the position error given that fault must be less than the HAL and VAL. It is impractical to calculate protection levels for every conceivable failure mode and fault magnitude. However, in practice, the HPL and VPL need only be determined for the signal or sensor for which faults are most difficult to observe, as this produces the largest position error at the fault detection threshold.

The protection limits are often expressed in terms of the largest value, across signals or sensors, of the ratio of the estimated position error to the test statistic, known as the *slope* [26, 27]. For GNSS and most other position-fixing systems, the measurement geometry is varying and not symmetric, so the slope varies between signals or environmental features. By reweighting the least-squares position solution or Kalman filter measurements so that faults on all measurements are equally observable, the maximum slope is reduced, resulting in a smaller HPL and VPL. Integrity performance is thus improved, albeit at the expense of navigation-solution accuracy [28, 29]. Another option is to select the measurement weighting that minimizes the HPL or the VPL [30], whichever is more critical.

An *average integrity risk* applies to the overall probability distribution. The error distribution of each measurement is averaged across the no-fault case and all failure modes, weighted according to their probabilities. The position error distribution for the case that no fault is detected is then calculated based on the measurement error distribution and the characteristics of the fault detection algorithms. Finally, the HPL and VPL are calculated from the position error distribution using the average integrity risk.

Comparing the two types of integrity risk, a given specific integrity risk is much more demanding than an average integrity risk of the same value because it applies to the specific case of a fault occurring as opposed to the overall behavior of the system. A HPL or VPL calculated using an average integrity risk will be two to five times smaller than one calculated using the same specific integrity risk and is thus much more likely to fall within a given HAL or VAL [25].

The final part of the integrity requirement is the time to alarm. This is the maximum time between the HAL or VAL being breached and the user being alerted of this.

Continuity is the ability to maintain a navigation solution within the HAL or VAL over a certain time period, known as the protection window, even if a fault occurs, noting that this requires FDE. The continuity requirement is specified in terms of the probability of a fault occurring within the protection window multiplied by the probability of the HAL or VAL being breached in the event of that fault. This is determined by calculating the HPL and VPL of each navigation solution excluding one signal or sensor, propagated over the protection window, and comparing them with the HAL or VAL.

Continuity is important for applications such as instrument or automated landing of an aircraft, whereby there is a window of a certain length between the pilot committing to a landing and completing that landing. The pilot is unable to respond to integrity alerts during that window. Therefore, for these applications, continuity calculations are performed in real time and a continuity alert is flagged to the user if any HPL or VPL is predicted to breach the HAL or VAL within the protection window in the event of a fault occurring.

There is a tradeoff between continuity and integrity performance. A lower test threshold will reduce the missed detection probability, improving integrity. However, it will also increase the false alarm rate, degrading continuity [18].

The availability requirement is the proportion of time that the navigation system must operate with the accuracy requirement met and the integrity and continuity alert limits protected. Note that if the false alarm rate is set high to minimize the HPL and VPL for a given missed detection probability, the availability will be low.

For systems in which the HPL and VPL vary, such as GNSS, the availability is calculated by determining the HPL and VPL across a grid of position and time, covering every possible signal geometry. The proportion of these HPLs and VPLs that fall within the HAL and VAL denotes the availability. This is easier for integrity monitoring methods in which the HPL and VPL may be determined analytically. For innovation-based methods, the HPL and VPL calculations are empirical and must be validated using Monte Carlo simulations, which is not always practical [31].

Table 17.1 lists some example RNPs for aircraft navigation [23]. Stand-alone GPS fails to meet the availability and continuity requirements of practical RNPs due to an insufficient number of satellites. Less demanding RNPs, such as aircraft en-route navigation and nonprecision approach and marine harbor entrance and approach, can be met by GPS through integration with other navigation technologies, such as DME, use of SBAS signals or addition of GLONASS or Galileo signals [32]. Meeting the RNP for aircraft landing additionally requires the use of differential GNSS.

Table 17.1 Example RNPs for Aircraft Navigation

Operation		En Route (Oceanic)	Nonprecision Approach	Category I Landing
Accuracy (95%)	Horizontal	3.7 km	220 m	16 m
	Vertical	—	—	—
Specific integrity risk		10^{-7} per hour	10^{-7} per hour	2×10^{-7} per approach
	HAL	7.4 km	556m	40m
	VAL	—	—	15–10m
Time to alert		5 minutes	10 seconds	6 seconds
Continuity failure rate		10^{-4} to 10^{-8} per hour	10^{-4} to 10^{-8} per hour	8×10^{-6} per 15 seconds
Availability		99–99.999%	99–99.999%	99–99.999%

17.7 Testing

To verify that a navigation system meets the required performance standards, it must be tested. This section discusses the different approaches to testing, including field trials, recorded data testing, laboratory testing, and software simulation.

17.7.1 Field Trials

Before a new design of navigation system is delivered, it must be tested in a realistic operational environment. Testing the whole system in this way requires field trials, whereby its navigation solution is compared against a truth reference over a range of scenarios covering the host vehicle's operational envelope.

Some authors determine dead-reckoning performance by conducting a trial over a closed-loop test circuit and measuring the difference between the reported positions at the start and finish. However, this can give misleading results as the heading initialization error has least impact at this point, as Figure 17.11 shows. Therefore, a position truth reference must be provided at multiple locations. One option is to survey a series of waypoints and record the time when the navigation system passes each waypoint. Another approach is to follow a predetermined route on a map (which must be converted to the same datum as the navigation system under test). The accuracy of the truth reference then depends on both the accuracy of the map and the accuracy with which the test vehicle (or person) can follow the route. The final option is to use an aviation-grade INS/GNSS system, located in the same test vehicle as the system under test. This has the advantage of providing a velocity and attitude reference as well as position.

A limitation of field trials is the difficulty of performing tests under unusual conditions. Extreme dynamics require a test range and an expensive test vehicle. Faults in publicly available radio navigation signals cannot be generated without disrupting other users. Jamming of GNSS and other radio navigation signals requires government permission, advanced notification of safety-critical users, and must be performed in remote areas.

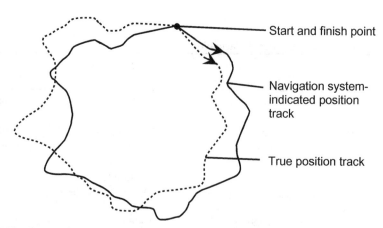

Figure 17.11 True and reported positions over a closed-loop test circuit in which the heading initialization error is the only error source.

17.7.2 Recorded Data Testing

Comparing the performance of different algorithms and software configurations using field trials is expensive. Recorded data testing enables one set of trials data to be used for multiple tests. As for conventional field trials, a truth reference is required. Typically, raw sensor measurements will be recorded from dead-reckoning systems (e.g., IMU specific force and angular rate measurements). For radio navigation systems, such as GNSS, two approaches may be adopted. Receiver measurements, such as pseudo-range, two-way range, RSS, or AOA, may be recorded. Alternatively, the radio signals themselves may be recorded, typically after digital sampling within the receiver front end. This enables baseband signal processing hardware to be tested as well as software [33]. GNSS record and replay devices are now widely available. Recorded data is also useful for investigating data synchronization issues, while, by manipulating the data, a navigation system's response to faults may be tested.

17.7.3 Laboratory Testing

Laboratory testing enables navigation equipment to be evaluated over a wider range of scenarios and under more controlled conditions than is feasible in field trials. After accounting for the costs of equipping the laboratory, the cost per test is also lower.

A number of tests may be applied to characterize the behavior of inertial sensors [34]. Static testing may be used to evaluate noise and biases, with a centrifuge used to generate angular rate and centripetal acceleration. Vibration tables and temperature-controlled chambers are also used. Similarly, odometry may be tested using a rolling road, while PDR may be tested using a treadmill.

For radio navigation, simulators are used to generate signals which are then input to the user equipment in place of the antenna signal. The time delay and Doppler shift of each signal may be adjusted to simulate user motion. Transmitter timing and position errors, signal propagation delays, multipath interference, RF noise, and faults may all be added to the signals. However, complex environments, such as urban areas, are difficult to simulate. Simulator testing is critical for testing the response of GNSS user equipment to new signal designs before they are widely available from satellites [35].

For user antenna testing, signal simulation equipment is connected to a set of transmit antennas installed, together with the user equipment, within an anechoic chamber. Interference may also be generated. Changes in the user orientation are simulated by rotating the antenna [36].

For integrated INS/GNSS testing, a GNSS signal simulator may be augmented with an IMU signal simulator that uses the same user motion model. Simulated IMU signals are input to the INS/GNSS system via a test interface that bypasses its internal IMU [37].

17.7.4 Software Simulation

Software simulation is a useful tool for predicting the performance of different navigation system configurations during the design process and for testing algorithms and software, particularly under unusual conditions. It has the advantages of cost-effectiveness, reproducibility, and the availability of accurate truth data to compare the navigation solution against. However, a software simulation is only as accurate as the sensor models on which it is based. Certain sources of error are often neglected, while misleading results can occur when scenarios outside the original design scope are simulated. Thus, software simulations must always be supported by other forms of testing.

Appendix J on the CD discusses the software simulation of all types of navigation system, with an emphasis on GNSS and inertial navigation. This is supported by some basic MATLAB simulation software, also on the accompanying CD.

Problems and exercises for this chapter are on the accompanying CD.

References

[1] Bhatti, U. I., and W. Y. Ochieng, "Failure Modes and Models for Integrated GPS/INS Systems," *Journal of Navigation*, Vol. 60, No. 2, 2007, pp. 327–348.

[2] Pullen, S., and J. Rife, "Differential GNSS: Accuracy and Integrity," in *GNSS Applications and Methods*, S. Gleason and D. Gebre-Egziabher, (eds.), Norwood, MA: Artech House, 2009, pp. 87–119.

[3] Chaffee, J., K. Kovach, and G. Robel, "Integrity and the Myth of Optimal Filtering," *Proc. ION NTM*, Santa Monica, CA, January 1997, pp. 453–461.

[4] Diesel, J., and S. Luu, "GPS/IRS AIME: Calculation of Thresholds and Protection Radius Using Chi-Square Methods," *Proc. ION GPS-95*, Palm Springs, CA, September 1995, pp. 1959–1964.

[5] Young, R. S. Y., and G. A. McGraw, "Fault Detection and Exclusion Using Normalised Solution Separation and Residual Monitoring Methods," *Navigation: JION*, Vol. 50, No. 3, 2003, pp. 151–169.

[6] Bhatti, U. I., W. Y. Ochieng, and S. Feng, "Integrity of an Integrated GPS/INS System in the Presence of Slowly Growing Errors. Part II: Analysis," *GPS Solutions*, Vol. 11, No. 3, 2007, pp. 183–192.

[7] Moore, S., G. Myers, and R. Hunt, "Future Integrated Navigation Guidance System," *Proc. ION NTM*, Long Beach, CA, January 2001, pp. 447–457.

[8] Groves, P. D., "Principles of Integrated Navigation," Course Notes, QinetiQ Ltd., 2002.

[9] Sukkarieh, S., et al., "A Low-Cost Redundant Inertial Measurement Unit for Unmanned Air Vehicles," *International Journal of Robotics Research*, Vol. 19, No. 11, 2000, pp. 1089–1103.

[10] Brown, R. G., "Receiver Autonomous Integrity Monitoring," in *Global Positioning System: Theory and Applications, Volume II*, B. W. Parkinson and J. J. Spilker, Jr., (eds.), Washington, D.C.: AIAA, 1996, pp. 143–165.

[11] Brown, R. G., and P. W. McBurney, "Self-Contained GPS Integrity Check Using Maximum Solution Separation as the Test Statistic," *Proc. ION GPS-87*, Colorado Springs, CO, September 1987, pp. 263–268.

[12] Lee, Y. C., "Analysis of Range and Position Comparison Methods as a Means to Provide GPS Integrity in the User Receiver," *Proc. ION 42nd AM*, Seattle, WA, June 1986, pp. 1–4.

[13] Parkinson, B. W., and P. Axelrad, "Autonomous GPS Integrity Monitoring Using the Pseudorange Residual," *Navigation: JION*, Vol. 35, No. 2, 1988, pp. 255–274.

[14] Sturza, M. A., "Navigation System Integrity Monitoring Using Redundant Measurements," *Navigation: JION*, Vol. 35, No. 4, 1988, pp. 483–501.

[15] Gratton, L., M. Joerger, and B. Pervan, "Carrier Phase Relative RAIM Algorithms and Protection Level Derivation," *Journal of Navigation*, Vol. 63, No. 2, 2010, pp. 215–231.

[16] Milner, C., et al., "A Holistic Approach to Carrier-Phase Receiver Autonomous Integrity Monitoring (CRAIM)," *Proc. ION GNSS 2011*, Portland, OR, September 2011, pp. 2689–2695.

[17] Brenner, M., "Integrated GPS/Inertial Fault Detection Availability," *Navigation: JION*, Vol. 43, No. 2, 1996, pp. 111–130.

[18] Rife, J., and S. Pullen, "Aviation Applications," in *GNSS Applications and Methods*, S. Gleason and D. Gebre-Egziabher, (eds.), Norwood, MA: Artech House, 2009, pp. 245–267.

[19] Xie, G., et al., "Integrity Design and Updated Test Results for the Stanford LAAS Integrity Monitor Testbed," *Proc. ION 57th AM*, Albuquerque, NM, June 2001, pp. 681–693.

[20] Fenton, P. C., and J. Jones, "The Theory and Performance of NovAtel Inc.'s Vision Correlator," *Proc. ION GNSS 2005*, Long Beach, CA, September 2005, pp. 2178–2186.

[21] Vidarsson, L., et al., "Satellite Autonomous Integrity Monitoring and Its Role in Enhancing GPS User Performance," *Proc. ION GPS 2001*, Salt Lake City, UT, September 2001, pp. 690–702.

[22] Xu, H., J. Wang, and X. Zhana, "GNSS Satellite Autonomous Integrity Monitoring (SAIM) Using Inter-Satellite Measurements," *Advances in Space Research*, Vol. 47, No. 7, 2011, pp. 1116–1126.

[23] Feng, S., et al., "A Measurement Domain Receiver Autonomous Integrity Monitoring Algorithm," *GPS Solutions*, Vol. 10, No. 2, 2006, pp. 85–96.

[24] DeCleene, B., "Defining Pseudorange Integrity-Overbounding," *Proc. ION GPS 2000*, Salt Lake City, UT, September 2000, pp. 1916–1924.

[25] Pullen, S., T. Walter, and P. Enge, "Integrity for Non-Aviation Users: Moving Away from Specific Risk," *GPS World*, July 2011, pp. 28–36.

[26] Conley, R., et al., "Performance of Stand-Alone GPS," in *Understanding GPS Principles and Applications*, 2nd ed., E. D. Kaplan and C. J. Hegarty, (eds.), Norwood, MA: Artech House, 2006, pp. 301–378.

[27] Powe, M., and J. Owen, "A Flexible RAIM Algorithm," *Proc. ION GPS-97*, Kansas, MO, September 1997, pp. 439–449.

[28] Hwang, P. Y., and R. G. Brown, "RAIM FDE Revisited: A New Breakthrough in Availability Performance with NioRAIM (Novel Integrity-Optimized RAIM)," *Navigation: JION*, Vol. 53, No. 1, 2006, pp. 41–51.

[29] Hwang, P. Y., "Applying NioRAIM to the Solution Separation Method for Inertially-Aided Aircraft Autonomous Integrity Monitoring," *Proc. ION NTM*, San Diego, CA, January 2005, pp. 992–1000.

[30] Milner, C. D., and W. Y. Ochieng, "Weighted RAIM for APV: The Ideal Protection Level," *Journal of Navigation*, Vol. 64, No. 1, 2011, pp. 61–73.

[31] Lee, Y. C., and D. G. O'Laughlin, "Performance Analysis of a Tightly Coupled GPS/Inertial System for Two Integrity Monitoring Methods," *Navigation: JION*, Vol. 47, No. 3, 2000, pp. 175–189.

[32] Ochieng, W. Y., et al., "Potential Performance Levels of a Combined Galileo/GPS Navigation System," *Journal of Navigation*, Vol. 54, No. 2, 2001, pp. 185–197.

[33] Vinande, E., et al., "GNSS Receiver Evaluation Record-and-Playback Test Methods," *GPS World*, January 2010, pp. 28–34.

[34] Titterton, D. H., and J. L. Weston, *Strapdown Inertial Navigation Technology*, 2nd ed., Stevenage, U.K.: IEE, 2004.

[35] Boulton, P., A. Read, and R. Wong, "Formal Verification Testing of Galileo RF Constellation Simulators," *Proc. ION GNSS 2007*, Fort Worth, TX, September 2007, pp. 1564–1575.

[36] Boasman, N. J., and P. Briggs, "The Development of an Anechoic GPS Test Facility," *Proc. ION 58th AM*, Albuquerque, NM, June 2002, pp. 483–494.

[37] Fedora, N., C. Ford, and P. Boulton, "A Versatile Solution for Testing GPS/Inertial Navigation Systems," *Proc. ION GNSS 2008*, Savannah, GA, September 2008, pp. 1227–1236.

CHAPTER 18
Applications and Future Trends

This chapter discusses how the technology described in the preceding chapters may be deployed to meet the requirements of a wide range of navigation applications. Section 18.1 discusses the design and development process in general terms, and the following sections each focus on one application area. Section 18.2 covers aviation; Section 18.3 discusses UAVs and guided weapons; Section 18.4 describes land vehicle applications; Section 18.5 discusses rail navigation; Section 18.6 covers marine navigation; Section 18.7 reviews underwater navigation; Section 18.8 discusses the navigation of spacecraft; Section 18.9 covers pedestrian navigation; and Section 18.10 discusses other applications of navigation technology. The U.S. Federal Radionavigation Plan [1] reviews the requirements for a wide range of navigation and positioning applications.

Section 18.11 concludes the book by discussing future trends in navigation and positioning technology.

18.1 Design and Development

Figure 18.1 shows a possible design and development process for a new navigation system. The first step is to determine the requirements, which may be divided into performance requirements, environmental constraints, engineering constraints, and economic constraints.

The most fundamental requirement is the outputs that the system must provide. This could be just position, or it may include velocity, acceleration, attitude, and angular rate. For each of these, key performance requirements may be expressed in terms of accuracy, integrity, continuity, and availability, as described in Section 17.6. For some applications, only some of these criteria will apply. There will also be an update-rate requirement.

Environmental constraints define where the navigation system is expected to operate. This could be on land, in the air, in space, on water, or underwater. For land applications, there are then significant differences between open, mountainous, woodland, urban, indoor, and underground environments. Also important are the host vehicle dynamics, such as the maximum speed, acceleration, jerk, angular rate, and angular acceleration, together with the vibration and temperature environments. Another consideration is whether the users are trained professionals or consumers.

The key engineering constraints are size, mass, and power consumption. These are much more limited for small-scale applications, such as pedestrian navigation, than for large vehicles, such as airliners and ships. A requirement for some military applications is stealth, whereby the navigation system must not output trackable signals.

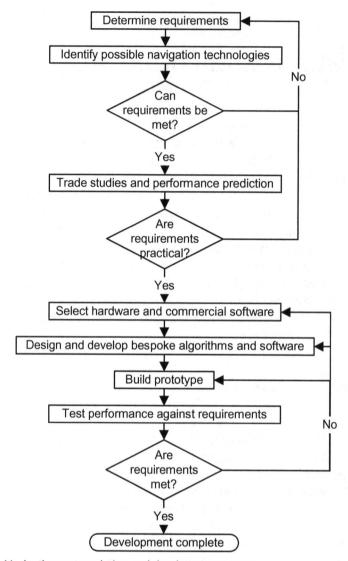

Figure 18.1 Navigation system design and development process.

Finally, the economic constraints include the delivery schedule, the number of units required, and the cost, which may be expressed per unit or for the project as a whole. Other issues are whether the market is regulated, whether there is a single customer or multiple customers, and whether customers are product buyers or service subscribers. These constraints help to determine the viability of research and development effort and deployment of new infrastructure.

The second stage of the design process is to identify possible navigation technologies. Some sets of requirements can be met by a single technology, while others will need a multisensor solution. For multisensor approaches, a main technology that meets most of the requirements should be selected first; this will often be GNSS. Next, the gaps in this technology's capability should be identified. These may include parts of the operating environment, such as indoors or underwater; the update rate and dynamics; and providing continuity through interruptions. One or more technologies to bridge this gap should then be identified, based on dead reckoning and/or position fixing as appropriate. When no suitable navigation sensor combination is identified, it will be necessary to revise the requirements.

Often, multiple solutions will be identified, involving different navigation technologies or different grades of the same technology, in which case trade studies should be conducted to determine the best option. Performance may be predicted over a range of operational scenarios using software simulation (Section 17.7.4 and Appendix J on the CD). At this stage, the practicality of the requirements should be considered and revised where necessary.

The next stages are to select the hardware and commercial software to be used and to design the bespoke algorithms and software needed to meet the requirements. This may include navigation sensor output processing (Chapters 4 to 13), navigation system integration and alignment (Chapters 14 to 16), fault detection and integrity monitoring (Chapter 17), and communications and user interfacing. Supporting research may also be required. This is followed by the assembly of a prototype system, which must be tested as discussed in Section 17.7. Modifications to the design and system implementation are made until it has been verified that all requirements have been met.

18.2 Aviation

Aviation may be divided into three categories: commercial air carriers, general aviation, and military aviation. Each has different navigation requirements and deploys a different navigation solution. Requirements also depend on the region of airspace used. However, for all aircraft, continuity is critical. It is not possible to simply stop an aircraft if the navigation system fails. Therefore, backup technologies are needed and duplicate user equipment is often deployed. Aircraft also require an airspeed measurement and roll- and pitch-axis attitude solutions for flight control.

Commercial air carriers are large aircraft that transport passengers and cargo. Instrument flight rules (IFR) apply, enabling flight to continue when visibility is limited. Safety is critical. The air space used is controlled with strict separation maintained between aircraft and a formal required navigation performance [1, 2]. An aircraft must not drift into the airspace allocated to another aircraft due to an undetected navigation system fault (known faults can be reported to air traffic control and the separation increased). Integrity is therefore a key requirement, more important than accuracy. Alert limits vary with the phase of flight, reflecting different aircraft separations. Figure 18.2 shows some examples, noting that the specific integrity risk is 10^{-7} per hour in all cases.

GPS augmented with SBAS is certified for use in the en-route, terminal, nonprecision approach (NPA), and precision approach phases of flight. Multiconstellation GNSS should also meet the requirements but had not been certified at the time of writing. DME/VOR and DME integrated with inertial navigation are certified for en-route, terminal, and NPA flight, but are only available over land. Stand-alone inertial navigation with redundant sensors meets the oceanic en-route requirements and also provides the attitude solution. At altitudes above 5.486 km, the flight levels

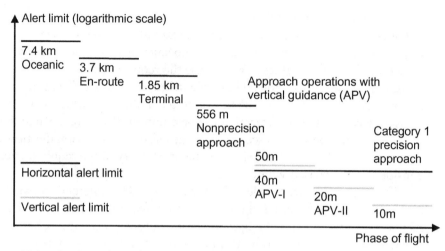

Figure 18.2 Alert limits for civil aviation.

allocated to aircraft in controlled airspace are defined in terms of barometric altitude computed using a standard atmospheric model, so a baro must be used for the aircraft's altitude solution.

The smallest alert limits apply to automated and instrument-based approach and landing. ILS is used in the vast majority of cases. However, GBAS systems are now being deployed at airports to enable use of GPS for category I landing, the landing category with the least demanding RNP.

General aviation (GA) includes leisure, emergency services, private business travel, and crop spraying. Aircraft must meet the relevant RNP to operate under IFR in controlled airspace. Otherwise, collisions must be avoided using visual flight rules (VFR) and airspace requiring IFR must not be used. Most GA aircraft use GNSS for horizontal position, but some rely on VOR and DME. Manual navigation using landmarks is the backup for aircraft operating under VFR. Attitude is typically provided by an AHRS, although older aircraft may combine a magnetic compass, directional gyro, and tilt sensors. Exceptions are airborne mapping, remote sensing, and precision agriculture, for which much higher position and attitude accuracy is required. Here, carrier-phase differential GNSS is typically integrated with inertial navigation.

Military aircraft perform many different roles [3]. Accuracy requirements are generally more demanding than for civil aviation and the dynamics are often much greater. Conversely, integrity is less important as an element of risk is accepted in military missions. Exceptions are use of civil airspace for which IFR rules apply and automated landing, particularly on ships [4]. Integrated INS/GNSS forms the core of most navigation systems used for military aviation. Military GNSS signals are used by those countries that have access to them. Jamming of GNSS signals is a major threat that must be countered as discussed in Section 10.3 and by including backups such as other radio signals, barometric altitude, and terrain referenced navigation, as well as INS.

Helicopters are used in both general and military aviation. Their relatively low speed and height of operation makes them suited to Doppler radar, used in conjunction with an AHRS, to provide cost-effective dead reckoning and aid hovering.

A radar altimeter is used for terrain collision avoidance of both helicopters and low-flying fixed-wing aircraft. It can also be used to aid automated landing.

18.3 Guided Weapons and Small UAVs

Guided weapons include missiles, guided bombs, and guided shells. They may seek their own target, be guided to a target at a particular location, or be guided towards a point near the target at which a seeker takes over. A navigation system is only required in the latter two cases. Missiles are self-propelled and require an IMU for flight control. Short-range missiles can use inertial navigation alone, while medium- and long-range missiles also use GNSS and sometimes terrain-referenced navigation or image matching. Guided bombs and shells use control surfaces to alter their trajectories; navigation may be inertial, GNSS, or INS/GNSS. However, for guided shells, which are gun-launched, the ability of the navigation sensors to withstand the enormous launch forces is critical. For all weapon applications, cost, size, weight, and power consumption are important design constraints.

Unmanned air vehicles range in mass from less than 1 kg to over 10 tonnes and in flight altitude from less than 250m to over 20 km [5]. Navigation requirements and solutions for larger UAVs, which are mostly used for military applications, are essentially the same as for manned aircraft (Section 18.2). However, for mini UAVs (<30 kg) and micro air vehicles (<1 kg), size and mass constraints significantly limit the choice of navigation sensors. For example, inertial sensors must be MEMS-based, limiting the achievable performance. Furthermore, these UAVs often operate around buildings (or even indoors in the case of MAVs), which limits GNSS signal availability. Therefore, GNSS and inertial navigation are typically augmented using the UAV's camera. Horizon sensing [6] may be used to improve attitude stability, while image matching and tracking techniques may improve positioning.

18.4 Land Vehicle Applications

Road vehicle navigation, commonly known as "satnav," provides the driver with route guidance as depicted in Figure 18.3 [7]. This is not safety-critical so there are no formal performance requirements. In practice, positioning must be sufficiently accurate to correctly determine the road segment occupied by the host vehicle. This typically varies from a 10-m accuracy in dense urban areas to a 100-m accuracy in rural areas. The system must be sufficiently reliable not to annoy the user. In practice, mapping limitations cause more problems than positioning errors.

Intelligent transportation system applications such as traffic monitoring, vehicle tracking, and emergency notification have similar requirements to route guidance. However, for other ITS applications, such as road-user charging, distance-based insurance, law enforcement, adaptive cruise control, and automatic collision avoidance, integrity and continuity are critical and the accuracy must always be known [1]. For some ITS applications, vehicle positions are broadcast to other users via V2V and V2I communications, which may also be used for network assistance and cooperative positioning.

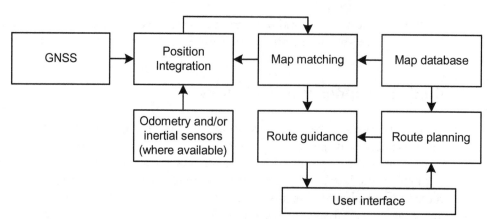

Figure 18.3 Components of a road vehicle navigation system (including route guidance).

GNSS and map matching are the core technologies for road vehicle navigation. Both must be used for reliable performance in urban areas. Factory-fitted navigation systems can also use odometry and sometimes incorporate a yaw-axis gyro. For after-market systems, interfacing to the vehicle odometry measurements is difficult and the use of MEMS partial IMUs for increased robustness has been investigated. Dead reckoning is also important for off-road applications where map matching cannot be used to supplement GNSS.

Autonomous vehicles and robots are equipped with cameras and sometimes a laser scanner or radar. These can be used for both dead-reckoning and position fixing by environmental feature matching.

For mobile mapping, either differential or RTK carrier-phase GNSS is used, depending on the mapping accuracy required. Mapping sensor orientation is provided by multi-antenna GNSS attitude, integrated with inertial navigation [8], which also boosts continuity. Precision INS/GNSS is used for autonomous vehicle applications, including precision agriculture, mining, construction, robotics, and unmanned ground vehicles (UGVs). Odometry is also used for these applications, noting that comparison with inertial navigation can be used for wheel slip detection [9].

18.5 Rail Navigation

Train position determination for fixed-block signaling schemes has traditionally used rail-side infrastructure to determine whether a particular section of track is occupied or whether a train has passed a particular location. However, implementation of more efficient moving-block signaling requires continuous position and velocity determination of all trains which, in turn, requires train-based navigation technology. This safety-critical application requires high continuity, integrity, and availability. Along-track accuracy must always be known so that train separation distances can be set appropriately. It must also be determined which set of parallel tracks each train is occupying.

Reception of GNSS and other radio signals is constrained by tunnels, covered stations, cuttings, and trackside buildings. Reception is generally better in the along-track

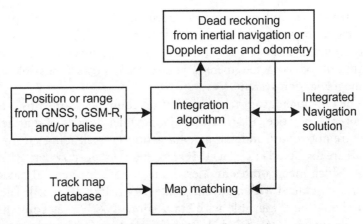

Figure 18.4 Error-state integration architecture for a rail navigation system.

direction than across the tracks, while GSM-Rail (GSM-R) base stations are being deployed to provide good coverage across rail networks. Trains are constrained to move on rails so that, except at switching points, rail navigation is essentially a one-dimensional problem. One solution is to use odometry (calibrated for wheel slip) and Doppler radar, interspersed with balise position fixes. The dead-reckoning performance requirement is an error of no more than 5m + 5% of distance traveled since the last balise [10]. However, integrating GNSS with a MEMS IMU and track map matching offers the prospect of higher performance at a lower infrastructure cost [11]. Figure 18.4 depicts a suitable rail integration architecture. GNSS is already used for nonsafety-critical rail applications such as real-time passenger information.

18.6 Marine Navigation

Navigation in the marine world encompasses not only positioning, heading and speed determination, and course guidance, but also obstacle avoidance and detection. This is the responsibility of the *navigator*, a ship's officer or boat crewmember. Each piece of navigation equipment onboard the vessel is known as a *navigation aid* (navaid), while an external device, such as a light, buoy, or radio transmitters, is known as an *aid to navigation*. Integrated navigation is a relatively new concept in the marine world as it has historically been the navigator's responsibility to compare and combine information from different sources. A marine integrated bridge system, sometimes known as an *integrated navigation system*, incorporates route guidance, collision avoidance, and chart display, as well as integrating position-fixing and dead-reckoning sensors.

Navigation requirements for merchant shipping are governed by the International Convention on Safety of Life at Sea (SOLAS) and are regulated by the International Maritime Organization (IMO). These are expressed in terms of the equipment that must be carried by vessels over a certain mass (including cargo). In mid-ocean, a position accuracy of several kilometers is sufficient for course determination. However, integrity is important for transmitting position to other vessels, via the Automatic Identification System (AIS), for collision avoidance. In coastal areas, ports and harbors, and inland waterway, accuracy is much more important as there are land, fixed obstacles, and other vessels to avoid [1]. There is also a significant risk of grounding if a vessel leaves a designated shipping channel. Thus, position errors should not exceed a few meters. Figure 18.5 shows a possible architecture for a marine navigation system.

The only mandatory radio navigation equipment for ships is a direction finder. However, in practice, all ships use GNSS. Older designs of marine GPS user equipment are not robust and can give hazardously misleading information when GPS signals are disrupted [12]. The IMO has therefore proposed an RNP for modernized GNSS which incorporates an alert limit of 25m for general navigation and 2.5m in ports, combined with an average integrity risk of 10^{-5} [13]. The general requirement should be achievable with stand-alone GNSS user equipment and the port requirement with DGNSS [14]. Back-up position-fixing technologies include other radio navigation systems, where signals are available; radar and visual fixes, where suitable landmarks or buoys are available; and celestial, or stellar, navigation [15].

Position fixing is supplemented by dead reckoning. Merchant vessels determine heading using a gyrocompass or AHRS (often known as a strapdown gyrocompass), and speed from an electromagnetic speed log, DVL, and/or CVL. Naval ships and hydrographic survey vessels use inertial navigation, often with an indexed IMU, which provides the high bandwidth and precision attitude required for sensor alignment.

Dynamic positioning is used to maintain the absolute or relative position of a vessel or offshore platform without using physical anchors. An accuracy of a few meters is required. However, integrity and continuity are critical, so multiple independent position-fixing systems are used, such as GNSS, acoustic ranging, typically integrated with inertial navigation. Systems based on lasers or taut wires may also be used.

Regulatory requirements for small boats and yachts are minimal. They typically use GNSS, a magnetic compass, and a basic speed log.

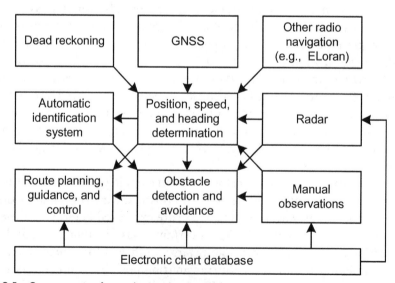

Figure 18.5 Components of a marine navigation system.

18.7 Underwater Navigation

Underwater navigation is used by the military, scientific, and oil and gas exploration communities. Vehicles include manned submarines [16] and submersibles, remotely operated vehicles, autonomous underwater vehicles [17], and individual divers [18]. A key problem is that radio signals do not significantly penetrate water, except for VLF systems (see Section K.2 of Appendix K on the CD). Consequently, to obtain a GNSS fix, an antenna must be raised above the surface. For military applications, this compromises stealth.

The core of an underwater navigation system is inertial navigation, integrated with a depth pressure sensor and either Doppler sonar or a sonar correlation velocity log. The INS provides high bandwidth and precision attitude, with sensor quality varying with the vehicle size and mission length. Indexed IMUs are often used in military submarines. The depth sensor stabilizes the vertical position solution and sonar constrains inertial error growth where the bottom can be tracked.

For localized operations, position fixes may be obtained from acoustic ranging. USBL or SBL ranging may be used close to the support ship. Otherwise, a network of LBL transponders is deployed. For long-range missions, underwater position fixes may be obtained using sonar TRN or gravity gradiometry and gravimetry. Figure 18.6 shows a suitable integration architecture for an ROV.

18.8 Spacecraft Navigation

Spacecraft navigation is partitioned into two problems: attitude determination and orbit determination. Attitude is determined with respect to the CIRS using a star tracker, with respect to the sun using a sun sensor, and with respect to the Earth using a horizon sensor. A gyro triad is used to stabilize the solution from the other sensors [19]. Low Earth orbit satellites may also use multi-antenna GNSS attitude.

Orbit determination is based on force modeling, supplemented by range measurements [20, 21]. The term applies to interplanetary craft as well as satellites. Figure 18.7 illustrates the main forces on a spacecraft. The gravitational forces due

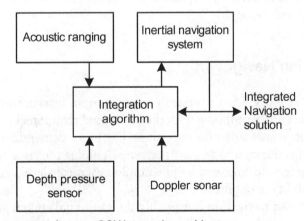

Figure 18.6 Error-state underwater ROV integration architecture.

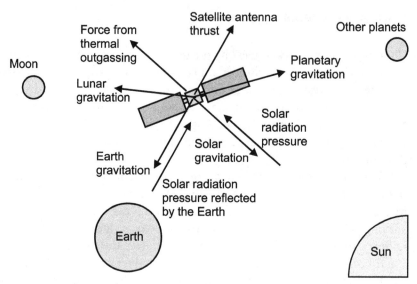

Figure 18.7 The main forces on a spacecraft in Earth orbit.

to multiple astronomical bodies must be accounted for. The effect of solar radiation pressure depends on the orientation of the spacecraft and the reflectivity and absorbtivity of all its surfaces across the solar spectrum. Reflections off the Earth, Moon, and other planetary bodies must be accounted for as must reflection from one part of the spacecraft to another. With a good model, satellite positions can be predicted to within a few meters over a complete orbit without new measurements.

In Earth orbit, ranges may be measured using GNSS, Doppler Orbitography and Radiopositioning Integrated by Satellite (DORIS), radar tracking, and satellite laser ranging (SLR), which uses Earth-bound lasers and a satellite-based retroreflector [22]. GNSS receivers for use in space must account for the higher Doppler shifts that occur at orbital velocities, greater variation in signal strengths, differences in atmospheric propagation errors, and the availability of signals at negative elevation angles [23, 24]. Like all space hardware, they must also be hardened against radiation and extreme temperatures.

Interplanetary craft can use two-way ranging over communication links with multiple Earth stations [25], pulsar navigation, and observations of the directions of planets and moons.

18.9 Pedestrian Navigation

Pedestrian route guidance typically requires room-level accuracy indoors and building-level accuracy outdoors with the additional requirement that the system should identify on which side of a street it is. For more demanding applications, such as positioning military and security personnel, guiding the visually impaired, and directing firefighters through smoke-filled buildings, meter-level accuracy combined with high reliability is required.

Pedestrian navigation is arguably the most challenging application of all. User equipment must be small, lightweight, low power, and, in many cases, low cost. Low

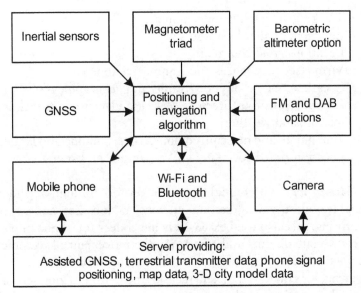

Figure 18.8 Potential smartphone navigation and positioning technology.

dynamics are combined with high vibration and there are many different routes a pedestrian may take through a given environment. Radio navigation is complex and signals are often highly attenuated.

Many different technologies have been proposed for pedestrian navigation [26, 27]. A combination of UWB ranging and foot-mounted inertial navigation is suitable for locating military, security, and firefighting personnel. However, this is unlikely to extend to mass-market applications due to the high cost of extensive UWB infrastructure and likely consumer resistance to foot-mounted equipment.

Room-level indoor positioning can be achieved using Wi-Fi RSS pattern matching, BLE, RFID, and IMES where the appropriate infrastructure is provided. Reliability can be enhanced with PDR using step detection and map matching. A group of companies have formed the In-Location Alliance to develop and standardize indoor positioning for smartphone users [28]. Outdoors, GNSS will easily provide building-level accuracy in open space, but performance is severely degraded in dense urban areas. New intelligent urban positioning technologies that combine multiconstellation GNSS with 3-D city models, PDR, and/or other radio signals may offer improved performance [29]. However, at the time of this writing, determination of the most effective techniques for pedestrian navigation was still a subject for research. Figure 18.8 depicts the navigation and positioning technology potentially available in a smartphone.

18.10 Other Applications

Technology originally designed for navigation has found many other applications. Location-based service (LBS) is the generic term for any service based on position information. GNSS, Wi-Fi, and mobile phone positioning are the main technologies used. Examples of LBS include [30, 31]:

- Tracking vehicles, people, animals, and inventory items;
- Emergency caller location;
- Location-based charging;
- Virtual security fences and fraud detection;
- Location tagging of photographs and geographic data;
- Location-sensitive directory services, advertising, travel news, parking information, and weather;
- Tour guides and museum commentaries; building and landmark identification;
- Location-based social networks, gaming, and sports.

Carrier-phase differential GNSS is a key tool for high-accuracy application such as engineering surveying, setting out construction sites, and structural deformation monitoring. It is also used extensively in geodesy for applications such as establishing coordinate datums and measuring the movement of tectonic plates, sea levels, and polar ice caps [32]. Inertial sensors are also used for underground and borehole surveying, for gravitational field measurement, for imaging sensor stabilization, and in vehicle autopilots [33].

GNSS is also used extensively for time synchronization. Telecommunications, digital broadcasting, and power distribution have all become critically dependent on it. Nations are increasingly considering GNSS and other technologies, such as ELoran, to be part of a critical positioning, navigation, and timing (PNT) infrastructure.

18.11 Future Trends

The navigation and positioning world has undergone a revolution over the past 25 years. This will continue as new application requirements drive the development of new technologies and new technologies pave the way for new applications. Based on current research trends, the following 10 areas are worth watching:

- *Sensor technology:* The cost, size, weight, and power consumption of sensors offering a given performance level are likely to drop. For example, the Defense Advanced Research Projects Agency (DARPA) is investing in new accelerometer, gyro, and clock technology [34].
- *GNSS:* New techniques that exploit the new GNSS constellations and signals to improve positioning performance are already under development.
- *Communications:* New communications standards are constantly under development and will be exploited for positioning wherever practicable.
- *Cameras:* Digital cameras are small, cheap, and likely to become increasingly ubiquitous. They can be used for navigation and positioning in many different ways and are the subject of a lot of research interest.
- *Mapping:* The increasing prevalence of 3-D mapping can be used to aid positioning and navigation in many different ways, including pedestrian map matching, signal shadowing prediction, height aiding, and terrain-referenced navigation.
- *Multisensor navigation:* Robust solutions to the most challenging navigation and positioning problems require a range of different technologies and methods

to be deployed. Consequently, multisensor integrated navigation systems are likely to become more complex, incorporating a greater number of subsystems.

- *Context:* Different navigation technologies and different configurations of those technologies are suited to different environments and different host vehicle or user behavior. Signal reception and environmental feature availability varies from place to place, while prior knowledge of motion characteristics can be used to aid navigation. An optimal navigation system should therefore detect its operating context and adapt accordingly.
- *Opportunism:* In most places, there are many different signals of opportunity and environmental features that could be used for positioning. Companies have mapped particular classes of signal (e.g., Wi-Fi or phone signals). However, using a heterogeneous mix of different signals and features offers greater flexibility and robustness, while users can build their own maps and databases.
- *Cooperation:* There is increasing interest in cooperation (or collaboration) between users. This enables rapid distribution of signal and feature information, extending to cooperative mapping, while relative positioning enables users to make use of signals and features that they cannot observe directly. Effective indoor positioning also requires cooperation between users and building operators.
- *Integrity:* As technologies mature, users expect greater reliability. Conversely, more reliable technology is trusted for use in safety-critical and mission-critical applications. Integrity is thus likely to become a driving factor for a much wider range of positioning applications in future.

Problems and exercises for this chapter are on the accompanying CD.

References

[1] *2010 Federal Radionavigation Plan*, U.S. Departments of Defense, Homeland Security, and Transportation, 2010.

[2] "Radio Navigation Aids," Annex 10, Volume 1 of *Standards and Recommended Practices (SARPS)*, 6th ed., International Civil Aviation Organization, 2012.

[3] Kayton, M., "Introduction," in *Avionics Navigation Systems*, 2nd ed., M. Kayton and W. R. Fried, (eds.), New York: Wiley, 1997, pp. 1–20.

[4] Mather, C., et al., "Performance of Integrity Monitoring Techniques for Shipboard Relative GPS Landing Systems," *Proc. ION GNSS 2005*, Long Beach, CA, September 2005, pp. 2882–2894.

[5] de Fátima Bento, M., "Unmanned Aerial Vehicles: An Overview," *Inside GNSS*, January/February 2008, pp. 54–61.

[6] Winkler, S., et al., "Improving Low-Cost GPS/MEMS-Based INS Integration for Autonomous MAV Navigation by Visual Aiding," *Proc. ION GNSS 2004*, Long Beach, CA, September 2004, pp. 1069–1075.

[7] Zhao, Y., *Vehicle Location and Navigation Systems*, Norwood, MA: Artech House, 1997.

[8] Piras, M., A. Cina, and A. Lingua, "Low Cost Mobile Mapping Systems: An Italian Experience," *Proc. IEEE/ION PLANS*, Monterey, CA, May 2008, pp. 1033–1045.

[9] Bevly, D., and S. Cobb, (eds.), *GNSS for Vehicle Control*, Norwood, MA: Artech House, 2010.

[10] *ECTS System Description*, Railway Group GE/GN8605, Rail Safety and Standards Board, February 2010.

[11] Jiang, Z., *Digital Route Model Aided Integrated Satellite Navigation and Low-Cost Inertial Sensors for High-Performance Positioning on the Railways*, Ph.D. Thesis, University College London, 2010.

[12] Grant, A., et al., "GPS Jamming and the Impact on Maritime Navigation," *Journal of Navigation*, Vol. 62, No. 2, 2009, pp. 173–187.

[13] International Maritime Organization, *Revised Maritime Policy and Requirements for a Future Global Navigation Satellite System (GNSS)*, Resolution A.915(22), November 2001.

[14] Parkins, A. J., *Performance of Precise Marine Positioning Using Future Modernised Global Satellite Positioning Systems and a Novel Partial Ambiguity Resolution Technique*, Ph.D. Thesis, University College London, 2009.

[15] Bowditch, N. A., et al., *The American Practical Navigator*, Bicentennial ed., Bethesda, MD: National Imagery and Mapping Agency, 2002.

[16] Vajda, S., and A. Zorn, "Survey of Existing and Emerging Technologies for Strategic Submarine Navigation," *Proc. IEEE PLANS*, Palm Springs, CA, April 1998, pp. 309–315.

[17] Butler, B., and R. Verrall, "Precision Hybrid Inertial/Acoustic Navigation System for a Long-Range Autonomous Underwater Vehicle," *Navigation: JION*, Vol. 48, No. 1, 2001, pp. 1–12.

[18] Hartman, R., W. Hawkinson, and K. Sweeney, "Tactical Underwater Navigation System (TUNS)," *Proc. IEEE/ION PLANS*, Monterey, CA, May 2008, pp. 898–911.

[19] Wertz, J. R., (ed.), *Spacecraft Attitude Determination and Control*, Dordecht, Netherlands: Reidel, 1980.

[20] Tapley, B. D., B. E. Schutz, and G. H. Born, *Statistical Orbit Determination*, Burlington, MA: Academic Press, 2004.

[21] Milani, A., and G. Gronchi, *Theory of Orbit Determination*, Cambridge, U.K.: Cambridge University Press, 2009.

[22] Seeber, G., *Satellite Geodesy*, New York: De Gruyter, 2003.

[23] Lightsey, E. G., "Space Applications," in *GNSS Applications and Methods*, S. Gleason and D. Gebre-Egziabher, (eds.), Norwood, MA: Artech House, 2009, pp. 329–346.

[24] Ebinuma, T., and M. Unwin, "GPS Receiver Demonstration on a Galileo Test Bed Satellite," *Journal of Navigation*, Vol. 60, No. 3, 2007, pp. 349–362.

[25] Lanyi, G. E., J. S. Border, and D. K. Shin, "Radiometric Spacecraft Tracking for Deep Space Navigation," *Proc. ION NTM*, San Diego, CA, January 2008, pp. 86–90.

[26] Grejner-Brzezinska, D. A., C. K. Toth, and S. Moafipoor, "A Step Ahead: Human Motion, Machine Learning Combine for Personal Navigation," *GPS World*, November 2008, pp. 34–41.

[27] Challamel, R., et al., "Performance Assessment of Indoor Location Technologies," *Proc. IEEE/ION PLANS*, Monterey, CA, May 2008, pp. 624–632.

[28] "Indoor Location: To Boldly Go ... with or without GNSS," *Inside GNSS*, www.insidegnss.com, September 2012.

[29] Groves, P. D., et al., "Intelligent Urban Positioning Using Multi-Constellation GNSS with 3D Mapping and NLOS Signal Detection," *Proc. ION GNSS 2012*, Nashville, TN, September 2012, pp. 458–472.

[30] Kolodziej, K. W., and J. Hjelm, *Local Positioning Systems: LBS Applications and Services*, Boca Raton, FL: CRC/Taylor and Francis, 2006.

[31] Mannings, R., *Ubiquitous Positioning*, Norwood, MA: Artech House, 2008.

[32] Rizos, C., and D. A. Grejner-Brzezinska, "Geodesy and Surveying," in *GNSS Applications and Methods*, S. Gleason and D. Gebre-Egziabher, (eds.), Norwood, MA: Artech House, 2009, pp. 347–380.

[33] Titterton, D. H., and J. L. Weston, *Strapdown Inertial Navigation Technology*, 2nd ed., Stevenage, U.K.: IEE, 2004.

[34] Shkel, A. M., "Microtechnology Comes of Age," *GPS World*, September 2011, pp. 43–50.

List of Key Symbols

Symbols appearing in multiple parts of the book are listed here. A full list may be found on the DVD, and all symbols are defined on first use. The symbols are divided into matrices, denoted by upper case bold; vectors, denoted by lower case bold; scalars, denoted by italics; subscripts and superscripts; and qualifiers. Subscripts and superscripts are only listed separately where they are used with more than one parent symbol; otherwise, the compound symbol is listed. Components of vectors and matrices are denoted by the equivalent scalar with subscript indices added. The magnitude of a vector is denoted by the equivalent scalar with no subscript index. Submatrices retain matrix notation, but have subscript indices added.

Matrices

C	coordinate transformation matrix
C	covariance matrix (general)
D	differencing matrix
F	system matrix
G_g	gyro g-dependent errors
H	measurement matrix
I_n	$n \times n$ identity matrix (diagonal elements = 1, off-diagonal elements = 0)
K	Kalman gain
M	scale-factor and cross-coupling errors
P	error covariance matrix
Q	system noise covariance matrix
Q'	approximated system noise covariance matrix
R	measurement noise covariance matrix
S	power spectral density matrix
T	position change transformation matrix
Π	cofactor matrix
Φ	transition matrix
Ω	skew-symmetric matrix of angular rate

Vectors

a	acceleration
b	bias errors
c	vector of calibration coefficients

f	specific force
f	system function
g	acceleration due to gravity
h	measurement function
k	Kalman gain for scalar measurement
l	lever arm
m	cross-coupling errors
m	magnetic flux density
m	quantities measured
p	curvilinear position (geodetic latitude, longitude, and geodetic height)
q	quaternion attitude
r	Cartesian position
s	scale-factor errors
$\mathbf{s}_{cg}$	satellite clock g-dependent error coefficients
u	unit vector and line-of-sight unit vector
v	velocity
w	vector of white noise sources
$\mathbf{w}_m$	measurement noise vector
$\mathbf{w}_s$	system noise vector
x	state vector
z	measurement vector
$\boldsymbol{\alpha}$	attitude increment
$\boldsymbol{\gamma}$	acceleration due to the gravitational force
$\Delta \mathbf{r}$	position displacement
$\delta \mathbf{x}$	state vector residual
$\delta \mathbf{z}^-$	measurement innovation
$\delta \mathbf{z}^+$	measurement residual
$\boldsymbol{\upsilon}$	integrated specific force
$\boldsymbol{\psi}$	Euler attitude {roll, pitch, yaw} (no superscript)
$\boldsymbol{\psi}$	Small-angle attitude (superscript indicates resolving axes)
$\boldsymbol{\omega}$	angular rate

Scalars

A	constant proportional to signal amplitude
a_f	satellite clock calibration coefficient
b	bias error
B	double-sided noise bandwidth
B_{L_CA}	carrier-phase tracking-loop bandwidth
B_{L_CF}	carrier-frequency tracking-loop bandwidth
B_{L_CO}	code tracking-loop bandwidth
B_{PC}	double-sided precorrelation bandwidth
C	spreading code
c	speed of light in free space or fiber-optic coil
c	calibration coefficient
C/N_0	$10\log_{10}$ carrier power to noise density

List of Key Symbols

c/n_0	carrier power to noise density
D	navigation data message
D	dilution of precision
D	code discriminator function
d	spacing of early and late correlation channels in code chips
d	depth
e	eccentricity of the ellipsoid
F	carrier-frequency discriminator function
f	frequency
f	probability density function
H	orthometric height
h	geodetic height
I	in-phase sample
k	discriminator gain
L	geodetic latitude
l	number of system-noise-vector components
m	number of measurement-vector components
m	quantity measured
m	number of transmitters
N	geoid height
N	number of samples
N	sample from Gaussian distribution
N	integer ambiguity
n	number of state-vector components
n	number of steps
P	carrier-phase discriminator function
p	pressure
p	hypothesis probability
Q	quadraphase sample
r	(geometric) range
r	iteration counter in summation
R_E	transverse radius of curvature
R_N	meridian radius of curvature
R_P	polar Earth radius
R_0	equatorial Earth radius
S	subcarrier function
S	power spectral density or cross spectral density
s	scale factor error
T	track width
t	time
t_{sa}	time of signal arrival
t_{st}	time of signal transmission
W	weighting factor
w	white noise source
w_ρ	pseudo-range tracking error
w_Φ	carrier phase tracking error expressed as a range
w_r	pseudo-range rate tracking error

Symbol	Description
x	first component of Cartesian position or a generic vector
x	code tracking error in code chips
y	second component of Cartesian position or a generic vector
z	third component of Cartesian position or a generic vector
α	magnetic declination angle/variation
β	magnitude of the projection of position onto the equatorial plane
γ	magnetic inclination/dip angle
Δf	Doppler frequency shift
Δr	distance traveled
δr_L	linearization error
δt	time increment
δt_c	clock offset (time)
$\delta \rho_c$	range error due to clock offset
$\Delta \rho_{dc}$	differential correction
$\delta \rho_e$	range error due to ephemeris data
$\delta \rho_I$	ionosphere propagation error
$\delta \rho_{ie}$	Sagnac correction
$\delta \rho_M$	range error due to multipath
$\delta \rho_T$	troposphere propagation error
$\delta \Phi_b$	phase bias, expressed as a range
$\delta \Phi_I$	phase error due to ionosphere propagation, expressed as a range
$\delta \Phi_p$	line-of-sight-dependent phase wind-up error, expressed as a range
θ	pitch or elevation angle
θ_{nu}	elevation angle of satellite line-of-sight vector
λ	longitude
Λ	likelihood
λ_{ca}	carrier wavelength
μ	magnitude of angle or rotation
μ	Earth's gravitational constant
ρ	pseudo-range
σ	standard deviation or error standard deviation
σ_{IQ}	noise standard deviation of accumulated correlator outputs
τ	correlation time
τ	propagation time
τ_a	correlator accumulation interval
τ_i	inertial navigation integration interval
τ_s	system propagation time
Φ	accumulated delta range
ϕ	roll or bank angle
ϕ	phase
ϕ_0	phase offset
ψ	yaw or heading angle
ψ_{bb}	boresight angle
ψ_{mu}	bearing of line-of-sight vector with respect to magnetic north
ψ_{nu}	azimuth angle of line-of-sight vector
ω	angular frequency

List of Key Symbols

Subscripts and Superscripts

A	denotes accelerometer indicated
A	denotes attitude-matching transfer alignment measurement
a	denotes accelerometer
a	denotes user antenna body coordinate frame
a	denotes user feature-matching sensor coordinate frame (in Section 16.3)
B	denotes barometric height measurement
b	denotes body or INS body coordinate frame
b	denotes barometric altimeter
bad	denotes accelerometer dynamic bias
bgd	denotes gyro dynamic bias
C	denotes corrected
c	denotes camera or feature-matching sensor body coordinate frame
ca	denotes carrier or carrier phase
cf	denotes carrier frequency
cf	denotes clock frequency drift
co	denotes code
$c\phi$	denotes clock phase drift
D	denotes down component
D	denotes Doppler measurement
D	denotes database-indicated
d	denotes dynamic
DC	denotes differentially corrected
E	denotes east component
E	denotes early correlation channel
E	denotes Earth's geomagnetic field
e	denotes Earth-centered Earth-fixed coordinate frame
ECD	denotes envelope-to-cycle difference
F	denotes feature-matching measurement
f	denotes front-wheel coordinate frame
f	denotes finish-point coordinate frame
f	denotes feature coordinate frame
G	denotes resultant position and time
G	denoted GNSS-derived
g	denotes gyro
$GNSS$	denotes GNSS partition
H	denotes horizontal
h	denotes height
h	denotes hard-iron
I	denotes ECI frame synchronized with ECEF at time of signal arrival
I	denotes in-phase
I	denotes INS-derived
i	generic index
i	denotes Earth-centered inertial coordinate frame
INS	denotes INS partition

j	satellite, signal, or tracking channel number
k	iteration index for Kalman filter or tracking loop
L	denotes late correlation channel
L	denotes latitude
L	denotes left (wheel)
L	denotes leveling measurement
L	denotes line-fix measurement
l	denotes local tangent plane coordinate frame
l	denotes a particular signal from a given satellite or other transmitter
$L\lambda$	denotes latitude and longitude
M	denotes magnetic heading measurement or error states
M	denotes due to multipath
m	denotes Markov process
m	denotes magnetometer-measured flux density and frame thereof
m	denotes magnetometer
m	denotes coordinate frame of nearest point on a line
N	denotes north component
N	denotes ambiguity states
n	denotes local navigation coordinate frame
Nav	denotes navigation solution
O	denotes odometry measurement
o	denotes orbital coordinate frame
o	denotes odometer
P	denotes position
P	denotes prompt correlation channel
P	denotes PDR measurement
P	denotes prefilter
p	denotes planar coordinate frame
p	denotes curvilinear position
Q	denotes quadraphase
Q	denotes quasi-stationary alignment measurement
q	denotes coordinate frame of intersection point on a line
R	denotes right (wheel)
R	denotes terrestrial radio navigation and measurement thereof
R	denotes feature-matching measurement (in Section 16.3)
R	denotes raw (i.e. uncorrected)
r	denotes pseudo-range rate
r	denotes rear-wheel coordinate frame
r	denotes receiver
r	denotes reference receiver or reference antenna body frame
r	denotes reference body coordinate frame
ra	denotes accelerometer random noise
Ref	denotes reference navigation system
rg	denotes gyro random noise
S	denotes a point on the Earth's ellipsoidal surface
s	denotes a point on the Earth's geoid surface or water surface
s	denotes static

s	denotes satellite body coordinate frame
s	denotes soft-iron
s	denotes scattering-surface coordinate frame
s	denotes start-point coordinate frame
Sensor	denotes sensor
T	denotes the transpose of a matrix
T	denotes TRN measurement
t	denotes transmitter or transmit antenna body frame
t	denotes feature or feature body frame (in Section 16.3)
u	denotes user receiver
V	denotes velocity-matching transfer alignment measurement
V	denotes vertical
v	denotes velocity
VC	denotes velocity constraint measurement
w	denotes wander-azimuth coordinate frame
x	denotes first component of a vector or axis
y	denotes second component of a vector or axis
z	denotes third component of a vector or axis
ZA	denotes ZARU measurement
ZV	denotes ZVU measurement
α	denotes a generic object frame
β	denotes a generic reference frame
γ	denotes a generic set of resolving axes or frame
Δ	denotes differenced measurement or subcarrier
δ	denotes a generic coordinate frame
ε	denotes measurement residual of single-epoch navigation solution
θ	denotes elevation measurement
λ	denotes longitude
ρ	denotes pseudo-range
Φ	denotes accumulated delta range
ψ	denotes attitude/ GNSS attitude/ heading/azimuth measurement
0	denotes value at the geoid
0	denotes initialization value
0	denotes a constant value
0	denotes at the reference time
$-$	denotes after state propagation and before measurement update
$-$	denotes prediction
$+$	denotes after measurement update
$\perp$	denotes perpendicular

Qualifiers

$E()$	expectation operator
δ	denotes a small increment or error
Δ	denotes an increment, change, or difference across transmitters
∇	denotes a difference across receivers

′	denotes alternative version
″	denotes alternative version
^	denotes estimate
ˇ	denotes ambiguity-fixed estimate
~	denotes a navigation system measurement
−	denotes average value
(−), −	denotes at the beginning of the navigation processing cycle
(+), +	denotes at the end of the navigation processing cycle

Acronyms and Abbreviations

Acronyms and abbreviations appearing in multiple parts of the book are listed here. A full list may be found on the accompanying DVD and all acronyms are defined on first use.

2-D	Two dimensional
3-D	Three dimensional
ADC	Analog-to-digital converter
ADR	Accumulated delta range
AGC	Automatic gain control
AGNSS	Assisted GNSS
AHRS	Attitude and heading reference system
AltBOC	Alternate binary offset carrier
AM	Amplitude modulation
ANN	Artificial neural network
AOA	Angle of arrival
ASF	Additional secondary factor
ATAN	Arctangent
ATAN2	Arctangent (four quadrant)
AUV	Autonomous underwater vehicle
baro	barometric altimeter
BLE	Bluetooth low energy
BOC	Binary offset code
BPSK	Biphase shift key
C/A	Coarse/acquisition
CD	Compact disc
CDMA	Code-division multiple access
CGCS	China Geodetic Coordinate System
CIRS	Conventional inertial reference system
CL	Civil long
CM	Civil moderate
CRPA	Controlled-reception-pattern antenna
CSAC	Chip-scale atomic clock
CVL	Correlation velocity log
CVN	Continuous visual navigation
DGNSS	Differential GNSS
DLL	Delay lock loop
DLoran	Differential long-range navigation
DME	Distance measuring equipment

DOP	Dilution of precision
DPP	Dot-product power
DR	Dead reckoning
DSRC	Dedicated short-range communication
DSSS	Direct-sequence spread spectrum
DVL	Doppler velocity log
ECEF	Earth-centered earth-fixed
ECI	Earth-centered inertial
EGNOS	European Geostationary Navigation Overlay System
EKF	Extended Kalman filter
ELE	Early-minus-late envelope
ELoran	Enhanced long-range navigation
ELP	Early-minus-late power
FDMA	Frequency-division multiple access
FDE	Fault detection and exclusion
FDI	Fault detection and isolation
FDR	Fault detection and recovery
FFR	Federated fusion-reset
FFT	Fast Fourier transform
FH	Frequency hopping
FLL	Frequency lock loop
FM	Frequency modulation
FOC	Full operational capability
FWHM	Full width at half maximum
GAGAN	GPS/GLONASS and Geo-augmented Navigation
GBAS	Ground-based augmentation system
GDOP	Geometric dilution of precision
GIS	Geographic information system
GLONASS	GLObal'naya NAvigatsionnaya Sputnikovaya Sistema
GNSS	Global navigation satellite systems
GPS	Global Positioning System
GSM	Global standard for mobile communications
GSM-R	Global standard for mobile communications—Rail
GTRF	Galileo Terrestrial Reference Frame
gyro	Gyroscope
HAIP	High-accuracy indoor positioning
HAL	Horizontal alert limit
HDOP	Horizontal dilution of precision
HPL	Horizontal protection level
IAE	Innovation-based adaptive estimation
ICD	Interface control document
ID	Identification
IERS	International Earth Rotation and reference systems Service
IF	Intermediate frequency
IFOG	Interferometric fiber-optic gyro
IGMAP	Iterative Gaussian mixture approximation of the posterior
ILS	Instrument landing system; iterated least squares

Acronyms and Abbreviations

IMES	Indoor Messaging System
IMU	Inertial measurement unit
INS	Inertial navigation system
IOC	Initial operational capability
IRM	IERS reference meridian
IRNSS	Indian Regional Navigation Satellite System
IRP	IERS reference pole
IS	Interface standard
ISM	Industrial, scientific, and medical
ITRF	International Terrestrial Reference Frame
ITS	Intelligent transportation systems
JTIDS	Joint tactical information distribution system
KF	Kalman filter
KNN	K nearest neighbors
laser	Light amplification by stimulated emission and radiation
LADAR	Laser detection and ranging
LADGNSS	Local area differential GNSS
LBL	Long baseline
LBS	Location-based services
LHCP	Left-handed circular polarization
LEO	Low Earth orbit
LF	Low frequency
LIDAR	Light detection and ranging
LOP	Line of position
Loran	Long-range navigation
LOS	Line of sight
L1/2/3/5	Link 1/2/3/5
L1C	Link 1 civil
L2C	Link 2 civil
L5I	Link 5 in-phase
L5Q	Link 5 quadraphase
M	Military
MAV	Micro air vehicle
MBOC	Multiplexed binary offset carrier
MEDLL	Multipath-estimating delay lock loop
MEMS	Microelectromechanical systems
MEO	Medium Earth orbit
MF	Medium frequency
MHKF	Multiple-hypothesis Kalman filter
MIDS	Multifunctional information distribution system
MMAE	Multiple-model adaptive estimation
MSAS	MTSat Satellite Augmentation System
NATO	North Atlantic Treaty Organization
NCO	Numerically controlled oscillator
NDB	Nondirectional beacon
NGA	National Geospatial-Intelligence Agency
NLOS	Nonline-of-sight

OFDM	Orthogonal frequency division multiplex
P	Precise
PDF	Probability density function
PDOP	Position dilution of precision
PDR	Pedestrian dead reckoning
PLL	Phase lock loop
PNT	Positioning, navigation, and timing
ppm	parts per million
PPP	Precise point positioning
PPS	Precise positioning service
PRN	Pseudo-random noise
PRS	Public regulated service
P(Y)	Precise (encrypted precise)
QPSK	Quadrature-phase shift key
QZSS	Quasi-Zenith Satellite System
radalt	radar altimeter
RAIM	Receiver autonomous integrity monitoring
RDSS	Radio determination satellite service
RF	Radio frequency
RFID	Radio-frequency identification
RHCP	Right-handed circular polarization
RLG	Ring laser gyro
RMS	Root mean square
RNP	Required navigation performance
ROV	Remotely operated vehicle
RSS	Received signal strength
RTK	Real-time kinematic
RTT	Round-trip time
SBAS	Space-based augmentation system
SBL	Short baseline
SD	Standard deviation
SDCM	System of Differential Corrections and Monitoring
SI	Système International d'unités
SLAM	Simultaneous localization and mapping
SNAS	Satellite Navigation Augmentation System
SOL	Safety of life
SOOP	Signal(s) of opportunity
SOP	Surface of position
SPS	Standard positioning service
TACAN	Tactical air navigation
TBD	To be determined
TCXO	Temperature-compensated crystal oscillator
TDM	Time-division multiplex
TDMA	Time-division multiple access
TDOA	Time difference of arrival
TDOP	Time dilution of precision
TOA	Time of arrival

TOF	Time of flight
TRN	Terrain-referenced navigation
TTFF	Time to first fix
VHF	Very high frequency
UAV	Unmanned air vehicle
UDA	Undefined acronym
UHF	Ultra-high frequency
UKF	Unscented Kalman filter
USBL	Ultra-short baseline
UTC	Coordinated Universal Time
UWB	Ultra-wideband
VAL	Vertical alert limit
VBA	Vibrating-beam accelerometer
VDOP	Vertical dilution of precision
VHF	Very high frequency
VOR	VHF omnidirectional radiorange
VPL	Vertical protection level
VRE	Vibration rectification error
V2I	Vehicle to infrastructure
V2V	Vehicle to vehicle
WAAS	Wide Area Augmentation System
WGS 84	World Geodetic System 1984
Wi-Fi	Wireless Fidelity
WLAN	Wireless local area network
WPAN	Wireless personal area network
WSS	Wheel speed sensor
Y	Encrypted precise
ZARU	Zero angular rate update
ZVU	Zero velocity update

About the Author

Paul D. Groves holds a B.A. (with first class honors) in physics, an M.A. (Oxon), and a D.Phil. (Doctor of Philosophy) in experimental atomic and laser physics, all from the University of Oxford. He has been active in navigation systems research and development since joining the Defence Evaluation and Research Agency in January 1997. He transferred to QinetiQ Ltd. when it was formed in July 2001. Since October 2009, he has been a lecturer (academic faculty member) at University College London, leading a program of research into robust positioning and navigation.

Dr. Groves is interested in all navigation and positioning technologies. He has worked on pedestrian, road vehicle, aircraft, guided-weapon, ship, and autonomous-underwater-vehicle applications. He is an inventor of the GNSS shadow-matching technique and has contributed to innovations in terrain-referenced navigation, visual navigation, positioning using AM radio broadcasts, heterogeneous feature-matching, and detection of GNSS NLOS reception and multipath interference. He has developed algorithms for all forms of INS/GNSS integration, multisensor integration, transfer alignment, quasi-stationary alignment, zero-velocity updates, pedestrian dead reckoning, GNSS C/N_0 measurement, and inertial navigation. He also has extensive experience in GNSS and IMU software simulation.

Dr. Groves is a Fellow of the Royal Institute of Navigation and has served on its Technical, R&D Group, and NAV Conference Committees, including periods as secretary and chair of the R&D committee. He is also an active member of the Institute of Navigation and a chartered member of the Institute of Physics. He has served as an associate editor of *Navigation: Journal of the ION* and the *IEEE Transactions on Aerospace and Electronic Systems* and has been a reviewer for many different journals.

DVD Contents

Appendix A. Vectors and Matrices

A.1	Introduction to Vectors	A-1
A.2	Introduction to Matrices	A-3
A.3	Special Matrix Types	A-5
A.4	Matrix Inversion	A-6
A.5	Calculus	A-7
A.6	Eigenvalues, Eigenvectors and Matrix Factorization	A-7
	References	A-10

Appendix B. Statistical Measures, Probability, and Random Processes

B.1	Statistical Measures of Data	B-1
B.2	Probability	B-2
B.3	Standard Distributions	B-11
B.4	Random Processes	B-15
B.5	Hypothesis Testing	B-20
	References	B-24

Appendix C. Position Representations, Transformations, and Conversions

C.1	Datums	C-1
C.2	Cartesian to Curvilinear Coordinate Conversion Methods	C-2
C.3	Normal Vector	C-5
C.4	Transverse Mercator Projection	C-7
	References	C-9

Appendix D. Additional Topics on State Estimation

D.1	Least-Squares Estimation	D-1
D.2	Schmidt-Kalman Filter	D-3
D.3	Particle Filtering Topics	D-4
	References	D-7

Appendix E. Additional Topics on Inertial Sensors and Inertial Navigation

E.1	New Inertial Sensor Technology	E-1
E.2	Spinning-Mass Gyroscopes	E-3
E.3	All-Accelerometer IMUs	E-6
E.4	History of Inertial Navigation	E-7
E.5	Platform INS	E-7
E.6	Quaternion Navigation Equations Implementation	E-8
E.7	Local Tangent-Plane Frame Navigation Equations	E-14
E.8	Navigation Equations Iteration Rate Issues	E-15
E.9	Effects of Timing Errors	E-16
E.10	Pseudo-measurement Generation using Motion Constraints	E-16
	References	E-18

Appendix F. Additional Topics on Radio Positioning

F.1	Angular Positioning using Iterated Least Squares	F-1
F.2	Radio Determination Satellite Service	F-3
F.3	Landing Guidance Systems	F-4
F.4	Radio Tracking Systems	F-5
F.5	Phone Positioning Terminology	F-7
F.6	Digital Television and Radio Broadcasts	F-8
	References	F-9

Appendix G. Additional Topics on GNSS

G.1	GNSS Space and Control Segments	G-2
G.2	GNSS Development History	G-5
G.3	GNSS Signal Design Notes	G-8
G.4	Satellite Position, Velocity, and Acceleration	G-11
G.5	Relativistic Frequency Shift	G-14
G.6	Receiver and Ranging Processor Design Notes	G-14
G.7	Range Error Corrections	G-19
G.8	Navigation Processor	G-23
G.9	Integer Wavelength Ambiguity Resolution	G-29
G.10	Vector Tracking and Acquisition	G-35
G.11	Multipath and NLOS Mitigation	G-36
G.12	Positioning using Ambiguous Pseudo-ranges	G-37
G.13	GNSS Repeaters	G-37
	References	G-37

Appendix H. Additional Topics on Environmental Feature Matching

H.1	Continuous Visual Navigation	H-1
H.2	Alternative Environmental Features	H-2
	References	H-3

Appendix I. Additional Topics on Integration and Alignment

I.1	Higher-Order INS State Transition Matrices	I-2
I.2	Local-Navigation-Frame Cartesian Position Error	I-4
I.3	Wander-Azimuth Implementation	I-9
I.4	Local Tangent-Plane Implementation	I-16
I.5	Body-Frame-Resolved Attitude Error	I-18
I.6	Estimation of the Time Synchronization Error	I-26
I.7	Integer Wavelength Ambiguity Resolution	I-32
I.8	Dead Reckoning, Attitude, and Height	I-35
I.9	Terrestrial Radio Navigation and Environmental Feature Matching	I-39
	References	I-42

Appendix J. Simulation

J.1	Reasons for Simulation	J-1
J.2	Simulation Architecture and Fidelity	J-2
J.3	Truth Model	J-6
J.4	Error Generation	J-9
J.5	Monte Carlo Simulation	J-12
	References	J-12

Appendix K. Historical Navigation Systems

K.1	From Navigation Aids to Navigation Systems	K-2
K.2	Very Low Frequency Navigation	K-3
K.3	Decca Navigator System	K-5
K.4	Loran	K-7
K.5	Transit and Tsikada	K-9
K.6	Other Radio Navigation Systems	K-11
K.7	Celestial Navigation, Dead Reckoning, and Attitude Determination	K-14
	References	K-16

Problems and Exercises

Full List of Symbols
Full List of Acronyms and Abbreviations
Guide to MATLAB Simulation Software
Guide for First Edition Readers

Example 2.1: Attitude Conversion
Example 2.2: Cartesian and Curvilinear Position Conversion
Example 2.3: Calculation of Gravity
Example 2.4: Transformations between ECEF and ECI Frames
Example 2.5: Transformations between ECEF and Local Navigation Frames
Example 2.6: Transformations between ECI and Local Navigation Frames

Example 3.1: Kalman filter estimating single-axis position and velocity
Example 3.2: Kalman filter estimating 2D position

Example 4.1: Accelerometer Error

Example 5.1: Inertial Navigation in Two Dimensions
Example 5.2: ECI-Frame Inertial Navigation Update Cycle
Example 5.3: Leveling and Direct Gyrocompassing
Example 5.4: Short-term INS Error Growth

Example 6.1: Magnetic Heading
Example 6.2: Rear-wheel Odometry

Example 7.1: Exact Positioning from Ranging in Two Dimensions
Example 7.2: Least-squares Positioning from Ranging in Two Dimensions

Example 8.1: Calculation of Range, Line of sight, Range-rate, and Elevation

Example 9.1: Least-squares Positioning from GNSS Pseudo-range Measurements
Example 9.2: Dilution of Precision Calculations

Index

Acceleration, 50–51
Accelerometers, 6, 139–142, Appendix E
  errors, 150–160, 203–214, 574–576, 581, 589–593
  leveling, 197–198, 226–227, 670–672
Accumulated delta range (ADR), 388
Accumulation of correlator outputs (GNSS), 355–359, 364–366
Accuracy (RNP), 720
Acoustic ranging, 509–512
Acquisition, 280
  aiding, 463
  GNSS, 367–372, 455–456, Appendix G
Adaptive Kalman filter, 124–125
Additional secondary factor, 485, 487
Aid to navigation, 4
Aiding, 18–19, 463–464, 597–571, 698–699
Air navigation, 731–733
Aircraft-based augmentation system (ABAS), 316
Aircraft directional gyro, Appendix E
Aircraft navigation, 473–481, 731–732
Air data, 249
Alignment
  coarse, 196–200, 634–637
  fine, 200–203, 637
  GNSS, 201–202
  ground, 201–202
  of INS, 196–202, 627–637
  quasi-stationary, 634–637
  rapid, 631–633
  self, 196–200
  transfer, 201–202, 627–634
Almanac, 327–328
Altimeter
  barometric, 212, 230–231
  radar, 232
  ultrasonic, 233
Altitude. *See* height
Ambient sound, Appendix H
Ambiguity, 260, 276
  GNSS carrier phase, 442–449, 541, 616–617, Appendix G
Ambiguous measurements, 125–129, 260, 695–696
Amplitude modulation (AM) radio, 492–493
Angle of arrival, 9, 269
Angular positioning, 9–10, 269–271, 690–694, Appendix F
Angular rate, 44–46, 142–149
  Earth, 66–67
Analog-to-digital converter (ADC), 280, 353, 355
Antennas, 279
  choke-ring (GNSS), 459
  dual polarization, 462
  controlled-reception-pattern, 271, 452–453, 460
  GNSS, 350–351, 452–453, 459–460
  synthetic aperture, 453, 460
Antispoofing, 321
Assistance, 19, 464
Assisted GNSS, 464
Astronomical latitude, 57
Atom interferometry, Appendix E

Atomic clock, 352
Attitude, 30–43, 217–229
   GNSS, 450–451, 618–619
   increment, 149, 183, 185
   interpolation, 43
   update matrix, 183–184, 192–193
Attitude and heading reference system (AHRS), 228–229
   integration, 672
Augmentation systems, 314–316, 327, 329, 332
Automatic Identification System (AIS), 492
Automatic gain control (AGC), 355
Availability, 723
Aviation, 731–733
Axis of rotation, 39, 40, 42
Azimuth, 34, 269, 344, 690–694

Bandwidth
   carrier-tracking, 383
   code-tracking, 375
   precorrelation, 353–354, 360–361, 404–405
   predetection, 366
Bank, 34
Barometric altimeter (baro), 212, 230–231
   integration, 673
   terrain-referenced navigation, 537
Baseband signal processor, 279–280
   GNSS, 355–366
Batch processing (GNSS), 454–455
Bearing, 9–10, 269, 690–694
Beidou, 314, 326, 330–332, Appendix F.
   See also Global Navigation Satellite Systems
Bias errors, 152–153, 205–206, 210, 581, 589–592
Binary offset carrier (BOC), 318–326, 353, 361–362, 371–372, 377, Appendix G
Bi-phase shift key (BPSK) modulation, 303, 318–327, 359–361
Bluetooth, 507–508
Bluetooth Low Energy, 508–509

Body frame, 28–29
Boeing Timing and Navigation (BTN), 491–492
Boresight error, 245, 677–680, Appendix I
Broadcasting, 492–494, Appendix F

C/A code, 303, 320–323
Camera, 539–542, 546–547
Carouseling IMU, 214–215
Carrier
   discriminators (GNSS), 379–382, 398–399
   INS/GNSS integration, 616–617
   phase measurements (GNSS), 388
   phase positioning, 276
   phase positioning (GNSS), 442–451
   tracking (GNSS), 377–384, 398–401, 454–458
   wipe-off (GNSS), 356–357
Carrier power to noise density (C/N_0), 362–363, 386–387, Appendix G
Cascaded integration, 648–653, 658
Celestial navigation, 548–550, Appendix K
Cell ID, 489
Cellphone, 463–464, 488–491, Appendix F
Centralized integration, 651, 654–655
Centrifugal force and acceleration, 51–53, 70–71, 175, 179
Centripetal force and acceleration, 51
Chayka, 482–483
Chi-square distribution, Appendix B
Chip-scale atomic clock (CSAC), 352, Appendix G
Circular error probable (CEP), Appendix B
Clock
   GNSS receiver, 305, 308–309, 351–352
   GNSS satellite, 390–391
   receiver, 262–264, 280
   synchronization, 262–264
   transmitter, 262–264
Code (GNSS), 303–305, 317–327, Appendix G

Index

correlation, 357–363
discriminators, 373–375, 396, 460
tracking, 372–377, 395–397, 454–458
Coherent and noncoherent integration, 365, 455
Cold-atom sensor, Appendix E
Cold start, 369
Collaborative positioning. *See* Cooperative positioning
Compass navigation system. *See* Beidou
Compass (magnetic), 218–222
integration, 668–669
Coning, 191
Consistency checks, 712–718
Containment intersection, 7–8, 259
Context, 17, 197, 215, 242–244, 517, 519, 529, 638–641, 665–666, 705
Continuity, 722–723
Continuous Visual Navigation (CVN), 545, 696, Appendix H
Controlled-reception-pattern antenna (CRPA), 271, 452–453, 460
Conventional Inertial Reference System (CIRS), 26
Conventional Terrestrial Pole (CTP), 26
Conventional Terrestrial reference System (CTRS), 27
Conventional Zero Meridian (CZM), 27, 58
Cooperative positioning, 19–20, 257–258, 463, 464, 673, 687, 697
Coordinate frame, 23–30
transformations, 36–37, 72–78
Coordinate system, 23–30
Coordinate transformation matrix, 35–40
normalization, 187
orthogonalization, 186–187
time derivative, 45–46
Coriolis force and acceleration, 51–53, 175, 179, 188, 211
Correlation of signals, 280–281, 290–291, 494

GNSS, 355–366
semi-codeless, 366
Correlation velocity log (CVL), 249, 250
Correlators (GNSS), 355–366
Covariance, Appendix B
Covariance Factorization, 114
Cross-coupling error, 154–155, 213–214, 589, 592
Crystal oscillator, 352
Cumulative distribution function, Appendix B
Cycle slip, 385, 464

Datum, 55–56, Appendix C
Data synchronization. *See* Time Synchronization
Day (sidereal and solar), 67
Dead reckoning, 3–7, 233–250, 666–667, 674–682
Decca, Appendix K
Dedicated short-range communication (DSRC), 509
Deeply coupled (deep) INS/GNSS integration, 561, 571–573, 606–614
Delay Lock Loop (DLL), 372–377
Delta range
GNSS, 388
TDOA, 264
Depth pressure sensor, 231–232
Differential GNSS, 437–441
integration with INS, 615
Differential Loran, 488
Differential positioning, 264–266
Diffraction, 290, 402
Digital Audio Broadcasting (DAB), 494, Appendix F
Digital road map, 520–521
Digital television, 493–494, Appendix F
Dilution of precision (DOP), 292–297, 424–429
Direction cosine matrix, 36
Direction finding, 9–10, 12, 270–271
Directional gyro, Appendix E

Discriminator functions, 373–375, 379–382, 396, 398–399, 460
Distance Measuring Equipment (DME), 8, 12, 474–480, 690
Doppler
  effect, 245
  integration, 680–681
  positioning, 11, 274–276
  radar, 6, 245–248, 702–703
  shift, 354, 376, 387–388, Appendix G
  sonar, 245–249, 702–703
  wipe-off, 356–357
Dynamic positioning, 2, 442

Earth, 53–72
  geomagnetic field, 218–219
  rotation, 66–67, 173, 177, 198–200
  surface, 54–56
  tides, 64–65
Earth-centered Earth-fixed (ECEF) frame, 26–27, 73–76, 172–176, 577–582
Earth-centered inertial (ECI) frame, 25–26, 73–76, 168–172, 582–584
Eccentricity of the ellipsoid, 54, 56
Electrolytic tilt sensor, 227
Elevation (angle), 34, 344, 690–694
Elevation (terrain), 64–65
Ellipsoid, 54–56
Enhanced Loran (ELoran), 481–488
Enhanced Position Location Reporting System (EPLRS), Appendix F
Environmental feature matching, 7–12, 14–15, 517–552
  failure modes, 704
  integration, 682–699
Ephemeris, 309, 327–329, 332–333
  errors, 390–391, 438
Equipotential surface, 64
Error, 87–88
  accelerometer, 151–160, 203–214, 574–576, 581, 589–593
  barometric altimeter (baro), 230, 673

covariance matrix, 89–90, 98–100, 109–110
diffraction, 290, 402
DME, 478
Doppler radar and sonar, 247–249, 680–681
ephemeris, 390–391, 438
generation, Appendix J
GNSS, 309–311, 342–343, 389–407, 415–418, 421–424, 593–594
gyroscope, 151–160, 203–214, 574–576, 581, 589–593
IMU, 151–160, 203–214, 574–576, 581, 589–593, 621–622
INS, 191–192, 203–214, 574–593
ionosphere, 287, 391–395, 438
Loran, 487–488
magnetic heading, 219–222, 668–669
multipath, 290, 401–407, 458–462
nonline-of-sight (NLOS), 401–402, 409, 458–462
odometer, 237–239, 674–677
pedestrian dead reckoning, 242–243, 680
radio navigation, 287–292
receiver clock, 352, 415–418, 576, 593–594
satellite clock, 390–391, 438
signal in space, 391
tracking, 290–291, 359, 395–401
transmitter location and clock, 292
troposphere, 287–288, 392, 394–395, 438
wheel speed sensor, 237–239, 674–677
Error-state integration, 652–659, 661–663
Euler attitude, 33–35
Euler force, 51, 53
Eurofix, 484
European Geostationary Navigation Overlay System (EGNOS), 315, 327, 329, 332
Extended Kalman filter (EKF), 118–121

GNSS, 419–421
Extended range tracking, 454–455

Failure modes, 702–705
Fault detection, 20, 701–723
Fault exclusion, 20, 701, 718, 722
Fault isolation, 20, 701, 715–718
Fault recovery, 701
Federated integration, 655–658
Feature matching. *See* Environmental feature matching
Fiber-optic gyro (FOG), 145–146
Field trials, 724
Filtered integration, 652–663
Fingerprinting, 271–274
Flattening of the ellipsoid, 54, 56
Float solution, 446–447
Force-feedback accelerometer, 141
Foucault frequency, 211
Frame transformations, 36–37, 72–78
Frequency lock loop (FLL), 377–384
Frequency modulation (FM) radio, 493
Front-end, 279–280
   GNSS receiver, 352–355

Galileo, 13, 313–314, 324–326, 329, Appendix G.
   See also Global Navigation Satellite Systems
Gaussian distribution, Appendix B
Gauss-Markov sequence, 88, Appendix B
G-dependent errors, 157, 589, 592
Geocentric frame, 29
Geocentric latitude, 57–58
Geodetic frame, 27.
   See also Local navigation frame
Geographic frame, 27.
   See also Local navigation frame
Geoid, 64
Geomagnetic field, 218–219
Geometric dilution of precision (GDOP), 424–427
Global Navigation Satellite Systems (GNSS), 12–13, 299–466, Appendix G

acquisition, 367–372, 455–456
aiding, 463–464, 569–571
antennas, 350–351, 452–453, 459–460
assisted, 464
attitude, 450–451, 618–619
attenuation, 452
carrier-phase positioning, 442–451
control segment, 301–302, Appendix G
differential, 437–441
error sources, 309–311, 342–343, 389–407, 415–418, 421–424, 593–594
failure modes, 703
integration, 201–202, 559–622
integrity monitoring, 701–703, 705–723
interference, 452
jamming, 452
navigation data message, 327–330, 365, 385–386
navigation solution, 307–309, 407–424, 456–458, 461–462, Appendix G
orbit prediction, 465
positioning, 307–309, 407–424, 456–458, 461–462, Appendix G
ranging, 304–309, 339–344, 367–389
receiver, 302–307, 350–366, 453–458, 460–461
receiver autonomous integrity monitoring (RAIM), 713–715
relative, 441–442
repeaters, Appendix G
satellites, 300–301, 330–339, Appendix G
shadow matching, 465–466
signals, 303–305, 317–327, Appendix G
signal geometry, 424–429
space segment, 300–301, 330–339, Appendix G
system architecture, 300–303
time to first fix (TTFF), 311, 464

Global Navigation Satellite Systems (GNSS) *(Cont.)*
  tracking, 372–384, 395–401, 454–458
  user equipment, 302–303
  vector tracking, 456–458
Global Positioning System (GPS), 12–13, 312, 320–323, 327–328, Appendix G.
  See also Global Navigation Satellite Systems
GLONASS, 13, 313, 323–324, 328, Appendix G.
  See also Global Navigation Satellite Systems
GNSS/INS integration. *See* INS/GNSS integration
Goniometer, 270
GPS/GLONASS and GEO-Augmented Navigation (GAGAN), 315, 327, 329, 332
Gravitation, 67–72
Gravity, 68–72, 209–212
  gradiometry, 551
  potential, 64
Ground-based augmentation system (GBAS), 315–316
Ground wave, 288, 481–488
Guided weapons, 733
Gyrocompass, 222–223
  strapdown, 228–229
Gyrocompassing, 198–200
Gyroscopes, 142–149, Appendix E
  errors, 151–160, 203–214, 574–576, 581, 589–593
  directional, Appendix E
  strapdown yaw-axis, 223–225

Heading, 34
  from trajectory, 225
  integrated, 226, 669–670
  magnetic, 218–222, 226, 228–229, 668–669
Heading and attitude reference system (HARS), 228–229
  integration, 672

Height
  above mean sea level, 64
  barometric, 230–231
  geodetic, 60–64
  orthometric, 64
Helicopter navigation, 732–733
High-accuracy indoor positioning (HAIP), 509
Horizon sensor, 227–228
Hot start, 369
Horizontal dilution of precision (HDOP), 292–297, 424–429
Hybridization, 2, 559, 647
Hyperbolic ranging, 264, 286
  integration, 689–690
Hypothesis testing, Appendix B
IERS Reference Meridian (IRM), 27, 58
IERS Reference Pole (IRP), 26

Image-based navigation, 14, 538–550
  integration, 682–699
Inclined geostationary orbit, 330–332
Inclinometer, 227
Indian Regional Navigation Satellite System (IRNSS), 314, 326, 332
Indexed IMU, 214–215
Indoor Messaging System (IMES), 501
Inertial frame, 25
  See also Earth-centered inertial frame
Inertial measurement unit (IMU), 6–7, 137–138, 149–151
  calibration, 149–150, 201
  errors, 151–160, 203–214, 574–576, 581, 589–593, 621–622
  indexed, 214–215
  measurement outputs, 139, 143, 149, 189–194
  size effect, 150–151
Inertial navigation, 6–7, 163–215, Appendix E
  alignment, 196–203, 622–637
  correction, 562–565

equations, 163–195
errors, 203–214, 574–593, 594–596, 621–622, Appendix E
failure modes, 702
initialization, 195–200
integrity monitoring, 701–702, 705–718, 720–723
platform, Appendix E
vibration, 159, 189–195
Inertial sensors, 137–160
errors, 151–160, 203–214, 574–576, 581, 589–593
failure modes, 702
Infrared positioning, 512–513
Initialization (INS), 195–200
Innovation-based adaptive estimation (IAE), 124–125
Innovation filtering, 706–709
Innovation sequence monitoring, 706–707, 709–711
INS/GNSS integration, 200–203, 559–622, Appendix I
carrier-phase GNSS, 616–617
deeply coupled, 561, 571–573, 606–614
differential GNSS, 615
GNSS aiding, 463–464, 569–571
GNSS attitude, 618–619
INS error modeling, 577–596, 621–622
integrity monitoring, 701–702, 706–712, 715–718
large heading errors, 619–621
loosely coupled, 561, 566–567, 598–602
measurement models, 596–615
observability, 574–577
smoothing, 622
system models, 577–596
tightly coupled, 561, 567–569, 602–606
ultra-tightly coupled, 561, 571
INS/GPS integration
See INS/GNSS integration
Instrument Landing System (ILS), 474, Appendix F

Integer wavelength ambiguity, 276, 442–449, 616–617, Appendix G
Integration, 18, 200–203, 226, 228–229, 559–622, 647–699, Appendix I
architectures, 647–666
Integrity (RNP), 720–723
Integrity monitoring, 20, 701–723
Interconstellation timing bias, 390, Appendix G
Interferometric fiber-optic gyro (IFOG), 145–146
International Terrestrial Reference Frame (ITRF), 56
Intersignal timing bias, 390, Appendix G
Ionosphere, 287, 391–395, 438, Appendix G
Iridium, 491–492

Joint Tactical Information Distribution System (JTIDS), 257, 481

Kalman filter, 81–136
adaptive, 124–125
algorithm, 84–86, 91–96, 111–113
closed-loop, 106–107
convergence, 103–104, 109–111
equilibrium, 103–104
extended, 118–121, 419–421, 602–606
failure modes, 704
GNSS, 413–424
innovation filtering, 706–709
innovation sequence monitoring, 706–707, 709–711
INS/GNSS, 577–615
integrity monitoring, 701–702, 706–712, 715–718
linearized, 121
measurement model, 84, 100–103, 107–108
multiple hypothesis, 125–129
observability, 104–106
open-loop, 96, 107

Kalman filter *(Cont.)*
  parallel filters integrity monitoring, 715–718
  range checks, 706
  sequential measurement update, 107–108
  sigma point, 121–123
  stability, 109–111, 113–114
  system model, 83–84, 96–100
  time synchronization, 114–117, 661, 663, Appendix I
  tuning, 109–111
  unscented, 121–123
Kalman gain matrix, 94, 101–103
Kalman smoother, 129–130
Keplerian orbit, 332–339
Kinematic carrier-phase positioning (KCPT), 442
Kinematic positioning, 2, 442

LADAR, 535–536, 540–541, 547–548
LAMBDA method, 448, Appendix G
Land navigation, 733–735, 738–739
Landing systems, 315–316, 474, Appendix F
Landmark tracking, 548
Laser
  gyro, 144–145
  image matching, 540–541, 543, 547–548
  rangefinder, 233
  scanner, 535–536, 540–541, 547–548
  terrain-referenced navigation, 535–536
Latitude (geodetic), 58–63
Least-squares estimation, 285–286, 410–413, 478, 486, 505, Appendix D
Least-squares integration, 648–651
Leveling, 197–198, 226–227, 670–672
Lever arm, 77–78
LIDAR, 535–536, 540–541
Light intensity, Appendix H
Lightning, Appendix H
Line fix, 527, 541–542, 694–695

Line of position, 7–8, 260–261, 269–270
Line of sight, 269, 341, 344
  frame, 30
Linearized Kalman filter, 121
Link 16 positioning, 481
Local Area Augmentation System, 315
Local navigation frame, 27–28, 74–76, 176–180, 584–589
Local tangent-plane frame, 28
Locata, 500
Localization, 3
Location, 3
Location-based services, 739–740
Location signature, 10, 518, 532
Lock detection (GNSS), 384–385
Long range navigation (Loran), 12, 481–488, Appendix K
Long-range radio navigation (non-GNSS), 473–488, 491–492, Appendices F and K
Longitude, 58–62
Loosely coupled INS/GNSS integration, 561, 566–567, 598–602
Loosely coupled integration (general), 683–685

Magnetic
  anomalies, 220–221
  compass, 218–222
  declination angle, 218
  dip, 218
  heading, 219–222, 226, 228–229
  heading integration, 668–670
  inclination, 218
  positioning, 513, 522
  variation, 218
Magnetometers, 220
  failure modes, 702
Map matching, 15, 519–530
  integration, 683–685, 694–696
Marine gyrocompass, 222–223
Marine navigation, 735–736
Marine radio beacons, 492

Markov sequence, 88, Appendix B
Matrices, Appendix A
M code, 321–323
Mean, Appendix B
Measurement
  GNSS, 387–389
  innovation, 90
  matrix, 93
  model, 84, 100–103, 107–108
  noise covariance matrix, 91, 109–110
  residual, 91
  vector, 90, 100–101
Medium-range radio navigation, 473–481, 488–491, 492–494, Appendices F and K
Meridian, 57
Microclimate, Appendix H
Micro-electro-mechanical systems (MEMS), 137–138, 140–142, 148–149, 152–159
Microwave Landing System (MLS), 474, Appendix F
Misalignment, 154
Mobile positioning, 2
Mobile phone, 463–464, 488–491, Appendix F
Modulation, 276–277
  correlation, 494
Moment arm, 77–78
Monte Carlo estimation, 131
Motion classification, 243–244, 529, 665–666
Motion constraints, 215, 641–644, 675–677, 679, Appendix E
MTSat Satellite Augmentation System (MSAS), 315, 327, 329, 332
Multi-Functional Information Distribution System (MIDS) navigation, 257, 481
Multilateral positioning, 255
Multipath, 290
  GNSS, 401–407, 458–462, Appendix G
Multiple-hypotheses filtering, 125–129, 695–696

Multiple-model adaptive estimation (MMAE), 125
Multiplexed Binary Offset Carrier (MBOC), 320–321, 324, 326, Appendix G
Multisensor integration, 647–699

Navigation
  definitions, 1
  frame (local), 27–28
  processor (GNSS), 407–424
  processor (inertial), 163–203
Navigation data message, 327–330, 365, 385–386
Network assistance, 19, 464
Noise sources, 87–88
  inertial sensors, 155–157, 207
  radio frequency, 291
  time-correlated, 123–124
Nondirectional beacon (NDB), 480–481
Nonholonomic constraint, 215, 641–644, 675–677, 679
Nonisotropic transmission, 271
Nonline-of-sight (NLOS) reception, 289, 401–402, 407, 458–462
Normal distribution, Appendix B
Numerically-controlled oscillator (NCO), 356–358, 372–373, 376–377, 384, 456–458, 571, 613–614

Object frame, 23–24, 44
Observability, 104–106
  deterministic, 104–106
  INS/GNSS integration, 574–577
  matrix, 105
  stochastic, 106
Odometer/ Odometry, 6, 233–240
  differential, 226, 238–239
  failure modes, 702
  integration, 674–677
Omega, 12, Appendix K
Optical gyro, 143–146
Optical positioning, 513, 538–550
Orbital coordinate frame, 333

Orbit prediction, 465
Orthogonal Frequency Division Multiplex (OFDM), 277, 489, 502–503
Oscillator (crystal), 352

Parallel, 57
Partial IMU, 215, 239–240, Appendix E
Particle filter, 131–134, Appendix D
Passive ranging, 8, 261–264, 282–286, 307–309, 407–424
Pattern matching, 10–11, 271–274, 465–466, 506–507, 530–537, 551–552, 648, 653, 663–664, 682, 695–696, 699
P code, 320–323
Pedestrian dead reckoning, 5, 240–245
  failure modes, 702
  integration, 677–680
Pedestrian navigation, 240–245, 638–641, 643–644, 738–739
  map matching, 528–530
Peer-to-peer positioning. *See* Cooperative positioning
Pendulous accelerometer, 140–142
Phase center, 351
Phase lock loop (PLL), 377–384
Phase wind-up (GNSS), 443
Pitch, 33–35
Platform INS, Appendix E
Position
  Cartesian, 46–48, 57–63
  conversion, 61–63, Appendix C
  curvilinear, 57–63
  normal vector, Appendix C
  projected, 65–66, Appendix C
Position dilution of precision (PDOP), 424–427
Position fixing, 3–4, 7–15
  integration, 682–699
Positioning
  categories, 1–2
  methods, 3–15, 255–276
Postprocessing, 1, 129–130, 442
Power spectral density (PSD), 88, 156, Appendix B

Precise point positioning (PPP), 440–442
Precise Positioning Service (PPS), 312, 321
Precorrelation bandwidth, 353–354, 360–361, 404–405
Prediction, 659–660
Prefilter, 572, 608–612
Primary mode/sensor, 663–664
Probabilistic data association filter (PDAF), 126–127
Probability, Appendix B
Probability density function (PDF,) Appendix B
Projected coordinates, 65–66, Appendix C
Proximity positioning, 7, 258–260, 489–490, 506, 507–509
Pseudo-force, 51–53
Pseudo-random noise (PRN), 304–305
Pseudo-range, 261–264, 304–306, 308–309, 342–343, 387, 407–409, 412, 419–421, 602–603
  carrier smoothed, 389, 394
Pseudo-range rate, 307, 342–363, 387–388, 408, 411–412, 419–421, 602–603
Pseudolites, 499–501
Public Regulated Service (PRS), 313, 324
Pulsar navigation, 552
P(Y) code, 320–322

Quantization, 157
Quasi-stationary alignment, 201–202, 634–637
Quasi-Zenith Satellite System (QZSS), 314, 326, 332
Quaternion attitude, 40–42, Appendix E

R mode, 492
Radar
  altimeter, 232–233, 530–535, 673
  Doppler, 245–248, 680–681, 702–703

imaging, 540, 543, 545
Radial error, Appendix B
Radiation, Appendix H
Radio broadcasts, 492–494, Appendix F
Radio determination satellite service (RDSS), 473, Appendix F
Radio engineers' system of short-range navigation (RSBN), 480
Radio-frequency identification (RFID) tags, 259, 508
Radio navigation, 7–14, 255–297
  GNSS, 299–466, Appendix G
  long-range (non-GNSS), 473–488, 491–492, Appendices F and K
  medium-range, 473–481, 488–491, 492–494, Appendices F and K
  short-range, 499–509, 512
  terrestrial, 473–494, 499–509, 512, Appendices F and K
Radio signpost, 512
Radio spectrum, 277–279
Radius
  geocentric, 54
  equatorial, 54, 56
  polar, 54, 56
Radius of curvature (meridian and transverse), 59–60
Rail navigation, 734–735
  map matching, 527–528
Random noise, 155–157
Random process, Appendix B
Random walk, 156, 660–661
Range, 260–261, 339–341, 343–344
Range checks, 705–706
Range-rate, 341–344
Ranging, 7–8, 260–269, 304–309
  differential, 264–266
  hyperbolic, 264, 286
  integration, 685–690
  passive, 8, 261–264, 282–286, 307–309, 407–424
  position determination, 282–287, 407–424, 476–478, 480, 485–486, 504–505
  resolution, 290–291
  two-way, 266–268, 282–285

Rapid alignment, 631–633
Real-time kinematic (RTK) positioning, 442–449
Real-time positioning, 1–2, 442
Received signal strength (RSS), 268, 271–274
Receiver
  baseband signal processor, 279–280, 355–366
  clock, 262, 304, 308, 351–352
  front-end, 279–280, 352–355, 453–454
  GNSS, 302–303, 350–366, 453–455
Receiver autonomous integrity monitoring (RAIM), 713–715
Reference frame, 23–24, 44
Reflection of signals, 289–290, 401–402
Relative GNSS, 441–442
Relative navigation and positioning, 257–258
Remote positioning, 2, 255–257
Required navigation performance (RNP), 720–723, 729
Resolving axes, 25, 44
Reversionary mode/sensor, 663–664
Ring laser gyro (RLG), 144–145
Road map, 520–521
  matching, 519–527
Road signs, 542
Road texture, Appendix H
Road-vehicle navigation, 733–734
Roll, 33–35
Rotation axis, 39–40, 42
Rotation vector, 42–43, 193

Sagnac effect (gyroscopes), 143–144
Sagnac correction (rotating coordinate frames), 339–340, 342
Satellite,
  attitude determination, 737
  GNSS, 300–301, 330–332, Appendix G
  orbit determination, 737–738
  position, velocity, and acceleration, 332–339, Appendix G

Satellite autonomous integrity monitoring (SAIM), 719
Satellite clocks, 308, 327–330, 390–391, 438
Satellite navigation
  See Global Navigation Satellite Systems
Satellite Navigation Augmentation System (SNAS), 315, 327, 329, 332
Scale-factor error, 154–155, 213–214, 589, 592
Scale-factor nonlinearity, 158
Scattering, 289
Schmidt-Kalman filter, Appendix D
Schuler frequency and oscillation, 209–211
Scintillation, 395
Sculling, 192
Sea navigation, 735–737
Self positioning, 2, 255–257
Semi-codeless correlation, 366
Semi-major axis, 54
Semi-minor axis, 54
Sequential measurement update, 107–108
Sferics, Appendix H
Shadow matching, 465–466
Ship navigation, 735–736
Ship's speed log, 5, 250, Appendix K
Short-range radio navigation, 499–509, 512
Sigma-point Kalman filter, 121–123
Signal
  acquisition, 280–281, 367–372, 455–456
  attenuation, 288–289, 452
  coherence, 364–365, 455
  conditioning, 352–354
  correlation, 280–281, 290–291, 355–366
  DME, 474–476
  geometry, 292–297, 424–429
  GNSS, 303–305, 317–327, Appendix G
  Loran, 482–484
  of opportunity, 257, 489–494, 506, 682, 687
  processing, 280–282, 290–291, 355–366
  propagation, 287–290
  reflection, 289–290, 401–402
  sampling, 355
  timing, 280–282
  tracking, 281, 372–384, 395–401, 454–458
  weak, 452–458
Signal-in-space error, 390–391
Signal to noise (GNSS), 362–364, 386–387, Appendix G
Signs, 542
Simultaneous localization and mapping (SLAM), 274, 518–519, 548, 698
Single-epoch integration, 648–651
Single-epoch position solution, 282–287, 409–413, 476–478, 480, 485–486, 504–505
Size effect, 150–151
Skew symmetric matrix, 45, Appendix A
Skyhook Wireless, 506
Sky wave, 287, 483
Smells, Appendix H
Smoothing
  fixed-gain, 226, 388–389, 394
  Kalman, 129–130
  INS/GNSS integration, 622
Snapshot position solution. *See* Single-epoch position solution
Snap to map. *See* Map matching
Software receiver, 356
Solid Earth tides, 64–65
Sonar
  altimeter, 233
  correlation velocity log (CVL), 249, 250
  Doppler/ Doppler velocity log (DVL), 245–250, 702–703
  terrain referenced navigation, 536
Sound. *See* acoustic positioning, ambient sound, sonar, ultrasound

Index

Space-based augmentation systems (SBAS), 315, 327, 329, 332, 440, 719
Spacecraft navigation, 737–738
Spatial decorrelation, 438
Specular reflection, 289
Specific force, 67–69, 139
  integrated, 149
Spherical error probable (SEP), Appendix B
Spinning-mass gyro, Appendix E
Spreading code, 303–305, 317–327
Spread spectrum, 277, 303–305
Standard deviation, Appendix B
Standard navigation unit (SNU), 138
Standard Positioning Service (SPS), 312, 320–321
Star imager, 14, 548–550
Star tracker, 548–550, Appendix K
State vector, 89
Static positioning, 2, 442
Stationary-condition detection, 638–639
Stellar navigation, 548–550, Appendix K
Step detection, 241–242
Stochastic process, Appendix B
Strapdown
  gyrocompass, 228–229
  IMU, 137
  yaw-axis gyro, 223–225
Submarine navigation, 737
Surface of position, 8, 307
Surveillance, 2
Synchronization. *See* Time synchronization
Synthetic aperture radar (SAR), 540
System
  matrix, 96–97
  model, 83–84, 96–100
  propagation, 92–93, 96–100
  noise covariance matrix, 92–93, 99–100, 109–110
System of Differential Corrections and Monitoring (SDCM), 315, 327, 329, 332

Tactical Air Navigation (TACAN), 479–480
Tangent-plane frame, 28
Television, 493–494
Temporal decorrelation, 438
Terrain-aided navigation (TAN). *See* Terrain-referenced navigation
Terrain-contour navigation (TCN). *See* Terrain-referenced navigation
Terrain-referenced navigation (TRN), 530–537
  barometric, 537
  integration, 683–685, 695–696
  laser, 535–536
  sonar, 536
Terralite XPS, 500
Terrestrial radio navigation, 473–494, 499–509, 512, Appendices F and K
  failure modes, 703–704
  integration, 682–690
Testing, 724–726
Tightly coupled INS/GNSS integration, 561, 567–569, 602–606
Tightly coupled integration (general), 685–690
Tilt sensor, 226–227
Time-correlated noise, 123–124
Time dilution of precision (TDOP), 292–297, 424–429
Time difference, 264
Time difference of arrival (TDOA), 264, 286, 689–690
Time of arrival, 261–264, 285–286
Time of flight, 8, 260
Time of validity, 114–117, 661, 663
Time synchronization, 114–117, 262–263, 661, 663, 740, Appendix I
Time to first fix (TTFF), 311, 463–464
Timing of signals, 280–282
  resolution, 290–291
Total-state integration, 652–654, 659–661
Tracking, 2, 256, 281
  carrier (GNSS), 377–384, 398–401, 454–458

Tracking *(Cont.)*
  code (GNSS), 372–377, 395–397, 454–458
  combined signal, 456
  combined with GNSS navigation, 456–458, 571, 573, 606–614
  errors, 290–291, 359, 395–401
  extended range, 454–455
  lock detection (GNSS), 384–385
  semi-codeless, 366
  system, 2, Appendix F
  vector, 456–458, Appendix G
Trajectory-derived heading, 225
Transect, 11, 532–534
Transfer alignment, 201–202, 627–634
Transit, 12, Appendix K
Transition matrix, 92–93, 97–98
Transport rate, 177–179, 181, 185, 188
Trilateration, 260
Troposphere, 287–288, 392, 394–395, 438, Appendix G
Tsikada, 12, Appendix K
Two-way ranging, 8, 266–268, 282–285

Ultrasonic altimeter, 233
Ultrasound, 512
Ultratightly-coupled INS/GNSS integration, 561, 571
Ultrawideband (UWB) positioning, 501–505
Unilateral positioning, 255
Underwater navigation, 737
Unmanned air vehicle (UAV), 733
Unscented Kalman filter (UKF), 121–123
Urban canyon, 310–311, 465–466
User, 3
User equipment range error (UERE), 424, 430

Variance, Appendix B
Vectors, Appendix A

Vector delay lock loop, 456–458
Vector tracking, 456–458, Appendix G
Velocity, 48–50
Vertical dilution of precision (VDOP), 424–429
VHF Ominidirectional Radiorange (VOR), 10, 12, 271, 479–480
Vibrating beam accelerometer, 142
Vibration
  of inertial sensors, 159
  in inertial navigation, 191–192, 194–196, 198, 200
Vibration rectification error, 159
Visual odometry, 546–548, 682
Visual navigation, 14, 539–542, 546–548
  integration, 682–698
VORTAC, 479

Wander angle, 181–182
Wander-azimuth frame, 29, 180–182, Appendix I
Warm start, 369
Wheel speed sensor, 233–240

White noise, 87–88, Appendix B
Wide Area Augmentation System (WAAS), 315, 327, 329, 332
Wi-Fi, 506–507
Wireless local area network (WLAN), 259, 506–507
Wireless personal area network (WPAN), 259, 507–508
World Geodetic System (WGS), 56, 64, 67, 71

X-ray navigation, 552

Yaw, 33–35

Zero angular rate update (ZARU), 640–641
Zero velocity update (ZVU), 201, 638–640
Zigbee, 508